Student's Solutions Manual

for use with

Elementary and Intermediate Algebra

Second Edition

Mark Dugopolski
Southeastern Louisiana University

Boston Burr Ridge, IL Dubuque, IA Madison, WI New York San Francisco St. Louis
Bangkok Bogotá Caracas Kuala Lumpur Lisbon London Madrid Mexico City
Milan Montreal New Delhi Santiago Seoul Singapore Sydney Taipei Toronto

The McGraw-Hill Companies

Student's Solutions Manual for use with
ELEMENTARY AND INTERMEDIATE ALGEBRA, SECOND EDITION
Mark Dugopolski

Published by McGraw-Hill Higher Education, an imprint of the The McGraw-Hill Companies, Inc., 1221 Avenue of the Americas, New York, NY 10020.

This book is printed on recycled, acid-free paper containing 10% postconsumer waste.

1 2 3 4 5 6 7 8 9 0 QPD/QPD 0 9 8 7 6 5

ISBN: 0-07-293692-4

www.mhhe.com

Student's Solutions Manual

Elementary and Intermediate Algebra

Second Edition

Table of Contents

1.1 WARM-UPS

1. True, because we refer to the numbers 1, 2, 3, 4, . . . as either the natural numbers or the counting numbers.
2. True, because of the definition of counting numbers.
3. False, the smallest counting number is 1.
4. False, because zero can be expressed as a ratio of two integers. For example $0 = 0/4$.
5. True, 3 and -3 are opposites of each other.
6. False, because 4 is positive and the absolute value of a positive number is the number itself. So the absolute value of 4 is 4.
7. True, because the opposite of -9 is 9.
8. True, because π is approximately 3.14 and 3.14 is between 3 and 4. The interval $(3, 4)$ consists of all real numbers between 3 and 4.
9. False, because -6 is to the left of -3 on the number line and so -6 is less than -3.
10. False, because -5 is to the left of 0 on the number line and 4 and 6 are to the right of 0 on the number line.

1.1 EXERCISES

1. The integers are the numbers in the set $\{\ldots, -3, -2, -1, 0, 1, 2, 3, \ldots\}$.
3. A rational number is a ratio of integers and an irrational number is not.
5. The number a is larger than b if a lies to the right of b on the number line.
7. The larger number is 6, because 6 is located to the right of -3 on the number line.
9. The larger number is 0, because 0 is located to the right of -6 on the number line.
11. The larger number is -2, because -2 is located to the right of -3 on the number line.
13. The larger number is -12, because -12 is located to the right of -15 on the number line.
15. The larger number is -2.1, because -2.1 is located to the right of -2.9 on the number line.
17. The counting numbers smaller than 6 are 1, 2, 3, 4, and 5. We show them on a number line as follows.

1 2 3 4 5

19. The whole numbers smaller than 5 are 0, 1, 2, 3, and 4. We show them on a number line as follows.

21. The whole numbers are 0, 1, 2, 3, and so on. The whole numbers between -5 and 5 are 0, 1, 2, 3, and 4. Note that there are no negative whole numbers.

0 1 2 3 4

23. The counting numbers larger than -4 are 1, 2, 3, 4, 5, and so on. All of the counting numbers are larger than -4.

1 2 3 4 5 . . .

25. The integers larger than $1/2$ are 1, 2, 3, 4, 5, and so on. We show them on a number line as follows.

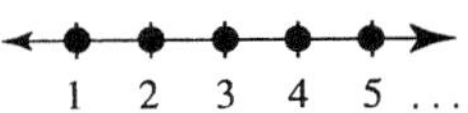

27. True, because every integer can be expressed as a ratio of two integers. For example, $-3 = -3/1$.
29. False, the smallest counting number is 1.
31. True, because the ratio the circumference and diameter of a circle is π and π is irrational.
33. True, because the ratio of a whole number and 1 is a ratio of integers. For example, 2 is a whole number and $2 = 2/1$ expresses 2 as a ratio of integers.
35. True, because the positive integers are natural numbers.
37. False, because 0 is rational $(0 = 0/1)$.
39. The set of real numbers between 0 and 1 is written in interval notation as $(0, 1)$ and graphed as follows.

−1 0 1 2

41. The set of real numbers between -2 and 2 inclusive is written in interval notation as $[-2, 2]$ and graphed as follows.

−4 −2 0 2 4

43. The set of real numbers greater than 0 and less than or equal to 5 is written in interval notation as $(0, 5]$ and graphed as follows.

−1 0 1 2 3 4 5 6

45. The set of real numbers greater than 4 is written in interval notation as $(4, \infty)$ and graphed as follows.

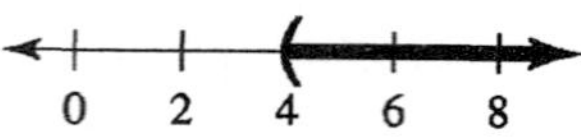

47. The set of real numbers less than or equal to -1 is written in interval notation as $(-\infty, -1]$ and graphed as follows.

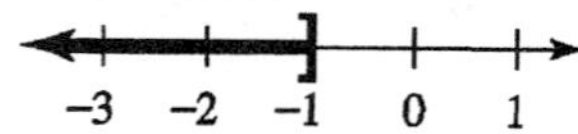

49. The set of real numbers greater than or equal to 0 is written in interval notation as $[0, \infty)$ and graphed as follows.

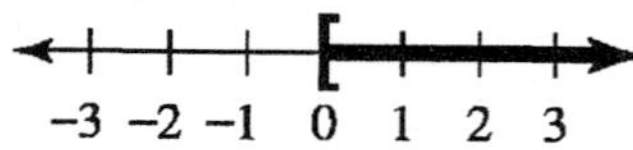

51. Because -6 is a negative number, $|-6| = -(-6) = 6$.

53. Because 0 is 0 units from 0 on the number line, $|0| = 0$.

55. Because 7 is positive $|7| = 7$.

57. Because -9 is negative, $|-9| = -(-9) = 9$.

59. Because -45 is negative, $|-45| = -(-45) = 45$.

61. Because $3/4$ is positive, $\left|\frac{3}{4}\right| = \frac{3}{4}$.

63. Because -5.09 is negative, $|-5.09| = -(-5.09) = 5.09$.

65. The smaller number is -16, because -16 is to the left of 9 on the number line.

67. First change the first fraction to fourths.

$$-\frac{5}{2} = -\frac{5 \cdot 2}{2 \cdot 2} = -\frac{10}{4}$$

On the number line $-10/4$ is to the left of $-9/4$. So the smaller number is $-10/4$ or $-5/2$.

69. Because $|-3| = 3$, the smaller number is 2.

71. Because $|-4| = 4$, the smaller number is 3.

73. Because $|-5| = 5$ and $|-9| = 9$, -9 has the larger absolute value.

75. Because $|16| = 16$ and $|-9| = 9$, 16 has the larger absolute value.

77. The absolute value of a number is the number's distance from 0 on the number line. The number with the smaller absolute value is closer to zero. Since $|-4| = 4$ and $|-5| = 5$, -4 is closer to zero.

79. The number with the smaller absolute value is closer to zero. Since $|-2.01| = 2.01$ and $|-1.99| = 1.99$, -1.99 is closer to zero.

81. The number with the smaller absolute value is closer to zero. Since $|-75| = 75$ and $|74| = 74$, 74 is closer to zero.

83. The distance on the number line between 0 and a number is the absolute value of the number. So the distance between 0 and 5.25 is 5.25.

85. The distance on the number line between 0 and a number is the absolute value of the number. So the distance between 0 and -40 is 40.

87. The distance on the number line between 0 and a number is the absolute value of the number. So the distance between 0 and $-1/2$ is $1/2$.

89. Since $|-3| = 3$ and $|3| = 3$, both -3 and 3 have an absolute value of 3.

91. Since the absolute values of $-4, -3, 3$, and 4 are $4, 3, 3$, and 4, respectively, the integers $-4, -3, 3$, and 4 each have an absolute value greater than 2.

93. Since the absolute values of $-1, 0$, and 1 are $1, 0$, and 1, respectively, the integers $-1, 0$, and 1 each have an absolute value less than 2.

95. The graph shows the real numbers between 3 and 8 inclusive, which is written in interval notation as $[3, 8]$.

97. The graph shows the real numbers between -30 and -20 including -20 but not -30, which is written in interval notation as $(-30, -20]$.

99. The graph shows the real numbers greater than or equal to 30, which is written in interval notation as $[30, \infty)$.

101. True, because $|-3| = 3$ and $|-5| = 5$, and if we add the absolute values we do get 8.

103. True, because the absolute value of any negative number is a positive number and any positive number is greater than 0.

105. True, because $|-9| = 9$ and $|6| = 6$.

107. What is the probability that a tossed coin turns up heads?

109. If a is negative, then $-a$ and $|-a|$ are positive. The rest are negative.

1.2 WARM-UPS

1. True, because we can multiply the numerator and denominator of a fraction by any nonzero number and get an equivalent fraction.

2. True, because they both reduce to $2/3$.

3. False, because $8/12$ in lowest terms is $2/3$.

4. True, because if we multiply we get $2/6$, which reduces to $1/3$.

5. True, because of the definition of multiplication of fractions.

6. True, because of the definition of multiplication of fractions.

7. True, because $\frac{1}{2} \div 3 = \frac{1}{2} \cdot \frac{1}{3} = \frac{1}{6}$.

8. True, because $5 \div \frac{1}{2} = 5 \cdot 2 = 10$.

9. False, because $\frac{1}{2} + \frac{1}{4} = \frac{2}{4} + \frac{1}{4} = \frac{3}{4}$.

10. True, because $2 - \frac{1}{2} = \frac{4}{2} - \frac{1}{2} = \frac{3}{2}$.

1.2 EXERCISES

1. If two fractions are identical when reduced to lowest terms, then they are equivalent fractions.

3. To reduce a fraction means to find an equivalent fraction that has no factor common to the numerator and denominator.

5. Convert a fraction to a decimal by dividing the denominator into the numerator.

7. $\frac{3}{4} = \frac{3 \cdot 2}{4 \cdot 2} = \frac{6}{8}$

9. $\frac{8}{3} = \frac{8 \cdot 4}{3 \cdot 4} = \frac{32}{12}$

11. $5 = 5 \cdot \frac{2}{2} = \frac{10}{2}$

13. $\frac{3}{4} = \frac{3 \cdot 25}{4 \cdot 25} = \frac{75}{100}$

15. $\frac{3}{10} = \frac{3 \cdot 10}{10 \cdot 10} = \frac{30}{100}$

17. $\frac{5}{3} = \frac{5 \cdot 14}{3 \cdot 14} = \frac{70}{42}$

19. $\frac{3}{6} = \frac{3 \cdot 1}{3 \cdot 2} = \frac{1}{2}$

21. $\frac{12}{18} = \frac{6 \cdot 2}{6 \cdot 3} = \frac{2}{3}$

23. $\frac{15}{5} = \frac{5 \cdot 3}{5} = 3$

25. $\frac{50}{100} = \frac{50 \cdot 1}{50 \cdot 2} = \frac{1}{2}$

27. $\frac{200}{100} = \frac{100 \cdot 2}{100} = 2$

29. $\frac{18}{48} = \frac{6 \cdot 3}{6 \cdot 8} = \frac{3}{8}$

31. $\frac{26}{42} = \frac{2 \cdot 13}{2 \cdot 21} = \frac{13}{21}$

33. $\frac{84}{91} = \frac{7 \cdot 12}{7 \cdot 13} = \frac{12}{13}$

35. $\frac{2}{3} \cdot \frac{5}{9} = \frac{10}{27}$

37. $\frac{1}{3} \cdot 15 = \frac{15}{3} = 5$

39. $\frac{3}{4} \cdot \frac{14}{15} = \frac{3}{2 \cdot 2} \cdot \frac{2 \cdot 7}{3 \cdot 5} = \frac{7}{10}$

41. $\frac{2}{5} \cdot \frac{35}{26} = \frac{2}{5} \cdot \frac{5 \cdot 7}{2 \cdot 13} = \frac{7}{13}$

43. $\frac{1}{2} \cdot \frac{6}{5} = \frac{1}{2} \cdot \frac{2 \cdot 3}{5} = \frac{3}{5}$

45. $\frac{1}{2} \cdot \frac{1}{3} = \frac{1}{6}$

47. $\frac{3}{4} \div \frac{1}{4} = \frac{3}{4} \cdot \frac{4}{1} = 3$

49. $\frac{1}{3} \div 5 = \frac{1}{3} \cdot \frac{1}{5} = \frac{1}{15}$

51. $5 \div \frac{5}{4} = 5 \cdot \frac{4}{5} = 4$

53. $\frac{6}{10} \div \frac{3}{4} = \frac{2 \cdot 3}{2 \cdot 5} \cdot \frac{2 \cdot 2}{3} = \frac{4}{5}$

55. $\frac{3}{16} \div \frac{5}{2} = \frac{3}{2 \cdot 2 \cdot 2 \cdot 2} \cdot \frac{2}{5} = \frac{3}{40}$

57. $\frac{1}{4} + \frac{1}{4} = \frac{2}{4} = \frac{2 \cdot 1}{2 \cdot 2} = \frac{1}{2}$

59. $\frac{5}{12} - \frac{1}{12} = \frac{4}{12} = \frac{1}{3}$

61. $\frac{1}{2} - \frac{1}{4} = \frac{1 \cdot 2}{2 \cdot 2} - \frac{1}{4} = \frac{2}{4} - \frac{1}{4} = \frac{1}{4}$

63. $\frac{1}{3} + \frac{1}{4} = \frac{1 \cdot 4}{3 \cdot 4} + \frac{1 \cdot 3}{4 \cdot 3} = \frac{4}{12} + \frac{3}{12} = \frac{7}{12}$

65. $\frac{3}{4} - \frac{2}{3} = \frac{3 \cdot 3}{4 \cdot 3} - \frac{2 \cdot 4}{3 \cdot 4} = \frac{9}{12} - \frac{8}{12} = \frac{1}{12}$

67. $\frac{1}{6} + \frac{5}{8} = \frac{1 \cdot 4}{6 \cdot 4} + \frac{5 \cdot 3}{8 \cdot 3} = \frac{4}{24} + \frac{15}{24} = \frac{19}{24}$

69. $\frac{5}{24} - \frac{1}{18} = \frac{5 \cdot 3}{24 \cdot 3} - \frac{1 \cdot 4}{18 \cdot 4} = \frac{15}{72} - \frac{4}{72}$
$= \frac{11}{72}$

71. $3\frac{5}{6} + \frac{5}{16} = \frac{23}{6} + \frac{5}{16} = \frac{23 \cdot 8}{6 \cdot 8} + \frac{5 \cdot 3}{16 \cdot 3}$
$= \frac{184}{48} + \frac{15}{48} = \frac{199}{48}$

73. $\frac{3}{5} = \frac{3 \cdot 20}{5 \cdot 20} = \frac{60}{100} = 60\%$,
$\frac{3}{5} = \frac{3 \cdot 2}{5 \cdot 2} = \frac{6}{10} = 0.6$

75. $9\% = \frac{9}{100} = 0.09$

77. $0.08 = \frac{8}{100} = 8\%$,
$0.08 = \frac{8}{100} = \frac{2 \cdot 4}{25 \cdot 4} = \frac{2}{25}$

79. $\frac{3}{4} = \frac{3 \cdot 25}{4 \cdot 25} = \frac{75}{100} = 0.75$,
$\frac{3}{4} = \frac{75}{100} = 75\%$

81. $2\% = \frac{2}{100} = \frac{2 \cdot 1}{2 \cdot 50} = \frac{1}{50}$,
$2\% = \frac{2}{100} = 0.02$

83. $0.01 = \frac{1}{100} = 1\%$

85. $\frac{3}{8} \div \frac{1}{8} = \frac{3}{8} \cdot \frac{8}{1} = 3$

87. $\frac{3}{4} \cdot \frac{28}{21} = \frac{3}{4} \cdot \frac{4 \cdot 7}{3 \cdot 7} = 1$

89. $\frac{7}{12} + \frac{5}{32} = \frac{7 \cdot 8}{12 \cdot 8} + \frac{5 \cdot 3}{32 \cdot 3} = \frac{56}{96} + \frac{15}{96}$
$= \frac{71}{96}$

91. $\frac{5}{24} - \frac{1}{15} = \frac{5 \cdot 5}{24 \cdot 5} - \frac{1 \cdot 8}{15 \cdot 8} = \frac{25}{120} - \frac{8}{120}$
$= \frac{17}{120}$

93. $3\frac{1}{8} + \frac{15}{16} = \frac{25}{8} + \frac{15}{16} = \frac{50}{16} + \frac{15}{16} = \frac{65}{16}$

95. $7\frac{2}{3} \cdot 2\frac{1}{4} = \frac{23}{3} \cdot \frac{9}{4} = \frac{69}{4}$

97. $\frac{1}{2} + \frac{1}{3} + \frac{1}{4} = \frac{1 \cdot 6}{2 \cdot 6} + \frac{1 \cdot 4}{3 \cdot 4} + \frac{1 \cdot 3}{4 \cdot 3}$
$= \frac{6}{12} + \frac{4}{12} + \frac{3}{12} = \frac{13}{12}$

99. $\frac{1}{2} \cdot \frac{1}{2} \cdot \frac{1}{2} = \frac{1}{8}$

101. Since $\frac{1}{4} = \frac{2}{8}$ we have $\frac{2}{8} + \frac{3}{8} = \frac{5}{8}$ or $\frac{1}{4} + \frac{3}{8} = \frac{5}{8}$

103. Since $\frac{1}{8} = \frac{2}{16}$ we have $\frac{5}{16} - \frac{3}{16} = \frac{2}{16}$ or $\frac{5}{16} - \frac{3}{16} = \frac{1}{8}$.

105. Since fractions are multiplied by multiplying their numerators and multiplying their denominators, we have $\frac{4}{9} \cdot \frac{2}{3} = \frac{8}{27}$.

107. Divide fractions by inverting and multiplying. So $\frac{2}{3} \div \frac{1}{2} = \frac{2}{3} \cdot 2 = \frac{4}{3}$ or $\frac{2}{3} \div \frac{1}{2} = \frac{4}{3}$.

109. $\frac{1}{6} + \frac{1}{32} = \frac{1 \cdot 16}{6 \cdot 16} + \frac{1 \cdot 3}{32 \cdot 3}$
$= \frac{16}{96} + \frac{3}{96} = \frac{19}{96}$

111. a) From the table, the amount needed is 1.3 yd^3.

b) $V = 12\frac{1}{2}\text{ft} \cdot 8\frac{3}{4}\text{ft} \cdot 4 \text{ in}$
$= \frac{25}{2}\text{ft} \cdot \frac{35}{4}\text{ft} \cdot \frac{1}{3}\text{ft} = \frac{875}{24}\text{ft}^3 = 36\frac{11}{24}\text{ft}^3$
$\frac{875}{24}\text{ft}^3 \cdot \frac{1 \text{ yd}^3}{27 \text{ ft}^3} = \frac{875}{648}\text{yd}^3 = 1\frac{227}{648}\text{yd}^3$

115. Each daughter gets $3 \text{ km}^2 \div 4$ or a $\frac{3}{4}\text{ km}^2$ piece of the farm. Divide the farm into 12 equal squares. Give each daughter an L-shaped piece consisting of 3 of those 12 squares.

1.3 WARM-UPS

1. True, because the difference between 9 and 8 is 1, and we use the negative sign because -9 has a larger absolute value than 8.
2. True, because to add two negative numbers we add their absolute values $(2 + 4 = 6)$ and then attach a negative sign to the result.
3. True, because $0 - 7 = 0 + (-7) = -7$.
4. False, because $5 - (-2) = 5 + 2 = 7$.
5. False, because $-5 - (-2) = -5 + 2 = -3$.
6. False, the additive inverse of -3 is 3.
7. True, because the variable b can represent a negative or a positive number, and when b is a negative number $-b$ is positive.
8. False, because the sum of a positive number and a negative number may be either positive or negative depending on which number has the larger absolute value.
9. True, because all subtraction can be expressed as addition of the opposite $(b - a = b + (-a))$.
10. False, because $-5 - (-7) = -5 + 7 = 2$.

1.3 EXERCISES

1. We studied addition and subtraction of signed numbers.
3. Two numbers are additive inverses of each other if their sum is zero.
5. To find the sum of two numbers with unlike signs subtract their absolute values. The answer is given the sign of the number with the larger absolute value.
7. $3 + 10 = 13$
9. $(-3) + (-10) = -(3 + 10) = -13$
11. $-0.25 + (-0.9) = -(0.25 + 0.90)$
$= -1.15$
13. $-\frac{1}{3} + \left(-\frac{1}{6}\right) = -\frac{1 \cdot 2}{3 \cdot 2} + \left(-\frac{1}{6}\right)$
$= -\frac{2}{6} + \left(-\frac{1}{6}\right) = -\frac{3}{6} = -\frac{1 \cdot 3}{2 \cdot 3} = -\frac{1}{2}$
15. Since -8 and 8 are additive inverses, their sum is 0, $-8 + 8 = 0$.
17. $-\frac{17}{50} + \frac{17}{50} = 0$
19. Since 9 has the larger absolute value, the result is positive, $-7 + 9 = 9 - 7 = 2$.

21. Since -13 has the larger absolute value, the result is negative,
$7+(-13)=-(13-7)=-6$.
23. Since 8.6 has the larger absolute value, the result is positive,
$8.6+(-3.0)=8.6-3.0=5.6$.
25. $3.9+(-6.8)=-(6.8-3.9)=-2.9$
27. Since $-1/2$ has the larger absolute value, the result is negative.
$\frac{1}{4}+\left(-\frac{1}{2}\right)=\frac{1}{4}+\left(-\frac{2}{4}\right)=-\frac{1}{4}$
29. Since subtraction is equivalent to addition of the opposite, $8-2=8+(-2)$.
31. Since subtraction is equivalent to addition of the opposite, $4-12=4+(-12)$.
33. Since subtraction is equivalent to addition of the opposite, $-3-(-8)=-3+8$.
35. Since subtraction is equivalent to addition of the opposite, $8.3-(-1.5)=8.3+1.5$.
37. $6-10=6+(-10)=-4$
39. $-3-7=-3+(-7)=-10$
41. $5-(-6)=5+6=11$
43. $-6-5=-6+(-5)=-11$
45. $\frac{1}{4}-\frac{1}{2}=\frac{1}{4}+\left(-\frac{2}{4}\right)=-\frac{1}{4}$
47. $\frac{1}{2}-\left(-\frac{1}{4}\right)=\frac{2}{4}+\frac{1}{4}=\frac{3}{4}$
49. $10-3=10+(-3)=7$
51. You must line up the decimal points when you subtract decimal numbers.
$1-0.07=1.00-0.07=0.93$
53. Remember to line up the decimal points.
$$7.3-(-2)=7.3+2.0=9.3$$
55. $-0.03-5=-0.03+(-5.00)=-5.03$
57. $-5+8=8-5=3$
59. $-6+(-3)=-(6+3)=-9$
61. $-80+(-40)=-(80+40)=-120$
63. $61-(-17)=61+17=78$
65. $(-12)+(-15)=-(12+15)=-27$
67. The result is negative because -20 has a larger absolute value than 13:
$$13+(-20)=-(20-13)=-7$$
69. $-102+(-99)=-(102+99)=-201$
71. $-161+(-161)=-(161+161)$
$=-322$
73. The result of this addition is negative because -16 has the larger absolute value:
$-16+0.03=-(16.00-0.03)=-15.97$
75. $0.08-3=0.08+(-3.00)$
$=-(3.00-0.08)=-2.92$
77. $-3.7+(-0.03)=-(3.70+0.03)$
$=-3.73$
79. $-2.3-(-6)=-2.3+6=6-2.3$
$=3.7$
81. $\frac{3}{4}+\left(-\frac{3}{5}\right)=\frac{3\cdot 5}{4\cdot 5}+\left(-\frac{3\cdot 4}{5\cdot 4}\right)$
$=\frac{15}{20}+\left(-\frac{12}{20}\right)=\frac{15}{20}-\frac{12}{20}=\frac{3}{20}$
83. $-\frac{1}{12}-\left(-\frac{3}{8}\right)=-\frac{2}{24}+\frac{9}{24}=\frac{7}{24}$
85. Since the sum is positive, the number in the blank must be positive, $-5+13=8$.
87. Since the sum is positive and smaller than 12, the number in the blank must be negative, $12+(-10)=2$.
89. Since the difference is negative, the number in the blank must be positive and larger than 10, $10-(14)=-4$.
91. Since the difference is positive and larger than 6, the number in the blank must be negative, $6-(-4)=6+4=10$.
93. Since the difference is negative, the number in the blank must be negative,
$-4-(-3)=-4+3=-1$.
95. $45.87+(-49.36)=-3.49$
Consult your calculator manual if you have trouble getting the answer with your calculator.
97. $0.6578+(-1)=-0.3422$
99. $-3.45-45.39=-48.84$
101. $-5.79-3.06=-8.85$
103. $97.86-27.89-42.32-25.00$
$-3.50-8=-\$8.85$
105. $5°-12°=-7°$ C
107. When adding signed numbers we add or subtract only positive numbers which are the absolute values of the original numbers. We then determine the appropriate sign for the answer. You could make up rules for adding without using the words absolute value, but the idea would have to be the same and the rules would probably be more complicated and less precise if absolute value is avoided.
109. The distance between x and y is given by either $|x-y|$ or $|y-x|$ for any numbers x and y. Try various numbers for x and y to see this.

1.4 WARM-UPS

1. True, in algebra the times symbol can be omitted between a number and a variable.
2. False, the product of -2 and 5 is -10.
3. True, the quotient is the result of division.
4. False, because $0 \div 6 = 0$.
5. True, because the quotient of two negative numbers is a positive number.
6. True, the quotient of a positive number and a negative number is a negative number.
7. True, the product of two negatives is positive.
8. False, because $(-0.2)(0.2) = -0.04$.
9. True, because
$$\left(-\tfrac{1}{2}\right) \div \left(-\tfrac{1}{2}\right) = \left(-\tfrac{1}{2}\right) \cdot \left(-\tfrac{2}{1}\right) = 1.$$
10. False, because division by zero is undefined.

1.4 EXERCISES

1. We learned to multiply and divide signed numbers.
3. To find the product of signed numbers, multiply their absolute values and then affix a negative sign if the two original numbers have opposite signs.
5. To find the quotient of nonzero numbers divide their absolute values and then affix a negative sign if the two original numbers have opposite signs.
7. $-3 \cdot 9 = -(3 \cdot 9) = -27$
9. $(-12)(-11) = 12 \cdot 11 = 132$
11. $-\frac{3}{4} \cdot \frac{4}{9} = \left(-\frac{3}{4}\right)\left(\frac{4}{3 \cdot 3}\right) = -\frac{1}{3}$
13. $0.5(-0.6) = -0.30 = -0.3$
15. $(-12)(-12) = 144$
17. $-3 \cdot 0 = 0$
19. $8 \div (-8) = -1$
21. $(-90) \div (-30) = 90 \div 30 = 3$
23. $\frac{44}{-66} = -\frac{2 \cdot 22}{3 \cdot 22} = -\frac{2}{3}$
25. $\left(-\frac{2}{3}\right) \div \left(-\frac{4}{5}\right) = \left(-\frac{2}{3}\right) \cdot \left(-\frac{5}{4}\right)$
$= \left(-\frac{2}{3}\right) \cdot \left(-\frac{5}{2 \cdot 2}\right) = \frac{5}{6}$
27. $\frac{-125}{0}$ is undefined.
29. $0 \div \left(-\frac{1}{3}\right) = 0$
31. $(40) \div (-0.5) = (400) \div (-5) = -80$
33. $-0.5 \div (-2) = 0.5(0.5) = 0.25$
35. $(25)(-4) = -(25 \cdot 4) = -100$
37. $(-3)(-9) = 27$
39. $-9 \div 3 = -3$
41. $20 \div (-5) = -4$
43. $(-6)(5) = -30$
45. $(-57) \div (-3) = 19$
47. $(0.6)(-0.3) = -0.18$
49. $(-0.03)(-10) = 0.30 = 0.3$
51. $(-0.6) \div (0.1) = (-6) \div (1) = -6$
53. $(-0.6) \div (-0.4) = 6 \div 4 = 1.5$
55. $-\frac{12}{5}\left(-\frac{55}{6}\right) = \frac{2 \cdot 6}{5} \cdot \frac{5 \cdot 11}{6} = 22$
57. $-2\frac{3}{4} \div 8\frac{1}{4} = -\frac{11}{4} \div \frac{33}{4}$
$= -\frac{11}{4} \cdot \frac{4}{33} = -\frac{1}{3}$
59. $(0.45)(-365) = -164.25$
61. $(-52) \div (-0.034) \approx 1529.41$
63. Since the product is positive, the missing number must be negative, $-5 \cdot (-12) = 60$.
65. Since the product is negative, the missing number must be negative, $12 \cdot (-8) = -96$.
67. Since the quotient is negative, the missing number must be negative, $24 \div (-6) = -4$.
69. Since the quotient is positive, the missing number must be negative, $-36 \div (-1) = 36$.
71. Since the quotient is negative, the missing number must be positive, $-40 \div (5) = -8$.
73. $(-4)(-4) = 16$
75. $-4 + (-4) = -(4 + 4) = -8$
77. $-4 + 4 = 0$
79. $-4 - (-4) = -4 + 4 = 0$
81. $0.1 - 4 = -(4.0 - 0.1) = -3.9$
83. $(-4) \div (0.1) = (-40) \div (1) = -40$
85. $(-0.1)(-4) = 0.4$
87. $|-0.4| = 0.4$
89. $\frac{-0.06}{0.3} = \frac{-6}{30} = -0.2$
91. $\frac{3}{-0.4} = -\frac{30}{4} = -7.5$
93. $-\frac{1}{5} + \frac{1}{6} = -\frac{1 \cdot 6}{5 \cdot 6} + \frac{1 \cdot 5}{6 \cdot 5} = -\frac{6}{30} + \frac{5}{30}$
$= -\frac{1}{30}$
95. $\left(-\frac{3}{4}\right)\left(\frac{2}{15}\right) = \left(-\frac{3}{2 \cdot 2}\right)\left(\frac{2}{3 \cdot 5}\right) = -\frac{1}{10}$
97. $\frac{45.37}{6} \approx 7.562$
99. $(-4.3)(-4.5) = 19.35$
101. $\frac{0}{6.345} = 0$
103. $199.4 \div 0$ is undefined.

105. If \$0 is divided among 5 people, then each person get \$0. But it is impossible to divide \$5 among no people. This illustrates why dividing 0 by a nonzero number makes sense, but division by zero is nonsense.
107. We learn multiplication before division because division is defined in terms of multiplication as $a \div b = a \cdot \frac{1}{b}$ provided $b \neq 0$.

1.5 WARM-UPS

1. False, because $(-3)^2 = (-3)(-3) = 9$.
2. False, because $5 - 3 \cdot 2 = 5 - 6 = -1$.
3. True, because we do operations within parentheses first.
4. False, because $|5 - 6| = |-1| = 1$ and $|5| - |6| = -1$.
5. False, because $5 + 6 \cdot 2 = 5 + 12 = 17$ and $(5 + 6) \cdot 2 = 11 \cdot 2 = 22$.
6. False, because $(2 + 3)^2 = 5^2 = 25$ and $2^2 + 3^2 = 4 + 9 = 13$.
7. False, because $5 - 3^3 = 5 - 27 = -22$.
8. True, because
$$(5 - 3)^3 = 2^3 = 2 \cdot 2 \cdot 2 = 8.$$
9. False, $6 - \frac{6}{2} = 6 - 3 = 3$ and $\frac{0}{2} = 0$.
10. True, because $\frac{6-6}{2} = \frac{0}{2} = 0$.

1.5 EXERCISES

1. An arithmetic expression is the result of writing numbers in a meaningful combination with the ordinary operations of arithmetic.
3. An exponential expression is an expression of the form a^n.
5. The order of operations tells us the order in which to perform operations when grouping symbols are omitted.
7. $(4 - 3)(5 - 9) = (1)(-4) = -4$
9. $|3 + 4| - |-2 - 4| = 7 - 6 = 1$
11. $\frac{7 - (-9)}{3 - 5} = \frac{16}{-2} = -8$
13. $(-6 + 5)(7) = (-1)(7) = -7$
15. $(-3 - 7) - 6 = (-10) - 6 = -16$
17. $-16 \div (8 \div 2) = -16 \div (4) = -4$
19. The exponent 4 indicates that 4 is used as a factor 4 times: $4 \cdot 4 \cdot 4 \cdot 4 = 4^4$.
21. To indicate that -5 is used as a factor 4 times we use the exponent 4.
$$(-5)(-5)(-5)(-5) = (-5)^4$$
23. The factor $-y$ appears 3 times so we use an exponent of 3: $(-y)(-y)(-y) = (-y)^3$.
25. Since $3/7$ appears as a factor 5 times,
$\frac{3}{7} \cdot \frac{3}{7} \cdot \frac{3}{7} \cdot \frac{3}{7} \cdot \frac{3}{7} = \left(\frac{3}{7}\right)^5$
27. The exponent 3 means that 5 is used as a factor 3 times, $5^3 = 5 \cdot 5 \cdot 5$.
29. The exponent 2 means that b is used as a factor 2 times, $b^2 = b \cdot b$.
31. The exponent 5 means that $-1/2$ is used as a factor 5 times,
$\left(-\frac{1}{2}\right)^5 = \left(-\frac{1}{2}\right)\left(-\frac{1}{2}\right)\left(-\frac{1}{2}\right)\left(-\frac{1}{2}\right)\left(-\frac{1}{2}\right)$.
33. The exponent 4 means that 0.22 is used as a factor 4 times,
$(0.22)^4 = (0.22)(0.22)(0.22)(0.22)$.
35. The exponent 4 means that 3 is used as a factor 4 times. So
$3^4 = 3 \cdot 3 \cdot 3 \cdot 3 = 9 \cdot 9 = 81$.
37. The exponent 9 indicates that 0 is used as a factor 9 times. So
$0^9 = 0 \cdot 0 \cdot 0 \cdot 0 \cdot 0 \cdot 0 \cdot 0 \cdot 0 \cdot 0 = 0$
39. $(-5)^4 = (-5)(-5)(-5)(-5) = 625$
41. $(-6)^3 = (-6)(-6)(-6) = -216$
43. $(10)^5 = 10 \cdot 10 \cdot 10 \cdot 10 \cdot 10 = 100{,}000$
45. $(-0.1)^3 = (-0.1)(-0.1)(-0.1) = -0.001$
47. $\left(\frac{1}{2}\right)^3 = \frac{1}{2} \cdot \frac{1}{2} \cdot \frac{1}{2} = \frac{1}{8}$
49. $\left(-\frac{1}{2}\right)^2 = \left(-\frac{1}{2}\right)\left(-\frac{1}{2}\right) = \frac{1}{4}$
51. $-8^2 = -(8 \cdot 8) = -64$
53. $-8^4 = -(8 \cdot 8 \cdot 8 \cdot 8) = -4096$
55. $-(7 - 10)^3 = -(-3)^3 = 27$
57. $(-2^2) - (3^2) = -4 - 9 = -13$
59. $3^2 \cdot 2^2 = 9 \cdot 4 = 36$
61. $-3 \cdot 2 + 4 \cdot 6 = -6 + 24 = 18$
63. $(-3)^3 + 2^3 = -27 + 8 = -19$
65. $-21 + 36 \div 3^2 = -21 + 36 \div 9$
$= -21 + 4 = -17$
67. $-3 \cdot 2^3 - 5 \cdot 2^2 = -3 \cdot 8 - 5 \cdot 4$
$= -24 - 20 = -44$
69. $\frac{-8}{2} + 2 \cdot 3 \cdot 5 - 2^3 = -4 + 30 - 8 = 18$
71. $(-3 + 4^2)(-6) = (13)(-6) = -78$
73. $(-3 \cdot 2 + 6)^3 = (-6 + 6)^3 = 0^3 = 0$

75. $2 - 5(3 - 4 \cdot 2) = 2 - 5(3 - 8)$
$= 2 - 5(-5) = 2 + 25 = 27$
77. $3 - 2 \cdot |5 - 6| = 3 - 2 \cdot |-1|$
$= 3 - 2 \cdot 1 = 1$
79. $(3^2 - 5) \cdot |3 \cdot 2 - 8|$
$= (9 - 5)|6 - 8| = (4)|-2| = 4 \cdot 2 = 8$
81. $\frac{3 - 4 \cdot 6}{7 - 10} = \frac{3 - 24}{-3} = \frac{-21}{-3} = 7$
83. $\frac{7 - 9 - 3^2}{9 - 7 - 3} = \frac{7 - 9 - 9}{2 - 3} = \frac{-11}{-1} = 11$
85. $3 + 4[9 - 6(2 - 5)] = 3 + 4[9 - 6(-3)]$
$= 3 + 4[9 - (-18)] = 3 + 4[27]$
$= 3 + 108 = 111$
87. $6^2 - [(2 + 3)^2 - 10] = 36 - [5^2 - 10]$
$= 36 - [15] = 21$
89. $4 - 5 \cdot |3 - (3^2 - 7)|$
$= 4 - 5 \cdot |3 - (2)|$
$= 4 - 5 \cdot |1| = 4 - 5 = -1$
91. $-2|3 - (7 - 3)| - |-9|$
$= -2|3 - (4)| - 9 = -2|-1| - 9$
$= -2 \cdot 1 - 9 = -2 - 9 = -11$
93. $1 + 2^3 = 1 + 8 = 9$
95. $(-2)^2 - 4(-1)(3) = 4 - (-12) = 16$
97. $4^2 - 4(1)(-3) = 16 - (-12) = 16 + 12$
$= 28$
99. $(-11)^2 - 4(5)(0) = 121 - 0 = 121$
101. $-5^2 - 3 \cdot 4^2 = -25 - 3 \cdot 16$
$= -25 - 48 = -73$
103. $[3 + 2(-4)]^2 = [3 + (-8)]^2 = [-5]^2$
$= 25$
105. $|-1| - |-1| = 1 - 1 = 0$
107. $\frac{4 - (-4)}{-2 - 2} = \frac{8}{-4} = -2$
109. $3(-1)^2 - 5(-1) + 4 = 3 \cdot 1 - (-5) + 4$
$= 3 + 5 + 4 = 12$
111. $5 - 2^2 + 3^4 = 5 - 4 + 81 = 82$
113. $-2 \cdot |9 - 6^2| = -2 \cdot |9 - 36|$
$= -2 \cdot 27 = -54$
115. $-3^2 - 5[4 - 2(4 - 9)]$
$= -9 - 5[4 - 2(-5)]$
$= -9 - 5[4 + 10] = -9 - 5[14] = -79$
117. $1 - 5|5 - (9 + 1)|$
$= 1 - 5|5 - (10)| = 1 - 5|-5|$
$= 1 - 5 \cdot 5 = 1 - 25 = -24$
119. $3.2^2 - 4(3.6)(-2.2) = 41.92$
121. $(5.63)^3 - [4.7 - (-3.3)^2]$
$= 184.643547$
123. $\frac{3.44 - (-8.32)}{6.89 - 5.43} \approx 8.0548$
125. a) $294.4(1.0105)^{11} \approx 330.2$
Population will be 330.2 million in 2015.
b) From the graph it appears that the population will be 350 million in around 2022.
127. The expressions $(-5)^3$, $-(5^3)$, -5^3, and $-1 \cdot 5^3$ have value -125. The value $-(-5)^3$ is 125.

1.6 WARM-UPS

1. True, because in $2x + 3y$ the last operation to be performed is addition.
2. False, because in $5(y - 9)$ the last operation to be performed is multiplication.
3. True, because the last operation to be performed is multiplication.
4. False, because the last operation to be performed is addition.
5. True, because the last operation to be performed is multiplication.
6. False, because $2(-2) + 4 = 0$.
7. False, $(-3)^3 - 5 = -27 - 5 = -32$.
8. False, because $2(5) - 3 = 13$ is incorrect.
9. True, because product means multiplication.
10. False, because $2(x + 7)$ is read as the product of 2 and $x + 7$.

1.6 EXERCISES

1. An algebraic expression is the result of combining numbers and variables with the operations of arithmetic in some meaningful way.
3. An algebraic expression is named according the last operation to be performed.
5. An equation is a sentence that expresses equality between two algebraic expressions.
7. Since subtraction is performed last in $a^3 - 1$, the expression is a difference.
9. In the expression $(w - 1)^3$, the subtraction is done first and the exponent is evaluated last. So the expression is called a cube.
11. In the expression $3x + 5y$, multiplication is done first and addition last. So the expression is called a sum.
13. Since the subtraction is performed after the division, the expression is a difference.

15. In the expression $3(x+5y)$ the last operation is multiplication. So the expression is a product.
17. In the expression squaring is done last. So the expression is a square.
19. The difference of x^2 and a^2
21. The square of $x-a$
23. The quotient of $x-4$ and 2
25. The difference of $x/2$ and 4
27. The cube of ab
29. The sum of 8 and y is written as $8+y$.
31. The product of $5x$ and z is written as $5x+z$.
33. The difference of 8 and $7x$ is written as $8-7x$.
35. The quotient of 6 and $x+4$ is written as $\frac{6}{x+4}$.
37. The square of $a+b$ is written as $(a+b)^2$.
39. The sum of the cube of x and the square of y is written as x^3+y^2.
41. The product of 5 and the square of m is written as $5m^2$.
43. The square of the sum of s and t is written as $(s+t)^2$.
45. Use $a=-1$ and $b=2$ in the expression.
$$-(a-b)=-(-1-2)=-(-3)=3$$
47. Use $b=2$ in the expression.
$$-2^2+7=-4+7=3$$
49. Use $c=-3$ in the expression.
$$c^2-2c+1=(-3)^2-2(-3)+1$$
$$=9-(-6)+1=9+6+1=16$$
51. Use $a=-1$ and $b=2$ in the expression.
$$a^3-b^3=(-1)^3-(2)^3=-1-8=-9$$
53. Use $a=-1$ and $b=2$ in the expression.
$$(a-b)(a+b)=(-1-2)(-1+2)$$
$$=(-3)(1)=-3$$
55. Use $a=-1$, $b=2$, and $c=-3$:
$$b^2-4ac=2^2-4(-1)(-3)$$
$$=4-12=-8$$
57. Use $a=-1$, $b=2$, and $c=-3$:
$$\frac{a-c}{a-b}=\frac{-1-(-3)}{-1-2}=\frac{2}{-3}=-\frac{2}{3}$$
59. Use $a=-1$, $b=2$, and $c=-3$:
$$\frac{2}{a}+\frac{6}{b}-\frac{9}{c}=\frac{2}{-1}+\frac{6}{2}-\frac{9}{-3}$$
$$=-2+3-(-3)=4$$
61. Use $a=-1$ in the expression.
$$a\div|-a|=-1\div|-(-1)|$$
$$=-1\div1=-1$$
63. Use $a=-1$ and $b=2$ in the expression.
$$|b|-|a|=|2|-|-1|$$
$$=2-1=1$$
65. Use $a=-1$ and $c=-3$ in the expression.
$$-|-a-c|=-|-(-1)-(-3)|$$
$$=-|1+3|=-|4|=-4$$
67. Use $a=-1$ and $b=2$ in the expression.
$$(3-|a-b|)^2=(3-|-1-2|)^2$$
$$=(3-|-3|)^2=(3-3)^2=(0)^2=0$$
69. Replace x by 2 in the equation to see if the equation is satisfied.
$$3x+7=13$$
$$3(2)+7=13$$
$$6+7=13 \quad \text{Correct}$$
So 2 is a solution to the equation.
71. Replace x by -2 in the equation.
$$\frac{3x-4}{2}=5$$
$$\frac{3(-2)-4}{2}=5$$
$$\frac{-10}{2}=5 \quad \text{Incorrect.}$$
So -2 is not a solution to the equation.
73. Replace x by -2 in the equation.
$$-x+4=6$$
$$-(-2)+4=6$$
$$2+4=6 \quad \text{Correct}$$
The number -2 satisfies the equation.
75. Replace x by 4 in the equation.
$$3x-7=x+1$$
$$3(4)-7=4+1$$
$$5=5 \quad \text{Correct}$$
The number 4 satisfies the equation.
77. Replace x by 3 in the equation.
$$-2(x-1)=2-2x$$
$$-2(3-1)=2-2(3)$$
$$-4=-4 \quad \text{Correct}$$
The number 3 satisfies the equation.
79. Replace x by 1 in the equation.
$$x^2+3x-4=0$$
$$1^2+3(1)-4=0$$
$$1+3-4=0 \quad \text{Correct}$$
The number 1 satisfies the equation.
81. Replace x by 8 in the equation.
$$\frac{x}{x-8}=0$$
$$\frac{8}{8-8}=0$$
$$\frac{8}{0}=0 \quad \text{Incorrect}$$

Since $8/0$ is an undefined expression, 8 does not satisfy this equation.

83. Replace x by -6 in the equation.

$$\frac{x+6}{x+6}=1$$
$$\frac{-6+6}{-6+6}=1$$
$$\frac{0}{0}=1 \text{ Incorrect}$$

Since $0/0$ is undefined, -6 is not a solution to the equation.

85. The sum of $5x$ and $3x$ is $5x+3x$. So we have $5x+3x=8x$.

87. Product means multiplication. So $3(x+2)=12$.

89. Quotient means division. So $\frac{x}{3}=5x$.

91. The sum of a and b is $a+b$. The square of the sum is $(a+b)^2$. So $(a+b)^2=9$.

93.

x	$2x-3$
-2	-7
-1	-5
0	-3
1	-1
2	1

95.

a	a^2	a^3	a^4
2	4	8	16
$\frac{1}{2}$	$\frac{1}{4}$	$\frac{1}{8}$	$\frac{1}{16}$
10	100	1000	10,000
0.1	0.01	0.001	0.0001

97. $b^2-4ac=(6.7)^2-4(4.2)(1.8)=14.65$

99. $b^2-4ac=(3.2)^2-4(-1.2)(5.6)$
$=37.12$

101. Replace T by 36.5:
$81.7+2.4(36.5)=169.3$
The height of the person is about 169.3 cm. From the graph, a male with height 180 cm has a tibia with a length of about 41 cm.

103. Boston
$$\text{GB}=\frac{(101-95)+(67-61)}{2}=6$$
Toronto
$$\text{GB}=\frac{(101-86)+(76-61)}{2}=15$$
Baltimore
$$\text{GB}=\frac{(101-71)+(91-61)}{2}=30$$
Tampa Bay
$$\text{GB}=\frac{(101-63)+(99-61)}{2}=38$$

105. Since 100 yds = 300 ft,
$2L+2W=2(300)+2(160)=920$ feet

107. Answers will vary. For the square of the sum consider $(2+3)^2=5^2=25$. For the sum of the squares consider $2^2+3^2=4+9=13$. So $(2+3)^2\neq 2^2+3^2$.

1.7 WARM-UPS

1. False, because $24\div(4\div 2)=24\div 2=12$ and $(24\div 4)\div 2=6\div 2=3$.

2. False, because $1\div 2=0.5$ and $2\div 1=2$.

3. True, because $6-5=1$ and $-5+6=1$.

4. False, because $9-(4-3)=9-1=8$ and $(9-4)-3=5-3=2$.

5. True, because of the commutative property of multiplication.

6. True, because of the distributive property.

7. True, because $(0.02)(50)=1$.

8. True, because of the distributive property.

9. True, because of the distributive property.

10. True, because $0+0=0$.

1.7 EXERCISES

1. The commutative property says that $a+b=b+a$ and the associative property says that $(a+b)+c=a+(b+c)$.

3. Factoring is the process of writing an expression or number as a product.

5. The properties help us to understand the operations and how they are related to each other.

7. $9+r=r+9$

9. $3(2+x)=3(x+2)$

11. Since $4-5x=4+(-5x)$, we can write $4-5x=-5x+4$.

13. $x\cdot 6=6x$.

15. $(x-4)(-2)=-2(x-4)$
Note that the commutative property of multiplication only allows us to rewrite the multiplication part of the expression.

17. By the commutative property of multiplication we have $4-y\cdot 8=4-8y$.

19. $(4w)(w)=4(w\cdot w)=4w^2$

21. $3a(ba)=3a(ab)=(3a\cdot a)b=(3a^2)b$
$=3a^2b$

23. $(x)(9x)(xz) = 9 \cdot x \cdot x \cdot x \cdot z = 9x^3z$
25. $8 - 4 + 3 - 10 = 11 - 14 = -3$
27. $8 - 10 + 7 - 8 - 7 = 15 - 25 = -10$
29. $-4 - 11 + 7 - 8 + 15 - 20$
$= -43 + 22 = -21$
31. $-3.2 + 2.4 - 2.8 + 5.8 - 1.6$
$= -7.6 + 8.2 = 0.6$
33. $3.26 - 13.41 + 5.1 - 12.35 - 5$
$= 8.36 - 30.76 = -22.4$
35. $3(x - 5) = 3x - 3 \cdot 5 = 3x - 15$
37. $a(2 + t) = a2 + at = 2a + at$
39. $-3(w - 6) = -3w - (-3)(6)$
$= -3w - (-18) = -3w + 18$
41. $-4(5 - y) = -4(5) - (-4y)$
$= -20 + 4y$
43. $-1(a - 7) = -1a - (-1)(7)$
$= -a - (-7) = -a + 7$
45. $-1(t + 4) = -1(t) + (-1)(4)$
$= -t + (-4) = -t - 4$
47. $2m + 12 = 2m + 2 \cdot 6 = 2(m + 6)$
49. $4x - 4 = 4x - 4 \cdot 1 = 4(x - 1)$
51. $4y - 16 = 4y - 4 \cdot 4 = 4(y - 4)$
53. $4a + 8 = 4a + 4 \cdot 2 = 4(a + 2)$
55. $x + xy = x \cdot 1 + xy = x(1 + y)$
57. $6a - 2b = 2 \cdot 3a - 2b = 2(3a - b)$
59. The multiplicative inverse of $\frac{1}{2}$ is 2 because $\frac{1}{2} \cdot 2 = 1$.
61. The multiplicative inverse of -5 is $-\frac{1}{5}$ because $-5 \cdot \left(-\frac{1}{5}\right) = 1$.
63. The multiplicative inverse of 7 is $\frac{1}{7}$ because $7 \cdot \frac{1}{7} = 1$.
65. The multiplicative inverse of 1 is 1 because $1 \cdot 1 = 1$.
67. The multiplicative inverse of -0.25 is -4 because $-0.25(-4) = 1$.
69. The multiplicative inverse of 2.5 is $\frac{2}{5}$ because $2.5 \cdot \frac{2}{5} = \frac{5}{2} \cdot \frac{2}{5} = 1$.
71. The statement $3 \cdot x = x \cdot 3$ is true because of the commutative property of multiplication.
73. The statement $2(x - 3) = 2x - 6$ is correct because of the distributive property.
75. The statement $-3(xy) = (-3x)y$ is correct because of the associative property of multiplication.
77. The statement $4 + (-4) = 0$ is true because of the additive inverse property.
79. The statement $x^2 \cdot 5 = 5x^2$ is correct because of the commutative property of multiplication.
81. The statement $1 \cdot 3y = 3y$ is correct because of the multiplicative identity property. One is the multiplicative identity.
83. The statement $2a + 5a = (2 + 5)a$ is correct because of the distributive property.
85. The statement $-7 + 7 = 0$ is correct because of the additive inverse property. The numbers 7 and -7 are additive inverses of each other.
87. The statement $(2346)0 = 0$ is correct because of the multiplication property of 0.
89. The statement $ay + y = y(a + 1)$ is correct because of the distributive property.
91. We can use the commutative property to reverse the addition and write $a + y = y + a$.
93. We can use the associative property to move the parentheses and write $5(aw) = (5a)w$.
95. Use the distributive property to write $\frac{1}{2}x + \frac{1}{2} = \frac{1}{2}x + \frac{1}{2} \cdot 1 = \frac{1}{2}(x + 1)$.
97. Use the distributive property to write $6x + 15 = 3 \cdot 2x + 3 \cdot 5 = 3(2x + 5)$.
99. Since 4 and 0.25 are multiplicative inverses of each other, we can write $4(0.25) = 1$.
101. By the multiplication property of 0, we can write $0 = 96(0)$.
103. The multiplicative inverse of 0.33 or $\frac{33}{100}$ is $\frac{100}{33}$. So we can write $0.33\left(\frac{100}{33}\right) = 1$.
105. a) $\frac{1}{0.04}$ bricks/hr + $\frac{1}{0.05}$ bricks/hr
= 45 bricks/hour
b) The bricklayer at 25 bricks/hour is working faster than the apprentice at 20 bricks/hour.
107. a) $\frac{1 \text{ person}}{0.4308 \text{ sec}} \approx 2.3213$ people/sec
b) $\frac{2.321262767 \text{ people}}{1 \text{ sec}} \times \frac{604{,}800 \text{ sec}}{\text{week}}$
$\approx 1{,}403{,}900$ people/week
109. Use the distributive property to write $P = 2L + 2W = 2(L + W)$. In words, the perimeter is twice the sum of the length and width.

111. a) Since you can put on your coat and hat in either order, the tasks are commutative.
b) Since you cannot put on your shirt and your coat in either order, the tasks are not commutative

1.8 WARM-UPS

1. True, because of the distributive property.
2. False, because if $x = 1$ then the equation $-3x + 9 = -3(x + 9)$ becomes $6 = -30$, which is incorrect.
3. True, because of the distributive property.
4. True, by combining like terms.
5. False, because if $a = 2$ then $(3a)(4a) = 12a$ becomes $48 = 24$.
6. False, $3(5 \cdot 2) = 30$ and $15 \cdot 6 = 90$.
7. False, because if $x = 1$ then $x + x = x^2$ becomes $2 = 1$.
8. False, because if $x = 1$ then $x \cdot x = 2x$ becomes $1 = 2$.
9. False, because if $x = 2$ then $3 + 2x = 5x$ becomes $7 = 10$.
10. True, because of the distributive property.

1.8 EXERCISES

1. Like terms are terms with the same variables and exponents.
3. We can add or subtract like terms.
5. If a negative sign precedes a set of parentheses, then signs for all terms in the parentheses are changed when the parentheses are removed.
7. $35(200) = (35 \cdot 2)100 = 70 \cdot 100 = 7000$
9. $\frac{4}{3}(0.75) = \frac{4}{3} \cdot \frac{3}{4} = 1$
11. $(256 + 78) + 22 = 256 + (78 + 22)$
$= 256 + 100 = 356$
13. $35 \cdot 3 + 35 \cdot 7 = 35(3 + 7) = 35(10)$
$= 350$
15. $18 \cdot 4 \cdot 2 \cdot \frac{1}{4} = 18 \cdot 2 \cdot 4 \cdot \frac{1}{4} = 18 \cdot 2 \cdot 1$
$= 36$
17. $(120)(300) = (120 \cdot 3)100 = 360 \cdot 100$
$= 36{,}000$
19. $12 \cdot 375(-6 + 6) = 12 \cdot 375(0) = 0$
21. $78 + 6 + 8 + 4 + 2$
$= 78 + (8 + 2) + (6 + 4)$
$= 78 + 10 + 10 = 98$
23. $5w + 6w = (5 + 6)w = 11w$
25. $4x - x = 4x - 1x = 3x$
27. $2x - (-3x) = 2x + 3x = 5x$
29. $-3a - (-2a) = -3a + 2a = -1a = -a$
31. $-a - a = -1a - 1a = (-1 - 1)a$
$= -2a$
33. The terms in $10 - 6t$ are not like terms. So they cannot be combined.
35. $3x^2 + 5x^2 = (3 + 5)x^2 = 8x^2$
37. The terms in $-4x + 2x^2$ are not like terms and so they cannot be combined.
39. $5mw^2 - 12mw^2 = (5 - 12)mw^2$
$= -7mw^2$
41. $\frac{1}{3}a + \frac{1}{2}a = \left(\frac{1}{3} + \frac{1}{2}\right)a = \left(\frac{2}{6} + \frac{3}{6}\right)a$
$= \frac{5}{6}a$
43. $3(4h) = (3 \cdot 4)h = 12h$
45. $6b(-3) = -18b$
47. $(-3m)(3m) = -3 \cdot 3 \cdot m \cdot m = -9m^2$
49. $(-3d)(-4d) = (-3)(-4)d \cdot d = 12d^2$
51. $(-y)(-y) = (-1y)(-1y) = (-1)(-1)yy$
$= y^2$
53. $-3a(5b) = (-3 \cdot 5)ab = -15ab$
55. $-3a(2 + b) = -3a(2) + (-3a)(b)$
$= -6a - 3ab$
57. $-k(1 - k) = -k(1) - (-k)(k)$
$= -k + k^2$
59. $\frac{3y}{3} = \frac{1}{3}(3y) = \left(\frac{1}{3} \cdot 3\right)y = 1 \cdot y = y$
61. $\frac{-15y}{5} = \frac{1}{5}(-15)y = -3y$
63. $2\left(\frac{y}{2}\right) = 2 \cdot \frac{1}{2} \cdot y = 1 \cdot y = y$
65. $8y\left(\frac{y}{4}\right) = \frac{8}{4}yy = 2y^2$
67. $\frac{6a - 3}{3} = \frac{1}{3}(6a - 3) = 2a - 1$
69. $\frac{-9x + 6}{-3} = -\frac{1}{3}(-9x + 6) = 3x - 2$
71. $x - (3x - 1) = x - 3x + 1 = -2x + 1$
73. $5 - (y - 3) = 5 - y + 3 = 8 - y$
75. $2m + 3 - (m + 9) = 2m + 3 - m - 9$
$= 2m - m + 3 - 9 = m - 6$
77. $-3 - (-w + 2) = -3 + w - 2 = w - 5$
79. $3x + 5x + 6 + 9 = 8x + 15$
81. $(-2x + 3) + (7x - 4) = 7x - 2x + 3 - 4$
$= 5x - 1$

83. $3a - 7 - (5a - 6) = 3a - 7 - 5a + 6$
$= 3a - 5a - 7 + 6 = -2a - 1$

85. $2(a - 4) - 3(-2 - a)$
$= 2a - 8 + 6 + 3a$
$= 2a + 3a - 8 + 6 = 5a - 2$

87. $3x(2x - 3) + 5(2x - 3)$
$= 6x^2 - 9x + 10x - 15 = 6x^2 + x - 15$

89. $-b(2b - 1) - 4(2b - 1)$
$= -2b^2 + b - 8b + 4 = -2b^2 - 7a + 4$

91. $-5m + 6(m - 3) + 2m$
$= -5m + 6m - 18 + 2m = 3m - 18$

93. $5 - 3(x + 2) - 6 = 5 - 3x - 6 - 6$
$= -3x - 7$

95. $x - 0.05(x + 10) = x - 0.05x - 0.5$
$= 1.00x - 0.05x - 0.5 = 0.95x - 0.5$

97. $3x - (4 - x) = 3x - 4 + x = 4x - 4$

99. $y - 5 - (-y - 9) = y - 5 + y + 9$
$= 2y + 4$

101. $7 - (8 - 2y - m) = 7 - 8 + 2y + m$
$= 2y + m - 1$

103. $\frac{1}{2}(10 - 2x) + \frac{1}{3}(3x - 6)$
$= 5 - x + x - 2 = 3$

105. $\frac{1}{2}(3a + 1) - \frac{1}{3}(a - 5)$
$= \frac{3}{2}a + \frac{1}{2} - \frac{1}{3}a + \frac{5}{3}$
$= \left(\frac{3}{2} - \frac{1}{3}\right)a + \frac{1}{2} + \frac{5}{3}$
$= \frac{7}{6}a + \frac{13}{6}$

107. $0.2(x + 3) - 0.05(x + 20)$
$= 0.2x + 0.6 - 0.05x - 1$
$= 0.15x - 0.4$

109. $2k + 1 - 3(5k - 6) - k + 4$
$= 2k + 1 - 15k + 18 - k + 4$
$= -14k + 23$

111. $-3m - 3[2m - 3(m + 5)]$
$= -3m - 3[-m - 15]$
$= -3m + 3m + 45 = 45$

113. a) $7820 + 0.25(x - 56{,}800)$
$= 0.25x - 6380$

b) $0.25(80{,}000) - 6380 = \$13{,}620$

c) From the graph, an income of \$200,000 pays approximately \$48,000 in tax.

d) A couple who paid \$80,000 in tax had an income of approximately \$300,000.

115. $2(x) + 2(x + 40) = 4x + 80$
If $x = 30$, then $4(30) + 80 = 200$ feet.

117. If $x = 5$, then $1/2 \cdot 5 = \frac{1}{2} \cdot 5 = 2.5$ because we do division and multiplication from left to right. To get the answer 0.1 you would have to multiply before dividing. If the expression was written $1/(2x)$ then the correct value would be 0.1.

Enriching Your Mathematical Word Power

1. c	**2.** b	**3.** a	**4.** d	**5.** b
6. d	**7.** a	**8.** d	**9.** c	**10.** a

CHAPTER 1 REVIEW

1. Of the numbers listed, the numbers 0, 1, 2, and 10 are whole numbers.

3. Of the numbers listed, the numbers -2, 0, 1, 2, and 10 are integers.

5. The only irrational numbers listed are $-\sqrt{5}$ and π. Note that any terminating decimal is rational.

7. True, because every whole number can be expressed as a ratio of integers. For example, $9 = 9/1$.

9. False, because the counting numbers between -4 and 4 are 1, 2, and 3.

11. False, because the national debt is finite.

13. True, because the integers greater than -1 are 0, 1, 2, 3, and so on.

15. The set of integers between -3 and 3 is $\{-2, -1, 0, 1, 2\}$. It is graphed as follows.

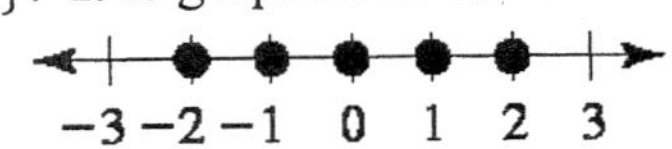

17. The set of real numbers between -1 and 4 is the interval $(-1, 4)$. It is graphed as follows.

−2 −1 0 1 2 3 4 5

19. The set of real numbers between 4 and 6 inclusive includes the endpoints 4 and 6. It is written in interval notation as $[4, 6]$.

21. The set of real numbers greater than or equal to -30 includes the endpoint -30. It is written in interval notation as $[-30, \infty)$.

23. $\frac{1}{3} + \frac{3}{8} = \frac{1 \cdot 8}{3 \cdot 8} + \frac{3 \cdot 3}{8 \cdot 3} = \frac{8}{24} + \frac{9}{24}$
$= \frac{17}{24}$

25. $\frac{3}{5} \cdot 10 = \frac{3}{5} \cdot 5 \cdot 2 = 6$

27. $\frac{2}{5} \cdot \frac{15}{14} = \frac{2}{5} \cdot \frac{3 \cdot 5}{2 \cdot 7} = \frac{3}{7}$

29. $4 + \frac{2}{3} = 4 \cdot \frac{3}{3} + \frac{2}{3} = \frac{12}{3} + \frac{2}{3} = \frac{14}{3}$

31. $\frac{1}{2} + \frac{1}{3} + \frac{1}{4} = \frac{1 \cdot 6}{2 \cdot 6} + \frac{1 \cdot 4}{3 \cdot 4} + \frac{1 \cdot 3}{4 \cdot 3}$

$= \frac{6}{12} + \frac{4}{12} + \frac{3}{12} = \frac{13}{12}$

33. $-5 + 7 = 7 - 5 = 2$

35. $35 - 48 = -(48 - 35) = -(13) = -13$

37. $-12 + 5 = -(12 - 5) = -(7) = -7$

39. $-12 - (-5) = -12 + 5 = -7$

41. $-0.05 + 12 = 12.00 - 0.05 = 11.95$

43. $-0.1 - (-0.05) = -0.1 + 0.05$

$= -0.10 + 0.05 = -0.05$

45. $\frac{1}{3} - \frac{1}{2} = \frac{1 \cdot 2}{3 \cdot 2} - \frac{1 \cdot 3}{2 \cdot 3} = \frac{2}{6} - \frac{3}{6} = -\frac{1}{6}$

47. $-\frac{1}{3} + \left(-\frac{2}{5}\right) = -\frac{1 \cdot 5}{3 \cdot 5} + \left(-\frac{2 \cdot 3}{5 \cdot 3}\right)$

$= -\frac{5}{15} + \left(-\frac{6}{15}\right) = -\frac{11}{15}$

49. $(-3)(5) = -15$

51. $(-8) \div (-2) = 4$

53. $\frac{-20}{-4} = 5$

55. $\left(-\frac{1}{2}\right)\left(-\frac{1}{3}\right) = \frac{1}{6}$

57. $-0.09 \div 0.3 = -0.9 \div 3 = -0.3$

59. $(0.3)(-0.8) = -0.24$

61. $(-5)(-0.2) = 1$

63. $3 + 7(9) = 3 + 63 = 66$

65. $(3 + 4)^2 = 7^2 = 49$

67. $3 + 2 \cdot |5 - 6 \cdot 4|$

$= 3 + 2 \cdot |5 - 24|$

$= 3 + 2 \cdot |-19| = 3 + 2 \cdot 19$

$= 3 + 38 = 41$

69. $(3 - 7) - (4 - 9) = (-4) - (-5)$

$= -4 + 5 = 1$

71. $-2 - 4(2 - 3 \cdot 5) = -2 - 4(2 - 15)$

$= -2 - 4(-13) = -2 + 52 = 50$

73. $3^2 - (7 + 5)^2 = 9 - 12^2 = -135$

75. $\frac{-3 - 5}{2 - (-2)} = \frac{-8}{2 + 2} = \frac{-8}{4} = -2$

77. $\frac{6 + 3}{3} - 5 \cdot 4 + 1 = 3 - 20 + 1 = -16$

79. $b^2 - 4ac = (-2)^2 - 4(-1)(3)$

$= 4 - (-12) = 4 + 12 = 16$

81. $(c - b)(c + b) = (3 - (-2))(3 + (-2))$

$= (3 + 2)(3 - 2) = (5)(1) = 5$

83. $a^2 + 2ab + b^2$

$= (-1)^2 + 2(-1)(-2) + (-2)^2$

$= 1 + 4 + 4 = 9$

85. $a^3 - b^3 = (-1)^3 - (-2)^3 = -1 - (-8)$

$= -1 + 8 = 7$

87. $\frac{b + c}{a + b} = \frac{-2 + 3}{-1 + (-2)} = \frac{1}{-3} = -\frac{1}{3}$

89. $|a - b| = |-1 - (-2)|$

$= |-1 + 2| = |1| = 1$

91. $(a + b)c = (-1 + (-2))(3) = -9$

93. Since $3(4) - 2 = 10$ is correct, 4 is a solution to the equation.

95. $\frac{3(-6)}{2} = 9$

$\frac{-18}{2} = 9$ Incorrect.

So -6 is not a solution to the equation.

97. $\frac{15 + 3}{2} = 9$

$\frac{18}{2} = 9$ Correct.

So 15 is a solution to the equation.

99. $-4 - 3 = 1$

$-7 = 1$ Incorrect.

So 4 is not a solution to the equation.

101. The statement $a(x + y) = ax + ay$ is correct because of the distributive property.

103. The statement $(0.001)(1000) = 1$ is correct because of the multiplicative inverse property.

105. The statement $0 + y = y$ is correct because of the additive identity property. Zero is the additive identity.

107. The statement $3 + (2 + x) = (3 + 2) + x$ is correct because of the associative property of addition.

109. The statement $5 \cdot 200 = 200 \cdot 5$ is correct because of the commutative property of multiplication.

111. The statement $-50 + 50 = 0$ is correct because of the additive inverse property. The numbers -50 and 50 are additive inverses of each other.

113. The statement $12 \cdot 1 = 12$ is correct because of the multiplicative identity property. One is the multiplicative identity.

115. $3a + 7 - (4a - 5) = 3a + 7 - 4a + 5$

$= -a + 12$

117. $2a(3a - 5) + 4a = 6a^2 - 10a + 4a$

$= 6a^2 - 6a$

119. $3(t - 2) - 5(3t - 9)$

$= 3t - 6 - 15t + 45 = -12t + 39$

121. $0.1(a + 0.3) - (a + 0.6)$

$= 0.1a + 0.03 - a - 0.6$

$= -0.9a - 0.57$

123. $0.05(x-20)-0.1(x+30)$
$= 0.05x - 1 - 0.1x - 3 = -0.05x - 4$

125. $5 - 3x(-5x-2) + 12x^2$
$= 5 + 15x^2 + 6x + 12x^2$
$= 27x^2 + 6x + 5$

127. $-(a-2) - 2 - a$
$= -a + 2 - 2 - a = -2a$

129. $x(x+1) + 3(x-1) = x^2 + x + 3x - 3$
$= x^2 + 4x - 3$

131. $752(-13) + 752(13) = 752(-13+13)$
$= 752(0) = 0$

133. $|15-23| = |-8| = 8$

135. $-6^2 + 3(5) = -36 + 15 = -21$

137. $\frac{2}{5} + \frac{1}{10} = \frac{2\cdot 2}{5\cdot 2} + \frac{1}{10} = \frac{4}{10} + \frac{1}{10}$
$= \frac{5}{10} = \frac{1}{2}$

139. $(0.05) \div (-0.1) = (0.5) \div (-1) = -0.5$

141. $2\left(-\frac{1}{2}\right)^2 + \left(-\frac{1}{2}\right) - 1 = 2\cdot\frac{1}{4} - \frac{1}{2} - 1$
$= \frac{2}{4} - \frac{1}{2} - 1 = \frac{1}{2} - \frac{1}{2} - 1 = -1$

143. $\frac{2x+4}{2} = \frac{1}{2}(2x+4) = x + 2$

145. The terms in $4 + 2x$ cannot be combined because they are not like terms.

147. $4 \cdot \frac{x}{2} = 2 \cdot 2 \cdot \frac{x}{2} = 2x$

149. $-4(x-2) = -4x - (-8) = -4x + 8$

151. $4x + 2x = (4+2)x = 6x$

153. $4 \cdot \frac{x}{4} = x$

155. $2 \cdot x \cdot 4 = 2 \cdot 4 \cdot x = 8x$

157. $2(x-4) - x(x-4) = 2x - 8 - x^2 + 4x$
$= -x^2 + 6x - 8$

159. $\frac{1}{2}(x-4) - \frac{1}{4}(x-2)$
$= \frac{1}{2}x - 2 - \frac{1}{4}x + \frac{1}{2} = \frac{1}{4}x - \frac{3}{2}$

161. Evaluate $-\frac{1}{3}x + 1$ for $x = -6, -3, 0, 3, 6$ to get the second column of the table:

x	$-\frac{1}{3}x + 1$
-6	3
-3	2
0	1
3	0
6	-1

163. Evaluate a^2, a^3, and a^4 for $a = 5$ and $a = -4$ to get the table values:

a	a^2	a^3	a^4
5	25	125	625
-4	16	-64	256

165. $\frac{1 \text{ membership}}{0.125 \text{ hours}} + \frac{1 \text{ membership}}{0.1 \text{ hours}}$
$= 8 \text{ mbs/hr} + 10 \text{ mbs/hr}$
$= 18 \text{ memberships/hr}$

CHAPTER 1 TEST

1. The only whole numbers listed are 0 and 8.

2. The only integers listed are -3, 0, and 8.

3. The rational numbers listed are -3, $-\frac{1}{4}$, 0, and 8.

4. The irrational numbers are the real numbers that are not rational. So the irrational numbers listed are $-\sqrt{3}$, $\sqrt{5}$, and π.

5. $6 + 3(-9) = 6 + (-27) = -21$

6. $(-2)^2 - 4(-2)(-1) = 4 - 8 = -4$

7. $\frac{-3^2 - 9}{3 - 5} = \frac{-9-9}{-2} = \frac{-18}{-2} = 9$

8. $-5 + 6 - 12 + 4 = -17 + 10 = -7$

9. $0.05 - 1 = -(1.00 - 0.05) = -0.95$

10. $(5-9)(5+9) = (-4)(14) = -56$

11. $(878 + 89) + 11 = 878 + (89 + 11)$
$= 878 + 100 = 978$

12. $6 + |3 - 5(2)| = 6 + |3 - 10|$
$= 6 + |-7| = 6 + 7 = 13$

13. $8 - 3|7 - 10| = 8 - 3|-3|$
$= 8 - 3\cdot 3 = 8 - 9 = -1$

14. $(839 + 974)[3(-4) + 12]$
$= (839 + 974)[0] = 0$

15. $974(7) + 974(3) = 974(10) = 9740$

16. $-\frac{2}{3} + \frac{3}{8} = -\frac{2\cdot 8}{3\cdot 8} + \frac{3\cdot 3}{8\cdot 3} = -\frac{16}{24} + \frac{9}{24}$
$= -\frac{7}{24}$

17. $(-0.05)(400) = -20$

18. $\left(-\frac{3}{4}\right)\left(\frac{2}{9}\right) = \left(-\frac{3}{2\cdot 2}\right)\left(\frac{2}{3\cdot 3}\right) = -\frac{1}{6}$

19. $13 \div \left(-\frac{1}{3}\right) = 13 \cdot (-3) = -39$

20. The set of whole numbers less than 5 is $\{0, 1, 2, 3, 4\}$ and it is graphed as follows:

21. The set of real numbers less than or equal to 4 is the interval $(-\infty, 4]$, which includes 4, and it is graphed as follows:

−1 0 1 2 3 4 5

22. The real numbers greater than 2 does not include 2. It is written in interval notation as $(2, \infty)$.

23. The real numbers greater than or equal to 3 and less than 9 includes 3 but not 9. It is written in interval notation as $[3, 9)$.

24. The statement $2(x+7) = 2x + 14$ is correct because of the distributive property.

25. $48 \cdot 1000 = 1000 \cdot 48$ is correct because of the commutative property of multiplication.

26. $2 + (6 + x) = (2 + 6) + x$ is correct because of the associative property of addition.

27. $-348 + 348 = 0$ is correct because of the additive inverse property.

28. $1 \cdot (-6) = -6$ is correct because of the multiplicative identity property.

29. The statement $0 \cdot 388 = 0$ is correct because of the multiplication property of 0.

30. $3x + 30 = 3x + 3 \cdot 10 = 3(x + 10)$

31. $7w - 7 = 7w - 7 \cdot 1 = 7(w - 1)$

32. $6 + 4x + 2x = 6x + 6$

33. $6 + 4(x - 2) = 6 + 4x - 8 = 4x - 2$

34. $5x - (3 - 2x) = 5x - 3 + 2x = 7x - 3$

35. $x + 10 - 0.1(x + 25)$
$= x + 10 - 0.1x - 2.5$
$= 0.9x + 7.5$

36. $2a(4a - 5) - 3a(-2a - 5)$
$= 8a^2 - 10a + 6a^2 + 15a = 14a^2 + 5a$

37. $\frac{6x + 12}{6} = \frac{1}{6}(6x + 12)$
$= \frac{1}{6} \cdot 6x + \frac{1}{6} \cdot 12 = x + 2$

38. $8 \cdot \frac{t}{2} = 4 \cdot 2 \cdot \frac{t}{2} = 4t$

39. $(-9xy)(-6xy) = 54x^2y^2$

40. $\frac{1}{2}(3x + 2) - \frac{1}{4}(3x - 2)$
$= \frac{3}{2}x + 1 - \frac{3}{4}x + \frac{1}{2}$
$= \frac{6}{4}x - \frac{3}{4}x + \frac{2}{2} + \frac{1}{2} = \frac{3}{4}x + \frac{3}{2}$

41. $b^2 - 4ac = 3^2 - 4(-2)(4) = 9 - (-32)$
$= 9 + 32 = 41$

42. $\frac{a - b}{b - c} = \frac{-2 - 3}{3 - 4} = \frac{-5}{-1} = 5$

43. $(a - c)(a + c) = (-2 - 4)(-2 + 4)$
$= (-6)(2) = -12$

44. Replace x in the equation with -2:
$3(-2) - 4 = 2$
$-10 = 2$ Incorrect.
So -2 is not a solution to the equation.

45. Replace x in the equation with 13:
$\frac{13 + 3}{8} = 2$
$\frac{16}{8} = 2$ Correct.
So 13 is a solution to the equation.

46. $-(-3) + 5 = 8$
$3 + 5 = 8$ Correct.
So -3 is a solution to the equation.

47. $\frac{1 \text{ delivery}}{0.25 \text{ hr}} + \frac{1 \text{ delivery}}{0.2 \text{ hr}}$
$= 4 \text{ del/hr} + 5 \text{ del/hr}$
$= 9 \text{ deliveries/hour}$

48. $80.405 + 3.66R - 0.06(A - 30)$
$= 3.66R - 0.06A + 82.205$
Replace R with 25 and A with 80:
$3.66(25) - 0.06(80) + 82.205 = 168.905$ cm

2.1 WARM-UPS

1. True, because $10 - 5 = 5$ is correct.
2. True, because 8 satisfies $\frac{x}{2} = 4$ and 8 satisfies $x = 8$.
3. False, because we should multiply by the reciprocal of $\frac{3}{4}$, which is $\frac{4}{3}$.
4. True, because dividing by 7 is equivalent to multiplying by $\frac{1}{7}$.
5. False, because the number must be a nonzero real number.
6. True, because subtracting t from each side yields $t = 7$.
7. True, because $\frac{3}{2}$ is the reciprocal of $\frac{2}{3}$.
8. True, because of the addition property of equality.
9. True, because $5 \cdot 0 = 0$ is correct.
10. True, because $2 \cdot 4 - 3 = 4 + 1$ is correct.

2.1 EXERCISES

1. The addition property of equality says that adding the same number to each side of an equation does not change the solution to the equation.
3. The multiplication property of equality says that multiplying both sides of an equation by the same nonzero number does not change the solution to the equation.
5. Replace the variable in the equation with your solution. If the resulting statement is correct, then the solution is correct.

7.
$$\begin{aligned} x - 6 &= -5 \\ x - 6 + 6 &= -5 + 6 \\ x &= 1 \end{aligned}$$
The solution set is $\{1\}$.

9.
$$\begin{aligned} -13 + x &= -4 \\ -13 + x + 13 &= -4 + 13 \\ x &= 9 \end{aligned}$$
The solution set is $\{9\}$.

11.
$$\begin{aligned} y - \frac{1}{2} &= \frac{1}{2} \\ y - \frac{1}{2} + \frac{1}{2} &= \frac{1}{2} + \frac{1}{2} \\ y &= 1 \end{aligned}$$
The solution set is $\{1\}$.

13.
$$\begin{aligned} w - \frac{1}{3} &= \frac{1}{3} \\ w - \frac{1}{3} + \frac{1}{3} &= \frac{1}{3} + \frac{1}{3} \\ w &= \frac{2}{3} \end{aligned}$$
The solution set is $\left\{\frac{2}{3}\right\}$.

15.
$$\begin{aligned} a - 0.2 &= -0.08 \\ a - 0.2 + 0.2 &= -0.08 + 0.2 \\ a &= 0.12 \end{aligned}$$
The solution set is $\{0.12\}$.

17.
$$\begin{aligned} x + 3 &= -6 \\ x + 3 - 3 &= -6 - 3 \\ x &= -9 \end{aligned}$$
The solution set is $\{-9\}$.

19.
$$\begin{aligned} 12 + x &= -7 \\ 12 + x - 12 &= -7 - 12 \\ x &= -19 \end{aligned}$$
The solution set is $\{-19\}$.

21.
$$\begin{aligned} t + \frac{1}{2} &= \frac{3}{4} \\ t + \frac{1}{2} - \frac{1}{2} &= \frac{3}{4} - \frac{1}{2} \\ t &= \frac{1}{4} \end{aligned}$$
The solution set is $\left\{\frac{1}{4}\right\}$.

23.
$$\begin{aligned} \frac{1}{19} + m &= \frac{1}{19} \\ \frac{1}{19} + m - \frac{1}{19} &= \frac{1}{19} - \frac{1}{19} \\ m &= 0 \end{aligned}$$
The solution set is $\{0\}$.

25.
$$\begin{aligned} a + 0.05 &= 6 \\ a + 0.05 - 0.05 &= 6 - 0.05 \\ a &= 5.95 \end{aligned}$$
The solution set is $\{5.95\}$.

27.
$$\begin{aligned} 2 &= x + 7 \\ 2 - 7 &= x + 7 - 7 \\ -5 &= x \end{aligned}$$
The solution set is $\{-5\}$.

29.
$$\begin{aligned} -13 &= y - 9 \\ -13 + 9 &= y - 9 + 9 \\ -4 &= y \end{aligned}$$
The solution set is $\{-4\}$.

31.
$$\begin{aligned} 0.5 &= -2.5 + x \\ 0.5 + 2.5 &= -2.5 + x + 2.5 \\ 3 &= x \end{aligned}$$
The solution set is $\{3\}$.

33.
$$\begin{aligned} \frac{1}{8} &= -\frac{1}{8} + r \\ \frac{1}{8} + \frac{1}{8} &= -\frac{1}{8} + r + \frac{1}{8} \\ \frac{1}{4} &= r \end{aligned}$$
The solution set is $\left\{\frac{1}{4}\right\}$.

35.
$$\begin{aligned} \frac{x}{2} &= -4 \\ 2 \cdot \frac{x}{2} &= 2 \cdot (-4) \\ x &= -8 \end{aligned}$$
The solution set is $\{-8\}$.

37.
$$\begin{aligned} 0.03 &= \frac{y}{60} \\ 60 \cdot 0.03 &= 60 \cdot \frac{y}{60} \\ 1.8 &= y \end{aligned}$$
The solution set is $\{1.8\}$.

39.
$$\begin{aligned} \frac{a}{2} &= \frac{1}{3} \\ 2 \cdot \frac{a}{2} &= 2 \cdot \frac{1}{3} \\ a &= \frac{2}{3} \end{aligned}$$
The solution set is $\left\{\frac{2}{3}\right\}$.

41.
$$\begin{aligned} \frac{1}{6} &= \frac{c}{3} \\ 3 \cdot \frac{1}{6} &= 3 \cdot \frac{c}{3} \\ \frac{1}{2} &= c \end{aligned}$$
The solution set is $\left\{\frac{1}{2}\right\}$.

43.
$$\begin{aligned} -3x &= 15 \\ \frac{-3x}{-3} &= \frac{15}{-3} \\ x &= -5 \end{aligned}$$
The solution set is $\{-5\}$.

45.
$$\begin{aligned} 20 &= 4y \\ \frac{20}{4} &= \frac{4y}{4} \\ 5 &= y \end{aligned}$$
The solution set is $\{5\}$.

47.
$$\begin{aligned} 2w &= 2.5 \\ \frac{2w}{2} &= \frac{2.5}{2} \\ w &= 1.25 \end{aligned}$$
The solution set is $\{1.25\}$.

49.
$$\begin{aligned} 5 &= 20x \\ \frac{5}{20} &= \frac{20x}{20} \\ \frac{1}{4} &= x \end{aligned}$$
The solution set is $\left\{\frac{1}{4}\right\}$.

51.
$$\begin{aligned} 5x &= \frac{3}{4} \\ \frac{1}{5} \cdot 5x &= \frac{1}{5} \cdot \frac{3}{4} \\ x &= \frac{3}{20} \end{aligned}$$
The solution set is $\left\{\frac{3}{20}\right\}$.

53.
$$\begin{aligned} \frac{3}{2}x &= -3 \\ \frac{2}{3} \cdot \frac{3}{2}x &= \frac{2}{3} \cdot (-3) \\ x &= -2 \end{aligned}$$
The solution set is $\{-2\}$.

55.
$$\begin{aligned} 90 &= \frac{3y}{4} \\ \frac{4}{3} \cdot 90 &= \frac{4}{3} \cdot \frac{3y}{4} \\ 120 &= y \end{aligned}$$
The solution set is $\{120\}$.

57.
$$\begin{aligned} -\frac{3}{5}w &= -\frac{1}{3} \\ -\frac{5}{3}\left(-\frac{3}{5}w\right) &= -\frac{5}{3}\left(-\frac{1}{3}\right) \\ w &= \frac{5}{9} \end{aligned}$$
The solution set is $\left\{\frac{5}{9}\right\}$.

59.
$$\frac{2}{3} = -\frac{4x}{3}$$
$$-\frac{3}{4}\left(\frac{2}{3}\right) = -\frac{3}{4}\left(-\frac{4x}{3}\right)$$
$$-\frac{1}{2} = x$$
The solution set is $\left\{-\frac{1}{2}\right\}$.

61.
$$-x = 8$$
$$-1(-x) = -1(8)$$
$$x = -8$$
The solution set is $\{-8\}$.

63.
$$-y = -\frac{1}{3}$$
$$-1(-y) = -1\left(-\frac{1}{3}\right)$$
$$y = \frac{1}{3}$$
The solution set is $\left\{\frac{1}{3}\right\}$.

65.
$$3.4 = -z$$
$$-1(3.4) = -1(-z)$$
$$-3.4 = z$$
The solution set is $\{-3.4\}$.

67.
$$-k = -99$$
$$-1(-k) = -1(-99)$$
$$k = 99$$
The solution set is $\{99\}$.

69.
$$4x = 3x - 7$$
$$4x - 3x = 3x - 7 - 3x$$
$$x = -7$$
The solution set is $\{-7\}$.

71.
$$9 - 6y = -5y$$
$$9 - 6y + 6y = -5y + 6y$$
$$9 = y$$
The solution set is $\{9\}$.

73.
$$-6x = 8 - 7x$$
$$-6x + 7x = 8 - 7x + 7x$$
$$x = 8$$
The solution set is $\{8\}$.

75.
$$\frac{1}{2}c = 5 - \frac{1}{2}c$$
$$\frac{1}{2}c + \frac{1}{2}c = 5 - \frac{1}{2}c + \frac{1}{2}c$$
$$c = 5$$
The solution set is $\{5\}$.

77.
$$12 = x + 17$$
$$12 - 17 = x + 17 - 17$$
$$-5 = x$$
The solution set is $\{-5\}$.

79.
$$\frac{3}{4}y = -6$$
$$\frac{4}{3} \cdot \frac{3}{4}y = \frac{4}{3}(-6)$$
$$y = -8$$
The solution set is $\{-8\}$.

81.
$$-3.2 + x = -1.2$$
$$-3.2 + x + 3.2 = -1.2 + 3.2$$
$$x = 2$$
The solution set is $\{2\}$.

83.
$$2a = \frac{1}{3}$$
$$\frac{1}{2} \cdot 2a = \frac{1}{2} \cdot \frac{1}{3}$$
$$a = \frac{1}{6}$$
The solution set is $\left\{\frac{1}{6}\right\}$.

85.
$$-9m = 3$$
$$\frac{-9m}{-9} = \frac{3}{-9}$$
$$m = -\frac{1}{3}$$
The solution set is $\left\{-\frac{1}{3}\right\}$.

87.
$$-b = -44$$
$$-1(-b) = -1(-44)$$
$$b = 44$$
The solution set is $\{44\}$.

89.
$$\frac{2}{3}x = \frac{1}{2}$$
$$\frac{3}{2} \cdot \frac{2}{3}x = \frac{3}{2} \cdot \frac{1}{2}$$
$$x = \frac{3}{4}$$
The solution set is $\left\{\frac{3}{4}\right\}$.

91.
$$-5x = 7 - 6x$$
$$-5x + 6x = 7 - 6x + 6x$$
$$x = 7$$
The solution set is $\{7\}$.

93. $$\frac{5a}{7} = -10$$
$$\frac{7}{5} \cdot \frac{5a}{7} = \frac{7}{5}(-10)$$
$$a = -14$$
The solution set is $\{-14\}$.

95. $$\frac{1}{2}v = -\frac{1}{2}v + \frac{3}{8}$$
$$\frac{1}{2}v + \frac{1}{2}v = -\frac{1}{2}v + \frac{3}{8} + \frac{1}{2}v$$
$$v = \frac{3}{8}$$
The solution set is $\left\{\frac{3}{8}\right\}$.

97. a) The 48.5 births per 1000 females in 2000 is $\frac{4}{5}$ of the birth rate in 1991. If x is the rate in 1991, we can write the following equation.
$$48.5 = \frac{4}{5}x$$
$$\frac{5}{4}(48.5) = \frac{5}{4} \cdot \frac{4}{5}x$$
$$60.625 = x$$
In 1991 the birth rate was about 60.6 births per 1000 females.

b) From the graph it appears that in 1996 the rate was about 54 births per 1000 females.

99. The number of advancers, 1918, was $\frac{2}{3}$of the number traded, t :
$$1918 = \frac{2}{3}t$$
$$\frac{3}{2}(1918) = \frac{3}{2} \cdot \frac{2}{3}t$$
$$2877 = t$$
So 2877 stocks were traded on that day.

2.2 WARM-UPS

1. True, because $4(3) - 3 = 3(3)$ is correct.
2. True, because subtracting 7 from both sides of $2x + 7 = 8$ yields $2x = 1$.
3. True, because that will get the like terms combined.
4. False, you should subtract 5 from each side and then subtract $7x$ from each side.
5. True, this is the multiplication property of equality.
6. False, add 7 to each side and then divide each side by 3.
7. True, because $\frac{4}{3}$ is the reciprocal of $\frac{3}{4}$.
8. True, because multiplying each side by -1 gives us $n = -9$.
9. True, because multiplying each side by -1 gives us $y = 7$.
10. True, because $7 \cdot 0 = 5 \cdot 0$ is correct.

2.2 EXERCISES

1. We can solve $ax + b = 0$ with the addition property and the multiplication property of equality.
3. Use the multiplication property of equality to solve $-x = 8$. Multiply each side by -1.

5. $$5a - 10 = 0$$
$$5a - 10 + 10 = 0 + 10$$
$$5a = 10$$
$$\frac{5a}{5} = \frac{10}{5}$$
$$a = 2$$
The solution set is $\{2\}$.

7. $$-3y - 6 = 0$$
$$-3y - 6 + 6 = 0 + 6$$
$$-3y = 6$$
$$\frac{-3y}{-3} = \frac{6}{-3}$$
$$y = -2$$
The solution set is $\{-2\}$.

9. $$3x - 2 = 0$$
$$3x - 2 + 2 = 0 + 2$$
$$3x = 2$$
$$\frac{3x}{3} = \frac{2}{3}$$
$$x = \frac{2}{3}$$
The solution set is $\left\{\frac{2}{3}\right\}$.

11. $$2p + 5 = 0$$
$$2p + 5 - 5 = 0 - 5$$
$$2p = -5$$
$$\frac{2p}{2} = \frac{-5}{2}$$
$$p = -\frac{5}{2}$$

The solution set is $\left\{-\frac{5}{2}\right\}$.

13.
$$\begin{aligned}\frac{1}{2}w-3&=0\\ \frac{1}{2}w-3+3&=0+3\\ \frac{1}{2}w&=3\\ 2\cdot\frac{1}{2}w&=2\cdot 3\\ w&=6\end{aligned}$$
The solution set is $\{6\}$.

15.
$$\begin{aligned}-\frac{2}{3}x+8&=0\\ -\frac{2}{3}x+8-8&=0-8\\ -\frac{2}{3}x&=-8\\ -\frac{3}{2}\left(-\frac{2}{3}x\right)&=-\frac{3}{2}(-8)\\ x&=12\end{aligned}$$
The solution set is $\{12\}$.

17.
$$\begin{aligned}-m+\frac{1}{2}&=0\\ -m+\frac{1}{2}-\frac{1}{2}&=0-\frac{1}{2}\\ -m&=-\frac{1}{2}\\ -1(-m)&=-1\left(-\frac{1}{2}\right)\\ m&=\frac{1}{2}\end{aligned}$$
The solution set is $\left\{\frac{1}{2}\right\}$.

19.
$$\begin{aligned}3p+\frac{1}{2}&=0\\ 3p+\frac{1}{2}-\frac{1}{2}&=0-\frac{1}{2}\\ 3p&=-\frac{1}{2}\\ \frac{1}{3}\cdot 3p&=\frac{1}{3}\left(-\frac{1}{2}\right)\\ p&=-\frac{1}{6}\end{aligned}$$
The solution set is $\left\{-\frac{1}{6}\right\}$.

21.
$$\begin{aligned}6x-8&=4x\\ 6x-8+8&=4x+8\\ 6x&=4x+8\\ 6x-4x&=4x+8-4x\\ 2x&=8\\ \frac{2x}{2}&=\frac{8}{2}\\ x&=4\end{aligned}$$
The solution set is $\{4\}$.

23.
$$\begin{aligned}4z&=5-2z\\ 4z+2z&=5-2z+2z\\ 6z&=5\\ \frac{6z}{6}&=\frac{5}{6}\\ z&=\frac{5}{6}\end{aligned}$$
The solution set is $\left\{\frac{5}{6}\right\}$.

25.
$$\begin{aligned}4a-9&=7\\ 4a-9+9&=7+9\\ 4a&=16\\ \frac{4a}{4}&=\frac{16}{4}\\ a&=4\end{aligned}$$
The solution set is $\{4\}$.

27.
$$\begin{aligned}9&=-6-3b\\ 9+6&=-6-3b+6\\ 15&=-3b\\ \frac{15}{-3}&=\frac{-3b}{-3}\\ -5&=b\end{aligned}$$
The solution set is -5.

29.
$$\begin{aligned}\frac{1}{2}w-4&=13\\ \frac{1}{2}w-4+4&=13+4\\ \frac{1}{2}w&=17\\ 2\cdot\frac{1}{2}w&=2(17)\\ w&=34\end{aligned}$$
The solution set is $\{34\}$.

31.
$$\begin{aligned} 6-\frac{1}{3}d &= \frac{1}{3}d \\ 6-\frac{1}{3}d+\frac{1}{3}d &= \frac{1}{3}d+\frac{1}{3}d \\ 6 &= \frac{2}{3}d \\ \frac{3}{2}\cdot 6 &= \frac{3}{2}\cdot\frac{2}{3}d \\ 9 &= d \end{aligned}$$
The solution set is $\{9\}$.

33.
$$\begin{aligned} 2w-0.4 &= 2 \\ 2w-0.4+0.4 &= 2+0.4 \\ 2w &= 2.4 \\ \frac{2w}{2} &= \frac{2.4}{2} \\ w &= 1.2 \end{aligned}$$
The solution set is $\{1.2\}$.

35.
$$\begin{aligned} x &= 3.3-0.1x \\ x+0.1x &= 3.3-0.1x+0.1x \\ 1.1x &= 3.3 \\ \frac{1.1x}{1.1} &= \frac{3.3}{1.1} \\ x &= 3 \end{aligned}$$
The solution set is $\{3\}$.

37.
$$\begin{aligned} 3x-3 &= x+5 \\ 3x &= x+8 \\ 2x &= 8 \\ \frac{2x}{2} &= \frac{8}{2} \\ x &= 4 \end{aligned}$$
The solution set is $\{4\}$.

39.
$$\begin{aligned} 4-7d &= 13-4d \\ -7d &= 9-4d \\ -3d &= 9 \\ \frac{-3d}{-3} &= \frac{9}{-3} \\ d &= -3 \end{aligned}$$
The solution set is $\{-3\}$.

41.
$$\begin{aligned} c+\frac{1}{2} &= 3c-\frac{1}{2} \\ c+1 &= 3c \\ 1 &= 2c \\ \frac{1}{2} &= \frac{2c}{2} \\ \frac{1}{2} &= c \end{aligned}$$
The solution set is $\left\{\frac{1}{2}\right\}$.

43.
$$\begin{aligned} \frac{2}{3}a-5 &= \frac{1}{3}a+5 \\ \frac{2}{3}a &= \frac{1}{3}a+10 \\ \frac{1}{3}a &= 10 \\ 3\cdot\frac{1}{3}a &= 3\cdot 10 \\ a &= 30 \end{aligned}$$
The solution set is $\{30\}$.

45.
$$\begin{aligned} 5(a-1)+3 &= 28 \\ 5a-5+3 &= 28 \\ 5a-2 &= 28 \\ 5a &= 30 \\ a &= 6 \end{aligned}$$
The solution set is $\{6\}$.

47.
$$\begin{aligned} 2-3(q-1) &= 10-(q+1) \\ 2-3q+3 &= 10-q-1 \\ -3q+5 &= 9-q \\ -2q &= 4 \\ q &= -2 \end{aligned}$$
The solution set is $\{-2\}$.

49.
$$\begin{aligned} 2(x-1)+3x &= 6x-20 \\ 2x-2+3x &= 6x-20 \\ 5x-2 &= 6x-20 \\ 5x+18 &= 6x \\ 18 &= x \end{aligned}$$
The solution set is $\{18\}$.

51.
$$\begin{aligned} 2\left(y-\frac{1}{2}\right) &= 4\left(y-\frac{1}{4}\right)+y \\ 2y-1 &= 4y-1+y \\ 2y-1 &= 5y-1 \\ 2y &= 5y \\ -3y &= 0 \\ y &= 0 \end{aligned}$$
The solution set is $\{0\}$.

53. Multiply each side by $1/2$:
$$\begin{aligned} 2x &= \frac{1}{3} \\ x &= \frac{1}{6} \end{aligned}$$
The solution set is $\left\{\frac{1}{6}\right\}$.

55.
$$\begin{aligned} 5t &= -2+4t \\ t &= -2 \end{aligned}$$

The solution set is $\{-2\}$.

57.
$$\begin{aligned} 3x - 7 &= 0 \\ 3x &= 7 \\ \frac{3x}{3} &= \frac{7}{3} \\ x &= \frac{7}{3} \end{aligned}$$

The solution set is $\left\{\frac{7}{3}\right\}$.

59.
$$\begin{aligned} -x + 6 &= 5 \\ -x &= -1 \\ -1(-x) &= -1(-1) \\ x &= 1 \end{aligned}$$

The solution set is $\{1\}$.

61.
$$\begin{aligned} -9 - a &= -3 \\ -9 - a + 9 &= -3 + 9 \\ -a &= 6 \\ -1(-a) &= -1(6) \\ a &= -6 \end{aligned}$$

The solution set is $\{-6\}$.

63.
$$\begin{aligned} 2q + 5 &= q - 7 \\ 2q &= q - 12 \\ q &= -12 \end{aligned}$$

The solution set is $\{-12\}$.

65.
$$\begin{aligned} -3x + 1 &= 5 - 2x \\ -3x + 1 - 1 &= 5 - 2x - 1 \\ -3x &= 4 - 2x \\ -x &= 4 \\ x &= -4 \end{aligned}$$

The solution set is $\{-4\}$.

67.
$$\begin{aligned} -12 - 5x &= -4x + 1 \\ -13 - 5x &= -4x \\ -13 - 5x + 5x &= -4x + 5x \\ -13 &= x \end{aligned}$$

The solution set is $\{-13\}$.

69.
$$\begin{aligned} 3x + 0.3 &= 2 + 2x \\ 3x + 0.3 - 0.3 &= 2 + 2x - 0.3 \\ 3x &= 1.7 + 2x \\ 3x - 2x &= 1.7 + 2x - 2x \\ x &= 1.7 \end{aligned}$$

The solution set is $\{1.7\}$.

71.
$$\begin{aligned} k - 0.6 &= 0.2k + 1 \\ k &= 0.2k + 1.6 \\ 0.8k &= 1.6 \\ \frac{0.8k}{0.8} &= \frac{1.6}{0.8} \\ k &= 2 \end{aligned}$$

The solution set is $\{2\}$.

73.
$$\begin{aligned} 0.2x - 4 &= 0.6 - 0.8x \\ 0.2x &= 4.6 - 0.8x \\ 0.2x + 0.8x &= 4.6 - 0.8x + 0.8x \\ x &= 4.6 \end{aligned}$$

The solution set is $\{4.6\}$.

75.
$$\begin{aligned} -3(k - 6) &= 2 - k \\ -3k + 18 &= 2 - k \\ -3k + 16 &= -k \\ 16 &= 2k \\ 8 &= k \end{aligned}$$

The solution set is $\{8\}$.

77.
$$\begin{aligned} 2(p + 1) - p &= 36 \\ 2p + 2 - p &= 36 \\ p + 2 &= 36 \\ p + 2 - 2 &= 36 - 2 \\ p &= 34 \end{aligned}$$

The solution set is $\{34\}$.

79.
$$\begin{aligned} 7 - 3(5 - u) &= 5(u - 4) \\ 7 - 15 + 3u &= 5u - 20 \\ -8 + 3u &= 5u - 20 \\ 12 &= 2u \\ 6 &= u \end{aligned}$$

The solution set is $\{6\}$.

81.
$$\begin{aligned} 4(x + 3) &= 12 \\ 4x + 12 &= 12 \\ 4x &= 0 \\ \frac{4x}{4} &= \frac{0}{4} \\ x &= 0 \end{aligned}$$

The solution set is $\{0\}$.

83.
$$\begin{aligned} \frac{w}{5} - 4 &= -6 \\ \frac{w}{5} - 4 + 4 &= -6 + 4 \\ \frac{w}{5} &= -2 \\ 5 \cdot \frac{w}{5} &= 5(-2) \\ w &= -10 \end{aligned}$$

The solution set is $\{-10\}$.

85. $$\frac{2}{3}y - 5 = 7$$
$$\frac{2}{3}y - 5 + 5 = 7 + 5$$
$$\frac{2}{3}y = 12$$
$$\frac{3}{2} \cdot \frac{2}{3}y = \frac{3}{2}(12)$$
$$y = 18$$
The solution set is $\{18\}$.

87. $$4 - \frac{2n}{5} = 12$$
$$4 - \frac{2n}{5} - 4 = 12 - 4$$
$$-\frac{2n}{5} = 8$$
$$-\frac{5}{2}\left(-\frac{2n}{5}\right) = -\frac{5}{2}(8)$$
$$n = -20$$
The solution set is $\{-20\}$.

89. $$-\frac{1}{3}p - \frac{1}{2} = \frac{1}{2}$$
$$-\frac{1}{3}p - \frac{1}{2} + \frac{1}{2} = \frac{1}{2} + \frac{1}{2}$$
$$-\frac{1}{3}p = 1$$
$$-3\left(-\frac{1}{3}p\right) = -3(1)$$
$$p = -3$$
The solution set is $\{-3\}$.

91. $$3.5x - 23.7 = -38.75$$
$$3.5x - 23.7 + 23.7 = -38.75 + 23.7$$
$$3.5x = -15.05$$
$$\frac{3.5x}{3.5} = \frac{-15.05}{3.5}$$
$$x = -4.3$$
The solution set is $\{-4.3\}$.

93. Let x represent the number of hours that the lawyer worked:
$$300 + 65x = 1405$$
$$65x = 1105$$
$$x = 17$$
The lawyer worked 17 hours on the case.

95. $$\frac{9}{5}C + 32 = 68$$
$$\frac{9}{5}C = 36$$
$$\frac{5}{9} \cdot \frac{9}{5}C = \frac{5}{9}(36)$$
$$C = 20$$
The temperature is 20° C.

97. $$2x + 2(x + 3) = 42$$
$$2x + 2x + 6 = 42$$
$$4x = 36$$
$$x = 9$$
The width is 9 feet.

99. $$x + 0.09x + 150 = 16,009.50$$
$$1.09x = 15,859.50$$
$$x = 14,550$$
The price of the car was \$14,550.

2.3 WARM-UPS

1. True, because 6 is the LCD.
2. True, because multiplying by 6 eliminates all of the fractions.
3. False, because multiplying by 100 yields $20x + 3x = 800$.
4. False, because $3(8/3) + 8 \neq 0$.
5. False, because $5a + 3 = 0$ is equivalent to $a = -3/5$ and the solution set is $\{-3/5\}$.
6. True, because $2t = t$ is equivalent to $t = 0$.
7. True, because if we simplify the left side we get $0.9w = 0.9w$.
8. False, because $1 \div 0$ is undefined.
9. True, because $x/x = 1$ is satisfied by every real number except 0 and x/x is undefined if $x = 0$.
10. True, because if we simplify the left side we get the inconsistent equation $0 = 99$.

2.3 EXERCISES

1. If an equation involves fractions we usually multiply each side by the LCD of all of the fractions.
3. An identity is an equation that is satisfied by all numbers for which both sides are defined.
5. An inconsistent equation has no solutions.

7. $$\begin{aligned} \frac{x}{4}-\frac{3}{10} &= 0 \\ 20\left(\frac{x}{4}-\frac{3}{10}\right) &= 20(0) \\ 5x-6 &= 0 \\ 5x-6+6 &= 0+6 \\ 5x &= 6 \\ x &= \frac{6}{5} \end{aligned}$$
The solution set is $\left\{\frac{6}{5}\right\}$.

9. $$\begin{aligned} 3x-\frac{1}{6} &= \frac{1}{2} \\ 6\left(3x-\frac{1}{6}\right) &= 6\cdot\frac{1}{2} \\ 18x-1 &= 3 \\ 18x-1+1 &= 3+1 \\ 18x &= 4 \\ x &= \frac{2}{9} \end{aligned}$$
The solution set is $\left\{\frac{2}{9}\right\}$.

11. $$\begin{aligned} \frac{x}{2}+3 &= x-\frac{1}{2} \\ 2\left(\frac{x}{2}+3\right) &= 2\left(x-\frac{1}{2}\right) \\ x+6 &= 2x-1 \\ x+6+1 &= 2x-1+1 \\ x+7 &= 2x \\ x+7-x &= 2x-x \\ 7 &= x \end{aligned}$$
The solution set is $\{7\}$.

13. $$\begin{aligned} \frac{x}{2}+\frac{x}{3} &= 20 \\ 6\left(\frac{x}{2}+\frac{x}{3}\right) &= 6\cdot 20 \\ 3x+2x &= 120 \\ 5x &= 120 \\ \frac{5x}{5} &= \frac{120}{5} \\ x &= 24 \end{aligned}$$
The solution set is $\{24\}$.

15. $$\begin{aligned} \frac{w}{2}+\frac{w}{4} &= 12 \\ 4\left(\frac{w}{2}+\frac{w}{4}\right) &= 4\cdot 12 \\ 2w+w &= 48 \\ 3w &= 48 \\ \frac{3w}{3} &= \frac{48}{3} \\ w &= 16 \end{aligned}$$
The solution set is $\{16\}$.

17. $$\begin{aligned} \frac{3z}{2}-\frac{2z}{3} &= -10 \\ 6\left(\frac{3z}{2}-\frac{2z}{3}\right) &= 6(-10) \\ 9z-4z &= -60 \\ 5z &= -60 \\ \frac{5z}{5} &= \frac{-60}{5} \\ z &= -12 \end{aligned}$$
The solution set is $\{-12\}$.

19. $$\begin{aligned} \frac{1}{3}p-5 &= \frac{1}{4}p \\ 12\left(\frac{1}{3}p-5\right) &= 12\cdot\frac{1}{4}p \\ 4p-60 &= 3p \\ 4p-60+60 &= 3p+60 \\ 4p &= 3p+60 \\ 4p-3p &= 3p+60-3p \\ p &= 60 \end{aligned}$$
The solution set is $\{60\}$.

21. $$\begin{aligned} \frac{1}{6}v+1 &= \frac{1}{4}v-1 \\ 12\left(\frac{1}{6}v+1\right) &= 12\left(\frac{1}{4}v-1\right) \\ 2v+12 &= 3v-12 \\ 2v+12+12 &= 3v-12+12 \\ 2v+24 &= 3v \\ 2v+24-2v &= 3v-2v \\ 24 &= v \end{aligned}$$
The solution set is $\{24\}$.

23. $$\begin{aligned} x-0.2x &= 72 \\ 10(x-0.2x) &= 10\cdot 72 \\ 10x-2x &= 720 \\ 8x &= 720 \\ \frac{8x}{8} &= \frac{720}{8} \\ x &= 90 \end{aligned}$$
The solution set is $\{90\}$.

25. $$\begin{aligned} 0.3x+1.2 &= 0.5x \\ 10(0.3x+1.2) &= 10(0.5x) \\ 3x+12 &= 5x \\ 3x+12-3x &= 5x-3x \\ 12 &= 2x \\ \frac{12}{2} &= \frac{2x}{2} \\ 6 &= x \end{aligned}$$
The solution set is $\{6\}$.

27. $$\begin{aligned} 0.02x-1.56 &= 0.8x \\ 100(0.02x-1.56) &= 100(0.8x) \\ 2x-156 &= 80x \\ 2x-156-2x &= 80x-2x \\ -156 &= 78x \\ \frac{-156}{78} &= \frac{78x}{78} \\ -2 &= x \end{aligned}$$
The solution set is $\{-2\}$.

29. $$\begin{aligned} 0.1a-0.3 &= 0.2a-8.3 \\ 10(0.1a-0.3) &= 10(0.2a-8.3) \end{aligned}$$

$$a-3=2a-83$$
$$a-3+83=2a-83+83$$
$$a+80=2a$$
$$a+80-a=2a-a$$
$$80=a$$
The solution set is $\{80\}$.

31. $$0.05r+0.4r=27$$
$$100(0.05r+0.4r)=100\cdot 27$$
$$5r+40r=2700$$
$$45r=2700$$
$$\frac{45r}{45}=\frac{2700}{45}$$
$$r=60$$
The solution set is $\{60\}$.

33. $$0.05y+0.03(y+50)=17.5$$
$$100[0.05y+0.03(y+50)]=100(17.5)$$
$$5y+3(y+50)=1750$$
$$5y+3y+150=1750$$
$$8y+150=1750$$
$$8y+150-150=1750-150$$
$$8y=1600$$
$$\frac{8y}{8}=\frac{1600}{8}$$
$$y=200$$
The solution set is $\{200\}$.

35. $$0.1x+0.05(x-300)=105$$
$$100[0.1x+0.05(x-300)]=100\cdot 105$$
$$10x+5(x-300)=10500$$
$$10x+5x-1500=10500$$
$$15x-1500=10500$$
$$15x-1500+1500=10500+1500$$
$$15x=12000$$
$$\frac{15x}{15}=\frac{12000}{15}$$
$$x=800$$
The solution set is $\{800\}$.

37. $$2x-9=0$$
$$2x=9$$
$$x=\frac{9}{2}$$
The solution set is $\left\{\frac{9}{2}\right\}$.

39. $$-2x+6=0$$
$$-2x=-6$$
$$x=3$$
The solution set is $\{3\}$.

41. $$\frac{z}{5}+1=6$$
$$\frac{z}{5}=5$$
$$z=25$$
The solution set is $\{25\}$.

43. $$\frac{c}{2}-3=-4$$
$$\frac{c}{2}=-1$$
$$c=-2$$
The solution set is $\{-2\}$.

45. $$3=t+6$$
$$-3=t$$
The solution set is $\{-3\}$.

47. $$5+2q=3q$$
$$5=q$$
The solution set is $\{5\}$.

49. $$8x-1=9+9x$$
$$-10=x$$
The solution set is $\{-10\}$.

51. $$-3x+1=-1-2x$$
$$2=x$$
The solution set is $\{2\}$.

53. $$x+x=2x$$
$$2x=2x$$
All real numbers satisfy the equation. The equation is an identity.

55. $$a-1=a+1$$
$$a-1-a=a+1-a$$
$$-1=1$$
The equation has no solution. It is an inconsistent equation.

57. $$3y+4y=12y$$
$$7y=12y$$
$$0=5y$$
$$0=y$$
The solution set is $\{0\}$. The equation is a conditional equation.

59. $$-4+3(w-1)=w+2(w-2)-1$$
$$-4+3w-3=w+2w-4-1$$
$$3w-7=3w-5$$
$$-7=-5$$
The equation has no solution. The solution set is $\emptyset$. It is an inconsistent equation.

61. $$3(m+1)=3(m+3)$$
$$3m+3=3m+9$$
$$3m+3-3m=3m+9-3m$$
$$3=9$$
The equation has no solution. The solution set is $\emptyset$. It is an inconsistent equation.

63. $$x+x=2$$
$$2x=2$$
$$\frac{2x}{2}=\frac{2}{2}$$

$$x = 1$$

The solution set is $\{1\}$. It is a conditional equation.

65. $2 - 3(5 - x) = 3x$

$$2 - 15 + 3x = 3x$$
$$-13 + 3x = 3x$$
$$-13 + 3x - 3x = 3x - 3x$$
$$-13 = 0$$

The equation has no solution. The solution set is $\emptyset$. It is an inconsistent equation.

67. $(3 - 3)(5 - z) = 0$

$$0(5 - z) = 0$$
$$0 = 0$$

All real numbers satisfy the equation. The equation is an identity.

69. $\frac{0}{x} = 0$

The equation is satisfied by every nonzero real number. The equation is an identity.

71. $x \cdot x = x^2$

$$x^2 = x^2$$

All real numbers satisfy this identity.

73. $3x - 5 = 2x - 9$

$$3x - 5 + 5 = 2x - 9 + 5$$
$$3x = 2x - 4$$
$$3x - 2x = 2x - 4 - 2x$$
$$x = -4$$

The solution set is $\{-4\}$.

75. $x + 2(x + 4) = 3(x + 3) - 1$

$$3x + 8 = 3x + 8$$

All real numbers satisfy the equation. The solution set is R.

77. $23 - 5(3 - n) = -4(n - 2) + 9n$

$$23 - 15 + 5n = -4n + 8 + 9n$$
$$8 + 5n = 5n + 8$$

All real numbers satisfy the equation. The solution set is R.

79. $0.05x + 30 = 0.4x - 5$

$$100(0.05x + 30) = 100(0.4x - 5)$$
$$5x + 3000 = 40x - 500$$
$$5x + 3000 + 500 = 40x - 500 + 500$$
$$5x + 3500 = 40x$$
$$5x + 3500 - 5x = 40x - 5x$$
$$3500 = 35x$$
$$\frac{3500}{35} = \frac{35x}{35}$$
$$100 = x$$

The solution set is $\{100\}$.

81. $-\frac{2}{3}a + 1 = 2$

$$3\left(-\frac{2}{3}a + 1\right) = 3 \cdot 2$$
$$-2a + 3 = 6$$
$$-2a + 3 - 3 = 6 - 3$$
$$-2a = 3$$
$$\frac{-2a}{-2} = \frac{3}{-2}$$
$$x = -\frac{3}{2}$$

The solution set is $\left\{-\frac{3}{2}\right\}$.

83. $\frac{y}{2} + \frac{y}{6} = 20$

$$6\left(\frac{y}{2} + \frac{y}{6}\right) = 6 \cdot 20$$
$$3y + y = 120$$
$$4y = 120$$
$$\frac{4y}{4} = \frac{120}{4}$$
$$y = 30$$

The solution set is $\{30\}$.

85. $0.09x - 0.2(x + 4) = -1.46$

$$0.09x - 0.2x - 0.8 = -1.46$$
$$9x - 20x - 80 = -146$$
$$-11x - 80 = -146$$
$$-11x = -66$$
$$x = 6$$

The solution set is $\{6\}$.

87. $436x - 789 = -571$

$$436x - 789 + 789 = -571 + 789$$
$$436x = 218$$
$$\frac{436x}{436} = \frac{218}{436}$$
$$x = 0.5$$

The solution set is $\{0.5\}$.

89. $\frac{x}{344} + 235 = 292$

$$\frac{x}{344} + 235 - 235 = 292 - 235$$
$$\frac{x}{344} = 57$$
$$344\left(\frac{x}{344}\right) = 344 \cdot 57$$
$$x = 19608$$

The solution set is $\{19{,}608\}$.

91. $x - 0.08x = 117{,}760$

$$0.92x = 117{,}760$$
$$\frac{0.92x}{0.92} = \frac{117{,}760}{0.92}$$
$$x = 128{,}000$$

The selling price was \$128,000.

93. a) From the graph it appears that the taxable income is approximately \$240,000.

b)

$$\begin{aligned}39{,}096.50 + 0.33(x - 174{,}700) &= 60{,}531\\ 39{,}096.50 + 0.33x - 57{,}651 &= 60{,}531\\ 0.33x - 18{,}554.5 &= 60{,}531\\ 0.33x &= 79{,}085.5\\ x &\approx 239{,}653\end{aligned}$$

Taxable income was \$239,653.

2.4 WARM-UPS

1. False, because $D = R \cdot T$ solved for T is $T = D/R$.
2. False, because a also appears on the right side of the equation.
3. False, because $A = L \cdot W$ solved for L is $L = A/W$.
4. False, because $D = R \cdot T$ solved for R is $R = D/T$. It is incorrect to use lower case letters in place of uppercase.
5. False, because the perimeter of a rectangle is $P = 2L + 2W$.
6. True, because for a rectangular solid, $V = LWH$.
7. True, because $P = 2(L + W)$.
8. True, because if we add x to each side of $y - x = 5$ we get $y = x + 5$.
9. False, because if $x = -1$, then $y = -3(-1) + 6 = 3 + 6 = 9$.
10. True, because $C = \pi D$.

2.4 EXERCISES

1. A formula is an equation with two or more variables.
3. To solve for a variable means to find an equivalent equation in which the variable is isolated.
5. To find the value of a variable in a formula we can solve for the variable then insert values for the other variables, or insert values for the other variables and then solve for the variable.

7.
$$\begin{aligned}D &= R \cdot T\\ \frac{D}{T} &= \frac{R \cdot T}{T}\\ \frac{D}{T} &= R\\ R &= \frac{D}{T}\end{aligned}$$

9.
$$\begin{aligned}C &= \pi D\\ \frac{C}{\pi} &= D\\ D &= \frac{C}{\pi}\end{aligned}$$

11.
$$\begin{aligned}I &= Prt\\ \frac{I}{rt} &= \frac{Prt}{rt}\\ P &= \frac{I}{rt}\end{aligned}$$

13.
$$\begin{aligned}F &= \tfrac{9}{5}C + 32\\ F - 32 &= \tfrac{9}{5}C\\ \tfrac{5}{9}(F - 32) &= \tfrac{5}{9} \cdot \tfrac{9}{5}C\\ \tfrac{5}{9}(F - 32) &= C\\ C &= \tfrac{5}{9}(F - 32)\end{aligned}$$

15.
$$\begin{aligned}A &= \tfrac{1}{2}bh\\ 2 \cdot A &= 2 \cdot \tfrac{1}{2}bh\\ 2A &= bh\\ \frac{2A}{b} &= \frac{bh}{b}\\ \frac{2A}{b} &= h\\ h &= \frac{2A}{b}\end{aligned}$$

17.
$$\begin{aligned}P &= 2L + 2W\\ P - 2W &= 2L + 2W - 2W\\ P - 2W &= 2L\\ \frac{P - 2W}{2} &= \frac{2L}{2}\\ \frac{P - 2W}{2} &= L\\ L &= \frac{P - 2W}{2}\end{aligned}$$

19.
$$\begin{aligned}A &= \tfrac{1}{2}(a + b)\\ 2A &= 2 \cdot \tfrac{1}{2}(a + b)\\ 2A &= a + b\\ 2A - b &= a + b - b\\ 2A - b &= a\\ a &= 2A - b\end{aligned}$$

21.
$$\begin{aligned}S &= P + Prt\\ S - P &= P + Prt - P\\ S - P &= Prt\\ \frac{S - P}{Pt} &= \frac{Prt}{Pt}\\ \frac{S - P}{Pt} &= r\\ r &= \frac{S - P}{Pt}\end{aligned}$$

23.
$$\begin{aligned}A &= \tfrac{1}{2}h(a + b)\\ 2A &= 2 \cdot \tfrac{1}{2}h(a + b)\\ 2A &= h(a + b)\end{aligned}$$

$$2A = ah + bh$$
$$2A - bh = ah + bh - bh$$
$$2A - bh = ah$$
$$\frac{2A - bh}{h} = \frac{ah}{h}$$
$$a = \frac{2A - hb}{h}$$

25. $5x + a = 3x + b$
$$5x = 3x + b - a$$
$$5x - 3x = b - a$$
$$2x = b - a$$
$$x = \frac{b-a}{2}$$

27. $4(a + x) - 3(x - a) = 0$
$$4a + 4x - 3x + 3a = 0$$
$$x + 7a = 0$$
$$x = -7a$$

29. $3x - 2(a - 3) = 4x - 6 - a$
$$3x - 2a + 6 = 4x - 6 - a$$
$$3x - 2a + a + 6 + 6 = 4x$$
$$3x - a + 12 = 4x$$
$$-a + 12 = 4x - 3x$$
$$12 - a = x$$
$$x = 12 - a$$

31. $3x + 2ab = 4x - 5ab$
$$3x + 2ab + 5ab = 4x$$
$$3x + 7ab = 4x$$
$$7ab = 4x - 3x$$
$$7ab = x$$
$$x = 7ab$$

33. $x + y = -9$
$$y = -x - 9$$

35. $x + y - 6 = 0$
$$y = -x + 6$$

37. $2x - y = 2$
$$2x = y + 2$$
$$2x - 2 = y$$
$$y = 2x - 2$$

39. $3x - y + 4 = 0$
$$3x + 4 = y$$
$$y = 3x + 4$$

41. $x + 2y = 4$
$$2y = -x + 4$$
$$\frac{1}{2} \cdot 2y = \frac{1}{2} \cdot (-x + 4)$$
$$y = -\frac{1}{2}x + 2$$

43. $2x - 2y = 1$
$$-2y = -2x + 1$$
$$-\frac{1}{2}(-2y) = -\frac{1}{2}(-2x + 1)$$
$$y = x - \frac{1}{2}$$

45. $y + 2 = 3(x - 4)$
$$y + 2 = 3x - 12$$
$$y = 3x - 12 - 2$$
$$y = 3x - 14$$

47. $y - 1 = \frac{1}{2}(x - 2)$
$$y - 1 = \frac{1}{2}x - 1$$
$$y = \frac{1}{2}x$$

49. $\frac{1}{2}x - \frac{1}{3}y = -2$
$$-\frac{1}{3}y = -\frac{1}{2}x - 2$$
$$-3\left(-\frac{1}{3}y\right) = -3\left(-\frac{1}{2}x - 2\right)$$
$$y = \frac{3}{2}x + 6$$

51. $y - 2 = \frac{3}{2}(x + 3)$
$$y - 2 = \frac{3}{2}x + \frac{9}{2}$$
$$y = \frac{3}{2}x + \frac{13}{2}$$

53. $y - \frac{1}{2} = -\frac{1}{4}\left(x - \frac{1}{2}\right)$
$$y - \frac{1}{2} = -\frac{1}{4}x + \frac{1}{8}$$
$$y = -\frac{1}{4}x + \frac{5}{8}$$

55. For each given value of x, find y using $y = -3x + 30$. For example, if $x = -10$, then $y = -3(-10) + 30 = 60$.

x	y
-10	60
0	30
10	0
20	-30
30	-60

57. $F = \frac{9}{5}C + 32$

C	F
-10	14
-5	23
0	32
40	104
100	212

59. $T = \frac{400}{R}$

R(mph)	T(hr)
10	40
20	20
40	10
80	5
100	4

61. $S = \dfrac{n(n+1)}{2}$

For $n = 1,\ S = \dfrac{1(1+1)}{2} = 1.$

For $n = 2,\ S = \dfrac{2(2+1)}{2} = 3.$

For $n = 3,\ S = \dfrac{3(3+1)}{2} = 6.$

For $n = 4,\ S = \dfrac{4(4+1)}{2} = 10.$

For $n = 5,\ S = \dfrac{5(5+1)}{2} = 15.$

n	S
1	1
2	3
3	6
4	10
5	15

63. Let $x = 2$ in the equation $y = 3x - 4$.

$y = 3(2) - 4 = 6 - 4 = 2$

65. Let $x = 2$ in the equation $3x - 2y = -8$.

$$\begin{aligned} 3(2) - 2y &= -8 \\ 6 - 2y &= -8 \\ -2y &= -14 \\ y &= 7 \end{aligned}$$

67. Let $x = 2$ in the equation $\frac{3x}{2} - \frac{5y}{3} = 6$.

$$\begin{aligned} \frac{3(2)}{2} - \frac{5y}{3} &= 6 \\ 3 - \frac{5y}{3} &= 6 \\ 9 - 5y &= 18 \\ -5y &= 9 \\ y &= -\frac{9}{5} \end{aligned}$$

69. Let $x = 2$ in the equation $y - 3 = \frac{1}{2}(x - 6)$.

$$\begin{aligned} y - 3 &= \tfrac{1}{2}(2 - 6) \\ y - 3 &= \tfrac{1}{2}(-4) \\ y - 3 &= -2 \\ y &= 1 \end{aligned}$$

71. Let $x = 2$ in $y - 4.3 = 0.45(x - 8.6)$.

$$\begin{aligned} y - 4.3 &= 0.45(2 - 8.6) \\ y - 4.3 &= 0.45(-6.6) \\ y - 4.3 &= -2.97 \\ y &= -2.97 + 4.3 \\ y &= 1.33 \end{aligned}$$

73. Use $P = 5000$, $t = 3$, and $I = 600$ in the formula for simple interest, $I = Prt$.

$$\begin{aligned} 600 &= 5000r(3) \\ 600 &= 15000r \\ \frac{600}{15000} &= r \\ 0.04 &= r \end{aligned}$$

The interest rate is 4%.

75. Use $I = 500$, $P = 2500$, and $r = 5\%$ in the formula for simple interest $I = Prt$.

$$\begin{aligned} 500 &= 2500(0.05)t \\ 500 &= 125t \\ \frac{500}{125} &= t \\ 4 &= t \end{aligned}$$

The time was 4 years.

77. The formula for the area of a rectangle is $A = LW$. Use $A = 28$ and $W = 4$ to find L.

$$\begin{aligned} 28 &= 4L \\ 7 &= L \end{aligned}$$

The length is 7 yards.

79. Use $P = 600$ and $W = 75$ in the formula for the perimeter of a rectangle, $P = 2L + 2W$.

$$\begin{aligned} 600 &= 2L + 2(75) \\ 600 &= 2L + 150 \\ 450 &= 2L \\ 225 &= L \end{aligned}$$

The length is 225 feet.

81. Use $S = 54{,}450$, and $r = 10\%$ in the formula for the sale price, $S = L - rL$.

$$\begin{aligned} 54{,}450 &= L - 0.10L \\ 54{,}450 &= 0.90L \\ \frac{54{,}450}{0.90} &= L \\ 60{,}500 &= L \end{aligned}$$

The MSRP is $60,500.

83. Use $S = 255$, and $r = 15\%$ in the formula for the sale price, $S = L - rL$.

$$\begin{aligned} 255 &= L - 0.15L \\ 255 &= 0.85L \end{aligned}$$

$\frac{255}{0.85} = L$
$300 = L$
The original price is $\$300$.
85. Use $d = 40$ and $b = 200$ in the formula for discount, $d = br$.
$40 = 200r$
$\frac{40}{200} = r$
$0.2 = r$
The rate of discount was 20%.
87. The length of a football field is 100 yards or 300 feet. Use $P = 920$ and $L = 300$ in the formula $P = 2L + 2W$.
$920 = 2(300) + 2W$
$920 = 600 + 2W$
$320 = 2W$
$160 = W$
The width is 160 feet.
89. Use $W = 2$, $L = 3$, and $H = 4$ in the formula for the volume of a rectangular solid, $V = LWH$.
$V = 3 \cdot 2 \cdot 4 = 24$
The volume is 24 cubic feet.
91. Use $C = 8\pi$ in the formula for the circumference of a circle, $C = 2\pi r$.

$8\pi = 2\pi r$
$\frac{8\pi}{2\pi} = \frac{2\pi r}{2\pi}$
$4 = r$

The radius is 4 inches.
93. Use $A = 16$ and $b = 4$ in the formula for the area of a triangle, $A = \frac{1}{2}bh$.
$16 = \frac{1}{2} \cdot 4h$
$16 = 2h$
$8 = h$
The height is 8 feet.
95. Use $A = 200$, $h = 20$, and $b_1 = 8$ in the formula for the area of a trapezoid,
$A = \frac{1}{2}h(b_1 + b_2)$.
$200 = \frac{1}{2} \cdot 20(8 + b_2)$
$200 = 10(8 + b_2)$
$200 = 80 + 10b_2$
$120 = 10b_2$
$12 = b_2$
The length of the upper base is 12 inches.

97. Let $D = 1000$ and $a = 8$ in $d = 0.08aD$:
$d = 0.08(8)(1000) = 640$
Child's dosage is 640 milligrams.
From the graph it appears that a child gets same dosage as an adult at about 13 years of age.
99. Amount $= \frac{750 \text{ mg}}{1 \text{ gram}} \times 5$ milliliters
$= \frac{750 \text{ mg}}{1000 \text{ mg}} \times 5 \text{ ml} = 3.75 \text{ ml}$
101.
$\frac{L + 2D - F\sqrt{S}}{2.37} = 2.4$
$L + 2D - F\sqrt{S} = 5.688$
$L = F\sqrt{S} - 2D + 5.688$

2.5 WARM-UPS

1. True, because $x + 6$ is 6 more than x.
2. True, because $a + (10 - a) = 10$.
3. True, because $D = RT$.
4. False, because $T = D/R$ means that her time is $10/x$ hours.
5. True, because the selling price minus the commission is what the owner gets.
6. False, because
$55{,}000 - 0.10(55{,}000) \neq 50{,}000$.
7. False, because if x is odd then $x + 1$ is even.
8. False, because n nickels at 5 cents each and d dimes at 10 cents each have total value in cents of $5n + 10d$.
9. True, because to get the total bill we add the amount of tax $0.05x$ to the amount of goods x, to get $x + 0.05x = 1.05x$.
10. False, because the perimeter is found by adding twice the length to twice the width.

2.5 EXERCISES

1. To express addition we use word such as plus, sum, increased by, and more than.
3. Complementary angles have degree measures with a sum of 90°.
5. Distance is the product of rate and time.
7. The sum of a number and 3 indicates addition and so the algebraic expression for that phrase is $x + 3$.

9. Three less than a number indicates subtraction. So the algebraic expression for the phrase is $x - 3$.
11. The product of a number and 5 indicates multiplication. So the algebraic expression is $5x$.
13. Ten percent of a number is found by multiplying the number by 0.10 or 0.1. So the algebraic expression is $0.1x$.
15. $x/3$
17. $\frac{1}{3}x$
19. If x is the smaller number, then x and $x + 15$ have a difference of 15. If x is the larger number, then x and $x - 15$ have a difference of 15 because $x-(x - 15) = 15$.
21. Two numbers with a sum of 6 are x and $6 - x$ because $x + (6 - x) = 6$.
23. Two numbers with a sum of -4 are x and $-4 - x$ because $x + (-4 - x) = -4$.
25. If x is the smaller number, then the numbers are x and $x + 3$. If x is the larger number, then the numbers are x and $x - 3$.
27. If x is one of the numbers, then 5% of x is $0.05x$. So the numbers are x and $0.05x$.
29. If one of the numbers is x, then 30% of x is $0.30x$ and a number that is 30% larger than x is $x + 0.30x$ or $1.30x$. So the numbers are x and $1.30x$.
31. Since the two angles are complementary we can use x and $90 - x$.
33. Since the sum of all angles of a triangle is 180°, the two unknown angles have a sum of 120°. So they are x and $120 - x$.
35. Since consecutive even integers differ by 2, two consecutive even integers are expressed as n and $n + 2$, where n is an even integer.
37. Consecutive integers differ by 1. So two consecutive integers are represented as x and $x + 1$, where x is an integer.
39. Three consecutive odd integers are expressed as x, $x + 2$, and $x + 4$, where x is an odd integer.
41. Four consecutive even integers are expressed as x, $x + 2$, $x + 4$, and $x + 6$, where x is an even integer.
43. If we use $R = x$ and $T = 3$ in the formula $D = RT$, we get $D = 3x$. So an expression for the distance is $3x$ miles.
45. Since the discount is 25% of the original price q, the discount is $0.25q$ dollars.
47. Use $D = x$ and $R = 20$ in the formula $T = D/R$, to get $T = x/20$. So an expression for the time is $x/20$ hr.
49. Use $D = x - 100$ and $T = 12$ in the formula $R = D/T$, to get $R = (x - 100)/12$. So an expression for the rate is $(x - 100)/12$ m/sec.
51. Since the area of a rectangle is the length times the width, the area is $5x$ m^2.
53. Since the perimeter of a rectangle is twice the length plus twice the width, the perimeter in this case is $2w + 2(w + 3)$ in.
55. If the perimeter of the rectangle is 300, then the total of the length and width is 150. If the length is x, then the width is expressed as $150 - x$ ft.
57. If the width is x and the length is 1 foot longer than twice the width, then the length is expressed as $2x + 1$ ft.
59. If the width is x and the length is 5 meters longer, then the length is $x + 5$. Since the area is length times width for a rectangle, we can express the area as $x(x + 5)$ m^2.
61. The simple interest is given by the formula $I = Prt$. So if P is $x + 1000$, r is 18%, and t is 1 year, then the simple interest is expressed as $0.18(x + 1000)$.
63. To find the price per pound we divide the total price by the number of pounds. So the price per pound for the peaches is expressed as $16.50/x$ dollars per pound.
65. Since the sum of complementary angles is 90°, the degree measure is $90 - x$ degrees.
67. If x represents the smaller number, then two numbers that differ by 5 are expressed as x and $x + 5$. If their product is 8, we can write the equation $x(x + 5) = 8$.
69. If x is the selling price, the agent gets $0.07x$. Since Herman receives the selling price less the commission, $x - 0.07x = 84{,}532$.
71. To find a percent of 500 we multiply the rate (or percent) x by 500: $500x = 100$.

73. The value in dollars of x nickels is $0.05x$. The value in dollars of $x+2$ dimes is $0.10(x+2)$. Since we know that the total value is \$3.80, we can write the equation $0.05x + 0.10(x+2) = 3.80$. We could express the total value in cents as $5x + 10(x+2) = 380$.
75. Sum indicates addition. The sum of a number (x) and 5 is 13 is written as the equation $x + 5 = 13$.
77. Three consecutive integers are represented as x, $x+1$, and $x+2$, where x is the smallest of the three integers. Since their sum is 42, we can write $x + (x+1) + (x+2) + 42$.
79. Two consecutive integers are represented as x and $x+1$, where x is an integer. Since their product is 182, we can write the equation $x(x+1) = 182$.
81. To find 12% of Harriet's income we multiply Harriet's income (x) by 0.12. Since we know that 12% of her income is \$3000, we can write the equation $0.12x = 3000$.
83. To find 5% of a number we multiply the number (x) by 0.05. Since we know that 5% of the number is 13, we can write the equation $0.05x = 13$.
85. Since the length is 5 feet longer than the width, let x represent the length and $x+5$ represent the width. Since the area is 126, we can write the equation $x(x+5) = 126$.
87. The number of cents in n nickels is $5n$ and the number of cents in $n-1$ dimes is $10(n-1)$. Since the total value is 95 cents, we can write the equation $5n + 10(n-1) = 95$.
89. The measures of the two angles are x and $x - 38$. Since the angles are supplementary, we have $x + x - 38 = 180$.
91. a) If r is the resting hear rate then we subtract the sum of the age and the resting heart rate from 220 to get $220 - (30 + r)$. Now take 60% of that result and add it to the resting heart rate to get the target heart rate of 144:

$$r + 0.60(220 - (30 + r)) = 144$$

b) As the resting heart rate increases, the target heart rate also increases.
93. The sum of 6 and x is $6 + x$ or $x + 6$.
95. To express m increased by 9, write $m + 9$.
97. To express t multiplied by 11 write $t \cdot 11$ or $11t$.
99. To express 5 times the difference between x and 2, write $5(x-2)$.
101. To express m decreased by the product of 3 and m, write $m - 3m$.
103. The ratio of 8 more than h and h is written as $(h+8)/h$ or $\frac{h+8}{h}$.
105. To express 5 divided by the difference between y and 9, write $5/(y-9)$ or $\frac{5}{y-9}$.
107. The quotient of 8 less then w and twice w is written as $(w-8)/(2w)$ or $\frac{w-8}{2w}$.
109. To express 9 less than the product of v and -3, write $-3v - 9$.
111. To express x decreased by the quotient of x and 7, write $x - x/7$ or $x - \frac{x}{7}$.
113. The difference between the square of m and the total of m and 7 is $m^2 - (m+7)$.
115. To express x increased by the difference between 9 times x and 8 write $x + (9x - 8)$ or $x + 9x - 8$.
117. To express 9 less than the product of 13 and n write $13n - 9$.
119. To express 6 increased by one-third of the sum of x and 2 write $6 + \frac{1}{3}(x+2)$.
121. The sum of x divided by 2 and x is written $x/2 + x$ or $\frac{x}{2} + x$.
123. Because the area of a rectangle is the length times the width, we have $x(x+3) = 24$.
125. Because the area of a parallelogram is the product of the base and height, we have $w(w-4) = 24$.

2.6 WARM-UPS

1. False, because before you write an equation you must read the problem and identify the variables.
2. True, because we must be certain what the variable represents in order to write an equation about it.
3. True, because diagrams and tables help us to understand the situation.

4. False, because consecutive odd integers differ by 2.
5. False, because if $5x$ is larger than $3(x+20)$ we can write the equation $5x-2=3(x+20)$.
6. True, because $x+6-x=6$ is true for any real number x.
7. True, because $x+7-x=7$ for any real number x.
8. True, because complementary angles have a sum of 90°.
9. False, because x and $x+180$ do not have a sum of 180 for all values of x.
10. True, because $x=\frac{1}{2}(x+50)$ or $2x=x+50$.

2.6 EXERCISES

1. In this section we studied number, geometric, and uniform motion problems.
3. Uniform motion is motion at a constant rate of speed.
5. Complementary angles are angles whose degree measures have a sum of 90°.

7. Let $x=$ the first integer, $x+1=$ the second integer, and $x+2=$ the third integer. Since their sum is 141, we can write the following equation.

$$x+x+1+x+2=141$$
$$3x+3=141$$
$$3x=138$$
$$x=46$$
$$x+1=47$$
$$x+2=48$$

The three integers are 46, 47, and 48.

9. Let $x=$ the first odd integer and $x+2=$ the second odd integer. Since their sum is 152, we can write the following equation.

$$x+x+2=152$$
$$2x+2=152$$
$$2x=150$$
$$x=75$$
$$x+2=77$$

The two consecutive odd integers are 75 and 77.

11. Let $x=$ the first integer, $x+1=$ the second integer, $x+2=$ the third integer, and $x+3=$ the fourth integer. Since their sum is 194, we can write the following equation.

$$x+x+1+x+2+x+3=194$$
$$4x+6=194$$
$$4x=188$$
$$x=47$$
$$x+1=48$$
$$x+2=49$$
$$x+3=50$$

The three integers are 47, 48, 49, and 50.

13. Let $x=$ the width and $2x=$ the length. Since the perimeter is 150, we can write the following.

$$2(x)+2(2x)=150$$
$$6x=150$$
$$x=25$$
$$2x=50$$

The length is 50 m and the width is 25 m.
15. Let $x=$ the width and $x+4=$ the length. Since the perimeter is 176, we can write the following equation.

$$2x+2(x+4)=176$$
$$2x+2x+8=176$$
$$4x+8=176$$
$$4x=168$$
$$x=42$$
$$x+4=46$$

The width is 42 inches and the length is 46 inches.
17. Let $x=$ the length of each of the equal sides and $x-5=$ the length of the base of the triangle. Since the perimeter is 34 inches, we can write the following equation.

$$x+x+x-5=34$$
$$3x=39$$
$$x=13$$

The length of each of the equal sides is 13 inches.
19. $2w+2w+40=180$

$$4w+40=180$$
$$4w=140$$
$$w=35$$

So the angle marked w is 35°.
21. $96\text{ ft}=96\text{ ft}\cdot\frac{12\text{ in.}}{1\text{ ft}}=1152\text{ in.}$

23. $14.22 \text{ mi} = 14.22 \text{ mi} \cdot \dfrac{1.609 \text{ km}}{1 \text{ mi}}$
$\approx 22.88 \text{ km}$

25. $13.5 \text{ cm} = 13.5 \text{ cm} \cdot \dfrac{0.3937 \text{ in.}}{1 \text{ cm}}$
$\approx 5.31 \text{ in.}$

27. $42.1 \text{ in.} = 42.1 \text{ in.} \cdot \dfrac{2.540 \text{ cm}}{1 \text{ in.}}$
$\approx 106.93 \text{ cm}$

29. 40 mph
$= \dfrac{40 \text{ mi}}{1 \text{ hr}} \cdot \dfrac{1 \text{hr}}{60 \text{ min}} \cdot \dfrac{1 \text{min}}{60 \text{ sec}} \cdot \dfrac{5280 \text{ ft}}{1 \text{ mi}}$

≈ 58.67 ft/sec

31. 500 ft/sec $=$
$\dfrac{500 \text{ ft}}{1 \text{ sec}} \cdot \dfrac{60 \text{ sec}}{1 \text{ min}} \cdot \dfrac{60 \text{ min}}{1 \text{ hr}} \cdot \dfrac{1 \text{ mi}}{5280 \text{ ft}} \cdot \dfrac{1 \text{ km}}{0.6215 \text{ mi}}$
≈ 548.53 km/hr

33. Let $x =$ his speed on the freeway and $x - 20 =$ his speed on the country road. Since $D = RT$, his distance on the freeway was $4x$ and his distance on the country road was $5(x - 20)$. Since his total distance was 485 miles, we can write the following equation.

$$4x + 5(x - 20) = 485$$
$$4x + 5x - 100 = 485$$
$$9x = 585$$
$$x = 65$$

He traveled 65 mph on the freeway.

35. Let $x =$ her speed after dawn and $x + 5 =$ her speed before dawn. Her distance after dawn was $6x$ and her distance before dawn was $5(x + 5)$. Since her total distance was 630 miles, we can write the following equation.

$$6x + 5(x + 5) = 630$$
$$6x + 5x + 25 = 630$$
$$11x + 25 = 630$$
$$11x = 605$$
$$x = 55$$

Her speed after dawn was 55 mph.

37. Let $x =$ the time in hours to L.A. and $x + 48/60 =$ the time in hours to Chicago. Since $D = RT$, we have

$$640x = 512(x + 0.8)$$
$$640x = 512x + 409.6$$
$$128x = 409.6$$
$$x = 3.2$$
$$640x = 2048$$

The trip from L.A. to Chicago was 3.2 hours and the trip from Chicago to L.A. was 4 hours. The distance from Chicago to L.A. is 2048 miles.

39. Let $x =$ the length and $x - 8 =$ the width of the frame. Use $P = 2L + 2W$.

$$2x + 2(x - 8) = 64$$
$$4x - 16 = 64$$
$$4x = 80$$
$$x = 20$$
$$x - 8 = 12$$

So the length is 20 in. and the width is 12 in.

41. Let $x =$ the length of each of the equal sides and $x - 2 =$ the length of the third (shortest) side. The perimeter is 13 feet.

$$x + x + x - 2 = 13$$
$$3x = 15$$
$$x = 5$$
$$x - 2 = 3$$

So the sides are 5 ft, 5 ft, and 3 ft.

43. Let $x =$ degree measure of the smallest angle, $6x =$ the degree measure of the largest angle, and $2x =$ the degree measure of the middle angle. The sum of the degree measures of all three angles in any triangle is 180°.

$$x + 6x + 2x = 180$$
$$9x = 180$$
$$x = 20$$
$$6x = 120$$
$$2x = 40$$

So the angles are $20°$, $40°$, and $120°$.

45. Let $x =$ degree measure of the smallest angle and $4x =$ the degree measure of each of the equal angles. The sum of the degree measures of all three angles in any triangle is 180°.

$$x + 4x + 4x = 180$$
$$9x = 180$$
$$x = 20$$
$$4x = 80$$

So the angles are $20°$, $80°$, and $80°$.

47. Let $x =$ the number of points scored by the Raiders and $x - 18 =$ the number scored by the Vikings.

$$x + x - 18 = 46$$
$$2x = 64$$
$$x = 32$$
$$x - 18 = 14$$

The scored was Raiders 32, Vikings 14.

49. Let $x =$ the driving time before lunch and $x - 1 =$ the driving time after lunch. Since $D = RT$,

$$50x + 53(x - 1) = 256$$
$$103x - 53 = 256$$
$$103x = 309$$
$$x = 3$$
$$x - 1 = 2$$

She drove for 3 hours before lunch. The distance from Ardmore to Lawton is 2(53) or 106 miles.

51. Let $x =$ Crawford's age in 1950, $x - 1 =$ John Wayne's age in 1950, and $x - 2 =$ James Stewart's age in 1950.

$$x + x - 1 + x - 2 = 129$$
$$3x = 132$$
$$x = 44$$
$$x - 1 = 43$$
$$x - 2 = 42$$

So Crawford was born in 1906, Wayne in 1907, and Stewart in 1908.

53. Let $x =$ the length of the shortest piece. The longest piece is $2x + 2$ feet. The total length for two short pieces and one long piece is 30 feet.

$$2x + 2x + 2 = 30$$
$$4x = 28$$
$$x = 7$$

The short pieces are 7 feet each and the long piece is $2(7) + 2$ or 16 feet.

2.7 WARM-UPS

1. True, because 12% of \$1000 is \$120.
2. False, because 5% of \$80,000 is \$4000.
3. True, because the discount is $0.20x$ and the price she pays is $x - 0.20x$ or $0.8x$.
4. False, because 6% as a decimal is 0.06 and t he discount is $0.06x$ dollars.
5. True, because selling price is list price minus the amount of discount.
6. True, to find the commission we multiply 8% and the selling price.
7. False, because $4{,}000 \neq 0.10(44{,}000)$.
8. True, because 20% of 10 plus 30% of x is written $0.20(10) + 0.30x$ or $2 + 0.3x$.
9. False, the percent of acid in the mixture will be between 10% and 14%.
10. True, because $x + 0.05x = 1.05x$.

2.7 EXERCISES

1. We studied discount, investment, and mixture problems in this section.

3. The product of the rate and the original price gives the amount of discount. The original price minus the discount is the sale price.

5. A table helps us to organize the information given in a problem.

7. Let $x =$ the original price of the television and $0.25x =$ the amount of the discount. Since the amount of the discount is \$80, we can write the following equation.

$$0.25x = 80$$
$$x = \frac{80}{0.25} = 320$$

The original price was \$320.

9. Let $x =$ the original price and $0.20x =$ the amount of the discount. Since the price after the discount was \$320, we can write the following equation.

$$x - 0.20x = 320$$
$$0.80x = 320$$
$$x = \frac{320}{0.80} = 400$$

The original price was \$400.

11. Let $x =$ the selling price, and $0.10x =$ the real estate commission. The selling price minus the commission is what Kirk receives.

$$x - 0.08x = 115{,}000$$
$$0.92x = 115{,}000$$
$$x = 125{,}000$$

The house should sell for \$125,000.

13. Let $x =$ the amount of her sales and $0.07x =$ the amount of sales ta$x$. Since her total receipts were \$462.24, we can write the following equation.

$$x + 0.07x = 462.24$$
$$1.07x = 462.24$$
$$x = 432$$

The sales tax was 0.07(432) or \$30.24.

15. Let $x =$ the amount invested in the 100 fund and $x + 3000 =$ the amount invested in the 101 fund.

$$\begin{aligned}0.18x + 0.15(x + 3000) &= 3750\\ 0.33x + 450 &= 3750\\ 0.33x &= 3300\\ x &= 10{,}000\\ x + 3000 &= 13{,}000\end{aligned}$$

He invested \$10,000 in the 100 fund and \$13,000 in the 101 fund.

17. Let $x =$ the amount invested at 5% and $25000 - x =$ the amount invested at 4%. His income on the first investment was $0.05x$ and his income from the second investment was $0.04(25000 - x)$. Since his total income was actually \$1140, we can write the following equation.

$$\begin{aligned}0.05x + 0.04(25000 - x) &= 1140\\ 0.05x + 1000 - 0.04x &= 1140\\ 0.01x &= 140\\ x &= 14{,}000\\ 25{,}000 - x &= 11{,}000\end{aligned}$$

He invested \$14,000 in Fidelity and \$11,000 in Price.

19. Let $x =$ the amount of 1% milk. The x gallons of 1% milk are mixed with 30 gallons of 3% milk to obtain $x + 30$ gallons of 2% milk. In the 1% milk there are $0.01x$ gallons of fat. In the 3% milk there are $0.03(30)$ gallons of fat. In the 2% milk there are $0.02(x + 30)$ gallons of fat. We can write an equation expressing the fact that the total of the fat in the two milks that are mixed is equal to the fat in the final mixture.

$$\begin{aligned}0.01x + 0.03(30) &= 0.02(x + 30)\\ 0.01x + 0.9 &= 0.02x + 0.6\\ 0.9 &= 0.01x + 0.6\\ 0.3 &= 0.01x\\ 100(0.3) &= 100(0.01x)\\ 30 &= x\end{aligned}$$

Use 30 gallons of 1% milk.

21. Let $x =$ the number of liters of 5% solution and $30 - x =$ the number of liters of 20% solution. The amount of alcohol in the 5% solution is $0.05x$. The amount of alcohol in the 20% solution is $0.20(30 - x)$. The amount of alcohol in the final 10% solution is $0.10(30)$. We can write an equation expressing the fact that the total of the alcohol in each of the two solutions mixed is equal to the alcohol in the final result.

$$\begin{aligned}0.05x + 0.20(30 - x) &= 0.10(30)\\ 0.05x + 6 - 0.20x &= 3\\ 6 - 0.15x &= 3\\ -0.15x &= -3\\ x &= \frac{-3}{-0.15} = 20\\ 30 - x &= 10\end{aligned}$$

He should use 20 liters of 5% alcohol and 10 liters of 20% alcohol.

23. Let $x =$ the number of registered voters. We can write the following equation.

$$\begin{aligned}0.60x &= 33420\\ x &= \frac{33420}{0.6} = 55{,}700\end{aligned}$$

There are 55,700 registered voters.

25. Let $x =$ the price of the car and $0.08x =$ the amount of sales tax. Since the amount of sales tax was \$1200, we can write the following equation.

$$\begin{aligned}0.08x &= 1200\\ x &= \frac{1200}{0.08} = 15000\end{aligned}$$

The price of the car was \$15,000.

27. Let $x =$ the percent increase and $8x =$ the amount of increase. Since the actual amount of increase is \$6, we can write the following equation.

$$\begin{aligned}8x &= 6\\ x &= \frac{6}{8} = 0.75\end{aligned}$$

The price of the shirts is increased 75%.

29. Let $x =$ the number of students at Jefferson and $x + 400 =$ the number of students in the combined school. The number of African American students at Jefferson is $0.60x$. The number of African American students at Wilson is $0.20(400)$. The number of African American students in the combined school will be $0.44(x + 400)$.

$$\begin{aligned}0.60x + 0.20(400) &= 0.44(x + 400)\\ 0.60x + 80 &= 0.44x + 176\\ 0.16x + 80 &= 176\\ 0.16x &= 96\\ x &= 600\end{aligned}$$

The number of students at Jefferson is 600.

31. Let $x =$ the number of people in private rooms and $x + 18 =$ the number of people in semiprivate rooms. The revenue from the private rooms is $200x$ dollars and the revenue from the semiprivate rooms is $150(x + 18)$

dollars. We can write an equation for the total receipts.

$$\begin{aligned} 200x + 150(x+18) &= 17{,}400 \\ 200x + 150x + 2700 &= 17{,}400 \\ 350x + 2700 &= 17{,}400 \\ 350x &= 14{,}700 \\ x &= 42 \\ x + 18 &= 60 \end{aligned}$$

They have 42 private rooms and 30 semi-private rooms (holding 60 people).

33. Let $x =$ the number of pounds of pistachios. We can write an equation expressing the total cost of the mixture.

$$\begin{aligned} 6.40x + 4.80(20) &= 5.40(x+20) \\ 6.4x + 96 &= 5.4x + 108 \\ x &= 12 \end{aligned}$$

We should mix 12 pounds of pistachios with 20 pounds of cashews to get a mix that sells for \$5.40 per pound.

35. Let $x =$ the number of nickels and $10 - x =$ the number of dimes. The value in cents of the nickels is $5x$ and the value in cents of the dimes is $10(10 - x)$. Since she has 80 cents altogether, we can write the following equation.

$$\begin{aligned} 5x + 10(10 - x) &= 80 \\ 5x + 100 - 10x &= 80 \\ -5x &= -20 \\ x &= 4 \\ 10 - x &= 6 \end{aligned}$$

She used 4 nickels and 6 dimes.

37. Let $x =$ the number of gallons of corn oil.

$$\begin{aligned} 0.14x + 0.07(600) &= 0.11(x+600) \\ 0.14x + 42 &= 0.11x + 66 \\ 0.03x &= 24 \\ x &= 800 \end{aligned}$$

Crisco should use 800 gallons of corn oil.

39. Let $x =$ the number of gallons of water. One gallon of Hawaiian Punch contains $0.10(1)$ or 0.10 gallon of fruit juice. The final mix will contain $0.06(x+1)$ gallons of fruit juice.

$$\begin{aligned} 0.10(1) &= 0.06(x+1) \\ 0.10 &= 0.06x + 0.06 \\ 0.04 &= 0.06x \\ x &= \frac{0.04}{0.06} = \frac{2}{3} \end{aligned}$$

So, mix 2/3 gal of water with one gallon of Hawaiian Punch.

41. Let $x =$ the price in dollars for a top and $2x =$ the price in dollars for a pair of shorts. The total price is \$108.

$$\begin{aligned} 5(2x) + 8(x) &= 108 \\ 18x &= 108 \\ x &= 6 \\ 2x &= 12 \end{aligned}$$

So shorts are \$12 and tops are \$6.

2.8 WARM-UPS

1. True, because -2 is equal to -2.
2. True, because $-5 < 4$ and $4 < 6$ are both correct.
3. False, because $0 < -1$ is false.
4. True, because any number to the right of 7 on the number line satisfies both $7 < x$ and $x > 7$.
5. False, because the graph of $x < -3$ includes only points to the left of -3.
6. True, because $5 > 2$ is correct.
7. False, because $-2 < -3$ is incorrect.
8. False, because $2(4) - 1 < 4$ is incorrect.
9. True, because $2(0) - 3 \leq 5(0) - 3$ is equivalent to $-3 \leq -3$ and that is correct.
10. False, because 0 satisfies $2x - 1 < x$ but does not satisfy $x < 2x - 1$.

2.8 EXERCISES

1. The inequality symbols are $<$, $\leq$, $>$, and $\geq$.
3. For $\leq$ and $\geq$ use a bracket and for $<$ and $>$ use a parenthesis.
5. The compound inequality $a < b < c$ means $b > a$ and $b < c$, or b is between a and c.
7. True, because -3 is to the left of 5 on the number line.
9. True, because $4 \leq 4$ is true if either $4 < 4$ or $4 = 4$ is correct.
11. False, because -6 is to the left of -5 on the number line.
13. True, because $-4 < -3$ is correct.
15. True, $(-3)(4) - 1 < 0 - 3$ is equivalent to $-13 < -3$.
17. True, because $-4(5) - 6 \geq 5(-6)$ is equivalent to $-26 \geq -30$.

19. True, because $7(4) - 12 \le 3(9) - 2$ is equivalent to $16 \le 25$.

21. The graph of $x \le 3$ consists of the numbers to the left of 3 including 3 on the number line. It is written in interval notation as $(-\infty, 3]$.

−1 0 1 2 3 4 5

23. The graph of $x > -2$ consists of the numbers to the right of -2 on the number line. It is written in interval notation as $(-2, \infty)$.

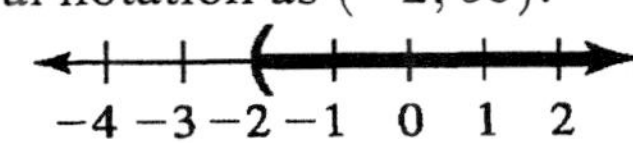

25. The inequality $-1 > x$ is the same as $x < -1$. It is written in interval notation as $(-\infty, -1)$.

−5 −4 −3 −2 −1 0 1

27. The graph of $-2 \le x$ is the same as the graph of $x \ge -2$, the numbers to the right of and including -2. It is written in interval notation as $[-2, \infty)$.

−4 −3 −2 −1 0 1 2

29. The graph of $x \ge \frac{1}{2}$ consists of the numbers to the right of and including $\frac{1}{2}$. It is written in interval notation as $[1/2, \infty)$.

$\frac{1}{2}$

−1 0 1 2 3

31. The graph of $x \le 5.3$ consists of the numbers to the left of and including 5.3.It is written in interval notation as $(-\infty, 5.3]$.

5.3

1 2 3 4 5 6 7

33. The graph of $-3 < x < 1$ consists of the numbers between -3 and 1. It is written in interval notation as $(-3, 1)$.

−4 −3 −2 −1 0 1 2

35. The graph of $3 \le x \le 7$ consists of the numbers between 3 and 7, including 3 and 7. It is written in interval notation as $[3, 7]$.

2 3 4 5 6 7 8

37. The graph of $-5 \le x < 0$ consists of the numbers between -5 and 0, including -5 but not including 0. It is written in interval notation as $[-5, 0)$.

−5 −4 −3 −2 −1 0 1

39. The graph of $40 < x \le 100$ consists of the numbers between 40 and 100, including 100 but not including 40. It is written in interval notation as $(40, 100]$.

20 40 60 80 100 120

41. The graph shows the numbers to the right of 3. The inequality $x > 3$ describes this graph. The solution set in interval notation is $(3, \infty)$.

43. The graph shows the numbers to the left of and including 2. The inequality $x \le 2$ describes this graph. The solution set in interval notation is $(-\infty, 2]$.

45. The graph shows the numbers between 0 and 2. The inequality $0 < x < 2$ describes this graph. The solution set in interval notation is $(0, 2)$.

47. The graph shows the numbers between -5 and 7, including 7 but not -5. The inequality $-5 < x \le 7$ describes this graph. The solution set in interval notation is $(-5, 7]$.

49. The graph shows the numbers to the right of -4. The inequality $x > -4$ describes this graph. The solution set in interval notation is $(-4, \infty)$.

51. Replace x by -9 in $-x > 3$.

$$-(-9) > 3$$
$$9 > 3$$

Since the last inequality is correct, -9 satisfies $-x > 3$.

53. Replace x by -2 in $5 \le x$.

$$5 \le -2$$

Since the inequality is incorrect, -2 does not satisfy $5 \le x$.

55. Replace x by -6 in $2x - 3 > -11$.

$$2(-6) - 3 > -11$$
$$-15 > -11$$

Since the last inequality is incorrect, -6 does not satisfy $2x - 3 > -11$.

57. Replace x by 3 in $-3x + 4 > -7$.

$$-3(3) + 4 > -7$$
$$-5 > -7$$

Since $-5 > -7$ is correct, 3 satisfies $-3x + 4 > -7$.

59. Replace x by 0 in $3x - 7 \le 5x - 7$.

$$3(0) - 7 \le 5(0) - 7$$
$$-7 \le -7$$

Since $-7 \le -7$ is correct, 0 satisfies $3x - 7 \le 5x - 7$.

61. Replace x by 2.5 in $-10x + 9 \le 3(x + 3)$.

$$-10(0) + 9 \le 3(0 + 3)$$
$$9 \le 9$$

Since the last inequality is correct, 0 satisfies the inequality.

63. Replace x by -7 in $-5 < x < 9$.

$$-5 < -7 < 9$$

Since -7 is not between -5 and 9, -7 does not satisfy $-5 < x < 9$.

65. Replace x by -2 in $-3 \le 2x + 5 \le 9$.

$$-3 \le 2(-2) + 5 \le 9$$
$$-3 \le 1 \le 9$$

Since 1 is between -3 and 9, -2 does satisfy $-3 \le 2x + 5 \le 9$.

67. Replace x by -3.4 in

$$-4.25x - 13.29 < 0.89.$$
$$-4.25(-3.4) - 13.29 < 0.89$$
$$1.16 < 0.89$$

Since $1.16 < 0.89$ is incorrect, -3.4 does not satisfy $-4.25x - 13.29 < 0.89$.

69. Since $-5.1 > -5$ is false, $0 > -5$ is true, and $5.1 > -5$ is true, only 0 and 5.1 satisfy $x > -5$.

71. Since $5 < -5.1$ is false, $5 < 0$ is false, and $5 < 5.1$ is true, only 5.1 satisfies $5 < x$.

73. Only 5.1 is between 5 and 7.

75. All three given numbers satisfy $-6 < -x < 6$.

77. Let $p =$ the sale price of the car and $0.08p =$ the amount of sales tax. The sales tax was more than \$1500 is expressed as $0.08p > 1500$.

79. Let $p =$ the price of an order of fries, $2p =$ the price of a hamburger, and $p + 0.25 =$ the price of a Coke. If the price of all three is under \$2.00, then we can write $p + 2p + p + 0.25 < 2.00$.

81. Let $s =$ his score on the remaining test. The average is found by adding the scores and then dividing by 3. Since the average must be at least 60, we can write $\frac{44 + 72 + s}{3} \ge 60$.

83. Let $R =$ his speed and $8R =$ his daily distance. His distance was between 396 and 453 is expressed as $396 < 8R < 453$.

85. The angle at the base of the ladder is $90 - x$. So $60 < 90 - x < 70$.

87. a) The girth is the sum of the length, twice the width and twice the height (h). So

$$45 + 2(30) + 2h \le 130.$$

b) From the graph you can see that 130 inches of girth corresponds to about 12 in. in height. So the maximum height is 12 in.

89. In the formula $r = \frac{Nw}{n}$ let $w = 27$, $N = 50$, and $n = 17$:

$$r = \frac{50 \cdot 27}{17} \approx 79$$

The gear ratio is approximately 79 and according to the chart it is used for moderate effort on level ground.

2.9 WARM-UPS

1. True, because dividing each side of $2x > 18$ by 2 gives $x > 9$.

2. False, because 0 satisfies $x < 5$ but 0 does not satisfy $x - 5 > 0$.

3. False, we cannot divide each side by 0.

4. True, because dividing each side of $-2x \le 6$ by 2 gives $-x \le 3$.

5. False, because x is at most 7 means that x is less than or equal to 7.

6. True, because at least means greater than or equal to.

7. False, because if x is not more than 85 then x is less than or equal to 85.

8. True, because in either inequality x is greater than -9 and x is less than -3.

9. True, because $1.08x$ gives the truck price plus the sales tax.

10. False, because at least means greater than or equal to.

2.9 EXERCISES

1. Equivalent inequalities are inequalities that have the same solutions.

3. According to the multiplication property of inequality, the inequality symbol is reversed when multiplying (or dividing) by a negative number and not reversed when multiplying (or dividing) by a positive number.

5. We solve compound inequalities using the properties of inequality as we do for simple inequalities.

7. Subtract 7 from each side to get $x > -7$.

9. Divide each side by 3 to get $3 \le w$, or $w \ge 3$.

11. Divide each side by -4 and reverse the inequality to get $k > 1$.

13. Multiply each side by -2 and reverse the inequality to get $y \le -8$.

15.
$$\begin{aligned} x + 3 &> 0 \\ x + 3 - 3 &> 0 - 3 \\ x &> -3 \end{aligned}$$
The solution set is the interval $(-3, \infty)$.

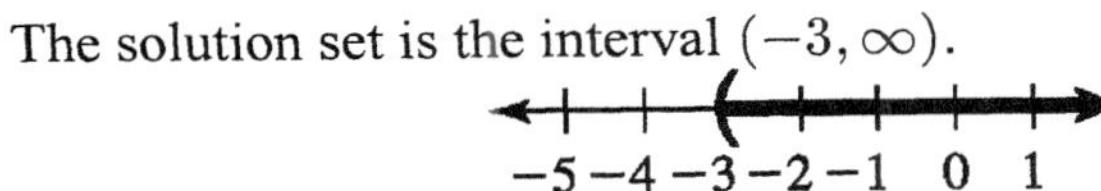

17.
$$\begin{aligned} -3 &< w - 1 \\ -2 &< w \\ w &> -2 \end{aligned}$$
The solution set is the interval $(-2, \infty)$.

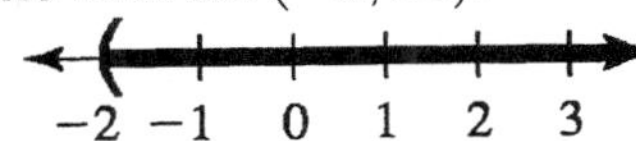

19.
$$\begin{aligned} 8 &> 2b \\ 4 &> b \\ b &< 4 \end{aligned}$$
The solution set is the interval $(-\infty, 4)$.

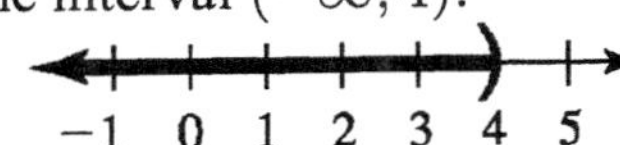

21.
$$\begin{aligned} -8z &\le 4 \\ z &\ge -\tfrac{1}{2} \end{aligned}$$
The solution set is the interval $[-1/2, \infty)$.

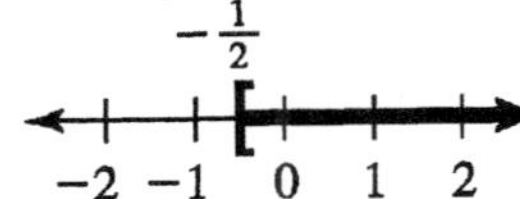

23.
$$\begin{aligned} 3y - 2 &< 7 \\ 3y &< 9 \\ y &< 3 \end{aligned}$$
The solution set is the interval $(-\infty, 3)$.

−1 0 1 2 3 4 5

25.
$$\begin{aligned} 3 - 9z &\le 6 \\ -9z &\le 3 \\ \frac{-9z}{-9} &\ge \frac{3}{-9} \\ z &\ge -\frac{1}{3} \end{aligned}$$
The solution set is the interval $[-1/3, \infty)$.

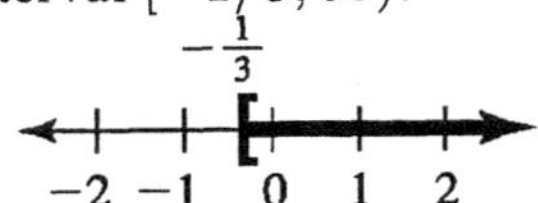

27.
$$\begin{aligned} 6 &> -r + 3 \\ r &> -3 \end{aligned}$$
The solution set is the interval $(-3, \infty)$.

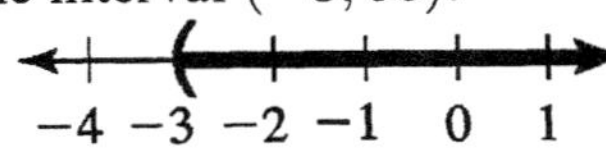

29.
$$\begin{aligned} 5 - 4p &> -8 - 3p \\ -p &> -13 \\ p &< 13 \end{aligned}$$
The solution set is the interval $(-\infty, 13)$.

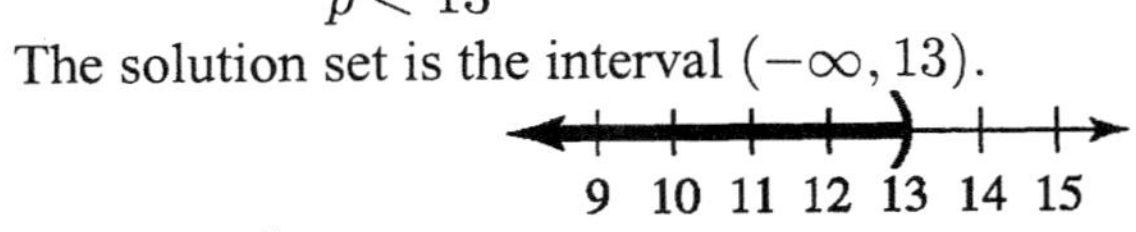

31.
$$\begin{aligned} -\frac{5}{6}q &\ge -20 \\ -\frac{6}{5}\left(-\frac{5}{6}\right)q &\le -\frac{6}{5}(-20) \\ q &\le 24 \end{aligned}$$
The solution set is the interval $(-\infty, 24]$.

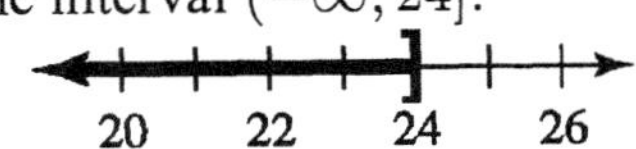

33.
$$\begin{aligned} 1 - \frac{1}{4}t &\ge \frac{1}{8} \\ -\frac{1}{4}t &\ge -\frac{7}{8} \\ -4\left(-\frac{1}{4}t\right) &\le -4\left(-\frac{7}{8}\right) \\ t &\le \frac{7}{2} \end{aligned}$$
The solution set is the interval $(-\infty, 7/2]$.

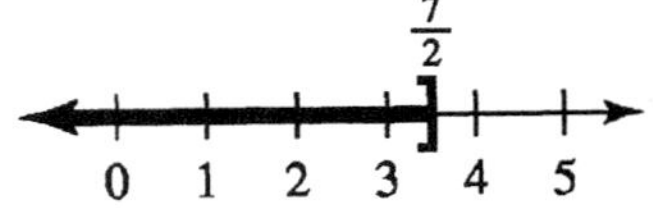

35.
$$\begin{aligned} 0.1x + 0.35 &> 0.2 \\ 10x + 35 &> 20 \\ 10x &> -15 \\ x &> -1.5 \end{aligned}$$
The solution set is the interval $(-1.5, \infty)$.

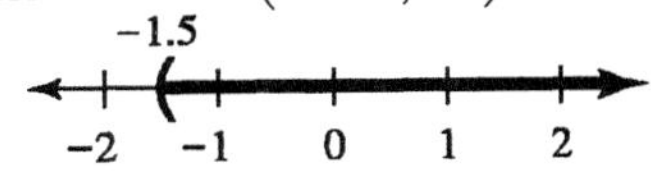

37.
$$\begin{aligned} 2x+5 &< x-6 \\ x+5 &< -6 \\ x &< -11 \end{aligned}$$
The solution set is the interval $(-\infty, -11)$.

−15 −14 −13 −12 −11 −10 −9

39.
$$\begin{aligned} x-4 &< 2(x+3) \\ x-4 &< 2x+6 \\ -4 &< x+6 \\ -10 &< x \\ x &> -10 \end{aligned}$$
The solution set is the interval $(-10, \infty)$.

−12 −11 −10 −9 −8 −7 −6

41.
$$\begin{aligned} 0.52x - 35 &< 0.45x + 8 \\ 0.52x &< 0.45x + 43 \\ 0.07x &< 43 \\ x &< \frac{43}{0.07} \\ x &< 614.3 \end{aligned}$$
The solution set is the interval $(-\infty, 614.3)$.

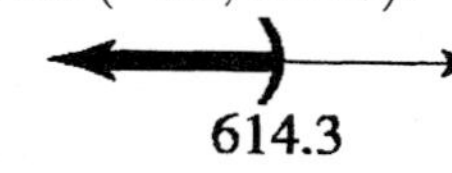

43.
$$\begin{aligned} 5 &< x-3 < 7 \\ 5+3 &< x-3+3 < 7+3 \\ 8 &< x < 10 \end{aligned}$$
The solution set is the interval $(8, 10)$.

6 7 8 9 10 11 12

45.
$$\begin{aligned} 3 &< 2v+1 < 10 \\ 2 &< 2v < 9 \\ 1 &< v < \frac{9}{2} \end{aligned}$$
The solution set is the interval $(1, 9/2)$.

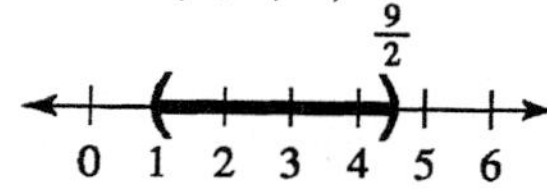

47.
$$\begin{aligned} -4 &\le 5-k \le 7 \\ -9 &\le -k \le 2 \\ (-1)(-9) &\ge (-1)(-k) \ge (-1)(2) \\ 9 &\ge k \ge -2 \\ -2 &\le k \le 9 \end{aligned}$$
The solution set is the interval $[-2, 9]$.

9

−2 0 2 4 6 8 10

49.
$$\begin{aligned} -2 &< 7-3y \le 22 \\ -9 &< -3y \le 15 \\ 3 &> y \ge -5 \\ -5 &\le y < 3 \end{aligned}$$
The solution set is the interval $[-5, 3)$.

−5 −4 −3 −2 −1 0 1 2 3

51.
$$\begin{aligned} 5 &< \frac{2u}{3} - 3 < 17 \\ 8 &< \frac{2u}{3} < 20 \\ 24 &< 2u < 60 \\ 12 &< u < 30 \end{aligned}$$
The solution set is the interval $(12, 30)$.

12 18 24 30

53.
$$\begin{aligned} -2 &< \frac{4m-4}{3} \le \frac{2}{3} \\ -6 &< 4m-4 \le 2 \\ -2 &< 4m \le 6 \\ -\frac{1}{2} &< m \le \frac{3}{2} \end{aligned}$$
The solution set is the interval $(-1/2, 3/2]$.

$-\frac{1}{2}$ $\frac{3}{2}$

−2 −1 0 1 2 3

55.
$$\begin{aligned} 0.02 &< 0.54 - 0.0048x < 0.05 \\ -0.52 &< -0.0048x < -0.49 \\ \frac{-0.52}{-0.0048} &> \frac{-0.0048x}{-0.0048} > \frac{-0.49}{-0.0048} \\ 108.3 &> x > 102.1 \\ 102.1 &< x < 108.3 \end{aligned}$$
The solution set is the interval $(102.1, 108.3)$.

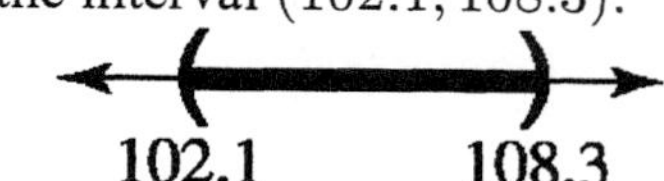

57.
$$\begin{aligned} \frac{1}{2}x - 1 &\le 4 - \frac{1}{3}x \\ 6\left(\frac{1}{2}x - 1\right) &\le 6\left(4 - \frac{1}{3}x\right) \\ 3x - 6 &\le 24 - 2x \\ 5x &\le 30 \\ x &\le 6 \end{aligned}$$
The solution set is the interval $(-\infty, 6]$.

1 2 3 4 5 6 7

59.
$$\begin{aligned} \frac{1}{2}\left(x - \frac{1}{4}\right) &> \frac{1}{4}\left(6x - \frac{1}{2}\right) \\ \frac{1}{2}x - \frac{1}{8} &> \frac{3}{2}x - \frac{1}{8} \\ 4x - 1 &> 12x - 1 \\ -8x &> 0 \\ x &< 0 \end{aligned}$$
The solution set is the interval $(-\infty, 0)$.

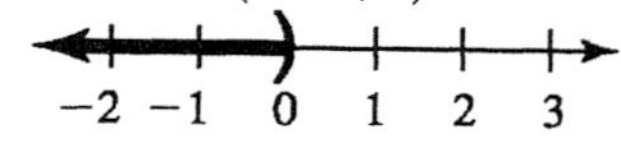

61. $\frac{1}{3} < \frac{1}{4}x - \frac{1}{6} < \frac{7}{12}$
$4 < 3x - 2 < 7$
$6 < 3x < 9$
$2 < x < 3$

The solution set is the interval $(2, 3)$.

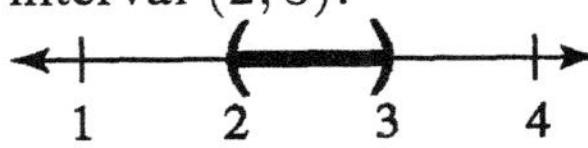

63. Let $w =$ the width and $w + 4 =$ the length. The perimeter is $2w + 2(w + 4)$. Since the perimeter is at least 120, we can write the following inequality.

$$2w + 2(w + 4) \geq 120$$
$$4w + 8 \geq 120$$
$$4w \geq 112$$
$$w \geq 28$$

The width must be at least 28meters.

65. Let $x =$ the price of the car. Since the tax is $0.05x$, we can write the following inequality.

$$x + 0.05x + 144 < 9970$$
$$1.05x < 9826$$
$$x < 9358$$

The price of the car must be less than \$9358.

67. Let $x =$ the price of the microwave. The cost of the microwave plus the tax is $1.08x$. Since she has at most \$594, we can write the following inequality.

$$1.08x \leq 594$$
$$x \leq \frac{594}{1.08}$$
$$x \leq 550$$

The price of the microwave is at most \$550.

69. Let $x =$ Tilak's score on the last test. His average for the three tests is $\frac{44 + 72 + x}{3}$. Since his test average must be at least 60, we can write the following inequality.

$$\frac{44 + 72 + x}{3} \geq 60$$
$$44 + 72 + x \geq 180$$
$$116 + x \geq 180$$
$$x \geq 64$$

He must score at least 64 on the last test to pass the course.

71. Let $x =$ the final exam score. Stacy's semester average is $\frac{1}{3}(48) + \frac{2}{3}x$. Since the semester average must be between 70 and 79 inclusive, we can write the following inequality.

$$70 \leq \frac{1}{3}(48) + \frac{2}{3}x \leq 79$$
$$70 \leq 16 + \frac{2}{3}x \leq 79$$
$$54 \leq \frac{2}{3}x \leq 63$$
$$\frac{3}{2} \cdot 54 \leq \frac{3}{2} \cdot \frac{2}{3}x \leq \frac{3}{2} \cdot 63$$
$$81 \leq x \leq 94.5$$

To get a C, Stacy must score between 81 and 94.5 inclusive on the final exam.

73. Let $x =$ her average speed for a day. Her distance each day was $8x$. Since her distance was between 396 and 453 we can write the following inequality.

$$396 < 8x < 453$$
$$49.5 < x < 56.625$$

Her average speed each day was between 49.5 and 56.625 miles per hour.

75. The supplement to the 85° angle is 95°. The angle at the lighthouse is $180 - 95 - x$ degrees. Since the angle at the lighthouse is less then 30° we have the following inequality.

$$180 - 95 - x < 30$$
$$85 - x < 30$$
$$-x < -55$$
$$x > 55$$

So x must be greater than 55°. From the diagram, x must be less than 85°. So x is between 55° and 85°.

77. a) Since $60 < r < 80$, we have the following inequality.

$$60 < \frac{N \cdot 27}{12} < 80$$
$$720 < 27N < 960$$
$$26.7 < N < 35.6$$

The gear ratio is between 60 and 80 if the number of teeth on the chain ring is between 27 and 35 inclusive.

b) $65 < \frac{48 \cdot w}{17} < 70$

$$1105 < 48w < 1190$$
$$23.02 < w < 24.79$$

The wheel diameter is between 23.02 in. and 24.79 in.

c) $\frac{40 \cdot 26}{n} < 75$

$$1040 < 75n$$
$$13.87 < n$$

Because n is a positive integer, we did not reverse the inequality when we multiplied each side by n. The number of teeth on the cog is greater than or equal to 14.

Enriching Your Mathematical Word Power

1. b	**2.** d	**3.** c	**4.** c	**5.** d
6. d	**7.** a	**8.** b	**9.** c	**10.** d

CHAPTER 2 REVIEW

1.
$$x - 23 = 12$$
$$x - 23 + 23 = 12 + 23$$
$$x = 35$$
The solution set is $\{35\}$.

3.
$$\frac{2}{3}u = -4$$
$$\frac{3}{2} \cdot \frac{2}{3}u = \frac{3}{2}(-4)$$
$$u = -6$$
The solution set is $\{-6\}$.

5.
$$-5y = 35$$
$$\frac{-5y}{-5} = \frac{35}{-5}$$
$$y = -7$$
The solution set is $\{-7\}$.

7.
$$6m = 13 + 5m$$
$$m = 13$$
The solution set is $\{13\}$.

9.
$$2x - 5 = 9$$
$$2x = 14$$
$$x = 7$$
The solution set is $\{7\}$.

11.
$$3p - 14 = -4p$$
$$3p = -4p + 14$$
$$7p = 14$$
$$p = 2$$
The solution set is $\{2\}$.

13.
$$2z + 12 = 5z - 9$$
$$2z = 5z - 21$$
$$-3z = -21$$
$$z = 7$$
The solution set is $\{7\}$.

15.
$$2(h - 7) = -14$$
$$2h - 14 = -14$$
$$2h = 0$$
$$h = 0$$
The solution set is $\{0\}$.

17.
$$3(w - 5) = 6(w + 2) - 3$$
$$3w - 15 = 6w + 12 - 3$$
$$-3w = 24$$
$$w = -8$$
The solution set is $\{-8\}$.

19.
$$2(x - 7) - 5 = 5 - (3 - 2x)$$
$$2x - 14 - 5 = 5 - 3 + 2x$$
$$2x - 19 = 2 + 2x$$
$$-19 = 2$$
There is no solution to this equation. It is an inconsistent equation. The solution set is $\emptyset$.

21.
$$2(w - w) = 0$$
$$2(0) = 0$$
$$0 = 0$$
All real numbers satisfy this equation. It is an identity.

23.
$$\frac{3r}{3r} = 1$$
$$\frac{r}{r} = 1$$
A number divided by itself is 1 except for $0/0$, which is undefined. The solution set is all real numbers except 0. It is an identity.

25.
$$\frac{1}{2}a - 5 = \frac{1}{3}a - 1$$
$$6\left(\frac{1}{2}a - 5\right) = 6\left(\frac{1}{3}a - 1\right)$$
$$3a - 30 = 2a - 6$$
$$a - 30 = -6$$
$$a = 24$$
The solution set is $\{24\}$. It is a conditional equation.

27.
$$0.06q + 14 = 0.3q - 5.2$$
$$0.06q = 0.3q - 19.2$$
$$-0.24q = -19.2$$
$$q = 80$$
The solution set is $\{80\}$. It is a conditional equation.

29.
$$0.05(x + 100) + 0.06x = 115$$
$$0.05x + 5 + 0.06x = 115$$
$$0.11x = 110$$
$$x = 1000$$
The solution set is $\{1000\}$. It is a conditional equation.

31.
$$2x + \frac{1}{2} = 3x + \frac{1}{4}$$
$$8x + 2 = 12x + 1$$
$$-4x = -1$$
$$x = \frac{1}{4}$$
The solution set is $\left\{\frac{1}{4}\right\}$.

33. $\frac{x}{2} - \frac{3}{4} = \frac{x}{6} + \frac{1}{8}$
$$12x - 18 = 4x + 3$$
$$8x = 21$$
$$x = \frac{21}{8}$$
The solution set is $\left\{\frac{21}{8}\right\}$.

35. $\frac{5}{6}x = -\frac{2}{3}$
$$x = \frac{6}{5}\left(-\frac{2}{3}\right) = -\frac{4}{5}$$
The solution set is $\left\{-\frac{4}{5}\right\}$.

37. $-\frac{1}{2}(x - 10) = \frac{3}{4}x$
$$-\frac{1}{2}x + 5 = \frac{3}{4}x$$
$$-2x + 20 = 3x$$
$$-5x = -20$$
$$x = 4$$
The solution set is $\{4\}$.

39. $3 - 4(x - 1) + 6 = -3(x + 2) - 5$
$$3 - 4x + 4 + 6 = -3x - 6 - 5$$
$$-4x + 13 = -3x - 11$$
$$-x = -24$$
$$x = 24$$
The solution set is $\{24\}$.

41. $5 - 0.1(x - 30) = 18 + 0.05(x + 100)$
$$5 - 0.1x + 3 = 18 + 0.05x + 5$$
$$-0.1x + 8 = 23 + 0.05x$$
$$-0.15x = 15$$
$$x = -100$$
The solution set is $\{-100\}$.

43. $ax + b = 0$
$$ax = -b$$
$$x = -\frac{b}{a}$$

45. $ax - 2 = b$
$$ax = b + 2$$
$$x = \frac{b + 2}{a}$$

47. $LWx = V$
$$x = \frac{V}{LW}$$

49. $2x - b = 5x$
$$-b = 3x$$
$$-\frac{b}{3} = x$$
$$x = -\frac{b}{3}$$

51. $5x + 2y = 6$
$$2y = -5x + 6$$
$$y = -\frac{5}{2}x + 3$$

53. $y - 1 = -\frac{1}{2}(x - 6)$
$$y - 1 = -\frac{1}{2}x + 3$$
$$y = -\frac{1}{2}x + 4$$

55. $\frac{1}{2}x + \frac{1}{4}y = 4$
$$\frac{1}{4}y = -\frac{1}{2}x + 4$$
$$4 \cdot \frac{1}{4}y = 4\left(-\frac{1}{2}x + 4\right)$$
$$y = -2x + 16$$

57. Use -3 for x in $y = 3x - 4$.
$$y = 3(-3) - 4 = -13$$

59. Use -3 for x in $5xy = 6$.
$$5(-3)y = 6$$
$$-15y = 6$$
$$y = -\frac{6}{15} = -\frac{2}{5}$$

61. Use -3 for x in $y - 3 = -2(x - 4)$.
$$y - 3 = -2(-3 - 4)$$
$$y - 3 = 14$$
$$y = 17$$

63. $y = -5x + 10$
If $x = -1$, then $y = -5(-1) + 10 = 15$.
If $x = 0$, then $y = -5(0) + 10 = 10$.
If $x = 1$, then $y = -5(1) + 10 = 5$.
If $x = 2$, then $y = -5(2) + 10 = 0$.
If $x = 3$, then $y = -5(3) + 10 = -5$.

x	y
-1	15
0	10
1	5
2	0
3	-5

65. For each value of x find the y-value by using the formula $y = \frac{2}{3}x - 1$ as in the previous exercise.

x	y
-3	-3
0	-1
3	1
6	3

67. Sum indicates addition. So the sum of a number (x) and 9 is $x + 9$.

69. If two numbers differ by 8, then one number is 8 larger than the other. So if x is the

smaller, then x and $x + 8$ are used to represent the numbers. We could also use x and $x - 8$ with x being the larger number.

71. Sixty-five percent of a number is 0.65 times the number (x) or $0.65x$.

73. If x represents the width, then $x + 5$ is the length. Since the area of the rectangle is 98, we can write $x(x + 5) = 98$.

75. Let $x - 10 =$ Barbara's speed and $x =$ Jim's speed. In 3 hours Barbara travels $3(x - 10)$ miles. In 2 hours Jim travels $2x$ miles. Since the distances are the same, we can write the equation $2x = 3(x - 10)$.

77. If x is the first even integer, then $x + 2$ and $x + 4$ represent the second and the third. Since their sum is 90, we can write the equation $x + x + 2 + x + 4 = 90$.

79. Since the sum of the measures of a triangle is 180°, we have $t + 2t + t - 10 = 180$.

81. Let $x =$ the first odd integer, $x + 2 =$ the second odd integer, and $x + 4 =$ the third odd integer. Since their sum is 237, we can write the following equation.

$$\begin{aligned} x + x + 2 + x + 4 &= 237 \\ 3x + 6 &= 237 \\ 3x &= 231 \\ x &= 77 \\ x + 2 &= 79 \\ x + 4 &= 81 \end{aligned}$$

The three consecutive odd integers are 77, 79, and 81.

83. Let $x =$ Betty's rate of speed and $x + 15 =$ Lawanda's rate of speed. Since $D = RT$, Betty's distance is $4x$ and Lawanda's distance is $3(x + 15)$. Since their distances are equal, we can write the following equation.

$$\begin{aligned} 4x &= 3(x + 15) \\ 4x &= 3x + 45 \\ x &= 45 \\ x + 15 &= 60 \end{aligned}$$

Betty drives 45 mph and Lawanda drives 60 mph.

85. Let $x =$ the husband's income and $x + 6000 =$ Wanda's income. Wanda saves $0.10(x + 6000)$ and her husband saves $0.06x$. Since they save \$5400 together, we can write the following equation.

$$\begin{aligned} 0.10(x + 6000) + 0.06x &= 5400 \\ 0.10x + 600 + 0.06x &= 5400 \\ 0.16x &= 4800 \\ x &= 30{,}000 \\ x + 6000 &= 36{,}000 \end{aligned}$$

Wanda makes \$36,000 and her husband makes \$30,000 per year.

87. Use 3 for x in $-2x + 5 \le x - 6$.

$$\begin{aligned} -2(3) + 5 &\le 3 - 6 \\ -1 &\le -3 \end{aligned}$$

Since this inequality is incorrect, 3 is not a solution to $-2x + 5 \le x - 6$.

89. Use -1 for x in $-2 \le 6 + 4x < 0$.

$$-2 \le 6 + 4(-1) < 0$$
$$-2 \le 2 < 0$$

Since this last inequality is incorrect, -1 is not a solution to $-2 \le 6 + 4x < 0$.

91. The graph shows the numbers to the right of 1 on the number line. This graph indicates the solution to $x > 1$. The solution set is the interval $(1, \infty)$.

93. The graph shows the number to the right of and including 2. This graph indicates the solution to $x \ge 2$. The solution set is the interval $[2, \infty)$.

95. The graph shows the numbers between -3 and 3, including -3 but not 3. This graph indicates the solution to $-3 \le x < 3$. The solution set is the interval $[-3, 3)$.

97. The graph shows the numbers to the left of -1 on the number line. This graph indicates the solution to $x < -1$. The solution set is the interval $(-\infty, -1)$.

99. $\quad x + 2 > 1$

$$x > -1$$

The solution set is the interval $(-1, \infty)$.

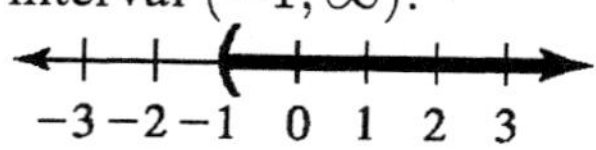

101. $3x - 5 < x + 1$

$$2x < 6$$
$$x < 3$$

The solution set is the interval $(-\infty, 3)$.

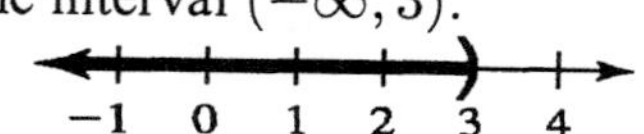

103. $\quad -\frac{3}{4}x \ge 3$

$$-\frac{4}{3}\left(-\frac{3}{4}x\right) \le -\frac{4}{3} \cdot 3$$
$$x \le -4$$

The solution set is the interval $(-\infty, -4]$.

$-8\ -7\ -6\ -5\ -4\ -3\ -2$

105. $3 - 2x < 11$

$-2x < 8$

$x > -4$

The solution set is the interval $(-4, \infty)$.

$-6\ -5\ -4\ -3\ -2\ -1\ \ 0$

107. $-3 < 2x - 1 < 9$

$-2 < 2x < 10$

$-1 < x < 5$

The solution set is the interval $(-1, 5)$.

$-1\ \ 0\ \ 1\ \ 2\ \ 3\ \ 4\ \ 5$

109. $0 \leq 1 - 2x < 5$

$-1 \leq -2x < 4$

$\frac{1}{2} \geq x > -2$

$-2 < x \leq \frac{1}{2}$

The solution set is the interval $(-2, 1/2]$.

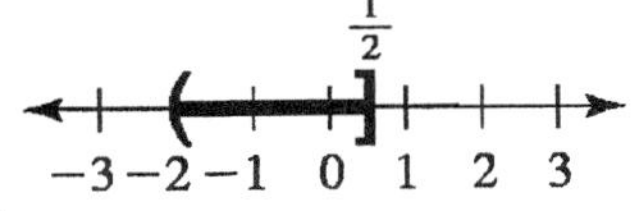

111. $-1 \leq \frac{2x-3}{3} \leq 1$

$-3 \leq 2x - 3 \leq 3$

$0 \leq 2x \leq 6$

$0 \leq x \leq 3$

The solution set is the interval $[0, 3]$.

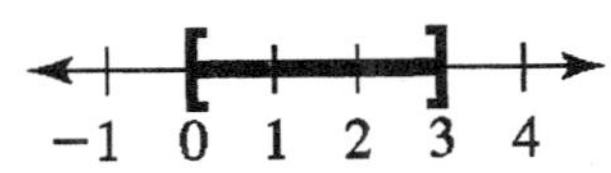

113. $\frac{1}{3} < \frac{1}{3} + \frac{x}{2} < \frac{5}{6}$

$2 < 2 + 3x < 5$

$0 < 3x < 3$

$0 < x < 1$

The solution set is the interval $(0, 1)$.

$-1\quad 0\quad 1\quad 2$

115. The interest rate for a 30-year bond was 5.375%. So the interest is 0.05375(10,000) or \$537.50.

117. Let $x =$ the number of movies at ABC. Since XYZ had 200 movies, the combined store has $x + 200$ movies. The number of children's movies at ABC was $0.60x$ and the number of children's movies after the merger is $0.40(x + 200)$. Since XYZ had no children's movies, these two amounts of children's movies are equal.

$0.60x = 0.40(x + 200)$

$0.60x = 0.40x + 80$

$0.20x = 80$

$x = 400$

So ABC had 400 movies before the merger.

119. Complementary angles have a sum of 90°.

$x + 2x - 3 = 90$

$3x = 93$

$x = 31$

The degree measure is 31°.

121. Let $x =$ the length of the shortest side, $x + 1 =$ the length of the second side, and $2x =$ the length of the third side. Since the perimeter is less than 25 feet we can write the following inequality.

$x + x + 1 + 2x < 25$

$4x + 1 < 25$

$4x < 24$

$x < 6$

The shortest side is less than 6 feet in length.

CHAPTER 2 TEST

1. $-10x - 6 + 4x = -4x + 8$

$-6x - 6 = -4x + 8$

$-2x - 6 = 8$

$-2x = 14$

$x = -7$

The solution set is $\{-7\}$.

2. $5(2x - 3) = x + 3$

$10x - 15 = x + 3$

$10x = x + 18$

$9x = 18$

$x = 2$

The solution set is $\{2\}$.

3. $-\frac{2}{3}x + 1 = 7$

$-\frac{2}{3}x = 6$

$-\frac{3}{2}\left(-\frac{2}{3}x\right) = -\frac{3}{2} \cdot 6$

$x = -9$

The solution set is $\{-9\}$.

4. $x + 0.06x = 742$

$1.06x = 742$

$x = 700$

The solution set is $\{700\}$.

5. $x - 0.03x = 0.97$

$0.97x = 0.97$

$x = 1$

The solution set is $\{1\}$.

6. $6x - 7 = 0$

$6x = 7$

$x = \frac{7}{6}$

The solution set is $\left\{\frac{7}{6}\right\}$.

7. $\frac{1}{2}x - \frac{1}{3} = \frac{1}{4}x + \frac{1}{6}$

$6x - 4 = 3x + 2$

$3x = 6$

$x = 2$

The solution set is $\{2\}$.

8. $2(x + 6) = 2x - 5$

$2x + 12 = 2x - 5$

$12 = -5$

The solution set is $\emptyset$.

9. $x + 7x = 8x$

$8x = 8x$

All real numbers satisfy the equation.

10. $2x - 3y = 9$

$-3y = -2x + 9$

$-\frac{1}{3}(-3y) = -\frac{1}{3}(-2x + 9)$

$y = \frac{2}{3}x - 3$

11. $m = aP - w$

$m + w = aP$

$\frac{m + w}{P} = a$

$a = \frac{m + w}{P}$

12. The graph shows the numbers between -3 and 2, including 2 but not including -3. This graph is the solution set to $-3 < x \leq 2$. The solution set is the interval $(-3, 2]$.

13. The graph shows the numbers to the right of 1 on the number line. This graph is the solution to the inequality $x > 1$. The solution set is the interval $(1, \infty)$.

14. $4 - 3(w - 5) < -2w$

$4 - 3w + 15 < -2w$

$19 - 3w < -2w$

$19 < w$

$w > 19$

The solution set is the interval $(19, \infty)$.

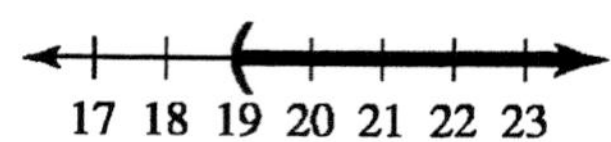

15. $1 < \frac{1 - 2x}{3} < 5$

$3 < 1 - 2x < 15$

$2 < -2x < 14$

$-1 > x > -7$

$-7 < x < -1$

The solution set is the interval $(-7, -1)$.

−7 −6 −5 −4 −3 −2 −1

16. $1 < 3x - 2 < 7$

$3 < 3x < 9$

$1 < x < 3$

The solution set is the interval $(1, 3)$.

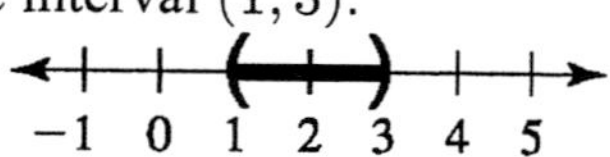

17. $-\frac{2}{3}y < 4$

$-\frac{3}{2}\left(-\frac{2}{3}y\right) > -\frac{3}{2} \cdot 4$

$y > -6$

The solution set is the interval $(-6, \infty)$.

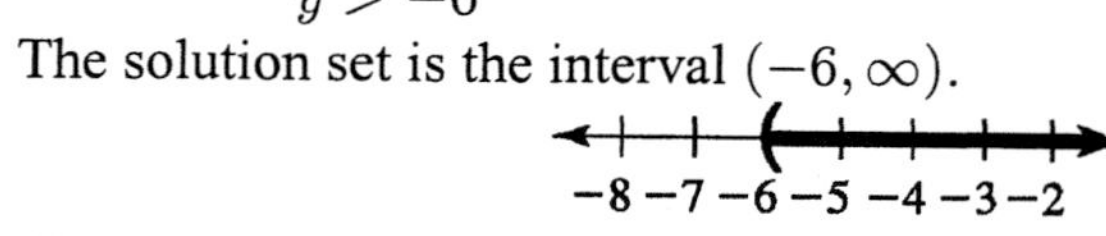

18. Let $x =$ the width and $x + 8 =$ the length. Since perimeter is 72, we can write the following equation.

$2x + 2(x + 8) = 72$

$4x + 16 = 72$

$4x = 56$

$x = 14$

$x + 8 = 22$

The width of the rectangle is 14 meters.

19. Use $A = 54$ and $b = 12$ in the formula for the area of a triangle, $A = \frac{1}{2}bh$.

$54 = \frac{1}{2} \cdot 12h$

$54 = 6h$

$9 = h$

The height is 9 inches.

20. Let $x =$ the number of liters of 20% solution. If she mixes the 20% solution with 50 liters of 60% solution she will obtain $x + 50$ liters of 30% solution. The amount of alcohol in the 20% solution is $0.20x$. The amount of alcohol in the 60% solution is $0.60(50)$. The amount of alcohol in the final 30% solution is $0.30(x + 50)$. The alcohol in the final solution is the total of the alcohol in the two solutions.

$$\begin{aligned} 0.20x + 0.60(50) &= 0.30(x + 50) \\ 0.20x + 30 &= 0.30x + 15 \\ 0.20x + 15 &= 0.30x \\ 15 &= 0.10x \\ 150 &= x \end{aligned}$$

She should use 150 liters of 20% solution.

21. Let $x =$ the original price of the diamonds. His discount is $0.40x$. The price he pays is $x - 0.40x + 250$.

$$\begin{aligned} x - 0.40x + 250 &\le 1450 \\ 0.60x &\le 1200 \\ x &\le 2000 \end{aligned}$$

The original price of the diamonds can be at most \$2000.

22. If $x =$ the degree measure of the smallest angle, then the degree measures of the other two are $2x$ and $3x$.

$$\begin{aligned} x + 2x + 3x &= 180 \\ 6x &= 180 \\ x &= 30 \end{aligned}$$

The angles are 30°, 60°, and 90°.

Making Connections

Chapters 1-2

1. $3x + 5x = (3 + 5)x = 8x$

2. $3x \cdot 5x = 3 \cdot 5 \cdot x \cdot x = 15x^2$

3. $\frac{4x+2}{2} = \frac{1}{2}(4x + 2) = \frac{1}{2} \cdot 4x + \frac{1}{2} \cdot 2$
$= 2x + 1$

4. $5 - 4(3 - x) = 5 - 12 + 4x = 4x - 7$

5. $3x + 8 - 5(x - 1) = 3x + 8 - 5x + 5$
$= -2x + 13$

6. $(-6)^2 - 4(-3)2 = 36 - (-24) = 36 + 24$
$= 60$

7. $3^2 \cdot 2^3 = 9 \cdot 8 = 72$

8. $4(-7) - (-6)(3) = -28 + 18 = -10$

9. $-2x \cdot x \cdot x = -2x^3$

10. $(-1)(-1)(-1)(-1)(-1) = -1$

11. $\frac{1}{2} + \frac{1}{3} = \frac{1}{2} \cdot \frac{3}{3} + \frac{1}{3} \cdot \frac{2}{2} = \frac{5}{6}$

12. $\frac{1}{2} - \frac{1}{3} = \frac{1}{2} \cdot \frac{3}{3} - \frac{1}{3} \cdot \frac{2}{2} = \frac{1}{6}$

13. $\frac{5}{3} \cdot \frac{1}{15} = \frac{5}{45} = \frac{1}{9}$

14. $\frac{2}{3} \cdot \frac{5}{6} = \frac{10}{18} = \frac{5}{9}$

15. $6 \cdot \left(\frac{5}{3} + \frac{1}{2}\right) = 10 + 3 = 13$

16. $15\left(\frac{2}{3} - \frac{2}{15}\right) = 10 - 2 = 8$

17. $4 \cdot \left(\frac{x}{2} + \frac{1}{4}\right) = 2x + 1$

18. $12\left(\frac{5}{6}x - \frac{3}{4}\right) = 10x - 9$

19.
$$\begin{aligned} x - \frac{1}{2} &= \frac{1}{6} \\ 6x - 3 &= 1 \\ 6x &= 4 \\ x &= \frac{2}{3} \end{aligned}$$

The solution set is $\left\{\frac{2}{3}\right\}$.

20.
$$\begin{aligned} x + \frac{1}{3} &= \frac{1}{2} \\ x &= \frac{1}{2} - \frac{1}{3} \\ x &= \frac{1}{6} \end{aligned}$$

The solution set is $\left\{\frac{1}{6}\right\}$.

21.
$$\begin{aligned} x - \frac{1}{2} &> \frac{1}{6} \\ x &> \frac{1}{6} + \frac{1}{2} \\ x &> \frac{2}{3} \end{aligned}$$

The solution set is the interval $(2/3, \infty)$.

22.
$$\begin{aligned} x + \frac{1}{3} &\le \frac{1}{2} \\ x &\le \frac{1}{2} - \frac{1}{3} \\ x &\le \frac{1}{6} \end{aligned}$$

The solution set is the interval $(-\infty, 1/6]$.

23.
$$\begin{aligned} \frac{3}{5}x &= \frac{1}{15} \\ 9x &= 1 \\ x &= \frac{1}{9} \end{aligned}$$

The solution set is $\left\{\frac{1}{9}\right\}$.

24.
$$\begin{aligned} \frac{3}{2}x &= \frac{5}{6} \\ 9x &= 5 \\ x &= \frac{5}{9} \end{aligned}$$

The solution set is $\left\{\frac{5}{9}\right\}$.

25.
$$\begin{aligned} -\frac{3}{5}x &\le \frac{1}{15} \\ x &\ge -\frac{5}{3} \cdot \frac{1}{15} \\ x &\ge -\frac{1}{9} \end{aligned}$$

The solution set is the interval $[-1/9, \infty)$.

26.
$$\begin{aligned} -\frac{3}{2}x &> \frac{5}{6} \\ x &< -\frac{2}{3} \cdot \frac{5}{6} \\ x &< -\frac{5}{9} \end{aligned}$$

The solution set is the interval $(-\infty, -5/9)$.

27. $\frac{5}{3}x + \frac{1}{2} = 1$
$10x + 3 = 6$
$10x = 3$
$x = \frac{3}{10}$
The solution set is $\left\{\frac{3}{10}\right\}$.

28. $\frac{2}{3}x - \frac{2}{15} = 2$
$10x - 2 = 30$
$x = \frac{32}{10} = \frac{16}{5}$
The solution set is $\left\{\frac{16}{5}\right\}$.

29. $\frac{x}{2} + \frac{1}{4} = \frac{1}{2}$
$2x + 1 = 2$
$x = \frac{1}{2}$
The solution set is $\left\{\frac{1}{2}\right\}$.

30. $\frac{5}{6}x - \frac{3}{4} = \frac{5}{12}$
$10x - 9 = 5$
$x = \frac{14}{10} = \frac{7}{5}$
The solution set is $\left\{\frac{7}{5}\right\}$.

31. $3x + 5x = 8$
$8x = 8$
$x = 1$
The solution set is $\{1\}$.

32. $3x + 5x = 8x$
$8x = 8x$
All real numbers satisfy this equation.

33. $3x + 5x = 7x$
$8x = 7x$
$8x - 7x = 7x - 7x$
$x = 0$
The solution set is $\{0\}$.

34. $3x + 5 = 8$
$3x = 3$
$x = 1$
The solution set is $\{1\}$.

35. $3x + 5x > 7x$
$8x > 7x$
$x > 0$
The solution set is the interval $(0, \infty)$.

36. $3x + 5x > 8x$
$8x > 8x$
$0 > 0$
Since $0 > 0$ is false, there is no solution to the inequality. The solution set is the empty set $\emptyset$.

37. $3x + 1 = 7$
$3x = 6$
$x = 2$
The solution set is $\{2\}$.

38. $5 - 4(3 - x) = 1$
$5 - 12 + 4x = 1$
$-7 + 4x = 1$
$4x = 8$
$x = 2$
The solution set is $\{2\}$.

39. $3x + 8 = 5(x - 1)$
$3x + 8 = 5x - 5$
$3x + 13 = 5x$
$13 = 2x$
$\frac{13}{2} = x$
The solution set is $\left\{\frac{13}{2}\right\}$.

40. $x - 0.05x = 190$
$0.95x = 190$
$x = \frac{190}{0.95} = 200$
The solution set is $\{200\}$.

41. a) $V = C - \frac{C - S}{5}t$
$V = 20{,}000 - \frac{20{,}000 - 4{,}000}{5} \cdot 2$
$= 13{,}600$
The value after 2 years is \$13,600.

b) $14{,}000 = 20{,}000 - \frac{20{,}000 - S}{5} \cdot 3$
$14{,}000 = 20{,}000 - \frac{60{,}000 - 3S}{5}$
$-6{,}000 = -\frac{60{,}000 - 3S}{5}$
$-30{,}000 = -60{,}000 + 3S$
$30{,}000 = 3S$
$S = 10{,}000$
The scrap value is \$10,000.

c) The scrap value is the value at time $t = 5$. From the graph it appears that the value at $t = 5$, the scrap value, is \$12,000.

3.1 WARM-UPS

1. False, because $2(4) - 3(2) \neq -8$.
2. False, because (1, 5) satisfies $y = x + 4$, but (5, 1) does not.
3. False, because the origin is not considered to be in any quadrant.
4. False, because the point (4, 0) is on the x-axis.
5. True, because we have agreed that when a variable is multiplied by 0, it may be omitted.
6. True, because all of the points with an x-coordinate of -5 form a vertical line.
7. True, because the graph consists of all points with a y-coordinate of 6.
8. False, because (0, 2.5) is the y-intercept.
9. False, because the point (5, -3) is in quadrant IV.
10. True, because every point with a y-coordinate of 0 is on the x-axis.

3.1 EXERCISES

1. An ordered pair is a pair of numbers in which there is a first number and a second number, usually written as (a, b).
3. The origin is the point of intersection of the x-axis and y-axis.
5. A linear equation in two variables is an equation of the form $Ax + By = C$ where A and B are not both zero.
7. If $x = 0$ in $y = 3x + 9$, then
$y = 3(0) + 9 = 9$.
If $y = 24$ in $y = 3x + 9$, then $24 = 3x + 9$, $15 = 3x$, or $x = 5$.
If $x = 2$ in $y = 3x + 9$, then
$y = 3(2) + 9 = 15$.
The ordered pairs are (0, 9), (5, 24), and (2, 15).
9. If $x = 0$ in $y = -3x - 7$, then
$y = -3(0) - 7 = -7$. If $x = \frac{1}{3}$ in
$y = -3x - 7$, then $y = -3\left(\frac{1}{3}\right) - 7 = -8$.
Replace y by -5 in $y = -3x - 7$:

$$-5 = -3x - 7$$
$$2 = -3x$$
$$-\frac{2}{3} = x$$

The ordered pairs are $(0, -7)$, $\left(\frac{1}{3}, -8\right)$, and $\left(-\frac{2}{3}, -5\right)$.
11. If $x = 0$ in $y = 1.2x + 54.3$, then
$y = 1.2(0) + 54.3 = 54.3$.
If $x = 10$ in $y = 1.2x + 54.3$, then
$y = 1.2(10) + 54.3 = 66.3$.
If $y = 54.9$ in $y = 1.2x + 54.3$, then
$54.9 = 1.2x + 54.3$, or $x = 0.5$.
The ordered pairs are (0, 54.3), (10, 66.3), and (0.5, 54.9).
13. Solve $2x - 3y = 6$ for y to get
$y = \frac{2}{3}x - 2$. Use this equation to calculate y for each given x as in the previous exercise.
The ordered pairs are (3, 0), (0, -2), and (12, 6).
15. Since the given equation simplifies to $x = 5$, the x coordinate in every ordered pair is 5. The ordered pairs are (5, -3), (5, 5), and (5, 0).
17. To plot (1, 5), start at the origin and move 1 unit to the right and 5 units up.
19. To plot (-2, 1) start at the origin and move 2 units to the left and then 1 unit up.
21. To plot (3, $-1/2$), start at the origin and move 3 units to the right and 1/2 unit down.
23. To plot (-2, -4), start at the origin and move 2 units to the left and 4 units down.
25. To plot (0, 3) move 3 units up from the origin.
27. To plot (-3, 0), move 3 units to the left of the origin.
29. To plot (π, 1) move approximately 3.14 units to the right of the origin, and then 1 unit up.
31. To plot (1.4, 4), start at the origin and move approximately 1.4 units to the right and then move 4 units up.

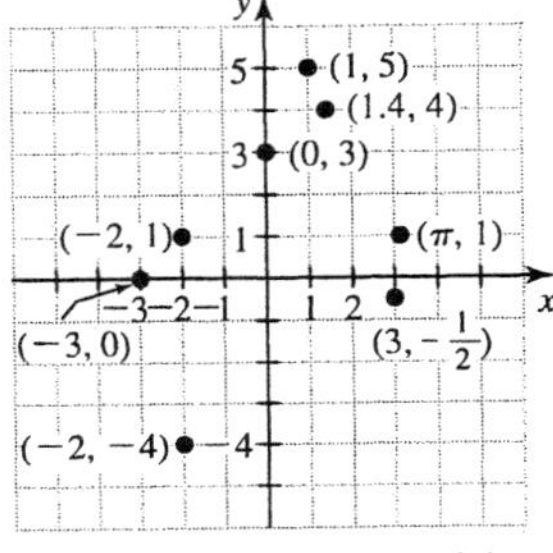

Graph for 17-31 odd.

33. $y = -2x + 5$
If $x = -2$, then $y = -2(-2) + 5 = 9$.
If $x = 0$, then $y = -2(0) + 5 = 5$.
If $x = 2$, then $y = -2(2) + 5 = 1$.
If $y = -3$, then $-3 = -2x + 5$. Solve for x to get $x = 4$.
If $y = -7$, then $-7 = -2x + 5$. Solve for x to get $x = 6$.

x	y
-2	9
0	5
2	1
4	-3
6	-7

35. Use the equation $y = \frac{1}{3}x + 2$ to find the missing coordinates as in the previous exercise.

x	y
-6	0
-3	1
0	2
3	3

37. $y - 20x = 400$ or $y = 20x + 400$
If $x = -30$, then
$y = 20(-30) + 400 = -200$.
If $y = 0$, then $0 - 20x = 400$ or $x = -20$.
If $x = -10$, then $y = 20(-10) + 400 = 200$.
If $x = 0$, then $y = 20(0) + 400 = 400$.
If $y = 600$, then $600 = 20x + 400$. Solve for x to get $x = 10$.

x	y
-30	-200
-20	0
-10	200
0	400
10	600

39. Select five different values for x and use $y = x + 1$ to find the y-coordinates. The ordered pairs (0, 1), (1, 2), (2, 3), (−1, 0), and (−2, −1) satisfy the equation. Plot these points and draw a line through them.

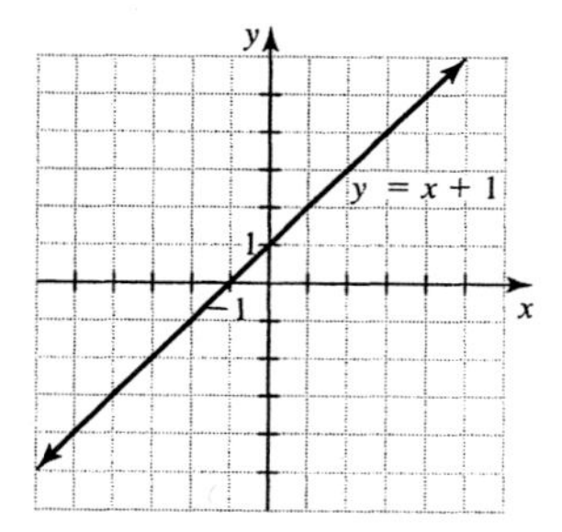

41. Select five different values for x and use $y = 2x + 1$ to find the y-coordinates. The ordered pairs (0, 1), (1, 3), (2, 5), (−1, −1), and (−2, −3) satisfy the equation. Plot these points and draw a line through them.

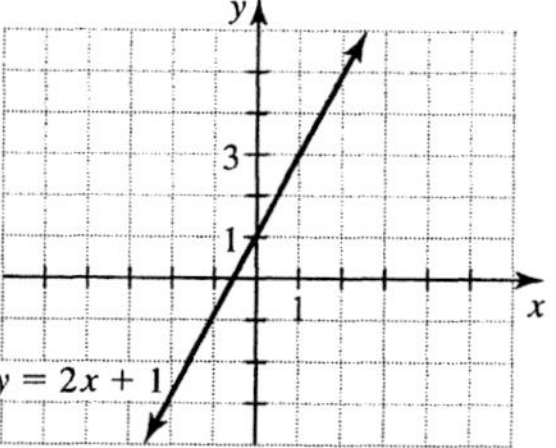

43. Select five different values for x and use $y = 3x - 2$ to find the y-coordinates. The ordered pairs (0, −2), (1, 1), (2, 4), (3, 7), and (−1, −5), satisfy the equation. Plot these points and draw a line through them.

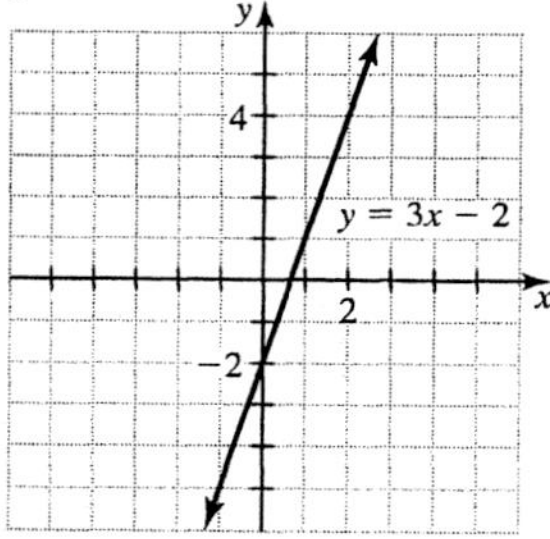

45. Select five different values for x and use $y = x$ to find the y-coordinates. The ordered pairs (0, 0), (1, 1), (2, 2), (−1, −1), and (−2, −2) satisfy the equation. Plot these points and draw a line through them.

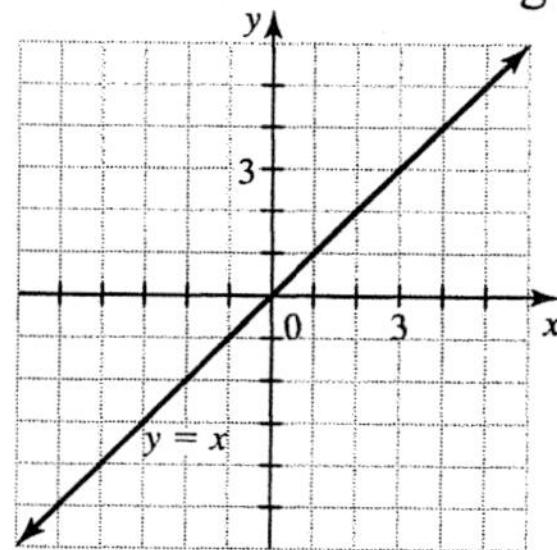

47. Select five different values for x and use $y = 1 - x$ to find the y-coordinates. The ordered pairs (0, 1), (1, 0), (2, −1), (−1, 2), and (−2, 3) satisfy the equation. Plot these points and draw a line through them.

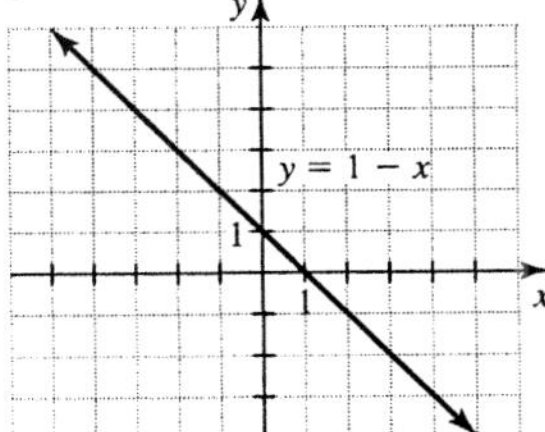

49. Select five different values for x and use $y = -2x + 3$ to find the y-coordinates. The ordered pairs (0, 3), (1, 1), (2, −1), (−1, 5), and (−2, 7) satisfy the equation. Plot these points and draw a line through them.

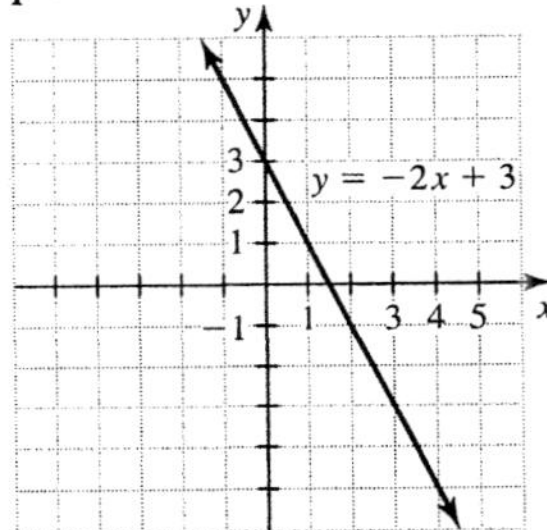

51. Select five different values for x and use $y = -3$ to find the y-coordinates. The ordered pairs (0, −3), (1, −3), (2, −3), (−1, −3), and (−2, −3) satisfy the equation. Plot these points and draw a line through them.

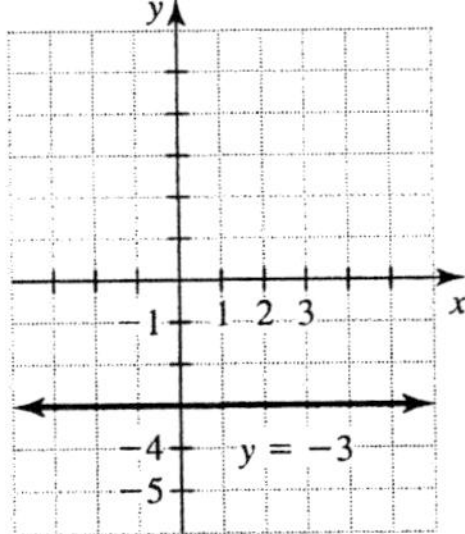

53. The equation $x = 2$ is equivalent to $0 \cdot y + x = 2$. The ordered pairs (2, 1), (2, 2), (2, 3), (2, 0), and (2, −1) satisfy the equation. Plot these points and draw a line through them.

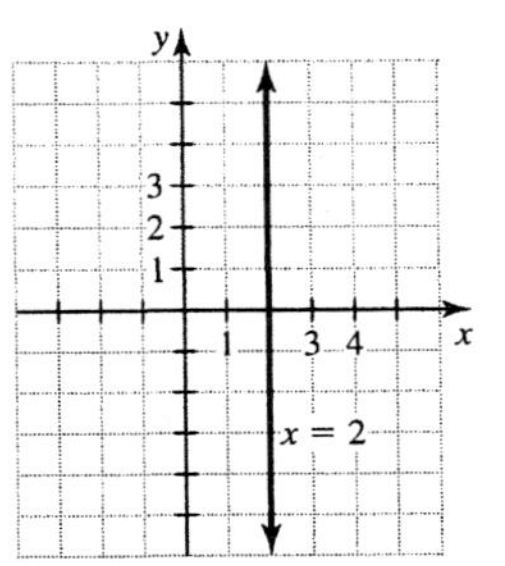

55. The equation $2x + y = 5$ is equivalent to $y = -2x + 5$. Select five different values for x and use $y = -2x + 5$ to find the y-coordinates. The ordered pairs (0, 5), (1, 3), (2, 1), (−1, 7), and (3, −1) satisfy the equation. Plot these points and draw a line through them.

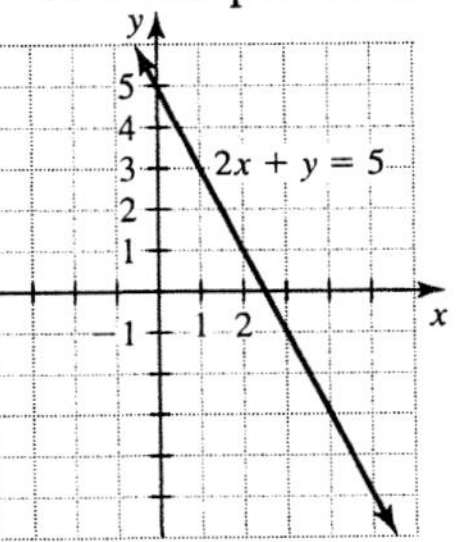

57. The equation $x + 2y = 4$ is equivalent to $y = -\frac{1}{2}x + 2$. The ordered pairs (0, 2), (2, 1), (4, 0), (−2, 3), and (−4, 4) satisfy the equation. Plot these points and draw a line through them.

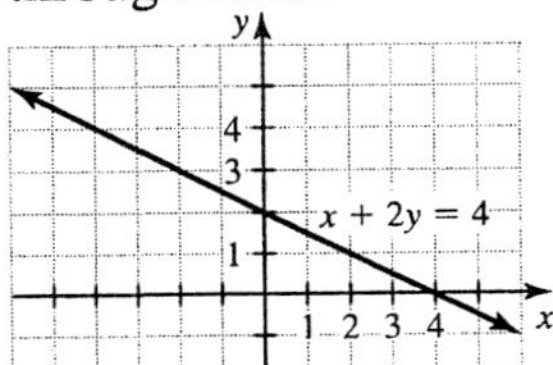

59. The equation $x - 3y = 6$ is equivalent to $y = \frac{1}{3}x - 2$. The ordered pairs (0, −2), (3, −1), (6, 0), (−3, −3), and (−6, −4) satisfy the equation. Plot these points and draw a line through them.

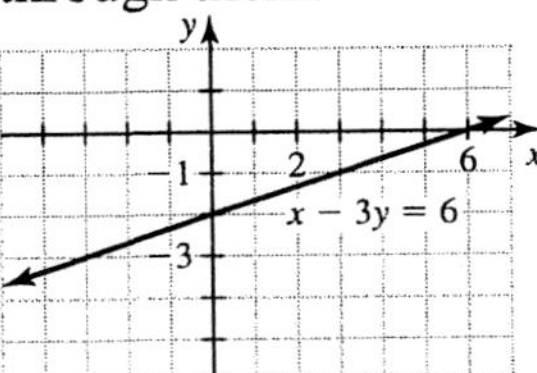

61. Select five different values for x and use $y = 0.36x + 0.4$ to find the y-coordinates. The ordered pairs (0, 0.4), (1, 0.76), (5, 2.2),

$(-1, 0.04)$, and $(-2, -0.32)$ satisfy the equation. Plot these points and draw a line through them.

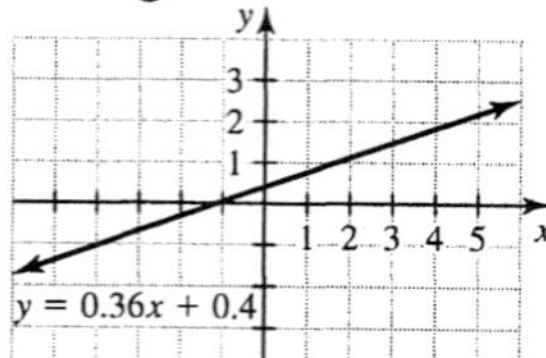

63. To get to $(-3, 45)$ we start at the origin and go to the left and then up. The point is in quadrant II.

65. Since the y-coordinate of $(-3, 0)$ is 0, the point is on the x-axis.

67. Since both coordinates of $(-2.36, -5)$ are negative, the point is in quadrant III.

69. Since both coordinates of $(3.4, 8.8)$ are positive, the point is in quadrant I.

71. Since the first coordinate is negative and the second coordinate positive, $(-1/2, 50)$ is in quadrant II.

73. Since the first coordinate is 0, $(0, -99)$ is on the y-axis.

75. Select five different values for x and use $y = x + 1200$ to find the y-coordinates. The ordered pairs (0, 1200), (600, 1800), $(1200, 2400)$, $(-600, 600)$, and $(-1200, 0)$ satisfy the equation. Plot these points and draw a line through them.

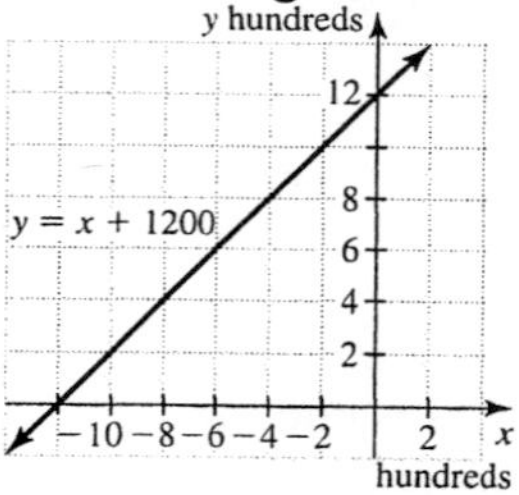

77. Select five different values for x and use $y = 50x - 2000$ to find the y-coordinates. The ordered pairs $(0, -2000)$, $(20, -1000)$, $(40, 0)$, $(-20, -3000)$, and $(-40, -4000)$ satisfy the equation. Plot these points and draw a line through them.

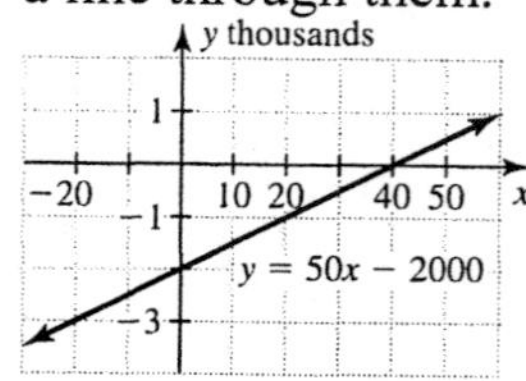

79. Select five different values for x and use $y = -400x + 2000$ to find the y-coordinates. The ordered pairs (0, 2000), (1, 1600), $(5, 0)$, $(6, -400)$, and $(-1, 2400)$ satisfy the equation. Plot these points and draw a line through them.

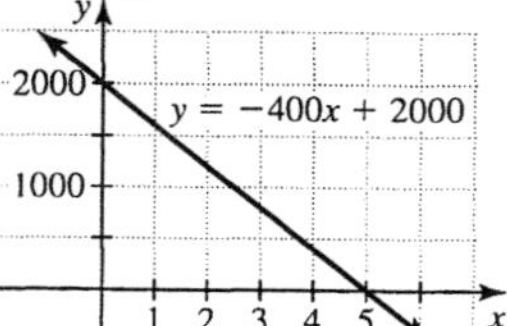

81. If $x = 0$ in $3x + 2y = 6$, then $0 + 2y = 6$, or $y = 3$. If $y = 0$ in $3x + 2y = 6$, then $3x + 0 = 6$, or $x = 2$. A third point is $(4, -3)$. Draw a line through $(0, 3)$, $(2, 0)$, and $(4, -3)$.

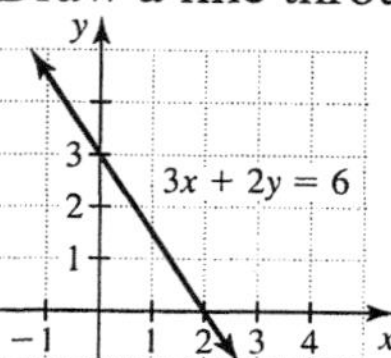

83. If $x = 0$ in $x - 4y = 4$, then $0 - 4y = 4$, or $y = -1$. If $y = 0$ in $x - 4y = 4$, then $x - 0 = 4$, or $x = 4$. A third point is $(8, 1)$. Draw a line through the intercepts $(0, -1)$ and $(4, 0)$, and through $(8, 1)$.

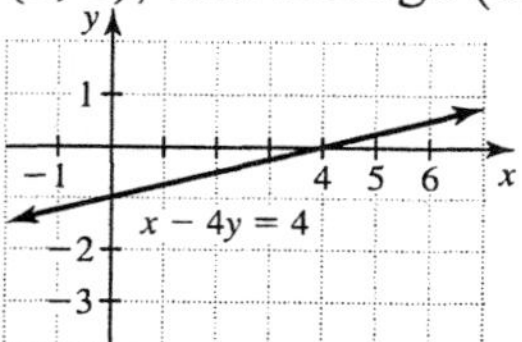

85. If $x = 0$ in $y = \frac{3}{4}x - 9$, then $y = -9$. If $y = 0$, then $0 = \frac{3}{4}x - 9$, or $x = 12$. A third point is $(4, -6)$. Draw a line through the intercepts $(0, -9)$ and $(12, 0)$, and through $(4, -6)$.

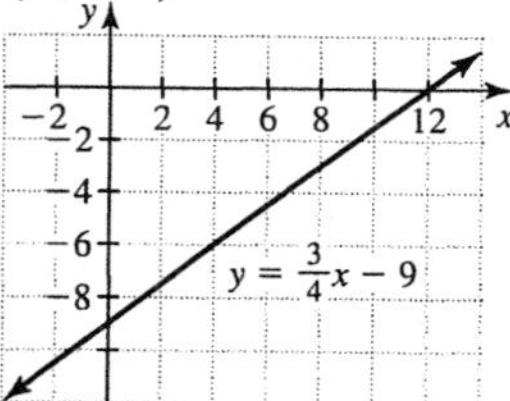

87. If $x = 0$ in $\frac{1}{2}x + \frac{1}{4}y = 1$, then $\frac{1}{2} \cdot 0 + \frac{1}{4}y = 1$, or $y = 4$. If $y = 0$, then

$\frac{1}{2}x + 0 = 1$, or $x = 2$. A third point is $(1, 2)$. Draw a line through the intercepts (0, 4) and (2, 0), and through $(1, 2)$.

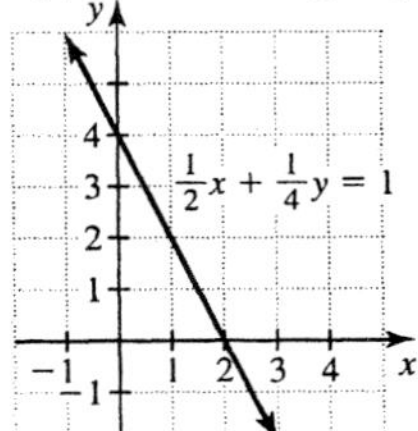

89. The benefit increases by 5% per year for the ages 62 through 64. So at age 63 the benefit is 75%. Because the graph goes through $(67, 100)$, full benefit is attained at age 67. More than full benefit is received for retiring at ages 68 and up.

91. a) The year 2000 is 10 years after 1990. So $n = 10$ and $C = 3.2(10) + 65.3 = 97.3$. So in 2000 the cost of Medicaid was \$97.3 billion.

b) Let $C = 150$ and solve for n:

$$3.2n + 65.3 = 150$$
$$3.2n = 84.7$$
$$n \approx 26.47$$

So approximately 26 years after 1990 or 2016 Medicaid spending will reach \$150 billion.

c) Draw a line through $(0, 65.3)$ and $(20, 129.3)$.

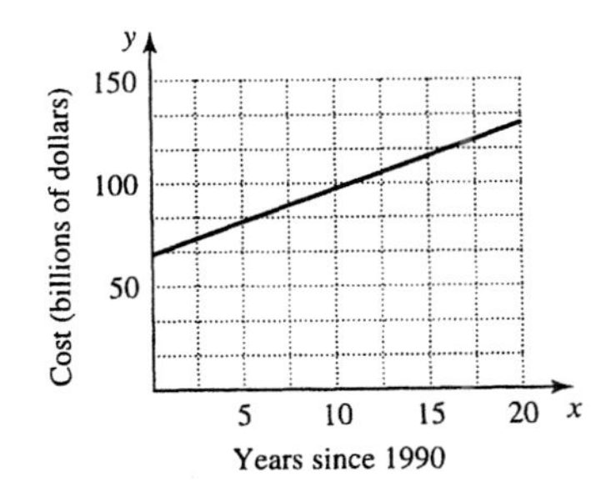

93. a) Nitrogen narcosis begins at depth 100 feet. If $d = 100$, then $A = 0.03(100) + 1 = 4$ atm.

b) At the maximum depth for intermediate divers $A = 4.9$.

$$0.03d + 1 = 4.9$$
$$0.03d = 3.9$$
$$d = 130 \text{ feet}$$

c) Draw a line through $(0, 1)$ and $(250, 8.5)$.

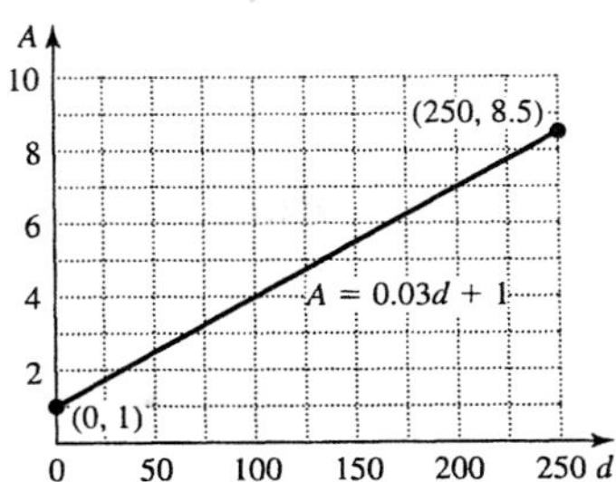

95. The variable x represents the number of radio ads and y represents the number of TV ads. The solutions to $300x + 400y = 24{,}000$ or $3x + 4y = 240$ that are whole numbers are (0, 60), (4, 57), (8, 54), (12, 51), (16, 48), (20, 45), (24, 42), ... (76, 3), (80, 0). Counting these ordered pairs, we get 21 solutions.

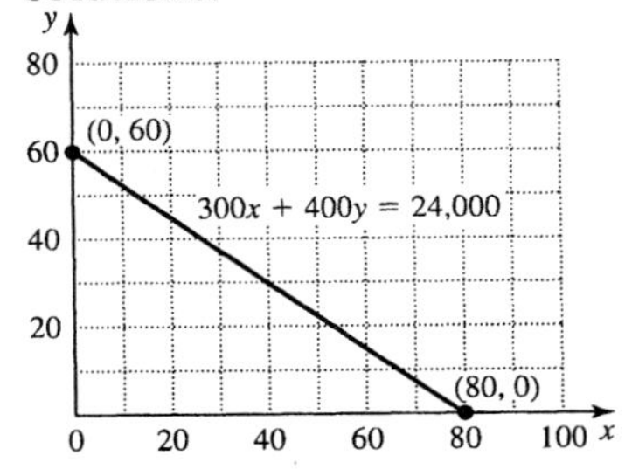

97. The intercepts are $(0, 400)$ and $(600, 0)$.

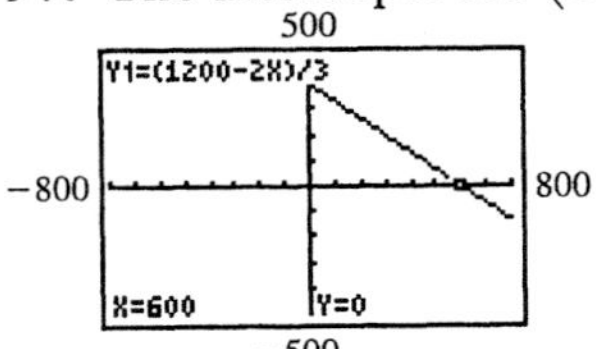

99. The intercepts are $(0, -0.02)$ and $(0.03, 0)$.

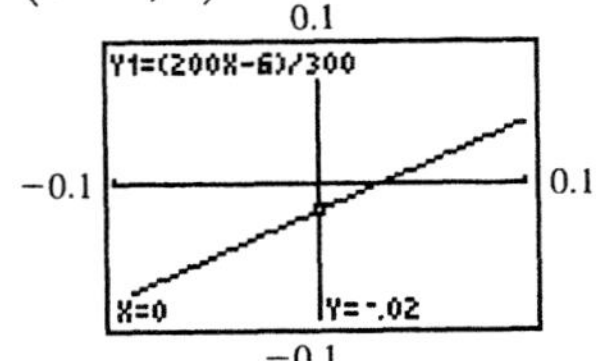

101. The intercepts are $(0, -1)$ and $(1/300, 0)$.

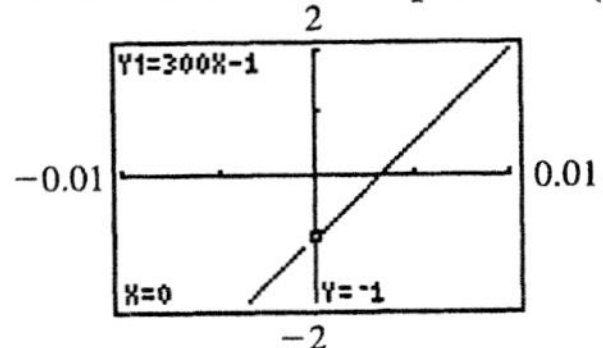

3.2 WARM-UPS

1. True, because of the definition of slope.
2. True, because that is the definition of slope.
3. False, because vertical lines do not have slope.
4. True, because to get to (1, 1) from the origin you must rise 1 and run 1.
5. False, because slope can be negative.
6. False, because a line perpendicular to a line with slope 2 must have slope $-1/2$.
7. False, because for (0, 3) and (4, 0), $m = \frac{3-0}{0-4} = -\frac{3}{4}$.
8. False, because different parallel lines have the same slope.
9. True, because for (1, 3) and (−5, 3), $m = \frac{3-3}{1-(-5)} = \frac{0}{6} = 0$.
10. True, because if y represents feet and x represents seconds, then slope is feet per second.

3.2 EXERCISES

1. The slope of a line is the ratio of its rise and run.
3. Slope is undefined for vertical lines.
5. Lines with positive slope are rising as you go from left to right, while lines with negative slope are falling as you go from left to right.
7. The line goes through the points (0, 2) and (3, 0). Its slope is $m = \frac{2-0}{0-3} = -\frac{2}{3}$.
9. The line goes through the points (−3, 0) and (0, 2). Its slope is

$$m = \frac{2-0}{0-(-3)} = \frac{2}{3}.$$

11. The line goes through the points (1, 1) and (−1, −2). Its slope is

$$m = \frac{-2-1}{-1-1} = \frac{-3}{-2} = \frac{3}{2}.$$

13. The line goes through the points (−2, 2) and (2, 2). Its slope is

$$m = \frac{2-2}{2-(-2)} = \frac{0}{4} = 0.$$

15. The line goes through the points (−3, −3) and (2, −1). Its slope is

$$m = \frac{-3-(-1)}{-3-2} = \frac{-2}{-5} = \frac{2}{5}.$$

17. The line goes through the points (2, −1) and (2, 2). This is a vertical line and it has no slope. Slope is undefined for this line.
19. $m = \frac{2-6}{1-3} = \frac{-4}{-2} = 2$
21. $m = \frac{4-(-1)}{2-5} = \frac{5}{-3} = -\frac{5}{3}$
23. $m = \frac{4-9}{-2-5} = \frac{-5}{-7} = \frac{5}{7}$
25. $m = \frac{-3-1}{-2-(-5)} = \frac{-4}{3} = -\frac{4}{3}$
27. $m = \frac{4-(-2)}{-3-3} = \frac{6}{-6} = -1$
29. $m = \frac{2-\frac{1}{2}}{\frac{1}{2}-(-1)} = \frac{\frac{3}{2}}{\frac{3}{2}} = 1$
31. Since $x_1 = x_2$, the slope is undefined.
33. $m = \frac{-5-(-5)}{-2-9} = \frac{0}{-11} = 0$
35. $m = \frac{0.9-(-0.3)}{0.3-(-0.1)} = \frac{1.2}{0.4} = 3$
37. Start at the point (1, 1) and rise 2 and go 3 units to the right to show a slope of $2/3$. Draw a line through the two points.

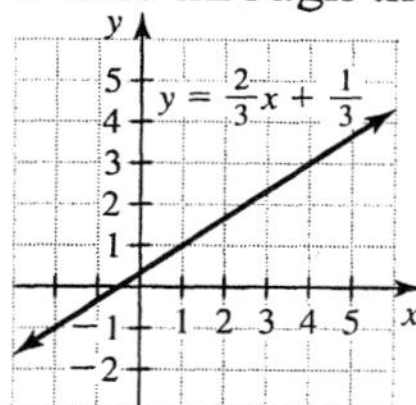

39. Start at the point (−2, 3) and go down 2 and then 1 unit to the right to show a slope of $-2/1$. Draw a line through the two points.

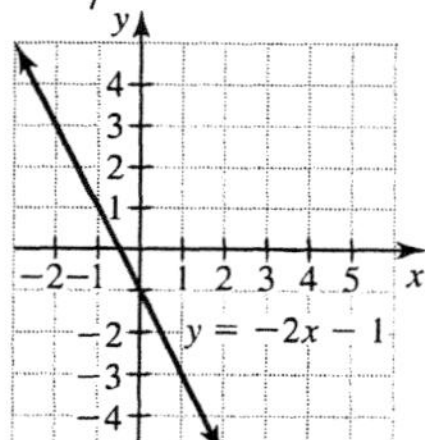

41. Start at the point (0, 0) and go down 2 units and then 5 units to the right to show a slope of $-2/5$. Draw a line through the two points.

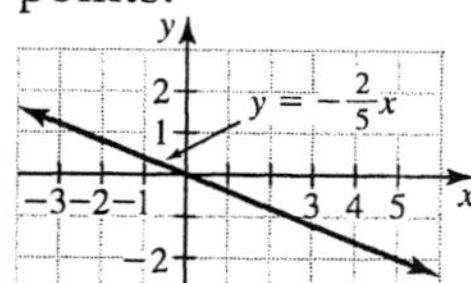

43. Start at the point (1, −2) and rise 1 and run 2 to get a second point on the line with slope

1/2. Draw the line. Start at $(-1, 1)$ and rise 1 and run 2 to get a second point on the line. Draw the line.

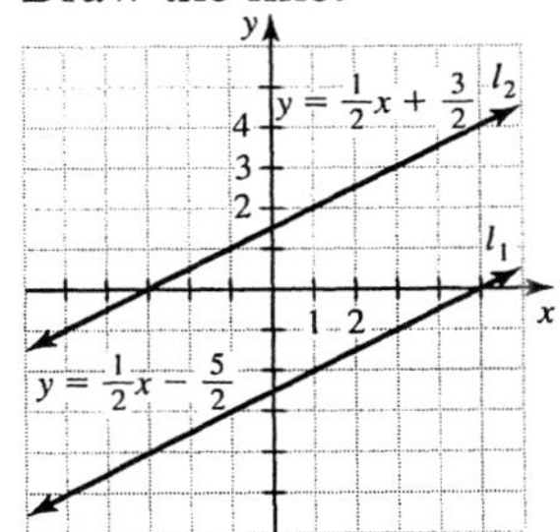

45. Start at (1, 2) and rise 1 and go 2 units to the right to indicate a slope of $1/2$. Draw a line through the two points. Now start at (1, 2) and go down 2 units and then 1 unit to the right to indicate a slope of $-2/1 = -2$. Draw a line through the two points.

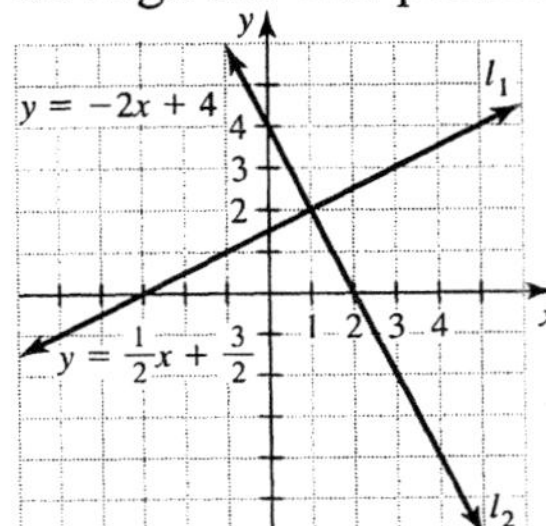

47. Draw the first line through $(0, 0)$ with slope $3/4$. The slope of a line perpendicular to a line with slope $3/4$ is the opposite of the reciprocal of $3/4$. The reciprocal of $3/4$ is $4/3$ and the opposite of that is $-4/3$. Draw the second line through $(0, 0)$ with slope $-4/3$. Graphs may vary.

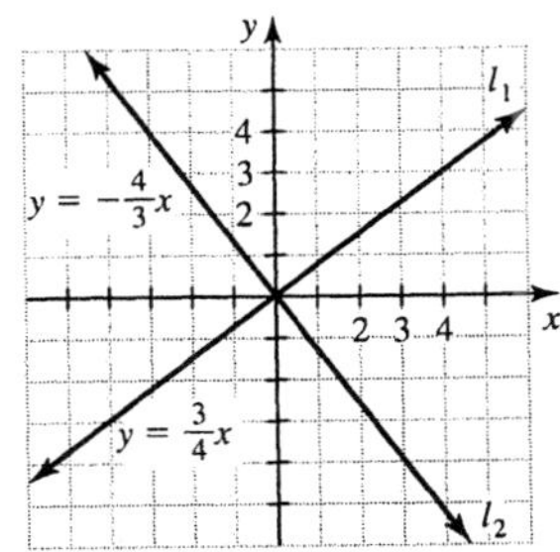

49. The line through $(-2, -3)$ and $(4, 0)$ has slope

$$m = \frac{-3-0}{-2-4} = \frac{-3}{-6} = \frac{1}{2}.$$

Any line parallel to it also has slope $\frac{1}{2}$. To draw the second line start at $(1, 2)$ and use the slope $1/2$.

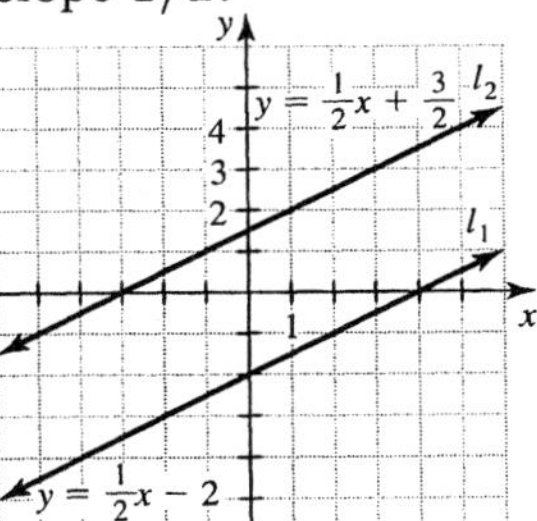

51. The line through $(-2, 4)$ and $(3, -1)$ has slope

$$m = \frac{4-(-1)}{-2-3} = \frac{5}{-5} = -1.$$

The slope of any line perpendicular to this line is the opposite of the reciprocal of -1. The reciprocal of -1 is -1. The opposite of -1 is 1. Draw the second line through $(1, 3)$ using slope 1.

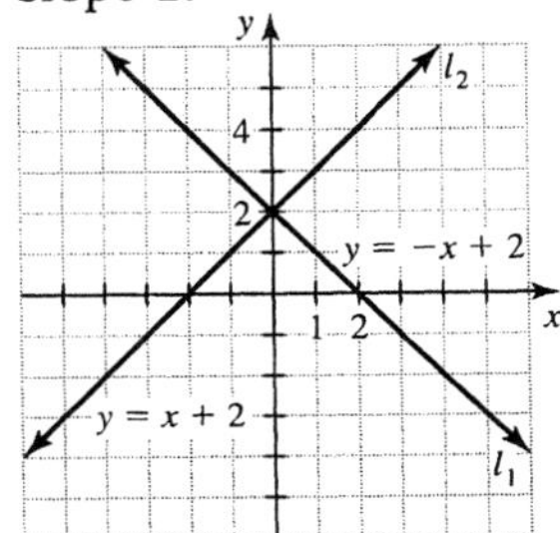

53. The slope of l_1 is

$$m_1 = \frac{7-5}{4-3} = \frac{2}{1} = 2.$$

The slope of l_2 is

$$m_2 = \frac{9-7}{12-11} = \frac{2}{1} = 2.$$

Since the slope are equal, the lines are parallel.

55. The slope of l_1 is

$$m_1 = \frac{6-(4)}{2-(-1)} = \frac{2}{3}.$$

The slope of l_2 is

$$m_2 = \frac{1-(-2)}{4-(2)} = \frac{3}{2}.$$

Since $m_1 \neq -1/m_2$ and $m_1 \neq m_2$ the lines are neither parallel nor perpendicular.

57. The slope of l_1 is

$$m_1 = \frac{6-4}{4-(-1)} = \frac{2}{5}.$$

The slope of l_2 is

$$m_2 = \frac{4-0}{3-(-7)} = \frac{4}{10} = 2/5.$$

Since the slope are equal, the lines are parallel.

59. The slope of l_1 is

$$m_1 = \frac{6-(5)}{3-(3)} = \frac{1}{0}.$$

So the line is vertical.
The slope of l_2 is

$$m_2 = \frac{4-4}{-3-(-2)} = \frac{0}{-1} = 0.$$

So the line is horizontal. Since any vertical line is perpendicular to any horizontal line, the lines are perpendicular.

61. a) $m = \frac{1.3-2.4}{1998-2004}$

$= \frac{-1}{-6} \approx 0.183$

The slope of 0.183 means that the cost is increasing on the average by about $183,000 per year.

b) The cost in 2002 was the 1998 cost plus $4(\$183,000)$ or $2.03 million. From the graph it appears that the cost in 2002 is about $2 million. So we are consistent with the graph.

c) If the cost keeps going up by $183,000 per year, then the cost in 2008 will be $2.03 million plus $6(\$183,000)$ or $3.13 million.

63. $m = \frac{25-5}{2002-1982} = \frac{20}{20} = 1$

The percentage increased 1% per year from 1982 to 2002.

65. Use the slope of $400 per year to fill in the table. The salary increases $400 per year.

Year	Salary (dollars)
2000	28,100
2002	28,900
2003	29,300
2012	32,900
2015	34,100

67. If the points in the table are on a straight line, then the slope must be the same for any two points in the table. In this table the slope is 3 for any two points. So the points are on a line.

69. If the points in the table are on a straight line, then the slope must be the same for any two points in the table. The slope for $(-2, 7)$ and $(0, 3)$ is -2. The slope for $(3, -3)$ and $(9, -16)$ is $-13/6$. So the points are not on a line.

3.3 WARM-UPS

1. True, because the slope determines a second point on the line and two points determine a line.

2. False, because (1, 2) does not satisfy $y = 3x + 2$.

3. True, because $x = -2$ runs parallel to the y-axis.

4. True, because any equation of the form $x = k$ for some real number k has a graph that is a vertical line.

5. True, because the slope of $y = x - 3$ is 1 and the slope of $y = 5 - x$ is -1.

6. False, because nonvertical parallel lines must have the same slope.

7. False, because $2y = 3x - 8$ is equivalent to $y = \frac{3}{2}x - 4$ and has slope $\frac{3}{2}$.

8. True, because even vertical lines that do not have slope can be written in standard form.

9. True, because $x = 2$ is vertical and $y = 5$ is horizontal.

10. False, because the y-intercept for $y = x$ is (0, 0).

3.3 EXERCISES

1. Slope-intercept form is $y = mx + b$.

3. The standard form is $Ax + By = C$, where A and B are not both zero.

5. The slope-intercept form allows us to write the equation from the y-intercept and the slope.

7. The line goes through the points (0, 1) and (2, 4). The slope of the line is $m = \frac{4-1}{2-0} = \frac{3}{2}$ and the y-intercept is (0, 1). The equation of the line is $y = \frac{3}{2}x + 1$.

9. The line goes through the points (0, 2) and (2, −2). The slope of the line is $m = \frac{-2-2}{2-0} = -\frac{4}{2} = -2$ and the y-intercept is (0, 2). The equation of the line is $y = -2x + 2$.

11. The line goes through the points (0, −2) and (3, 1). The slope of the line is $m = \frac{-2-1}{0-3} = \frac{-3}{-3} = 1$ and the y-intercept is (0, −2). The equation of the line is $y = x - 2$.

13. The line goes through the points (0, 0) and (2, −2). The slope of the line is $m = \frac{-2-0}{2-0} = \frac{-2}{2} = -1$ and the y-intercept is (0, 0). The equation of the line is $y = -x$.
15. The line goes through the points (0, −1) and (−3, −1). The slope of the line is $m = \frac{-1-(-1)}{0-(-3)} = \frac{0}{3} = 0$ and the y-intercept is (0, −1). The equation of the line is $y = -1$.
17. The line goes through the points (−2, 0) and (−2, 2). The line is vertical and so it has no slope. There is no slope-intercept form for this line. The equation of the line is $x = -2$.
19. In the form $y = mx + b$ the slope is m and the y-intercept is $(0, b)$. For the line $y = 3x - 9$, the slope is 3 and the y-intercept is (0, −9).
21. In the form $y = mx + b$ the slope is m and the y-intercept is $(0, b)$. For the line $y = -\frac{1}{2}x + 3$, the slope is $-1/2$ and the y-intercept is $(0, 3)$.
23. For the line $y = 4$, the slope is 0 and the y-intercept is (0, 4).
25. For the line $y = -3x$, the slope is −3 and the y-intercept is (0, 0).
27. The equation $x + y = 5$ is equivalent to $y = -x + 5$ or $y = -1x + 5$. The slope is −1 and the y-intercept is (0, 5).
29. The equation $x - 2y = 4$, is equivalent to $y = \frac{1}{2}x - 2$. The slope is $\frac{1}{2}$ and the y-intercept is (0, −2).
31. The equation $2x - 5y = 10$ is equivalent to $y = \frac{2}{5}x - 2$. The slope is $\frac{2}{5}$ and the y-intercept is (0, −2).
33. The equation $2x - y + 3 = 0$ is equivalent to $y = 2x + 3$. The slope is 2 and the y-intercept is (0, 3).
35. For the line $x = -3$, there is no y-intercept because the line is vertical. The slope is undefined for a vertical line.

37.
$$y = -x + 2$$
$$x + y = 2$$

39.
$$y = \frac{1}{2}x + 3$$
$$2y = x + 6$$
$$-x + 2y = 6$$
$$x - 2y = -6$$

Either of the last two equations is in standard form with only integers.

41.
$$y = \frac{3}{2}x - \frac{1}{3}$$
$$6y = 6\left(\frac{3}{2}x - \frac{1}{3}\right)$$
$$6y = 9x - 2$$
$$-9x + 6y = -2$$
$$9x - 6y = 2$$

43.
$$y = -\frac{3}{5}x + \frac{7}{10}$$
$$10y = 10\left(-\frac{3}{5}x + \frac{7}{10}\right)$$
$$10y = -6x + 7$$
$$6x + 10y = 7$$

45.
$$\frac{3}{5}x + 6 = 0$$
$$3x + 30 = 0$$
$$x = -10$$

47.
$$\frac{3}{4}y = \frac{5}{2}$$
$$4 \cdot \frac{3}{4}y = 4 \cdot \frac{5}{2}$$
$$3y = 10$$

49.
$$\frac{x}{2} = \frac{3y}{5}$$
$$5x = 6y$$
$$5x - 6y = 0$$

51.
$$y = 0.02x + 0.5$$
$$100y = 2x + 50$$
$$-2x + 100y = 50$$
$$x - 50y = -25$$

53. The line $y = 2x - 1$ has a y-intercept of (0, −1) and a slope of $2 = 2/1$. Start at (0, −1), rise 2 and then go 1 unit to the right to locate a second point on the line. Draw a line through the two points as shown.

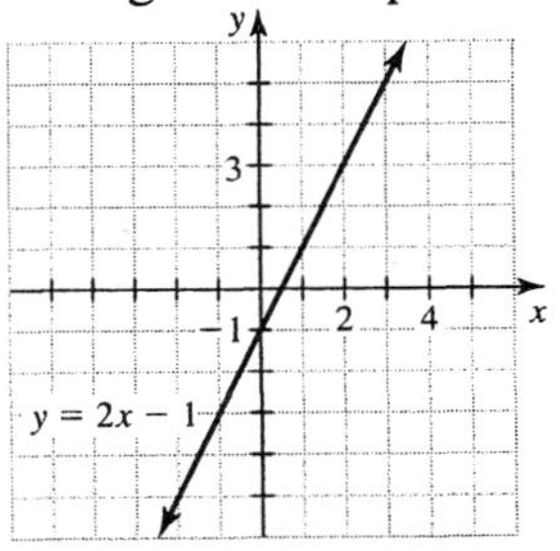

55. The line $y = -3x + 5$ has a y-intercept of (0, 5) and a slope of $-3 = -3/1$. Start at (0, 5), go down 3 units, and then 1 unit to the right to locate a second point on the line. Draw a line through the two points as shown.

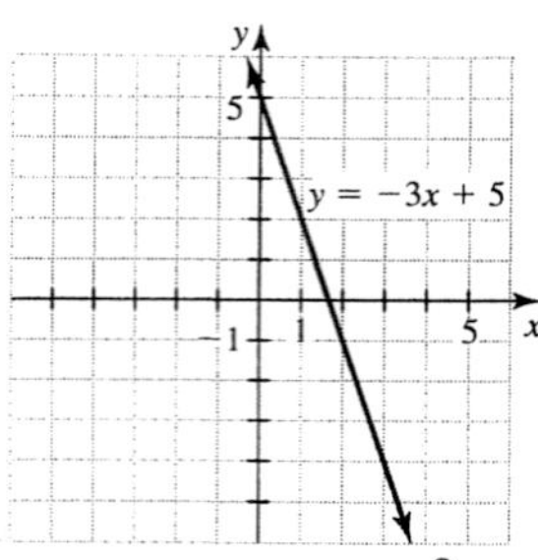

57. The line $y = \frac{3}{4}x - 2$ has a y-intercept of $(0, -2)$ and a slope of $3/4$. Start at $(0, -2)$, go up 3 units, and then 4 units to the right to locate a second point on the line. Draw a line as shown.

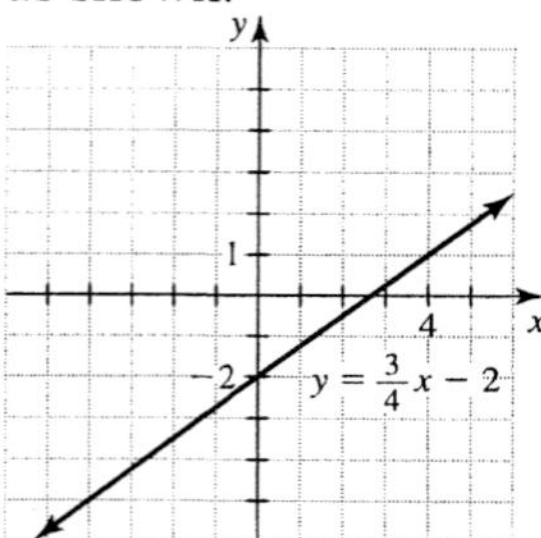

59. The equation $2y + x = 0$ is equivalent to $y = -\frac{1}{2}x$. The y-intercept is $(0, 0)$ and the slope is $-1/2$. Start at $(0, 0)$, go down 1 unit, and then go 2 units to the right to locate a second point on the line. Draw the line as shown.

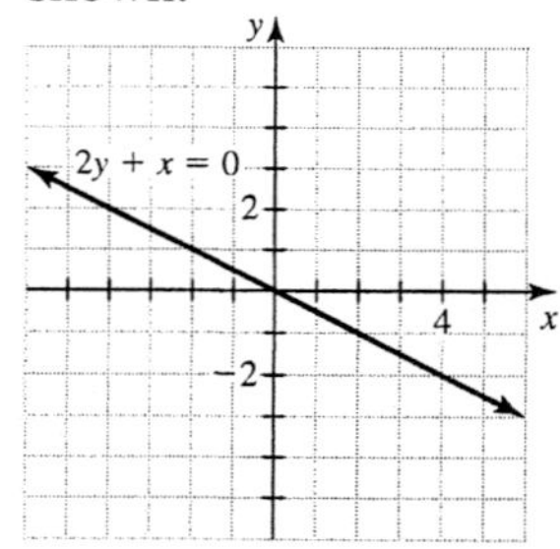

61. The equation $3x - 2y = 10$ is equivalent to $y = (3/2)x - 5$. The y-intercept is $(0, -5)$ and the slope is $3/2$. Start at $(0, -5)$, go up 3 units, and then 2 units to the right to locate a second point on the line. Draw a line through the two points as shown.

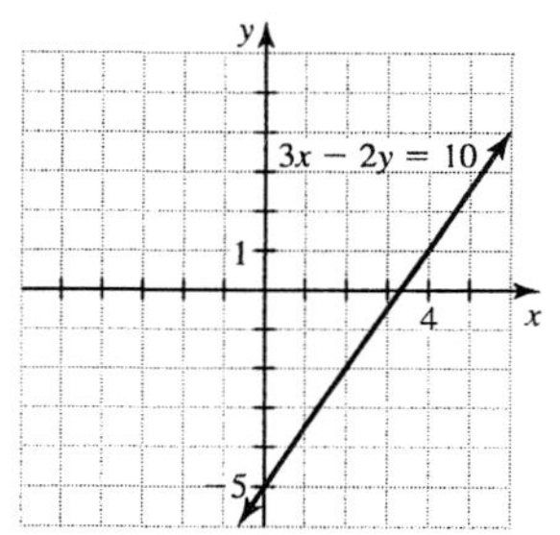

63. The equation $4y + x = 8$ is equivalent to $y = -\frac{1}{4}x + 2$. The y-intercept is $(0, 2)$ and the slope is $-1/4$. Start at $(0, 2)$, go 1 unit down, and then 4 units to the right to locate a second point on the line. Draw a line through the two points as shown.

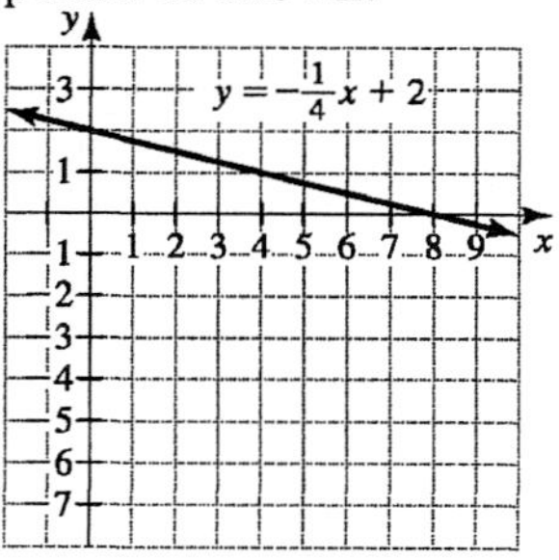

65. The equation $y - 2 = 0$ is equivalent to $y = 2$. The y-intercept is $(0, 2)$ and the slope is 0. Draw a horizontal line through $(0, 2)$.

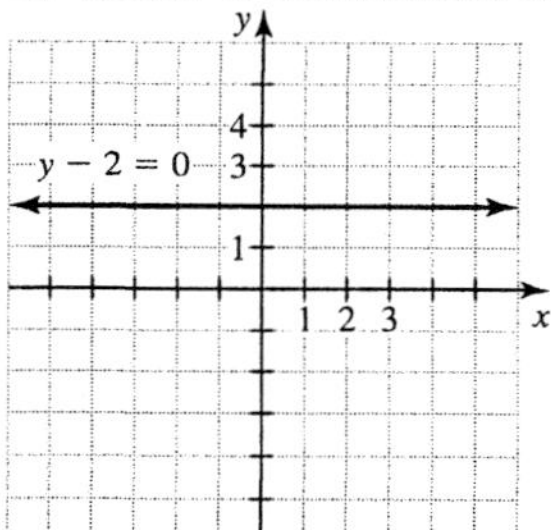

67. Since the lines both have slope 3, the lines are parallel.

69. Since the slopes are 2 and -2 the lines are neither parallel nor perpendicular.

71. Since the slopes are both zero, the lines are parallel.

73. Since the slopes are -4 and $1/4$, opposite signs and reciprocals, the lines are perpendicular.

75. The line through $(0, -4)$ with slope $\frac{1}{2}$ has equation $y = \frac{1}{2}x - 4$.

77. The line through $(0, 3)$ parallel to $y = 2x - 1$ has slope 2. So its equation is $y = 2x + 3$.
79. The line $y = 3x - 5$ has slope 3. Any line perpendicular to it has slope $-1/3$. The line through (0, 6) with slope $-1/3$ has equation $y = -\frac{1}{3}x + 6$.
81. The equation $2x + y = 5$ is equivalent to $y = -2x + 5$. Any line parallel to it has slope -2. The line through (0, 3) with slope -2 has equation $y = -2x + 3$.
83. Any line parallel to the x-axis has 0 slope. A line through (2, 3) with 0 slope has a y-intercept of (0, 3). The equation of the line through (0, 3) with 0 slope is $y = 3$.
85. The line $2x - 3y = 6$ is equivalent to $y = \frac{2}{3}x - 2$. Any line perpendicular to it has slope $-\frac{3}{2}$. The line through $(0, 4)$ with slope $-\frac{3}{2}$ is $y = -\frac{3}{2}x + 4$.
87. The line through (0, 4) and (5, 0) has slope $m = \frac{4-0}{0-5} = -\frac{4}{5}$. The line through $(0, 4)$ with slope $-4/5$ has equation $y = -\frac{4}{5}x + 4$.
89. If $n = 5000$, then
$C = 200(5000) + 150{,}000 = \$1{,}150{,}000$.
If $n = 5001$, then
$C = 200(5001) + 150{,}000 = \$1{,}150{,}200$.
The extra mower increased the cost by \$200.
91. a) $m = \frac{25-5}{2002-1982} = 1$
The slope is 1% per year. The percentage of workers receiving training is increasing by 1% per year.
b) The slope is 1 and the y-intercept is $(0, 5)$. The equation is $y = x + 5$ where x is the number of years since 1982.
c) The y-intercept of $(0, 5)$ means that 5% of the workers received training in 1982.
d) The year 2010 is 28 years after 1982. So $y = 28 + 5 = 33$. So 33% of workers should be receiving training in 2010.
93. a) The variable x represents the number of packs of snapdragons and y represents the number of packs of pansies.
b) The intercepts are $(0, 400)$ and $(200, 0)$.

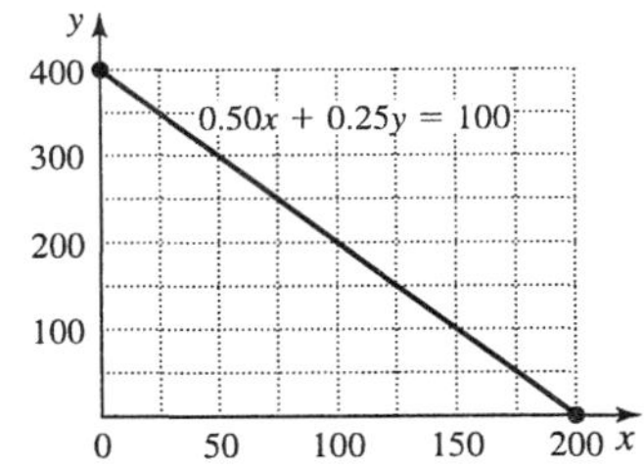

c)
$$0.50x + 0.25y = 100$$
$$0.25y = -0.50x + 100$$
$$y = -2x + 400$$
d) The slope is -2.
e) The slope -2 means increasing the number of packs of snapdragons by 1 causes the number of packs of pansies to decrease by 2. Which makes sense because the snapdragons cost twice as much as the pansies.
95. The lines will look perpendicular only if the unit length on each axis is the same. Use the square feature of your calculator or experiment with different viewing windows until the lines look perpendicular.

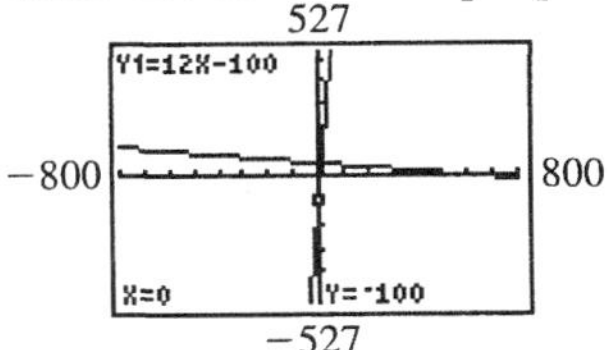

3.4 WARM-UPS

1. False, because the point-slope formula is $y - y_1 = m(x - x_1)$.
2. False, because we can find the equation of a line through any two points.
3. True, because the line is vertical and its slope is undefined.
4. True, because the line through the origin with slope 1 is $y = 1 \cdot x + 0$, or $y = x$.
5. False, because $5x + y = 4$ is equivalent to $y = -5x + 4$, which has slope -5.
6. True, because the line $y = 4x - 3$ has slope 4 and any line perpendicular to it has slope $-1/4$.
7. True, because $x + y = 1$ is equivalent to $y = -x + 1$, which has slope -1.
8. True, because $2(-2) - (-3) = -1$ is correct.
9. True, because $2x + y = 4$ is equivalent to $y = -2x + 4$ and both lines have slope -2.

10. True, because the slope of $y = x$ is 1 and the slope of $y = -x$ is -1.

3.4 EXERCISES

1. Point-slope form is $y - y_1 = m(x - x_1)$.
3. If you know two points on a line, find the slope. Then use it along with a point in point-slope form to write the equation of the line.
5. Nonvertical parallel lines have equal slopes.

7.
$$y - 1 = 5(x + 2)$$
$$y - 1 = 5x + 10$$
$$y = 5x + 11$$

9.
$$3x - 4y = 80$$
$$-4y = -3x + 80$$
$$y = \frac{3}{4}x - 20$$

11.
$$y - \frac{1}{2} = \frac{2}{3}\left(x - \frac{1}{4}\right)$$
$$y - \frac{1}{2} = \frac{2}{3}x - \frac{1}{6}$$
$$y = \frac{2}{3}x - \frac{1}{6} + \frac{1}{2}$$
$$y = \frac{2}{3}x + \frac{1}{3}$$

13. Use the point $(1, 2)$ and slope 3 in point-slope form $y - y_1 = m(x - x_1)$.
$$y - 2 = 3(x - 1)$$
$$y - 2 = 3x - 3$$
$$y = 3x - 1$$

15. Use the point $(2, 4)$ and slope $1/2$ in point-slope form $y - y_1 = m(x - x_1)$.
$$y - 4 = \frac{1}{2}(x - 2)$$
$$y - 4 = \frac{1}{2}x - 1$$
$$y = \frac{1}{2}x + 3$$

17. Use the point (2, 3) and slope $1/3$ in point-slope form $y - y_1 = m(x - x_1)$.
$$y - 3 = \frac{1}{3}(x - 2)$$
$$y - 3 = \frac{1}{3}x - \frac{2}{3}$$
$$y = \frac{1}{3}x - \frac{2}{3} + 3$$
$$y = \frac{1}{3}x + \frac{7}{3}$$

19. Use the point $(-2, 5)$ and slope $-1/2$ in point-slope form $y - y_1 = m(x - x_1)$.
$$y - 5 = -\frac{1}{2}(x - (-2))$$
$$y - 5 = -\frac{1}{2}x - 1$$
$$y = -\frac{1}{2}x + 4$$

21. Use the point $(-1, -7)$ and slope -6 in point-slope form $y - y_1 = m(x - x_1)$.
$$y - (-7) = -6(x - (-1))$$
$$y + 7 = -6x - 6$$
$$y = -6x - 13$$

23.
$$y - 3 = 2(x - 5)$$
$$y - 3 = 2x - 10$$
$$-2x + y = -7$$
$$2x - y = 7$$
Either of the last two equations is standard form with integers.

25.
$$y = \frac{1}{2}x - 3$$
$$2y = x - 6$$
$$-x + 2y = -6$$
$$x - 2y = 6$$
Either of the last two equations is standard form with integers.

27.
$$y - 2 = \frac{2}{3}(x - 4)$$
$$3y - 6 = 2x - 8$$
$$-2x + 3y = -2$$
$$2x - 3y = 2$$
Either of the last two equations is standard form with integers.

29. Find the slope of the line through the points (1, 3) and (2, 5).
$$m = \frac{3 - 5}{1 - 2} = \frac{-2}{-1} = 2$$
Use the point (1, 3) and slope 2 in the point-slope form $y - y_1 = m(x - x_1)$.
$$y - 3 = 2(x - 1)$$
$$y - 3 = 2x - 2$$
$$-2x + y = 1$$
$$2x - y = -1$$

31. Find the slope of the line through the points (1, 1) and (2, 2).
$$m = \frac{1 - 2}{1 - 2} = \frac{-1}{-1} = 1$$
Use the point (1, 1) and slope 1 in the point-slope form $y - y_1 = m(x - x_1)$.
$$y - 1 = 1(x - 1)$$
$$y - 1 = x - 1$$
$$-x + y = 0$$
$$x - y = 0$$

33. Find the slope of the line through the points (1, 2) and (5, 8).

$$m = \frac{2-8}{1-5} = \frac{-6}{-4} = \frac{3}{2}$$

Use the point (1, 2) and slope 3/2 in the point-slope form $y - y_1 = m(x - x_1)$.

$$y - 2 = \frac{3}{2}(x-1)$$
$$2y - 4 = 3(x-1)$$
$$2y - 4 = 3x - 3$$
$$-3x + 2y = 1$$
$$3x - 2y = -1$$

35. Find the slope of the line through the points (−2, −1) and (3, −4).

$$m = \frac{-1-(-4)}{-2-3} = \frac{3}{-5} = -\frac{3}{5}$$

Use the point (3, −4) and slope −3/5 in the point-slope form $y - y_1 = m(x - x_1)$.

$$y - (-4) = -\frac{3}{5}(x-3)$$
$$y + 4 = -\frac{3}{5}x + \frac{9}{5}$$
$$5y + 20 = -3x + 9$$
$$3x + 5y = -11$$

37. Find the slope of the line through the points (−2, 0) and (0, 2).

$$m = \frac{0-2}{-2-0} = \frac{-2}{-2} = 1$$

Use the point (0, 2) and slope 1 in the slope-intercept form $y = mx + b$.

$$y = 1x + 2$$
$$-x + y = 2$$
$$x - y = -2$$

39. Find the slope of the line through the points (2, 4) and (2, 6).

$$m = \frac{6-4}{2-2} = \frac{2}{0}$$

Since the slope is undefined, the line is vertical. Since it goes through $(2, 4)$ the equation is $x = 2$.

41. Find the slope of the line through the points (−3, 9) and (3, 9).

$$m = \frac{9-9}{3-(-3)} = \frac{0}{6} = 0$$

Since the slope is 0, the line is horizontal. Since it goes through $(-3, 9)$ the equation is $y = 9$.

43. The slope of the dashed line is 1. The slope of the solid line is −1 and it goes through (5, −1).

$$y - (-1) = -1(x-5)$$
$$y + 1 = -x + 5$$
$$y = -x + 4$$

45. The slope of the dashed line is $-3/5$. The solid line has slope $5/3$ and goes through $(3, 4)$.

$$y - 4 = \frac{5}{3}(x-3)$$
$$y - 4 = \frac{5}{3}x - 5$$
$$y = \frac{5}{3}x - 1$$

47. Any line perpendicular to $y = 3x - 1$ has slope $-1/3$. Use the point (3, 4) and the slope $-1/3$ in the point-slope form $y - y_1 = m(x - x_1)$.

$$y - 4 = -\frac{1}{3}(x-3)$$
$$y - 4 = -\frac{1}{3}x + 1$$
$$y = -\frac{1}{3}x + 5$$

49. Any line parallel to $y = x - 9$ has slope 1. Use the point (7, 10) and slope 1 in the point-slope form.

$$y - 10 = 1(x-7)$$
$$y = x + 3$$

51. The line $3x - 2y = 10$ is equivalent to $y = \frac{3}{2}x - 5$, and so its slope is $3/2$. Any line perpendicular to $3x - 2y = 10$ has slope $-2/3$. Use the point (1, 1) and slope $-2/3$ in the point-slope form.

$$y - 1 = -\frac{2}{3}(x-1)$$
$$y - 1 = -\frac{2}{3}x + \frac{2}{3}$$
$$y = -\frac{2}{3}x + \frac{5}{3}$$

53. The equation $2x + y = 8$ is equivalent to $y = -2x + 8$, which has slope -2. Any line parallel to $2x + y = 8$ also has slope -2. Use the point (−1, −3) and slope −2 in the point-slope form.

$$y - (-3) = -2(x - (-1))$$
$$y + 3 = -2x - 2$$
$$y = -2x - 5$$

55. The equation $3x + y = 5$ is equivalent to $y = -3x + 5$. It has slope of -3. Any line perpendicular to it has slope $1/3$. Use the point (−1, 2) and slope $1/3$ in the point-slope form.

$$y - 2 = \frac{1}{3}(x - (-1))$$
$$y - 2 = \frac{1}{3}x + \frac{1}{3}$$
$$y = \frac{1}{3}x + \frac{7}{3}$$

57. The equation $-2x + y = 6$ is equivalent to $y = 2x + 6$. It has slope 2. Any line parallel to $-2x + y = 6$ also has slope 2. Use the point (2, 3) and slope 2 in the point-slope form.

$$y - 3 = 2(x - 2)$$
$$y - 3 = 2x - 4$$
$$y = 2x - 1$$

59. The line through $(3, 2)$ with slope 0 is a horizontal line and has y-intercept $(0, 2)$. So its equation is $y = 0x + 2$ or $y = 2$.

61. The line through $(3, 2)$ and the origin has slope $2/3$ and y-intercept $(0, 0)$. So its equation is $y = \frac{2}{3}x$.

63. The line through $(5, 0)$ and $(0, 5)$ has slope -1 and any line parallel to it has slope -1. So the equation is $y = -x$.

65. A line perpendicular to $x = 400$ is a horizontal line with slope 0. If it goes through $(-30, 50)$ its equation is $y = 50$.

67. The line through $(0, 0)$ and $(3, 5)$ has slope $5/3$. A line perpendicular to it has slope $-3/5$. Use the point-slope form:

$$y - (-1) = -\frac{3}{5}(x - (-5))$$
$$y + 1 = -\frac{3}{5}x - 3$$
$$y = -\frac{3}{5}x - 4$$

69. The line through $(1, 3)$ and $(2, 5)$ has slope 2. Only choice (e) has slope 2.

71. A line with no x-intercept is a horizontal line. Since it goes through $(1, 3)$ its equation is $y = 3$. So the answer is (f).

73. A line through $(1, 3)$ and $(5, 0)$ has slope $\frac{0 - 3}{5 - 1}$ or $-3/4$. Using point-slope form we have $y - 0 = -\frac{3}{4}(x - 5)$ or $y + \frac{3}{4}x = \frac{15}{4}$ or $4y + 3x = 15$ or $3x + 4y = 15$, which is choice (h).

75. A line through $(1, 3)$ with slope -2 has equation $y - 3 = -2(x - 1)$ or $y - 3 = -2x + 2$ or $2x + y = 5$ which is choice (g).

77. a) $m = \dfrac{10.6 - 14.2}{0 - 4} = \dfrac{-3.6}{-4} = 0.9$

The slope 0.9 means that the number of ATM transactions is increasing by 0.9 billion per year.

b) Use slope 0.9 and point (0, 10.6) to get

$$y = 0.9x + 10.6,$$

where x is the number of years after 1996.

c) For the year 2010, x would be 14:

$$y = 0.9(14) + 10.6 = 23.2$$

So in 2010 there will be 23.2 billion transactions.

79. a) $m = \dfrac{62.7 - 71.6}{6 - 12} = \dfrac{-8.9}{-6} \approx 1.5$

$$y - 62.7 = \frac{8.9}{6}(x - 6)$$
$$y - 62.7 = 1.5x - 8.9$$
$$y = 1.5x + 53.8$$

b) x represents the number of years after 1990 and y represents the gross domestic product per employed person in thousands of dollars.

c) In 2010 we have $x = 20$ and $y = 1.5(20) + 53.8 = 83.8$ or \$83,800

d) Draw a line through (0, 53.8) and $(20, 83.8)$.

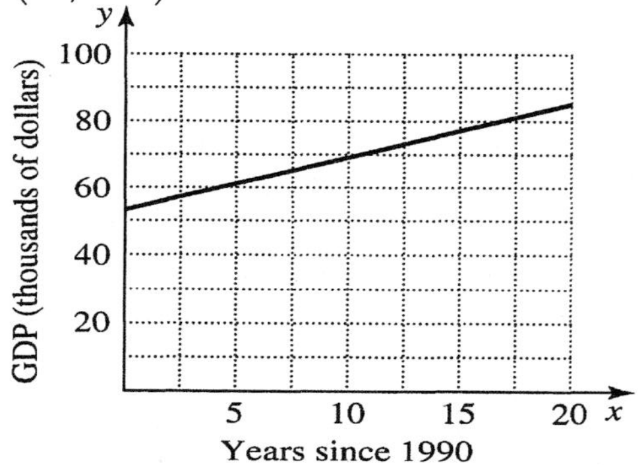

81. We want the equation of the line through the points (2, 70) and (4, 110). The slope is

$$m = \frac{110 - 70}{4 - 2} = \frac{40}{2} = 20.$$

Use the point (2, 70) and slope 20 in the point-slope form, where C is the charge and n is the number of hours worked.

$$C - 70 = 20(n - 2)$$
$$C - 70 = 20n - 40$$
$$C = 20n + 30$$

If $n = 7$ at Fred's house, then $C = 20(7) + 30 = \$170$.

83. We want the equation of the line through the points (7.75, 13) and (5.75, 7) where the first coordinate is the length of the foot L and the second coordinate is the shoe size S. First find the slope.

$$m = \frac{13-7}{7.75-5.75} = \frac{6}{2} = 3$$

Now use the point (5.75, 7) and the slope 3 in the point-slope form of the line.

$$S - 7 = 3(L - 5.75)$$
$$S - 7 = 3L - 17.25$$
$$S = 3L - 10.25$$
$$S = 3L - \frac{41}{4}$$

If $L = 6.25$, then $S = 3(6.25) - 10.25 = 8.5$. A child with a 6.25 inch foot wears a size 8.5.

85. We want the equation of the line through the points (1, 42) and (2, 74) where the first coordinate is time t in seconds, and the second coordinate is velocity v in feet per second. The slope of the line is

$$m = \frac{74-42}{2-1} = 32.$$

Use a slope of 32 and the point (1, 42) in the point-slope form.

$$v - 42 = 32(t - 1)$$
$$v - 42 = 32t - 32$$
$$v = 32t + 10$$

If $t = 3.5$ seconds, then the velocity is $v = 32(3.5) + 10 = 122$ feet per second.

87. a) We want the equation of the line through the points (90, 0.75) and (30, 1.25) where the first coordinate is temperature t and the second coordinate is width w in inches. The slope is

$$m = \frac{1.25-0.75}{30-90} = \frac{0.5}{-60} = -\frac{1}{120}.$$

Use a slope of $-1/120$ and the point $(90, 0.75)$ in the point-slope form.

$$w - 0.75 = -\frac{1}{120}(t - 90)$$
$$w - 0.75 = -\frac{1}{120}t + \frac{3}{4}$$
$$w = -\frac{1}{120}t + \frac{3}{2}$$

b) If $t = 80$, then the width is

$$w = -\frac{1}{120}(80) + \frac{3}{2} = -\frac{2}{3} + \frac{3}{2} = \frac{5}{6}.$$

When the temperature is 80° F, the width is $\frac{5}{6}$ inch.

c)
$$1 = -\frac{1}{120}t + \frac{3}{2}$$
$$120 = -t + 180$$
$$t = 60$$

When the width is 1 inch, the temperature is 60°F.

89. We want the equation of the line through the two points (3, 1.8) and (5, 3) where the first coordinate is weight w in pounds and the second coordinate is amount A in inches that the spring stretches. The slope is

$$m = \frac{3-1.8}{5-3} = \frac{1.2}{2} = 0.6.$$

Use the slope 0.6 and the point (5, 3) in the point-slope form.

$$A - 3 = 0.6(w - 5)$$
$$A - 3 = 0.6w - 3$$
$$A = 0.6w$$

If $w = 6$, then $A = 0.6(6) = 3.6$. With a weight of 6 pounds the spring will stretch 3.6 in.

91. a) Find the equation of the line through $(2, 0.16)$ and $(5, 0.40)$.

$$m = \frac{0.40-0.16}{5-2} = \frac{0.24}{3} = 0.08$$
$$a - 0.16 = 0.08(c - 2)$$
$$a - 0.16 = 0.08c - 0.16$$
$$a = 0.08c$$

b) If $c = 3$, then $a = 0.08(3) = 0.24$.

c) From the graph, it appears that if the absorption is 0.50, then the concentration is about 6.25 mg/ml.

93. Solve $2x + 3y = 9$ for y to get $y = -\frac{2}{3}x + 3$. The slope is $-2/3$. Solve $4x - 5y = 6$ for y to get $y = \frac{4}{5}x - \frac{6}{5}$. The slope is $4/5$. Fill in the table.

$Ax + By = C$	A	B	slope
$2x + 3y = 9$	2	3	$-\frac{2}{3}$
$4x - 5y = 6$	4	-5	$\frac{4}{5}$
$\frac{1}{2}x + 3y = 1$	$\frac{1}{2}$	3	$-\frac{1}{6}$
$2x - \frac{1}{3}y = 7$	2	$-\frac{1}{3}$	6

95. a) The intercepts are $(0, -300)$ and $(15, 0)$.

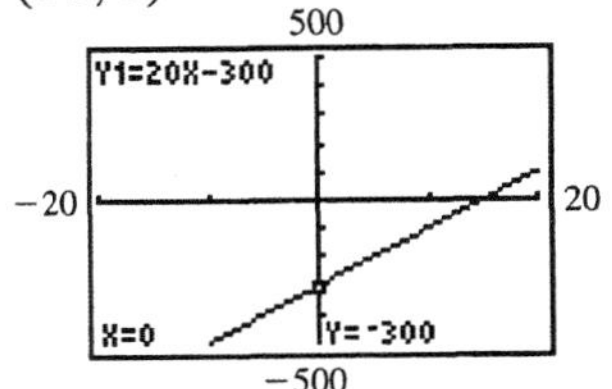

b) The intercepts are (0, 500) and (50/3, 0).

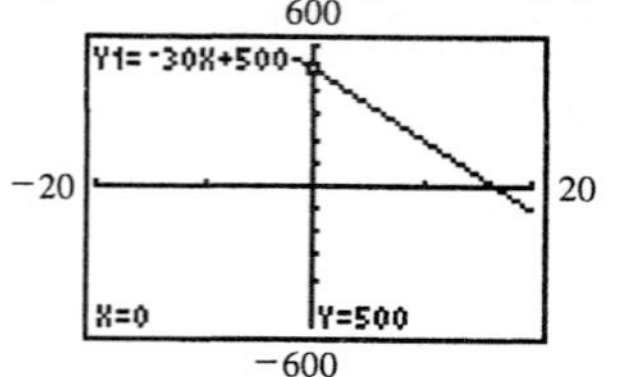

c) The intercepts are $(0, -2000)$ and $(3000, 0)$.

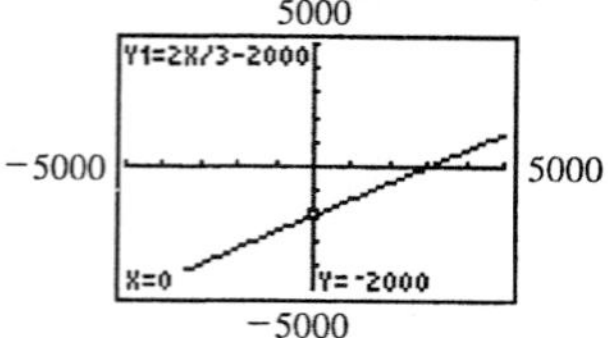

97. Since these are parallel lines that are very close together, use a small viewing window. Use the viewing window $-1 \le x \le 1$ and $-1 \le y \le 1$.

3.5 WARM-UPS

1. True, because of the definition of varies directly.
2. False, because if a varies inversely as b, then $a = k/b$ for some constant k.
3. True, because if y varies directly as x, then $y = kx$. If $y = 8$ when $x = 2$, then $8 = k \cdot 2$. So $k = 4$.
4. False, because if y varies inversely as x, then $y = k/x$. If $y = 8$ when $x = 2$, $8 = k/2$, or $k = 16$.
5. False, because if C varies jointly as h and t, then $C = kht$ for some constant k.
6. True, because if k is the sales tax rate, then the amount of tax is kP where P is the price of the car.
7. True, because if z varies inversely as w, then $z = k/w$. If $z = 10$ when $w = 2$, then $10 = k/2$, or $k = 20$. So $z = 20/w$.
8. True, because $T = D/R$ and if D is fixed, then T varies inversely with the rate.
9. False, because if m varies directly as w, then $m = kw$ for some constant k.
10. False, because if y varies jointly as x and z, then $y = kxz$.

3.5 EXERCISES

1. If y varies directly as x, then there is a constant k such that $y = kx$.
3. If y is inversely proportional to x, then there is a constant k such that $y = \frac{k}{x}$
5. T varies directly as h means T is a constant multiple of h, $T = kh$.
7. If y varies inversely as r, then y is a constant divided by r, $y = \frac{k}{r}$.
9. If R is jointly proportional to t and s, then R is a constant multiple of the product of t and s, $R = kts$.
11. If i is directly proportional to b, then i is a constant multiple of b, $i = kb$.
13. If A is jointly proportional to y and m, then A is a constant multiple of the product of y and m, $A = kym$.
15. If y varies directly as x, then $y = kx$. If $y = 5$ when $x = 3$, then $5 = k \cdot 3$, or $k = \frac{5}{3}$. So $y = \frac{5}{3}x$.
17. If A varies inversely as B, then $A = \frac{k}{B}$. If $A = 3$ when $B = 2$, then $3 = \frac{k}{2}$. So $k = 6$ and $A = \frac{6}{B}$.
19. If m varies inversely as p, then $m = \frac{k}{p}$. If $m = 22$ when $p = 9$, then $22 = \frac{k}{9}$. So $k = 198$ and $m = \frac{198}{p}$.
21. If A varies jointly as t and u, then $A = ktu$. If $A = 24$ when $t = 6$ and $u = 2$, then $24 = k \cdot 6 \cdot 2$. So $k = 2$ and $A = 2tu$.
23. If T varies directly as u, then $T = ku$. If $T = 9$ when $u = 2$, then $9 = k \cdot 2$. So $k = \frac{9}{2}$ and $T = \frac{9}{2}u$.
25. If Y varies directly as x, then $Y = kx$. If $Y = 100$ when $x = 20$, then $100 = k \cdot 20$ and $k = 5$. So the formula is $Y = 5x$. If $x = 5$, then $Y = 5(5) = 25$.
27. If a varies inversely as b, then $a = \frac{k}{b}$. If $a = 3$ when $b = 4$, then $3 = \frac{k}{4}$ and $k = 12$. So the formula is $a = \frac{12}{b}$. If $b = 12$, then $a = \frac{12}{12} = 1$.

29. If P varies jointly as s and t, then $P = kst$. If $P = 56$ when $s = 2$ and $t = 4$, then $56 = k \cdot 2 \cdot 4$ and $k = 7$. So the formula is $P = 7st$. If $s = 5$ and $t = 3$, then $P = 7 \cdot 5 \cdot 3 = 105$.

31. If $b = 300/a$, then b varies inversely as a. Use the formula to find missing values in the table. If $a = \frac{1}{2}$, then $b = \dfrac{300}{\frac{1}{2}} = 600$. If $a = 1$, then $b = 300$. If $b = 1$, then $1 = 300/a$ or $a = 300$.

a	b
$\frac{1}{2}$	600
1	300
30	10
900	$\frac{1}{3}$

33. If $b = \frac{3}{4}a$, then b varies directly as a. Use the formula to find missing values in the table.

a	b
$\frac{1}{3}$	$\frac{1}{4}$
8	6
12	9
20	15

35. As x increases, y increases and y is 3.5 times x in every entry in the table. So y varies directly as x and $y = 3.5x$.

37. As x increases, y decreases and y is 20 divided by x in every entry in the table. So y varies inversely as x and $y = \dfrac{20}{x}$.

39. Since $D = RT$ and $R = 65$, $D = 65T$.
If $T = 1$, then $D = 65(1) = 65$.
If $T = 2$, then $D = 65(2) = 130$.
If $T = 3$, then $D = 65(3) = 195$.
If $T = 4$, then $D = 65(1) = 260$.

Time (hr)	1	2	3	4
Distance (mi)	65	130	195	260

41. Since $T = D/R$ and $D = 400$, $T = 400/R$. Use this formula to fill in the table:

Speed (mph)	20	40	50	200
Time (hr)	20	10	8	2

43. We have $C = 20LW$ where L is the length and W is the width. Use this formula to fill in the table.

Length (yd)	Width (yd)	Cost ($)
8	10	1600
10	12	2400
12	14	3360

45. If the weight w varies directly as the length L, then $w = kL$. If a 12-foot boat weighs 86 pounds, then $86 = k \cdot 12$ and $k = \frac{86}{12} = \frac{43}{6}$. So the formula is $w = \frac{43}{6}L$. If the length is 14 feet, then $w = \frac{43}{6} \cdot 14 = \frac{301}{3} \approx 100.3$ pounds.

47. If the time t varies inversely as the number of elephants n, then $t = \frac{k}{n}$. If $t = 75$ when $n = 4$, then $75 = \frac{k}{4}$ and $k = 300$. So the formula is $t = \dfrac{300}{n}$. If $n = 6$, then $t = \dfrac{300}{6} = 50$ minutes.

49. If the cost C is jointly proportional to the length L and the diameter D, then $C = kLD$. If $C = \$5.80$ when $L = 10$ feet and $D = 1$ inch, then $5.80 = k \cdot 10 \cdot 1$ and $k = 0.58$. So the formula is $C = 0.58LD$. If $L = 15$ and $D = 2$, then $C = 0.58 \cdot 15 \cdot 2 = \$17.40$.

51. Since the approach speed A is directly proportional to the landing speed L, we have $A = kL$. Use $A = 90$ and $L = 75$ to find k:

$$k75 = 90$$
$$k = \frac{90}{75} = 1.2$$

So $A = 1.2L$. If $A = 96$, then $96 = 1.2L$ or $L = \dfrac{96}{1.2} = 80$ mph.

53. Since the number of days n varies inversely with the size of the family s, we have $n = k/s$. If $n = 7$ when $s = 3$, then $7 = k/3$ or $k = 21$. If $s = 7$, then $n = 21/7 = 3$. So a family of 7 can eat a box of cereal in 3 days.

55. If y varies directly as x, then $y = kx$ and the slope of the line is k. The y-intercept is $(0, 0)$. If $y = 3x + 2$, then y does not vary directly as x. Straight lines through the origin ($y = kx$) correspond to direct variations.

3.6 WARM-UPS

1. True, because $4 > 3(-1) + 1$ is correct.
2. True, because $3(2) - 2(-3) \geq 12$ is correct.
3. True, because the inequality $y > x + 9$ indicates the region above the line $y = x + 9$.
4. False, because the inequality $x < y + 2$ is equivalent to $y > x - 2$ and that indicates the region above the line.
5. False, because the graph of $x = 3$ is a vertical line.
6. False, because the graph of $y \leq 5$ is the region on or below the line $y = 5$.
7. True, because $x < 3$ indicates all points that have an x-coordinate smaller than 3.
8. False, because for the inequality $\geq$ we use a solid boundary line.
9. True, because $(0, 0)$ satisfies $y \geq x$.
10. False, because any point above $y = 2x + 1$ satisfies $y > 2x + 1$ and $(0, 0)$ does not satisfy $y > 2x + 1$.

3.6 EXERCISES

1. A linear inequality has the same form as a linear equation except that an inequality symbol is used.
3. If the inequality symbol includes equality then the boundary line is solid. Otherwise it is dashed.
5. In the test point method we test a point to see which side of the boundary line satisfies the inequality.
7. Check each point $(2, 3)$, $(-3, -9)$, and $(8, 3)$ in the inequality $x - y > 5$.

$2 - 3 > 5$ Incorrect.
$-3 - (-9) > 5$ Correct.
$8 - 3 > 5$ Incorrect.

Only $(-3, -9)$ satisfies the inequality.
9. Check each point $(3, 0)$, $(1, 3)$, and $(-2, 5)$ in the inequality $y \geq -2x + 5$.

$0 \geq -2(3) + 5$ Correct.
$3 \geq -2(1) + 5$ Correct.
$5 \geq -2(-2) + 5$ Incorrect.

The points $(3, 0)$ and $(1, 3)$ satisfy the inequality.
11. Check each point $(2, 3)$, $(7, -1)$, and $(0, 5)$ in the inequality $x > -3y + 4$.

$2 > -3(3) + 4$ Correct.
$7 > -3(-1) + 4$ Incorrect.
$0 > -3(5) + 4$ Correct.

The points $(2, 3)$ and $(0, 5)$ satisfy the inequality.
13. To graph $y < x + 4$ first graph the equation $y = x + 4$. The graph of $y = x + 4$ is a line with y-intercept $(0, 4)$ and slope 1. Draw the line dashed because of the inequality symbol. Shade the region below the line to indicate points that satisfy $y < x + 4$.

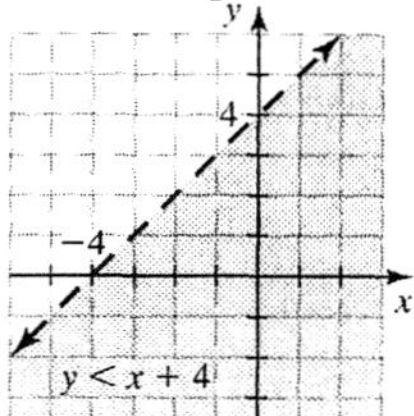

15. To graph $y > -x + 3$ first graph the equation $y = -x + 3$. The graph of $y = -x + 3$ is a line with y-intercept $(0, 3)$ and slope -1. Draw the line dashed because of the inequality symbol. Shade the region above the line to indicate points that satisfy $y > -x + 3$.

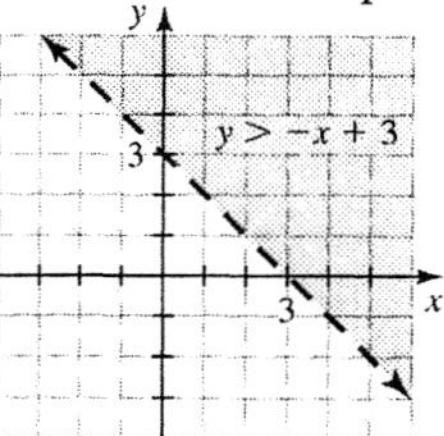

17. To graph $y > \frac{2}{3}x - 3$ first graph the equation $y = \frac{2}{3}x - 3$. The graph of $y = \frac{2}{3}x - 3$ is a line with y-intercept $(0, -3)$ and slope $2/3$. Draw the line dashed because of the inequality symbol. Shade the region above the line to indicate points that satisfy $y > \frac{2}{3}x - 3$.

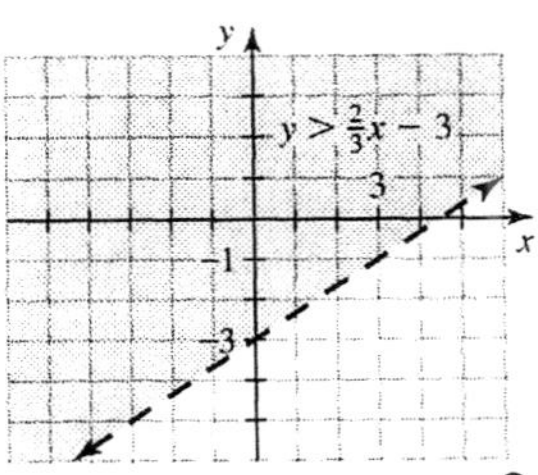

19. To graph $y \leq -\frac{2}{5}x + 2$ first graph the equation $y = -\frac{2}{5}x + 2$. The graph of $y = -\frac{2}{5}x + 2$ is a line with y-intercept $(0, 2)$ and slope $-2/5$. Draw a solid line because of the inequality symbol. Shade the region below the line to indicate the points that satisfy $y \leq -\frac{2}{5}x + 2$.

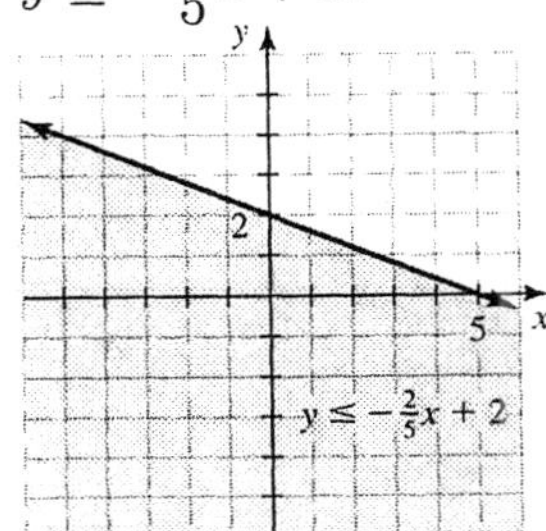

21. To graph the inequality $y - x \geq 0$ first graph the equation $y - x = 0$, or $y = x$. The graph of $y = x$ is a line through the origin with slope 1. Draw the line solid because of the inequality symbol. The inequality $y - x \geq 0$ is equivalent to $y \geq x$. So shade the region above the line to indicate the points that satisfy the inequality.

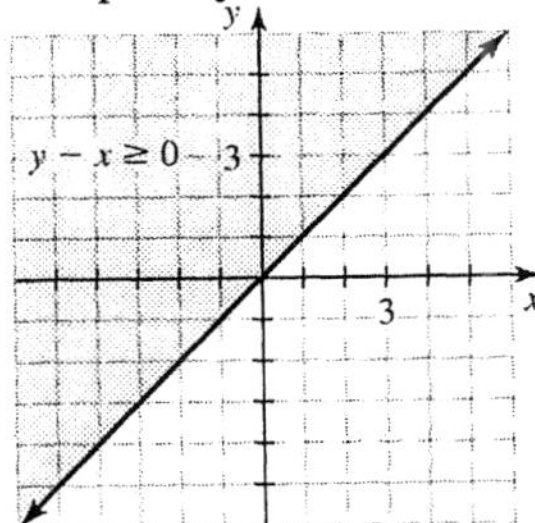

23. The inequality $x > y - 5$ is equivalent to $y < x + 5$. First graph the line $y = x + 5$ with y-intercept $(0, 5)$ and slope 1. Draw the line dashed because of the inequality symbol. Shade the region below the line to indicate the points that satisfy $y < x + 5$.

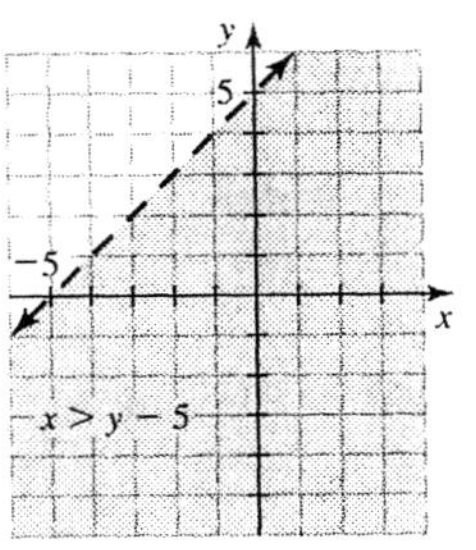

25. Solve the inequality $x - 2y + 4 \leq 0$ for y.

$$-2y \leq -x - 4$$
$$y \geq \frac{1}{2}x + 2$$

First graph the line $y = \frac{1}{2}x + 2$ with y-intercept $(0, 2)$ and slope $1/2$. Draw the line solid because of the inequality symbol. Shade the region above the line to indicate the points that satisfy $y \geq \frac{1}{2}x + 2$.

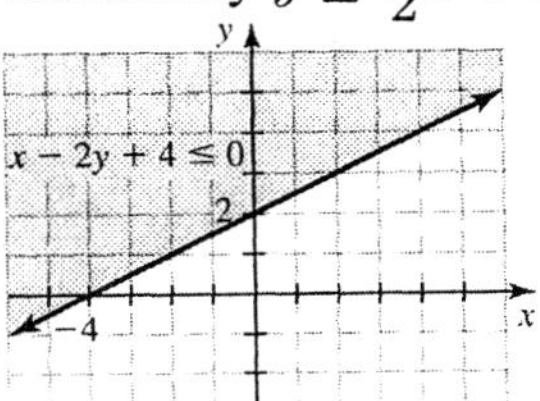

27. To graph $y \geq 2$ first graph $y = 2$. The graph of $y = 2$ is a horizontal line with y-intercept $(0, 2)$. Draw the line solid because of the inequality symbol. Shade the region above the line to indicate the points that satisfy $y \geq 2$.

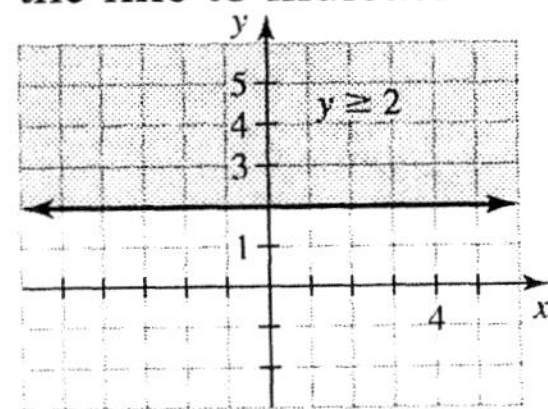

29. To graph $x > 9$ first graph $x = 9$. The graph of $x = 9$ is a vertical line with x-intercept at $(9, 0)$. Draw the line dashed because of the inequality symbol. Shade the region to the right to indicate the points that satisfy the inequality $x > 9$.

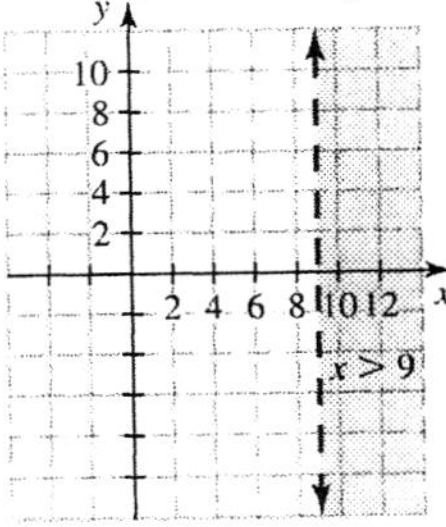

31. The inequality $x + y \leq 60$ is equivalent to $y \leq -x + 60$. First graph the equation $y = -x + 60$. The graph of $y = -x + 60$ is a line with y-intercept $(0, 60)$ and slope -1. Draw the line solid because the inequality symbol includes equality. Shade the region below the line to indicate the points that satisfy $y \leq -x + 60$.

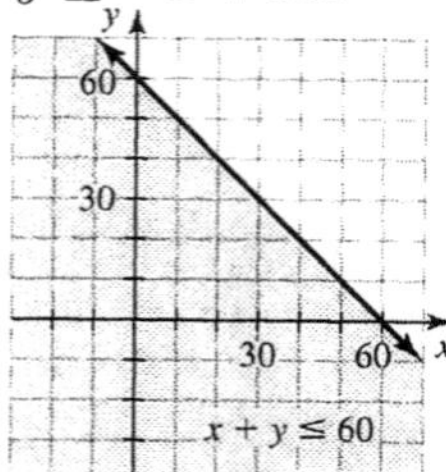

33. To graph $x \leq 100y$, or $y \geq \frac{1}{100}x$, first graph the equation $y = \frac{1}{100}x$. The graph of $y = \frac{1}{100}x$ goes through $(0, 0)$ and has slope $\frac{1}{100}$. Draw the line solid because the inequality symbol includes equality. Shade the region above the line to indicate the points that satisfy $y \geq \frac{1}{100}x$.

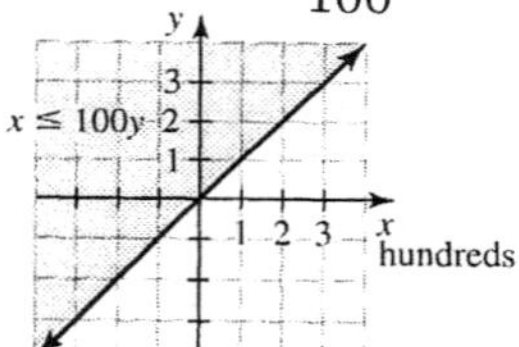

35. First solve the inequality for y.

$$3x - 4y \leq 8$$
$$-4y \leq -3x + 8$$
$$y \geq \frac{3}{4}x - 2$$

First graph the equation $y = \frac{3}{4}x - 2$. The graph of $y = \frac{3}{4}x - 2$ is a line with y-intercept $(0, -2)$ and slope $3/4$. Draw the line solid because $\geq$ includes the line. Shade the region above the line to indicate the points that satisfy $y \geq \frac{3}{4}x - 2$.

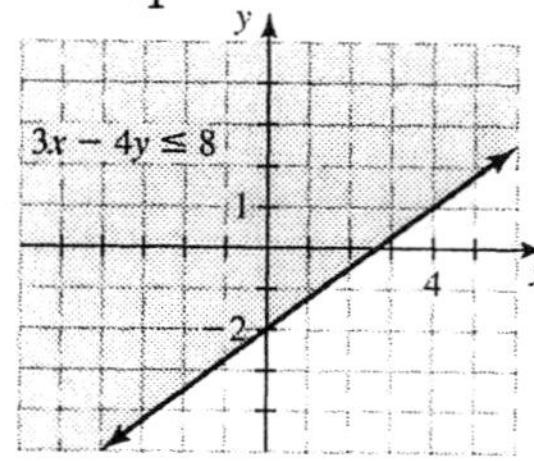

37. To graph $2x - 3y < 6$ first graph $2x - 3y = 6$. The line $2x - 3y = 6$ goes through $(0, -2)$ and $(3, 0)$. Draw the line dashed because of the inequality symbol. Use $(0, 0)$ as a test point. Since $2(0) - 3(0) < 6$ is correct, shade the side of the line that contains the point $(0, 0)$.

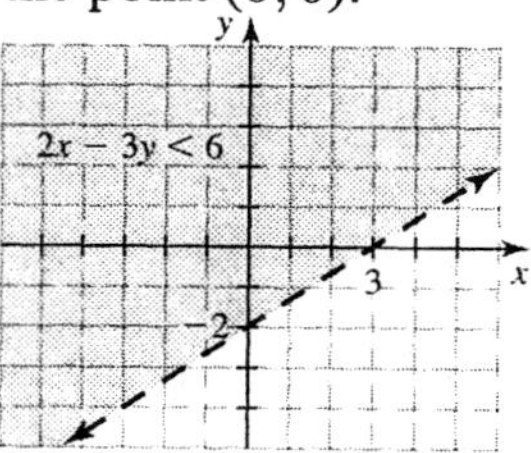

39. To graph $x - 4y \leq 8$ first graph the line $x - 4y = 8$ through $(0, -2)$ and $(8, 0)$. Draw the line solid because of the inequality symbol $\leq$. Test $(0, 0)$. Since $0 - 4 \cdot 0 \leq 8$ is correct, shade the region containing $(0, 0)$.

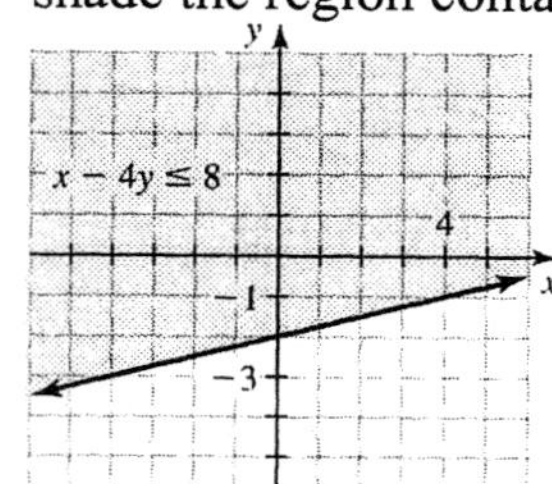

41. To graph $y - \frac{7}{2}x \leq 7$ first graph $2y - 7x = 14$ through $(0, 7)$ and $(-2, 0)$. Draw the line solid because of the inequality symbol $\leq$. Test $(0, 0)$ in the original inequality. Since $0 - \frac{7}{2} \cdot 0 \leq 7$ is correct, shade the region containing $(0, 0)$.

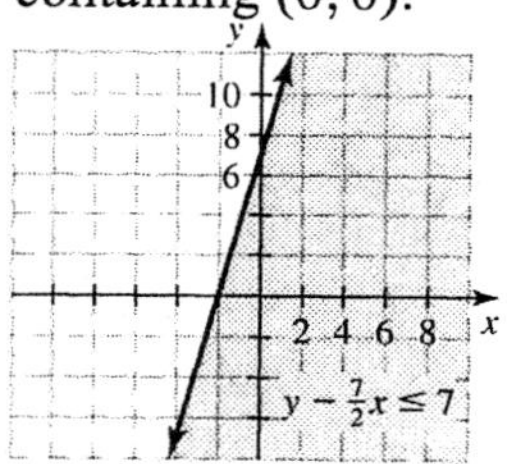

43. To graph $x - y < 5$ first graph $x - y = 5$. The line $x - y = 5$ goes through $(0, -5)$ and $(5, 0)$. Test $(0, 0)$ in the original inequality. Since $0 - 0 < 5$ is correct, shade the region containing $(0, 0)$.

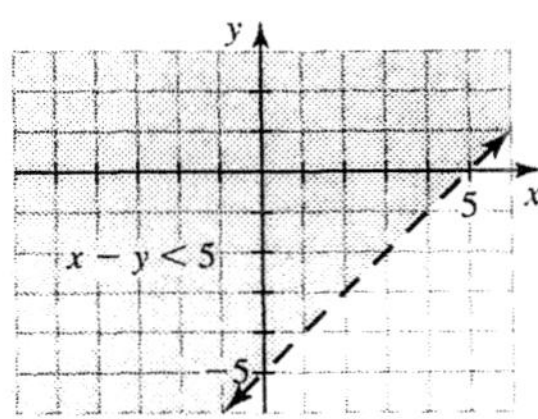

45. To graph $3x - 4y < -12$ first graph $3x - 4y = -12$ through $(0, 3)$ and $(-4, 0)$. Draw the line dashed because of the inequality symbol $<$. Test $(0, 0)$ in the original inequality. Since $3 \cdot 0 - 4 \cdot 0 < -12$ is incorrect, shade the region that does not contain $(0, 0)$.

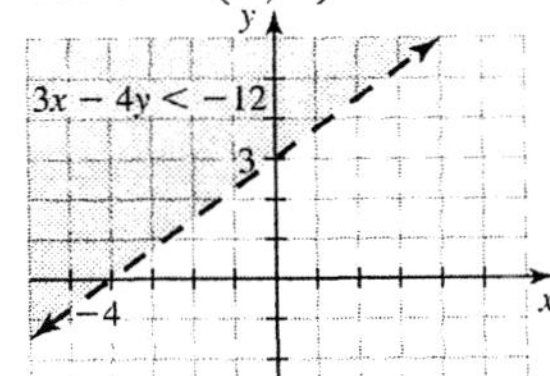

47. To graph $x < 5y - 100$ first graph $x = 5y - 100$ through $(0, 20)$ and $(-100, 0)$. Draw the line dashed because of the inequality symbol $<$. Test $(0, 0)$ in the original inequality. Since $0 < 5 \cdot 0 - 100$ is incorrect, we shade the region that does not contain $(0, 0)$.

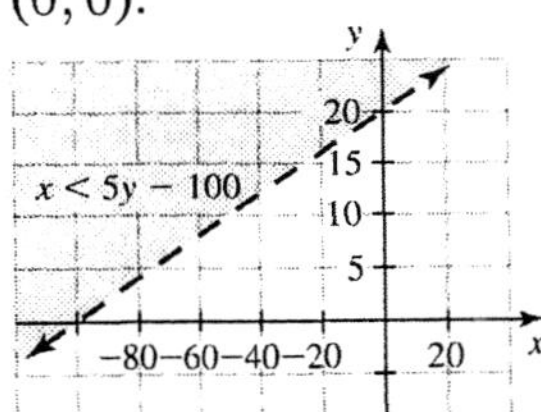

49. Let $x =$ the number of round tables and $y =$ the number of rectangular tables. Since there is at most 3850 cubic feet of storage space available, $25x + 35y \leq 3850$, or $5x + 7y \leq 770$. Graph $5x + 7y = 770$ through $(0, 110)$ and $(154, 0)$. Shade the region below this line in the first quadrant, because x and y cannot be negative.

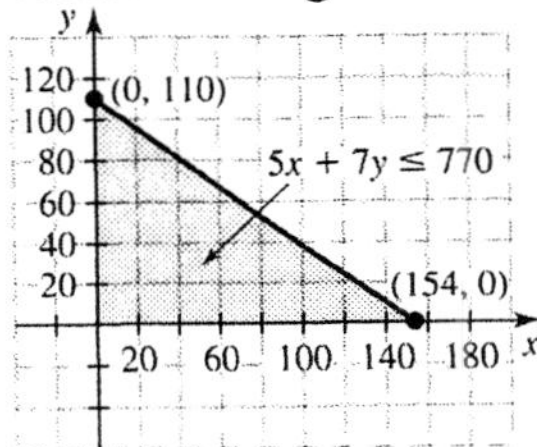

51. Let $x =$ the number of pens and $y =$ the number of notebooks. We have $0.25x + 0.40y \leq 4.00$, or $25x + 40y \leq 400$, or $5x + 8y \leq 80$. Graph $5x + 8y = 80$ through $(0, 10)$ and $(16, 0)$. Shade in the first quadrant below this line.

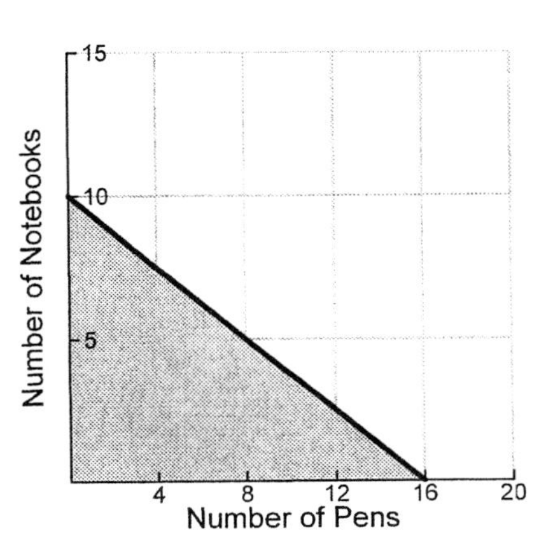

Enriching Your Mathematical Word Power

1. d **2.** a **3.** b **4.** c **5.** b
6. a **7.** c **8.** c **9.** d **10.** b
11. c **12.** d **13.** c

CHAPTER 3 REVIEW

1. To locate the point (−2, 5) we start at the origin and move 2 units to the left and then 5 units up. This point is in quadrant II.
3. Since the y-coordinate is zero in (3, 0), the point is on the x-axis.
5. Since the x-coordinate is zero in (0, −6), the point is on the y-axis.
7. To locate the point (1.414, −3) we start at the origin and move approximately 1.414 units to the right and then 3 units down. This point is in quadrant IV.
9. If $x = 0$, then $y = 3(0) - 5 = -5$.
If $x = -3$, then $y = 3(-3) - 5 = -14$.
If $x = 4$, then $y = 3(4) - 5 = 7$.
The points are (0, −5), (−3, −14), and (4, 7).
11. Solve $2x - 3y = 8$ for y.

$$y = \frac{2}{3}x - \frac{8}{3}$$

If $x = 0$, then $y = \frac{2}{3}(0) - \frac{8}{3} = -\frac{8}{3}$.
If $x = 3$, then $y = \frac{2}{3}(3) - \frac{8}{3} = -\frac{2}{3}$.
If $x = -6$, then $y = \frac{2}{3}(-6) - \frac{8}{3} = -\frac{20}{3}$.

The points are $\left(0, -\frac{8}{3}\right)$, $\left(3, -\frac{2}{3}\right)$, and $\left(-6, -\frac{20}{3}\right)$.

13. Three ordered pairs that satisfy $y = -3x + 4$ are $(0, 4)$, $(1, 1)$, and $(2, -2)$. Draw a line through these points as shown.

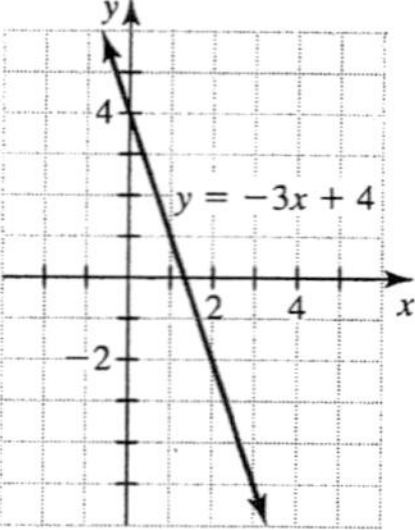

15. Three ordered pairs that satisfy $x + y = 7$ are $(1, 6)$, $(2, 5)$, and $(3, 4)$. Draw a line through these points as shown.

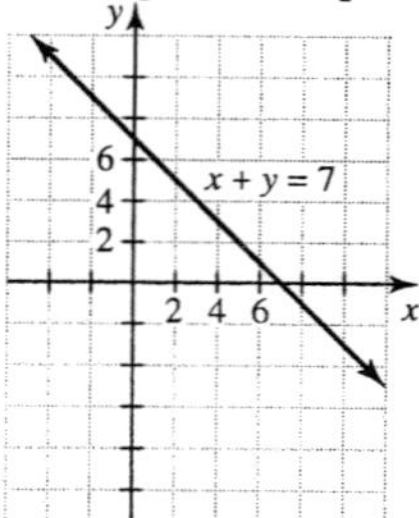

17. $m = \dfrac{1-0}{1-0} = 1$

19. $m = \dfrac{-3-0}{-2-0} = \dfrac{-3}{-2} = \dfrac{3}{2}$

21. $m = \dfrac{-2-1}{-4-3} = \dfrac{-3}{-7} = \dfrac{3}{7}$

23. The equation $y = 3x - 18$ is in slope-intercept form. So the slope is 3 and the y-intercept is $(0, -18)$.

25. Rewrite $2x - y = 3$ in slope-intercept form.

$$-y = -2x + 3$$
$$y = 2x - 3$$

The slope is 2 and the y-intercept is $(0, -3)$.

27. Rewrite $4x - 2y - 8 = 0$ in slope-intercept form.

$$-2y = -4x + 8$$
$$y = 2x - 4$$

The slope is 2 and the y-intercept is $(0, -4)$.

29. To sketch $y = \frac{2}{3}x - 5$, start at the y-intercept $(0, -5)$ and use the slope $\frac{2}{3}$ to locate a second point on the line. From $(0, -5)$ rise 2 and run 3. Draw a line through the two points.

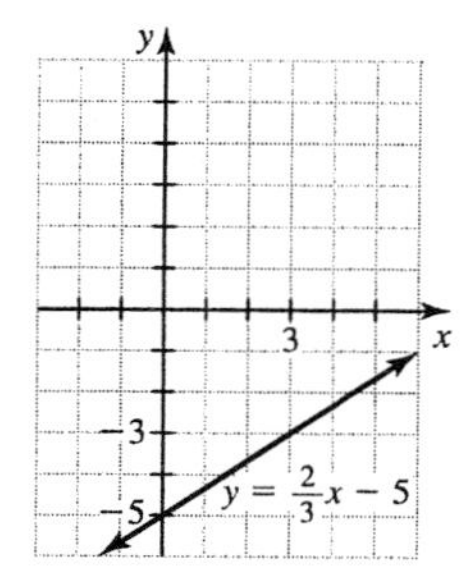

31. Rewrite $2x + y = -6$ in slope-intercept form.

$$y = -2x - 6$$

Start at the y-intercept $(0, -6)$ and use the slope $-2 = -2/1$ to locate a second point on the line. From $(0, -6)$ rise -2 and run 1, or rise 2 and run -1. Draw a line through the two points.

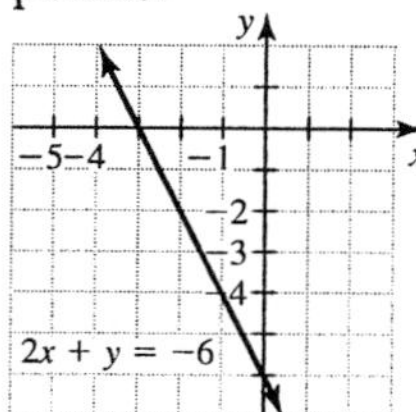

33. The line $y = -4$ has slope 0 and y-intercept $(0, -4)$. Draw a horizontal line through $(0, -4)$.

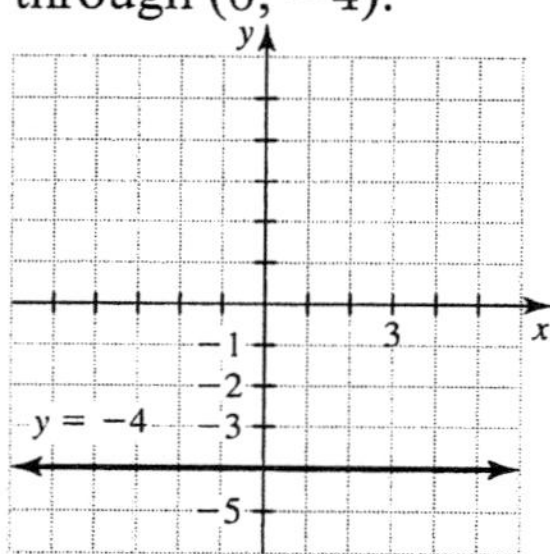

35. Use slope-intercept form with slope $1/3$ and y-intercept $(0, 4)$.

$$y = \frac{1}{3}x + 4$$
$$3y = x + 12$$
$$-x + 3y = 12$$
$$x - 3y = -12$$

37. Any line perpendicular to $y = 2x - 1$ has slope $-\frac{1}{2}$. The line through $(0, 0)$ with slope $-\frac{1}{2}$ has the following equation.

$$y = -\frac{1}{2}x$$
$$2y = -x$$
$$x + 2y = 0$$

39. Any line parallel to the x-axis has slope 0. The line through (3, 5) with slope 0 has y-intercept (0, 5).

$$y = 0 \cdot x + 5$$
$$y = 5$$

41.
$$y - 3 = \frac{2}{3}(x + 6)$$
$$y - 3 = \frac{2}{3}x + 4$$
$$y = \frac{2}{3}x + 7$$

43.
$$3x - 7y - 14 = 0$$
$$-7y = -3x + 14$$
$$y = \frac{3}{7}x - 2$$

45.
$$y - 5 = -\frac{3}{4}(x + 1)$$
$$y - 5 = -\frac{3}{4}x - \frac{3}{4}$$
$$y = -\frac{3}{4}x + \frac{17}{4}$$

47. Use the point-slope form with (−4, 7) and slope −2.

$$y - 7 = -2(x - (-4))$$
$$y - 7 = -2x - 8$$
$$y = -2x - 1$$

49. Find the slope of the line through (−2, 1) and (3, 7).

$$m = \frac{7 - 1}{3 - (-2)} = \frac{6}{5}$$

Use the slope 6/5 and the point (3, 7) in point-slope form.

$$y - 7 = \frac{6}{5}(x - 3)$$
$$y - 7 = \frac{6}{5}x - \frac{18}{5}$$
$$y = \frac{6}{5}x + \frac{17}{5}$$

51. Any line parallel to $y = 3x - 1$ has slope 3. Use point-slope form with slope 3 and the point (3, −5).

$$y - (-5) = 3(x - 3)$$
$$y + 5 = 3x - 9$$
$$y = 3x - 14$$

53. We want the equation of a line through the points (2, 113) and (5, 209). The slope is $m = \frac{209 - 113}{5 - 2} = \frac{96}{3} = 32$. Use slope 32 and the point $(2, 113)$ in the point-slope form.

$$C - 113 = 32(n - 2)$$
$$C - 113 = 32n - 64$$
$$C = 32n + 49$$

For 4 days the charge would be $C = 32(4) + 49 = \$177$.

55. a) Find the equation of the line through (0.9, 0.1) and (0.8, 0.2).

$$m = \frac{0.2 - 0.1}{0.8 - 0.9} = \frac{0.1}{-0.1} = -1$$
$$q - 0.1 = -1(p - 0.9)$$
$$q - 0.1 = -p + 0.9$$
$$q = -p + 1$$
$$q = 1 - p$$

b) From the graph, if the probability that it does not rain is 0, then the probability that it rains is 1.

57. Find the equation of the line through $(4, 1)$ and $(14, 2)$.

$$m = \frac{2 - 1}{14 - 4} = 0.1$$
$$y - 2 = 0.1(x - 14)$$
$$y - 2 = 0.1x - 1.4$$
$$y = 0.1x + 0.6$$

59. If y varies directly as w, then $y = kw$. If $y = 48$ when $w = 4$, then $48 = k \cdot 4$, or $k = 12$. The formula is $y = 12w$. If $w = 11$, then $y = 12 \cdot 11 = 132$.

61. If y varies inversely as v, then $y = k/v$. If $y = 8$ when $v = 6$, then $8 = k/6$, or $k = 48$. So the formula is $y = 48/v$. If $v = 24$, then $y = 48/24 = 2$.

63. If y varies jointly as u and v, then $y = kuv$. If $y = 72$ when $u = 3$ and $v = 4$, then $72 = k \cdot 3 \cdot 4$, or $k = 6$. So the equation is $y = 6uv$. If $u = 5$ and $v = 2$, then $y = 6 \cdot 5 \cdot 2 = 60$.

65. a) If the cost C (in dollars) varies directly with the length T of the ride (in minutes), then $C = kT$. If a 12-minute ride costs \$9.00, then $9 = k \cdot 12$, or $k = 0.75$. So the equation is $C = 0.75T$.

b) If $T = 20$ minutes, then $C = 0.75(20) = \$15$.

c) The cost increases as the length of the ride increases.

67. First graph the line $y = \frac{1}{3}x - 5$, which has y-intercept $(0, -5)$ and slope $1/3$. Use a dashed line because the inequality symbol is $>$. Shade the region above the line to indicate the points that satisfy $y > \frac{1}{3}x - 5$.

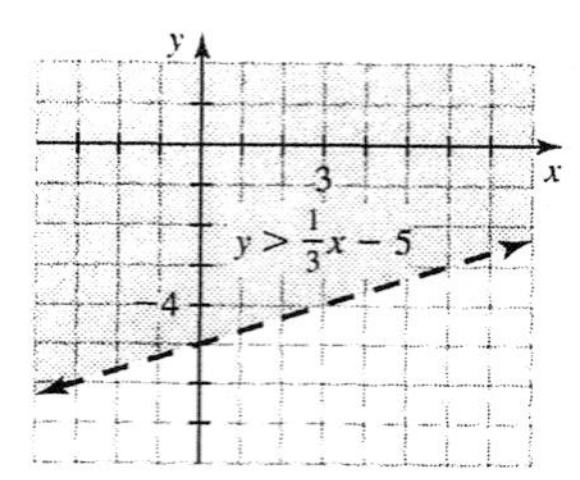

69. First graph the line $y = -2x + 7$, which has y-intercept $(0, 7)$ and slope -2. Use a solid line because of the inequality symbol $\leq$. Shade the region below the line to indicate the points that satisfy $y \leq -2x + 7$.

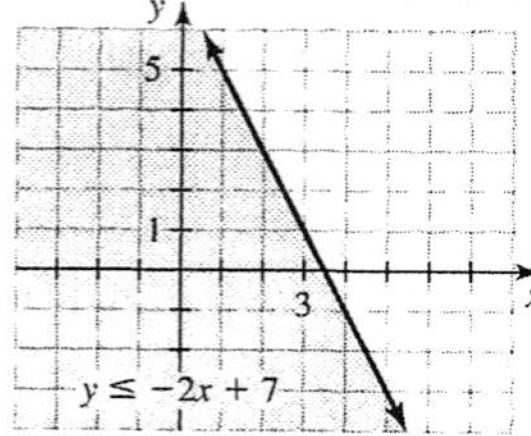

71. First graph the line $y = 8$. The line $y = 8$ is a horizontal line with y-intercept $(0, 8)$. Use a solid line because of the inequality symbol $\leq$. Shade the region below the line to indicate the points that satisfy $y \leq 8$.

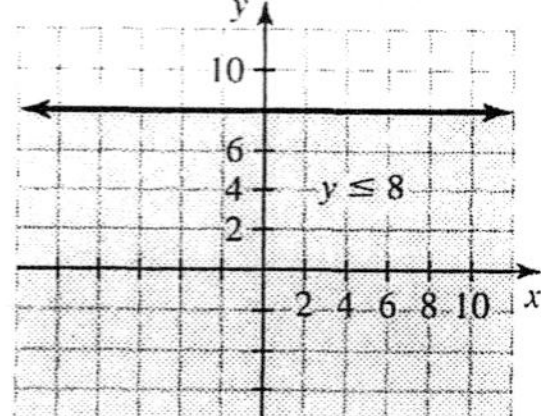

73. First graph the line $2x + 3y = -12$. The line goes through the points $(0, -4)$ and $(-6, 0)$. Draw a solid line because of the inequality symbol $\leq$. Test $(0, 0)$ in the inequality. Since $2(0) + 3(0) \leq -12$ is incorrect, we shade the region that does not contain $(0, 0)$ to indicate the points that satisfy $2x + 3y \leq -12$.

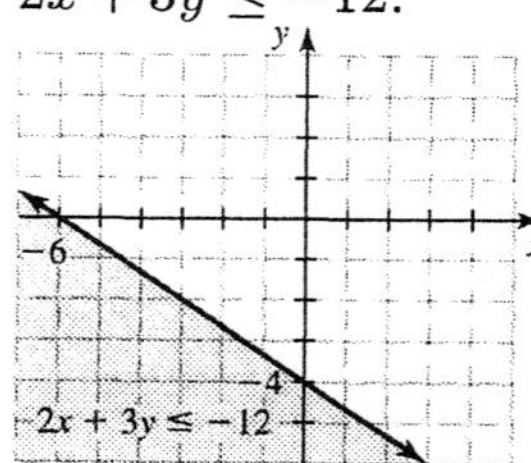

CHAPTER 3 TEST

1. Since the x-coordinate is negative and the y-coordinate is positive, the point $(-2, 7)$ lies in quadrant II.
2. Since the y-coordinate is 0, $(-\pi, 0)$ lies on the x-axis.
3. Since the x-coordinate is positive and the y-coordinate is negative, $(3, -6)$ lies in quadrant IV.
4. Since the x-coordinate is 0, $(0, 1785)$ lies on the y-axis.
5. $m = \frac{4-3}{4-3} = \frac{1}{1} = 1$
6. $m = \frac{-3-(-8)}{-2-4} = \frac{5}{-6} = -\frac{5}{6}$
7. The line $y = 3x - 5$ has slope 3.
8. The line $y = 3$ has slope 0.
9. The line $x = 5$ is a vertical line and slope is undefined for vertical lines.
10. Solving $2x - 3y = 4$ for y yields $y = \frac{2}{3}x - \frac{4}{3}$. So the slope is $\frac{2}{3}$.
11. Use the slope $-1/2$ and y-intercept $(0, 3)$ in the slope-intercept form: $y = -\frac{1}{2}x + 3$.

12. Use slope $3/7$ and the point $(-1, -2)$ in point-slope form.

$$\begin{aligned} y - (-2) &= \frac{3}{7}(x - (-1)) \\ y + 2 &= \frac{3}{7}x + \frac{3}{7} \\ y &= \frac{3}{7}x - \frac{11}{7} \end{aligned}$$

13. The line $y = -3x + 12$ has slope -3. Any line perpendicular to it has slope $1/3$. Use slope $1/3$ and the point $(2, -3)$ in the point slope form.

$$\begin{aligned} y - (-3) &= \frac{1}{3}(x - 2) \\ y + 3 &= \frac{1}{3}x - \frac{2}{3} \\ 3y + 9 &= x - 2 \\ -x + 3y &= -11 \\ x - 3y &= 11 \end{aligned}$$

14. The equation $5x + 3y = 9$ is equivalent to $y = -\frac{5}{3}x + 3$. Its slope is $-\frac{5}{3}$ and any line parallel to it has slope $-\frac{5}{3}$. Use slope $-\frac{5}{3}$ and the point (3, 4) in the point-slope form.

$$y - 4 = -\frac{5}{3}(x - 3)$$
$$y - 4 = -\frac{5}{3}x + 5$$
$$3y - 12 = -5x + 15$$
$$5x + 3y = 27$$

15. The graph of $y = \frac{1}{2}x - 3$ has y-intercept of $(0, -3)$ and a slope of $1/2$. Start at $(0, -3)$ and rise 1 and run 2 to locate a second point. Draw a line through the two points.

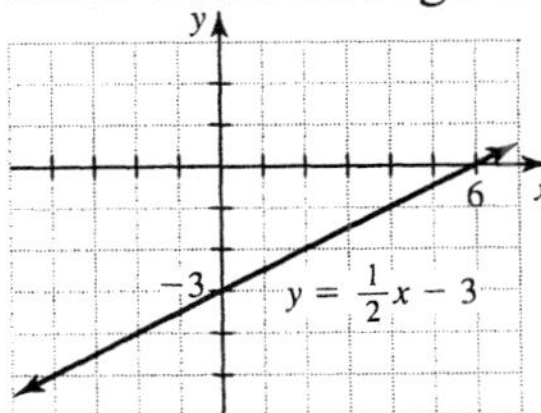

16. If $x = 0$ in $2x - 3y = 6$ then $y = -2$, and if $y = 0$ in $2x - 3y = 6$, then $x = 3$. So the x-intercept is $(3, 0)$ and the y-intercept is $(0, -2)$. Draw a line through the two intercepts.

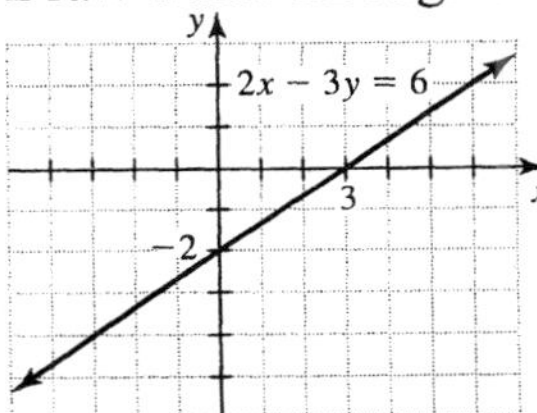

17. The graph of $y = 4$ is a horizontal line with y-intercept of $(0, 4)$.

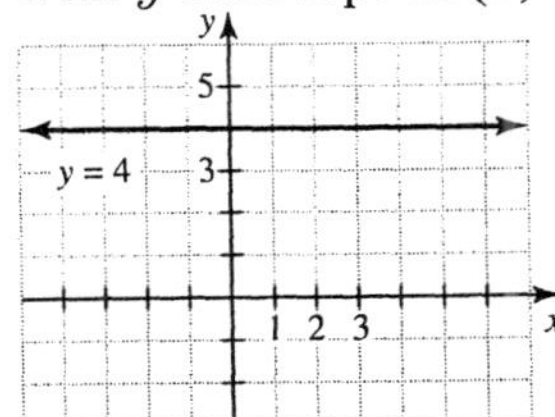

18. The graph of $x = -2$ is a vertical line with x-intercept $(-2, 0)$.

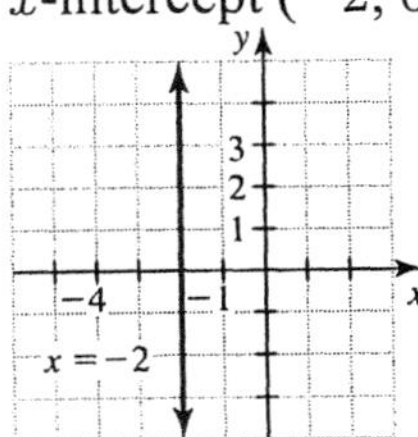

19. First graph the line $y = 3x - 5$, which has a y-intercept of $(0, -5)$ and a slope of 3. Use a dashed line. Shade the region above the line to indicate the points that satisfy $y > 3x - 5$.

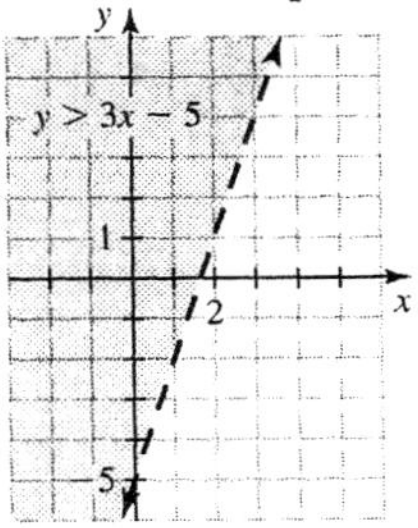

20. First graph $x - y = 3$. The line goes through the points $(3, 0)$ and $(0, -3)$. Test $(0, 0)$ in the inequality. Since $0 - 0 < 3$ is correct, we shade the region containing $(0, 0)$ to indicate the points that satisfy $x - y < 3$.

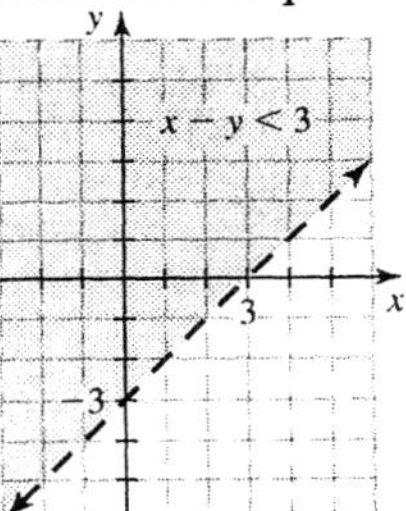

21. First graph $x - 2y = 4$ through the points $(4, 0)$ and $(0, -2)$. Test $(0, 0)$ in the inequality. Since $0 - 2(0) \geq 4$ is incorrect, we shade the region that does not contain $(0, 0)$ to indicate the points that satisfy $x - 2y \geq 4$.

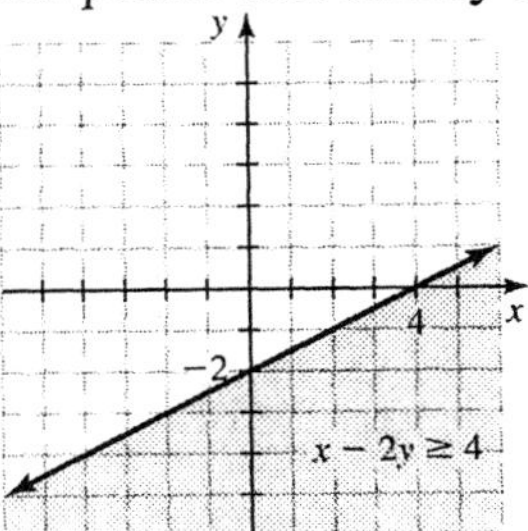

22. To find the total shipping and handling fee S, we multiply the number of CDs n by \$0.75 and add on the \$2.50 fee: $S = 0.75n + 2.50$.

23. Find the equation of the line through $(10, 50)$ and $(16, 68)$. First find the slope.

$$m = \frac{68 - 50}{16 - 10} = \frac{18}{6} = 3$$

Use slope 3 and the point (10, 50) in the point-slope form where P is the price in cents and v is the volume of the cup in ounces.

$$P - 50 = 3(v - 10)$$
$$P - 50 = 3v - 30$$
$$P = 3v + 20$$

If $v = 20$ ounces, then $P = 3(20) + 20 = 80$ cents.

24. If the price P (in dollars) varies directly with the weight w (in pounds), then $P = kw$. If a 30-pound watermelon sells for \$4.20, then $4.20 = k \cdot 30$, or $k = 0.14$. So the formula is $P = 0.14w$. If $w = 20$ pounds, then $P = 0.14(20) = \$2.80$.

25. Since the number of days n is inversely proportional to the sales s, $n = k/s$. If $n = 15$ when $s = 75{,}000$, then $15 = k/75{,}000$, or $k = 1{,}125{,}000$. The formula is $n = 1{,}125{,}000/s$. If $s = 60{,}000$, then $n = 1{,}125{,}000/60{,}000 = 18.75$ days. His road time decreases as his sales increase.

26. If the labor cost varies jointly with the length a width, then $C = kLW$. If the cost for an 8 by 10 room is \$400, then $400 = k \cdot 8 \cdot 10$ and $k = 5$. So $C = 5LW$. If $W = 11$ and $L = 14$, then $C = 5 \cdot 14 \cdot 11 = 770$. The cost for an 11 by 14 room is \$770.

Making Connections

Chapters 1-3

1. $9 - 5 \cdot 2 = 9 - 10 = -1$

2. $-4 \cdot 5 - 7 \cdot 2 = -20 - 14 = -34$

3. $3^2 - 2^3 = 9 - 8 = 1$

4. $3^2 \cdot 2^3 = 9 \cdot 8 = 72$

5. $(-4)^2 - 4(1)(5) = 16 - 20 = -4$

6. $-4^2 - 4 \cdot 3 = -16 - 12 = -28$

7. $\frac{-5 - 9}{2 - (-2)} = \frac{-14}{4} = -\frac{7}{2}$

8. $\frac{6 - 3.6}{6} = \frac{2.4}{6} = 0.4$

9. $\frac{1 - \frac{1}{2}}{4 - (-1)} = \frac{\frac{1}{2}}{5} = \frac{1}{2} \cdot \frac{1}{5} = \frac{1}{10}$

10. $\frac{4 - (-6)}{1 - \frac{1}{3}} = \frac{10}{\frac{2}{3}} = 10 \cdot \frac{3}{2} = 15$

11. $4x - (-9x) = 13x$

12. $4(x - 9) - x = 4x - 36 - x$
$= 3x - 36$

13.
$$5(x - 3) + x = 0$$
$$5x - 15 + x = 0$$
$$6x = 15$$
$$x = \frac{5}{2}$$

The solution set is $\left\{\frac{5}{2}\right\}$.

14.
$$5 - 2(x - 1) = x$$
$$5 - 2x + 2 = x$$
$$-3x = -7$$
$$x = \frac{7}{3}$$

The solution set is $\left\{\frac{7}{3}\right\}$.

15. $\frac{1}{2} - \frac{1}{3} = \frac{3}{6} - \frac{2}{6} = \frac{1}{6}$

16. $\frac{1}{4} + \frac{1}{6} = \frac{3}{12} + \frac{2}{12} = \frac{5}{12}$

17.
$$\frac{1}{2}x - \frac{1}{3} = \frac{1}{4}x + \frac{1}{6}$$
$$6x - 4 = 3x + 2$$
$$3x = 6$$
$$x = 2$$

The solution set is $\{2\}$.

18.
$$\frac{2}{3}x + \frac{1}{5} = \frac{3}{5}x - \frac{1}{15}$$
$$10x + 3 = 9x - 1$$
$$x = -4$$

The solution set is $\{-4\}$.

19. $\frac{4x - 8}{2} = \frac{1}{2}(4x - 8) = 2x - 4$

20. $\frac{-5x - 10}{-5} = -\frac{1}{5}(-5x - 10)$
$= x + 2$

21.
$$\frac{6 - 2(x - 3)}{2} = 1$$
$$\frac{12 - 2x}{2} = 1$$
$$6 - x = 1$$
$$5 = x$$

The solution set is $\{5\}$.

22.
$$\frac{20 - 5(x - 5)}{5} = 6$$
$$\frac{45 - 5x}{5} = 6$$
$$9 - x = 6$$
$$3 = x$$

The solution set is $\{3\}$.

23.
$$-4(x - 9) - 4 = -4x$$
$$-4x + 36 - 4 = -4x$$
$$32 = 0$$

The solution set is the empty set, $\emptyset$.

24. $4(x-6) = -4(6-x)$
$$4x - 24 = -24 + 4x$$

The solution set is the set of all real numbers.

25.
$$\begin{aligned} 2x - 3 &> 6 \\ 2x &> 9 \\ x &> 4.5 \end{aligned}$$
The solution set is the interval $(4.5, \infty)$.

26.
$$\begin{aligned} 5 - 3x &< 7 \\ -3x &< 2 \\ x &> -\frac{2}{3} \end{aligned}$$
The solution set is the interval $(-2/3, \infty)$.

27.
$$\begin{aligned} 51 - 2x &\leq 3x + 1 \\ -5x &\leq -50 \\ x &\geq 10 \end{aligned}$$
The solution set is the interval $[10, \infty)$.

28.
$$\begin{aligned} 4x - 80 &\geq 60 - 3x \\ 7x &\geq 140 \\ x &\geq 20 \end{aligned}$$
The solution set is the interval $[20, \infty)$.

29.
$$\begin{aligned} -1 < 4 - 2x &\leq 5 \\ -5 < -2x &\leq 1 \\ \frac{5}{2} > x &\geq -\frac{1}{2} \\ -\frac{1}{2} \leq x &< \frac{5}{2} \end{aligned}$$
The solution set is the interval $[-1/2, 5/2)$.

30.
$$\begin{aligned} 1 - 2x \leq x + 1 &< 3 - 2x \\ -2x \leq x &< 2 - 2x \\ 0 \leq 3x &< 2 \\ 0 \leq x &< \frac{2}{3} \end{aligned}$$
The solution set is the interval $[0, 2/3)$.

31.
$$\begin{aligned} 3\pi y + 2 &= t \\ 3\pi y &= t - 2 \\ y &= \frac{t-2}{3\pi} \end{aligned}$$

32.
$$\begin{aligned} x &= \frac{y-b}{m} \\ mx &= y - b \\ mx + b &= y \\ y &= mx + b \end{aligned}$$

33.
$$\begin{aligned} 3x - 3y - 12 &= 0 \\ -3y &= -3x + 12 \\ y &= x - 4 \end{aligned}$$

34.
$$\begin{aligned} 2y - 3 &= 9 \\ 2y &= 12 \\ y &= 6 \end{aligned}$$

35.
$$\begin{aligned} \frac{y}{2} - \frac{y}{4} &= \frac{1}{5} \\ 20\left(\frac{y}{2} - \frac{y}{4}\right) &= 20\left(\frac{1}{5}\right) \\ 10y - 5y &= 4 \\ 5y &= 4 \\ y &= \frac{4}{5} \end{aligned}$$

36.
$$\begin{aligned} 0.6y - 0.06y &= 108 \\ 60y - 6y &= 10800 \\ 54y &= 10800 \\ y &= 200 \end{aligned}$$

37. **a)** $m = \dfrac{3-1}{50-35} = \dfrac{2}{15}$

b) $m = \dfrac{6-3}{65-50} = \dfrac{3}{15} = \dfrac{1}{5}$

c) A 38 year-old man should save twice his income during the years 35 to 50, or \$80,000. To do this in 15 years he would have to save \$80,000 $\div$ 15 or \$5333 dollars per year.
$$\frac{5333}{40{,}000} \approx 0.133$$
So he should save about 13% of his income. Since this calculation does not include interest it is very approximate and other answers might also be reasonable.

d) The 58-year old women needs to get from \$180,000 to \$360,000 in 15 years, or she needs to save about \$12,000 per year. So at age 58 she should have saved

$$3(60{,}000) + 8(12{,}000) = \$276{,}000.$$

These figures do not include interest and are approximate.

4.1 WARM-UPS

1. False, because the coefficient of x is -4.
2. False, because the degree is 3, the highest power of the variable x.
3. True, because $x^2 - x$ could be written as $x^2 + (-1)x$.
4. True, because the degree is the highest power of x.
5. False, $x^4 + x$ is a binomial with degree 4.
6. True, because $P(5) = 3(5) - 1 = 14$.
7. False, $x^5 - 3x - 8$ is a trinomial with degree 5.
8. True, because the monomials are subtracted correctly.
9. True, because the polynomials are added correctly.
10. False, because the polynomials are not subtracted correctly.

4.1 EXERCISES

1. A term is a single number or the product of a number and one or more variables raised to powers.
3. The degree of a polynomial in one variable is the highest power of the variable in the polynomial.
5. Polynomials are added by adding the like terms.
7. The coefficient of x^3 is -3 and the coefficient of x^2 is 7.
9. Since x^3 does not appear, its coefficient is 0. The coefficient of x^2 is 6.
11. The polynomial can be written as $\frac{1}{3}x^3 + \frac{7}{2}x^2 - 4$. The coefficient of x^3 is $\frac{1}{3}$ and the coefficient of x^2 is $\frac{7}{2}$.
13. Since -1 has only one term, it is a monomial. Since there is no variable, the degree is 0.
15. Since m^3 has one term only, it is a monomial. The degree is 3, because that is the highest power of m.
17. Since $4x + 7$ has two terms, it is a binomial. The degree is 1, because that is the highest power of x in the polynomial.
19. Since the polynomial has three terms, it is a trinomial. The degree is 10, because that is the highest power of the variable x.
21. Since $x^6 + 1$ has two terms, it is a binomial. The degree is 6, because that is the highest power of x in the binomial.
23. Since the polynomial has 3 terms it is a trinomial. Since the highest power of a is 3, the degree is 3.
25. $2x^2 - 3x + 1 = 2(-1)^2 - 3(-1) + 1$
$= 2 + 3 + 1 = 6$
27. $\frac{1}{2}x^2 - x + 1 = \frac{1}{2}\left(\frac{1}{2}\right)^2 - \frac{1}{2} + 1$
$= \frac{1}{8} - \frac{4}{8} + \frac{8}{8} = \frac{5}{8}$
29. $-3x^3 - x^2 + 3x - 4$
$= -3(3)^3 - 3^2 + 3(3) - 4$
$= -81 - 9 + 9 - 4 = -85$
31. Replace x with 3 in $P(x) = x^2 - 4$ to get $P(3) = (3)^2 - 4 = 5$.
33. $P(-2) = 3(-2)^4 - 2(-2)^3 + 7$
$= 48 + 16 + 7 = 71$
35. $P(1.45)$
$= 1.2(1.45)^3 - 4.3(1.45) - 2.4$
$= -4.97665$
37. $(x - 3) + (3x - 5)$
$= x + 3x - 3 - 5 = 4x - 8$
39. $(q - 3) + (q + 3) = 2q$
41. $(3x + 2) + (x^2 - 4) = x^2 + 3x + 2 - 4$
$= x^2 + 3x - 2$
43. $(4x - 1) + (x^3 + 5x - 6)$
$= x^3 + 4x + 5x - 1 - 6$
$= x^3 + 9x - 7$
45. $(a^2 - 3a + 1) + (2a^2 - 4a - 5)$
$= a^2 + 2a^2 - 3a - 4a + 1 - 5$
$= 3a^2 - 7a - 4$
47. $(w^2 - 9w - 3) + (w - 4w^2 + 8)$
$= w^2 - 4w^2 - 9w + w - 3 + 8$
$= -3w^2 - 8w + 5$
49. $(5.76x^2 - 3.14x - 7.09)$
$+ (3.9x^2 + 1.21x + 5.6)$
$= 9.66x^2 - 1.93x - 1.49$
51. $(x - 2) - (5x - 8) = x - 2 - 5x + 8$
$= -4x + 6$
53. $(m - 2) - (m + 3) = m - 2 - m - 3$
$= -5$

55. $(2z^2 - 3z) - (3z^2 - 5z)$
$= 2z^2 - 3z^2 - 3z + 5z$
$= -z^2 + 2z$

57. $(w^5 - w^3) - (-w^4 + w^2)$
$= w^5 - w^3 + w^4 - w^2$
$= w^5 + w^4 - w^3 - w^2$

59. $(t^2 - 3t + 4) - (t^2 - 5t - 9)$
$= t^2 - t^2 - 3t + 5t + 4 + 9$
$= 2t + 13$

61. $(9 - 3y - y^2) - (2 + 5y - y^2)$
$= 9 - 3y - y^2 - 2 - 5y + y^2$
$= -8y + 7$

63. $(3.55x - 879) - (26.4x - 455.8)$
$= 3.55x - 879 - 26.4x + 455.8$
$= -22.85x - 423.2$

65.
$$\begin{array}{r} 3a - 4 \\ \underline{a + 6} \\ 4a + 2 \end{array}$$

67.
$$\begin{array}{r} 3x + 11 \\ \underline{5x + 7} \\ -2x + 4 \end{array} \quad 3x - 5x = -2x$$

69.
$$\begin{array}{r} a - b \\ \underline{a + b} \\ 2a \end{array}$$

71.
$$\begin{array}{r} -3m + 1 \\ \underline{2m - 6} \\ -5m + 7 \end{array} \quad 1 - (-6) = 7$$

73.
$$\begin{array}{r} 2x^2 - x - 3 \\ \underline{2x^2 + x + 4} \\ 4x^2 \qquad + 1 \end{array}$$

75.
$$\begin{array}{r} 3a^3 - 5a^2 \qquad + 7 \\ \underline{2a^3 + 4a^2 - 2a \quad} \\ a^3 - 9a^2 + 2a + 7 \end{array} \quad \begin{array}{r} -5 - 4 = -9 \\ 0 - (-2) = 2 \end{array}$$

77.
$$\begin{array}{r} x^2 - 3x + 6 \\ \underline{x^2 \qquad - 3} \\ -3x + 9 \end{array} \quad 6 - (-3) = 9$$

79.
$$\begin{array}{r} y^3 + 4y^2 - 6y - 5 \\ \underline{y^3 + 3y^2 + 2y - 9} \\ 2y^3 + 7y^2 - 4y - 14 \end{array}$$

81. $(4m - 2) + (2m + 4) - (9m - 1)$
$= 4m - 2 + 2m + 4 - 9m + 1$
$= -3m + 3$

83. $(6y - 2) - (8y + 3) - (9y - 2)$
$= 6y - 2 - 8y - 3 - 9y + 2$
$= -11y - 3$

85. $(-x^2 - 5x + 4) + (6x^2 - 8x + 9) - (3x^2 - 7x + 1)$
$= -x^2 - 5x + 4 + 6x^2 - 8x + 9 - 3x^2 + 7x - 1$
$= 2x^2 - 6x + 12$

87. $(-6z^4 - 3z^3 + 7z^2) - (5z^3 + 3z^2 - 2) + (z^4 - z^2 + 5)$
$= -6z^4 - 3z^3 + 7z^2 - 5z^3 - 3z^2 + 2 + z^4 - z^2 + 5$
$= -5z^4 - 8z^3 + 3z^2 + 7$

89. Since profit P is equal to revenue minus cost, we have
$P(x) = x^2 + 400x + 300 - (x^2 + 300x - 200)$
$= 100x + 500$ dollars
If $x = 50$, then
$P(50) = 100(50) + 500 = \$5500$

91. The perimeter is the total of the lengths of the three sides of the triangle.
$P(x) = x + (3x - 1) + (2x + 4) = 6x + 3$
The perimeter polynomial is $P(x) = 6x + 3$. If $x = 4$, then $P(4) = 6(4) + 3 = 27$ and the perimeter is 27 meters.

93. To find the total distance that she traveled, add the two binomials.
$$(2x + 50) + (3x - 10) = 5x + 40$$
She traveled a total of $5x + 40$ miles. If $x = 20$, then $5x + 40 = 5(20) + 40 = 140$ miles.

95. $D = (-16t^2 + 7400) - (-16t^2 + 6600)$
$= 800$
So $D(t) = 800$. If $t = 3$, then $D(3) = 800$ feet.

97. Add the two polynomials to find the total interest.
$0.08(x + 554) + 0.09(x + 335)$
$= 0.08x + 44.32 + 0.09x + 30.15$
$= 0.17x + 74.47$
His total interest was $0.17x + 74.47$ dollars. If $x = 1000$, then $0.17(1000) + 74.47$
$= \$244.47$

99. $655.1 + 9.56w + 1.85h - 4.68a$
$= 655.1 + 9.56(54) + 1.85(157) - 4.68(30)$
$= 1321.39$ calories

101. The sum of two natural numbers is a natural number, the sum of two integers is an integer, and the sum of two polynomials is a polynomial.

103. Because the highest power of x is 3, the degree of the polynomial is 3. The term 2^4 is actually the constant term 16.

4.2 WARM-UPS

1. False, because if x is replaced by -1, the two sides of the equation have different values. A corrected version could be written as $3x^3 \cdot 5x^4 = 15x^7$ for any value of x.
2. False, because if we replace x by 1, the two sides of the equation have different values. A corrected version could be written as $3x^2 \cdot 2x^7 = \; = 6x^9$ for any value of x.
3. True, because $(3y^3)^2 = 3y^3 \cdot 3y^3 = 9y^6$ for any value of y.
4. False, because if x is replaced by -1, the equation is incorrect. A corrected version could be written as $-3x(5x - 7x^2) = -15x^2 + 21x^3$ for any value of x.
5. True, because the monomial and the trinomial are correctly multiplied.
6. True, because $-2(3 - x) = -6 - (-2x)$ $= -6 + 2x = 2x - 6$ for any value of x.
7. True, because $(a + b)(c + d)$
$$= (a + b)c + (a + b)d$$
$$= ac + bc + ad + bd$$
for any values of $a, b, c,$ and d.
8. True, because $-(x - 7) = -1(x - 7)$ $= -x - (-7) = -x + 7 = 7 - x$ for any value of x.
9. True, because $-(a - b) = b - a$ for any values of a and b.
10. False, because if we replace x by 1, then $x + 3$ and $x - 3$ have values 4 and -2, which are not opposites. The opposite of $x + 3$ is $-x - 3$ for any number x.

4.2 EXERCISES

1. The product rule for exponents says that $a^m \cdot a^n = a^{m+n}$.
3. To multiply a monomial and a polynomial we use the distributive property.
5. To multiply any two polynomials we multiply each term of the first polynomial by every term of the second polynomial.
7. Multiply the coefficients and add the exponents: $3x^2 \cdot 9x^3 = 27x^5$.
9. Multiply the coefficients and add the exponents: $2a^3 \cdot 7a^8 = 14a^{11}$.
11. $-6x^2 \cdot 5x^2 = -30x^4$
13. $(-9x^{10})(-3x^7) = (-9)(-3)x^{10+7} = 27x^{17}$
15. $-6st \cdot 9st = -54s^2t^2$
17. $3wt \cdot 8w^7t^6 = 3 \cdot 8w^{1+7}t^{1+6} = 24t^7w^8$
19. $(5y)^2 = 5y \cdot 5y = 25y^2$
21. $(2x^3)^2 = 2x^3 \cdot 2x^3 = 4x^6$
23. $x(x + y^2) = x^2 + xy^2$
25. $4y^2(y^5 - 2y) = 4y^7 - 8y^3$
27. $-3y(6y - 4) = -18y^2 + 12y$
29. $(y^2 - 5y + 6)(-3y)$
$$= -3y^3 - (-3)5y^2 + (-3)6y$$
$$= -3y^3 + 15y^2 - 18y$$
31. $-x(y^2 - x^2) = -xy^2 - (-x)x^2$
$$= -xy^2 + x^3$$
33. $(3ab^3 - a^2b^2 - 2a^3b)(5a^3)$
$$= 15a^4b^3 - 5a^5b^2 - 10a^6b$$
35. $-\frac{1}{2}t^2v(4t^3v^2 - 6tv - 4v)$
$$= -2t^5v^3 + 3t^3v^2 + 2t^2v^2$$
37. $(x + 1)(x + 2) = (x + 1)x + (x + 1)2$
$$= x^2 + x + 2x + 2$$
$$= x^2 + 3x + 2$$
39. $(x - 3)(x + 5) = (x - 3)x + (x - 3)5$
$$= x^2 - 3x + 5x - 15$$
$$= x^2 + 2x - 15$$
41. $(t - 4)(t - 9) = (t - 4)t - (t - 4)9$
$$= t^2 - 4t - (9t - 36)$$
$$= t^2 - 4t - 9t + 36$$
$$= t^2 - 13t + 36$$
43. $(x + 1)(x^2 + 2x + 2)$
$$= (x + 1)x^2 + (x + 1)2x + (x + 1)2$$
$$= x^3 + x^2 + 2x^2 + 2x + 2x + 2$$
$$= x^3 + 3x^2 + 4x + 2$$
45. $(3y + 2)(2y^2 - y + 3)$
$$= (3y + 2)2y^2 - (3y + 2)y + (3y + 2)3$$
$$= 6y^3 + 4y^2 - 3y^2 - 2y + 9y + 6$$
$$= 6y^3 + y^2 + 7y + 6$$
47. $(y^2z - 2y^4)(y^2z + 3z^2 - y^4)$
$$= (y^2z - 2y^4)y^2z + (y^2z - 2y^4)3z^2 - (y^2z - 2y^4)y^4$$
$$= y^4z^2 - 2y^6z + 3y^2z^3 - 6y^4z^2 - y^6z + 2y^8$$
$$= 2y^8 - 3y^6z - 5y^4z^2 + 3y^2z^3$$

49.
$$\begin{array}{r} 2a-3 \\ a+5 \\ \hline 10a-15 \\ 2a^2-3a \quad \\ \hline 2a^2+7a-15 \end{array}$$

51.
$$\begin{array}{r} 7x+30 \\ 2x+5 \\ \hline 35x+150 \\ 14x^2+60x \quad \\ \hline 14x^2+95x+150 \end{array}$$

53.
$$\begin{array}{r} 5x+2 \\ 4x-3 \\ \hline -15x-6 \\ 20x^2+8x \quad \\ \hline 20x^2-7x-6 \end{array}$$

55.
$$\begin{array}{r} m-3n \\ 2a+b \\ \hline mb-3nb \\ 2am-6an \quad \\ \hline 2am-6an+bm-3bn \end{array}$$

57.
$$\begin{array}{r} x^2+3x-2 \\ x+6 \\ \hline 6x^2+18x-12 \\ x^3+3x^2-2x \quad \\ \hline x^3+9x^2+16x-12 \end{array}$$

59.
$$\begin{array}{r} 2a^3-3a^2+4 \\ -2a-3 \\ \hline -6a^3+9a^2-12 \\ -4a^4+6a^3-8a \quad \\ \hline -4a^4+9a^2-8a-12 \end{array}$$

61.
$$\begin{array}{r} x-y \\ x+y \\ \hline xy-y^2 \\ x^2-xy \quad \\ \hline x^2 \quad -y^2 \end{array}$$

63.
$$\begin{array}{r} x^2-xy+y^2 \\ x+y \\ \hline x^2y-xy^2+y^3 \\ x^3-x^2y+xy^2 \quad \\ \hline x^3 \quad +y^3 \end{array}$$

65. The opposite of $3t-u$ is $-3t+u$ or $u-3t$.

67. The opposite of $3x+y$ is $-3x-y$.

69. The opposite of $-3a^2-a+6$ is $3a^2+a-6$.

71. The opposite of $3v^2+v-6$ is $-3v^2-v+6$.

73. $-3x(2x-9) = -3x(2x)-(-3x)(9)$
$= -6x^2+27x$

75. $2-3x(2x-9) = 2-6x^2+27x$
$= -6x^2+27x+2$

77. $(2-3x)+(2x-9) = 2-9-3x+2x$
$= -x-7$

79. $(6x^6)^2 = 6x^6 \cdot 6x^6 = 36x^{12}$

81. $3ab^3 \cdot (-2a^2b^7) = -2 \cdot 3aa^2b^3b^7$
$= -6a^3b^{10}$

83. $(5x+6)(5x+6)$
$= (5x+6)5x+(5x+6)6$
$= 25x^2+30x+30x+36$
$= 25x^2+60x+36$

85. $(5x-6)(5x+6)$
$= (5x-6)5x+(5x-6)6$
$= 25x^2-30x+30x-36$
$= 25x^2-36$

87. $2x^2(3x^5-4x^2) = 2x^2 \cdot 3x^5 - 2x^2 \cdot 4x^2$
$= 6x^7-8x^4$

89. $(m-1)(m^2+m+1)$
$= (m-1)m^2+(m-1)m+(m-1)1$
$= m^3-m^2+m^2-m+m-1$
$= m^3-1$

91. $(3x-2)(x^2-x-9)$
$= x^2(3x-2)-x(3x-2)-9(3x-2)$
$= 3x^3-2x^2-3x^2+2x-27x+18$
$= 3x^3-5x^2-25x+18$

93. Since area of a rectangle is length times width, the area is $x(x+4)$ or x^2+4x square feet. If $x=10$, then the area is $10^2+4(10)$ or 140 ft^2.

95. For a triangle $A = \frac{1}{2}bh$. So $A(x) = \frac{1}{2}x(2x+1) = x^2+\frac{1}{2}x$ square feet. If $x=5$, then $A(5) = 5^2+\frac{1}{2}\cdot 5 = 27.5$. The area is 27.5 ft^2.

97. Two numbers that differ by 5 can be represented as x and $x+5$. Their product is x^2+5x.

99. The area for a rectangle is length times width. So we find the product:
$(2.3x+1.2)(3.5x+5.1)$
$= (2.3x+1.2)3.5x+(2.3x+1.2)5.1$
$= 8.05x^2+4.2x+11.73x+6.12$
$= 8.05x^2 + 15.93x + 6.12$ square meters

101. If $p = 10$, then
$40{,}000 - 1000(10) = 30{,}000$.
At \$10 each 30,000 tickets will be sold.
The revenue is $30{,}000 \cdot 10$, or \$300,000.
The total revenue is $p(40{,}000 - 1000p)$, or
$40{,}000p - 1000p^2$.
103. The values at the end of each year follow:
$10x$
$(10x + 10)x = 10x^2 + 10x$
$(10x^2 + 10x + 10)x = 10x^3 + 10x^2 + 10x$
$(10x^3 + 10x^2 + 10x + 10)x$
$= 10x^4 + 10x^3 + 10x^2 + 10x$
$(10x^4 + 10x^3 + 10x^2 + 10x + 10)x$
$= 10x^5 + 10x^4 + 10x^3 + 10x^2 + 10x$
If $r = 0.10$ then $x = 1.10$:
$10(1.10)^5 + 10(1.10)^4 + 10(1.10)^3 + 10(1.10)^2$
$+ 10(1.10) \approx 67.16$
So the value of the investment is \$67.16.

4.3 WARM-UPS

1. False, because if we replace x by 1 we get $(1 + 3)(1 + 2) = 1^2 + 6$, which is incorrect.
2. True, because the binomials are correctly multiplied.
3. True, because the binomials are correctly multiplied.
4. True, because the binomials are correctly multiplied.
5. True, because the binomials are correctly multiplied.
6. False, because $3a^2 \cdot 3a^2 = 9a^4$.
7. True, because the binomials are correctly multiplied.
8. False, because $(-9)(-2) = 18$.
9. False, because $4x - 7x = -3x$.
10. False, because we must be able to multiply binomials quickly and FOIL enables us to do that.

4.3 EXERCISES

1. We use the distributive property to find the product of two binomials.
3. The purpose of FOIL is to provide a faster method for finding the product of two binomials.
5. $(x + 2)(x + 4) = x^2 + 2x + 4x + 8$
$= x^2 + 6x + 8$
7. $(a + 1)(a + 4) = a^2 + 1a + 4a + 4$
$= a^2 + 5x + 4$
9. $(x + 9)(x + 10) = x^2 + 9x + 10x + 90$
$= x^2 + 19x + 90$
11. $(2x + 1)(x + 3) = 2x^2 + 1x + 6x + 3$
$= 2x^2 + 7x + 3$
13. $(a - 3)(a + 2) = a^2 - 3a + 2a - 6$
$= a^2 - a - 6$
15. $(2x - 1)(x - 2) = 2x^2 - x - 4x + 2$
$= 2x^2 - 5x + 2$
17. $(2a - 3)(a + 1) = 2a^2 - 3a + 2a - 3$
$= 2a^2 - a - 3$
19. $(w - 50)(w - 10)$
$= w^2 - 50w - 10w + 500$
$= w^2 - 60w + 500$
21. $(y - a)(y + 5) = y^2 + 5y - ay - 5a$
23. $(5 - w)(w + m) = 5w + 5m - w^2 - mw$
25. $(2m - 3t)(5m + 3t)$
$= 10m^2 - 15mt + 6mt - 9t^2$
$= 10m^2 - 9mt - 9t^2$
27. $(5a + 2b)(9a + 7b)$
$= 45a^2 + 18ab + 35ab + 14b^2$
$= 45a^2 + 53ab + 14b^2$
29. $(x^2 - 5)(x^2 + 2) = x^4 - 5x^2 + 2x^2 - 10$
$= x^4 - 3x^2 - 10$
31. $(h^3 + 5)(h^3 + 5) = h^6 + 5h^3 + 5h^3 + 25$
$= h^6 + 10h^3 + 25$
33. $(3b^3 + 2)(b^3 + 4) = 3b^6 + 2b^3 + 12b^3 + 8$
$= 3b^6 + 14b^3 + 8$
35. $(y^2 - 3)(y - 2) = y^3 - 2y^2 - 3y + 6$
37. $(3m^3 - n^2)(2m^3 + 3n^2)$
$= 6m^6 + 7m^3n^2 - 3n^4$
39. $(3u^2v - 2)(4u^2v + 6)$
$= 12u^4v^2 - 8u^2v + 18u^2v - 12$
$= 12u^4v^2 + 10u^2v - 12$
41. $(b + 4)(b + 5) = b^2 + 4b + 5b + 20$
$= b^2 + 9b + 20$
43. $(x - 3)(x + 9) = x^2 - 3x + 9x - 27$
$= x^2 + 6x - 27$
45. $(a + 5)(a + 5) = a^2 + 5a + 5a + 25$
$= a^2 + 10a + 25$
47. $(2x - 1)(2x - 1) = 4x^2 - 2x - 2x + 1$
$= 4x^2 - 4x + 1$
49. $(z - 10)(z + 10) = z^2 - 10z + 10z - 100$
$= z^2 - 100$

51. $(a+b)(a+b) = a^2 + ab + ab + b^2$
$= a^2 + 2ab + b^2$
53. $(a-1)(a-2) = a^2 - a - 2a + 2$
$= a^2 - 3a + 2$
55. $(2x-1)(x+3) = 2x^2 - x + 6x - 3$
$= 2x^2 + 5x - 3$
57. $(5t-2)(t-1) = 5t^2 - 2t - 5t + 2$
$= 5t^2 - 7t + 2$
59. $(h-7)(h-9) = h^2 - 7h - 9h + 63$
$= h^2 - 16h + 63$
61. $(h+7w)(h+7w)$
$= h^2 + 7hw + 7hw + 49w^2$
$= h^2 + 14hw + 49w^2$
63. $(2h^2-1)(2h^2-1)$
$= 4h^4 - 2h^2 - 2h^2 + 1$
$= 4h^4 - 4h^2 + 1$
65. $(a+1)(a-2)(a+5)$
$= (a^2 - a - 2)(a+5)$
$= a^2(a+5) - a(a+5) - 2(a+5)$
$= a^3 + 5a^2 - a^2 - 5a - 2a - 10$
$= a^3 + 4a^2 - 7a - 10$
67. $(h+2)(h+3)(h+4)$
$= (h^2 + 5h + 6)(h+4)$
$= h^2(h+4) + 5h(h+4) + 6(h+4)$
$= h^3 + 4h^2 + 5h^2 + 20h + 6h + 24$
$= h^3 + 9h^2 + 26h + 24$
69. $\left(\frac{1}{2}x+4\right)\left(\frac{1}{2}x-4\right)(4x-8)$
$= \left(\frac{1}{4}x^2 - 16\right)(4x-8)$
$= x^3 - 2x^2 - 64x + 128$
71. $\left(x+\frac{1}{2}\right)\left(x-\frac{1}{2}\right)(x+8)$
$= \left(x^2 - \frac{1}{4}\right)(x+8)$
$= x^3 + 8x^2 - \frac{1}{4}x - 2$
73. $(x+10)(x+5) = x^2 + 10x + 5x + 50$
$= x^2 + 15x + 50$
75. $\left(x+\frac{1}{2}\right)\left(x+\frac{1}{2}\right)$
$= x^2 + \frac{1}{2}x + \frac{1}{2}x + \frac{1}{4}$
$= x^2 + x + \frac{1}{4}$
77. $\left(4x+\frac{1}{2}\right)\left(2x+\frac{1}{4}\right)$
$= 8x^2 + x + x + \frac{1}{8}$
$= 8x^2 + 2x + \frac{1}{8}$
79. $\left(2a+\frac{1}{2}\right)\left(4a-\frac{1}{2}\right)$
$= 8a^2 + 2a - a - \frac{1}{4}$
$= 8a^2 + a - \frac{1}{4}$
81. $\left(\frac{1}{2}x-\frac{1}{3}\right)\left(\frac{1}{4}x+\frac{1}{2}\right)$
$= \frac{1}{8}x^2 - \frac{1}{12}x + \frac{1}{4}x - \frac{1}{6}$
$= \frac{1}{8}x^2 + \frac{1}{6}x - \frac{1}{6}$
83. $a(a+3)(a+4)$
$= a(a^2 + 7a + 12)$
$= a^3 + 7a^2 + 12a$
85. $x^3(x+6)(x+7)$
$= x^3(x^2 + 13x + 42)$
$= x^5 + 13x^4 + 42x^3$
87. $-2x^4(3x-1)(2x+5)$
$= -2x^4(6x^2 + 13x - 5)$
$= -12x^6 - 26x^5 + 10x^4$
89. $(x-1)(x+1)(x+3)$
$= (x-1)(x^2 + 4x + 3)$
$= x(x^2 + 4x + 3) - 1(x^2 + 4x + 3)$
$= x^3 + 4x^2 + 3x - x^2 - 4x - 3$
$= x^3 + 3x^2 - x - 3$
91. $(3x-2)(3x+2)(x+5)$
$= (9x^2 - 4)(x+5)$
$= 9x^3 + 45x^2 - 4x - 20$
93. $(x-1)(x+2) - (x+3)(x-4)$
$= (x^2 + x - 2) - (x^2 - x - 12)$
$= x^2 + x - 2 - x^2 + x + 12$
$= 2x + 10$
95. The area of a rectangle is length times width. So the area of the rug is $(x+3)(2x-1) = 2x^2 + 5x - 3$ square feet.
97. The area of a triangle is $\frac{1}{2}bh$. So the area is $\frac{1}{2}(4.57x+3)(2.3x-1.33)$
$= 5.2555x^2 + 0.41095x - 1.995$ square meters.
99. The areas of the four regions are 12 ft^2, $3h$ ft^2, $4h$ ft^2, and h^2 ft^2. The total area is $h^2 + 7h + 12$ ft^2. The total area is also $(h+4)(h+3)$ ft^2. So $(h+4)(h+3) = h^2 + 7h + 12$.

4.4 WARM-UPS

1. False, because $(2+3)^2 = 5^2 = 25$ and $2^2 + 3^2 = 4 + 9 = 13$.
2. True, because the binomial is squared correctly.
3. True, because $(3+5)^2 = 3^2 + 2 \cdot 3 \cdot 5 + 5^2$
$= 9 + 30 + 25$.

4. True, because $(2x+7)^2$
$= 4x^2 + 2 \cdot 2x \cdot 7 + 7^2 = 4x^2 + 28x + 49$.
5. False, because if y is replaced by 1 we get $(1+8)^2 = 1^2 + 64$, which is incorrect.
6. True, because $(a-b)(a+b) = a^2 - b^2$.
7. True, because $(40-1)(40+1) = 40^2 - 1^2$
$= 1600 - 1 = 1599$.
8. True, because $49 \cdot 51 = (50-1)(50+1)$
$= 2500 - 1 = 2499$.
9. False, because if we replace x by 3 we get $(3-3)^2 = 3^2 - 3(3) + 9$, which is incorrect.
10. False, because $(a+b)^2 = a^2 + 2ab + b^2$.

4.4 EXERCISES

1. The special products are $(a+b)^2$, $(a-b)^2$, and $(a+b)(a-b)$.
3. It is faster to square a sum using the new rule than with FOIL.
5. $(a+b)(a-b) = a^2 - b^2$
7. $(x+1)^2 = x^2 + 2 \cdot x \cdot 1 + 1^2$
$= x^2 + 2x + 1$
9. $(y+4)^2 = y^2 + 2 \cdot y \cdot 4 + 4^2$
$= y^2 + 8y + 16$
11. $(m+6)^2 = m^2 + 2 \cdot m \cdot 6 + 6^2$
$= m^2 + 12m + 36$
13. $(3x+8)^2 = (3x)^2 + 2 \cdot 3x \cdot 8 + 8^2$
$= 9x^2 + 48x + 64$
15. $(s+t)^2 = s^2 + 2st + t^2$
17. $(2x+y)^2 = (2x)^2 + 2 \cdot 2x(y) + (y)^2$
$= 4x^2 + 4xy + y^2$
19. $(2t+3h)^2 = (2t)^2 + 2 \cdot 2t \cdot 3h + (3h)^2$
$= 4t^2 + 12ht + 9h^2$
21. $(p-2)^2 = p^2 - 2 \cdot p \cdot 2 + 2^2$
$= p^2 - 4p + 4$
23. $(a-3)^2 = a^2 - 2 \cdot a \cdot 3 + 3^2$
$= a^2 - 6a + 9$
25. $(t-1)^2 = (t)^2 - 2 \cdot t \cdot 1 + 1^2$
$= t^2 - 2t + 1$
27. $(3t-2)^2 = (3t)^2 - 2 \cdot 3t \cdot 2 + 2^2$
$= 9t^2 - 12t + 4$
29. $(s-t)^2 = s^2 - 2st + t^2$
31. $(3a-b)^2 = (3a)^2 - 2 \cdot 3a(b) + (b)^2$
$= 9a^2 - 6ab + b^2$
33. $(3z-5y)^2 = (3z)^2 - 2(3z)(5y) + (5y)^2$
$= 9z^2 - 30yz + 25y^2$
35. $(a-5)(a+5) = a^2 - 5^2 = a^2 - 25$
37. $(y-1)(y+1) = y^2 - 1^2 = y^2 - 1$
39. $(3x-8)(3x+8) = (3x)^2 - 8^2$
$= 9x^2 - 64$
41. $(r+s)(r-s) = r^2 - s^2$
43. $(8y-3a)(8y+3a) = 64y^2 - 9a^2$
45. $(5x^2-2)(5x^2+2) = (5x^2)^2 - 2^2$
$= 25x^4 - 4$
47. $(x+1)^3 = (x+1)(x^2+2x+1)$
$= (x+1)x^2 + (x+1)2x + (x+1)1$
$= x^3 + x^2 + 2x^2 + 2x + x + 1$
$= x^3 + 3x^2 + 3x + 1$
49. $(2a-3)^3 = (2a-3)(4a^2-12a+9)$
$= 2a(4a^2-12a+9) - 3(4a^2-12a+9)$
$= 8a^3 - 24a^2 + 18a - 12a^2 + 36a - 27$
$= 8a^3 - 36a^2 + 54a - 27$
51. $(a-3)^4 = (a-3)^2(a-3)^2$
$= (a^2-6a+9)(a^2-6a+9)$
$= a^2(a^2-6a+9) - 6a(a^2-6a+9)$
$+ 9(a^2-6a+9)$
$= a^4 - 12a^3 + 54a^2 - 108a + 81$
53. $(a+b)^4 = (a^2+2ab+b^2)(a^2+2ab+b^2)$
$= a^4 + 4a^3b + 6a^2b^2 + 4ab^3 + b^4$
55. $(a-20)(a+20) = a^2 - 20^2 = a^2 - 400$
57. $(x+8)(x+7) = x^2 + 8x + 7x + 56$
$= x^2 + 15x + 56$
59. $(4x-1)(4x+1) = (4x)^2 - 1^2$
$= 16x^2 - 1$
61. $(9y-1)^2 = 81y^2 - 2 \cdot 9y \cdot 1 + 1^2$
$= 81y^2 - 18y + 1$
63. $(2t-5)(3t+4) = 6t^2 - 15t + 8t - 20$
$= 6t^2 - 7t - 20$
65. $(2t-5)^2 = 4t^2 - 2 \cdot 2t \cdot 5 + 5^2$
$= 4t^2 - 20t + 25$
67. $(2t+5)(2t-5) = (2t)^2 - 5^2 = 4t^2 - 25$
69. $(x^2-1)(x^2+1) = (x^2)^2 - 1^2 = x^4 - 1$
71. $(2y^3-9)^2 = (2y^3)^2 - 2(2y^3)(9) + 9^2$
$= 4y^6 - 36y^3 + 81$

73. $(2x^3+3y^2)^2$
$= (2x^3)^2 + 2(2x^3)(3y^2) + (3y^2)^2$
$= 4x^6 + 12x^3y^2 + 9y^4$

75. $\left(\frac{1}{2}x + \frac{1}{3}\right)^2$
$= \left(\frac{1}{2}x\right)^2 + 2 \cdot \frac{1}{2}x \cdot \frac{1}{3} + \left(\frac{1}{3}\right)^2$
$= \frac{1}{4}x^2 + \frac{1}{3}x + \frac{1}{9}$

77. $(0.2x - 0.1)^2$
$= (0.2x)^2 - 2(0.2x)(0.1) + (0.1)^2$
$= 0.04x^2 - 0.04x + 0.01$

79. $(a + b)^3 = (a + b)(a^2 + 2ab + b^2)$
$= a(a^2 + 2ab + b^2) + b(a^2 + 2ab + b^2)$
$= a^3 + 3a^2b + 3ab^2 + b^3$

81. $(1.5x + 3.8)^2$
$= 2.25x^2 + 2(1.5x)(3.8) + (3.8)^2$
$= 2.25x^2 + 11.4x + 14.44$

83. $(3.5t - 2.5)(3.5t + 2.5) = (3.5t)^2 - (2.5)^2$
$= 12.25t^2 - 6.25$

85. Let $x =$ the length of a side of the square garden. The new length will be $x + 5$ feet and the new width will be $x - 5$ feet. The old area is x^2 square feet, and the new area is $(x + 5)(x - 5) = x^2 - 25$ square feet. So the new garden is 25 square feet smaller than the old square garden.

87. The area of a circle is πr^2. The area of a circle with radius $b + 1$ is
$3.14(b + 1)^2 = 3.14(b^2 + 2b + 1)$
$= 3.14b^2 + 6.28b + 3.14$
square meters.

89. $v = k(R - r)(R + r)$
$v = k(R^2 - r^2)$

91. $P(1 + r)^2 = P(1 + 2r + r^2)$
$= P + 2Pr + Pr^2$
If $P = 200$ and $r = 0.10$, then
$200 + 2 \cdot 200 \cdot 0.10 + 200 \cdot (0.10)^2 = \$242.$

93. At 7.3%, investing in treasury bills would yield
$P(1 + r)^{10} = 10{,}000(1 + 0.073)^{10}$
$= \$20{,}230.06.$

95. The equation $(x + 5)^2 = x^2 + 10x + 25$ is an identity, because the binomial is squared correctly. The equation $(x + 5)^2 = x^2 + 25$ is a conditional equation that is satisfied only if $x = 0$.

4.5 WARM-UPS

1. False, because $2^{10} \div 2^2 = 1024 \div 4 = 256$ and $2^5 = 32$. By the quotient rule $y^{10} \div y^2 = y^8$ for any nonzero value of y.
2. False, because if $x = 0$, the left side is $2/7$ and the right side is 2.
3. True, because $7/7 = 1$.
4. False, because the quotient is $x^2 + 2$.
5. True, because $2y(2y - 3) = 4y^2 - 6y$.
6. False, because the quotient times the divisor plus the remainder is equal to the dividend.
7. True, because $(x + 2)(x + 1) + 3$
$= x^2 + 2x + x + 2 + 3$
$= x^2 + 3x + 5.$
8. True, because of the fact stated in Warm-up number 7.
9. True, because of the fact stated in Warm-up number 7.
10. True, because of the fact that dividend is equal to the divisor times the quotient plus the remainder.

4.5 EXERCISES

1. The quotient rule is used for dividing monomials.

3. When dividing a polynomial by a monomial the quotient should have the same number of terms as the polynomial.

5. The long division process stops when the degree of the remainder is less than the degree of the divisor.

7. $9^0 = 1$

9. $(-2x^3)^0 = 1$

11. $2 \cdot 5^0 - 3^0 = 2 \cdot 1 - 1 = 1$

13. $(2x - y)^0 = 1$

15. $\frac{x^8}{x^2} = x^{8-2} = x^6$

17. $\frac{a^5}{a^{14}} = \frac{1}{a^{14-5}} = \frac{1}{a^9}$

19. $\frac{6a^7}{2a^{12}} = \frac{3}{a^{12-7}} = \frac{3}{a^5}$

21. $a^9 \div a^3 = \frac{a^9}{a^3} = a^6$

23. $-12x^5 \div (3x^9) = \frac{-12x^5}{3x^9} = \frac{-4}{x^4}$

25. $-6y^2 \div (6y) = \frac{-6y^2}{6y} = -1y^{2-1} = -y$

27. $\frac{-6x^3y^2}{2x^2y^2} = -3x^{3-2}y^{2-2} = -3x$

29. $\frac{-9x^2y^2}{3x^5y^2} = \frac{-3y^0}{x^3} = \frac{-3}{x^3}$

31. $\frac{3x - 6}{3} = \frac{3x}{3} - \frac{6}{3} = x - 2$

33. $\frac{x^5 + 3x^4 - x^3}{x^2} = x^{5-2} + 3x^{4-2} - x^{3-2}$
$= x^3 + 3x^2 - x$

35. $\dfrac{-8x^2y^2+4x^2y-2xy^2}{-2xy}$
$= \dfrac{-8x^2y^2}{-2xy} + \dfrac{4x^2y}{-2xy} - \dfrac{2xy^2}{-2xy}$
$= 4xy - 2x + y$

37. $(x^2y^3 - 3x^3y^2) \div (x^2y) = \dfrac{x^2y^3}{x^2y} - \dfrac{3x^3y^2}{x^2y}$
$= y^2 - 3xy$

39.
$$\begin{array}{r} 2 \\ x-1\,\overline{)\,2x-3} \\ \underline{2x-2} \\ -1 \end{array}$$
The quotient is 2 and the remainder is -1.

41.
$$\begin{array}{r} x+5 \\ x-3\,\overline{)\,x^2+2x+1} \\ \underline{x^2-3x} \\ 5x+1 \\ \underline{5x-15} \\ 16 \end{array}$$
The quotient is $x+5$ and the remainder is 16.

43.
$$\begin{array}{r} x+2 \\ x+3\,\overline{)\,x^2+5x+13} \\ \underline{x^2+3x} \\ 2x+13 \\ \underline{2x+6} \\ 7 \end{array}$$
The quotient is $x+2$ and the remainder is 7.

45.
$$\begin{array}{r} 2 \\ x+5\,\overline{)\,2x+0} \\ \underline{2x+10} \\ -10 \end{array}$$
The quotient is 2 and the remainder is -10.

47.
$$\begin{array}{r} a^2+2a+8 \\ a-2\,\overline{)\,a^3+0a^2+4a-3} \\ \underline{a^3-2a^2} \\ 2a^2+4a \\ \underline{2a^2-4a} \\ 8a-3 \\ \underline{8a-16} \\ 13 \end{array}$$
The quotient is a^2+2a+8 and remainder is 13.

49.
$$\begin{array}{r} x-4 \\ x+1\,\overline{)\,x^2-3x+0} \\ \underline{x^2+x} \\ -4x+0 \\ \underline{-4x-4} \\ 4 \end{array}$$
The quotient is $x-4$ and the remainder is 4.

51.
$$\begin{array}{r} h^2+3h+9 \\ h-3\,\overline{)\,h^3+0h^2+0h-27} \\ \underline{h^3-3h^2} \\ 3h^2+0h \\ \underline{3h^2-9h} \\ 9h-27 \\ \underline{9h-27} \\ 0 \end{array}$$
The quotient is h^2+3h+9 and the remainder is 0.

53.
$$\begin{array}{r} 2x-3 \\ 3x-2\,\overline{)\,6x^2-13x+7} \\ \underline{6x^2-4x} \\ -9x+7 \\ \underline{-9x+6} \\ 1 \end{array}$$
The quotient is $2x-3$ and the remainder is 1.

55.
$$\begin{array}{r} x^2+0x+1 \\ x-1\,\overline{)\,x^3-x^2+x-2} \\ \underline{x^3-x^2} \\ 0x^2+x \\ \underline{0x^2+0x} \\ x-2 \\ \underline{x-1} \\ -1 \end{array}$$
The quotient is x^2+1 and the remainder is -1.

57.
$$\begin{array}{r} 3 \\ x-5\,\overline{)\,3x+0} \\ \underline{3x-15} \\ 15 \end{array}$$
$$\frac{3x}{x-5} = 3 + \frac{15}{x-5}$$

59. $$\begin{array}{r} -1 \\ x+3\,\overline{)\,-x+0} \\ \underline{-x-3} \\ 3 \end{array}$$

$$\frac{-x}{x+3} = -1 + \frac{3}{x+3}$$

61. $\dfrac{x-1}{x} = \dfrac{x}{x} - \dfrac{1}{x} = 1 - \dfrac{1}{x}$

63. $\dfrac{3x+1}{x} = \dfrac{3x}{x} + \dfrac{1}{x} = 3 + \dfrac{1}{x}$

65. $$\begin{array}{r} x-1 \\ x+1\,\overline{)\,x^2+0x+0} \\ \underline{x^2+x} \\ -x+0 \\ \underline{-x-1} \\ 1 \end{array}$$

$$\frac{x^2}{x+1} = x - 1 + \frac{1}{x+1}$$

67. $$\begin{array}{r} x-2 \\ x+2\,\overline{)\,x^2+0x+4} \\ \underline{x^2+2x} \\ -2x+4 \\ \underline{-2x-4} \\ 8 \end{array}$$

$$\frac{x^2+4}{x+2} = x - 2 + \frac{8}{x+2}$$

69. $$\begin{array}{r} x^2+2x+4 \\ x-2\,\overline{)\,x^3+0x^2+0x+0} \\ \underline{x^3-2x^2} \\ 2x^2+0x \\ \underline{2x^2-4x} \\ 4x+0 \\ \underline{4x-8} \\ 8 \end{array}$$

$$\frac{x^3}{x-2} = x^2 + 2x + 4 + \frac{8}{x-2}$$

71. $\dfrac{x^3+3}{x} = \dfrac{x^3}{x} + \dfrac{3}{x} = x^2 + \dfrac{3}{x}$

73. $-6a^3b \div (2a^2b) = \dfrac{-6a^3b}{2a^2b} = -3a^{3-2}b^0$

$= -3a$

75. $-8w^4t^7 \div (-2w^9t^3) = \dfrac{-8w^4t^7}{-2w^9t^3} = \dfrac{4t^4}{w^5}$

77. $(3a - 12) \div (-3) = \dfrac{3a}{-3} - \dfrac{12}{-3}$

$= -a + 4$

79. $(3x^2 - 9x) \div (3x) = \dfrac{3x^2}{3x} - \dfrac{9x}{3x} = x - 3$

81. $(12x^4 - 4x^3 + 6x^2) \div (-2x^2)$

$= \dfrac{12x^4}{-2x^2} - \dfrac{4x^3}{-2x^2} + \dfrac{6x^2}{-2x^2}$

$= -6x^2 + 2x - 3$

83. $$\begin{array}{r} t+4 \\ t-9\,\overline{)\,t^2-5t-36} \\ \underline{t^2-9t} \\ 4t-36 \\ \underline{4t-36} \\ 0 \end{array}$$

The quotient is $t + 4$.

85. $$\begin{array}{r} 2w+1 \\ 3w-5\,\overline{)\,6w^2-7w-5} \\ \underline{6w^2-10w} \\ 3w-5 \\ \underline{3w-5} \\ 0 \end{array}$$

The quotient is $2w + 1$.

87. $$\begin{array}{r} 4x^2-6x+9 \\ 2x+3\,\overline{)\,8x^3+0x^2+0x+27} \\ \underline{8x^3+12x^2} \\ -12x^2+0x \\ \underline{-12x^2-18x} \\ 18x+27 \\ \underline{18x+27} \\ 0 \end{array}$$

The quotient is $4x^2 - 6x + 9$.

89. $$\begin{array}{r} t^2-t+3 \\ t-2\,\overline{)\,t^3-3t^2+5t-6} \\ \underline{t^3-2t^2} \\ -t^2+5t \\ \underline{-t^2+2t} \\ 3t-6 \\ \underline{3t-6} \\ 0 \end{array}$$

The quotient is $t^2 - t + 3$.

91. $$\begin{array}{r} v^2-2v+1 \\ v-4\,\overline{)\,v^3-6v^2+9v-4} \\ \underline{v^3-4v^2} \\ -2v^2+9v \\ \underline{-2v^2+8v} \\ v-4 \\ \underline{v-4} \\ 0 \end{array}$$

The quotient is $v^2 - 2v + 1$.

93. Since $A = LW$ for a rectangle, we can write $W = A/L$. To find the width divide the area by the length.

$$\begin{array}{r|l} & x - 5 \\ \hline x+6 & x^2 + x - 30 \\ & x^2 + 6x \\ \hline & \quad -5x - 30 \\ & \quad -5x - 30 \\ \hline & \qquad 0 \end{array}$$

The width is $x - 5$ meters.

95. $(x^9 - 1) \div (x - 1) =$
$x^8 + x^7 + x^6 + x^5 + x^4 + x^3 + x^2 + x + 1$

97. In $10x \div 5x$ there are no grouping symbols. So according to the order of operations, $10x$ is divided by 5 and then that result is multiplied by x. If $x = 3$, $10 \cdot 3 \div 5 \cdot 3 = 18$ whereas the other two expressions have value 2 when $x = 3$.

4.6 WARM-UPS

1. False, because $-3^0 = -(3^0) = -1$.
2. False, because $2^5 \cdot 2^8 = 2^{13}$.
3. False, because $2^3 \cdot 3^2 = 8 \cdot 9 = 72$ and $6^5 \neq 72$.
4. False, because $(2x)^4 = 16x^4$ for any real number x.
5. False, because $(q^3)^5 = q^{15}$.
6. False, because $(-3x^2)^3 = -27x^6$.
7. True, because $(ab^3)^4 = a^4(b^3)^4 = a^4b^{12}$.
8. False, because $a^{12} \div a^4 = a^8$.
9. False, because $\frac{6w^4}{3w^9} = \frac{2}{w^5}$.
10. True, because $\left(\frac{2y^3}{9}\right)^2 = \frac{2^2(y^3)^2}{9^2} = \frac{4y^6}{81}$.

4.6 EXERCISES

1. The product rule states that $a^m a^n = a^{m+n}$.
3. These rules do not make sense without identical bases.
5. The power of a product rule states that $(ab)^n = a^n b^n$.
7. $2^2 \cdot 2^5 = 2^7 = 128$
9. $(-3u^8)(-2u^2) = 6u^{10}$
11. $a^3b^4 \cdot ab^6(ab)^0 = a^4b^{10} \cdot 1 = a^4b^{10}$
13. $\frac{-2a^3}{4a^7} = \frac{-2}{2 \cdot 2a^{7-3}} = \frac{-1}{2a^4}$
15. $\frac{2a^5b \cdot 3a^7b^3}{15a^6b^8} = \frac{2 \cdot 3a^{12}b^4}{3 \cdot 5a^6b^8} = \frac{2a^6}{5b^4}$
17. $2^3 \cdot 5^2 = 8 \cdot 25 = 200$
19. $(x^2)^3 = x^{2 \cdot 3} = x^6$
21. $2x^2 \cdot (x^2)^5 = 2x^2 \cdot x^{10} = 2x^{12}$
23. $\frac{(t^2)^5}{(t^3)^4} = \frac{t^{10}}{t^{12}} = \frac{1}{t^2}$
25. $\frac{3x(x^5)^2}{6x^3(x^2)^4} = \frac{3xx^{10}}{6x^3x^8} = \frac{3x^{11}}{6x^{11}} = \frac{1}{2}$
27. $(xy^2)^3 = x^3(y^2)^3 = x^3y^6$
29. $(-2t^5)^3 = (-2)^3(t^5)^3 = -8t^{15}$
31. $(-2x^2y^5)^3 = (-2)^3(x^2)^3(y^5)^3 = -8x^6y^{15}$
33. $\frac{(a^4b^2c^5)^3}{a^3b^4c} = \frac{a^{12}b^6c^{15}}{a^3b^4c} = a^9b^2c^{14}$
35. $\left(\frac{x^4}{4}\right)^3 = \frac{(x^4)^3}{4^3} = \frac{x^{12}}{64}$
37. $\left(\frac{-2a^2}{b^3}\right)^4 = \frac{(-2)^4(a^2)^4}{(b^3)^4} = \frac{16a^8}{b^{12}}$
39. $\left(\frac{2x^2y}{-4y^2}\right)^3 = \frac{2^3(x^2)^3y^3}{(-4)^3(y^2)^3} = \frac{8x^6y^3}{-64y^6}$
$= \frac{x^6}{-8y^3} = -\frac{x^6}{8y^3}$
41. $\left(\frac{-6x^2y^4z^9}{3x^6y^4z^3}\right)^2 = \frac{(-6)^2(x^2)^2(y^4)^2(z^9)^2}{3^2(x^6)^2(y^4)^2(z^3)^2}$
$= \frac{36x^4y^8z^{18}}{9x^{12}y^8z^6} = \frac{4z^{12}}{x^8}$
43. $3^2 + 6^2 = 9 + 36 = 45$
45. $(3 + 6)^2 = 9^2 = 81$
47. $2^3 - 3^3 = 8 - 27 = -19$
49. $(2 - 3)^3 = (-1)^3 = -1$
51. $\left(\frac{2}{5}\right)^3 = \frac{2^3}{5^3} = \frac{8}{125}$
53. $5^2 \cdot 2^3 = 25 \cdot 8 = 200$
55. $2^3 \cdot 2^4 = 2^7 = 128$
57. $\left(\frac{2^3}{2^5}\right)^2 = \frac{2^6}{2^{10}} = \frac{1}{2^4} = \frac{1}{16}$
59. $x^4 \cdot x^3 = x^{4+3} = x^7$
61. $(a^8)^4 = a^{8 \cdot 4} = a^{32}$
63. $(a^4b^2)^3 = a^{4 \cdot 3}b^{2 \cdot 3} = a^{12}b^6$
65. $\frac{x^4}{x^7} = \frac{1}{x^{7-4}} = \frac{1}{x^3}$
67. $\frac{a^{13}}{a^9} = a^{13-9} = a^4$
69. $\left(\frac{a^3}{b^4}\right)^3 = \frac{a^{3 \cdot 3}}{b^{4 \cdot 3}} = \frac{a^9}{b^{12}}$

71. $\left(\frac{x^3}{x^4}\right)^5 = \frac{x^{15}}{x^{20}} = \frac{1}{x^5}$
73. $3x^4 \cdot 5x^7 = 15x^{11}$
75. $(-5x^4)^3 = -125x^{12}$
77. $-3y^5z^{12} \cdot 9yz^7 = -27y^6z^{19}$
79. $\frac{-9u^4v^9}{-3u^5v^8} = \frac{3v}{u}$
81. $(-xt^2)(-2x^2t)^4 = -xt^2 \cdot 16x^8t^4$
$= -16x^9t^6$
83. $\left(\frac{2x^2}{x^4}\right)^3 = \frac{2^3(x^2)^3}{(x^4)^3} = \frac{8x^6}{x^{12}} = \frac{8}{x^6}$
85. $\left(\frac{-8a^3b^4}{4c^5}\right)^5 = \frac{(-2)^5(a^3)^5(b^4)^5}{(c^5)^5}$
$= \frac{-32a^{15}b^{20}}{c^{25}}$
87. $\left(\frac{-8x^4y^7}{-16x^5y^6}\right)^5 = \left(\frac{y}{2x}\right)^5 = \frac{y^5}{32x^5}$
89. By the product rule,
$P(1+r)^{10}(1+r)^5 = P(1+r)^{15}$.
91. Because $(a+b)^2 = a^2 + 2ab + b^2$, the square of a sum cannot be found by simply squaring each term of the sum.

4.7 WARM-UPS

1. True, because $10^{-2} = \frac{1}{10^2} = \frac{1}{100}$.
2. False, because $\left(-\frac{1}{5}\right)^{-1} = -5$.
3. False, $3^{-2} \cdot 2^{-1} = \frac{1}{9} \cdot \frac{1}{2} = \frac{1}{18} \neq 6^{-3}$.
4. True, $\frac{3^{-2}}{3^{-1}} = 3^{-2-(-1)} = 3^{-1} = \frac{1}{3}$.
5. False, $2.37 \times 10^{-1} = 2.37 \times \frac{1}{10} = 0.237$.
6. True, because multiplying by 10^{-5} moves the decimal point 5 places to the left.
7. True, because $25 \times 10^7 = 2.5 \times 10^1 \times 10^7$
$= 2.5 \times 10^8$.
8. True, because 0.442×10^{-3}
$= 4.42 \times 10^{-1} \times 10^{-3} = 4.42 \times 10^{-4}$.
9. True, because $(3 \times 10^{-9})^2 = 3^2 \times (10^{-9})^2$
$= 9 \times 10^{-18}$.
10. False, because $(2 \times 10^{-5})(4 \times 10^4)$
$= 8 \times 10^{-5+4} = 8 \times 10^{-1}$.

4.7 EXERCISES

1. A negative exponent means "reciprocal," as in $a^{-n} = \frac{1}{a^n}$.
3. The new quotient rule is $a^m/a^n = a^{m-n}$ for any integers m and n.
5. Convert from standard notation by counting the number of places the decimal must move so that there is one nonzero digit to the left of the decimal point.
7. $3^{-1} = \frac{1}{3}$
9. $(-2)^{-4} = \frac{1}{(-2)^4} = \frac{1}{16}$
11. $-4^{-2} = -\frac{1}{4^2} = -\frac{1}{16}$
13. $\frac{5^{-2}}{10^{-2}} = \frac{10^2}{5^2} = \frac{100}{25} = 4$
15. $\left(\frac{5}{2}\right)^{-3} = \left(\frac{2}{5}\right)^3 = \frac{8}{125}$
17. $6^{-1} + 6^{-1} = \frac{1}{6} + \frac{1}{6} = \frac{1}{3}$
19. $\frac{10}{5^{-3}} = 10 \cdot 5^3 = 10 \cdot 125 = 1250$
21. $\frac{1}{4^{-3}} + \frac{3^2}{2^{-1}} = 4^3 + 9 \cdot 2 = 64 + 18 = 82$
23. $x^{-1}x^2 = x^1 = x$
25. $-2x^2 \cdot 8x^{-6} = -16x^{-4} = -\frac{16}{x^4}$
27. $-3a^{-2}(-2a^{-3}) = 6a^{-5} = \frac{6}{a^5}$
29. $\frac{u^{-5}}{u^3} = u^{-8} = \frac{1}{u^8}$
31. $\frac{8t^{-3}}{-2t^{-5}} = -4t^{-3-(-5)} = -4t^2$
33. $\frac{-6x^5}{-3x^{-6}} = 2x^{5-(-6)} = 2x^{11}$
35. $(x^2)^{-5} = x^{-10} = \frac{1}{x^{10}}$
37. $(a^{-3})^{-3} = a^{(-3)(-3)} = a^9$
39. $(2x^{-3})^{-4} = 2^{-4}x^{12} = \frac{x^{12}}{2^4} = \frac{x^{12}}{16}$
41. $(4x^2y^{-3})^{-2} = 4^{-2}x^{-4}y^6 = \frac{y^6}{16x^4}$
43. $\left(\frac{2x^{-1}}{y^{-3}}\right)^{-2} = \frac{2^{-2}(x^{-1})^{-2}}{(y^{-3})^{-2}} = \frac{x^2}{4y^6}$
45. $\left(\frac{2a^{-3}}{ac^{-2}}\right)^{-4} = \frac{2^{-4}a^{12}}{a^{-4}c^8} = \frac{a^{16}}{2^4c^8} = \frac{a^{16}}{16c^8}$
47. $2 \cdot 3w^{-5} = 2 \cdot 3 \cdot \frac{1}{w^5} = \frac{6}{w^5}$
49. $(2h)^{-3} = 2^{-3} \cdot h^{-3} = \frac{1}{2^3} \cdot \frac{1}{h^3} = \frac{1}{8h^3}$
51. $(x^{-4})^{-3}(x^{-5})^6 = x^{12} \cdot x^{-30}$
$= x^{-18} = \frac{1}{x^{18}}$
53. $\frac{(b^3)^{-5}}{(b^{-7})^4} = \frac{b^{-15}}{b^{-28}} = b^{-15-(-28)} = b^{13}$

55. $\dfrac{(v^{-3})^6(v^{-5})^{-4}}{(v^{-7})^3} = \dfrac{v^{-18}v^{20}}{v^{-21}}$

$= \dfrac{v^2}{v^{-21}} = v^{2-(-21)} = v^{23}$

57. $\dfrac{(c^{-1})^{-12}(c^{-5})^6}{(c^{-4})^0(c^3)^{-3}} = \dfrac{c^{12}c^{-30}}{1 \cdot c^{-9}}$

$= \dfrac{c^{-18}}{c^{-9}} = c^{-18-(-9)} = c^{-9} = \dfrac{1}{c^9}$

59. $2^{-1} \cdot 3^{-1} = \frac{1}{2} \cdot \frac{1}{3} = \frac{1}{6}$

61. $(2 \cdot 3^{-1})^{-1} = 2^{-1} \cdot 3^1 = \frac{3}{2}$

63. $(x^{-2})^{-3} + 3x^7(-5x^{-1}) = x^6 - 15x^6$

$= -14x^6$

65. $\dfrac{a^3b^{-2}}{a^{-1}} + \left(\dfrac{b^6a^{-2}}{b^5}\right)^{-2} = \dfrac{a^4}{b^2} + \dfrac{b^{-12}a^4}{b^{-10}}$

$= \dfrac{a^4}{b^2} + \dfrac{a^4}{b^2} = \dfrac{2a^4}{b^2}$

67. The exponent 9 indicates that the decimal point moves 9 places to the right.
$9.86 \times 10^9 = 9{,}860{,}000{,}000$

69. The exponent -3 indicates that the decimal point moves 3 places to the left.
$1.37 \times 10^{-3} = 0.00137$

71. The exponent -6 indicates that the decimal point moves 6 places to the left.
$1 \times 10^{-6} = 0.000001$

73. The exponent 5 indicates that the decimal point moves 5 places to the right.
$6 \times 10^5 = 600{,}000$

75. Move the decimal point 3 places. Since 9000 is larger than 10 the exponent is positive.
$9000 = 9 \times 10^3$

77. Move the decimal point 4 places. Since 0.00078 is smaller than 1, the exponent is negative. $0.00078 = 7.8 \times 10^{-4}$

79. Move the decimal point 6 places. Since 0.0000085 is smaller than 1, the exponent is negative. $0.0000085 = 8.5 \times 10^{-6}$

81. $525 \times 10^9 = 5.25 \times 10^2 \times 10^9$

$= 5.25 \times 10^{11}$

83. $(3 \times 10^5)(2 \times 10^{-15}) = 6 \times 10^{5+(-15)}$

$= 6 \times 10^{-10}$

85. $\dfrac{4 \times 10^{-8}}{2 \times 10^{30}} = 2 \times 10^{-8-30} = 2 \times 10^{-38}$

87. $\dfrac{3 \times 10^{20}}{6 \times 10^{-8}} = 0.5 \times 10^{28}$

$= 5 \times 10^{-1} \times 10^{28}$

$= 5 \times 10^{27}$

89. $(3 \times 10^{12})^2 = 3^2 \times (10^{12})^2 = 9 \times 10^{24}$

91. $(5 \times 10^4)^3 = 125 \times 10^{12}$

$= 1.25 \times 10^2 \times 10^{12}$

$= 1.25 \times 10^{14}$

93. $(4 \times 10^{32})^{-1} = 4^{-1} \times 10^{-32} = \frac{1}{4} \times 10^{-32}$

$= 0.25 \times 10^{-32}$

$= 2.5 \times 10^{-1} \times 10^{-32}$

$= 2.5 \times 10^{-33}$

95. $(4300)(2{,}000{,}000) = 4.3 \times 10^3 \cdot 2 \times 10^6$

$= 8.6 \times 10^9$

97. $(4{,}200{,}000)(0.00005)$

$= 4.2 \times 10^6 \cdot 5 \times 10^{-5}$

$= 21 \times 10^1$

$= 2.1 \times 10^1 \times 10^1$

$= 2.1 \times 10^2$

99. $(300)^3(0.000001)^5$

$= (3 \times 10^2)^3(1 \times 10^{-6})^5$

$= 27 \times 10^6 \cdot 1 \times 10^{-30}$

$= 27 \times 10^{-24}$

$= 2.7 \times 10^1 \times 10^{-24} = 2.7 \times 10^{-23}$

101. $\dfrac{(4000)(90{,}000)}{(0.00000012)} = \dfrac{4 \times 10^3 \cdot 9 \times 10^4}{12 \times 10^{-8}}$

$= \dfrac{36 \times 10^7}{12 \times 10^{-8}}$

$= 3 \times 10^{15}$

103. $(6.3 \times 10^6)(1.45 \times 10^{-4}) = 9.135 \times 10^2$

105. $(5.36 \times 10^{-4}) + (3.55 \times 10^{-5})$

$= 5.715 \times 10^{-4}$

107. $\dfrac{(3.5 \times 10^5)(4.3 \times 10^{-6})}{3.4 \times 10^{-8}} \approx 4.426 \times 10^7$

109. $(3.56 \times 10^{85})(4.43 \times 10^{96})$

$\approx 15.77 \times 10^{181}$

$\approx 1.577 \times 10^{182}$

111. Multiply 93 million miles by 5,280 feet per mile.

$9.3 \times 10^7 \text{ miles} \cdot \dfrac{5280 \text{ feet}}{1 \text{ mile}}$

$\approx 4.910 \times 10^{11}$ feet

113. Since $T = D/R$, we divide 93 million miles by 2×10^{35} miles per hour.

$\dfrac{9.3 \times 10^7 \text{ miles}}{2 \times 10^{35} \text{ miles per hour}} = 4.65 \times 10^{-28}$ hours

115. Since $C = 2\pi r$, we can divide C by 2π to get the radius.

$r = \dfrac{5.68 \times 10^9 \text{ feet}}{2\pi} \approx 9.040 \times 10^8$ feet

117. $P = 50{,}000(1 + 0.08)^{-20} \approx \$10{,}727.41$

119. a) If $w^{-3} < 0$, then $\frac{1}{w^3} < 0$ and $w^3 < 0$. Because 3 is odd, $w < 0$.
b) If $(-5)^m < 0$ and m is an integer, then m must be odd.
c) If $w^m < 0$ and m is an integer, then m must be odd and $w < 0$.

Enriching Your Mathematical Word Power

1. a **2.** d **3.** b **4.** c **5.** d
6. b **7.** a **8.** b **9.** c **10.** a
11. a **12.** c

CHAPTER 4 REVIEW

1. $(2w - 6) + (3w + 4) = 5w - 2$

3. $(x^2 - 2x - 5) - (x^2 + 4x - 9)$
$= x^2 - 2x - 5 - x^2 - 4x + 9$
$= -6x + 4$

5. $(5 - 3w + w^2) + (w^2 - 4w - 9)$
$= w^2 + w^2 - 3w - 4w - 9 + 5$
$= 2w^2 - 7w - 4$

7. $(4 - 3m - m^2) - (m^2 - 6m + 5)$
$= 4 - 3m - m^2 - m^2 + 6m - 5$
$= -2m^2 + 3m - 1$

9. $5x^2 \cdot (-10x^9) = -50x^{2+9}$
$= -50x^{11}$

11. $(-11a^7)^2 = (-11a^7)(-11a^7) = 121a^{14}$

13. $x - 5(x - 3) = x - 5x + 15$
$= -4x + 15$

15. $5x + 3(x^2 - 5x + 4)$
$= 5x + 3x^2 - 15x + 12$
$= 3x^2 - 10x + 12$

17. $3m^2(5m^3 - m + 2)$
$= 3m^2 \cdot 5m^3 - 3m^2 \cdot m + 3m^2 \cdot 2$
$= 15m^5 - 3m^3 + 6m^2$

19. $(x - 5)(x^2 - 2x + 10)$
$= (x - 5)x^2 - (x - 5)(2x) + (x - 5)(10)$
$= x^3 - 5x^2 - 2x^2 + 10x + 10x - 50$
$= x^3 - 7x^2 + 20x - 50$

21. $(x^2 - 2x + 4)(3x - 2)$
$= (x^2 - 2x + 4)3x - (x^2 - 2x + 4)2$
$= 3x^3 - 6x^2 + 12x - 2x^2 + 4x - 8$
$= 3x^3 - 8x^2 + 16x - 8$

23. $(q - 6)(q + 8) = q^2 - 6q + 8q - 48$
$= q^2 + 2q - 48$

25. $(2t - 3)(t - 9) = 2t^2 - 3t - 18t + 27$
$= 2t^2 - 21t + 27$

27. $(4y - 3)(5y + 2) = 20y^2 - 15y + 8y - 6$
$= 20y^2 - 7y - 6$

29. $(3x^2 + 5)(2x^2 + 1)$
$= 6x^4 + 10x^2 + 3x^2 + 5$
$= 6x^4 + 13x^2 + 5$

31. $(z - 7)(z + 7) = z^2 - 7z + 7z - 49$
$= z^2 - 49$

33. $(y + 7)^2 = y^2 + 2 \cdot y \cdot 7 + 7^2$
$= y^2 + 14y + 49$

35. $(w - 3)^2 = w^2 - 2 \cdot 3w + 3^2$
$= w^2 - 6w + 9$

37. $(x^2 - 3)(x^2 + 3) = x^4 - 3x^2 + 3x^2 - 9$
$= x^4 - 9$

39. $(3a + 1)^2 = 9a^2 + 2 \cdot 3a \cdot 1 + 1^2$
$= 9a^2 + 6a + 1$

41. $(4 - y)^2 = 4^2 - 2 \cdot 4 \cdot y + y^2$
$= 16 - 8y + y^2$

43. $-10x^5 \div (2x^3) = \frac{-10x^5}{2x^3} = -5x^2$

45. $\frac{6a^5b^7c^6}{-3a^3b^9c^6} = \frac{-2a^2}{b^2}$

47. $\frac{3x - 9}{-3} = \frac{3x}{-3} - \frac{9}{-3} = -x + 3$

49. $\frac{9x^3 - 6x^2 + 3x}{-3x} = \frac{9x^3}{-3x} - \frac{6x^2}{-3x} + \frac{3x}{-3x}$
$= -3x^2 + 2x - 1$

51. $(a - 1) \div (1 - a) = \frac{a - 1}{1 - a}$
$= \frac{-1(1 - a)}{(1 - a)} = -1$

53.
$$\begin{array}{r|l} & m^3 + 2m^2 + 4m + 8 \\ \hline m - 2 & m^4 + 0m^3 + 0m^2 + 0m - 16 \\ & \underline{m^4 - 2m^3} \\ & 2m^3 + 0m^2 \\ & \underline{2m^3 - 4m^2} \\ & 4m^2 + 0m \\ & \underline{4m^2 - 8m} \\ & 8m - 16 \\ & \underline{8m - 16} \\ & 0 \end{array}$$

The quotient is $m^3 + 2m^2 + 4m + 8$.

55. $\frac{3m^3 - 9m^2 + 18m}{3m}$
$= \frac{3m^3}{3m} - \frac{9m^2}{3m} + \frac{18m}{3m}$
$= m^2 - 3m + 6$

The quotient is $m^2 - 3m + 6$ and the remainder is 0.

57.
$$\begin{array}{r} b-5 \\ b+2\,\overline{)\,b^2-3b+5} \\ \underline{b^2+2b} \\ -5b+5 \\ \underline{-5b-10} \\ 15 \end{array}$$

The quotient is $b-5$ and the remainder is 15.

59.
$$\begin{array}{r} 2x-1 \\ 2x+1\,\overline{)\,4x^2+0x-9} \\ \underline{4x^2+2x} \\ -2x-9 \\ \underline{-2x-1} \\ -8 \end{array}$$

The quotient is $2x-1$ and the remainder is -8.

61.
$$\begin{array}{r} x^2+2x-9 \\ x-1\,\overline{)\,x^3+x^2-11x+10} \\ \underline{x^3-x^2} \\ 2x^2-11x \\ \underline{2x^2-2x} \\ -9x+10 \\ \underline{-9x+9} \\ 1 \end{array}$$

The quotient is x^2+2x-9 and remainder is 1.

63.
$$\begin{array}{r} 2 \\ x-3\,\overline{)\,2x+0} \\ \underline{2x-6} \\ 6 \end{array}$$

$$\frac{2x}{x-3} = 2 + \frac{6}{x-3}$$

65.
$$\begin{array}{r} -2 \\ 1-x\,\overline{)\,2x+0} \\ \underline{2x-2} \\ 2 \end{array}$$

$$\frac{2x}{1-x} = -2 + \frac{2}{1-x}$$

67.
$$\begin{array}{r} x-1 \\ x+1\,\overline{)\,x^2+0x-3} \\ \underline{x^2+x} \\ -x-3 \\ \underline{-x-1} \\ -2 \end{array}$$

$$\frac{x^2-3}{x+1} = x-1-\frac{2}{x+1}$$

69.
$$\begin{array}{r} x-1 \\ x+1\,\overline{)\,x^2+0x+0} \\ \underline{x^2+x} \\ -x+0 \\ \underline{-x-1} \\ 1 \end{array}$$

$$\frac{x^2}{x+1} = x-1+\frac{1}{x+1}$$

71. $2y^{10} \cdot 3y^{20} = 6y^{30}$

73. $\dfrac{-10b^5c^3}{2b^5c^9} = \dfrac{-5}{c^6}$

75. $(b^5)^6 = b^{30}$

77. $(-2x^3y^2)^3 = -8x^9y^6$

79. $\left(\dfrac{2a}{b}\right)^3 = \dfrac{8a^3}{b^3}$

81. $\left(\dfrac{-6x^2y^5}{-3z^6}\right)^3 = \left(\dfrac{2x^2y^5}{z^6}\right)^3 = \dfrac{8x^6y^{15}}{z^{18}}$

83. $2^{-5} = \dfrac{1}{2^5} = \dfrac{1}{32}$

85. $10^{-3} = \dfrac{1}{10^3} = \dfrac{1}{1000}$

87. $x^5x^{-8} = x^{5-8} = x^{-3} = \dfrac{1}{x^3}$

89. $\dfrac{a^{-8}}{a^{-12}} = a^{-8-(-12)} = a^4$

91. $\dfrac{a^3}{a^{-7}} = a^{3-(-7)} = a^{10}$

93. $(x^{-3})^4 = x^{-3\cdot 4} = x^{-12} = \dfrac{1}{x^{12}}$

95. $(2x^{-3})^{-3} = 2^{-3}x^9 = \dfrac{x^9}{8}$

97. $\left(\dfrac{a}{3b^{-3}}\right)^{-2} = \dfrac{a^{-2}}{(3b^{-3})^{-2}} = \dfrac{a^{-2}}{3^{-2}b^6} = \dfrac{9}{a^2b^6}$

99. Move the decimal point 3 places to the left. Since 5000 is larger than 10 the exponent is positive. $5000 = 5 \times 10^3$

101. Move the decimal point 5 places to the right. $3.4 \times 10^5 = 340{,}000$

103. Move the decimal point 5 places to the right. Since 0.0000461 is smaller than 1, the exponent is negative.
$0.0000461 = 4.61 \times 10^{-5}$

105. Move the decimal point 6 places to the left. $5.69 \times 10^{-6} = 0.00000569$

107. $(3.5 \times 10^8)(2.0 \times 10^{-12}) = 7 \times 10^{-4}$

109.
$$\begin{aligned}(2 \times 10^{-4})^4 &= 2^4 \times 10^{-16} = 16 \times 10^{-16} \\ &= 1.6 \times 10^1 \times 10^{-16} \\ &= 1.6 \times 10^{-15}\end{aligned}$$

111. $(0.00000004)(2{,}000{,}000{,}000)$
$= 4 \times 10^{-8} \cdot 2 \times 10^{9}$
$= 8 \times 10^{1}$

113. $(0.0000002)^5 = (2 \times 10^{-7})^5$
$= 32 \times 10^{-35}$
$= 3.2 \times 10^{1} \times 10^{-35} = 3.2 \times 10^{-34}$

115. $(x+3)(x+7) = x^2 + 3x + 7x + 21$
$= x^2 + 10x + 21$

117. $(t - 3y)(t - 4y)$
$= t^2 - 3ty - 4ty + 12y^2 = t^2 - 7ty + 12y^2$

119. $(2x^3)^0 + (2y)^0 = 1 + 1 = 2$

121. $(-3ht^6)^3 = -27h^3t^{18}$

123. $(2w+3)(w-6)$
$= 2w^2 + 3w - 12w - 18$
$= 2w^2 - 9w - 18$

125. $(3u - 5v)(3u + 5v) = 9u^2 - 25v^2$

127. $(3h+5)^2 = 9h^2 + 30h + 25$

129. $(x+3)^3 = (x+3)(x+3)^2$
$= (x+3)(x^2+6x+9)$
$= x(x^2+6x+9) + 3(x^2+6x+9)$
$= x^3 + 9x^2 + 27x + 27$

131. $(-7s^2t)(-2s^3t^5) = 14s^5t^6$

133. $\left(\dfrac{k^4m^2}{2k^2m^2}\right)^4 = \left(\dfrac{k^2}{2}\right)^4 = \dfrac{k^8}{16}$

135. $(5x^2 - 8x - 8) - (4x^2 + x - 3)$
$= 5x^2 - 8x - 8 - 4x^2 - x + 3$
$= x^2 - 9x - 5$

137. $(2x^2 - 2x - 3) + (3x^2 + x - 9)$
$= 5x^2 - x - 12$

139. $(x+4)(x^2 - 5x + 1)$
$= x(x^2 - 5x + 1) + 4(x^2 - 5x + 1)$
$= x^3 - 5x^2 + x + 4x^2 - 20x + 4$
$= x^3 - x^2 - 19x + 4$

141.

$$
\begin{array}{r|l}
 & x + 6 \\
\hline
x - 2 & x^2 + 4x - 12 \\
 & \underline{x^2 - 2x} \\
 & \quad 6x - 12 \\
 & \quad \underline{6x - 12} \\
 & \qquad 0
\end{array}
$$

The quotient is $x + 6$.

143. If w is the width, then the length is $w + 44$. Since $P = 2L + 2W$,
$P(w) = 2(w+44) + 2(w)$ or
$P(w) = 4w + 88$ and $A(w) = w(w+44)$ or
$A(w) = w^2 + 44w$. If $w = 50$, then
$P(50) = 4(50) + 88 = 288$ ft and
$A(50) = 50^2 + 44(50) = 4700$ ft^2.

145. $R = p(600 - 15p) = -15p^2 + 600p$. If $p = 12$, then
$R = -15 \cdot 12^2 + 600 \cdot 12 = 5040$.
The revenue is \$5040 when the price is \$12 each. From the graph, the maximum weekly revenue occurs when the price is \$20 each.

CHAPTER 4 TEST

1. $(7x^3 - x^2 - 6) + (5x^2 + 2x - 5)$
$= 7x^3 + 4x^2 + 2x - 11$

2. $(x^2 - 3x - 5) - (2x^2 + 6x - 7)$
$= x^2 - 3x - 5 - 2x^2 - 6x + 7$
$= -x^2 - 9x + 2$

3. $\dfrac{6y^3 - 9y^2}{-3y} = \dfrac{6y^3}{-3y} - \dfrac{9y^2}{-3y} = -2y^2 + 3y$

4. $(x-2) \div (2-x) = \dfrac{x-2}{2-x}$
$= \dfrac{-1(2-x)}{2-x} = -1$

5.

$$
\begin{array}{r|l}
 & x^2 + x - 1 \\
\hline
x - 3 & x^3 - 2x^2 - 4x + 3 \\
 & \underline{x^3 - 3x^2} \\
 & \quad x^2 - 4x \\
 & \quad \underline{x^2 - 3x} \\
 & \qquad -x + 3 \\
 & \qquad \underline{-x + 3} \\
 & \qquad\quad 0
\end{array}
$$

$(x^3 - 2x^2 - 4x + 3) \div (x - 3) = x^2 + x - 1$

6. $3x^2(5x^3 - 7x^2 + 4x - 1)$
$= 15x^5 - 21x^4 + 12x^3 - 3x^2$

7. $(x+5)(x-2) = x^2 + 5x - 2x - 10$
$= x^2 + 3x - 10$

8. $(3a - 7)(2a + 5) = 6a^2 + a - 35$

9. $(a-7)^2 = a^2 - 2 \cdot a \cdot 7 + 7^2$
$= a^2 - 14a + 49$

10. $(4x + 3y)^2 = 16x^2 + 2 \cdot 4x \cdot 3y + (3y)^2$
$= 16x^2 + 24xy + 9y^2$

11. $(b-3)(b+3) = b^2 - 3b + 3b - 9$
$= b^2 - 9$

12. $(3t^2 - 7)(3t^2 + 7)$
$= 9t^4 - 21t^2 + 21t^2 - 49$
$= 9t^4 - 49$

13. $(4x^2 - 3)(x^2 + 2) = 4x^4 - 3x^2 + 8x^2 - 6$
$= 4x^4 + 5x^2 - 6$

14. $(x-2)(x+3)(x-4)$
$= (x-2)(x^2-x-12)$
$= x(x^2-x-12) - 2(x^2-x-12)$
$= x^3 - 3x^2 - 10x + 24$

15.
$$\begin{array}{r|l} & 2 \\ \hline x-3 & 2x+0 \\ & \underline{2x-6} \\ & 6 \end{array}$$

$\frac{2x}{x-3} = 2 + \frac{6}{x-3}$

16.
$$\begin{array}{r|l} & x-5 \\ \hline x+2 & x^2-3x+5 \\ & \underline{x^2+2x} \\ & -5x+5 \\ & \underline{-5x-10} \\ & 15 \end{array}$$

$\frac{x^2-3x+5}{x+2} = x - 5 + \frac{15}{x+2}$

17. $-5x^3 \cdot 7x^5 = -35x^{3+5} = -35x^8$
18. $3x^3y(2xy^4)^2 = 3x^3y \cdot 4x^2y^8 = 12x^5y^9$
19. $-4a^6b^5 \div (2a^5b) = -2a^{6-5}b^{5-1} = -2ab^4$
20. $3x^{-2} \cdot 5x^7 = 15x^{-2+7} = 15x^5$
21. $\left(\frac{-2a}{b^2}\right)^5 = \frac{-32a^5}{b^{10}}$
22. $\frac{-6a^7b^6c^2}{-2a^3b^8c^2} = \frac{3a^{7-3}}{b^{8-6}} = \frac{3a^4}{b^2}$
23. $\frac{6t^{-7}}{2t^9} = 3t^{-7-9} = 3t^{-16} = \frac{3}{t^{16}}$
24. $\frac{w^{-6}}{w^{-4}} = w^{-6-(-4)} = w^{-2} = \frac{1}{w^2}$
25. $(-3s^{-3}t^2)^{-2} = (-3)^{-2}s^6t^{-4} = \frac{s^6}{9t^4}$
26. $(-2x^{-6}y)^3 = (-2)^3x^{-18}y^3 = \frac{-8y^3}{x^{18}}$
27. Move the decimal point 6 places to the left. Since 5,433,000 is larger than 10, the exponent is positive. $5{,}433{,}000 = 5.433 \times 10^6$
28. Move the decimal point 6 places to the right. Since 0.0000065 is smaller than 1, the exponent is negative. $0.0000065 = 6.5 \times 10^{-6}$
29. $(80{,}000)(0.000006) = (8 \times 10^4)(6 \times 10^{-6})$
$= 48 \times 10^{-2}$
$= 4.8 \times 10^1 \times 10^{-2}$
$= 4.8 \times 10^{-1}$
30. $(0.0000003)^4 = (3 \times 10^{-7})^4$
$= 81 \times 10^{-28}$
$= 8.1 \times 10^1 \times 10^{-28}$
$= 8.1 \times 10^{-27}$

31.
$$\begin{array}{r|l} & x-2 \\ \hline x-3 & x^2-5x+9 \\ & \underline{x^2-3x} \\ & -2x+9 \\ & \underline{-2x+6} \\ & 3 \end{array}$$

The quotient is $x-2$ and the remainder is 3.
32. $(x^2-3x+6) - (3x^2-4x-9)$
$= -2x^2 + x + 15$
33. The width is x and the length is $x+4$. The area is $A(x) = x(x+4)$ or $A(x) = x^2 + 4x$. The perimeter is $P(x) = 2x + 2(x+4)$ or $P(x) = 4x + 8$.
If $x = 4$, then
$A(x) = 4^2 + 4 \cdot 4 = 32$ ft^2.
If $x = 4$, then
$P(4) = 4(4) + 8 = 24$ ft.

34. $R = q(3000 - 150q) = -150q^2 + 3000q$
If $q = 8$, then
$R = -150 \cdot 8^2 + 3000 \cdot 8 = 14{,}400$.
At \$8 each the weekly revenue is \$14,400.

Making Connections

Chapters 1 - 4

1. $-16 \div (-2) = 8$
2. $-16 \div \left(-\frac{1}{2}\right) = -16(-2) = 32$
3. $(-5)^2 - 3(-5) + 1 = 25 + 15 + 1 = 41$
4. $-5^2 - 4(-5) + 3 = -25 + 20 + 3 = -2$
5. $2^{15} \div 2^{10} = 2^{15-10} = 2^5 = 32$
6. $2^6 - 2^5 = 64 - 32 = 32$
7. $-3^2 \cdot 4^2 = -9 \cdot 16 = -144$
8. $(-3 \cdot 4)^2 = (-12)^2 = 144$

9. $\left(\frac{1}{2}\right)^3 + \frac{1}{2} = \frac{1}{8} + \frac{1}{2} = \frac{5}{8}$

10. $\left(\frac{2}{3}\right)^2 - \frac{1}{3} = \frac{4}{9} - \frac{3}{9} = \frac{1}{9}$

11. $(5+3)^2 = 8^2 = 64$

12. $5^2 + 3^2 = 25 + 9 = 34$

13. $3^{-1} + 2^{-1} = \frac{1}{3} + \frac{1}{2} = \frac{5}{6}$

14. $2^{-2} - 3^{-2} = \frac{1}{4} - \frac{1}{9} = \frac{5}{36}$

15. $(30-1)(30+1) = 30^2 - 1^2 = 900 - 1$
$= 899$

16. $(30-1) \div (1-30) = 29 \div (-29) = -1$

17. $(x+3)(x+5) = x^2 + 3x + 5x + 15$
$= x^2 + 8x + 15$

18. $x + 3(x+5) = x + 3x + 15$
$= 4x + 15$

19. $-5t^3v \cdot 3t^2v^6 = -15t^{3+2}v^{1+6} = -15t^5v^7$

20. $\dfrac{-10t^3v^2}{-2t^2v} = 5t^{3-2}v^{2-1} = 5tv$

21. $(x^2 + 8x + 15) + (x+5) = x^2 + 9x + 20$

22. $(x^2 + 8x + 15) - (x+5) = x^2 + 7x + 10$

23. Because of the Exercise 17
$(x^2 + 8x + 15) \div (x+5) = x + 3.$

24. $(x^2 + 8x + 15)(x+5)$
$= (x^2 + 8x + 15)x + (x^2 + 8x + 15)5$
$= x^3 + 8x^2 + 15x + 5x^2 + 40x + 75$
$= x^3 + 13x^2 + 55x + 75$

25. $\dfrac{-6y^3 + 8y^2}{-2y^2} = \dfrac{-6y^3}{-2y^2} + \dfrac{8y^2}{-2y^2} = 3y - 4$

26. $\dfrac{18y^4 - 12y^3 + 3y^2}{3y^2}$
$= \dfrac{18y^4}{3y^2} - \dfrac{12y^3}{3y^2} + \dfrac{3y^2}{3y^2} = 6y^2 - 4y + 1$

27. $2x + 1 = 0$
$2x = -1$
$x = -\frac{1}{2}$

The solution set to the equation is $\left\{-\frac{1}{2}\right\}$.

28. $x - 7 = 0$
$x = 7$

The solution set to the equation is $\{7\}$.

29. $\frac{3}{4}x - 3 = \frac{1}{2}$
$3x - 12 = 2$
$3x = 14$
$x = \frac{14}{3}$

The solution set to the equation is $\left\{\frac{14}{3}\right\}$.

30. $\frac{x}{2} - \frac{3}{4} = \frac{1}{8}$
$4x - 6 = 1$
$4x = 7$
$x = \frac{7}{4}$

The solution set to the equation is $\left\{\frac{7}{4}\right\}$.

31. $2(x-3) = 3(x-2)$
$2x - 6 = 3x - 6$
$0 = x$

The solution set to the equation is $\{0\}$.

32. $2(3x-3) = 3(2x-2)$
$6x - 6 = 6x - 6$

The solution set to the equation is the set of all real numbers.

33. To find the x-intercept, let $y = 0$:
$0 = 2x + 1$
$-2x = 1$
$x = -\frac{1}{2}$

The x-intercept is $\left(-\frac{1}{2}, 0\right)$.

34. To find the y-intercept, let $x = 0$:
$y = 0 - 7 = -7$

The y-intercept is $(0, -7)$.

35. The slope of $y = 2x + 1$ is 2.

36. $m = \dfrac{\frac{1}{3} - 0}{\frac{1}{2} - 0} = \frac{1}{3} \cdot 2 = \frac{2}{3}$

37. If $y = \frac{1}{2}$, then $\frac{3}{4}x - 3 = \frac{1}{2}$. Solve for x:
$4 \cdot \frac{3}{4}x - 4 \cdot 3 = 4 \cdot \frac{1}{2}$
$3x - 12 = 2$
$3x = 14$
$x = \frac{14}{3}$

38. If $x = \frac{1}{2}$ and $y = \frac{x}{2} - \frac{3}{4}$, then
$y = \dfrac{\frac{1}{2}}{2} - \frac{3}{4} = \frac{1}{4} - \frac{3}{4} = -\frac{2}{4} = -\frac{1}{2}.$

39. The average cost is $\dfrac{2.25n + 100{,}000}{n}$.

If $n = 1000$, then
$\dfrac{2.25(1000) + 100{,}000}{1000} = \$102.25.$

If $n = 100{,}000$, then the average cost is
$\dfrac{2.25(100{,}000) + 100{,}000}{100{,}000} = \$3.25$

If $n = 1{,}000{,}000$, then the average cost is
$\dfrac{2.25(1{,}000{,}000) + 100{,}000}{1{,}000{,}000} = \$2.35$

At one million CDs the initial investment is actually only \$0.10 per CD. The initial investment is spread over all of the 1,000,000 CDs.

5.1 WARM-UPS

1. False, because there are infinitely many prime numbers.
2. False, because $32 = 2^5$.
3. False, because $51 = 3 \cdot 17$.
4. True, because $12 = 3 \cdot 4$ and $16 = 4 \cdot 4$.
5. True, because $10 = 2 \cdot 5$ and $21 = 3 \cdot 7$.
6. True, because $x^5y^3 - x^4y^7 = x^4y^3(x - y^4)$.
7. True, because $2x^2y - 6xy^2 = 2xy(x - 3y)$ or $2x^2y - 6xy^2 = -2xy(-x + 3y)$.
8. False, because
$8a^3b - 12a^2b = 4a^2b(2a - 3)$.
9. False, because if $x = 0$ then $x - 7 = 7 - x$ becomes $-7 = 7$.
10. True, because of the distributive property.

5.1 EXERCISES

1. To factor means to write as a product.
3. You can find the prime factorization by dividing by prime factors until the result is prime.
5. The GCF for two monomials consists of the GCF of their coefficients and every variable that they have in common raised to the lowest power that appears on the variable.
7. $18 = 2 \cdot 9 = 2 \cdot 3 \cdot 3 = 2 \cdot 3^2$
9. $52 = 2 \cdot 26 = 2 \cdot 2 \cdot 13 = 2^2 \cdot 13$
11. $98 = 2 \cdot 49 = 2 \cdot 7 \cdot 7 = 2 \cdot 7^2$
13. $460 = 2 \cdot 230 = 2 \cdot 10 \cdot 23$
$= 2 \cdot 2 \cdot 5 \cdot 23 = 2^2 \cdot 5 \cdot 23$
15. $924 = 2 \cdot 462 = 2 \cdot 2 \cdot 231$
$= 2 \cdot 2 \cdot 3 \cdot 77 = 2 \cdot 2 \cdot 3 \cdot 7 \cdot 11$
$= 2^2 \cdot 3 \cdot 7 \cdot 11$
17. Since $8 = 2^3$ and $20 = 2^2 \cdot 5$, the GCF is 2^2 or 4.
19. Since $36 = 2^2 \cdot 3^2$ and $60 = 2^2 \cdot 3 \cdot 5$, the GCF is $2^2 \cdot 3$ or 12.
21. Since $40 = 2^3 \cdot 5$, $48 = 2^4 \cdot 3$, and $88 = 2^3 \cdot 11$, the GCF is 2^3 or 8.
23. Since $76 = 2^2 \cdot 19$, $84 = 2^2 \cdot 3 \cdot 7$, and $100 = 2^2 \cdot 5^2$, the GCF is 2^2 or 4.
25. Since $39 = 3 \cdot 13$, $68 = 2^2 \cdot 17$, and $77 = 7 \cdot 11$, the GCF is 1.
27. Since $6x = 2 \cdot 3 \cdot x$ and $8x^3 = 2^3x^3$, the GCF is $2x$.
29. Since $12x^3 = 2^2 \cdot 3x^3$, $4x^2 = 2^2x^2$, and $6x = 2 \cdot 3x$, the GCF is $2x$.
31. The GCF for $3x^2y$ and $2xy^2$ is xy.
33. Since $24a^2bc = 2^3 \cdot 3a^2bc$ and $60ab^2 = 2^2 \cdot 3 \cdot 5ab^2$, the GCF is $2^2 \cdot 3ab$ or $12ab$.
35. Since $12u^3v^2 = 2^2 \cdot 3u^3v^2$ and $25s^2t^4 = 5^2s^2t^4$, the GCF is 1.
37. Since $18a^3b = 2 \cdot 3^2a^3b$, $30a^2b^2 = 2 \cdot 3 \cdot 5a^2b^2$, and $54ab^3 = 2 \cdot 3^3ab^3$, the GCF is $2 \cdot 3ab = 6ab$.
39. $27x = 9(3x)$
41. $24t^2 = 8t(3t)$
43. $36y^5 = 4y^2(9y^3)$
45. $u^4v^3 = uv(u^3v^2)$
47. $-14m^4n^3 = 2m^4(-7n^3)$
49. $-33x^4y^3z^2 = -3x^3yz(11xy^2z)$
51. The GCF for $2w$ and $4t$ is 2. So $2w + 4t = 2 \cdot w + 2 \cdot 2t = 2(w + 2t)$.
53. The GCF for $12x$ and $18y$ is 6. So $12x - 18y = 6 \cdot 2x - 6 \cdot 3y = 6(2x - 3y)$.
55. The GCF for x^3 and $6x$ is x. So $x^3 - 6x = x(x^2 - 6)$.
57. The GCF for $5ax$ and $5ay$ is $5a$. So $5ax + 5ay = 5a(x + y)$.
59. The GCF for h^5 and h^3 is h^3. So $h^5 + h^3 = h^3(h^2 + 1)$.
61. The GCF for $2k^7m^4$ and $4k^3m^6$ is $2k^3m^4$. So $-2k^7m^4 + 4k^3m^6 = 2k^3m^4(-k^4 + 2m^2)$.
63. The GCF for $2x^3$, $6x^2 = 2 \cdot 3x^2$, and $8x = 2^3x$ is $2x$. So we can factor out $2x$: $2x^3 - 6x^2 + 8x = 2x(x^2 - 3x + 4)$.
65. Since $12x^4t = 2^2 \cdot 3x^4t$, $30x^3t = 2 \cdot 3 \cdot 5x^3t$, and $24x^2t = 2^3 \cdot 3x^2t^2$, the GCF is $2 \cdot 3x^2t = 6x^2t$. So
$12x^4t + 30x^3t - 24x^2t^2$
$= 6x^2t(2x^2 + 5x - 4t)$.
67. The GCF of $(x - 3)a$ and $(x - 3)b$ is $x - 3$. So
$(x - 3)a + (x - 3)b = (x - 3)(a + b)$.
69. The GCF for $x(x - 1)$ and $5(x - 1)$ is $(x - 1)$. So
$x(x - 1) - 5(x - 1) = (x - 5)(x - 1)$.
71. The GCF for $m(m + 9)$ and $(m + 9)$ is $(m + 9)$. So
$m(m + 9) + (m + 9) = (m + 1)(m + 9)$.

73. The GCF of $a(y+1)^2$ and $b(y+1)^2$ is $(y+1)^2$. So
$a(y+1)^2 + b(y+1)^2 = (a+b)(y+1)^2$.
75. The GCF of $8x$ and $8y$ is 8. However, we can factor out either 8 or -8. So
$8x - 8y = 8(x - y)$ or
$8x - 8y = -8(-x + y)$.
77. The GCF of $4x$ and $8x^2$ is $4x$. We can factor out either $4x$ or $-4x$. So
$-4x + 8x^2 = 4x(-1 + 2x)$ or
$-4x + 8x^2 = -4x(1 - 2x)$.
79. $x - 5 = 1(x - 5)$
$x - 5 = -1(-x + 5)$
81. $4 - 7a = 1(4 - 7a)$
$4 - 7a = -1(-4 + 7a)$
83. $-24a^3 + 16a^2 = 8a^2(-3a + 2)$
$-24a^3 + 16a^2 = -8a^2(3a - 2)$
85. $-12x^2 - 18x = 6x(-2x - 3)$
$-12x^2 - 18x = -6x(2x + 3)$
87. $-2x^3 - 6x^2 + 14x$
$= 2x(-x^2 - 3x + 7)$
$-2x^3 - 6x^2 + 14x = -2x(x^2 + 3x - 7)$
89. $4a^3b - 6a^2b^2 - 4ab^3$
$= 2ab(2a^2 - 3ab - 2b^2)$
$4a^3b - 6a^2b^2 - 4ab^3$
$= -2ab(-2a^2 + 3ab + 2b^2)$
91. Note that we can write her distance as $20x + 40 = 20(x + 2)$ and that $D = RT$. Since her rate was 20 mph, her time is $x + 2$ hours.
93. a) $S = 2\pi r^2 + 2\pi rh$
$S = 2\pi r(r + h)$
b) If $h = 5$, then $S = 2\pi r^2 + 2\pi r(5)$ or $S = 2\pi r^2 + 10\pi r$.
c) From the graph we see that the maximum surface area occurs when the radius is 3 in.
95. The greatest common factor of $-6x^2 + 3x$ is $3x$, which is an algebraic expression. We cannot determine whether $3x$ is positive or negative unless we have a value for x.

5.2 WARM-UPS

1. False, because $x^2 + 16$ is a sum of two squares.
2. True, because $x^2 - 8x + 16 = (x - 4)^2$.
3. False, because the first term is $(3x)^2$, the last terms is $(7)^2$, but the middle term is not $2 \cdot 3x \cdot 7$ or $42x$.
4. False, because if $x = 1$, then $4 \cdot 1^2 + 4 = 8$ and $(2 \cdot 1 + 2)^2 = 16$.
5. True, because $a^2 - b^2 = (a + b)(a - b)$.
6. True, because $16y + 1$ cannot be factored.
7. False, because
$(x + 3)(x + 3) = x^2 + 6x + 9$.
8. False, because $x^2 - 1$ is not prime.
9. True, because $y^2 - 2y + 1$ is a perfect square trinomial and it is correctly factored as $(y - 1)^2$.
10. True, because $2x^2 - 18 = 2(x^2 - 9)$
$= 2(x - 3)(x + 3)$ is the correct factorization.

5.2 EXERCISES

1. A perfect square is a square of an integer or an algebraic expression.
3. A perfect square trinomial is of the form $a^2 + 2ab + b^2$ or $a^2 - 2ab + b^2$.
5. A polynomial is factored completely when it is a product of prime polynomials.
7. The polynomial $a^2 - 4$ is the difference of two squares, $a^2 - 2^2$. So
$a^2 - 4 = (a - 2)(a + 2)$.
9. The polynomial $x^2 - 49$ is a difference of two squares $x^2 - 7^2$. So
$x^2 - 49 = (x - 7)(x + 7)$.
11. The polynomial $y^2 - 9x^2$ is a difference of two squares, $y^2 - (3x)^2$. So
$y^2 - 9x^2 = (y + 3x)(y - 3x)$.
13. The polynomial $25a^2 - 49b^2$ is a difference of two squares, $(5a)^2 - (7b)^2$. So
$25a^2 - 49b^2 = (5a + 7b)(5a - 7b)$.
15. The polynomial $121m^2 - 1$ is a difference of two squares, $(11m)^2 - 1^2$.
So $121m^2 - 1 = (11m + 1)(11m - 1)$.
17. The polynomial $9w^2 - 25c^2$ is a difference of two squares, $(3w)^2 - (5c)^2$. So
$9w^2 - 25c^2 = (3w - 5c)(3w + 5c)$.
19. Since the first and last terms of $x^2 - 20x + 100$ are perfect squares and $-20x = -2 \cdot x \cdot 10$, the polynomial is a perfect square trinomial.
21. Since $y^2 - 40$ has only two terms and 40 is not a perfect square, this polynomial is neither

a perfect square trinomial nor a difference of two squares.

23. Since the first and last terms of $4y^2 + 12y + 9$ are perfect squares and $12y = 2 \cdot 2y \cdot 3$, the polynomial is a perfect square trinomial.

25. The first and last terms of $x^2 - 8x + 64$ are perfect squares, but $8x$ is not equal to $2 \cdot x \cdot 8$. So the polynomial is not a perfect square trinomial. Since it has three terms it cannot be a difference of two squares, so it is neither a perfect square trinomial nor a difference of two squares.

27. Since $9y^2 - 25c^2 = (3y)^2 - (5c)^2$, the polynomial is a difference of two squares.

29. The first and last terms of $9a^2 + 6ab + b^2$ are perfect squares. Since $6ab = 2 \cdot 3a \cdot b$, it is a perfect square trinomial.

31. Since x^2 and 1 are perfect squares and $2x = 2 \cdot x \cdot 1$, $x^2 + 2x + 1$ is a perfect square trinomial and

$$x^2 + 2x + 1 = (x + 1)^2.$$

33. Since a^2 and 9 are perfect squares and $6a = 2 \cdot a \cdot 3$, $a^2 + 6a + 9$ is a perfect square trinomial and

$$a^2 + 6a + 9 = (a + 3)^2.$$

35. Since x^2 and 36 are perfect squares and $12x = 2 \cdot x \cdot 6$, $x^2 + 12x + 36$ is a perfect square trinomial and

$x^2 + 12x + 36 = (x + 6)^2$.

37. Since a^2 and 4 are perfect squares and $-4a = -2 \cdot a \cdot 2$, $a^2 - 4a + 4$ is a perfect square trinomial and $a^2 - 4a + 4 = (a - 2)^2$.

39. $4w^2 + 4w + 1 = (2w)^2 + 2 \cdot 2w \cdot 1 + 1^2$
$= (2w + 1)^2$

41. $16x^2 - 8x + 1 = (4x)^2 - 2 \cdot 4x \cdot 1 + 1^2$
$= (4x - 1)^2$

43. $4t^2 + 20t + 25 = (2t)^2 + 2 \cdot 2t \cdot 5 + 5^2$
$= (2t + 5)^2$

45. $9w^2 + 42w + 49$
$= (3w)^2 + 2 \cdot 3w \cdot 7 + 7^2$
$= (3w + 7)^2$

47. $n^2 + 2nt + t^2 = (n + t)^2$

49. $5x^2 - 125 = 5(x^2 - 25)$
$= 5(x - 5)(x + 5)$

51. We could factor out 2 or -2. If we factor out -2, then we have a difference of two squares and we can factor again.

$-2x^2 + 18 = -2(x^2 - 9)$
$= -2(x - 3)(x + 3)$

53. $a^3 - ab^2 = a(a^2 - b^2) = a(a - b)(a + b)$

55. $3x^2 + 6x + 3 = 3(x^2 + 2x + 1)$
$= 3(x + 1)^2$

57. $-5y^2 + 50y - 125$
$= -5(y^2 - 10y + 25)$
$= -5(y - 5)^2$

59. $x^3 - 2x^2y + xy^2 = x(x^2 - 2xy + y^2)$
$= x(x - y)^2$

61. We could factor out 3 or -3. If we factor out -3, then we get a difference of two squares and we can factor completely.

$-3x^2 + 3y^2 = -3(x^2 - y^2)$
$= -3(x - y)(x + y)$

63. $2ax^2 - 98a = 2a(x^2 - 49)$
$= 2a(x - 7)(x + 7)$

65. $3ab^2 - 18ab + 27a = 3a(b^2 - 6b + 9)$
$= 3a(b - 3)^2$

67. $-4m^3 + 24m^2n - 36mn^2$
$= -4m(m^2 - 6mn + 9n^2)$
$= -4m(m - 3n)^2$

69. $bx + by + cx + cy = b(x + y) + c(x + y)$
$= (b + c)(x + y)$

71. Note that we must factor out -4 from the last two terms to get the common factor $(x + 1)$.

$x^3 + x^2 - 4x - 4 = x^2(x + 1) - 4(x + 1)$
$= (x^2 - 4)(x + 1)$
$= (x - 2)(x + 2)(x + 1)$

73. Note that we must factor out $-x$ from the last two terms to get the common factor $(a - b)$.

$3a - 3b - xa + xb = 3(a - b) - x(a - b)$
$= (3 - x)(a - b)$

75. Note that the GCF for the last two terms is 1. So we factor out 1.

$a^3 + 3a^2 + a + 3 = a^2(a + 3) + 1(a + 3)$
$= (a^2 + 1)(a + 3)$

77. $xa + ay + 3y + 3x$
$= a(x + y) + 3(x + y)$
$= (a + 3)(x + y)$

79. $abc + c - 3 - 3ab$
$= c(ab + 1) - 3(1 + ab)$
$= c(ab + 1) - 3(ab + 1)$
$= (c - 3)(ab + 1)$

81. $x^2a - a + bx^2 - b$
$= a(x^2 - 1) + b(x^2 - 1)$

$= (a+b)(x^2-1)$
$= (a+b)(x-1)(x+1)$

83. $y^2+y+by+b = y(y+1)+b(y+1)$
$= (y+b)(y+1)$

85. $6a^3y+24a^2y^2+24ay^3$
$= 6ay(a^2+4ay+4y^2)$
$= 6ay(a+2y)^2$

87. $24a^3y-6ay^3 = 6ay(4a^2-y^2)$
$= 6ay(2a-y)(2a+y)$

89. $2a^3y^2-6a^2y = 2a^2y(ay-3)$

91. $ab+2bw-4aw-8w^2$
$= b(a+2w)-4w(a+2w)$
$= (b-4w)(a+2w)$

93. $h = -16t^2+6400 = -16(t^2-400)$
$h = -16(t-20)(t+20)$
If $t = 2$, then
$h = -16(2-20)(2+20) = -16(-18)(22)$
$= 6336$ feet

95. $V = y^3-6y^2+9y = y(y^2-6y+9)$
$V = y(y-3)^2$
Since $V = LWH$ and the base is square, the side of the base is $y-3$ inches.

97. To factor a four-term polynomial by grouping, factor out a common factor out of the first two terms and a common factor out of the last two terms, then factor out a common factor from the remaining two terms. The terms might have to be rearranged to accomplish the factoring.

5.3 WARM-UPS

1. True, because
$(x-3)^2 = x^2-2\cdot3\cdot x+3^2 = x^2-6x+9.$

2. True, because
$(x+3)^2 = x^2+2\cdot3\cdot x+3^2 = x^2+6x+9.$

3. False, because
$(x-9)(x-1) = x^2-10x+9.$

4. False, because
$(x-8)(x-9) = x^2-17x+72.$

5. True, because
$(x+9)(x-1) = x^2+9x-x-9$
$= x^2+8x-9.$

6. False, because $(x+3)^2 = x^2+6x+9.$

7. True, because
$(x-y)(x-9y) = x^2-xy-9xy+9y^2$
$= x^2-10xy+9y^2.$

8. False, because
$(x+1)(x+1) = x^2+2x+1.$

9. False, because
$(x+5y)(x-4y) = x^2+xy-20y^2.$

10. False, because
$(x+1)(x+1) = x^2+2x+1.$

5.3 EXERCISES

1. We factored ax^2+bx+c with $a = 1$.

3. If there are no two integers that have product of c and a sum of b, then x^2+bx+c is prime.

5. A polynomial is factored completely when all of the factors are prime polynomials.

7. Two numbers that have a product of 3 and a sum of 4 are 3 and 1. So
$x^2+4x+3 = (x+3)(x+1).$

9. Two numbers that have a product of 18 and a sum of 9 are 3 and 6. So
$x^2+9x+18 = (x+3)(x+6).$

11. Two numbers that have a product of 10 and a sum of 7 are 2 and 5. So
$a^2+7a+10 = (a+2)(a+5).$

13. Two numbers that have a product of 12 and a sum of -7 are -3 and -4. So
$a^2-7a+12 = (a-3)(a-4).$

15. Two numbers that have a product of -6 and a sum of -5 are -6 and 1. So
$b^2-5b-6 = (b-6)(b+1).$

17. Since -2 and 5 have a product of -10 and a sum of 3, we have
$x^2+3x-10 = (x-2)(x+5).$

19. Since 2 and 5 have a product of 10 and a sum of 7, we have
$y^2+7y+10 = (y+2)(y+5).$

21. Two numbers with a product of 8 and a sum of -6 are -2 and -4. So we can write
$a^2-6a+8 = (a-2)(a-4)$

23. Two numbers that have a product of 16 and a sum of -10 are -8 and -2. So we can write $m^2-10m+16 = (m-8)(m-2).$

25. Two numbers that have a product of -10 and a sum of 9 are 10 and -1. So we can write
$w^2+9w-10 = (w+10)(w-1).$

27. Two numbers with a product of -8 and a sum of -2 are -4 and 2. So
$w^2 - 8 - 2w = w^2 - 2w - 8$
$= (w-4)(w+2).$
29. The only numbers that have a product of -12 are 1 and -12, -1 and 12, 2 and -6, -2 and 6, 3 and -4, -3 and 4. None of these pairs of numbers have a sum of -2. So the polynomial is a prime polynomial.
31. Two numbers that have a product of -16 and a sum of 15 are 16 and -1. So we can write
$15m - 16 + m^2 = m^2 + 15m - 16$
$= (m+16)(m-1).$
33. The only numbers that have a product of 12 are 1 and 12, -1 and -12, 2 and 6, -2 and -6, 3 and 4, -3 and -4. None of these pairs of numbers have a sum of -4. So the polynomial is a prime polynomial.
35. $z^2 - 25 = z^2 - 5^2 = (z-5)(z+5)$
37. A sum of two squares is prime.
39. $m^2 + 12m + 20 = (m+2)(m+10)$
41. The only numbers that have a product of 10 are 1 and 10, -1 and -10, 2 and 5, -2 and -5. Since none of these pairs have a sum of -3, the polynomial is prime.
43. $m^2 - 18 - 17m = m^2 - 17m - 18$
$= (m-18)(m+1)$
45. Two numbers that have a product of 24 are 1 and 24, -1 and -24, 2 and 12, -2 and -12, 3 and 8, -3 and -8, 4 and 6, -4 and -6. Since none of these pairs has a sum of -23, the polynomial is prime.
47. Two numbers that have a product of -24 and a sum of 5 are 8 and -3. So
$5t - 24 + t^2 = t^2 + 5t - 24 = (t+8)(t-3).$
49. Two numbers that have a product of -24 and a sum of -2 are -6 and 4. So
$t^2 - 2t - 24 = (t-6)(t+4).$
51. Two numbers that have a product of -200 and a sum of -10 are -20 and 10. So
$t^2 - 10t - 200 = (t-20)(t+10).$
53. Two numbers that have a product of -150 and a sum of -5 are -15 and 10. So
$x^2 - 5x - 150 = (x-15)(x+10).$
55. Two numbers that have a product of 30 and a sum of 13 are 3 and 10. So
$13y + 30 + y^2 = y^2 + 13y + 30$
$= (y+3)(y+10).$
57. Two number that have a product of 6 and a sum of 5 are 3 and 2. So
$x^2 + 5ax + 6a^2 = (x+2a)(x+3a).$
59. Two numbers that have a product of -12 and a sum of -4 are -6 and 2. So
$x^2 - 4xy - 12y^2 = (x-6y)(x+2y).$
61. Two numbers that have a product of 12 and a sum of -13 are -1 and -12. So
$x^2 - 13xy + 12y^2 = (x-12y)(x-y).$
63. There is no pair of numbers with a product of -33 and a sum of 4. So $x^2 + 4xz - 33z^2$ is prime.
65. $5x^3 + 5x = 5x(x^2+1)$
Note that $x^2 + 1$ is prime.
67. $w^2 - 8w = w(w-8)$
69. $2w^2 - 162 = 2(w^2 - 81)$
$= 2(w-9)(w+9)$
71. $-2b^2 - 98 = -2(b^2 + 49)$
Note that $b^2 + 49$ is prime.
73. $x^3 - 2x^2 - 9x + 18$
$= x^2(x-2) - 9(x-2)$
$= (x^2-9)(x-2)$
$= (x+3)(x-3)(x-2)$
75. $4r^2 + 9$ is a sum of two squares and is a prime polynomial.
77. $x^2w^2 + 9x^2 = x^2(w^2+9)$
79. $w^2 - 18w + 81 = w^2 - 2 \cdot 9w + 9^2$
$= (w-9)^2$
81. $6w^2 - 12w - 18 = 6(w^2 - 2w - 3)$
$= 6(w-3)(w+1)$
83. $3y^2 + 75 = 3(y^2 + 25)$
Note that $y^2 + 25$ is prime.
85. $ax + ay + cx + cy$
$= a(x+y) + c(x+y)$
$= (a+c)(x+y)$
87. $-2x^2 - 10x - 12$
$= -2(x^2 + 5x + 6)$
$= -2(x+2)(x+3)$
89. $32x^2 - 2x^4 = 2x^2(16 - x^2)$
$= 2x^2(4-x)(4+x)$
91. $3w^2 + 27w + 54 = 3(w^2 + 9w + 18)$
$= 3(w+3)(w+6)$
93. $18w^2 + w^3 + 36w = w^3 + 18w^2 + 36w$
$= w(w^2 + 18w + 36)$
95. $9y^2 + 1 + 6y = 9y^2 + 6y + 1$
$= (3y+1)^2$

97. $8vw^2 + 32vw + 32v = 8v(w^2 + 4w + 4)$
$= 8v(w + 2)^2$
99. $6x^3y + 30x^2y^2 + 36xy^3$
$= 6xy(x^2 + 5xy + 6y^2)$
$= 6xy(x + 3y)(x + 2y)$
101. $5 + 8w + 3w^2 = 3w^2 + 8w + 5$
$= (3w + 5)(w + 1)$
103. $-3y^3 + 6y^2 - 3y$
$= -3y(y^2 - 2y + 1)$
$= -3y(y - 1)^2$
105. $a^3 + ab + 3b + 3a^2$
$= a(a^2 + b) + 3(b + a^2)$
$= a(a^2 + b) + 3(a^2 + b)$
$= (a + 3)(a^2 + b)$
107. The area of a rectangle is given by the formula $A = LW$ and the area of this rectangle is $x^2 + 6x + 8 = (x + 4)(x + 2)$. If the width is $x + 2$ feet, then the length must be $x + 4$ feet.
109. $V = x^3 + 8x^2 + 15x = x(x^2 + 8x + 15)$
$= x(x + 3)(x + 5)$
Since $V = LWH$, the increases to the length and width are 3 feet and 5 feet.
111. Find each product. The products in a, b, and c are $2x^2 + 2x - 12$. The product in d is $4x^2 + 4x - 24$.

5.4 WARM-UPS

1. True, because $(2x + 1)(x + 1)$
$= 2x^2 + x + 2x + 1 = 2x^2 + 3x + 1$.
2. False, because $(2x + 1)(x + 3)$
$= 2x^2 + x + 6x + 3 = 2x^2 + 7x + 3$.
3. True, because $(3x + 1)(x + 3)$
$= 3x^2 + x + 9x + 3 = 3x^2 + 10x + 3$.
4. False, because $(3x + 7)(5x + 2)$
$= 15x^2 + 35x + 6x + 14 = 15x^2 + 41x + 14$
5. True, because $(2x - 9)(x + 1)$
$= 2x^2 - 9x + 2x - 9 = 2x^2 - 7x - 9$.
6. False, because $(2x + 3)(x - 3)$
$= 2x^2 + 3x - 6x - 9 = 2x^2 - 3x - 9$.
7. False, because $(2x - 9)(2x + 1)$
$= 4x^2 - 18x + 2x - 9 = 4x^2 - 16x - 9$.
8. False, because $(4x - 1)(2x + 5)$
$= 8x^2 - 2x + 20x - 5 = 8x^2 + 18x - 5$.
9. False, because $(5x - 1)(4x + 1)$
$= 20x^2 - 4x + 5x - 1 = 20x^2 + x - 1$.
10. True, because $(3x - 1)(4x - 3)$
$= 12x^2 - 4x - 9x + 3 = 12x^2 - 13x + 3$.

5.4 EXERCISES

1. We factored $ax^2 + bx + c$ with $a \neq 1$.
b, and then we use factoring by grouping,
3. If there are no two integers whose product is ac and whose sum is b, then $ax^2 + bx + c$ is prime.
5. Two integers with a product of 20 and a sum of 12 are 2 and 10.
7. Two integers with a product of -12 and a sum of -4 are -6 and 2.
9. Since $ac = 12$, we need two numbers with a product of 12 and a sum of 7. The numbers are 3 and 4.
11. Since $ac = 18$, we need two numbers with a product of 18 and a sum of -11. The numbers are -2 and -9.
13. Since $ac = -12$, we need two numbers with a product of -12 and sum of 1. The numbers are -3 and 4.
15. Since $ac = 2 \cdot 1 = 2$, we need two numbers that have a product of 2 and a sum of 3. The numbers are 2 and 1.
$2x^2 + 3x + 1 = 2x^2 + 2x + x + 1$
$= 2x(x + 1) + 1(x + 1)$
$= (2x + 1)(x + 1)$
17. Since $ac = 2 \cdot 4 = 8$, we need two numbers that have a product of 8 and a sum of 9. The numbers are 8 and 1.
$2x^2 + 9x + 4 = 2x^2 + 8x + x + 4$
$= 2x(x + 4) + 1(x + 4)$
$= (2x + 1)(x + 4)$
19. Since $ac = 3 \cdot 2 = 6$, we need two numbers that have a product of 6 and a sum of 7. The numbers are 6 and 1.
$3t^2 + 7t + 2 = 3t^2 + 6t + t + 2$
$= 3t(t + 2) + 1(t + 2)$
$= (3t + 1)(t + 2)$
21. Since $ac = 2 \cdot (-3) = -6$, we need two numbers that have a product of -6 and a sum of 5. The numbers are 6 and -1.
$2x^2 + 5x - 3 = 2x^2 + 6x - 1x - 3$
$= 2x(x + 3) - 1(x + 3)$
$= (2x - 1)(x + 3)$
23. Since $ac = 6 \cdot (-3) = -18$, we need two numbers that have a product of -18 and a sum

of 7. The numbers are 9 and -2.

$6x^2 + 7x - 3 = 6x^2 + 9x - 2x - 3$
$= 3x(2x + 3) - 1(2x + 3)$
$= (3x - 1)(2x + 3)$

25. Since $ac = 3 \cdot 4 = 12$, we need two numbers that have a product of 12 and a sum of -5. There are no such integers. So the polynomial is prime.

27. Since $ac = 2 \cdot 6 = 12$, we need two numbers that have a product of 12 and a sum of -7. The numbers are -3 and -4.

$2x^2 - 7x + 6 = 2x^2 - 4x - 3x + 6$
$= 2x(x - 2) - 3(x - 2)$
$= (2x - 3)(x - 2)$

29. Since $ac = 5 \cdot 6 = 30$, we need two numbers that have a product of 30 and a sum of -13. The numbers are -10 and -3.

$5b^2 - 13b + 6 = 5b^2 - 10b - 3b + 6$
$= 5b(b - 2) - 3(b - 2)$
$= (5b - 3)(b - 2)$

31. Since $ac = 4 \cdot (-3) = -12$, we need two numbers that have a product of -12 and a sum of -11. The numbers are -12 and 1.

$4y^2 - 11y - 3 = 4y^2 - 12y + 1y - 3$
$= 4y(y - 3) + 1(y - 3)$
$= (4y + 1)(y - 3)$

33. Since $ac = 3 \cdot 1 = 3$, we need two numbers that have a product of 3 and a sum of 2. There is no pair of numbers that has a product of 3 and a sum of 2. So the polynomial is prime.

35. Since $ac = 8 \cdot (-1) = -8$, we need two numbers that have a product of -8 and a sum of -2. The numbers are -4 and 2.

$8x^2 - 2x - 1 = 8x^2 - 4x + 2x - 1$
$= 4x(2x - 1) + 1(2x - 1)$
$= (4x + 1)(2x - 1)$

37. Since $ac = 9 \cdot 2 = 18$, we need two numbers that have a product of 18 and a sum of -9. The numbers are -6 and -3.

$9t^2 - 9t + 2 = 9t^2 - 6t - 3t + 2$
$= 3t(3t - 2) - 1(3t - 2)$
$= (3t - 1)(3t - 2)$

39. Since $ac = 15 \cdot 2 = 30$, we need two numbers that have a product of 30 and a sum of 13. The numbers are 10 and 3.

$15x^2 + 13x + 2 = 15x^2 + 10x + 3x + 2$
$= 5x(3x + 2) + 1(3x + 2)$
$= (5x + 1)(3x + 2)$

41. Since $ac = 4 \cdot 15 = 60$, we need two numbers that have a product of 60 and a sum of 16. The numbers are 10 and 6.

$4a^2 + 16ab + 15b^2$
$= 4a^2 + 10ab + 6ab + 15b^2$
$= 2a(2a + 5b) + 3b(2a + 5b)$
$= (2a + 3b)(2a + 5b)$

43. Since $ac = 6(-5) = -30$, we need two numbers that have a product of -30 and a sum of -7. The numbers are -10 and 3.

$6m^2 - 7mn - 5n^2$
$= 6m^2 - 10mn + 3mn - 5n^2$
$= 2m(3m - 5n) + n(3m - 5n)$
$= (3m - 5n)(2m + n)$

45. Since $ac = 3(5) = 15$, we need two numbers that have a product of 15 and a sum of -8. The numbers are -3 and -5.

$3x^2 - 8xy + 5y^2$
$= 3x^2 - 3xy - 5xy + 5y^2$
$= 3x(x - y) - 5y(x - y)$
$= (x - y)(3x - 5y)$

47. There is only one way to factor $5a^2$ (as $5a \cdot a$) and only one way to factor 1 (as $1 \cdot 1$). So $5a^2 + 6a + 1 = (5a + 1)(a + 1)$.

49. There are two ways to factor $6x^2$ ($6x \cdot x$ or $3x \cdot 2x$). So try $(6x + 1)(x + 1)$ and $(2x + 1)(3x + 1)$. So

$6x^2 + 5x + 1 = (2x + 1)(3x + 1)$.

51. $5a^2 + 11a + 2 = (5a + 1)(a + 2)$

53. $4w^2 + 8w + 3 = (2w + 3)(2w + 1)$

55. $15x^2 - x - 2 = (5x - 2)(3x + 1)$

57. $8x^2 - 6x + 1 = (4x - 1)(2x - 1)$

59. $15x^2 - 31x + 2 = (15x - 1)(x - 2)$

61. For $4x^2 - 4x + 3$ we have $ac = 12$. Since there are no integers with a product 12 and a sum of -4 the polynomial is prime.

63. $2x^2 + 18x - 90 = 2(x^2 + 9x - 45)$

65. $3x^2 + x - 10 = (3x - 5)(x + 2)$

67. $10x^2 - 3xy - y^2 = (5x + y)(2x - y)$

69. $42a^2 - 13ab + b^2 = (6a - b)(7a - b)$

71. $3x^2 + 7x + 2 = (x + 2)(3x + 1)$

73. $5x^2 + 11x + 2 = (5x + 1)(x + 2)$

75. $6a^2 - 17a + 5 = (3a - 1)(2a - 5)$

77. $81w^3 - w = w(81w^2 - 1)$
$= w(9w - 1)(9w + 1)$

79. $4w^2 + 2w - 30 = 2(2w^2 + w - 15)$
$= 2(2w - 5)(w + 3)$
81. $12x^2 + 36x + 27 = 3(4x^2 + 12x + 9)$
$= 3(2x + 3)^2$
83. $6w^2 - 11w - 35$
$= 6w^2 - 21w + 10w - 35$
$= 3w(2w - 7) + 5(2w - 7)$
$= (3w + 5)(2w - 7)$
85. $3x^2z - 3zx - 18z = 3z(x^2 - x - 6)$
$= 3z(x - 3)(x + 2)$
87. $9x^3 - 21x^2 + 18x = 3x(3x^2 - 7x + 6)$
89. Two numbers that have a product of -15 and a sum of 2 are 5 and -3.
$a^2 + 2ab - 15b^2 = (a + 5b)(a - 3b)$
91. $2x^2y^2 + xy^2 + 3y^2 = y^2(2x^2 + x + 3)$
93. $-6t^3 - t^2 + 2t = -t(6t^2 + t - 2)$
$= -t(3t + 2)(2t - 1)$
95. $12t^4 - 2t^3 - 4t^2 = 2t^2(6t^2 - t - 2)$
$= 2t^2(3t - 2)(2t + 1)$
97. $4x^2y - 8xy^2 + 3y^3$
$= y(4x^2 - 8xy + 3y^2)$
$= y(2x - y)(2x - 3y)$
99. $-4w^2 + 7w - 3 = -1(4w^2 - 7w + 3)$
$= -1(w - 1)(4w - 3)$
101. $-12a^3 + 22a^2b - 6ab^2$
$= -2a(6a^2 - 11ab + 3b^2)$
$= -2a(2a - 3b)(3a - b)$
103. $h = -16t^2 + 40t + 24$
$h = -8(2t^2 - 5t - 3)$
$h = -8(2t + 1)(t - 3)$
If $t = 3$, then $h = -8(2 \cdot 3 + 1)(3 - 3) = 0$.
If $t = 3$ seconds, the height is 0 feet.
105. a) To factor $x^2 + bx + 3$ we need two integers that have a product of 3 and a sum of b. The only integers with a product of 3 are 3 and 1 or -3 and -1. The sums of these pairs of integers are 4 and -4. So b must be either 4 or -4 for the trinomial to factor.
b) To factor $3x^2 + bx + 5$ we need two integers that have a product of 15 and a sum of b. So in this case b could be ± 8 or ± 16.
c) To factor $2x^2 + bx - 15$ we need two integers that have a product of -30 and a sum of b. So b could be $\pm 1, \pm 7, \pm 13$, or ± 29.
107. Since $2x^2 + 3x + 1 = (2x + 1)(x + 1)$, you can make a rectangle with length $2x + 1$ and width $x + 1$.

5.5 WARM-UPS

1. False, because if $x = 3$, $3^2 - 4 = 5$ and $(3 - 2)^2 = 1$.
2. False, because although the first term is $(2x)^2$ and the last term is 3^2, the middle term is not $2 \cdot 2x \cdot 3 = 12x$.
3. True, because $4y^2 + 25$ there is no pair of integers with a product of 100 and a sum of 0.
4. True, because
$(x + y)(3 + a) = (x + y)3 + (x + y)a$
$= 3x + 3y + ax + ay$.
5. False, because $3x^2 + 51 = 3(x^2 + 17)$.
6. True, because if more factoring is necessary it will be easier with the GCF factored out.
7. False, because if $x = 1$, $1^2 + 9 = 10$ and $(1 + 3)^2 = 16$.
8. True, because no pair of integers has a product of -5 and a sum of -3.
9. True, because
$y^2 - 5y - my + 5m = y(y - 5) - m(y - 5)$
$= (y - m)(y - 5)$.
10. False, because $x^2 + ax = x(x + a)$ but $-3x + 3a = -3(x - a) = 3(-x + a)$. We cannot produce the factor $x + a$ out of the last two terms.

5.5 EXERCISES

1. If there is no remainder, then the dividend factors as the divisor times the quotient.
3. If you divide $a^3 + b^3$ by $a + b$ there will be no remainder.
5. $a^3 + b^3 = (a + b)(a^2 - ab + b^2)$

7.
$$\begin{array}{r} x^2 - x - 6 \\ x + 4 \overline{\smash{)}\, x^3 + 3x^2 - 10x - 24} \\ \underline{x^3 + 4x^2} \\ -x^2 - 10x \\ \underline{-x^2 - 4x} \\ -6x - 24 \\ \underline{-6x - 24} \\ 0 \end{array}$$

$x^3 + 3x^2 - 10x - 24 = (x + 4)(x^2 - x - 6)$
$= (x + 4)(x - 3)(x + 2)$

9.

$$\begin{array}{r} x^2 + 5x + 6 \\ x - 1 \overline{) x^3 + 4x^2 + x - 6} \\ \underline{x^3 - x^2} \\ 5x^2 + x \\ \underline{5x^2 - 5x} \\ 6x - 6 \\ \underline{6x - 6} \\ 0 \end{array}$$

$x^3 + 4x^2 + x - 6 = (x - 1)(x^2 + 5x + 6)$
$= (x - 1)(x + 3)(x + 2)$

11.

$$\begin{array}{r} x^2 + 2x + 4 \\ x - 2 \overline{) x^3 + 0x^2 + 0x - 8} \\ \underline{x^3 - 2x^2} \\ 2x^2 + 0x \\ \underline{2x^2 - 4x} \\ 4x - 8 \\ \underline{4x - 8} \\ 0 \end{array}$$

$x^3 - 8 = (x - 2)(x^2 + 2x + 4)$
There is no pair of integers that has a product of 4 and a sum of 2. So $x^2 + 2x + 4$ is prime.

13.

$$\begin{array}{r} x^2 - x + 2 \\ x + 5 \overline{) x^3 + 4x^2 - 3x + 10} \\ \underline{x^3 + 5x^2} \\ -x^2 - 3x \\ \underline{-x^2 - 5x} \\ 2x + 10 \\ \underline{2x + 10} \\ 0 \end{array}$$

$x^3 + 4x^2 - 3x + 10 = (x + 5)(x^2 - x + 2)$
There is no pair of integers that has a product of 2 and a sum of -1. So $x^2 - x + 2$ is prime.

15.

$$\begin{array}{r} x^2 + x + 1 \\ x + 1 \overline{) x^3 + 2x^2 + 2x + 1} \\ \underline{x^3 + x^2} \\ x^2 + 2x \\ \underline{x^2 + x} \\ x + 1 \\ \underline{x + 1} \\ 0 \end{array}$$

$x^3 + 2x^2 + 2x + 1 = (x + 1)(x^2 + x + 1)$
There is no pair of integers that has a product of 1 and a sum of 1. So $x^2 + x + 1$ is prime.

17. $m^3 - 1 = (m - 1)(m^2 + m + 1)$

19. $x^3 + 8 = (x + 2)(x^2 - 2x + 4)$

21. $a^3 + 125 = (a + 5)(a^2 - 5a + 25)$

23. $c^3 - 343 = (c - 7)(c^2 + 7c + 49)$

25. $8w^3 + 1 = (2w)^3 + 1^3$
$= (2w + 1)(4w^2 - 2w + 1)$

27. $8t^3 - 27 = (2t)^3 - 3^3$
$= (2t - 3)(4t^2 + 6t + 9)$

29. $x^3 - y^3 = (x - y)(x^2 + xy + y^2)$

31. $8t^3 + y^3 = (2t)^3 + y^3$
$= (2t + y)(4t^2 - 2ty + y^2)$

33. $x^4 - y^4 = (x^2 - y^2)(x^2 + y^2)$
$= (x - y)(x + y)(x^2 + y^2)$

35. $x^4 - 1 = (x^2 - 1)(x^2 + 1)$
$= (x - 1)(x + 1)(x^2 + 1)$

37. $16b^4 - 1 = (4b^2 - 1)(4b^2 + 1)$
$= (2b - 1)(2b + 1)(4b^2 + 1)$

39. $a^4 - 81b^4 = (a^2 - 9b^2)(a^2 + 9b^2)$
$= (a - 3b)(a + 3b)(a^2 + 9b^2)$

41. $2x^2 - 18 = 2(x^2 - 9)$
$= 2(x - 3)(x + 3)$

43. Since $a^2 + 4$ is a sum of two squares of the form $a^2 + b^2$ it is prime.

45. $4x^2 + 8x - 60 = 4(x^2 + 2x - 15)$
$= 4(x + 5)(x - 3)$

47. $x^3 + 4x^2 + 4x = x(x^2 + 4x + 4)$
$= x(x + 2)^2$

49. $5max^2 + 20ma = 5am(x^2 + 4)$

51. Since there are no integers with a product of -2 and a sum of -3, $2x^2 - 3x - 1$ is prime.

53. $9x^2 + 6x + 1 = (3x)^2 + 2 \cdot 3x \cdot 1 + 1^2$
$= (3x + 1)^2$

55. $9m^2 + 1$ is a sum of two squares and it is a prime polynomial.

57. $w^4 - z^4 = (w^2 - z^2)(w^2 + z^2)$
$= (w - z)(w + z)(w^2 + z^2)$

59. $6x^2y + xy - 2y = y(6x^2 + x - 2)$
$= y(3x + 2)(2x - 1)$

61. There is no pair of integers that has a product of -25 and a sum of 10. So the polynomial is prime.

63. $48a^2 - 24a + 3 = 3(16a^2 - 8a + 1)$
$= 3(4a - 1)^2$

65. $16m^2 - 4m - 2 = 2(8m^2 - 2m - 1)$
$= 2(4m + 1)(2m - 1)$

67. $s^4 - 16t^4$
$= (s^2 - 4t^2)(s^2 + 4t^2)$
$= (s - 2t)(s + 2t)(s^2 + 4t^2)$

69. $9a^2 + 24a + 16 = (3a)^2 + 2 \cdot 3a \cdot 4 + 4^2$
$= (3a + 4)^2$
71. $24x^2 - 26x + 6 = 2(12x^2 - 13x + 3)$
$= 2(3x - 1)(4x - 3)$
73. $3m^2 + 27 = 3(m^2 + 9)$
75. $3a^2 - 27a = 3a(a - 9)$
77. $8 - 2x^2 = 2(4 - x^2) = 2(2 - x)(2 + x)$

79. Since there is no pair of integers with a product of 4 and a sum of 0, $w^2 + 4t^2$ is prime.

81. $6x^3 - 5x^2 + 12x = x(6x^2 - 5x + 12)$
There is no pair of integers that has a product of 72 $(6 \cdot 12)$ and a sum of -5. So $6x^2 - 5x + 12$ is prime.
83. $a^3b - 4ab = ab(a^2 - 4)$
$= ab(a - 2)(a + 2)$
85. $x^3 + 2x^2 - 4x - 8$
$= x^2(x + 2) - 4(x + 2)$
$= (x^2 - 4)(x + 2)$
$= (x - 2)(x + 2)(x + 2)$
$= (x - 2)(x + 2)^2$
87. $-7m^3n - 28mn^3 = -7mn(m^2 + 4n^2)$
89. $2x^3 + 16 = 2(x^3 + 8)$
$= 2(x + 2)(x^2 - 2x + 4)$
91. $2w^4 - 16w = 2w(w^3 - 8)$
$= 2w(w - 2)(w^2 + 2w + 4)$
93. $3a^2w - 18aw + 27w = 3w(a^2 - 6a + 9)$
$= 3w(a - 3)^2$
95. $5x^2 - 500 = 5(x^2 - 100)$
$= 5(x - 10)(x + 10)$
97. $2m + 2n - wm - wn$
$= 2(m + n) - w(m + n)$
$= (2 - w)(m + n)$
99. $3x^4 + 3x = 3x(x^3 + 1)$
$= 3x(x + 1)(x^2 - x + 1)$
101. $4w^2 + 4w - 4 = 4(w^2 + w - 1)$
There is no pair of integers that has a product of -1 and a sum of 1. So $w^2 + w - 1$ is prime.
103. $a^4 + 7a^3 - 30a^2 = a^2(a^2 + 7a - 30)$
$= a^2(a + 10)(a - 3)$
105. $4aw^3 - 12aw^2 + 9aw$
$= aw(4w^2 - 12w + 9)$
$= aw(2w - 3)^2$
107. $t^2 + 6t + 9 = t^2 + 2 \cdot t \cdot 3 + 3^2$
$= (t + 3)^2$

109.
$$\begin{array}{r} x^2 + 8x + 15 \\ x + 2 \overline{\big) x^3 + 10x^2 + 31x + 30} \\ \underline{x^3 + 2x^2} \qquad\qquad\quad \\ 8x^2 + 31x \qquad \\ \underline{8x^2 + 16x} \qquad \\ 15x + 30 \\ \underline{15x + 30} \\ 0 \end{array}$$
$V = (x + 2)(x^2 + 8x + 15)$
$V = (x + 2)(x + 3)(x + 5)$
The new length is $x + 5$ centimeters and the new width is $x + 3$ centimeters.
111. If $a = 0$ and $b = 0$, then $(a + b)^3 = a^3 + b^3$. There are infinitely many other possibilities. If $a = 2$ and $b = 1$, then $(a + b)^3 \neq a^3 + b^3$. So $(a + b)^3$ is not equivalent to $a^3 + b^3$.

5.6 WARM-UPS

1. False, because the zero factor property only works for a product that is equal to zero.
2. False, because the equation $(x - 3)(x - 3) = 0$ has only one solution, 3.
3. True, because of the zero factor property.
4. True, because the area of a rectangle is the product of the length and the width.
5. True, because $(1 - 1)(1 + 4) = 0$ and $(-4 - 1)(-4 + 4) = 0$ are both correct.
6. False, because the Pythagorean theorem applies to right triangles only.
7. True, because $P = 2L + 2W = 2(L + W)$, the length plus the width is equal to one-half of the perimeter.
8. True, because $x(x - 1)(x - 2) = 0$ has three solutions: 0, 1, and 2.
9. True, because $0(0 - 2) = 0$ and $2(2 - 2) = 0$ are both correct.
10. False, because 3 is not a solution: $3(3 - 2)(3 + 5) = 0$ is not correct.

5.6 EXERCISES

1. A quadratic equation has the form $ax^2 + bx + c = 0$ with $a \neq 0$.
3. The zero factor property says that if $ab = 0$ then $a = 0$ or $b = 0$.

5. Dividing each side by a variable is not usually done because the variable might have a value of zero.

7. $(x+5)(x+4)=0$

$x+5=0$ or $x+4=0$

$x=-5$ or $x=-4$

The solutions to the equation are -4 and -5.

9. $(2x+5)(3x-4)=0$

$2x+5=0$ or $3x-4=0$

$2x=-5$ or $3x=4$

$x=-\frac{5}{2}$ or $x=\frac{4}{3}$

The solutions to the equation are $-\frac{5}{2}$ and $\frac{4}{3}$.

11. $x^2+3x+2=0$

$(x+2)(x+1)=0$

$x+2=0$ or $x+1=0$

$x=-2$ or $x=-1$

The solutions to the equation are -2 and -1.

13. $w^2-9w+14=0$

$(w-2)(w-7)=0$

$w-2=0$ or $w-7=0$

$w=2$ or $w=7$

The solutions to the equation are 2 and 7.

15. $y^2-2y-24=0$

$(y-6)(y+4)=0$

$y-6=0$ or $y+4=0$

$y=6$ or $y=-4$

The solutions to the equation are -4 and 6.

17. $2m^2+m-1=0$

$(2m-1)(m+1)=0$

$2m-1=0$ or $m+1=0$

$m=\frac{1}{2}$ or $m=-1$

The solutions to the equation are -1 and $\frac{1}{2}$.

19. $x^2=x$

$x^2-x=0$

$x(x-1)=0$

$x=0$ or $x-1=0$

$x=0$ or $x=1$

The solutions to the equation are 0 and 1.

21. $m^2=-7m$

$m^2+7m=0$

$m(m+7)=0$

$m=0$ or $m+7=0$

$m=0$ or $m=-7$

The solutions to the equation are 0 and -7.

23. $a^2+a=20$

$a^2+a-20=0$

$(a+5)(a-4)=0$

$a+5=0$ or $a-4=0$

$a=-5$ or $a=4$

The solutions to the equation are -5 and 4.

25. $2x^2+5x=3$

$2x^2+5x-3=0$

$(2x-1)(x+3)=0$

$2x-1=0$ or $x+3=0$

$2x=1$ or $x=-3$

$x=\frac{1}{2}$

The solutions to the equation are $\frac{1}{2}$ and -3.

27. $(x+2)(x+6)=12$

$x^2+8x+12=12$

$x^2+8x=0$

$x(x+8)=0$

$x=0$ or $x+8=0$

$x=-8$

The solutions to the equation are 0 and -8.

29. $(a+3)(2a-1)=15$

$2a^2+5a-3=15$

$2a^2+5a-18=0$

$(2a+9)(a-2)=0$

$2a+9=0$ or $a-2=0$

$2a=-9$ or $a=2$

$a=-\frac{9}{2}$ or $a=2$

The solutions to the equation are $-\frac{9}{2}$ and 2.

31. $2(4-5h)=3h^2$

$8-10h-3h^2=0$

$3h^2+10h-8=0$

$(3h-2)(h+4)=0$

$3h-2=0$ or $h+4=0$

$3h=2$ or $h=-4$

$h=\frac{2}{3}$ or $h=-4$

The solutions to the equation are $\frac{2}{3}$ and -4.

33. $2x^2+50=20x$

$2x^2-20x+50=0$

$x^2-10x+25=0$

$(x-5)^2=0$

$x-5=0$

$x=5$

The solution to the equation is 5.

35. $4m^2-12m+9=0$

$(2m-3)^2=0$

$2m-3=0$

$$m = \frac{3}{2}$$

The solution to the equation is $\frac{3}{2}$.

37. $x^3 - 9x = 0$

$x(x^2 - 9) = 0$

$x(x-3)(x+3) = 0$

$x = 0$ or $x - 3 = 0$ or $x + 3 = 0$

$x = 0$ or $x = 3$ or $x = -3$

The solutions to the equation are 0, −3, and 3.

39. $w^3 + 4w^2 - 4w = 16$

$w^3 + 4w^2 - 4w - 16 = 0$

$w^2(w+4) - 4(w+4) = 0$

$(w^2 - 4)(w+4) = 0$

$(w-2)(w+2)(w+4) = 0$

$w - 2 = 0$ or $w + 2 = 0$ or $w + 4 = 0$

$w = 2$ or $w = -2$ or $w = -4$

The solutions are −4, −2, and 2.

41. $n^3 - 3n^2 + 3 = n$

$n^3 - 3n^2 - n + 3 = 0$

$n^2(n-3) - 1(n-3) = 0$

$(n^2 - 1)(n-3) = 0$

$(n-1)(n+1)(n-3) = 0$

$n - 1 = 0$ or $n + 1 = 0$ or $n - 3 = 0$

$n = 1$ or $n = -1$ or $n = 3$

The solutions to the equation are −1, 1, and 3.

43. $y^3 - 9y^2 + 20y = 0$

$y(y^2 - 9y + 20) = 0$

$y(y-4)(y-5) = 0$

$y = 0$ or $y - 4 = 0$ or $y - 5 = 0$

$y = 0$ or $y = 4$ or $y = 5$

The solutions to the equation are 0, 4, and 5.

45. $x^2 - 16 = 0$

$(x-4)(x+4) = 0$

$x - 4 = 0$ or $x + 4 = 0$

$x = 4$ or $x = -4$

The solutions to the equation are −4 and 4.

47. $x^2 = 9$

$x^2 - 9 = 0$

$(x-3)(x+3) = 0$

$x - 3 = 0$ or $x + 3 = 0$

$x = 3$ or $x = -3$

The solutions to the equation are −3 and 3.

49. $a^3 = a$

$a^3 - a = 0$

$a(a^2 - 1) = 0$

$a(a-1)(a+1) = 0$

$a = 0$ or $a - 1 = 0$ or $a + 1 = 0$

$a = 0$ or $a = 1$ or $a = -1$

The solutions to the equation are 0, −1, and 1.

51. $3x^2 + 15x + 18 = 0$

$3(x^2 + 5x + 6) = 0$

$3(x+3)(x+2) = 0$

$x + 3 = 0$ or $x + 2 = 0$

$x = -3$ or $x = -2$

The solutions to the equation are −3 and −2.

53. $z^2 + \frac{11}{2}z = -6$

$2z^2 + 11z = -12$

$2z^2 + 11z + 12 = 0$

$(2z+3)(z+4) = 0$

$2z + 3 = 0$ or $z + 4 = 0$

$2z = -3$ or $z = -4$

$z = -\frac{3}{2}$

The solutions to the equation are $-\frac{3}{2}$ and −4.

55. $(t-3)(t+5) = 9$

$t^2 + 2t - 15 = 9$

$t^2 + 2t - 24 = 0$

$(t+6)(t-4) = 0$

$t + 6 = 0$ or $t - 4 = 0$

$t = -6$ or $t = 4$

The solutions to the equation are −6 and 4.

57. $(x-2)^2 + x^2 = 10$

$x^2 - 4x + 4 + x^2 = 10$

$2x^2 - 4x - 6 = 0$

$x^2 - 2x - 3 = 0$

$(x-3)(x+1) = 0$

$x - 3 = 0$ or $x + 1 = 0$

$x = 3$ or $x = -1$

The solutions to the equation are −1 and 3.

59. $\frac{1}{16}x^2 + \frac{1}{8}x = \frac{1}{2}$

$x^2 + 2x = 8$

$x^2 + 2x - 8 = 0$

$(x+4)(x-2) = 0$

$x + 4 = 0$ or $x - 2 = 0$

$x = -4$ or $x = 2$

The solutions to the equation are −4 and 2.

61. $a^3 + 3a^2 - 25a = 75$

$a^3 + 3a^2 - 25a - 75 = 0$

$a^2(a+3) - 25(a+3) = 0$

$(a^2 - 25)(a+3) = 0$

$(a-5)(a+5)(a+3) = 0$

$a - 5 = 0$ or $a + 5 = 0$ or $a + 3 = 0$

$a = 5$ or $a = -5$ or $a = -3$

The solutions are −5, −3 and 5.

63. If the perimeter is 34 feet, then the sum of the length and width is 17 feet. Let $x =$ the length and $17 - x =$ the width. Since the diagonal is the hypotenuse of a right triangle, we can use the Pythagorean theorem to write the following equation.

$$x^2 + (17 - x)^2 = 13^2$$
$$x^2 + 289 - 34x + x^2 = 169$$
$$2x^2 - 34x + 120 = 0$$
$$2(x^2 - 17x + 60) = 0$$
$$2(x - 12)(x - 5) = 0$$
$$x - 12 = 0 \quad \text{or} \quad x - 5 = 0$$
$$x = 12 \quad \text{or} \quad x = 5$$
$$17 - x = 5 \qquad 17 - x = 12$$

The length is 12 feet and the width is 5 feet.

65. Let $x =$ the width and $2x + 2 =$ the length. Since the diagonal is the hypotenuse of a right triangle, we can use the Pythagorean theorem to write the following equation.

$$x^2 + (2x + 2)^2 = 13^2$$
$$x^2 + 4x^2 + 8x + 4 = 169$$
$$5x^2 + 8x - 165 = 0$$
$$(5x + 33)(x - 5) = 0$$
$$5x + 33 = 0 \quad \text{or} \quad x - 5 = 0$$
$$5x = -33 \quad \text{or} \quad x = 5$$
$$x = -\frac{33}{5}$$

Since the width is a positive number, the width is 5 feet and the length $(2x + 2)$ is 12 feet.

67. Let $x =$ the first integer and $x + 1 =$ the second integer. The sum of their squares is 13, is expressed with the following equation.

$$x^2 + (x + 1)^2 = 13$$
$$x^2 + x^2 + 2x + 1 = 13$$
$$2x^2 + 2x - 12 = 0$$
$$x^2 + x - 6 = 0$$
$$(x + 3)(x - 2) = 0$$
$$x + 3 = 0 \quad \text{or} \quad x - 2 = 0$$
$$x = -3 \quad \text{or} \quad x = 2$$
$$x + 1 = -2 \qquad x + 1 = 3$$

The consecutive integers are 2 and 3, or -3 and -2.

69. Since the sum of the two numbers is 11, we can let $x =$ one number and $11 - x =$ the other number. From the fact that their product is 30 we can write the following equation.

$$x(11 - x) = 30$$
$$11x - x^2 = 30$$
$$-x^2 + 11x - 30 = 0$$
$$x^2 - 11x + 30 = 0$$
$$(x - 5)(x - 6) = 0$$
$$x - 5 = 0 \quad \text{or} \quad x - 6 = 0$$
$$x = 5 \quad \text{or} \quad x = 6$$
$$11 - x = 6 \qquad 11 - x = 5$$

The numbers are 5 and 6.

71. Three consecutive even integers are represented by x, $x + 2$, and $x + 4$, where x is the smallest of the three even integers.

$$x^2 + (x + 2)^2 + (x + 4)^2 = 116$$
$$x^2 + x^2 + 4x + 4 + x^2 + 8x + 16 = 116$$
$$3x^2 + 12x - 96 = 0$$
$$x^2 + 4x - 32 = 0$$
$$(x + 8)(x - 4) = 0$$
$$x + 8 = 0 \quad \text{or} \quad x - 4 = 0$$
$$x = -8 \quad \text{or} \quad x = 4$$
$$x + 2 = -6 \qquad x + 2 = 6$$
$$x + 4 = -4 \qquad x + 4 = 8$$

The integers are $-8, -6,$ and -4 or 4, 6, and 8.

73. Two consecutive integers are represented by x and $x + 1$, where x is the smaller integer.

$$x(x + 1) = x + x + 1 + 5$$
$$x^2 + x = 2x + 6$$
$$x^2 - x - 6 = 0$$
$$(x - 3)(x + 2) = 0$$
$$x - 3 = 0 \quad \text{or} \quad x + 2 = 0$$
$$x = 3 \quad \text{or} \quad x = -2$$
$$x + 1 = 4 \qquad x + 1 = -1$$

So the two integers are -2 and -1 or 3 and 4.

75. Two integers that differ by 5 are represented by x and $x + 5$, where x is the smaller integer.

$$x^2 + (x + 5)^2 = 53$$
$$x^2 + x^2 + 10x + 25 = 53$$
$$2x^2 + 10x - 28 = 0$$
$$x^2 + 5x - 14 = 0$$
$$(x - 2)(x + 7) = 0$$
$$x - 2 = 0 \quad \text{or} \quad x + 7 = 0$$
$$x = 2 \quad \text{or} \quad x = -7$$
$$x + 5 = 7 \qquad x + 5 = -2$$

So the two integers are -7 and -2 or 2 and 7.

77. Let x represent the length and $x + 6$ represent the width of the rectangle.
Since $A = LW$ for a rectangle, we can write the following equation.

$$x(x + 6) = 72$$

$$x^2 + 6x - 72 = 0$$
$$(x + 12)(x - 6) = 0$$
$$x + 12 = 0 \text{ or } \quad x - 6 = 0$$
$$x = -12 \text{ or } \quad x = 6$$
$$x + 6 = 12$$

So the length is 12 feet and the width is 6 feet.

79. Let x represent the length of the shorter leg and $x + 3$ represent the length of the longer leg. Use the Pythagorean theorem to write the following equation.

$$x^2 + (x + 3)^2 = 15^2$$
$$x^2 + x^2 + 6x + 9 = 225$$
$$2x^2 + 6x - 216 = 0$$
$$x^2 + 3x - 108 = 0$$
$$(x + 12)(x - 9) = 0$$
$$x + 12 = 0 \text{ or } \quad x - 9 = 0$$
$$x = -12 \text{ or } \quad x = 9$$
$$x + 3 = 12$$

So the legs are 9 meters and 12 meters in length.

81. a) $-16t^2 + 10{,}000 = 0$

$$t^2 - 625 = 0$$
$$(t - 25)(t + 25) = 0$$
$$t - 25 = 0 \text{ or } \quad t + 25 = 0$$
$$t = 25 \text{ or } \quad t = -25$$

It would take 25 sec to reach the ground.

b) The sky diver falls approximately 1000 feet in the first 5 seconds and approximately 3000 feet in the last 5 seconds. So the sky diver falls further in the last 5 seconds.

c) The velocity of the sky diver is increasing as she falls.

83. To find the time that it takes for the sandbag to reach the ground, we must solve the equation $0 = -16t^2 - 24t + 720$.

$$-16t^2 - 24t + 720 = 0$$
$$2t^2 + 3t - 90 = 0$$
$$(2t + 15)(t - 6) = 0$$
$$2t + 15 = 0 \text{ or } t - 6 = 0$$
$$t = -\frac{15}{2} \text{ or } \quad t = 6$$

Since the answer must be positive, the time it takes to reach the ground is 6 seconds.

85. Let $x =$ the length of the base and $2x + 1 =$ the height. Since the area of the triangle is 39 square inches, we can write the following equation.

$$\frac{1}{2}x(2x + 1) = 39$$
$$x^2 + \frac{1}{2}x = 39$$
$$2x^2 + x = 78$$
$$2x^2 + x - 78 = 0$$
$$(2x + 13)(x - 6) = 0$$
$$2x + 13 = 0 \quad \text{or} \quad x - 6 = 0$$
$$2x = -13 \text{ or } \quad x = 6$$
$$x = -\frac{13}{2}$$

Since the base must have a positive length, the length of the base is 6 inches and the height $(2x + 1)$ is 13 inches.

87. Let $x =$ the length of a side of the original square garden. The new garden will be $x - 5$ feet by $x - 8$ feet. Since the area of the new garden is 180 square feet, we can write the following equation.

$$(x - 5)(x - 8) = 180$$
$$x^2 - 13x + 40 = 180$$
$$x^2 - 13x - 140 = 0$$
$$(x - 20)(x + 7) = 0$$
$$x - 20 = 0 \text{ or } x + 7 = 0$$
$$x = 20 \text{ or } \quad x = -7$$

Since the length of a side of the garden must be positive, the original garden was 20 ft by 20 ft.

89. Let $x =$ the distance between Imelda and Gordon and $x + 20 =$ the altitude of the kite. The distance between the kite and Imelda is the length of the hypotenuse of a right triangle.

$$x^2 + (x + 20)^2 = 100^2$$
$$x^2 + x^2 + 40x + 400 = 10{,}000$$
$$2x^2 + 40x - 9600 = 0$$
$$x^2 + 20x - 4800 = 0$$
$$(x + 80)(x - 60) = 0$$
$$x + 80 = 0 \quad \text{or } x - 60 = 0$$
$$x = -80 \text{ or } \quad x = 60$$
$$x + 20 = 80$$

Since x must be positive, x is 60 and $x + 20$ is 80. The altitude is 80 feet.

91. Let $x =$ the width of one square room and $x + 3 =$ the width of the other square room. Since the total area of the two rooms is 45 square yards, we can write the following equation.

$$x^2 + (x + 3)^2 = 45$$
$$x^2 + x^2 + 6x + 9 = 45$$
$$2x^2 + 6x - 36 = 0$$
$$x^2 + 3x - 18 = 0$$
$$(x + 6)(x - 3) = 0$$

$x + 6 = 0$ or $x - 3 = 0$
$x = -6$ or $x = 3$
Since the width of each room is a positive number, one room is 3 yards by 3 yards and the other is 3 yards wider, 6 yards by 6 yards.
93. Let x and $17 - x$ represent the two legs of the right triangle.
$$x^2 + (17 - x)^2 = 13^2$$
$$x^2 + 289 - 34x + x^2 = 169$$
$$2x^2 - 34x + 120 = 0$$
$$x^2 - 17x + 60 = 0$$
$$(x - 5)(x - 12) = 0$$
$$x - 5 = 0 \text{ or } x - 12 = 0$$
$$x = 5 \text{ or } x = 12$$
The distance from A to B is 12 miles.
95.
$$16{,}000(1 + r)^2 = 25000$$
$$16000(1 + 2r + r^2) = 25000$$
$$16000r^2 + 32000r - 9000 = 0$$
$$16r^2 + 32r - 9 = 0$$
$$(4r + 9)(4r - 1) = 0$$
$$4r + 9 = 0 \text{ or } 4r - 1 = 0$$
$$r = -\frac{9}{4} \text{ or } r = \frac{1}{4} = 25\%$$
So the average annual return was 25%.

Enriching Your Mathematical Word Power
1. a **2.** d **3.** c **4.** a **5.** c
6. b **7.** c **8.** a **9.** d **10.** c

CHAPTER 5 REVIEW

1. $144 = 2 \cdot 72 = 2 \cdot 2 \cdot 36 = 2 \cdot 2 \cdot 2 \cdot 18$
$= 2 \cdot 2 \cdot 2 \cdot 2 \cdot 9 = 2 \cdot 2 \cdot 2 \cdot 2 \cdot 3 \cdot 3 = 2^4 \cdot 3^2$
3. $58 = 2 \cdot 29$
5. $150 = 2 \cdot 75 = 2 \cdot 3 \cdot 25 = 2 \cdot 3 \cdot 5 \cdot 5$
$= 2 \cdot 3 \cdot 5^2$
7. Since $36 = 2^2 \cdot 3^2$ and $90 = 2 \cdot 3^2 \cdot 5$, the GCF is $2 \cdot 3^2$ or 18.
9. Since the GCF for 8 and 12 is 4, the GCF is $4x$.
11. $3x + 6 = 3(x + 2)$
13. $2a - 20 = -2(-a + 10)$
15. $2a - a^2 = a(2 - a)$
17. $6x^2y^2 - 9x^5y = 3x^2y(2y - 3x^3)$
19. $3x^2y - 12xy - 9y^2 = 3y(x^2 - 4x - 3y)$
21. $y^2 - 400 = y^2 - 20^2 = (y - 20)(y + 20)$
23. $w^2 - 8w + 16 = w^2 - 2 \cdot w \cdot 4 + 4^2$
$= (w - 4)^2$
25. $4y^2 + 20y + 25 = (2y)^2 + 2 \cdot 2y \cdot 5 + 5^2$
$= (2y + 5)^2$
27. $r^2 - 4r + 4 = r^2 - 2 \cdot r \cdot 2 + 2^2$
$= (r - 2)^2$
29. $8t^3 - 24t^2 + 18t = 2t(4t^2 - 12t + 9)$
$= 2t(2t - 3)^2$
31. $x^2 + 12xy + 36y^2$
$= x^2 + 2 \cdot x \cdot 6y + (6y)^2$
$= (x + 6y)^2$
33. $x^2 + 5x - xy - 5y = x(x + 5) - y(x + 5)$
$= (x - y)(x + 5)$
35. Two numbers that have a product of -24 and a sum of 5 are 8 and -3.
$b^2 + 5b - 24 = (b + 8)(b - 3)$
37. Two numbers that have product of -60 and a sum of -4 are -10 and 6.
$r^2 - 4r - 60 = (r - 10)(r + 6)$
39. Two numbers that have a product of -55 and a sum of -6 are -11 and 5.
$y^2 - 6y - 55 = (y - 11)(y + 5)$
41. $u^2 + 26u + 120 = (u + 20)(u + 6)$
43. $3t^3 + 12t^2 = 3t^2(t + 4)$
45. $5w^3 + 25w^2 + 25w = 5w(w^2 + 5w + 5)$
47. $2a^3b + 3a^2b^2 + ab^3 = ab(2a^2 + 3ab + b^2)$
$= ab(2a + b)(a + b)$
49. $9x^3 - xy^2 = x(9x^2 - y^2)$
$= x(3x - y)(3x + y)$
51. Since $ac = 14 \cdot (-3) = -42$, we need two integers that have a product of -42 and a sum of 1. The integers are 7 and -6.
$14t^2 + t - 3 = 14t^2 + 7t - 6t - 3$
$= 7t(2t + 1) - 3(2t + 1)$
$= (7t - 3)(2t + 1)$
53. Since $ac = 6 \cdot (-7) = -42$, we need two integers that have a product of -42 and a sum of -19. The integers are 2 and -21.
$6x^2 - 19x - 7 = 6x^2 - 21x + 2x - 7$
$= 3x(2x - 7) + 1(2x - 7)$
$= (3x + 1)(2x - 7)$
55. Since $ac = 6 \cdot (-4) = -24$, we need two integers that have a product of -24 and a sum of 5. The numbers are 8 and -3.
$6p^2 + 5p - 4 = 6p^2 - 3p + 8p - 4$
$= 3p(2p - 1) + 4(2p - 1)$
$= (3p + 4)(2p - 1)$
57. $-30p^3 + 8p^2 + 8p = -2p(15p^2 - 4p - 4)$
$= -2p(5p + 2)(3p - 2)$
59. $6x^2 - 29xy - 5y^2 = (6x + y)(x - 5y)$

61. $32x^2 + 16xy + 2y^2 = 2(16x^2 + 8xy + y^2)$
$= 2(4x + y)^2$
63. $5x^3 + 40x = 5x(x^2 + 8)$
65. Since $ac = 9(-2) = -18$, we need two integers that have a product of -18 and a sum of 3. The integers are 6 and -3.
$9x^2 + 3x - 2 = 9x^2 + 6x - 3x - 2$
$= 3x(3x + 2) - 1(3x + 2)$
$= (3x - 1)(3x + 2)$
67. $n^2 + 64$ is a sum of two squares of the form $a^2 + b^2$ and so it is a prime polynomial.
69. $x^3 + 2x^2 - x - 2 = x^2(x + 2) - 1(x + 2)$
$= (x + 2)(x^2 - 1)$
$= (x + 2)(x - 1)(x + 1)$
71. $x^2y - 16xy^2 = xy(x - 16y)$
73. There are no integers that have a product of 5 and a sum of 4. So $w^2 + 4w + 5$ is a prime polynomial.
75. $a^2 + 2a + 1 = (a + 1)^2$
77. $x^3 - x^2 + x - 1 = x^2(x - 1) + 1(x - 1)$
$= (x^2 + 1)(x - 1)$
79. $a^2 + ab + 2a + 2b = a(a + b) + 2(a + b)$
$= (a + 2)(a + b)$

81. $-2x^2 + 16x - 24 = -2(x^2 - 8x + 12)$
$= -2(x - 6)(x - 2)$

83. $m^3 - 1000 = m^3 - 10^3$
$= (m - 10)(m^2 + 10m + 100)$

85. $p^4 - q^4 = (p^2 - q^2)(p^2 + q^2)$
$= (p - q)(p + q)(p^2 + q^2)$

87.
$$\begin{array}{r|l} & x^2 - 2x + 5 \\ \hline x + 2 & x^3 + 0x^2 + x + 10 \\ & \underline{x^3 + 2x^2} \\ & -2x^2 + x \\ & \underline{-2x^2 - 4x} \\ & 5x + 10 \\ & \underline{5x + 10} \\ & 0 \end{array}$$

$x^3 + x + 10 = (x + 2)(x^2 - 2x + 5)$

Since no pair of integers has a product of 5 and a sum of -2, $x^2 - 2x + 5$ is prime.

89.
$$\begin{array}{r|l} & x^2 + 2x - 15 \\ \hline x + 4 & x^3 + 6x^2 - 7x - 60 \\ & \underline{x^3 + 4x^2} \\ & 2x^2 - 7x \\ & \underline{2x^2 + 8x} \\ & -15x - 60 \\ & \underline{-15x - 60} \\ & 0 \end{array}$$

$x^3 + 6x^2 - 7x - 60 = (x + 4)(x^2 + 2x - 15)$
$= (x + 4)(x + 5)(x - 3)$

91. $x^3 = 5x^2$
$x^3 - 5x^2 = 0$
$x^2(x - 5) = 0$
$x^2 = 0$ or $x - 5 = 0$
$x = 0$ or $x = 5$
The solutions to the equation are 0 and 5.

93. $(a - 2)(a - 3) = 6$
$a^2 - 5a + 6 = 6$
$a^2 - 5a = 0$
$a(a - 5) = 0$
$a = 0$ or $a - 5 = 0$
$a = 5$
The solutions to the equation are 0 and 5.

95. $2m^2 - 9m - 5 = 0$
$(2m + 1)(m - 5) = 0$
$2m + 1 = 0$ or $m - 5 = 0$
$2m = -1$ or $m = 5$
$m = -\frac{1}{2}$
The solutions to the equation are $-\frac{1}{2}$ and 5.

97. $m^3 + 4m^2 - 9m = 36$
$m^3 + 4m^2 - 9m - 36 = 0$
$m^2(m + 4) - 9(m + 4) = 0$
$(m^2 - 9)(m + 4) = 0$
$(m - 3)(m + 3)(m + 4) = 0$
$m - 3 = 0$ or $m + 3 = 0$ or $m + 4 = 0$
$m = 3$ or $m = -3$ or $m = -4$
The solutions are -4, -3, and 3.

99. $(x + 3)^2 + x^2 = 5$
$x^2 + 6x + 9 + x^2 - 5 = 0$
$2x^2 + 6x + 4 = 0$
$x^2 + 3x + 2 = 0$
$(x + 2)(x + 1) = 0$
$x + 2 = 0$ or $x + 1 = 0$
$x = -2$ or $x = -1$
The solutions to the equation are -2 and -1.

101. $p^2 + \frac{1}{4}p - \frac{1}{8} = 0$

$$8p^2 + 2p - 1 = 0$$
$$(4p - 1)(2p + 1) = 0$$
$$4p - 1 = 0 \text{ or } 2p + 1 = 0$$
$$p = \frac{1}{4} \text{ or } \quad p = -\frac{1}{2}$$

The solutions to the equation are $-\frac{1}{2}$ and $\frac{1}{4}$.

103. Since the numbers differ by 6, we can let $x =$ one number and $x + 6 =$ the other. Since their squares differ by 96, we can write the following equation.

$$(x + 6)^2 - x^2 = 96$$
$$x^2 + 12x + 36 - x^2 = 96$$
$$12x + 36 = 96$$
$$12x = 60$$
$$x = 5$$
$$x + 6 = 11$$

The numbers are 5 and 11.

105. If the perimeter is 28 inches, then the sum of the length and width is 14. Let $x =$ the width and $14 - x =$ the length. The diagonal is the hypotenuse of a right triangle.

$$x^2 + (14 - x)^2 = 10^2$$
$$x^2 + 196 - 28x + x^2 = 100$$
$$2x^2 - 28x + 96 = 0$$
$$x^2 - 14x + 48 = 0$$
$$(x - 6)(x - 8) = 0$$
$$x - 6 = 0 \quad \text{or} \quad x - 8 = 0$$
$$x = 6 \quad \text{or} \quad x = 8$$
$$14 - x = 8 \qquad 14 - x = 6$$

The notebook is 6 inches by 8 inches.

107. $v = kR^2 - kr^2$

$$v = k(R^2 - r^2)$$
$$v = k(R - r)(R + r)$$

109. Let x be the distance from the top of the ladder to the ground and $x - 2$ be the distance from the bottom of the ladder to the building.

$$x^2 + (x - 2)^2 = 10^2$$
$$x^2 + x^2 - 4x + 4 = 100$$
$$2x^2 - 4x - 96 = 0$$
$$x^2 - 2x - 48 = 0$$
$$(x - 8)(x + 6) = 0$$
$$x - 8 = 0 \text{ or } x + 6 = 0$$
$$x = 8 \text{ or } \quad x = -6$$

If $x = 8$, then $x - 2 = 6$. The distance from the bottom of the ladder to the building is 6 feet.

CHAPTER 5 TEST

1. $66 = 6 \cdot 11 = 2 \cdot 3 \cdot 11$

2. $336 = 2 \cdot 168 = 2 \cdot 2 \cdot 84 = 2 \cdot 2 \cdot 2 \cdot 42$
$= 2 \cdot 2 \cdot 2 \cdot 2 \cdot 21 = 2^4 \cdot 3 \cdot 7$

3. Since $48 = 2^4 \cdot 3$ and $80 = 2^4 \cdot 5$, the GCF is 2^4 or 16.

4. Since $42 = 2 \cdot 3 \cdot 7$, $66 = 2 \cdot 3 \cdot 11$, and $78 = 2 \cdot 3 \cdot 13$, the GCF of 42, 66, and 78 is $2 \cdot 3$ or 6.

5. The GCF for $6y^2$ and $15y^3$ is $3y^2$.

6. The GCF for $12a^2b$, $18ab^2$, and $24a^3b^3$ is $6ab$.

7. $5x^2 - 10x = 5x(x - 2)$

8. $6x^2y^2 + 12xy^2 + 12y^2 = 6y^2(x^2 + 2x + 2)$
Since no pair of integers has a product of 2 and a sum of 2, $x^2 + 2x + 2$ is prime.

9. $3a^3b - 3ab^3 = 3ab(a^2 - b^2)$
$= 3ab(a - b)(a + b)$

10. Two numbers that have a product of -24 and a sum of 2 are 6 and -4.
$a^2 + 2a - 24 = (a + 6)(a - 4)$

11. $4b^2 - 28b + 49 = (2b)^2 - 2 \cdot 2b \cdot 7 + 7^2$
$= (2b - 7)^2$

12. $3m^3 + 27m = 3m(m^2 + 9)$ Since $m^2 + 9$ is a sum of two squares, it is prime.

13. $ax - ay + bx - by = a(x - y) + b(x - y)$
$= (a + b)(x - y)$

14. $ax - 2a - 5x + 10 = a(x - 2) - 5(x - 2)$
$= (a - 5)(x - 2)$

15. Since $ac = 6(-5) = -30$, we need two integers that have a product of -30 and a sum of -7. The integers are -10 and 3.
$6b^2 - 7b - 5 = 6b^2 + 3b - 10b - 5$
$= 3b(2b + 1) - 5(2b + 1) = (3b - 5)(2b + 1)$

16. $m^2 + 4mn + 4n^2$
$= m^2 + 2 \cdot m \cdot 2n + (2n)^2 = (m + 2n)^2$

17. Since $ac = 2(15) = 30$, we need two integers that have a product of 30 and a sum of -13. The integers are -10 and -3.
$2a^2 - 13a + 15 = 2a^2 - 10a - 3a + 15$
$= 2a(a - 5) - 3(a - 5)$
$= (2a - 3)(a - 5)$

18. $z^3 + 9z^2 + 18z = z(z^2 + 9z + 18)$
$= z(z + 3)(z + 6)$

19. $x^3 + 125 = (x + 5)(x^2 - 5x + 25)$

20. $a^4 - ab^3 = a(a^3 - b^3)$
$= a(a - b)(a^2 + ab + b^2)$

21. $$\begin{array}{r} x^2 - 5x + 6 \\ x - 1 \overline{) x^3 - 6x^2 + 11x - 6} \\ \underline{x^3 - x^2} \\ -5x^2 + 11x \\ \underline{-5x^2 + 5x} \\ 6x - 6 \\ \underline{6x - 6} \\ 0 \end{array}$$

$$x^3 - 6x^2 + 11x - 6 = (x-1)(x^2 - 5x + 6) = (x-1)(x-2)(x-3)$$

22. $$\begin{aligned} x^2 + 6x + 9 &= 0 \\ (x+3)^2 &= 0 \\ x + 3 &= 0 \\ x &= -3 \end{aligned}$$

The solution to the equation is -3.

23. $$\begin{aligned} 2x^2 + 5x - 12 &= 0 \\ 2x^2 + 8x - 3x - 12 &= 0 \\ 2x(x+4) - 3(x+4) &= 0 \\ (2x-3)(x+4) &= 0 \\ 2x - 3 = 0 \text{ or } x + 4 &= 0 \\ 2x = 3 \text{ or } \quad x &= -4 \\ x = \tfrac{3}{2} \end{aligned}$$

The solutions to the equation are $\frac{3}{2}$ and -4.

24. $$\begin{aligned} 3x^3 &= 12x \\ 3x^3 - 12x &= 0 \\ 3x(x^2 - 4) &= 0 \\ 3x(x-2)(x+2) &= 0 \\ 3x = 0 \text{ or } x - 2 = 0 \text{ or } x + 2 &= 0 \\ x = 0 \text{ or } \quad x = 2 \text{ or } \quad x &= -2 \end{aligned}$$

The solutions to the equation are 0, -2, and 2.

25. $$\begin{aligned} (2x-1)(3x+5) &= 5 \\ 6x^2 + 7x - 5 &= 5 \\ 6x^2 + 7x - 10 &= 0 \\ (6x-5)(x+2) &= 0 \\ 6x - 5 = 0 \text{ or } x + 2 &= 0 \\ x = \tfrac{5}{6} \text{ or } \quad x &= -2 \end{aligned}$$

The solutions to the equation are -2 and $\frac{5}{6}$.

26. Let $x =$ the width and $x + 3 =$ the length. Since the diagonal is the hypotenuse of a right triangle, we can write the following equation.

$$\begin{aligned} x^2 + (x+3)^2 &= 15^2 \\ x^2 + x^2 + 6x + 9 &= 225 \\ 2x^2 + 6x - 216 &= 0 \\ x^2 + 3x - 108 &= 0 \\ (x-9)(x+12) &= 0 \\ x - 9 = 0 \quad \text{or } x + 12 &= 0 \\ x = 9 \quad \text{or} \qquad x &= -12 \\ x + 3 = 12 \end{aligned}$$

The length is 12 feet and the width is 9 feet.

27. Let x and $4 - x$ represent the numbers.

$$\begin{aligned} x(4-x) &= -32 \\ -x^2 + 4x + 32 &= 0 \\ x^2 - 4x - 32 &= 0 \\ (x-8)(x+4) &= 0 \\ x - 8 = 0 \text{ or } x + 4 &= 0 \\ x = 8 \text{ or } \quad x &= -4 \end{aligned}$$

If $x = 8$, then $4 - x = 4 - 8 = -4$. If $x = -4$, then $4 - x = 4 - (-4) = 8$. The numbers are -4 and 8.

Making Connections

Chapters 1 - 5

1. $\dfrac{91 - 17}{17 - 91} = \dfrac{-1(17 - 91)}{17 - 91} = -1$

2. $\dfrac{4 - 18}{-6 - 1} = \dfrac{-14}{-7} = 2$

3. $5 - 2(7 - 3) = 5 - 2(4) = 5 - 8 = -3$

4. $3^2 - 4(6)(-2) = 9 - (-48) = 9 + 48 = 57$

5. $2^5 - 2^4 = 32 - 16 = 16$

6. $0.07(37) + 0.07(63) = 0.07(37 + 63) = 0.07(100) = 7$

7. $x \cdot 2x = 2 \cdot x \cdot x = 2x^2$

8. $x + 2x = (1 + 2)x = 3x$

9. $\dfrac{6 + 2x}{2} = \dfrac{6}{2} + \dfrac{2x}{2} = 3 + x$

10. $\dfrac{6 \cdot 2x}{2} = 6x \cdot \dfrac{2}{2} = 6x \cdot 1 = 6x$

11. $2 \cdot 3y \cdot 4z = 2 \cdot 3 \cdot 4yz = 24yz$

12. $2(3y + 4z) = 2 \cdot 3y + 2 \cdot 4z = 6y + 8z$

13. $2 - (3 - 4z) = 2 - 3 + 4z = 4z - 1$

14. $t^8 \div t^2 = t^{8-2} = t^6$

15. $t^8 \cdot t^2 = t^{8+2} = t^{10}$

16. $\dfrac{8t^8}{2t^2} = 4t^{8-2} = 4t^6$

17. $$\begin{aligned} 2x - 5 &> 3x + 4 \\ -5 &> x + 4 \\ -9 &> x \\ x &< -9 \end{aligned}$$

18. $$\begin{aligned} 4 - 5x &\le -11 \\ -5x &\le -15 \\ \frac{-5x}{-5} &\ge \frac{-15}{-5} \\ x &\ge 3 \end{aligned}$$

1 2 3 4 5 6 7

19. $-\frac{2}{3}x + 3 < -5$

$-2x + 9 < -15$

$-2x < -24$

$x > 12$

10 11 12 13 14 15 16

20. $0.05(x - 120) - 24 < 0$

$0.05x - 6 - 24 < 0$

$0.05x < 30$

$x < 600$

0 200 400 600 800

21. $2x - 3 = 0$

$2x = 3$

$x = \frac{3}{2}$

The solution set to the equation is $\left\{\frac{3}{2}\right\}$.

22. $2x + 1 = 0$

$2x = -1$

$x = -\frac{1}{2}$

The solution set to the equation is $\left\{-\frac{1}{2}\right\}$.

23. $(x - 3)(x + 5) = 0$

$x - 3 = 0$ or $x + 5 = 0$

$x = 3$ or $x = -5$

The solution set to the equation is $\{3, -5\}$.

24. $(2x - 3)(2x + 1) = 0$

$2x - 3 = 0$ or $2x + 1 = 0$

$2x = 3$ or $2x = -1$

$x = \frac{3}{2}$ or $x = -\frac{1}{2}$

The solution to the equation is $\left\{\frac{3}{2}, -\frac{1}{2}\right\}$.

25. $3x(x - 3) = 0$

$3x = 0$ or $x - 3 = 0$

$x = 0$ or $x = 3$

The solution set to the equation is $\{0, 3\}$.

26. $x^2 = x$

$x^2 - x = 0$

$x(x - 1) = 0$

$x = 0$ or $x - 1 = 0$

$x = 0$ or $x = 1$

The solution set to the equation is $\{0, 1\}$.

27. Since $3x - 3x = 0$ is equivalent to $0 = 0$, the solution set is $(-\infty, \infty)$, or R.

28. Since $3x - 3x = 1$ is equivalent to $0 = 1$, the solution set is the empty set, $\emptyset$.

29. $0.01x - x + 14.9 = 0.5x$

$x - 100x + 1490 = 50x$

$-149x = -1490$

$x = 10$

The solution set to the equation is $\{10\}$.

30. $0.05x + 0.04(x - 40) = 2$

$0.05x + 0.04x - 1.6 = 2$

$0.09x = 3.6$

$9x = 360$

$x = 40$

The solution set to the equation is $\{40\}$.

31. $2x^2 - 18 = 0$

$2(x^2 - 9) = 0$

$2(x - 3)(x + 3) = 0$

$x - 3 = 0$ or $x + 3 = 0$

$x = 3$ or $x = -3$

The solution set to the equation is $\{-3, 3\}$.

32. $2x^2 + 7x - 15 = 0$

$(2x - 3)(x + 5) = 0$

$2x - 3 = 0$ or $x + 5 = 0$

$2x = 3$ or $x = -5$

$x = \frac{3}{2}$

The solution set to the equation is $\left\{-5, \frac{3}{2}\right\}$.

33. If the perimeter is 69 feet, then the sum of the length and width is 34.5 feet. So if w is the width, then the length is $34.5 - w$.

$w(34.5 - w) = 283.5$

$34.5w - w^2 - 283.5 = 0$

$w^2 - 34.5w + 283.5 = 0$

$2w^2 - 69w + 567 = 0$

To factor by the ac method we need two numbers that have a product of $2 \cdot 567 = 1134$ and a sum of -69. Use a calculator to try possible divisors of 1134 until you find that the numbers are -42 and -27.

$2w^2 - 42w - 27w + 567 = 0$

$2w(w - 21) - 27(w - 21) = 0$

$(2w - 27)(w - 21) = 0$

$2w - 27 = 0$ or $w - 21 = 0$

$w = 13.5$ or $w = 21$

If $w = 13.5$, then the length is $34.5 - 13.5 = 21$. If $w = 21$, then the length is $34.5 - 21 = 13.5$. So the length is 21 feet and the width is 13.5 feet.

6.1 WARM-UPS

1. False, because 3003 is an odd number and it does not have a factor of 2.
2. True, because 2, 3, and 5 are prime numbers and $2^3 \cdot 3 \cdot 5 = 8 \cdot 15 = 120$.
3. True, because we cannot get zero in the denominator in the expression $\frac{x-2}{5}$.
4. False, because replacing x by -1 gives us $\frac{x+1}{x-3} = \frac{-1+1}{-1-3} = 0$.
5. False, because the two's cannot be divided out since 2 is not a factor of the numerator.
6. True, because 2 is a factor of both the numerator and denominator.
7. True, because of the quotient rule.
8. False, because the expression is already in lowest terms.
9. False, because $a - b = -1(b - a)$ and their quotient is -1.
10. True, because

$$\frac{-3x-6}{x+2} = \frac{-3(x+2)}{x+2} = -3.$$

6.1 EXERCISES

1. A rational number is a ratio of two integers with the denominator not 0.
3. A rational number is reduced to lowest terms by dividing the numerator and denominator by the GCF.
5. The quotient rule is used in reducing ratios of monomials.
7. Replace x with -2:

$$\frac{3(-2)-3}{-2+5} = \frac{-9}{3} = -3$$

9. Replace x with 3:

$$R(3) = \frac{2(3)+9}{3} = \frac{15}{3} = 5$$

11. Replace x with 2, -4, -3.02, and -2.96:

$$R(2) = \frac{2-5}{2+3} = \frac{-3}{5} = -0.6$$
$$R(-4) = \frac{-4-5}{-4+3} = \frac{-9}{-1} = 9$$
$$R(-3.02) = \frac{-3.02-5}{-3.02+3} = 401$$
$$R(-2.96) = \frac{-2.96-5}{-2.96+3} = -199$$

13. We cannot use numbers that cause the denominator to have a value of 0 and $x + 1 = 0$ if $x = -1$. We cannot use -1 in place of x.
15. The denominator has a value of 0, if $3a - 5 = 0$ or $a = 5/3$. So we cannot use $5/3$ in place of a.
17. We cannot use numbers that cause the denominator to have a value of 0.

$$x^2 - 16 = 0$$
$$(x-4)(x+4) = 0$$
$$x - 4 = 0 \text{ or } x + 4 = 0$$
$$x = 4 \text{ or } x = -4$$

We cannot use -4 or 4 in place of x.
19. Since the denominator is 2, the denominator is never zero. So any number can be used in place of p in the expression.

21. $\frac{6}{27} = \frac{2 \cdot 3}{3 \cdot 3 \cdot 3} = \frac{2}{9}$

23. $\frac{42}{90} = \frac{2 \cdot 3 \cdot 7}{2 \cdot 3 \cdot 3 \cdot 5} = \frac{7}{15}$

25. $\frac{36a}{90} = \frac{2 \cdot 2 \cdot 3 \cdot 3a}{2 \cdot 3 \cdot 3 \cdot 5} = \frac{2a}{5}$

27. $\frac{78}{30w} = \frac{2 \cdot 3 \cdot 13}{2 \cdot 3 \cdot 5w} = \frac{13}{5w}$

29. $\frac{6x+2}{6} = \frac{2(3x+1)}{2 \cdot 3} = \frac{3x+1}{3}$

31. $\frac{2x+4y}{6y+3x} = \frac{2(x+2y)}{3(x+2y)} = \frac{2}{3}$

33. $\frac{w^2-49}{w+7} = \frac{(w-7)(w+7)}{w+7} = w - 7$

35. $\frac{a^2-1}{a^2+2a+1} = \frac{(a-1)(a+1)}{(a+1)^2} = \frac{a-1}{a+1}$

37. $\frac{2x^2+4x+2}{4x^2-4} = \frac{2(x+1)^2}{4(x-1)(x+1)}$

$$= \frac{x+1}{2(x-1)}$$

39. $\frac{3x^2+18x+27}{21x+63} = \frac{3(x+3)^2}{21(x+3)} = \frac{x+3}{7}$

41. $\frac{x^{10}}{x^7} = x^{10-7} = x^3$

43. $\frac{z^3}{z^8} = \frac{1}{z^{8-3}} = \frac{1}{z^5}$

45. $\frac{4x^7}{-2x^5} = -2x^{7-5} = -2x^2$

47. $\frac{-12m^9n^{18}}{8m^6n^{16}} = \frac{-3 \cdot 4m^{9-6}n^{18-16}}{2 \cdot 4}$

$$= \frac{-3m^3n^2}{2}$$

49. $\frac{6b^{10}c^4}{-8b^{10}c^7} = \frac{2 \cdot 3b^{10-10}}{-2 \cdot 4c^{7-4}} = \frac{-3}{4c^3}$

51. $\frac{30a^3bc}{18a^7b^{17}} = \frac{6 \cdot 5c}{6 \cdot 3a^{7-3}b^{17-1}} = \frac{5c}{3a^4b^{16}}$

53. $\frac{210}{264} = \frac{2 \cdot 3 \cdot 5 \cdot 7}{2 \cdot 2 \cdot 2 \cdot 3 \cdot 11} = \frac{35}{44}$

55. $\frac{231}{168}=\frac{3\cdot7\cdot11}{2\cdot2\cdot2\cdot3\cdot7}=\frac{11}{8}$

57. $\frac{630x^5}{300x^9}=\frac{30\cdot21}{30\cdot10x^{9-5}}=\frac{21}{10x^4}$

59. $\frac{924a^{23}}{448a^{19}}=\frac{2\cdot2\cdot3\cdot7\cdot11a^{23-19}}{2\cdot2\cdot2\cdot2\cdot2\cdot2\cdot7}=\frac{33a^4}{16}$

61. $\frac{3a-2b}{2b-3a}=\frac{-1(2b-3a)}{2b-3a}=-1$

63. $\frac{h^2-t^2}{t-h}=\frac{(h-t)(h+t)}{-1(h-t)}=\frac{h+t}{-1}$
$=-h-t$

65. $\frac{2g-6h}{9h^2-g^2}=\frac{-2(3h-g)}{(3h-g)(3h+g)}=\frac{-2}{3h+g}$

67. $\frac{x^2-x-6}{9-x^2}=\frac{(x-3)(x+2)}{(3-x)(3+x)}$
$=\frac{-1(x+2)}{3+x}=\frac{-x-2}{x+3}$

69. $\frac{-x-6}{x+6}=\frac{-1(x+6)}{x+6}=-1$

71. $\frac{-2y-6y^2}{3+9y}=\frac{-2y(1+3y)}{3(1+3y)}=\frac{-2y}{3}$

73. $\frac{-3x-6}{3x-6}=\frac{-3(x+2)}{3(x-2)}=\frac{x+2}{-1(x-2)}$
$=\frac{x+2}{2-x}$

75. $\frac{-12a-6}{2a^2+7a+3}=\frac{-6(2a+1)}{(2a+1)(a+3)}$
$=\frac{-6}{a+3}$

77. $\frac{2x^{12}}{4x^8}=\frac{2x^{12-8}}{2\cdot2}=\frac{x^4}{2}$

79. $\frac{2x+4}{4x}=\frac{2(x+2)}{2(2x)}=\frac{x+2}{2x}$

81. $\frac{a-4}{4-a}=\frac{-1(4-a)}{4-a}=-1$

83. $\frac{2c-4}{4-c^2}=\frac{-2(2-c)}{(2-c)(2+c)}=\frac{-2}{c+2}$

85. $\frac{x^2+4x+4}{x^2-4}=\frac{(x+2)^2}{(x-2)(x+2)}=\frac{x+2}{x-2}$

87. $\frac{-2x-4}{x^2+5x+6}=\frac{-2(x+2)}{(x+2)(x+3)}=\frac{-2}{x+3}$

89. $\frac{2q^8+q^7}{2q^6+q^5}=\frac{q^7(2q+1)}{q^5(2q+1)}=q^{7-5}=q^2$

91. $\frac{u^2-6u-16}{u^2-16u+64}=\frac{(u-8)(u+2)}{(u-8)^2}$
$=\frac{u+2}{u-8}$

93. $\frac{a^3-8}{2a-4}=\frac{(a-2)(a^2+2a+4)}{2(a-2)}$
$=\frac{a^2+2a+4}{2}$

95. $\frac{y^3-2y^2-4y+8}{y^2-4y+4}=\frac{(y^2-4)(y-2)}{(y-2)^2}$
$=\frac{(y-2)(y+2)(y-2)}{(y-2)^2}=y+2$

97. Since $D=RT$, we have $T=D/R$. So Sergio's time was $\frac{300}{x+10}$ hours.

99. The cost per pound is found by dividing the total cost by the number of pounds. So if $x+4$ pounds cost \$4.50, then the cost per pound is $\frac{4.50}{x+4}$ dollars per pound.

101. If she could clean the whole pool in 3 hours, she would be cleaning 1/3 of the pool per hour. So if she cleans the whole pool in x hours, she cleans $\frac{1}{x}$ of the pool per hour.

103. a) From the graph it appears that the cost per report for printing 1000 reports is \$0.75

b) $C(1000)=\frac{150+0.60(1000)}{1000}=\$0.75$

$C(5000)=\frac{150+0.60(5000)}{1000}=\$0.63$

$C(10{,}000)=\frac{150+0.60(10000)}{10000}=\$0.615$

c) It approaches \$0.60

6.2 WARM-UPS

1. True, because $\frac{a}{b}\cdot\frac{c}{d}=\frac{ac}{bd}$ if $b\neq0$ and $d\neq0$.

2. True, because
$\frac{x-7}{3}\cdot\frac{6}{7-x}=\frac{-1(7-x)}{3}\cdot\frac{2\cdot3}{7-x}=-2.$

3. True, because of the definition of division of fractions.

4. False, because if $x=1$ we get $3\div1=\frac{1}{3}\cdot1$, which is incorrect.

5. True, because we must factor before we can divide out the common factors.

6. False, because $\frac{1}{2}\cdot\frac{1}{4}=\frac{1}{8}$.

7. False, because $\frac{1}{2}\div3=\frac{1}{2}\cdot\frac{1}{3}=\frac{1}{6}$.

8. True, because $\frac{839-487}{487-839}$
$=\frac{-1(487-839)}{487-839}=-1.$

9. True, because $\frac{a}{3}\div3=\frac{a}{3}\cdot\frac{1}{3}=\frac{a}{9}$.

10. True, because $\frac{a}{b}\cdot\frac{b}{a}=\frac{ab}{ab}=1$.

6.2 EXERCISES

1. Rational numbers are multiplied by multiplying their numerators and their denominators.

3. Reducing can be done before multiplying rational numbers or expressions.

5. $\frac{2}{3} \cdot \frac{5}{6} = \frac{2}{3} \cdot \frac{5}{2 \cdot 3} = \frac{5}{9}$

7. $\frac{8}{15} \cdot \frac{35}{24} = \frac{2 \cdot 2 \cdot 2}{3 \cdot 5} \cdot \frac{5 \cdot 7}{2 \cdot 2 \cdot 2 \cdot 3} = \frac{7}{9}$

9. $\frac{12}{17} \cdot \frac{51}{10} = \frac{2 \cdot 6}{17} \cdot \frac{3 \cdot 17}{2 \cdot 5} = \frac{18}{5}$

11. $24 \cdot \frac{7}{20} = 4 \cdot 6 \cdot \frac{7}{4 \cdot 5} = \frac{42}{5}$

13. $\frac{2x}{3} \cdot \frac{5}{4x} = \frac{2x}{3} \cdot \frac{5}{2 \cdot 2x} = \frac{5}{6}$

15. $\frac{5a}{12b} \cdot \frac{3ab}{55a} = \frac{5a}{2 \cdot 2 \cdot 3b} \cdot \frac{3ab}{5 \cdot 11a} = \frac{a}{44}$

17. $\frac{-2x^6}{7a^5} \cdot \frac{21a^2}{6x} = \frac{-2x^6}{7a^5} \cdot \frac{3 \cdot 7a^2}{2 \cdot 3x} = \frac{-x^5}{a^3}$

19. $\frac{15t^3y^5}{20w^7} \cdot 24t^5w^3y^2$

$= \frac{3 \cdot 5t^3y^5}{2 \cdot 2 \cdot 5w^7} \cdot 2^3 \cdot 3t^5w^3y^2$

$= \frac{18t^8y^7}{w^4}$

21. $\frac{2x+2y}{7} \cdot \frac{15}{6x+6y}$

$= \frac{2(x+y)}{7} \cdot \frac{3 \cdot 5}{2 \cdot 3(x+y)} = \frac{5}{7}$

23. $\frac{3a+3b}{15} \cdot \frac{10a}{a^2-b^2}$

$= \frac{3(a+b)}{3 \cdot 5} \cdot \frac{2 \cdot 5a}{(a-b)(a+b)} = \frac{2a}{a-b}$

25. $(x^2-6x+9) \cdot \frac{3}{x-3} = (x-3)^2 \cdot \frac{3}{x-3}$

$= 3(x-3) = 3x-9$

27. $\frac{16a+8}{5a^2+5} \cdot \frac{2a^2+a-1}{4a^2-1}$

$= \frac{8(2a+1)}{5(a^2+1)} \cdot \frac{(2a-1)(a+1)}{(2a-1)(2a+1)}$

$= \frac{8(a+1)}{5(a^2+1)} = \frac{8a+8}{5(a^2+1)}$

29. $\frac{1}{4} \div \frac{1}{2} = \frac{1}{2 \cdot 2} \cdot \frac{2}{1} = \frac{1}{2}$

31. $12 \div \frac{2}{5} = 12 \cdot \frac{5}{2} = 30$

33. $\frac{5}{7} \div \frac{15}{14} = \frac{5}{7} \cdot \frac{14}{15} = \frac{2}{3}$

35. $\frac{40}{3} \div 12 = \frac{40}{3} \cdot \frac{1}{12} = \frac{2^3 \cdot 5}{3 \cdot 2^2 \cdot 3} = \frac{10}{9}$

37. $\frac{x^2}{4} \div \frac{x}{2} = \frac{x^2}{2 \cdot 2} \cdot \frac{2}{x} = \frac{x}{2}$

39. $\frac{5x^2}{3} \div \frac{10x}{21} = \frac{5x^2}{3} \cdot \frac{3 \cdot 7}{2 \cdot 5x} = \frac{7x}{2}$

41. $\frac{8m^3}{n^4} \div (12mn^2) = \frac{2^3m^3}{n^4} \cdot \frac{1}{2^2 \cdot 3mn^2}$

$= \frac{2m^2}{3n^6}$

43. $\frac{y-6}{2} \div \frac{6-y}{6} = \frac{y-6}{2} \cdot \frac{6}{6-y}$

$= \frac{-1(6-y)}{2} \cdot \frac{2 \cdot 3}{6-y} = -3$

45. $\frac{x^2+4x+4}{8} \div \frac{(x+2)^3}{16}$

$= \frac{(x+2)^2}{8} \cdot \frac{16}{(x+2)^3} = \frac{2}{x+2}$

47. $\frac{t^2+3t-10}{t^2-25} \div (4t-8)$

$= \frac{t^2+3t-10}{t^2-25} \cdot \frac{1}{4t-8}$

$= \frac{(t+5)(t-2)}{(t-5)(t+5)} \cdot \frac{1}{4(t-2)}$

$= \frac{1}{4(t-5)} = \frac{1}{4(t-5)}$

49. $(2x^2-3x-5) \div \frac{2x-5}{x-1}$

$= (2x-5)(x+1) \cdot \frac{x-1}{2x-5} = x^2-1$

51. $\frac{\frac{x-2y}{5}}{\frac{1}{10}} = \frac{x-2y}{5} \cdot \frac{10}{1}$

$= 2(x-2y) = 2x-4y$

53. $\frac{\frac{x^2-4}{12}}{\frac{x-2}{6}} = \frac{x^2-4}{12} \cdot \frac{6}{x-2}$

$= \frac{(x-2)(x+2)}{2 \cdot 6} \cdot \frac{6}{x-2}$

$= \frac{x+2}{2}$

55. $\frac{\frac{x^2+9}{3}}{5} = \frac{x^2+9}{3} \cdot \frac{1}{5} = \frac{x^2+9}{15}$

57. $\frac{\frac{x^2-y^2}{\frac{x-y}{9}}}{} = (x^2-y^2)\cdot\frac{9}{x-y}$
$= (x-y)(x+y)\cdot\frac{9}{x-y} = 9(x+y)$
$= 9x + 9y$

59. $\frac{x-1}{3}\cdot\frac{9}{1-x} = \frac{-1(1-x)3\cdot 3}{3(1-x)} = -3$

61. $\frac{3a+3b}{a}\cdot\frac{1}{3} = \frac{3(a+b)}{a}\cdot\frac{1}{3} = \frac{a+b}{a}$

63. $\frac{\frac{b}{a}}{\frac{1}{2}} = \frac{b}{a}\cdot\frac{2}{1} = \frac{2b}{a}$

65. $\frac{6y}{3}\div(2x) = \frac{6y}{3}\cdot\frac{1}{2x} = \frac{6y}{6x} = \frac{y}{x}$

67. $\frac{a^3b^4}{-2ab^2}\cdot\frac{a^5b^7}{ab} = \frac{a^8b^{11}}{-2a^2b^3} = \frac{-a^6b^8}{2}$

69. $\frac{2mn^4}{6mn^2}\div\frac{3m^5n^7}{m^2n^4} = \frac{2mn^4}{6mn^2}\cdot\frac{m^2n^4}{3m^5n^7}$

$= \frac{2m^3n^8}{18m^6n^9} = \frac{1}{9m^3n}$

71. $\frac{3x^2+16x+5}{x}\cdot\frac{x^2}{9x^2-1}$
$= \frac{(3x+1)(x+5)}{x}\cdot\frac{x^2}{(3x-1)(3x+1)}$
$= \frac{(x+5)x}{3x-1} = \frac{x^2+5x}{3x-1}$

73. $\frac{a^2-2a+4}{a^2-4}\cdot\frac{(a+2)^3}{2a+4}$
$= \frac{a^2-2a+4}{(a-2)(a+2)}\cdot\frac{(a+2)^3}{2(a+2)}$
$= \frac{(a^2-2a+4)(a+2)}{2(a-2)}$
$= \frac{a^3+8}{2(a-2)}$

75. $\frac{2x^2+19x-10}{x^2-100}\div\frac{4x^2-1}{2x^2-19x-10}$
$= \frac{2x^2+19x-10}{x^2-100}\cdot\frac{2x^2-19x-10}{4x^2-1}$
$= \frac{(2x-1)(x+10)}{(x-10)(x+10)}\cdot\frac{(2x+1)(x-10)}{(2x-1)(2x+1)}$
$= 1$

77. $\frac{m^2+6m+9}{m^2-6m+9}\cdot\frac{m^2-9}{m^2+mk+3m+3k}$
$= \frac{(m+3)^2}{(m-3)^2}\cdot\frac{(m-3)(m+3)}{(m+3)(m+k)}$
$= \frac{(m+3)^2}{(m-3)(m+k)}$
$= \frac{m^2+6m+9}{(m-3)(m+k)}$

79. $D = RT = \frac{26.2}{x}\text{mph}\cdot\frac{1}{2}\text{hr} = \frac{13.1}{x}\text{ mi}$

81. $A = LW = x\cdot\frac{5}{x} = 5\text{ m}^2$

83. **a)** $\frac{1}{2}\cdot\frac{1}{4} = \frac{1}{8}$ **b)** $\frac{1}{3}\cdot 4 = \frac{4}{3}$
c) $\frac{1}{2}\cdot\frac{4x}{3} = \frac{2x}{3}$ **d)** $\frac{1}{2}\cdot\frac{3x}{2} = \frac{3x}{4}$

6.3 WARM-UPS

1. False, because to get an equivalent fraction with a denominator of 18 we must multiply both numerator and denominator by 6.
2. False, because we must factor each denominator to determine the LCD.
3. True, because
$\frac{3}{2ab^2} = \frac{3\cdot 5a^2b^2}{2ab^2\cdot 5a^2b^2} = \frac{15a^2b^2}{10a^3b^4}$.
4. True, because $2^5\cdot 3^2$ is a multiple of both $2^5\cdot 3$ and $2^4\cdot 3^2$, and it is the smallest number that is a multiple of both.
5. False, because 30 is a multiple of both 6 and 10.
6. False, because the LCD for $6a^2b$ and $4ab^3$ is $12a^2b^3$.
7. False, because a^2+1 is not a multiple of $a+1$.
8. False, because if $x = 1$ then $\frac{1}{2}\neq\frac{1+7}{2+7}$.
9. True, because $x^2-4 = (x-2)(x+2)$.
10. True, because $x = x\cdot\frac{3}{3} = \frac{3x}{3}$.

6.3 EXERCISES

1. We can build up a denominator by multiplying the numerator and denominator of a fraction by the same nonzero number.
3. For fractions the LCD is the smallest number that is a multiple of all of the denominators.
5. $\frac{1}{3} = \frac{1\cdot 9}{3\cdot 9} = \frac{9}{27}$

7. $\frac{3}{4} = \frac{3 \cdot 4}{4 \cdot 4} = \frac{12}{16}$

9. $2 = \frac{2}{1} = \frac{2 \cdot 6}{1 \cdot 6} = \frac{12}{6}$

11. $\frac{5}{x} = \frac{5 \cdot a}{x \cdot a} = \frac{5a}{ax}$

13. $7 = 7 \cdot \frac{2x}{2x} = \frac{14x}{2x}$

15. $\frac{5}{b} = \frac{5 \cdot 3t}{b \cdot 3t} = \frac{15t}{3bt}$

17. $\frac{-9z}{2aw} = \frac{-9z \cdot 4z}{2aw \cdot 4z} = \frac{-36z^2}{8awz}$

19. $\frac{2}{3a} = \frac{2 \cdot 5a^2}{3a \cdot 5a^2} = \frac{10a^2}{15a^3}$

21. $\frac{4}{5xy^2} = \frac{4 \cdot 2xy^3}{5xy^2 \cdot 2xy^3} = \frac{8xy^3}{10x^2y^5}$

23. $\frac{5}{x+3} = \frac{5(2)}{(x+3)(2)} = \frac{10}{2x+6}$

25. $\frac{5}{2x+2} = \frac{5(-4)}{(2x+2)(-4)} = \frac{-20}{-8x-8}$

27. $\frac{8a}{5b^2-5b} = \frac{8a(-4b)}{(5b^2-5b)(-4b)}$
$= \frac{-32ab}{20b^2-20b^3}$

29. $\frac{3}{x+2} = \frac{3(x-2)}{(x+2)(x-2)} = \frac{3x-6}{x^2-4}$

31. $\frac{3x}{x+1} = \frac{3x(x+1)}{(x+1)(x+1)} = \frac{3x^2+3x}{x^2+2x+1}$

33. $\frac{y-6}{y-4} = \frac{(y-6)(y+5)}{(y-4)(y+5)} = \frac{y^2-y-30}{y^2+y-20}$

35. Since $12 = 2^2 \cdot 3$ and $16 = 2^4$, the LCD is $2^4 \cdot 3$ or 48.
37. Since $12 = 2^2 \cdot 3$, $18 = 2 \cdot 3^2$, and $20 = 2^2 \cdot 5$, the LCD is $2^2 \cdot 3^2 \cdot 5$ or 180.
39. Since $6a^2 = 2 \cdot 3a^2$ and $15a = 3 \cdot 5a$, the LCD is $2 \cdot 3 \cdot 5a^2$ or $30a^2$.
41. The LCD for $2a^4b$, $3ab^6$ and $4a^3b^2$ is $12a^4b^6$.

43. Since $x^2 - 16 = (x-4)(x+4)$ and $x^2 + 8x + 16 = (x+4)^2$, the LCD is $(x-4)(x+4)^2$.
45. The LCD for x, $x+2$, and $x-2$ is $x(x+2)(x-2)$.
47. Since $x^2 - 4x = x(x-4)$, $x^2 - 16 = (x-4)(x+4)$, and $2x = 2x$, the LCD is $2x(x-4)(x+4)$.
49. The LCD for 6 and 8 is 24:
$\frac{1}{6} = \frac{1 \cdot 4}{6 \cdot 4} = \frac{4}{24}$, $\frac{3}{8} = \frac{3 \cdot 3}{8 \cdot 3} = \frac{9}{24}$
51. The LCD for $2x$ and $6x$ is $6x$:
$\frac{1}{2x} = \frac{1 \cdot 3}{2x \cdot 3} = \frac{3}{6x}$, $\frac{5}{6x}$
53. The LCD for $3a$ and $2b$ is $6ab$:
$\frac{2}{3a} = \frac{2 \cdot 2b}{3a \cdot 2b} = \frac{4b}{6ab}$, $\frac{1}{2b} = \frac{1 \cdot 3a}{2b \cdot 3a} = \frac{3a}{6ab}$
55. Since $84 = 2^2 \cdot 3 \cdot 7$ and $63 = 3^2 \cdot 7$, the LCD is $2^2 \cdot 3^2 \cdot 7ab$ or $252ab$.

$\frac{3}{84a} = \frac{3 \cdot 3b}{84a \cdot 3b} = \frac{9b}{252ab}$,

$\frac{5}{63b} = \frac{5 \cdot 4a}{63b \cdot 4a} = \frac{20a}{252ab}$

57. The LCD for $3x^2$ and $2x^5$ is $3 \cdot 2x^5$ or $6x^5$.

$\frac{1}{3x^2} = \frac{1 \cdot 2x^3}{3x^2 \cdot 2x^3} = \frac{2x^3}{6x^5}$,

$\frac{3}{2x^5} = \frac{3 \cdot 3}{2x^5 \cdot 3} = \frac{9}{6x^5}$

59. Since $9y^5z = 3 \cdot 3y^5z$, $12x^3 = 2^2 \cdot 3x^3$, and $6x^2y = 2 \cdot 3x^2y$, the LCD is $36x^3y^5z$.
$\frac{x}{9y^5z} = \frac{x \cdot 4x^3}{9y^5z \cdot 4x^3} = \frac{4x^4}{36x^3y^5z}$,

$\frac{y}{12x^3} = \frac{y \cdot 3y^5z}{12x^3 \cdot 3y^5z} = \frac{3y^6z}{36x^3y^5z}$

$\frac{1}{6x^2y} = \frac{1 \cdot 6xy^4z}{6x^2y \cdot 6xy^4z} = \frac{6xy^4z}{36x^3y^5z}$

61. The LCD for $x-3$ and $x+2$ is $(x-3)(x+2)$.
$\frac{2x}{x-3} = \frac{2x(x+2)}{(x-3)(x+2)} = \frac{2x^2+4x}{(x-3)(x+2)}$

$\frac{5x}{x+2} = \frac{5x(x-3)}{(x+2)(x-3)} = \frac{5x^2-15x}{(x-3)(x+2)}$

63. Since $6 - a = -1(a - 6)$, the LCD is $a - 6$. The expression $\frac{4}{a-6}$ already has the required denominator.

$\frac{5}{6-a} = \frac{5(-1)}{(6-a)(-1)} = \frac{-5}{a-6}$

65. Since $x^2 - 9 = (x-3)(x+3)$ and $x^2 - 6x + 9 = (x-3)^2$, the LCD is $(x+3)(x-3)^2$.

$\frac{x}{x^2-9} = \frac{x(x-3)}{(x-3)(x+3)(x-3)}$
$= \frac{x^2-3x}{(x-3)^2(x+3)}$

$\frac{5x}{x^2-6x+9} = \frac{5x(x+3)}{(x-3)^2(x+3)}$
$= \frac{5x^2+15x}{(x-3)^2(x+3)}$

67. Since $w^2 - 2w - 15 = (w-5)(w+3)$ and $w^2 - 4w - 5 = (w-5)(w+1)$, the LCD is $(w-5)(w+1)(w+3)$.

$\frac{w+2}{(w-5)(w+3)} = \frac{(w+2)(w+1)}{(w-5)(w+3)(w+1)}$
$= \frac{w^2+3w+2}{(w-5)(w+3)(w+1)}$

$\frac{-2w}{(w-5)(w+1)} = \frac{-2w(w+3)}{(w-5)(w+1)(w+3)}$
$= \frac{-2w^2-6w}{(w-5)(w+3)(w+1)}$

69. $\frac{-5}{6(x-2)} = \frac{-5(x+2)}{6(x-2)(x+2)}$
$= \frac{-5x-10}{6(x-2)(x+2)}$

$\frac{x}{(x-2)(x+2)} = \frac{6x}{6(x-2)(x+2)}$

$\frac{3}{2(x+2)} = \frac{3 \cdot 3(x-2)}{2(x+2) \cdot 3(x-2)}$
$= \frac{9x-18}{6(x-2)(x+2)}$

71. $\frac{2}{(2q+1)(q-3)}$
$= \frac{2(q+4)}{(2q+1)(q-3)(q+4)}$
$= \frac{2q+8}{(2q+1)(q-3)(q+4)}$

$\frac{3}{(2q+1)(q+4)} = \frac{3(q-3)}{(2q+1)(q+4)(q-3)}$
$= \frac{3q-9}{(2q+1)(q-3)(q+4)}$

$\frac{4}{(q+4)(q-3)} = \frac{4(2q+1)}{(q+4)(q-3)(2q+1)}$
$= \frac{8q+4}{(2q+1)(q-3)(q+4)}$

73. Identical denominators are need for addition and subtraction.

6.4 WARM-UPS

1. False, because $\frac{1}{2} + \frac{1}{3} = \frac{3}{6} + \frac{2}{6} = \frac{5}{6}$.
2. True, because $\frac{7}{12} - \frac{1}{12} = \frac{6}{12} = \frac{1}{2}$.
3. True, because $\frac{3}{5} + \frac{4}{3} = \frac{3 \cdot 3 + 4 \cdot 5}{3 \cdot 5} = \frac{29}{15}$.
4. True, because $\frac{4}{5} - \frac{5}{7} = \frac{4 \cdot 7 - 5 \cdot 5}{5 \cdot 7} = \frac{3}{35}$.
5. True, because $\frac{5}{20} + \frac{3}{4} = \frac{1}{4} + \frac{3}{4} = \frac{4}{4} = 1$.
6. False, because if $x = 2$ the equation becomes $\frac{2}{2} + 1 = \frac{3}{2}$, which is incorrect.
7. True, because $1 + \frac{1}{a} = \frac{a}{a} + \frac{1}{a} = \frac{a+1}{a}$.
8. False, because if $a = 0$ we get $0 - \frac{1}{4} = \frac{3}{4} \cdot 0$, which is incorrect.
9. True, because $\frac{a}{2} + \frac{b}{3} = \frac{3a}{6} + \frac{2b}{6} = \frac{3a+2b}{6}$.
10. False, because the LCD for x and $x - 1$ is $x(x-1)$ or $x^2 - x$.

6.4 EXERCISES

1. We can add or subtract rational numbers with identical denominators as follows: $\frac{a}{c} + \frac{b}{c} = \frac{a+b}{c}$ and $\frac{a}{c} - \frac{b}{c} = \frac{a-b}{c}$.
3. The LCD is the smallest number that is a multiple of all denominators.

5. $\frac{1}{10} + \frac{1}{10} = \frac{2}{10} = \frac{2 \cdot 1}{2 \cdot 5} = \frac{1}{5}$

7. $\frac{7}{8} - \frac{1}{8} = \frac{6}{8} = \frac{3 \cdot 2}{4 \cdot 2} = \frac{3}{4}$

9. $\frac{1}{6} - \frac{5}{6} = \frac{-4}{6} = \frac{-2 \cdot 2}{3 \cdot 2} = -\frac{2}{3}$

11. $-\frac{7}{8} + \frac{1}{8} = \frac{-6}{8} = \frac{-3 \cdot 2}{4 \cdot 2} = -\frac{3}{4}$

13. $\frac{1}{3} + \frac{2}{9} = \frac{1 \cdot 3}{3 \cdot 3} + \frac{2}{9} = \frac{3}{9} + \frac{2}{9} = \frac{5}{9}$

15. The LCD for $16 = 2^4$ and $18 = 2 \cdot 3^2$ is $2^4 \cdot 3^2$ or 144.

$\frac{7}{16} + \frac{5}{18} = \frac{7 \cdot 9}{16 \cdot 9} + \frac{5 \cdot 8}{18 \cdot 8} = \frac{63}{144} + \frac{40}{144} = \frac{103}{144}$

17. The LCD for $8 = 2^3$ and $10 = 2 \cdot 5$ is $2^3 \cdot 5$ or 40.

$\frac{1}{8} - \frac{9}{10} = \frac{1 \cdot 5}{8 \cdot 5} - \frac{9 \cdot 4}{10 \cdot 4} = \frac{5}{40} - \frac{36}{40} = -\frac{31}{40}$

19. The LCD for $6 = 2 \cdot 3$ and $8 = 2^3$ is $2^3 \cdot 3$ or 24.

$-\frac{1}{6} - \left(-\frac{3}{8}\right) = -\frac{1}{6} + \frac{3}{8} = -\frac{1 \cdot 4}{6 \cdot 4} + \frac{3 \cdot 3}{8 \cdot 3}$
$= -\frac{4}{24} + \frac{9}{24} = \frac{5}{24}$

21. $\frac{1}{2x} + \frac{1}{2x} = \frac{2}{2x} = \frac{1}{x}$

23. $\frac{3}{2w} + \frac{7}{2w} = \frac{10}{2w} = \frac{2 \cdot 5}{2w} = \frac{5}{w}$

25. $\frac{3a}{a+5} + \frac{15}{a+5} = \frac{3a+15}{a+5} = \frac{3(a+5)}{a+5}$
$= 3$

27. $\frac{q-1}{q-4} - \frac{3q-9}{q-4} = \frac{q-1-3q+9}{q-4}$

$= \frac{-2q+8}{q-4} = \frac{-2(q-4)}{q-4} = -2$

29. $\frac{4h-3}{h(h+1)} - \frac{h-6}{h(h+1)} = \frac{4h-3-h+6}{h(h+1)}$
$= \frac{3h+3}{h(h+1)} = \frac{3(h+1)}{h(h+1)} = \frac{3}{h}$

31. $\frac{x^2-x-5}{(x+1)(x+2)} + \frac{1-2x}{(x+1)(x+2)}$
$= \frac{x^2-3x-4}{(x+1)(x+2)} = \frac{(x-4)(x+1)}{(x+1)(x+2)} = \frac{x-4}{x+2}$

33. The LCD for a and $2a$ is $2a$.

$\frac{1}{a} + \frac{1}{2a} = \frac{1 \cdot 2}{a \cdot 2} + \frac{1}{2a} = \frac{2}{2a} + \frac{1}{2a} = \frac{3}{2a}$

35. The LCD for 3 and 2 is 6.

$\frac{x}{3} + \frac{x}{2} = \frac{x \cdot 2}{3 \cdot 2} + \frac{x \cdot 3}{2 \cdot 3} = \frac{2x}{6} + \frac{3x}{6} = \frac{5x}{6}$

37. The LCD for 5 and 1 is 5.

$\frac{m}{5} + m = \frac{m}{5} + \frac{m \cdot 5}{1 \cdot 5} = \frac{m}{5} + \frac{5m}{5} = \frac{6m}{5}$

39. The LCD for x and y is xy.

$\frac{1}{x} + \frac{2}{y} = \frac{1 \cdot y}{x \cdot y} + \frac{2 \cdot x}{y \cdot x} = \frac{y}{xy} + \frac{2x}{xy} = \frac{2x+y}{xy}$

41. The LCD for $2a$ and $5a$ is $2 \cdot 5a$ or $10a$.

$\frac{3}{2a} + \frac{1}{5a} = \frac{3 \cdot 5}{2a \cdot 5} + \frac{1 \cdot 2}{5a \cdot 2} = \frac{15}{10a} + \frac{2}{10a} = \frac{17}{10a}$

43. $\frac{w-3}{9} - \frac{w-4}{12} = \frac{(w-3)4}{9 \cdot 4} - \frac{(w-4)3}{12 \cdot 3}$

$= \frac{4w-12}{36} - \frac{3w-12}{36} = \frac{w}{36}$

45. $\frac{b^2}{4a} - c = \frac{b^2}{4a} - c \cdot \frac{4a}{4a} = \frac{b^2}{4a} - \frac{4ac}{4a}$
$= \frac{b^2-4ac}{4a}$

47. $\frac{2}{wz^2} + \frac{3}{w^2z} = \frac{2 \cdot w}{wz^2 \cdot w} + \frac{3 \cdot z}{w^2z \cdot z}$
$= \frac{2w}{w^2z^2} + \frac{3z}{w^2z^2} = \frac{2w+3z}{w^2z^2}$

49. $\frac{1}{x} + \frac{1}{x+2} = \frac{1(x+2)}{x(x+2)} + \frac{1(x)}{(x+2)(x)}$
$= \frac{x+2}{x(x+2)} + \frac{x}{x(x+2)} = \frac{2x+2}{x(x+2)}$

51. $\frac{2}{x+1} - \frac{3}{x} = \frac{2(x)}{(x+1)(x)} - \frac{3(x+1)}{x(x+1)}$
$= \frac{2x}{x(x+1)} - \frac{3x+3}{x(x+1)} = \frac{2x-3x-3}{x(x+1)} = \frac{-x-3}{x(x+1)}$

53. $\frac{2}{a-b} + \frac{1}{a+b}$
$= \frac{2(a+b)}{(a-b)(a+b)} + \frac{1(a-b)}{(a+b)(a-b)}$
$= \frac{2a+2b}{(a-b)(a+b)} + \frac{a-b}{(a+b)(a-b)}$
$= \frac{3a+b}{(a-b)(a+b)}$

55. $\frac{3}{x^2+x} - \frac{4}{5x+5}$
$= \frac{3}{x(x+1)} - \frac{4}{5(x+1)}$
$= \frac{3 \cdot 5}{x(x+1) \cdot 5} - \frac{4 \cdot x}{5(x+1) \cdot x}$
$= \frac{15}{5x(x+1)} - \frac{4x}{5x(x+1)}$
$= \frac{15-4x}{5x(x+1)}$

57. $\frac{2a}{a^2-9} + \frac{a}{a-3}$
$= \frac{2a}{(a-3)(a+3)} + \frac{a}{a-3}$
$= \frac{2a}{(a-3)(a+3)} + \frac{a(a+3)}{(a-3)(a+3)}$
$= \frac{2a+a^2+3a}{(a-3)(a+3)} = \frac{a^2+5a}{(a-3)(a+3)}$

59. $\frac{4}{a-b}+\frac{4}{b-a}=\frac{4}{a-b}+\frac{4(-1)}{(b-a)(-1)}$

$=\frac{4}{a-b}+\frac{-4}{a-b}=\frac{0}{a-b}=0$

61. $\frac{3}{2a-2}-\frac{2}{1-a}=\frac{3}{2(a-1)}-\frac{2}{1-a}$

$=\frac{3}{2(a-1)}-\frac{2(-2)}{(1-a)(-2)}$

$=\frac{3}{2(a-1)}-\frac{-4}{2(a-1)}=\frac{7}{2(a-1)}$

63. $\frac{1}{x^2-4}-\frac{3}{x^2-3x-10}$

$=\frac{1}{(x-2)(x+2)}-\frac{3}{(x-5)(x+2)}$

$=\frac{1(x-5)}{(x-2)(x+2)(x-5)}-\frac{3(x-2)}{(x-5)(x+2)(x-2)}$

$=\frac{x-5}{(x-5)(x+2)(x-2)}-\frac{3x-6}{(x-5)(x+2)(x-2)}$

$=\frac{x-5-3x+6}{(x-5)(x+2)(x-2)}$

$=\frac{-2x+1}{(x-5)(x+2)(x-2)}$

65. $\frac{3}{x^2+x-2}+\frac{4}{x^2+2x-3}$

$=\frac{3}{(x+2)(x-1)}+\frac{4}{(x+3)(x-1)}$

$=\frac{3(x+3)}{(x+2)(x-1)(x+3)}+\frac{4(x+2)}{(x+3)(x-1)(x+2)}$

$=\frac{3x+9+4x+8}{(x+2)(x-1)(x+3)}$

$=\frac{7x+17}{(x+2)(x-1)(x+3)}$

67. $\frac{1}{a}+\frac{1}{b}+\frac{1}{c}=\frac{1(bc)}{a(bc)}+\frac{1(ac)}{b(ac)}+\frac{1(ab)}{c(ab)}$

$=\frac{bc}{abc}+\frac{ac}{abc}+\frac{ab}{abc}=\frac{bc+ac+ab}{abc}$

69. $\frac{2}{x}-\frac{1}{x-1}+\frac{1}{x+2}$

$=\frac{2(x-1)(x+2)}{x(x-1)(x+2)}-\frac{1(x)(x+2)}{(x-1)(x)(x+2)}$

$+\frac{1(x)(x-1)}{(x+2)(x)(x-1)}$

$=\frac{2x^2+2x-4}{x(x-1)(x+2)}-\frac{x^2+2x}{x(x-1)(x+2)}$

$+\frac{x^2-x}{x(x-1)(x+2)}$

$=\frac{2x^2+2x-4-x^2-2x+x^2-x}{x(x-1)(x+2)}$

$=\frac{2x^2-x-4}{x(x-1)(x+2)}$

71. $\frac{5}{3(a-3)}-\frac{3}{2a}+\frac{4}{a(a-3)}$

$=\frac{5\cdot 2a}{3(a-3)\cdot 2a}-\frac{3(3)(a-3)}{2a(3)(a-3)}+\frac{4\cdot 6}{a(a-3)6}$

$=\frac{10a-9a+27+24}{6a(a-3)}$

$=\frac{a+51}{6a(a-3)}$

73. a) $\frac{1}{y}+2=\frac{1}{y}+\frac{2y}{y}=\frac{2y+1}{y}$ F

b) $\frac{1}{y}+\frac{2}{y}=\frac{3}{y}$ A

c) $\frac{1}{y}+\frac{1}{2}=\frac{2}{2y}+\frac{y}{2y}=\frac{y+2}{2y}$ E

d) $\frac{1}{y}+\frac{1}{2y}=\frac{2}{2y}+\frac{1}{2y}=\frac{3}{2y}$ B

e) $\frac{2}{y}+1=\frac{2}{y}+\frac{y}{y}=\frac{y+2}{y}$ D

f) $\frac{y}{2}+1=\frac{y}{2}+\frac{2}{2}=\frac{y+2}{2}$ C

75. $\frac{3}{2p}-\frac{1}{2p+8}=\frac{3}{2p}-\frac{1}{2(p+4)}$

$=\frac{3(p+4)}{2p(p+4)}-\frac{1p}{2(p+4)p}=\frac{2p+12}{2p(p+4)}$

$=\frac{2(p+6)}{2p(p+4)}=\frac{p+6}{p(p+4)}$

77. $\frac{3}{a^2+3a+2}+\frac{3}{a^2+5a+6}$

$=\frac{3}{(a+1)(a+2)}+\frac{3}{(a+2)(a+3)}$

$=\frac{3(a+3)}{(a+1)(a+2)(a+3)}$

$+\frac{3(a+1)}{(a+1)(a+2)(a+3)}$

$=\frac{6a+12}{(a+1)(a+2)(a+3)}$

$=\frac{6(a+2)}{(a+1)(a+2)(a+3)}=\frac{6}{(a+1)(a+3)}$

79. $\frac{2}{b^2+4b+3}-\frac{1}{b^2+5b+6}$

$=\frac{2}{(b+1)(b+3)}-\frac{1}{(b+2)(b+3)}$

$=\frac{2(b+2)}{(b+1)(b+3)(b+2)}$

$-\frac{1(b+1)}{(b+2)(b+3)(b+1)}$

$=\frac{b+3}{(b+1)(b+2)(b+3)}=\frac{1}{(b+1)(b+2)}$

81. $\frac{3}{2t}-\frac{2}{t+2}-\frac{3}{t^2+2t}$

$=\frac{3(t+2)}{2t(t+2)}-\frac{2\cdot 2t}{(t+2)2t}-\frac{3\cdot 2}{t(t+2)2}$

$=\frac{3t+6}{2t(t+2)}-\frac{4t}{2t(t+2)}-\frac{6}{2t(t+2)}$

$=\frac{-t}{2t(t+2)}=\frac{-1}{2(t+2)}$

83. Since the perimeter of a rectangle is given by $P = 2L + 2W$, we have
$P = 2\left(\frac{3}{x}\right) + 2\left(\frac{5}{2x}\right) = \frac{6}{x} + \frac{5}{x} = \frac{11}{x}$ feet.

85. Since $T = D/R$, we have time before $= \frac{120}{x}$ hr, time after $= \frac{195}{x+5}$ hr. Total time is

$$\frac{120}{x} + \frac{195}{x+5} = \frac{120(x+5) + 195x}{x(x+5)} = \frac{315x + 600}{x(x+5)}$$

If $x = 60$, then total time is $\frac{315 \cdot 60 + 600}{60(60+5)}$ or 5 hours.

87. Use the fact that the product of rate and time is the amount of work completed. Work completed by Kent $= \frac{1}{x} \cdot 2 = \frac{2}{x}$ job, work completed by Keith $= \frac{1}{x+2} \cdot 2 = \frac{2}{x+3}$ job.
Work completed in 2 days together is

$$\frac{2}{x} + \frac{2}{x+3} = \frac{2(x+3)}{x(x+3)} + \frac{2x}{(x+3)x}$$
$$= \frac{4x+6}{x(x+3)} \text{ job.}$$

If $x = 6$, then $\frac{4 \cdot 6 + 6}{6(6+3)} = \frac{30}{54} = \frac{5}{9}$ job.

6.5 WARM-UPS

1. False, because the LCD for 4, x, 6, and x^2 is $12x^2$.
2. True, because $2b - 2a = -2(a - b)$.
3. False, because a complex fraction has fractions in its numerator, denominator, or both.
4. False, because $3 - a = -1(a - 3)$.
5. False, because there is no largest common denominator.
6. False, because the LCD for the denominators 2, 3, 4, and 5 is 60.
7. False, because the LCD of the denominators b and a is ab.
8. True, because this complex fraction is simplified correctly and $-3/2$ is the only value that causes the denominator to be zero.
9. True, because $-3/2$ is the only value that causes the denominator to be zero.
10. True, because $\frac{1}{2} + \frac{1}{3} = \frac{5}{6}$ and $1 + \frac{1}{2} = \frac{3}{2}$.

6.5 EXERCISES

1. A complex fraction is a fraction that has fractions in its numerator, denominator, or both.

3. $$\frac{\frac{1}{2} + \frac{1}{4}}{\frac{1}{2} + \frac{3}{4}} = \frac{\left(\frac{1}{2} + \frac{1}{4}\right)(4)}{\left(\frac{1}{2} + \frac{3}{4}\right)(4)} = \frac{2+1}{2+3} = \frac{3}{5}$$

5. $$\frac{\frac{1}{2} + \frac{1}{3}}{\frac{1}{4} - \frac{1}{2}} = \frac{\left(\frac{1}{2} + \frac{1}{3}\right)(12)}{\left(\frac{1}{4} - \frac{1}{2}\right)(12)} = \frac{6+4}{3-6} = \frac{10}{-3} = -\frac{10}{3}$$

7. $$\frac{\frac{2}{5} + \frac{5}{6} - \frac{1}{2}}{\frac{1}{2} - \frac{1}{3} + \frac{1}{15}} = \frac{\frac{2}{5} \cdot 30 + \frac{5}{6} \cdot 30 - \frac{1}{2} \cdot 30}{\frac{1}{2} \cdot 30 - \frac{1}{3} \cdot 30 + \frac{1}{15} \cdot 30}$$
$$= \frac{12 + 25 - 15}{15 - 10 + 2} = \frac{22}{7}$$

9. $$\frac{1 + \frac{1}{2}}{2 + \frac{1}{4}} = \frac{\left(1 + \frac{1}{2}\right)4}{\left(2 + \frac{1}{4}\right)4} = \frac{4+2}{8+1} = \frac{6}{9} = \frac{2}{3}$$

11. $$\frac{3 + \frac{1}{2}}{5 - \frac{3}{4}} = \frac{\left(3 + \frac{1}{2}\right)4}{\left(5 - \frac{3}{4}\right)4} = \frac{12+2}{20-3} = \frac{14}{17}$$

13. $$\frac{1 - \frac{1}{6} + \frac{2}{3}}{1 + \frac{1}{15} - \frac{3}{10}} = \frac{\left(1 - \frac{1}{6} + \frac{2}{3}\right)30}{\left(1 + \frac{1}{15} - \frac{3}{10}\right)30}$$
$$= \frac{30 - 5 + 20}{30 + 2 - 9} = \frac{45}{23}$$

15. $$\frac{\frac{1}{a} + \frac{1}{b}}{\frac{2}{a} + \frac{2}{b}} = \frac{\frac{1}{a} \cdot ab + \frac{1}{b} \cdot ab}{\frac{2}{a} \cdot ab + \frac{2}{b} \cdot ab} = \frac{b+a}{2b+2a}$$
$$= \frac{1(b+a)}{2(b+a)} = \frac{1}{2}$$

17. $$\frac{\frac{1}{a} + \frac{3}{b}}{\frac{1}{b} - \frac{3}{a}} = \frac{\frac{1}{a} \cdot ab + \frac{3}{b} \cdot ab}{\frac{1}{b} \cdot ab - \frac{3}{a} \cdot ab} = \frac{b+3a}{a-3b}$$
$$= \frac{3a+b}{a-3b}$$

19. $\dfrac{5-\frac{3}{a}}{3+\frac{1}{a}} = \dfrac{\left(5-\frac{3}{a}\right)(a)}{\left(3+\frac{1}{a}\right)(a)} = \dfrac{5a-3}{3a+1}$

21. $\dfrac{\frac{1}{2}-\frac{2}{x}}{3-\frac{1}{x^2}} = \dfrac{\frac{1}{2}\cdot 2x^2-\frac{2}{x}\cdot 2x^2}{3\cdot 2x^2-\frac{1}{x^2}\cdot 2x^2}$

$= \dfrac{x^2-4x}{6x^2-2} = \dfrac{x^2-4x}{2(3x^2-1)}$

23. $\dfrac{\frac{3}{2b}+\frac{1}{b}}{\frac{3}{4}-\frac{1}{b^2}} = \dfrac{\frac{3}{2b}\cdot 4b^2+\frac{1}{b}\cdot 4b^2}{\frac{3}{4}\cdot 4b^2-\frac{1}{b^2}\cdot 4b^2} = \dfrac{6b+4b}{3b^2-4}$

$= \dfrac{10b}{3b^2-4}$

25. $\dfrac{\frac{1}{x+1}+1}{\frac{3}{x+1}+3} = \dfrac{\frac{1}{x+1}(x+1)+1(x+1)}{\frac{3}{x+1}(x+1)+3(x+1)}$

$= \dfrac{1+x+1}{3+3x+3} = \dfrac{x+2}{3x+6} = \dfrac{x+2}{3(x+2)} = \dfrac{1}{3}$

27. $\dfrac{1-\frac{3}{y+1}}{3+\frac{1}{y+1}} = \dfrac{1(y+1)-\frac{3}{y+1}(y+1)}{3(y+1)+\frac{1}{y+1}(y+1)}$

$= \dfrac{y+1-3}{3y+3+1} = \dfrac{y-2}{3y+4}$

29. $\dfrac{x+\frac{4}{x-2}}{x-\frac{x+1}{x-2}} = \dfrac{x(x-2)+\frac{4}{x-2}(x-2)}{x(x-2)-\frac{x+1}{x-2}(x-2)}$

$= \dfrac{x(x-2)+4}{x(x-2)-(x+1)} = \dfrac{x^2-2x+4}{x^2-3x-1}$

31. $\dfrac{\frac{1}{3-x}-5}{\frac{1}{x-3}-2} = \dfrac{\frac{1}{3-x}(x-3)-5(x-3)}{\frac{1}{x-3}(x-3)-2(x-3)}$

$= \dfrac{-1-5x+15}{1-2x+6}$

$= \dfrac{14-5x}{7-2x} = \dfrac{5x-14}{2x-7}$

33. $\dfrac{1-\frac{5}{a-1}}{3-\frac{2}{1-a}} = \dfrac{1(a-1)-\frac{5}{a-1}(a-1)}{3(a-1)-\frac{2}{1-a}(a-1)}$

$= \dfrac{a-1-5}{3a-3-2(-1)} = \dfrac{a-6}{3a-1}$

35. $\dfrac{\frac{1}{m-3}-\frac{4}{m}}{\frac{3}{m-3}+\frac{1}{m}}$

$= \dfrac{\frac{1}{m-3}m(m-3)-\frac{4}{m}m(m-3)}{\frac{3}{m-3}m(m-3)+\frac{1}{m}m(m-3)}$

$= \dfrac{m-4(m-3)}{3m+m-3} = \dfrac{-3m+12}{4m-3}$

37. $\dfrac{\frac{2}{w-1}-\frac{3}{w+1}}{\frac{4}{w+1}+\frac{5}{w-1}} =$

$= \dfrac{\frac{2}{w-1}(w+1)(w-1)-\frac{3}{w+1}(w+1)(w-1)}{\frac{4}{w+1}(w+1)(w-1)+\frac{5}{w-1}(w+1)(w-1)}$

$= \dfrac{2(w+1)-3(w-1)}{4(w-1)+5(w+1)} = \dfrac{-w+5}{9w+1}$

39. $\dfrac{\frac{1}{a-b}-\frac{1}{a+b}}{\frac{1}{b-a}+\frac{1}{b+a}}$

$= \dfrac{\frac{1}{a-b}(a-b)(a+b)-\frac{1}{a+b}(a-b)(a+b)}{\frac{1}{b-a}(a-b)(a+b)+\frac{1}{b+a}(a-b)(a+b)}$

$= \dfrac{a+b-(a-b)}{-1(a+b)+a-b} = \dfrac{a+b-a+b}{-a-b+a-b}$

$= \dfrac{2b}{-2b} = -1$

41. $\dfrac{1-\frac{4}{a^2}}{1+\frac{2}{a}-\frac{6}{a^2}}$

$= \dfrac{1\cdot a^2-\frac{4}{a^2}\cdot a^2}{1\cdot a^2+\frac{2}{a}\cdot a^2-\frac{6}{a^2}\cdot a^2}$

$= \dfrac{a^2-4}{a^2+2a-8} = \dfrac{(a+2)(a-2)}{(a+4)(a-2)} = \dfrac{a+2}{a+4}$

43. $\dfrac{\frac{1}{2}+\frac{1}{4x}}{\frac{x}{3}-\frac{1}{12x}} = \dfrac{\frac{1}{2}12x+\frac{1}{4x}12x}{\frac{x}{3}12x-\frac{1}{12x}12x}$

$= \dfrac{6x+3}{4x^2-1} = \dfrac{3(2x+1)}{(2x+1)(2x-1)} = \dfrac{3}{2x+1}$

45. $\dfrac{\frac{1}{3}-\frac{5}{3x}+\frac{2}{x^2}}{\frac{1}{3}-\frac{3}{x^2}}$

$= \dfrac{\frac{1}{3}3x^2-\frac{5}{3x}3x^2+\frac{2}{x^2}3x^2}{\frac{1}{3}3x^2-\frac{3}{x^2}3x^2}$

$= \dfrac{x^2-5x+6}{x^2-9} = \dfrac{(x-2)(x-3)}{(x+3)(x-3)} = \dfrac{x-2}{x+3}$

47. $\dfrac{2x-9}{6}\cdot\dfrac{9}{2x-3} = \dfrac{2x-9}{2\cdot 3}\cdot\dfrac{3\cdot 3}{2x-3}$

$= \dfrac{6x-27}{4x-6} = \dfrac{6x-27}{2(2x-3)}$

49. $\dfrac{2(x-2y)}{xy^2}\cdot\dfrac{x^3y}{3(x-2y)} = \dfrac{2x^2}{3y}$

51. $\dfrac{(a+6)(a-4)}{a+1}\cdot\dfrac{(a+1)^2}{(a-4)(a+3)}$

$= \dfrac{(a+6)(a+1)}{a+3} = \dfrac{a^2+7a+6}{a+3}$

53. $\dfrac{\frac{x}{x+1}}{\frac{1}{x^2-1}-\frac{1}{x-1}}$

$= \dfrac{\frac{x}{x+1}\cdot(x-1)(x+1)}{\frac{1}{x^2-1}(x-1)(x+1)-\frac{1}{x-1}(x-1)(x+1)}$

$= \dfrac{x(x-1)}{1-1(x+1)} = \dfrac{x(x-1)}{-x}$

$= -1(x-1) = 1-x$

55. Let $x =$ the number of males and $x =$ the number of females.

$\dfrac{\text{Calculus students}}{\text{Mathematics students}} = \dfrac{\frac{1}{3}x+\frac{1}{5}x}{\frac{5}{6}x+\frac{3}{4}x}$

$= \dfrac{\left(\frac{1}{3}x+\frac{1}{5}x\right)60}{\left(\frac{5}{6}x+\frac{3}{4}x\right)60} = \dfrac{20x+12x}{50x+45x} = \dfrac{32x}{95x} = \dfrac{32}{95}$

If $x =$ the number of males and $2x =$ the number of females.

$\dfrac{\text{Calculus students}}{\text{Mathematics students}} = \dfrac{\frac{1}{3}x+\frac{1}{5}2x}{\frac{5}{6}x+\frac{3}{4}2x}$

$= \dfrac{\left(\frac{1}{3}x+\frac{1}{5}2x\right)60}{\left(\frac{5}{6}x+\frac{3}{4}2x\right)60} = \dfrac{20x+24x}{50x+90x} = \dfrac{44x}{140x}$

$= \dfrac{11}{35}$

57. The values of the expressions are $2/3$, $3/5$, and $5/8$.

a) The fractions are alternating larger and smaller. From the third one on, each fraction lies between the previous two fractions. That is, $3/5$ is smaller than $2/3$, $5/8$ is larger than $3/5$ but smaller than $2/3$.

b) The next two expressions have values $8/13$ and $13/21$.

c) The fractions appear to be getting closer and closer to approximately 0.61803.

6.6 WARM-UPS

1. False, because the first step is to multiply each side by the LCD.

2. False, because we should not divide each side of an equation by a variable.

3. False, an extraneous solution could be rational.

4. False, because the first step is to multiply each side by the LCD of the denominators in the equation.

5. False, because if $x = 2$ then the denominator $x - 2$ has a value of zero.

6. True, because $3x^2 - 6x$ is the LCD.

7. True, because $1/4 + 1/2 = 3/4$.

8. True, $4 + 2x = 3x$ is obtained by multiplying each side by $4x$.

9. True, because $x^2 - 1$ is the LCD for the denominators $x - 1$ and $x + 1$.

10. True, because -1 and 1 cause the denominators to have a value of zero.

6.6 EXERCISES

1. The first step is usually to multiply each side by the LCD.

3. An extraneous solution is a number that appears when we solve an equation, but it does not check in the original equation.

5. $\frac{x}{2} + 1 = \frac{x}{4}$
$$4\left(\frac{x}{2} + 1\right) = 4\left(\frac{x}{4}\right)$$
$$2x + 4 = x$$
$$x = -4$$
The solution to the equation is -4.

7. $\frac{x}{3} - 5 = \frac{x}{2} - 7$
$$6\left(\frac{x}{3} - 5\right) = 6\left(\frac{x}{2} - 7\right)$$
$$2x - 30 = 3x - 42$$
$$-30 = x - 42$$
$$12 = x$$
The solution to the equation is 12.

9. $\frac{y}{5} - \frac{2}{3} = \frac{y}{6} + \frac{1}{3}$
$$30\left(\frac{y}{5} - \frac{2}{3}\right) = 30\left(\frac{y}{6} + \frac{1}{3}\right)$$
$$6y - 20 = 5y + 10$$
$$y - 20 = 10$$
$$y = 30$$
The solution to the equation is 30.

11. $\frac{3}{4} - \frac{t-4}{3} = \frac{t}{12}$
$$12 \cdot \frac{3}{4} - 12 \cdot \frac{t-4}{3} = 12 \cdot \frac{t}{12}$$
$$9 - 4t + 16 = t$$
$$-5t = -25$$
$$t = 5$$
The solution to the equation is 5.

13. $\frac{1}{5} - \frac{w+10}{15} = \frac{1}{10} - \frac{w+1}{6}$
$$30 \cdot \frac{1}{5} - 30 \cdot \frac{w+10}{15} = 30 \cdot \frac{1}{10} - 30 \cdot \frac{w+1}{6}$$
$$6 - 2w - 20 = 3 - 5w - 5$$
$$3w = 12$$
$$w = 4$$
The solution to the equation is 4.

15. $\frac{1}{x} + \frac{1}{2} = 3$
$$2x\left(\frac{1}{x} + \frac{1}{2}\right) = 2x(3)$$
$$2 + x = 6x$$
$$2 = 5x$$
$$\frac{2}{5} = x$$
The solution to the equation is $2/5$.

17. $\frac{1}{x} + \frac{2}{x} = 7$
$$x\left(\frac{1}{x} + \frac{2}{x}\right) = x(7)$$
$$1 + 2 = 7x$$
$$3 = 7x$$
$$\frac{3}{7} = x$$
The solution to the equation is $3/7$.

19. $\frac{1}{x} + \frac{1}{2} = \frac{3}{4}$
$$4x\left(\frac{1}{x} + \frac{1}{2}\right) = 4x\left(\frac{3}{4}\right)$$
$$4 + 2x = 3x$$
$$4 = x$$
The solution to the equation is 4.

21. $\frac{2}{3x} + \frac{1}{2x} = \frac{7}{24}$
$$24x\frac{2}{3x} + 24x\frac{1}{2x} = 24x\frac{7}{24}$$
$$16 + 12 = 7x$$
$$28 = 7x$$
$$4 = x$$
The solution to the equation is 4.

23. $\frac{1}{2} + \frac{a-2}{a} = \frac{a+2}{2a}$
$$2a \cdot \frac{1}{2} + 2a \cdot \frac{a-2}{a} = 2a \cdot \frac{a+2}{2a}$$
$$a + 2a - 4 = a + 2$$
$$2a = 6$$
$$a = 3$$
The solution to the equation is 3.

25. $\frac{1}{3} - \frac{k+3}{6k} = \frac{1}{3k} - \frac{k-1}{2k}$
$$6k \cdot \frac{1}{3} - 6k \cdot \frac{k+3}{6k} = 6k \cdot \frac{1}{3k} - 6k \cdot \frac{k-1}{2k}$$
$$2k - k - 3 = 2 - 3k + 3$$
$$4k = 8$$
$$k = 2$$
The solution to the equation is 2.

27. $\frac{x}{2} = \frac{5}{x+3}$
$$2(x+3)\frac{x}{2} = 2(x+3)\frac{5}{x+3}$$
$$x^2 + 3x = 10$$
$$x^2 + 3x - 10 = 0$$
$$(x+5)(x-2) = 0$$
$$x + 5 = 0 \text{ or } x - 2 = 0$$
$$x = -5 \text{ or } x = 2$$
The solutions to the equation are -5 and 2.

29. $\frac{x}{x+1} = \frac{6}{x+7}$
$$(x+1)(x+7)\frac{x}{x+1} = \frac{6}{x+7}(x+1)(x+7)$$
$$x^2 + 7x = 6x + 6$$
$$x^2 + x - 6 = 0$$
$$(x+3)(x-2) = 0$$
$$x - 2 = 0 \text{ or } x + 3 = 0$$
$$x = 2 \text{ or } x = -3$$
The solutions to the equation are -3 and 2.

31. $$\frac{2}{x+1} = \frac{1}{x} + \frac{1}{6}$$
$$6x(x+1)\frac{2}{x+1} = 6x(x+1)\frac{1}{x} + 6x(x+1)\frac{1}{6}$$
$$12x = 6x + 6 + x^2 + x$$
$$0 = x^2 - 5x + 6$$
$$0 = (x-2)(x-3)$$
$$x - 2 = 0 \text{ or } x - 3 = 0$$
$$x = 2 \text{ or } x = 3$$
The solutions to the equation are 2 and 3.

33. $$\frac{a-1}{a^2-4} + \frac{1}{a-2} = \frac{a+4}{a+2}$$
$$(a-2)(a+2)\left(\frac{a-1}{a^2-4} + \frac{1}{a-2}\right) = (a-2)(a+2)\frac{a+4}{a+2}$$
$$a - 1 + a + 2 = (a-2)(a+4)$$
$$2a + 1 = a^2 + 2a - 8$$
$$0 = a^2 - 9$$
$$(a-3)(a+3) = 0$$
$$a - 3 = 0 \text{ or } a + 3 = 0$$
$$a = 3 \text{ or } a = -3$$
The solution to the equation is -3 and 3.

35. $$\frac{1}{x-1} + \frac{2}{x} = \frac{x}{x-1}$$
$$x(x-1)\frac{1}{x-1} + x(x-1)\frac{2}{x} = x(x-1)\frac{x}{x-1}$$
$$x + 2x - 2 = x^2$$
$$0 = x^2 - 3x + 2$$
$$0 = (x-2)(x-1)$$
$$x - 2 = 0 \text{ or } x - 1 = 0$$
$$x = 2 \text{ or } x = 1$$
Since the denominator $x - 1$ has a value of 0 for $x = 1$, 1 is not a solution to the equation. The solution to the equation is 2.

37. $$\frac{5}{x+2} + \frac{2}{x-3} = \frac{x-1}{x-3}$$
$$(x+2)(x-3)\frac{5}{x+2} + (x+2)(x-3)\frac{2}{x-3} = (x+2)(x-3)\frac{x-1}{x-3}$$
$$5x - 15 + 2x + 4 = (x+2)(x-1)$$
$$7x - 11 = x^2 + x - 2$$
$$0 = x^2 - 6x + 9$$
$$0 = (x-3)^2$$
$$x - 3 = 0$$
$$x = 3$$
Since the value of the denominator $x - 3$ is 0 if $x = 3$, there is no solution to this equation.

39. $$1 + \frac{3y}{y-2} = \frac{6}{y-2}$$
$$(y-2)1 + (y-2)\frac{3y}{y-2} = (y-2)\frac{6}{y-2}$$
$$y - 2 + 3y = 6$$
$$4y = 8$$
$$y = 2$$
Since the value of the denominator $y - 2$ is 0 if $y = 2$, there is no solution to this equation.

41. $$\frac{z}{z+1} - \frac{1}{z+2} = \frac{2z+5}{z^2+3z+2}$$
$$(z+1)(z+2)\left(\frac{z}{z+1} - \frac{1}{z+2}\right) = (z+1)(z+2)\frac{2z+5}{z^2+3z+2}$$
$$z^2 + 2z - z - 1 = 2z + 5$$
$$z^2 - z - 6 = 0$$
$$(z-3)(z+2) = 0$$
$$z - 3 = 0 \text{ or } z + 2 = 0$$
$$z = 3 \text{ or } z = -2$$
Since the value of $z + 2$ is 0 if $z = -2$, the only solution to the equation is 3.

43. $$\frac{a}{4} = \frac{5}{2}$$
$$4 \cdot \frac{a}{4} = 4 \cdot \frac{5}{2}$$
$$a = 10$$
The solution to the equation is 10.

45. $$\frac{w}{6} = \frac{3w}{11}$$
$$66 \cdot \frac{w}{6} = 66 \cdot \frac{3w}{11}$$
$$11w = 18w$$
$$0 = 7w$$
$$0 = w$$
The solution to the equation is 0.

47. $$\frac{5}{x} = \frac{x}{5}$$
$$5x \cdot \frac{5}{x} = 5x \cdot \frac{x}{5}$$
$$25 = x^2$$
$$0 = x^2 - 25$$
$$0 = (x-5)(x+5)$$
$$x - 5 = 0 \text{ or } x + 5 = 0$$
$$x = 5 \text{ or } x = -5$$
The solutions to the equation are -5 and 5.

49. $$\frac{x-3}{5} = \frac{x-3}{x}$$
$$5x \cdot \frac{x-3}{5} = 5x \cdot \frac{x-3}{x}$$
$$x^2 - 3x = 5x - 15$$
$$x^2 - 8x + 15 = 0$$

$(x-3)(x-5)=0$
$x-3=0$ or $x-5=0$
$x=3$ or $x=5$
The solutions to the equation are 3 and 5.

51. $\frac{1}{x+2}=\frac{x}{x+2}$
$(x+2)\frac{1}{x+2}=(x+2)\frac{x}{x+2}$
$1=x$
The solution to the equation is 1.

53. $\frac{1}{2x-4}+\frac{1}{x-2}=\frac{3}{2}$

$2(x-2)\frac{1}{2x-4}+2(x-2)\frac{1}{x-2}$
$=2(x-2)\frac{3}{2}$
$1+2=3x-6$
$9=3x$
$3=x$
The solution to the equation is 3.

55. $\frac{3}{a^2-a-6}=\frac{2}{a^2-4}$
$\frac{3}{(a-3)(a+2)}=\frac{2}{(a-2)(a+2)}$
$(a-3)(a+2)(a-2)\frac{3}{(a-3)(a+2)}$
$=(a-3)(a+2)(a-2)\frac{2}{(a-2)(a+2)}$
$(a-2)3=(a-3)2$
$3a-6=2a-6$
$a=0$
The solution to the equation is 0.

57. $\frac{4}{c-2}-\frac{1}{2-c}=\frac{25}{c+6}$
$(c-2)(c+6)\frac{4}{c-2}-(c-2)(c+6)\frac{1}{2-c}$
$=(c-2)(c+6)\frac{25}{c+6}$
$4(c+6)-(c+6)(-1)=(c-2)25$
$4c+24+c+6=25c-50$
$5c+30=25c-50$
$5c+80=25c$
$80=20c$
$4=c$
The solution to the equation is 4.

59. $\frac{1}{x^2-9}+\frac{3}{x+3}=\frac{4}{x-3}$

$(x+3)(x-3)\frac{1}{x^2-9}+(x+3)(x-3)\frac{3}{x+3}$
$=(x+3)(x-3)\frac{4}{x-3}$
$1+3(x-3)=4(x+3)$
$1+3x-9=4x+12$
$-20=x$
The solution to the equation is -20.

61. $\frac{3}{2x+4}-\frac{1}{x+2}=\frac{1}{3x+1}$

$2(x+2)(3x+1)(\frac{3}{2x+4}-\frac{1}{x+2})$
$=2(x+2)(3x+1)\frac{1}{3x+1}$
$(3x+1)3-2(3x+1)=2(x+2)$
$9x+3-6x-2=2x+4$
$3x+1=2x+4$
$x=3$
The solution to the equation is 3.

63. $\frac{2t-1}{3(t+1)}+\frac{3t-1}{6(t+1)}=\frac{t}{t+1}$
$6(t+1)\frac{2t-1}{3(t+1)}+6(t+1)\frac{3t-1}{6(t+1)}$
$=6(t+1)\frac{t}{t+1}$
$4t-2+3t-1=6t$
$t=3$
The solution to the equation is 3.

65. $\frac{1}{50}=\frac{1}{600}+\frac{1}{i}$
$600i\frac{1}{50}=600i\frac{1}{600}+600i\frac{1}{i}$
$12i=i+600$
$11i=600$

$i=\frac{600}{11}=54\frac{6}{11}$ mm

6.7 WARM-UPS

1. True, because the fraction $40/30$ reduces to $4/3$.
2. False, because 2 yards = 6 feet and the ratio should be expressed as 3 to 6 or 1 to 2.
3. False, because the ratio of men to women being 3 to 2 means there would be 30 men to 20 women.
4. True, because if we multiply the numerator and denominator of $1.5/2$ by 2, we get the ratio $3/4$.
5. True, because of the definition of proportion.
6. True, because of the extremes-means property.
7. False, because the correct application of the extremes-means property gives us $3x=10$.

8. False, because the ratio should be given with the same units. One foot to 3 feet is a ratio of 1 to 3.
9. False, because the number who prefer aspirin is 30 and the number who do not is 70, giving a ratio of 3 to 7.
10. True, because of the extremes-means property.

6.7 EXERCISES

1. A ratio is a comparison of two numbers.
3. Equivalent ratios are ratios that are equivalent as fractions.
5. In the proportion $\frac{a}{b} = \frac{c}{d}$ the means are b and c and the extremes are a and d.
7. $\frac{4}{6} = \frac{(2)2}{(3)2} = \frac{2}{3}$
9. $\frac{200}{150} = \frac{(50)4}{(50)3} = \frac{4}{3}$
11. $\frac{2.5}{3.5} = \frac{(2.5)2}{(3.5)2} = \frac{5}{7}$
13. $\frac{0.32}{0.6} = \frac{0.32(100)}{0.6(100)} = \frac{32}{60} = \frac{8 \cdot 4}{15 \cdot 4} = \frac{8}{15}$
15. $\frac{35}{10} = \frac{7 \cdot 5}{2 \cdot 5} = \frac{7}{2}$
17. $\frac{4.5}{7} = \frac{(4.5)(2)}{(7)(2)} = \frac{9}{14}$

19. $\frac{\frac{1}{2}}{\frac{1}{5}} = \frac{\left(\frac{1}{2}\right)10}{\left(\frac{1}{5}\right)10} = \frac{5}{2}$

21. $\frac{5}{\frac{1}{3}} = \frac{5(3)}{\frac{1}{3}(3)} = \frac{15}{1}$

23. The ratio of men to women is 12/8 or 3/2.
25. The ratio of smokers to nonsmokers is 72/128, or 9/16, or 9 to 16.
27. The ratio of violence to kindness is 1240/40 or 31/1.
29. The ratio of rise to run is 8 to 12 or 2 to 3.
31.
$$\frac{4}{x} = \frac{2}{3}$$
$$2x = 12$$
$$x = 6$$
The solution to the proportion is 6.

33.
$$\frac{a}{2} = \frac{-1}{5}$$
$$5a = -2$$
$$a = -\frac{2}{5}$$
The solution to the proportion is $-\frac{2}{5}$.

35.
$$-\frac{5}{9} = \frac{3}{x}$$
$$-5x = 27$$
$$x = -\frac{27}{5}$$
The solution to the proportion is $-\frac{27}{5}$.

37.
$$\frac{10}{x} = \frac{34}{x+12}$$
$$34x = 10x + 120$$
$$24x = 120$$
$$x = 5$$
The solution to the proportion is 5.

39.
$$\frac{a}{a+1} = \frac{a+3}{a}$$
$$a^2 = (a+1)(a+3)$$
$$a^2 = a^2 + 4a + 3$$
$$0 = 4a + 3$$
$$-4a = 3$$
$$a = -\frac{3}{4}$$
The solution to the proportion is $-\frac{3}{4}$.

41.
$$\frac{m-1}{m-2} = \frac{m-3}{m+4}$$
$$(m-1)(m+4) = (m-2)(m-3)$$
$$m^2 + 3m - 4 = m^2 - 5m + 6$$
$$3m - 4 = -5m + 6$$
$$8m - 4 = 6$$
$$8m = 10$$
$$m = \frac{10}{8} = \frac{5}{4}$$
The solution to the proportion is $\frac{5}{4}$.

43. Let $x =$ the number of reruns. Since the ratio of new shows to reruns is 2 to 27, we can write the following proportion.
$$\frac{8}{x} = \frac{2}{27}$$
$$2x = 216$$
$$x = 108$$
There were 108 reruns.

45. Let $x =$ the number of votes for the incumbent. Since 220 out of 500 voters said they would vote for the incumbent, we can write the following proportion.
$$\frac{220}{500} = \frac{x}{400{,}000}$$
$$500x = 88{,}000{,}000$$
$$x = 176{,}000$$
The expected number of votes for the incumbent is 176,000.

47. Let $x =$ the number of points scored by the Tigers and $x + 34 =$ the number scored by the Lions.

$$\frac{x + 34}{x} = \frac{5}{3}$$
$$3x + 102 = 5x$$
$$102 = 2x$$
$$51 = x$$
$$x + 34 = 85$$

The score was Lions 85, Tigers 51.

49. Let $x =$ the number of luxury cars and $x + 20 =$ the number of sports cars. Since the ratio of sports cars to luxury cars sold is 3 to 2, we can write the following equation.

$$\frac{x + 20}{x} = \frac{3}{2}$$
$$3x = 2x + 40$$
$$x = 40$$
$$x + 20 = 60$$

There were 40 luxury cars and 60 sports cars sold.

51. Let $x =$ the number of inches in 7 feet. We can write the following proportion.

$$\frac{12}{1} = \frac{x}{7}$$
$$x = 84$$

There are 84 inches in 7 feet.

53. Let $x =$ the number of minutes in 0.25 hour.

$$\frac{x \text{ min}}{0.25 \text{ hr}} = \frac{60 \text{ min}}{1 \text{ hr}}$$
$$x = (0.25)(60) = 15 \text{ minutes}$$

55. Let $x =$ the number of miles traveled in 7 hours. Since he travels 230 miles in 3 hours, we can write the following proportion.

$$\frac{230}{3} = \frac{x}{7}$$
$$3x = 1610$$
$$x = \frac{1610}{3} \approx 536.7$$

He travels 536.7 miles in 7 hours.

57. Let $x =$ the force of a 280 lb player.

$$\frac{x}{280} = \frac{980}{70}$$
$$x = 280 \cdot \frac{980}{70} = 3920 \text{ pounds}$$

Using the graph to estimate the force for a 150-pound player we get approximately 2000 lbs.

59. Let $x =$ the number of trout in Trout Lake.

$$\frac{200}{x} = \frac{5}{150}$$
$$5x = 30000$$
$$x = 6000$$

So there are 6000 trout in the lake.

61. a) The ratio of customer waste to food waste is 6% to 34%, or 6 to 34, or 3 to 17.

b) Let $x =$ the number of pounds of customer waste, and $x + 67 =$ the number of pounds of food waste. Solve the following proportion.

$$\frac{x}{x + 67} = \frac{3}{17}$$
$$17x = 3(x + 67)$$
$$14x = 201$$
$$x = \frac{210}{14} \approx 14.4$$

So the customer waste is 14.4 pounds.

63. $k = 1 + 0 + 0.26 + 0 - 0.29 = 0.97$

If $B = 4200$ units and $k = 0.97$, we have

$$\frac{A}{4200} = 0.97$$

$$A = 4200(0.97) = 4074$$

65. We cannot apply the extremes-means property unless we have a proportion. The distributive property is used incorrectly. The number 1 disappeared in the fourth equation.

6.8 WARM-UPS

1. True, because we can obtain the second formula from the first by multiplying each side by m and then dividing by t.

2. True, because $2mn$ is the LCD of the denominators 2, m, and n.

3. False, her average speed is $\frac{300}{x}$ mph.

4. True, because 20 hard bargains divided by x hours is $20/x$ hard bargains per hour.

5. True, because if he paints the whole house in y days then he paints $1/y$ of the house per day.

6. False, because if $1/x$ is the smaller of the two quantities then we must add 1 to $1/x$ to get an equation.

7. False, because if $a = m/b$ then $b = m/a$.

8. True, because if we divide each side of $D = RT$ by R, we get $T = D/R$.

9. False, because if the equation is solved for P then P must not appear on the other side of the equation.

10. True, because the only way to get R isolated is to factor $3R + yR$ as $R(3 + y)$.

6.8 EXERCISES

1. $\frac{y-1}{x-3} = 2$

$(x-3)\frac{y-1}{x-3} = (x-3)2$

$y - 1 = 2x - 6$

$y = 2x - 5$

3. $\frac{y-1}{x+6} = -\frac{1}{2}$

$(x+6)\frac{y-1}{x+6} = -\frac{1}{2}(x+6)$

$y - 1 = -\frac{1}{2}x - 3$

$y = -\frac{1}{2}x - 2$

5. $\frac{y+a}{x-b} = m$

$(x-b)\frac{y+a}{x-b} = m(x-b)$

$y + a = mx - mb$

$y = mx - mb - a$

7. $\frac{y-1}{x+4} = -\frac{1}{3}$

$(x+4)\frac{y-1}{x+4} = -\frac{1}{3}(x+4)$

$y - 1 = -\frac{1}{3}x - \frac{4}{3}$

$y = -\frac{1}{3}x - \frac{1}{3}$

9. $A = \frac{B}{C}$

$C \cdot A = C \cdot \frac{B}{C}$

$AC = B$

$C = \frac{B}{A}$

11. $\frac{1}{a} + m = \frac{1}{p}$

$ap \cdot \frac{1}{a} + ap \cdot m = ap \cdot \frac{1}{p}$

$p + apm = a$

$p(1 + am) = a$

$p = \frac{a}{1 + am}$

13. $F = k\frac{m_1 m_2}{r^2}$

$r^2 \cdot F = k\frac{m_1 m_2}{r^2} \cdot r^2$

$Fr^2 = km_1 m_2$

$\frac{Fr^2}{km_2} = \frac{km_1 m_2}{km_2}$

$m_1 = \frac{r^2 F}{km_2}$

15. $\frac{1}{a} + \frac{1}{b} = \frac{1}{f}$

$abf \cdot \frac{1}{a} + abf \cdot \frac{1}{b} = abf \cdot \frac{1}{f}$

$bf + af = ab$

$bf = ab - af$

$bf = a(b - f)$

$\frac{bf}{b-f} = a$

$a = \frac{bf}{b-f}$

17. $S = \frac{a}{1-r}$

$(1-r)S = (1-r)\frac{a}{1-r}$

$S - rS = a$

$S = a + rS$

$S - a = rS$

$\frac{S-a}{S} = r$

$r = \frac{S-a}{S}$

19. $\frac{P_1 V_1}{T_1} = \frac{P_2 V_2}{T_2}$

$T_1 T_2 \cdot \frac{P_1 V_1}{T_1} = T_1 T_2 \cdot \frac{P_2 V_2}{T_2}$

$T_2 P_1 V_1 = T_1 P_2 V_2$

$\frac{P_1 V_1 T_2}{T_1 V_2} = P_2$

21. $V = \frac{4}{3}\pi r^2 h$

$3V = 3 \cdot \frac{4}{3}\pi r^2 h$

$3V = 4\pi r^2 h$

$\frac{3V}{4\pi r^2} = h$

23. Use $A = 12$ and $B = 5$ in the formula $A = \frac{B}{C}$.

$12 = \frac{5}{C}$

$12C = 5$

$C = \frac{5}{12}$

25. Use $p = 6$ and $m = 4$ in the formula $\frac{1}{a} + m = \frac{1}{p}$.

$\frac{1}{a} + 4 = \frac{1}{6}$

$6a \cdot \frac{1}{a} + 6a \cdot 4 = 6a \cdot \frac{1}{6}$

$6 + 24a = a$

$23a = -6$

$a = -\frac{6}{23}$

27. Use $F = 32$, $r = 4$, $m_1 = 2$, and $m_2 = 6$ in the formula $F = k\frac{m_1 m_2}{r^2}$.

$32 = k\frac{2 \cdot 6}{4^2}$

$32 = k\frac{12}{16}$

$$32 = k\frac{3}{4}$$
$$k = \frac{4}{3} \cdot 32 = \frac{128}{3}$$

29. Use $f = 3$ and $a = 2$ in the formula $\frac{1}{a} + \frac{1}{b} = \frac{1}{f}$.

$$\frac{1}{2} + \frac{1}{b} = \frac{1}{3}$$
$$6b \cdot \frac{1}{2} + 6b \cdot \frac{1}{b} = 6b \cdot \frac{1}{3}$$
$$3b + 6 = 2b$$
$$b = -6$$

31. Use $S = 3/2$ and $r = 1/5$ in the formula $S = \frac{a}{1-r}$.

$$\frac{3}{2} = \frac{a}{1 - \frac{1}{5}}$$
$$\frac{3}{2} = \frac{a}{\frac{4}{5}}$$
$$a = \frac{3}{2} \cdot \frac{4}{5} = \frac{12}{10} = \frac{6}{5}$$

33. Let $x =$ Frank's rate and $x + 1 =$ Marcie's rate. Since $T = DR$, Marcie's time is $8/(x + 1)$ and Frank's time is $6/x$. Since their times are equal, we can write the following equation.

$$\frac{6}{x} = \frac{8}{x+1}$$
$$6(x + 1) = 8x$$
$$6x + 6 = 8x$$
$$6 = 2x$$
$$3 = x$$
$$4 = x + 1$$

Marcie walks 4 mph and Frank walks 3 mph.

35. Let $x =$ Bob's rate and $x - 5 =$ Pat's rate. Since $T = D/R$, Bob's time is $75/x$ and Pat's time is $70/(x - 5)$. Since Pat takes $1/2$ of an hour longer than Bob, we can write the following equation.

$$\frac{75}{x} + \frac{1}{2} = \frac{70}{x-5}$$

$$2x(x-5)\frac{75}{x} + 2x(x-5)\frac{1}{2} = 2x(x-5)\frac{70}{x-5}$$
$$150(x - 5) + x(x - 5) = 140x$$
$$150x - 750 + x^2 - 5x = 140x$$
$$x^2 + 5x - 750 = 0$$
$$(x - 25)(x + 30) = 0$$
$$x - 25 = 0 \text{ or } x + 30 = 0$$
$$x = 25 \text{ or } x = -30$$
$$x - 5 = 20$$

Bob's rate is 25 mph and Pat's rate is 20 mph.

37. Let $x =$ his walking rate and $x + 5 =$ his running rate. Since $T = D/R$, his time walking was $6/x$ and his time running was $8/(x + 5)$. Since his total time was 2 hours, we can write the following equation.

$$\frac{6}{x} + \frac{8}{x+5} = 2$$
$$x(x+5)\frac{6}{x} + x(x+5)\frac{8}{x+5} = 2x(x+5)$$
$$6x + 30 + 8x = 2x^2 + 10x$$
$$-2x^2 + 4x + 30 = 0$$
$$x^2 - 2x - 15 = 0$$
$$(x - 5)(x + 3) = 0$$
$$x - 5 = 0 \text{ or } x + 3 = 0$$
$$x = 5 \text{ or } x = -3$$

His rate walking was 5 mph.

39. Let $x =$ the number of hours for Red to paint the fence by himself. Red paints $1/x$ of the fence per hour and Kiyoshi paints $1/3$ of the fence per hour. Together they paint $1/2$ of the fence per hour. We can write the following equation.

$$\frac{1}{x} + \frac{1}{3} = \frac{1}{2}$$
$$6x \cdot \frac{1}{x} + 6x \cdot \frac{1}{3} = 6x \cdot \frac{1}{2}$$
$$6 + 2x = 3x$$
$$6 = x$$

It would take Red 6 hours to paint the fence by himself.

41. Let $x =$ the number of hours for the dog and the boy working together. The dog destroys $1/2$ of the garden per hour and the boy destroys the whole garden in 1 hour. Since together they destroy $1/x$ of the garden per hour, we can write the following equation.

$$\frac{1}{2} + 1 = \frac{1}{x}$$
$$\frac{3}{2} = \frac{1}{x}$$
$$3x = 2$$
$$x = \frac{2}{3}$$

Working together they can destroy the whole garden in $2/3$ of an hour or 40 minutes.

43. Let $x =$ their time together. Edgar does $1/2$ of the job per hour and Ellen does $1/8$ of the job per hour. Since together they do $1/x$ of the job per hour, we can write the following equation.

$$\frac{1}{2} + \frac{1}{8} = \frac{1}{x}$$
$$8x \cdot \frac{1}{2} + 8x \cdot \frac{1}{8} = 8x \cdot \frac{1}{x}$$
$$4x + x = 8$$
$$5x = 8$$
$$x = \frac{8}{5} \text{ hours}$$

Together they can do the job in 1 hour 36 minutes.

45. Let $x =$ the number of pounds of apples and $18 - x =$ the number of pounds of bananas. The price per pound for the apples was $9/x$ and the price per pound for the bananas was $2.40/(18 - x)$. Since the price per pound for the apples was 3 times the price per pound for the bananas, we can write the following equation.

$$\frac{9}{x} = 3 \cdot \frac{2.40}{18 - x}$$

$$\frac{9}{x} = \frac{7.20}{18 - x}$$
$$9(18 - x) = 7.20x$$
$$162 - 9x = 7.2x$$
$$162 = 16.2x$$
$$10 = x$$
$$8 = 18 - x$$

She bought 8 pounds of bananas and 10 pounds of apples.

47. Let $x =$ the number of gallons used in the small truck and $110 - x =$ the number of gallons used in the large truck. The small truck got $600/x$ miles per gallon and the large truck got $800/(110 - x)$ miles per gallon. Since the small truck got twice as many miles per gallon as the large truck, we can write the following equation.

$$\frac{600}{x} = 2 \cdot \frac{800}{110 - x}$$

$$\frac{600}{x} = \frac{1600}{110 - x}$$
$$1600x = 66000 - 600x$$
$$2200x = 66000$$
$$x = 30$$
$$110 - x = 80$$

The large truck used 80 gallons of gasoline.

49. Let $x =$ the speed of the plane in still air. The speed with the wind is $x + 20$ and against the wind is $x - 20$. Since $T = D/R$ the time with the wind was $480/(x + 20)$ and against the wind was $480/(x - 20)$. Since the time against the wind was 1 hour longer, we have the following equation.

$$\frac{480}{x + 20} + 1 = \frac{480}{x - 20}$$
$$(x + 20)(x - 20)\left(\frac{480}{x + 20} + 1\right)$$
$$= (x + 20)(x - 20)\frac{480}{x - 20}$$
$$(x - 20)480 + (x + 20)(x - 20)$$
$$= (x + 20)480$$
$$480x - 9600 + x^2 - 400$$
$$= 480x + 9600$$
$$x^2 - 19600 = 0$$
$$(x - 140)(x + 140) = 0$$
$$x - 140 = 0 \quad \text{or} \quad x + 140 = 0$$
$$x = 140 \quad \text{or} \quad x = -140$$

So the speed of the plane in calm air is 140 mph.

51. Let $x =$ the speed of the wind. The speed with the wind is $120 + x$ and against the wind is $120 - x$. Since $T = D/R$ the time with the wind was $325/(120 + x)$ and against the wind was $275/(120 - x)$. Since the time against the is equal to the time with the wind, we have the following equation.

$$\frac{325}{120 + x} = \frac{275}{120 - x}$$
$$(120 - x)325 = 275(120 + x)$$
$$39000 - 325x = 33000 + 275x$$
$$-600x = -6000$$
$$x = 10$$

So the wind speed is 10 mph.

53. Let $x =$ the distance for Ben and $90 - x =$ the distance for Jerry.
The speed for Ben is $x/4$ and the speed for Jerry is $(90 - x)/4$. Since Ben rides twice as fast as Jerry, we have the following equation.

$$\frac{x}{4} = 2\left(\frac{90 - x}{4}\right)$$
$$\frac{x}{4} = \frac{90 - x}{2}$$
$$2x = 4(90 - x)$$
$$2x = 360 - 4x$$
$$6x = 360$$
$$x = 60$$

So Ben's distance is 60 miles and his time is 4 hours, his speed is 15 mph. Since Ben rides twice as fast as Jerry, Jerry's speed is 7.5 mph.

55. Let $x =$ the distance from their home to Las Vegas. The time to Las Vegas was $x/45$ and their time going back was $x/60$. Since the total time was 70 hours, we have the following equation.

$$\frac{x}{45} + \frac{x}{60} = 70$$
$$180 \cdot \frac{x}{45} + 180 \cdot \frac{x}{60} = 180 \cdot 70$$
$$4x + 3x = 12600$$
$$7x = 12600$$
$$x = 1800$$

So the distance from their home to Las Vegas is 1800 miles.

57. Let $x =$ the time to fill the fountain. The rate for the pipe is $\frac{1}{6}$ of the fountain per hour and the rate for the hose is $\frac{1}{12}$ of the fountain per hour. Since the work is the product of the rate and time, the work done by the pipe is $\frac{1}{6}x$ fountain and the work done by the hose is $\frac{1}{12}x$ fountain. The total work is 1 fountain.

$$\frac{1}{6}x + \frac{1}{12}x = 1$$
$$12 \cdot \frac{1}{6}x + 12 \cdot \frac{1}{12}x = 12 \cdot 1$$
$$2x + x = 12$$
$$3x = 12$$
$$x = 4$$

So it takes 4 hours to fill the fountain. The pipe does $\frac{2}{3}$ of the work and the hose does $\frac{1}{3}$ of the work.

59. Let $x =$ the time to print the report together. The old computer prints $\frac{1}{3}$ of the report per hour and the new computer prints $\frac{1}{2}$ of the computer per hour. The work done by the old computer is $\frac{1}{3}x$ report and the work done by the new computer is $\frac{1}{2}x$ report. Since the total work is 1 report, we have the following equation.

$$\frac{1}{3}x + \frac{1}{2}x = 1$$
$$6 \cdot \frac{1}{3}x + 6 \cdot \frac{1}{2}x = 6 \cdot 1$$
$$2x + 3x = 6$$
$$5x = 6$$
$$x = 1.2$$

So working together the report is printed in 1.2 hours of 1 hour 12 minutes.

61. Let $x =$ the time in minutes to fill the tub with the cold water faucet. The hot water faucet fills at the rate of $\frac{1}{12}$ tub/min and the cold water faucet fills at the rate of $\frac{1}{x}$ tub/min. Working together for 8 minutes, the hot water faucet fills $\frac{1}{12} \cdot 8 = \frac{2}{3}$ tub and the cold water faucet fills $\frac{1}{x} \cdot 8 = \frac{8}{x}$ tub. Since total work done is 1 tub, we have the following equation.

$$\frac{2}{3} + \frac{8}{x} = 1$$
$$3x \cdot \frac{2}{3} + 3x \cdot \frac{8}{x} = 3x \cdot 1$$
$$2x + 24 = 3x$$
$$24 = x$$

So the cold water faucet working alone will fill the tub in 24 minutes.

Enriching Your Mathematical Word Power

1. b **2.** a **3.** a **4.** d **5.** a
6. b **7.** d **8.** a **9.** c **10.** d

CHAPTER 6 REVIEW

1. $\frac{24}{28} = \frac{2 \cdot 2 \cdot 2 \cdot 3}{2 \cdot 2 \cdot 7} = \frac{6}{7}$

3. $\frac{2a^3c^3}{8a^5c} = \frac{c^{3-1}}{4a^{5-3}} = \frac{c^2}{4a^2}$

5. $\frac{6w - 9}{9w - 12} = \frac{3(2w - 3)}{3(3w - 4)} = \frac{2w - 3}{3w - 4}$

7. $\frac{x^2 - 1}{3 - 3x} = \frac{(x - 1)(x + 1)}{-3(x - 1)} = -\frac{x + 1}{3}$

9. $\frac{1}{6k} \cdot 3k^2 = \frac{1}{2}k^{2-1} = \frac{1}{2}k$

11. $\frac{2xy}{3} \div y^2 = \frac{2xy}{3} \cdot \frac{1}{y^2} = \frac{2x}{3y}$

13. $\frac{a^2 - 9}{a - 2} \cdot \frac{a^2 - 4}{a + 3}$
$= \frac{(a - 3)(a + 3)}{a - 2} \cdot \frac{(a - 2)(a + 2)}{a + 3}$
$= (a - 3)(a + 2) = a^2 - a - 6$

15. $\frac{w - 2}{3w} \div \frac{4w - 8}{6w} = \frac{w - 2}{3w} \cdot \frac{6w}{4w - 8}$
$= \frac{w - 2}{3w} \cdot \frac{2 \cdot 3w}{4(w - 2)} = \frac{2}{4} = \frac{1}{2}$

17. Since $36 = 2^2 \cdot 3^2$ and $54 = 2 \cdot 3^3$, the LCD is $2^2 \cdot 3^3$ or 108.

19. Since $6ab^3 = 2 \cdot 3ab^3$ and $8a^7b^2 = 2^3a^7b^2$, the LCD is $2^3 \cdot 3a^7b^3 = 24a^7b^3$.

21. Since $4x = 2^2 \cdot x$ and $6x - 6 = 2 \cdot 3(x - 1)$, the LCD is $2^2 \cdot 3x(x - 1) = 12x(x - 1)$.

23. Since $x^2 - 4 = (x - 2)(x + 2)$ and $x^2 - x - 2 = (x - 2)(x + 1)$, the LCD is $(x + 1)(x - 2)(x + 2)$.

25. $\frac{5}{12} = \frac{5 \cdot 3}{12 \cdot 3} = \frac{15}{36}$

27. $\frac{2}{3xy} = \frac{2(5x)}{3xy(5x)} = \frac{10x}{15x^2y}$

29. $\frac{5}{y-6} = \frac{5(-2)}{(y-6)(-2)} = \frac{-10}{12-2y}$

31. $\frac{x}{x-1} = \frac{x(x+1)}{(x-1)(x+1)} = \frac{x^2+x}{x^2-1}$

33. $\frac{5}{36} + \frac{9}{28} = \frac{5}{2^2 3^2} + \frac{9}{2^2 7}$

$= \frac{5 \cdot 7}{2^2 3^2 \cdot 7} + \frac{9 \cdot 3^2}{2^2 7 \cdot 3^2} = \frac{35}{252} + \frac{81}{252} = \frac{116}{252}$

$= \frac{29}{63}$

35. $3 - \frac{4}{x} = \frac{3x}{x} - \frac{4}{x} = \frac{3x-4}{x}$

37. $\frac{2}{ab^2} - \frac{1}{a^2b} = \frac{2(a)}{ab^2(a)} - \frac{1(b)}{a^2b(b)}$

$= \frac{2a}{a^2b^2} - \frac{b}{a^2b^2} = \frac{2a-b}{a^2b^2}$

39. $\frac{9a}{2a-3} + \frac{5}{3a-2}$

$= \frac{9a(3a-2)}{(2a-3)(3a-2)} + \frac{5(2a-3)}{(3a-2)(2a-3)}$

$= \frac{27a^2-18a}{(2a-3)(3a-2)} + \frac{10a-15}{(2a-3)(3a-2)}$

$= \frac{27a^2-8a-15}{(2a-3)(3a-2)}$

41. $\frac{1}{a-8} - \frac{2}{8-a} = \frac{1}{a-8} - \frac{2(-1)}{(8-a)(-1)}$

$= \frac{1}{a-8} - \frac{-2}{a-8} = \frac{3}{a-8}$

43. $\frac{3}{2x-4} + \frac{1}{x^2-4}$

$= \frac{3}{2(x-2)} + \frac{1}{(x-2)(x+2)}$

$= \frac{3(x+2)}{2(x-2)(x+2)} + \frac{1(2)}{(x-2)(x+2)(2)}$

$= \frac{3x+6+2}{2(x+2)(x-2)} = \frac{3x+8}{2(x+2)(x-2)}$

45. $\frac{\frac{1}{2}-\frac{3}{4}}{\frac{2}{3}+\frac{1}{2}} = \frac{\left(\frac{1}{2}-\frac{3}{4}\right)(12)}{\left(\frac{2}{3}+\frac{1}{2}\right)(12)}$

$= \frac{6-9}{8+6} = \frac{-3}{14} = -\frac{3}{14}$

47. $\frac{\frac{1}{a}+\frac{2}{3b}}{\frac{1}{2b}-\frac{3}{a}} = \frac{\left(\frac{1}{a}+\frac{2}{3b}\right)(6ab)}{\left(\frac{1}{2b}-\frac{3}{a}\right)(6ab)}$

$= \frac{6b+4a}{3a-18b} = \frac{6b+4a}{3(a-6b)}$

49. $\frac{\left(\frac{1}{x-2}-\frac{3}{x+3}\right)(x-2)(x+3)}{\left(\frac{2}{x+3}+\frac{1}{x-2}\right)(x-2)(x+3)}$

$= \frac{x+3-3(x-2)}{2(x-2)+1(x+3)} = \frac{-2x+9}{3x-1}$

51. $\frac{\frac{x-1}{x-3}(x-3)(x+2)}{\left(\frac{1}{x^2-x-6}-\frac{4}{x+2}\right)(x-3)(x+2)}$

$= \frac{(x-1)(x+2)}{1-4(x-3)} = \frac{x^2+x-2}{-4x+13}$

53. $\frac{-2}{5} = \frac{3}{x}$

$-2x = 15$

$x = -\frac{15}{2}$

The solution to the equation is $-\frac{15}{2}$.

55. $\frac{14}{a^2-1} + \frac{1}{a-1} = \frac{3}{a+1}$

$(a-1)(a+1)\frac{14}{a^2-1} + (a-1)(a+1)\frac{1}{a-1}$

$= (a-1)(a+1)\frac{3}{a+1}$

$14 + a + 1 = 3(a-1)$

$15 + a = 3a - 3$

$18 = 2a$

$9 = a$

The solution to the equation is 9.

57. $z - \frac{3z}{2-z} = \frac{6}{z-2}$

$(z-2)z - (z-2)\frac{3z}{2-z} = (z-2)\frac{6}{z-2}$

$z^2 - 2z - (-1)3z = 6$

$z^2 - 2z + 3z = 6$

$z^2 + z - 6 = 0$

$(z+3)(z-2) = 0$

$z+3=0$ or $z-2=0$

$z=-3$ or $z=2$

Since the value of $z-2$ is 0 when $z=2$, 2 is not a solution to the equation. The solution to the equation is -3.

59. $\frac{3}{x} = \frac{2}{7}$

$2x = 21$

$x = \frac{21}{2}$

The solution to the equation is $\frac{21}{2}$.

61. $\frac{2}{w-3} = \frac{5}{w}$

$5w - 15 = 2w$

$3w = 15$

$w = 5$

The solution to the proportion is 5.

63. Let $x =$ the number of private automobiles. Since the ratio of taxis to private automobiles was 15 to 2, we can write the following proportion.

$\frac{15}{2} = \frac{60}{x}$

$15x = 120$

$x = 8$

There were 8 private automobiles.

65. Let $x =$ the number of cups of rice and $x + 28 =$ the number of cups of water. Since the ratio of water to rice is 2 to 1, we can write the following proportion.

$$\frac{2}{1} = \frac{x+28}{x}$$
$$2x = x + 28$$
$$x = 28$$
$$x + 28 = 56$$

He used 56 cups of water and 28 cups of rice.

67. $$\frac{y-b}{m} = x$$
$$y - b = mx$$
$$y = mx + b$$

69. $$F = \frac{mv+1}{m}$$
$$Fm = mv + 1$$
$$Fm - mv = 1$$
$$m(F - v) = 1$$
$$m = \frac{1}{F-v}$$

71. $$\frac{y+1}{x-3} = 4$$
$$y + 1 = 4x - 12$$
$$y = 4x - 13$$

73. Let $x =$ the number of hours for Stacy or Tracy to assemble the puzzle working alone. Since Stacy works twice as fast as Fred, it takes Fred $2x$ hours to assemble the puzzle by himself. Since Stacy does $1/x$ of the puzzle per hour, Tracy does $1/x$ of the puzzle per hour and Fred does $1/(2x)$ of the puzzle per hour, we can write the following equation.

$$\frac{1}{x} + \frac{1}{x} + \frac{1}{2x} = \frac{1}{40}$$
$$40x \cdot \frac{1}{x} + 40x \cdot \frac{1}{x} + 40x \cdot \frac{1}{2x} = 40x \cdot \frac{1}{40}$$
$$40 + 40 + 20 = x$$
$$100 = x$$
$$200 = 2x$$

It would take Fred 200 hours to assemble the puzzle by himself.

75. Let $x =$ the number of cars owned by Ernie and $x + 10 =$ the number of cars owned by Bert. Ernie had $0.36x$ new cars and Bert had $0.25(x + 10)$ new cars. Since the total number of new cars was 33, we can write the following equation.

$$0.36x + 0.25(x + 10) = 33$$
$$0.36x + 0.25x + 2.5 = 33$$
$$0.61x = 30.5$$
$$x = \frac{30.5}{0.61} = 50$$
$$x + 10 = 60$$

Bert had 60 cars and Ernie had 50 cars before the merger.

77. Let $x =$ the amount of yard waste and $x + 59.8 =$ the amount of paper waste.

$$\frac{x}{x+59.8} = \frac{12.1\%}{38.1\%}$$
$$\frac{x}{x+59.8} = \frac{121}{381}$$
$$381x = 121x + 59.8(121)$$
$$260x = 7235.8$$
$$x = 27.83$$

The amount of yard waste was 27.83 million tons.

79. $\frac{5}{x} = \frac{5 \cdot 2}{x \cdot 2} = \frac{10}{2x}$

81. $\frac{2}{a-5} = \frac{2(-1)}{(a-5)(-1)} = \frac{-2}{5-a}$

83. $3 = 3 \cdot \frac{x}{x} = \frac{3x}{x}$

85. $m \div \frac{1}{2} = m \cdot 2 = 2m$

87. $2a \div \frac{1}{6} = 2a \cdot 6 = 12a$

89. $\frac{a-1}{a^2-1} = \frac{(a-1)1}{(a-1)(a+1)} = \frac{1}{a+1}$

91. $\frac{1}{a} - \frac{1}{5} = \frac{1 \cdot 5}{a \cdot 5} - \frac{1 \cdot a}{5 \cdot a}$
$= \frac{5}{5a} - \frac{a}{5a} = \frac{5-a}{5a}$

93. $\frac{a}{2} - 1 = \frac{a}{2} - \frac{2}{2} = \frac{a-2}{2}$

95. $(a - b) \div (-1) = (a - b)(-1) = b - a$

97. $\frac{\frac{1}{5a}}{2} = \frac{1}{5a} \cdot \frac{1}{2} = \frac{1}{10a}$

99. $\frac{1}{x} + \frac{1}{2x} = \frac{1 \cdot 2}{x \cdot 2} + \frac{1}{2x}$
$= \frac{2}{2x} + \frac{1}{2x} = \frac{3}{2x}$

101. $\frac{2}{3xy} + \frac{1}{6x}$
$= \frac{2 \cdot 2}{3xy \cdot 2} + \frac{1 \cdot y}{6x \cdot y} = \frac{4}{6xy} + \frac{y}{6xy}$
$= \frac{4+y}{6xy}$

103. $\frac{5}{a-5} - \frac{3}{5-a} = \frac{5}{a-5} - \frac{3(-1)}{(5-a)(-1)}$
$= \frac{5}{a-5} - \frac{-3}{a-5} = \frac{8}{a-5}$

105. $$\frac{2}{x-1} - \frac{2}{x} = 1$$
$$x(x-1)\frac{2}{x-1} - x(x-1)\frac{2}{x} = x(x-1)1$$
$$2x - 2(x-1) = x^2 - x$$
$$2 = x^2 - x$$
$$0 = x^2 - x - 2$$
$$0 = (x-2)(x+1)$$

$$x - 2 = 0 \text{ or } x + 1 = 0$$
$$x = 2 \text{ or } \quad x = -1$$

The solutions to the equation are -1 and 2.

107. $\frac{-3}{x+2} \cdot \frac{5x+10}{9} = \frac{-3}{x+2} \cdot \frac{5(x+2)}{3 \cdot 3}$
$$= -\frac{5}{3}$$

109. $\frac{1}{-3} = \frac{-2}{x}$
$$1x = (-2)(-3)$$
$$x = 6$$

The solution to the equation is 6.

111. $\frac{ax + am + 3x + 3m}{a^2 - 9} \div \frac{2x + 2m}{a - 3}$
$$= \frac{(a+3)(x+m)}{(a-3)(a+3)} \cdot \frac{a-3}{2(x+m)}$$
$$= \frac{1}{2}$$

113. $\frac{2}{x^2 - 25} + \frac{1}{x^2 - 4x - 5}$
$$= \frac{2}{(x-5)(x+5)} + \frac{1}{(x-5)(x+1)}$$
$$= \frac{2(x+1)}{(x-5)(x+5)(x+1)} + \frac{1(x+5)}{(x-5)(x+1)(x+5)}$$
$$= \frac{3x+7}{(x-5)(x+5)(x+1)}$$

115. $\frac{-3}{a^2 - 9} - \frac{2}{a^2 + 5a + 6}$
$$= \frac{-3}{(a-3)(a+3)} - \frac{2}{(a+2)(a+3)}$$
$$= \frac{-3(a+2)}{(a-3)(a+3)(a+2)} - \frac{2(a-3)}{(a+2)(a+3)(a-3)}$$
$$= \frac{-3a-6}{(a-3)(a+3)(a+2)} - \frac{2a-6}{(a-3)(a+3)(a+2)}$$
$$= \frac{-5a}{(a-3)(a+3)(a+2)}$$

117. $\frac{1}{a^2 - 1} + \frac{2}{1 - a} = \frac{3}{a+1}$
$$(a+1)(a-1)\frac{1}{a^2-1} + (a+1)(a-1)\frac{2}{1-a} = (a+1)(a-1)\frac{3}{a+1}$$
$$1 + (a+1)(-1)2 = (a-1)3$$
$$1 - 2a - 2 = 3a - 3$$
$$-1 - 2a = 3a - 3$$
$$2 = 5a$$
$$\frac{2}{5} = a$$

The solution to the equation is $\frac{2}{5}$.

CHAPTER 6 TEST

1. We cannot use any number for which $x^2 - 1 = 0$.
$$x^2 - 1 = 0$$
$$(x-1)(x+1) = 0$$
$$x - 1 = 0 \text{ or } x + 1 = 0$$
$$x = 1 \text{ or } \quad x = -1$$

We cannot use -1 or 1 in place of x.

2. We cannot use any number for which $2 - 3x = 0$.
$$2 - 3x = 0$$
$$-3x = -2$$
$$x = \frac{2}{3}$$

We cannot use $\frac{2}{3}$ in place of x.

3. We cannot use 0 for x because if $x = 0$, then the denominator of $1/x$ is 0.

4. $\frac{2}{15} - \frac{4}{9} = \frac{2 \cdot 3}{15 \cdot 3} - \frac{4 \cdot 5}{9 \cdot 5}$
$$= \frac{6}{45} - \frac{20}{45} = -\frac{14}{45}$$

5. $\frac{1}{y} + 3 = \frac{1}{y} + \frac{3y}{y} = \frac{1 + 3y}{y}$

6. $\frac{3}{a-2} - \frac{1}{2-a} = \frac{3}{a-2} - \frac{1(-1)}{(2-a)(-1)}$
$$= \frac{3}{a-2} - \frac{-1}{a-2} = \frac{4}{a-2}$$

7. $\frac{2}{x^2 - 4} - \frac{3}{x^2 + x - 2}$
$$= \frac{2}{(x-2)(x+2)} - \frac{3}{(x+2)(x-1)}$$
$$= \frac{2(x-1)}{(x-2)(x+2)(x-1)} - \frac{3(x-2)}{(x+2)(x-1)(x-2)}$$
$$= \frac{2x-2}{(x+2)(x-2)(x-1)} - \frac{3x-6}{(x+2)(x-2)(x-1)}$$
$$= \frac{-x+4}{(x+2)(x-2)(x-1)}$$

8. $\frac{m^2 - 1}{(m-1)^2} \cdot \frac{2m-2}{3m+3}$
$$= \frac{(m-1)(m+1)}{(m-1)^2} \cdot \frac{2(m-1)}{3(m+1)}$$
$$= \frac{2}{3}$$

9. $\frac{a-b}{3} \div \frac{b^2-a^2}{6} = \frac{a-b}{3} \cdot \frac{2 \cdot 3}{(b-a)(b+a)}$
$= \frac{-2}{a+b}$

10. $\frac{5a^2b}{12a} \cdot \frac{2a^3b}{15ab^6} = \frac{5 \cdot 2a^5b^2}{2 \cdot 2 \cdot 3 \cdot 3 \cdot 5a^2b^6}$
$= \frac{a^3}{18b^4}$

11. $\frac{\frac{2}{3}+\frac{4}{5}}{\frac{2}{5}-\frac{3}{2}} = \frac{\left(\frac{2}{3}+\frac{4}{5}\right)(30)}{\left(\frac{2}{5}-\frac{3}{2}\right)(30)}$
$= \frac{20+24}{12-45} = \frac{44}{-33} = -\frac{4}{3}$

12. $\frac{\frac{2}{x}+\frac{1}{x-2}}{\frac{1}{x-2}-\frac{3}{x}} = \frac{\left(\frac{2}{x}+\frac{1}{x-2}\right)(x)(x-2)}{\left(\frac{1}{x-2}-\frac{3}{x}\right)(x)(x-2)}$
$= \frac{2(x-2)+x}{x-3(x-2)} = \frac{3x-4}{-2x+6}$
$= \frac{-3x+4}{2x-6} = \frac{-3x+4}{2(x-3)}$

13.
$$\frac{3}{x} = \frac{7}{5}$$
$$7x = 15$$
$$x = \frac{15}{7}$$

The solution to the equation is $\frac{15}{7}$.

14.
$$\frac{x}{x-1} - \frac{3}{x} = \frac{1}{2}$$
$$2x(x-1)\frac{x}{x-1} - 2x(x-1)\frac{3}{x} = 2x(x-1)\frac{1}{2}$$
$$2x^2 - 6(x-1) = x(x-1)$$
$$2x^2 - 6x + 6 = x^2 - x$$
$$x^2 - 5x + 6 = 0$$
$$(x-2)(x-3) = 0$$
$$x-2=0 \text{ or } x-3=0$$
$$x=2 \text{ or } x=3$$

The solutions to the equation are 2 and 3.

15.
$$\frac{1}{x} + \frac{1}{6} = \frac{1}{4}$$
$$12x \cdot \frac{1}{x} + 12x \cdot \frac{1}{6} = 12x \cdot \frac{1}{4}$$
$$12 + 2x = 3x$$
$$12 = x$$

The solution to the equation is 12.

16.
$$\frac{y-3}{x+2} = \frac{-1}{5}$$
$$y - 3 = -\frac{1}{5}(x+2)$$
$$y - 3 = -\frac{1}{5}x - \frac{2}{5}$$
$$y = -\frac{1}{5}x - \frac{2}{5} + 3$$
$$y = -\frac{1}{5}x + \frac{13}{5}$$

17.
$$M = \frac{1}{3}b(c+d)$$
$$3M = b(c+d)$$
$$3M = bc + bd$$
$$3M - bd = bc$$
$$\frac{3M-bd}{b} = c$$
$$c = \frac{3M-bd}{b}$$

18. $R(0.9) = \frac{0.9+2}{1-0.9} = \frac{2.9}{0.1} = 29$

19. Let $x =$ the number of minutes it takes for them to do the job together. Reginald does $1/12$ of the job per minute and Norman does $1/18$ of the job per minute. Since together they get $1/x$ of the job done per minute, we can write the following equation.
$$\frac{1}{12} + \frac{1}{18} = \frac{1}{x}$$
$$36x \cdot \frac{1}{12} + 36x \cdot \frac{1}{18} = 36x \cdot \frac{1}{x}$$
$$3x + 2x = 36$$
$$5x = 36$$
$$x = \frac{36}{5} = 7.2$$

It takes them 7.2 minutes to complete the job when working together.

20. Let $x =$ Brenda's rate and $x + 5 =$ Randy's rate. Since Brenda rode 30 miles, her time was $30/x$ hours. Since Randy rode 60 miles, his time was $60/(x+5)$ hours. Since Randy traveled one hour longer, we can write the following equation.
$$\frac{30}{x} + 1 = \frac{60}{x+5}$$
$$x(x+5)\frac{30}{x} + x(x+5)1 = x(x+5)\frac{60}{x+5}$$
$$30x + 150 + x^2 + 5x = 60x$$
$$x^2 - 25x + 150 = 0$$
$$(x-15)(x-10) = 0$$
$$x-15=0 \text{ or } x-10=0$$
$$x=15 \text{ or } x=10$$
$$x+5=20 \text{ or } x+5=15$$

There are two possible answers to this problem. It could be that Brenda rode at 15 mph and Randy 20 mph, or it could be that Brenda rode at 10 mph and Randy 15 mph.

21. Let $x =$ the dollar value of imports. Since the ratio of exports to imports was 2 to 3, we can write the following equation.
$$\frac{2}{3} = \frac{48}{x}$$
$$2x = 144$$
$$x = 72$$

The value of imports was 72 billion dollars.

Making Connections

Chapters 1- 6

1. $3x - 2 = 5$
$$3x = 7$$
$$x = \frac{7}{3}$$
The solution to the equation is $\frac{7}{3}$.

2. $\frac{3}{5}x = -2$
$$\frac{5}{3} \cdot \frac{3}{5}x = -2 \cdot \frac{5}{3}$$
$$x = -\frac{10}{3}$$
The solution to the equation is $-\frac{10}{3}$.

3. $2(x - 2) = 4x$
$$2x - 4 = 4x$$
$$-4 = 2x$$
$$-2 = x$$
The solution to the equation is -2.

4. $2(x - 2) = 2x$
$$2x - 4 = 2x$$
$$-4 = 0$$
There is no solution to the equation.

5. $2(x + 3) = 6x + 6$
$$2x + 6 = 6x + 6$$
$$2x = 6x$$
$$0 = 4x$$
$$0 = x$$
The solution to the equation is 0.

6. $2(3x + 4) + x^2 = 0$
$$6x + 8 + x^2 = 0$$
$$x^2 + 6x + 8 = 0$$
$$(x + 2)(x + 4) = 0$$
$$x + 2 = 0 \quad \text{or} \quad x + 4 = 0$$
$$x = -2 \quad \text{or} \quad x = -4$$
The solutions to the equation are -4 and -2.

7. $4x - 4x^3 = 0$
$$4x(1 - x^2) = 0$$
$$4x(1 - x)(1 + x) = 0$$
$$4x = 0 \text{ or } 1 - x = 0 \text{ or } 1 + x = 0$$
$$x = 0 \text{ or } \quad 1 = x \text{ or } \quad x = -1$$
The solutions to the equation are -1, 0, and 1.

8. $\frac{3}{x} = \frac{-2}{5}$
$$-2x = 15$$
$$x = -\frac{15}{2}$$
The solution to the equation is $-\frac{15}{2}$.

9. $\frac{3}{x} = \frac{x}{12}$
$$x^2 = 36$$
$$x^2 - 36 = 0$$
$$(x - 6)(x + 6) = 0$$
$$x - 6 = 0 \text{ or } x + 6 = 0$$
$$x = 6 \text{ or } \quad x = -6$$
The solutions to the equation are -6 and 6.

10. $\frac{x}{2} = \frac{4}{x - 2}$
$$x(x - 2) = 8$$
$$x^2 - 2x = 8$$
$$x^2 - 2x - 8 = 0$$
$$(x - 4)(x + 2) = 0$$
$$x - 4 = 0 \text{ or } x + 2 = 0$$
$$x = 4 \text{ or } \quad x = -2$$
The solutions to the equation are -2 and 4.

11. $18 \cdot \frac{w}{18} - 18 \cdot \frac{w - 1}{9} = 18 \cdot \frac{4 - w}{6}$
$$w - 2(w - 1) = 3(4 - w)$$
$$-w + 2 = 12 - 3w$$
$$2w = 10$$
$$w = 5$$
The solution to the equation is 5.

12. $\frac{x}{x + 1} + \frac{1}{2(x + 1)} = \frac{7}{8}$
$$8(x + 1)\left(\frac{x}{x + 1} + \frac{1}{2(x + 1)}\right) = 8(x + 1)\frac{7}{8}$$
$$8x + 4 = 7x + 7$$
$$x = 3$$
The solution to the equation is 3.

13. $2x + 3y = c$
$$3y = -2x + c$$
$$y = \frac{-2x + c}{3}$$
$$y = \frac{c - 2x}{3}$$

14. $\frac{y - 3}{x - 5} = \frac{1}{2}$
$$y - 3 = \frac{1}{2}(x - 5)$$
$$y - 3 = \frac{1}{2}x - \frac{5}{2}$$
$$y = \frac{1}{2}x + \frac{1}{2}$$

15. $2y = ay + c$
$$2y - ay = c$$
$$y(2 - a) = c$$
$$y = \frac{c}{2 - a}$$

16. $\frac{A}{y} = \frac{C}{B}$
$$Cy = AB$$
$$y = \frac{AB}{C}$$

17. $\frac{A}{y} + \frac{1}{3} = \frac{B}{y}$

$3y \cdot \frac{A}{y} + 3y \cdot \frac{1}{3} = 3y \cdot \frac{B}{y}$

$3A + y = 3B$

$y = 3B - 3A$

18. $\frac{A}{y} - \frac{1}{2} = \frac{1}{3}$

$6y \cdot \frac{A}{y} - 6y \cdot \frac{1}{2} = 6y \cdot \frac{1}{3}$

$6A - 3y = 2y$

$6A = 5y$

$\frac{6A}{5} = y$

$y = \frac{6A}{5}$

19. $3y - 5ay = 8$

$y(3 - 5a) = 8$

$y = \frac{8}{3 - 5a}$

20. $y^2 - By = 0$

$y(y - B) = 0$

$y = 0$ or $y - B = 0$

$y = 0$ or $y = B$

21. $A = \frac{1}{2}h(b + y)$

$2A = h(b + y)$

$2A = hb + hy$

$2A - hb = hy$

$\frac{2A - hb}{h} = y$

$y = \frac{2A - hb}{h}$

22. $2(b + y) = b$

$2b + 2y = b$

$2y = -b$

$y = -\frac{b}{2}$

23. $b^2 - 4ac = 2^2 - 4(1)(-15)$
$= 4 - (-60) = 64$

24. $b^2 - 4ac = 8^2 - 4(1)(12)$
$= 64 - 48 = 16$

25. $b^2 - 4ac = 5^2 - 4(2)(-3)$
$= 25 - (-24) = 49$

26. $b^2 - 4ac = 7^2 - 4(6)(-3)$
$= 49 - (-72) = 121$

27. $(3x - 5) - (5x - 3) = 3x - 5 - 5x + 3$
$= -2x - 2$

28. $(2a - 5)(a - 3) = 2a^2 - 5a - 6a + 15$
$= 2a^2 - 11a + 15$

29. $x^7 \div x^3 = x^{7-3} = x^4$

30. $\frac{x-3}{5} + \frac{x+4}{5} = \frac{x - 3 + x + 4}{5}$
$= \frac{2x+1}{5}$

31. $\frac{1}{2} \cdot \frac{1}{x} = \frac{1 \cdot 1}{2 \cdot x} = \frac{1}{2x}$

32. $\frac{1}{2} + \frac{1}{x} = \frac{1 \cdot x + 2 \cdot 1}{2 \cdot x} = \frac{x+2}{2x}$

33. $\frac{1}{2} \div \frac{1}{x} = \frac{1}{2} \cdot x = \frac{x}{2}$

34. $\frac{1}{2} - \frac{1}{x} = \frac{1 \cdot x - 2 \cdot 1}{2x} = \frac{x-2}{2x}$

35. $\frac{x-3}{5} - \frac{x+4}{5} = \frac{x - 3 - x - 4}{5} = -\frac{7}{5}$

36. $\frac{3a}{2} \div 2 = \frac{3a}{2} \cdot \frac{1}{2} = \frac{3a}{4}$

37. $(x - 8)(x + 8) = x^2 - 8^2 = x^2 - 64$

38. $3x(x^2 - 7) = 3x^3 - 21x$

39. $2a^5 \cdot 5a^9 = 2 \cdot 5 \cdot a^{5+9} = 10a^{14}$

40. $x^2 \cdot x^8 = x^{2+8} = x^{10}$

41. $(k - 6)^2 = k^2 - 2 \cdot k \cdot 6 + 6^2$
$= k^2 - 12k + 36$

42. $(j + 5)^2 = j^2 + 2 \cdot j \cdot 5 + 5^2$
$= j^2 + 10j + 25$

43. $(g - 3) \div (3 - g) = \frac{g-3}{3-g}$
$= \frac{-1(3-g)}{3-g} = -1$

44. $(6x^3 - 8x^2) \div (2x) = \frac{6x^3}{2x} - \frac{8x^2}{2x}$
$= 3x^2 - 4x$

45. **a)** $P = \frac{1}{1+r} + \frac{1}{(1+r)^2}$
$= \frac{1(1+r)}{(1+r)(1+r)} + \frac{1}{(1+r)^2} = \frac{r+2}{(1+r)^2}$

b) If $r = 7\%$, then $P = \frac{0.07 + 2}{(1 + 0.07)^2} \approx 1.8080$ or \$1.81.

c) $P = \frac{1}{1 + 0.05} + \frac{1}{(1 + 0.05)^2}$
$+ \frac{1}{(1 + 0.05)^3} + \frac{1}{(1 + 0.05)^4} + \frac{1}{(1 + 0.05)^5}$
$+ \frac{1}{(1 + 0.05)^6} + \frac{1}{(1 + 0.05)^7} + \frac{1}{(1 + 0.05)^8}$
$+ \frac{1}{(1 + 0.05)^9} + \frac{1}{(1 + 0.05)^{10}}$
$\approx \$7.72$

7.1 WARM-UPS

1. True, because if $x = 1$ and $y = 2$, then $2(1) + 2 = 4$ is correct.
2. False, because if $x = 1$ and $y = 2$, then $3(1) - 2 = 6$ is incorrect. Both equations must be satisfied for the compound statement using "and" to be true.
3. False, because (2, 3) does not satisfy $4x - y = -5$.
4. True.
5. True, because when we substitute, one of the variables is eliminated.
6. True, because each of these lines has slope 3 and they are parallel.
7. True, because the lines are not parallel and they have different y-intercepts.
8. True, because the lines are parallel and have different y-intercepts.
9. True, because dependent equations have the same solution sets.
10. True, because a system of independent equations has one point in its solution set.

7.1 EXERCISES

1. The intersection point of the graphs is the solution to an independent system.
3. The graphing method can be very inaccurate.
5. If the equation you get after substituting turns out to be incorrect, such as $0 = 9$, then the system has no solution.
7. The graph of $y = 2x$ is a straight line with y-intercept (0, 0) and slope 2. The graph of $y = -x + 3$ is a straight line with y-intercept (0, 3) and slope -1. The graphs appear to intersect at (1, 2). Check that (1, 2) satisfies both equations. The solution set is $\{(1, 2)\}$.
9. The graph of $y = 2x - 1$ is a straight line with y-intercept $(0, -1)$ and slope 2. The graph of $2y = x - 2$ or $y = \frac{1}{2}x - 1$ is a straight line with y-intercept $(0, -1)$ and slope $\frac{1}{2}$. The graphs intersect at $(0, -1)$. The solution set to the system is $\{(0, -1)\}$.
11. The graph of $y = x - 3$ is a line with y-intercept $(0, -3)$ and slope 1. The graph of $x - 2y = 4$ or $y = \frac{1}{2}x - 2$ is a line with y-intercept $(0, -2)$ and slope $\frac{1}{2}$. The lines appear to intersect at $(2, -1)$. To be sure, check that $(2, -1)$ satisfies both of the original equations. The solution set is $\{(2, -1)\}$.
13. If we rewrite these equations in slope-intercept form, we get $y = x + 1$ and $y = x + 3$. The graphs are parallel lines with slopes of 1 and y-intercepts of (0, 1) and (0, 3). Since the lines do not intersect, the solution set is the empty set, $\emptyset$.
15. Solving $x + 2y = 8$ for y yields $y = -\frac{1}{2}x + 4$, which is the same as the first equation. So the equations have the same graph. All points on the line satisfy both equations. The solution set is $\{(x, y) \mid x + 2y = 8\}$.
17. The graph of $y = 2x + 4$ is a line with slope 2 and y-intercept $(0, 4)$. The graph of $3x + y = -1$ or $y = -3x - 1$ is a line with slope -3 and y-intercept $(0, -1)$. The lines appear to intersect at $(-1, 2)$. Check that $(-1, 2)$ satisfies both equations. The solution set is $\{(-1, 2)\}$
19. The graph of $y = -\frac{1}{4}x$ is a line with slope $-1/4$ and y-intercept $(0, 0)$. The graph of $x + 4y = 8$ or $y = -\frac{1}{4}x + 2$ is a line with slope $-1/4$ and y-intercept $(0, 2)$. The lines have the same slope and do not intersect. The solution set is the empty set, $\emptyset$.
21. The lines in graph (c) intersect at $(3, -2)$ and $(3, -2)$ satisfies both equations of the given system.
23. The lines in graph (b) intersect at $(-3, -2)$ and $(-3, -2)$ satisfies both equations.
25. Substitute $y = x - 5$ into $2x - 5y = 1$.

$$2x - 5(x - 5) = 1$$
$$2x - 5x + 25 = 1$$
$$-3x = -24$$
$$x = 8$$

Use $x = 8$ in $y = x - 5$ to find y.

$$y = 8 - 5$$
$$y = 3$$

The solution set is $\{(8, 3)\}$. The equations are independent.

27. Substitute $x = 2y - 7$ into $3x + 2y = -5$.

$$3(2y-7)+2y = -5$$
$$6y - 21 + 2y = -5$$
$$8y = 16$$
$$y = 2$$

Use $y = 2$ in $x = 2y - 7$ to find x.

$$x = 2(2) - 7 = -3$$

The solution set is $\{(-3, 2)\}$. The equations are independent.

29. Substitute $y = 2x - 30$ into the other equation.

$$\frac{1}{5}x - \frac{1}{2}(2x - 30) = -1$$
$$\frac{1}{5}x - x + 15 = -1$$
$$5\left(\frac{1}{5}x - x + 15\right) = 5(-1)$$
$$x - 5x + 75 = -5$$
$$-4x = -80$$
$$x = 20$$

If $x = 20$, then

$$y = 2x - 30 = 2(20) - 30 = 10.$$

The solution set is $\{(20, 10)\}$ and the equations are independent.

31. Write $x - y = 5$ as $x = y + 5$ and substitute into $2x = 2y + 14$.

$$2(y+5) = 2y + 14$$
$$2y + 10 = 2y + 14$$
$$10 = 14$$

Since this last equation is incorrect no matter what values are used for x and y, the equations are inconsistent and the solution set is $\emptyset$.

33. Substitute $y = 2x - 5$ into $y + 1 = 2(x - 2)$.

$$2x - 5 + 1 = 2(x-2)$$
$$2x - 4 = 2x - 4$$

Since the last equation is an identity, the equations are dependent. Any ordered pair that satisfies one equation will satisfy the other. The solution set is $\{(x, y) \mid y = 2x - 5\}$.

35. Write $2x + y = 9$ as $y = -2x + 9$ and substitute into $2x - 5y = 15$.

$$2x - 5(-2x + 9) = 15$$
$$2x + 10x - 45 = 15$$
$$12x - 45 = 15$$
$$12x = 60$$
$$x = 5$$

Use $x = 5$ in $y = -2x + 9$ to find y.

$$y = -2(5) + 9 = -1$$

The solution set is $\{(5, -1)\}$. The equations are independent.

37. Write $x - y = 0$ as $y = x$ and substitute into $2x + 3y = 35$.

$$2x + 3x = 35$$
$$5x = 35$$
$$x = 7$$

Since $y = x$, $y = 7$ also. The solution set is $\{(7, 7)\}$ and the equations are independent.

39. Write $x + y = 40$ as $x = 40 - y$ and substitute into $0.2x + 0.8y = 23$.

$$0.2(40 - y) + 0.8y = 23$$
$$8 - 0.2y + 0.8y = 23$$
$$0.6y = 15$$
$$y = 25$$

Since $x = 40 - y$, we get $x = 40 - 25 = 15$. The solution set is $\{(15, 25)\}$ and the equations are independent.

41. Substitute $y = \frac{5}{7}x$ into $x = -\frac{2}{3}y$:

$$x = -\frac{2}{3}\left(\frac{5}{7}x\right)$$
$$x = -\frac{10}{21}x$$
$$21x = -10x$$
$$31x = 0$$
$$x = 0$$
$$y = \frac{5}{7}x$$
$$y = \frac{5}{7} \cdot 0 = 0$$

The solution set is $\{(0, 0)\}$. The system is independent.

43. First simplify.

$$3(y-1) = 2(x-3)$$
$$3y - 3 = 2x - 6$$
$$3y - 2x = -3$$

By simplifying the first equation we see that it is the same as the second equation. There is no need to substitute now. The equations are dependent and the solution set is $\{(x, y) \mid 3y - 2x = -3\}$.

45. Substitute $y = 3x$ into $y = 3x + 1$.

$$3x = 3x + 1$$
$$0 = 1$$

Since the last equation is false, the equations are inconsistent. The solution set to the system is the empty set, $\emptyset$.

47. Substitute $y = \frac{5}{2}x$ into $x + 3y = 3$:

$$x + 3\left(\frac{5}{2}x\right) = 3$$
$$x + \frac{15}{2}x = 3$$
$$2x + 15x = 6$$
$$17x = 6$$
$$x = \frac{6}{17}$$
$$y = \frac{5}{2}x$$
$$y = \frac{5}{2} \cdot \frac{6}{17} = \frac{15}{17}$$

The solution set is $\left\{\left(\frac{6}{17}, \frac{15}{17}\right)\right\}$. The system is independent.

49. Write $x - y = 5$ as $x = y + 5$ and substitute into $x + y = 4$.

$$y + 5 + y = 4$$
$$2y = -1$$
$$y = -1/2$$

If $y = -1/2$, then

$$x = y + 5 = -1/2 + 5 = 9/2.$$

The solution set is $\left\{\left(\frac{9}{2}, -\frac{1}{2}\right)\right\}$ and the equations are independent.

51. Write $2x - 4y = 0$ as $y = \frac{1}{2}x$. Substitute $y = \frac{1}{2}x$ into $6x + 8y = 5$.

$$6x + 8\left(\frac{1}{2}x\right) = 5$$
$$6x + 4x = 5$$
$$10x = 5$$
$$x = \frac{1}{2}$$
$$y = \frac{1}{2}\left(\frac{1}{2}\right) = \frac{1}{4}$$

The solution set is $\left\{\left(\frac{1}{2}, \frac{1}{4}\right)\right\}$.

53. Write $3x + y = 2$ as $y = -3x + 2$. Substitute $y = -3x + 2$ into $-x - 3y = 6$.

$$-x - 3(-3x + 2) = 6$$
$$-x + 9x - 6 = 6$$
$$8x = 12$$
$$x = \frac{3}{2}$$
$$y = -3\left(\frac{3}{2}\right) + 2 = -\frac{9}{2} + \frac{4}{2} = -\frac{5}{2}$$

The solution set is $\left\{\left(\frac{3}{2}, -\frac{5}{2}\right)\right\}$.

55. Write $-9x + 6y = 3$ as $y = \frac{3}{2}x + \frac{1}{2}$. Substitute $y = \frac{3}{2}x + \frac{1}{2}$ into $18x + 30y = 1$.

$$18x + 30\left(\frac{3}{2}x + \frac{1}{2}\right) = 1$$
$$18x + 45x + 15 = 1$$
$$63x = -14$$
$$x = -\frac{2}{9}$$
$$y = \frac{3}{2}\left(-\frac{2}{9}\right) + \frac{1}{2} = \frac{1}{6}$$

The solution set is $\left\{\left(-\frac{2}{9}, \frac{1}{6}\right)\right\}$.

57. Substitute $y = -2x$ into $3y - x = 1$.

$$3(-2x) - x = 1$$
$$-7x = 1$$
$$x = -\frac{1}{7}$$
$$y = -2\left(-\frac{1}{7}\right) = \frac{2}{7}$$

The solution set is $\left\{\left(-\frac{1}{7}, \frac{2}{7}\right)\right\}$.

59. Substitute $x = -6y + 1$ into $2y = -5x$.

$$2y = -5(-6y + 1)$$
$$2y = 30y - 5$$
$$-28y = -5$$
$$y = \frac{5}{28}$$
$$x = -6\left(\frac{5}{28}\right) + 1 = -\frac{1}{14}$$

The solution set is $\left\{\left(-\frac{1}{14}, \frac{5}{28}\right)\right\}$.

61. Write $x - y = 0.1$ as $x = y + 0.1$ and substitute into $2x - 3y = -0.5$.

$$2(y + 0.1) - 3y = -0.5$$
$$2y + 0.2 - 3y = -0.5$$
$$-y = -0.7$$
$$y = 0.7$$

If $y = 0.7$, then $x = 0.7 + 0.1 = 0.8$. The solution set is $\{(0.8, 0.7)\}$.

63. Let x represent the width and y represent the length, then $2x + 2y = 84$ and $y = x + 12$.

$$2x + 2(x + 12) = 84$$
$$4x + 24 = 84$$
$$4x = 60$$
$$x = 15$$
$$y = 15 + 12 = 27$$

The length is 27 feet and the width is 15 feet.

65. Let x represent the width and y represent the length, then $2x + 2y = 28$ and $x = \frac{1}{2}y - 1$.

$$2\left(\frac{1}{2}y - 1\right) + 2y = 28$$
$$y - 2 + 2y = 28$$
$$3y = 30$$
$$y = 10$$
$$x = \frac{1}{2}(10) - 1 = 4$$

The length is 10 ft and the width is 4 ft.

67. Let x and y represent the numbers.

$$x + y = 10$$
$$x - y = 3$$

Solve $x - y = 3$ for x to get $x = y + 3$.
Substitute into $x + y = 10$.

$$y + 3 + y = 10$$
$$2y = 7$$
$$y = \frac{7}{2}$$
$$x = \frac{7}{2} + 3 = \frac{13}{2}$$

The numbers are $7/2$ and $13/2$ or as decimals 3.5 and 6.5.

69. Let x and y represent the numbers.

$$x + y = 1$$
$$x - y = 20$$

Solve $x - y = 20$ for x to get $x = y + 20$.
Substitute into $x + y = 1$.

$$y + 20 + y = 1$$
$$2y = -19$$
$$y = -9.5$$
$$x = -9.5 + 20 = 10.5$$

The numbers are -9.5 and 10.5.

71. Let x represent the number of \$200 tickets and y represent the number of \$250 tickets.

$$x + y = 200$$
$$200x + 250y = 44{,}000$$

Solve $x + y = 200$ for x to get $x = 200 - y$.
Substitute into $200x + 250y = 44,000$.

$$200(200 - y) + 250y = 44,000$$
$$40000 - 200y + 250y = 44,000$$
$$50y = 4000$$
$$y = 80$$
$$x = 200 - 80 = 120$$

There were 120 tickets for \$200 each and 80 tickets for \$250 each.

73. Let x represent the number of student tickets (\$6) and y represent the number of nonstudent tickets (\$11).

$$y = 2x$$
$$6x + 11y = 1540$$

Substitute $y = 2x$ into $6x + 11y = 1540$.

$$6x + 11(2x) = 1540$$
$$28x = 1540$$
$$x = 55$$
$$y = 2(55) = 110$$

There were 55 tickets for \$6 each and 110 tickets for \$11 each.

75. Let x represent the amount invested at 5% and y represent the amount invested at 8%.

$$x + y = 40,000$$
$$0.05x + 0.08y = 2300$$

Substitute $y = 40,000 - x$ into $0.05x + 0.08y = 2300$.

$$0.05x + 0.08(40,000 - x) = 2300$$
$$0.05x + 3200 - 0.08x = 2300$$
$$-0.03x = -900$$
$$x = 30,000$$
$$y = 40,000 - 30,000 = 10,000$$

She invested \$30,000 at 5% and \$10,000 at 8%.

77. Let $x =$ the amount invested at 5% and $y =$ the amount invested at 10%. The interest earned on the investments is $0.05x$ and $0.10y$ respectively. We can write two equations, one expressing the total amount invested, and the other expressing the total of the interest earned.

$$x + y = 30{,}000$$
$$0.05x + 0.10y = 2{,}300$$

Write the first equation as $y = 30{,}000 - x$ and substitute into the second.

$$0.05x + 0.10(30{,}000 - x) = 2{,}300$$
$$0.05x + 3{,}000 - 0.10x = 2{,}300$$
$$-0.05x = -700$$
$$x = 14{,}000$$

If $x = 14{,}000$, then $y = 30{,}000 - 14{,}000 = 16{,}000$. She invested \$14,000 at 5% and \$16,000 at 10%.

79. Let x and y be the numbers. The fact that their sum is 2 and their difference is 26 gives us a system of equations:

$$x + y = 2$$
$$x - y = 26$$

Write $x + y = 2$ as $x = 2 - y$ and substitute.

$$2 - y - y = 26$$
$$-2y = 24$$
$$y = -12$$

If $y = -12$, then $x = 2 - y = 2 - (-12) = 14$. The numbers are -12 and 14.

81. Let $x =$ the number of toasters and $y =$ the number of vacation coupons. We write one equation expressing the fact that the total number of prizes is 100 and the other expressing the total bill of \$708.

$$x + y = 100$$
$$6x + 24y = 708$$

Write $x + y = 100$ as $x = 100 - y$ and substitute.

$$6(100 - y) + 24y = 708$$
$$600 - 6y + 24y = 708$$
$$18y = 108$$
$$y = 6$$

If $y = 6$, then $x = 100 - y = 100 - 6 = 94$. He gave away 94 toasters and 6 vacation coupons.

83. $s = 0.05(100{,}000 - f)$
$f = 0.30(100{,}000 - s)$

Simplify:

$$s = 5000 - 0.05f$$
$$f = 30{,}000 - 0.30s$$

Substitute:

$$f = 30{,}000 - 0.30(5000 - 0.05f)$$
$$f = 30{,}000 - 1500 + 0.015f$$
$$0.985f = 28{,}500$$
$$f \approx 28{,}934$$
$$s = 5{,}000 - 0.05(28{,}934)$$
$$\approx 3553$$

Rounding to the nearest dollar, the state tax is \$3,553 and federal tax is \$28,934.

85. $B = 0.20N$
$N = 120{,}000 - B$

Substitute:

$$N = 120{,}000 - 0.20N$$
$$1.2N = 120{,}000$$
$$N = 100{,}000$$
$$B = 0.20 \cdot 100{,}000 = 20{,}000$$

The bonus is \$20,000.

87. a) Use $x = 10{,}000$ in $C = 10x + 400{,}000$ to get the cost of 10,000 textbooks. The cost is \$500,000.

b) Use $x = 10.000$ in $F = 30x$ to get the revenue for 10,000 textbooks. The revenue for 10,000 textbooks is \$300,000.

c) The cost and revenue are equal when

$$30x = 10x + 400{,}000$$
$$20x = 400{,}000$$
$$x = 20{,}000.$$

The cost is equal to the revenue for 20,000 textbooks.

d) Let $x = 0$ in $C = 10x + 400{,}000$ to get $C = 400{,}000$. The fixed cost is \$400,000.

89. Solve each equation for y to get $y = \frac{2}{3}x - 2$ in every case except (a), which is $y = \frac{2}{3}x + 2$.

91. a) Graph the equations and use the intersect feature of the calculator to find the point of intersection: (2.8, 2.6).

b) Solve the equations for y and then graph them on a calculator. Use the intersect feature to find the point of intersection: (1.0, −0.2).

7.2 WARM-UPS

1. True, because when we add the equations, the y-terms add up to 0 and one variable is eliminated.

2. False, multiply the first by −2 and the second by 3 to eliminate x.

3. True, multiplying the second by 2 and adding will eliminate both x and y.

4. True, because both ordered pairs satisfy both equations.

5. False, solution set is $\{(x, y) \mid 4x - 2y = 20\}$.

6. True, because the equations are inconsistent. **7.** True.

8. False, as long as we add the left sides and the right sides, the form does not matter.

9. True. **10.** True.

7.2 EXERCISES

1. In this section we learned the addition method.

3. In some cases we multiply one or both of the equations on each side to change the coefficients of the variable that we are trying to eliminate.

5. If an identity (such as $0 = 0$) results form addition of the equations, then the equations are dependent.

7.
$$\begin{array}{r} x + y = 7 \\ x - y = 9 \\ \hline 2x \quad = 16 \end{array}$$

$x = 8$

Use $x = 8$ in $x + y = 7$ to find y.

$$8 + y = 7$$
$$y = -1$$

The solution set is $\{(8, -1)\}$.

9.
$$\begin{array}{l} x - y = 12 \\ \underline{2x + y = 3} \\ 3x \quad\;\; = 15 \\ \quad x = 5 \end{array}$$

Use $x = 5$ in $x - y = 12$.

$$5 - y = 12$$
$$-7 = y$$

The solution set is $\{(5, -7)\}$.

11.
$$\begin{array}{l} 3x - y = 5 \\ \underline{5x + y = -2} \\ 8x \quad\;\; = 3 \\ \quad x = \frac{3}{8} \end{array}$$

Use $x = \frac{3}{8}$ in $5x + y = -2$.

$$5 \cdot \frac{3}{8} + y = -2$$
$$y = -2 - \frac{15}{8} = -\frac{31}{8}$$

The solution set is $\left\{\left(\frac{3}{8}, -\frac{31}{8}\right)\right\}$.

13. If we multiply $2x - y = -5$ by 2, we get $4x - 2y = -10$. Add this equation to the second equation.

$$\begin{array}{l} 4x - 2y = -10 \\ \underline{3x + 2y = 3} \\ 7x \quad\;\; = -7 \\ \quad x = -1 \end{array}$$

Use $x = -1$ in $2x - y = -5$.

$$2(-1) - y = -5$$
$$-2 - y = -5$$
$$3 = y$$

The solution set is $\{(-1, 3)\}$.

15. If we multiply the first equation $-3x + 5y = 1$ by 3, we get $-9x + 15y = 3$. Add this equation to the second equation.

$$\begin{array}{l} -9x + 15y = 3 \\ \underline{9x - 3y = 5} \\ \quad 12y = 8 \\ \quad y = \frac{2}{3} \end{array}$$

Use $y = 2/3$ in $9x - 3y = 5$:

$$9x - 3 \cdot \frac{2}{3} = 5$$
$$9x - 2 = 5$$
$$9x = 7$$
$$x = \frac{7}{9}$$

The solution set is $\left\{\left(\frac{7}{9}, \frac{2}{3}\right)\right\}$.

17. Multiply the first equation by 4 and the second by 5.

$$4(2x - 5y) = 4(13)$$
$$5(3x + 4y) = 5(-15)$$

$$\begin{array}{l} 8x - 20y = 52 \\ \underline{15x + 20y = -75} \\ 23x \qquad = -23 \\ \quad x = -1 \end{array}$$

Use $x = -1$ in $2x - 5y = 13$.

$$2(-1) - 5y = 13$$
$$-2 - 5y = 13$$
$$-5y = 15$$
$$y = -3$$

The solution set is $\{(-1, -3)\}$.

19. Rewrite the first equation to match the form of the second.

$$2x - 3y = 11$$
$$7x - 4y = 6$$

Multiply the first equation by -7 and the second by 2.

$$-7(2x - 3y) = -7(11)$$
$$2(7x - 4y) = 2(6)$$

$$\begin{array}{l} -14x + 21y = -77 \\ \underline{14x - 8y = 12} \\ \quad 13y = -65 \\ \quad y = -5 \end{array}$$

Use $y = -5$ in $2x = 3y + 11$.

$$2x = 3(-5) + 11$$
$$2x = -4$$
$$x = -2$$

The solution set is $\{(-2, -5)\}$.

21.
$$-12(x + y) = -12(48)$$
$$12x + 14y = 628$$

$$\begin{array}{l} -12x - 12y = -576 \\ \underline{12x + 14y = 628} \\ \quad 2y = 52 \\ \quad y = 26 \end{array}$$

$$x + 26 = 48$$
$$x = 22$$

The solution set is $\{(22, 26)\}$.

23. $$\begin{aligned} 3x - 4y &= 9 \\ -3x + 4y &= 12 \\ \hline 0 &= 21 \end{aligned}$$

Since we obtained an incorrect equation by addition, the equations are inconsistent and the solution set is $\emptyset$.

25. Multiply $5x - y = 1$ by -2, to get $-10x + 2y = -2$. Add this equation to the second equation.

$$\begin{aligned} -10x + 2y &= -2 \\ 10x - 2y &= 2 \\ \hline 0 &= 0 \end{aligned}$$

Since we obtained an identity by adding the equations, the equations are dependent and the solution set is $\{(x, y) \mid 5x - y = 1\}$.

27. $$\begin{aligned} 2x - y &= 5 \\ 2x + y &= 5 \\ \hline 4x \quad &= 10 \\ x &= 10/4 = 5/2 \end{aligned}$$

Use $x = 5/2$ in $2x + y = 5$ to find y.

$$\begin{aligned} 2(5/2) + y &= 5 \\ 5 + y &= 5 \\ y &= 0 \end{aligned}$$

The solution set is $\{(5/2, 0)\}$ and the equations are independent.

29. Multiplying the first equation by 12 to eliminate the fractions gives us the following system.

$$\begin{aligned} 3x + 4y &= 60 \\ x - y &= 6 \end{aligned}$$

Multiply the second equation by 4 and then add.

$$\begin{aligned} 3x + 4y &= 60 \\ 4x - 4y &= 24 \\ \hline 7x \quad &= 84 \\ x &= 12 \end{aligned}$$

Use $x = 12$ in $x - y = 6$ to find y.

$$\begin{aligned} 12 - y &= 6 \\ 6 &= y \end{aligned}$$

The solution set is $\{(12, 6)\}$.

31. If we multiply the first equation by 12 and the second by 24 to eliminate fractions, we get the following system.

$$\begin{aligned} 3x - 4y &= -48 \\ 3x + 4y &= 0 \\ \hline 6x \quad &= -48 \\ x &= -8 \end{aligned}$$

Use $x = -8$ in $3x + 4y = 0$ to find y.

$$\begin{aligned} 3(-8) + 4y &= 0 \\ 4y &= 24 \\ y &= 6 \end{aligned}$$

The solution set is $\{(-8, 6)\}$.

33. Multiply the first equation by 8 and the second by -16 to get the following system.

$$\begin{aligned} x + 2y &= 40 \\ -x - 8y &= -112 \\ \hline -6y &= -72 \\ y &= 12 \\ x + 2(12) &= 40 \\ x &= 16 \end{aligned}$$

The solution set is $\{(16, 12)\}$.

35. Multiply the first equation by 6 and the second by 12 to get the following system.

$$\begin{aligned} 2x + 3y &= 2 \\ 10x - 9y &= 2 \end{aligned}$$

Now multiply the first equation by -5 to get the following system:

$$\begin{aligned} -10x - 15y &= -10 \\ 10x - 9y &= 2 \\ \hline -24y &= -8 \\ y &= \frac{1}{3} \end{aligned}$$

Use $y = 1/3$ in $\frac{1}{3}x + \frac{1}{2}y = \frac{1}{3}$

$$\begin{aligned} \frac{1}{3}x + \frac{1}{2} \cdot \frac{1}{3} &= \frac{1}{3} \\ 2x + 1 &= 2 \\ 2x &= 1 \\ x &= \frac{1}{2} \end{aligned}$$

The solution set is $\left\{\left(\frac{1}{2}, \frac{1}{3}\right)\right\}$.

37. Multiply the first equation by 100 and the second by -10 to get the following system.

$$\begin{aligned} 5x + 10y &= 130 \\ -10x - 10y &= -190 \\ \hline -5x \quad &= -60 \\ x &= 12 \end{aligned}$$

Use $x = 12$ in $x + y = 19$ to get $y = 7$. The solution set is $\{(12, 7)\}$.

39. Multiply the first equation by -9 and the second by 100 to get the following system.

$$\begin{aligned} -9x - 9y &= -10{,}800 \\ 12x + 9y &= 12{,}000 \\ \hline 3x \quad &= 1200 \\ x &= 400 \end{aligned}$$

Use $x = 400$ in $x + y = 1200$ to get $y = 800$. The solution set is $\{(400, 800)\}$.

41. Multiply the first equation by -2 to get the following system.

$$\begin{aligned} -3x + \quad 4y &= \quad 0.5 \\ 3x + 1.5y &= 6.375 \\ \hline 5.5y &= 6.875 \\ y &= 1.25 \end{aligned}$$

Use $y = 1.25$ in $3x + 1.5y = 6.375$ to find x.

$$\begin{aligned} 3x + 1.5(1.25) &= 6.375 \\ 3x + 1.875 &= 6.375 \\ 3x &= 4.5 \\ x &= 1.5 \end{aligned}$$

The solution set is $\{(1.5, 1.25)\}$.

43. Multiply both equations by 100 to get the following system.

$$\begin{aligned} 24x + 60y &= 58 \\ 80x - 12y &= 52 \end{aligned}$$

Multiply the second equation by 5 and add:

$$\begin{aligned} 24x + 60y &= 58 \\ 400x - 60y &= 260 \\ \hline 424x &= 318 \\ x &= \frac{318}{424} = \frac{3}{4} \end{aligned}$$

Use $x = 3/4$ in $24x + 60y = 58$ to find y.

$$\begin{aligned} 24 \cdot \frac{3}{4} + 60y &= 58 \\ 60y &= 40 \\ y &= \frac{40}{60} = \frac{2}{3} \end{aligned}$$

The solution set is $\left\{\left(\frac{3}{4}, \frac{2}{3}\right)\right\}$.

45. Substitute $y = x + 1$ into $2x - 5y = -20$:

$$\begin{aligned} 2x - 5(x + 1) &= -20 \\ -3x - 5 &= -20 \\ -3x &= -15 \\ x &= 5 \end{aligned}$$

If $x = 5$, then $y = 5 + 1 = 6$. The solution set is $\{(5, 6)\}$.

47.
$$\begin{aligned} x - y &= 19 \\ 2x + y &= -13 \\ \hline 3x \quad &= 6 \\ x &= 2 \end{aligned}$$

Use $x = 2$ in $x - y = 19$.

$$\begin{aligned} 2 - y &= 19 \\ -17 &= y \end{aligned}$$

The solution set is $\{(2, -17)\}$.

49. Substitute $x = y - 1$ into $2y = x + 2$:

$$\begin{aligned} 2y &= y - 1 + 2 \\ y &= 1 \end{aligned}$$

If $y = 1$, then $x = 1 - 1 = 0$. The solution set is $\{(0, 1)\}$.

51.
$$\begin{aligned} 2y - 3x &= -1 \\ 5y + 3x &= 29 \\ \hline 7y \quad &= 28 \\ y &= 4 \end{aligned}$$

Use $y = 4$ in $5y + 3x = 29$.

$$\begin{aligned} 5(4) + 3x &= 29 \\ 3x &= 9 \\ x &= 3 \end{aligned}$$

The solution set is $\{(3, 4)\}$.

53. Substitute $y = \frac{2}{3}x$ into $6x + 3y = 4$:

$$\begin{aligned} 6x + 3\left(\frac{2}{3}x\right) &= 4 \\ 6x + 2x &= 4 \\ 8x &= 4 \\ x &= \frac{1}{2} \end{aligned}$$

$$\begin{aligned} y &= \frac{2}{3}x \\ y &= \frac{2}{3} \cdot \frac{1}{2} = \frac{1}{3} \end{aligned}$$

The solution set is $\left\{\left(\frac{1}{2}, \frac{1}{3}\right)\right\}$.

55. Substitute $y = 3x + 1$ into $x = \frac{1}{3}y + 5$:

$$\begin{aligned} x &= \frac{1}{3}(3x + 1) + 5 \\ x &= x + \frac{1}{3} + 5 \\ 0 &= \frac{16}{3} \end{aligned}$$

The solution set is the empty set, $\emptyset$.

57.
$$\begin{aligned} x - y &= 0 \\ x + y &= 2x \\ \hline 2x \quad &= 2x \end{aligned}$$

The system is dependent and the solution set is $\{(x, y) | \, y = x\}$.

59. If $(2, 3)$ satisfies the system then it satisfies $x + y = 5$ and $x - y = a$. So $2 + 3 = 5$ and $2 - 3 = a$. For these equations to be correct, a must be -1.

61. If $(5, 12)$ satisfies the system then it satisfies $y = ax + 2$ and $y = bx + 17$. So $12 = a(5) + 2$ and $12 = b(5) + 17$. For the first equation to be correct, we must have $5a + 2 = 12$, $5a = 10$, and $a = 2$. For the second equation to be correct, we must have $5b + 17 = 12, 5b = -5$, and $b = -1$.

63. Let $x =$ the price of one doughnut and $y =$ the price of one cup of coffee. His bills for Monday and Tuesday give us the following system of equations.

$$3x + 2y = 3.40$$
$$2x + 3y = 3.60$$
Multiply the first equation by -2 and the second by 3.
$$-6x - 4y = -6.80$$
$$6x + 9y = 10.80$$
$$5y = 4.00$$
$$y = 0.80$$
Use $y = 0.80$ in $3x + 2y = 3.40$ to find x.
$$3x + 2(0.80) = 3.40$$
$$3x + 1.60 = 3.40$$
$$3x = 1.80$$
$$x = 0.60$$
Doughnuts are $0.60 each and a cup of coffee is $0.80. So his total bill on Wednesday was $1.40.

65. Let $x =$ the number of boys and $y =$ the number of girls at Freemont High. We get a system of equations from the number of boys and girls who attended the dance and game.
$$\frac{1}{2}x + \frac{1}{3}y = 570$$
$$\frac{1}{3}x + \frac{1}{2}y = 580$$
Multiply the first equation by 12 and the second equation by -18.
$$6x + 4y = 6840$$
$$-6x - 9y = -10440$$
$$-5y = -3600$$
$$y = 720$$
Use $y = 720$ in $6x + 4y = 6840$ to find x.
$$6x + 4(720) = 6840$$
$$6x + 2880 = 6840$$
$$6x = 3960$$
$$x = 660$$
There are 660 boys and 720 girls at Freemont High, or 1380 students.

67. Let $x =$ the number of dimes and $y =$ the number of nickels. We write one equation about the total number of the coins and the other about the value of the coins (in cents).
$$x + y = 35$$
$$10x + 5y = 330$$
Multiply the first equation by -5.
$$-5x - 5y = -175$$
$$10x + 5y = 330$$
$$5x \qquad = 155$$
$$x = 31$$
Use $x = 31$ in $x + y = 35$ to find that $y = 4$. He has 31 dimes and 4 nickels.

69. a) From the graph it appears that there should be about 20 pounds of Chocolate fudge and 30 pounds of Peanut Butter fudge.

b) Let $x =$ the number of pounds of Chocolate fudge and $y =$ the number of pounds of Peanut Butter fudge.
$$x + y = 50$$
$$0.35x + 0.25y = 0.29(50)$$

$$y = 50 - x$$
$$0.35x + 0.25(50 - x) = 14.5$$
$$0.10x = 2$$
$$x = 20$$
$$20 + y = 50$$
$$y = 30$$
Use 20 pounds of Chocolate fudge and 30 pounds of Peanut Butter fudge.

71. Let $a =$ the time from Allentown to Harrisburg and $h =$ the time from Harrisburg to Pittsburgh.
$$a + h = 6$$
$$42a + 51h = 288$$

$$-51a - 51h = -306$$
$$42a + 51h = 288$$
$$-9a \qquad = -18$$
$$a = 2$$
$$h = 6 - 2 = 4$$
It took 4 hours to go from Harrisburg to Pittsburgh.

73. Let $p =$ the probability of rain and $q =$ the probability that it doesn't rain:
$$p + q = 1$$
$$p = 4q$$

$$4q + q = 1$$
$$5q = 1$$
$$q = 0.20 = 20\%$$
$$p = 80\%$$
The probability of rain is 80%.

75. Let $W =$ the width and $L =$ the length.
$$W = 0.75L$$
$$2W + 2L = 700$$

$$2(0.75L) + 2L = 700$$
$$1.5L + 2L = 700$$

$$3.5L = 700$$
$$L = 200$$
$$W = 0.75(200) = 150$$

The width is 150 m and the length is 200 m.

77. If one equation is already solved for x or y then substitution is usually easier. If both equations are in standard form then addition is usually easier.

79. a) Find the equation of the line through the two points. A system satisfied by both $(-1, 2)$ and $(4, 5)$ is $y = \frac{3}{5}x + \frac{13}{5}$ and $5y = 3x + 13$.

b) The equations are dependent.

c) There is not independent system satisfied by both lines, because two points determines a unique line.

7.3 WARM-UPS

1. False, because $1 + (-2) - 3 = 4$ is incorrect.

2. False, because there are infinitely many ordered triples that satisfy $x + y - z = 4$.

3. True, because if $x = 1$, $y = -1$, and $z = 2$ then each of the equations is satisfied.

4. False, we can use substitution and addition to eliminate variables.

5. False, because two planes never intersect in a single point.

6. True, because if we multiply the second one by -1 and add, then we get $0 = 2$.

7. True, because -2 times the first equation is the same as the second equation.

8. False, because the graph is a plane.

9. False, because the value of x nickels, y dimes, and z quarters is $5x + 10y + 25z$ cents.

10. False, because $x = -2$, $z = 3$, and $-2 + y + 3 = 6$ implies that $y = 5$.

7.3 EXERCISES

1. A linear equation in three variables is an equation of the form $Ax + By + Cz = D$ where A, B, and C cannot all be zero.

3. A solution to a system of linear equations in three variable is an ordered triple that satisfies all of the equations in the system.

5. The graph of a linear equation in three variables is a plane in a three-dimensional coordinate system.

7. Substituting $z = 4$ into the first two equations yields the following system.

$$x + y + 4 = 9$$
$$y + 4 = 7$$

Simplify.

$$x + y = 5$$
$$y = 3$$

Substitute.

$$x + 3 = 5$$
$$x = 2$$

The solution set is $\{(2, 3, 4)\}$.

9. Add the second and third equations to eliminate y.

$$\begin{array}{r} x - y = -1 \\ \underline{x + y = 5} \\ 2x \quad = 4 \end{array}$$
$$x = 2$$

Substitute $x = 2$ into $x + y = 5$.

$$2 + y = 5$$
$$y = 3$$

Substitute $x = 2$ and $y = 3$ into $x + y + z = 10$.

$$2 + 3 + z = 10$$
$$z = 5$$

The solution set is $\{(2, 3, 5)\}$.

11. Add the first and last equations.

$$\begin{array}{r} x + y + z = 6 \\ \underline{x - y - z = -4} \\ 2x \qquad = 2 \end{array}$$
$$x = 1$$

Replace x with 1 in the first two equations.

$$1 + y + z = 6$$
$$1 - y + z = 2$$

Simplify and add.

$$\begin{array}{r} y + z = 5 \\ \underline{-y + z = 1} \\ 2z = 6 \end{array}$$
$$z = 3$$

If $z = 3$ and $y + z = 5$, then $y = 2$. The solution set is $\{(1, 2, 3)\}$.

13. Add the first two equations.

$$\begin{array}{r} x + y + z = 2 \\ \underline{x + 2y - z = 6} \\ 2x + 3y \qquad = 8 \end{array} \quad \text{A}$$

Add the first and third equations.

$$\begin{array}{l} x + y + z = 2 \\ \underline{2x + y - z = 5} \\ 3x + 2y \quad = 7 \quad \text{B} \end{array}$$

Multiply equation A by 3 and equation B by -2.

$$\begin{array}{ll} 6x + 9y = 24 & 3 \times \text{A} \\ \underline{-6x - 4y = -14} & -2 \times \text{B} \\ 5y = 10 & \\ y = 2 & \end{array}$$

Use $y = 2$ in the equation $3x + 2y = 7$.

$$\begin{aligned} 3x + 2(2) &= 7 \\ 3x &= 3 \\ x &= 1 \end{aligned}$$

Use $x = 1$ and $y = 2$ in $x + y + z = 2$.

$$\begin{aligned} 1 + 2 + z &= 2 \\ z &= -1 \end{aligned}$$

The solution set is $\{(1, 2, -1)\}$.

15. Multiply the first equation by -1 to get $-x + 2y - 4z = -3$. Now add this equation and the second, and this equation and the third.

$$\begin{array}{ll} -x + 2y - 4z = -3 & -x + 2y - 4z = -3 \\ \underline{x + 3y - 2z = 6} & \underline{x - 4y + 3z = -5} \\ 5y - 6z = 3 & -2y - z = -8 \end{array}$$

Multiply $-2y - z = -8$ by -6 and add the result to $5y - 6z = 3$.

$$\begin{array}{l} 12y + 6z = 48 \\ \underline{5y - 6z = 3} \\ 17y \quad = 51 \\ y = 3 \end{array}$$

Use $y = 3$ in $5y - 6z = 3$.

$$\begin{aligned} 5(3) - 6z &= 3 \\ -6z &= -12 \\ z &= 2 \end{aligned}$$

Use $y = 3$ and $z = 2$ in $x + 3y - 2z = 6$.

$$\begin{aligned} x + 3(3) - 2(2) &= 6 \\ x &= 1 \end{aligned}$$

The solution set is $\{(1, 3, 2)\}$.

17. Multiply $2x - y + z = 10$ by 2 to get $4x - 2y + 2z = 20$. Add this equation and the second and this equation and the third.

$$\begin{array}{ll} 4x - 2y + 2z = 20 & 4x - 2y + 2z = 20 \\ \underline{3x - 2y - 2z = 7} & \underline{x - 3y - 2z = 10} \\ 7x - 4y \quad = 27 & 5x - 5y \quad = 30 \\ & x - y \quad = 6 \end{array}$$

Multiply $x - y = 6$ by -4 to get $-4x + 4y = -24$. Add this to $7x - 4y = 27$.

$$\begin{array}{l} -4x + 4y = -24 \\ \underline{7x - 4y = 27} \\ 3x \quad = 3 \\ x = 1 \end{array}$$

Use $x = 1$ in $7x - 4y = 27$ to find y.

$$\begin{aligned} 7(1) - 4y &= 27 \\ -4y &= 20 \\ y &= -5 \end{aligned}$$

Use $x = 1$ and $y = -5$ in $2x - y + z = 10$.

$$\begin{aligned} 2(1) - (-5) + z &= 10 \\ z &= 3 \end{aligned}$$

The solution set is $\{(1, -5, 3)\}$.

19. Multiply $x - y + 2z = -5$ by 2 to get $2x - 2y + 4z = -10$. Add this equation and the second equation, and add the first and second equations.

$$\begin{array}{ll} 2x - 2y + 4z = -10 & 2x - 3y + z = -9 \\ \underline{-2x + y - 3z = 7} & \underline{-2x + y - 3z = 7} \\ -y + z = -3 & -2y - 2z = -2 \\ & y + z = 1 \end{array}$$

Add the last two equations to eliminate y.

$$\begin{array}{l} -y + z = -3 \\ \underline{y + z = 1} \\ 2z = -2 \\ z = -1 \end{array}$$

Use $z = -1$ in $y + z = 1$ to get $y = 2$. Now use $z = -1$ and $y = 2$ in $x - y + 2z = -5$.

$$\begin{aligned} x - 2 + 2(-1) &= -5 \\ x &= -1 \end{aligned}$$

The solution set is $\{(-1, 2, -1)\}$.

21. Multiply the first equation by -2 and add the result to the last equation.

$$\begin{array}{l} -4x + 10y - 4z = -32 \\ \underline{4x - 3y + 4z = 18} \\ 7y \quad = -14 \\ y \quad = -2 \end{array}$$

Multiply the first equation by 3 and the second by 2 and add the results.

$$\begin{array}{l} 6x - 15y + 6z = 48 \\ \underline{6x + 4y - 6z = -38} \\ 12x - 11y \quad = 10 \end{array}$$

Use $y = -2$ in the last equation to find x.

$$\begin{aligned} 12x - 11(-2) &= 10 \\ 12x &= -12 \\ x &= -1 \end{aligned}$$

Use $x = -1$ and $y = -2$ in
$2x - 5y + 2z = 16.$

$$2(-1) - 5(-2) + 2z = 16$$
$$2z = 8$$
$$z = 4$$

The solution set is $\{(-1, -2, 4)\}$.

23. If we add the last two equations we get $x + 2y = 7$. Multiply the first equation by -1 to get $-x - y = -4$. Add these two equations.

$$\begin{array}{r} x + 2y = 7 \\ -x - y = -4 \\ \hline y = 3 \end{array}$$

Use $y = 3$ in $x + y = 4$ to get $x = 1$. Use $y = 3$ in $y - z = -2$.

$$3 - z = -2$$
$$5 = z$$

The solution set is $\{(1, 3, 5)\}$.

25. Multiply the first equation by -1 and add the result to the second equation.

$$\begin{array}{r} -x - y \quad = -7 \\ y - z = -1 \\ \hline -x \quad - z = -8 \end{array}$$

Add this result to the last equation.

$$\begin{array}{r} -x - z = -8 \\ x + 3z = 18 \\ \hline 2z = 10 \\ z = 5 \end{array}$$

Use $z = 5$ in $x + 3z = 18$ to get $x + 15 = 18$ or $x = 3$. Use $x = 3$ in $x + y = 7$ to get $3 + y = 7$ or $y = 4$. The solution set is $\{(3, 4, 5)\}$.

27. Use $z = 1$ two write the first equation as $x + y = 8$. Now the second equation indicates that $x + y = 5$. By substitution we have $8 = 5$, which is false. So there is no solution to the system. The solution set is the empty set $\emptyset$.

29. Note that the second equation is obtained from multiplying the first by -1 and the third is obtained from multiplying the first by 2. So all three equations are equivalent. So any point that satisfies one of them satisfies all of them. The solution set is $\{(x, y, z) | x + y - z = 2\}$.

31. Add the first two equations.

$$\begin{array}{r} x - y + 2z = 3 \\ 2x + y - z = 5 \\ \hline 3x \quad + z = 8 \end{array} \quad \text{A}$$

Multiply the second equation by 3 and add the result to the last equation.

$$\begin{array}{r} 6x + 3y - 3z = 15 \\ 3x - 3y + 6z = 4 \\ \hline 9x \quad + 3z = 19 \end{array} \quad \text{B}$$

Multiply A by -3 and add the result to B.

$$\begin{array}{r} -9x - 3z = -24 \\ 9x + 3z = 19 \\ \hline 0 = -5 \end{array}$$

Since the last equation is false no matter what values the variables have, the solution set is $\emptyset$.

33. Multiply the last equation $-x + 2y - 3z = -6$ by -2 to get $2x - 4y + 6z = 12$, the first equation. Multiply the last equation $-x + 2y - 3z = -6$ by -6 to get $6x - 12y + 18z = 36$, the second equation. Since the first and second equations are multiples of the last equation, the system is dependent. The solution set is $\{(x, y, z) \mid -x + 2y - 3z = -6\}$.

35. Add the first and second equation to get $x + z = 11$. Multiply this equation by -2 and then add the result to the last equation.

$$\begin{array}{r} -2x - 2z = -22 \\ 2x + 2z = 7 \\ \hline 0 = -15 \end{array}$$

Since the last result is false, the solution set is $\emptyset$.

37. Multiply the first equation by 300 to get $30x + 24y - 12z = 900$. Multiply the second equation by 6 to get $30x + 24y - 12z = 900$. Multiply the third equation by 100 to get $30x + 24y - 12z = 900$. Since all three of these equations are different forms of the same equation, the system is dependent. The solution set is $\{(x, y, z) \mid 5x + 4y - 2z = 150\}$.

39. Multiply the second equation by 10 and add the result to the first equation.

$$\begin{array}{r} 37x - 2y + 0.5z = 4.1 \\ 3x + 2y - 0.4z = 0.1 \\ \hline 40x \quad + 0.1z = 4.2 \end{array} \quad \text{A}$$

Multiply the second equation by 19 and add the result to the last equation.

$$\begin{aligned} 70.3x - 3.8y + 0.95z &= 7.79 \\ -2x + 3.8y - 2.1z &= -3.26 \\ \hline 68.3x \quad - 1.15z &= 4.53 \quad \text{B} \end{aligned}$$

Multiply equation A by 11.5 and add the result to equation B.

$$\begin{aligned} 460x + 1.15z &= 48.3 \\ 68.3x - 1.15z &= 4.53 \\ \hline 528.3x \quad &= 52.83 \\ x &= 0.1 \end{aligned}$$

Use $x = 0.1$ in $40x + 0.1z = 4.2$.

$$\begin{aligned} 40(0.1) + 0.1z &= 4.2 \\ 0.1z &= 0.2 \\ z &= 2 \end{aligned}$$

Use $x = 0.1$ and $z = 2$ in $3x + 2y - 0.4z = 0.1$.

$$\begin{aligned} 3(0.1) + 2y - 0.4(2) &= 0.1 \\ 0.3 + 2y - 0.8 &= 0.1 \\ 2y &= 0.6 \\ y &= 0.3 \end{aligned}$$

The solution set is $\{(0.1, 0.3, 2)\}$.

41. Let $x =$ the price of the Chevy, $y =$ the price of the Ford, and $z =$ the price of the Toyota. Write the following system.

$$\begin{aligned} x + y + z &= 66{,}000 \\ y &= x + 2000 \\ z &= y + 2000 \end{aligned}$$

By substitution $z = x + 4000$ and

$$\begin{aligned} x + (x + 2000) + (x + 4000) &= 66{,}000 \\ 3x &= 60{,}000 \\ x &= 20{,}000 \\ y = x + 2{,}000 &= 22{,}000 \\ z = y + 2{,}000 &= 24{,}000 \end{aligned}$$

The Chevrolet was \$20,000, the Ford was \$22,000, and the Toyota was \$24,000.

43. Let $x =$ the number of hours driven on the first day, $y =$ the number of hours for the second day, and $z =$ the number of hours for the third day. His distance each day was $64x$, $62y$, and $58z$.

$$\begin{aligned} x + y + z &= 36 \\ 64x + 62y + 58z &= 2196 \\ z &= x + 4 \end{aligned}$$

Substitute:

$$\begin{aligned} x + y + (x + 4) &= 36 \\ 64x + 62y + 58(x + 4) &= 2196 \end{aligned}$$

Simplify:

$$\begin{aligned} 2x + y &= 32 \\ 122x + 62y &= 1964 \end{aligned}$$

Multiply the first equation by -62:

$$\begin{aligned} -124x - 62y &= -1984 \\ 122x + 62y &= 1964 \\ \hline -2x \quad &= -20 \\ x &= 10 \end{aligned}$$

$$\begin{aligned} z = x + 4 = 10 + 4 &= 14 \\ x + y + z &= 36 \\ 10 + y + 14 &= 36 \\ y &= 12 \end{aligned}$$

So he drove 10 hours on the first day, 12 hours on the second day, and 14 hours on the third day.

45. Let $x =$ her investment in stocks, $y =$ her investment in bonds, and $z =$ her investment in a mutual funds. We can write 3 equations concerning x, y, and z.

$$\begin{aligned} x + y + z &= 12{,}000 \\ 0.10x + 0.08y + 0.12z &= 1230 \\ x + y &= z \end{aligned}$$

Substitute $z = x + y$ into the first equation.

$$\begin{aligned} x + y + x + y &= 12000 \\ 2x + 2y &= 12000 \\ x + y &= 6000 \quad \text{A} \end{aligned}$$

Substitute $z = x + y$ into the second equation.

$$\begin{aligned} 0.10x + 0.08y + 0.12(x + y) &= 1230 \\ 0.22x + 0.20y &= 1230 \\ -5(0.22x + 0.20y) &= -5(1230) \\ -1.1x - y &= -6150 \quad \text{B} \end{aligned}$$

Add equations A and B.

$$\begin{aligned} x + y &= 6000 \\ -1.1x - y &= -6150 \\ \hline -0.1x \quad &= -150 \\ x &= 1500 \end{aligned}$$

Use $x = 1500$ in $x + y = 6000$ to get $y = 4500$. Since $z = x + y$, we must have $z = 6000$. So Ann invested \$1500 in stocks, \$4500 in bonds, and \$6000 in a mutual fund.

47. Let $x =$ Anna's weight, $y =$ Bob's weight, and $z =$ Chris's weight. We are given three equations.

$$\begin{aligned} x + y \quad &= 226 \\ y + z &= 210 \\ x \quad + z &= 200 \end{aligned}$$

From the first equation we get $y = 226 - x$. Substitute this equation into the second equation to eliminate y.

$$226 - x + z = 210$$
$$-x + z = -16$$

Add this last equation to $x + z = 200$ to eliminate x.

$$\begin{aligned} -x + z &= -16 \\ x + z &= 200 \\ \hline 2z &= 184 \\ z &= 92 \end{aligned}$$

If $z = 92$ and $x + z = 200$, we get $x = 108$. If $z = 92$ and $y + z = 210$, we get $y = 118$. So Anna weighs 108 pounds, Bob weighs 118 pounds, and Chris weighs 92 pounds.

49. Let $x =$ the number of nickels, $y =$ the number of dimes, and $z =$ the number of quarters. Since he has 13 coins altogether, $x + y + z = 13$. Since the value of the coins in cents is 175, we have $5x + 10y + 25z = 175$ or $x + 2y + 5z = 35$. Since the number of dimes is twice the number of nickels, $y = 2x$.

$$x + y + z = 13$$
$$x + 2y + 5z = 35$$
$$y = 2x$$

Substitute $y = 2x$ into the other two equations to get the following system.

$$3x + z = 13$$
$$5x + 5z = 35$$

Divide the second equation by -5 and add:

$$\begin{aligned} 3x + z &= 13 \\ -x - z &= -7 \\ \hline 2x \quad &= 6 \\ x &= 3 \end{aligned}$$

If $x = 3$, then $y = 6$. If $x = 3$ and $y = 6$, then $3 + 6 + z = 13$, or $z = 4$. So he used 3 nickels, 6 dimes, and 4 quarters.

51. Let $x =$ her income from teaching, $y =$ her income from house painting, and $z =$ her royalties.

$$x + y + z = 48{,}000$$
$$x - y = 6000$$
$$z = \tfrac{1}{7}(x + y)$$

The last equation can be written as $x + y = 7z$. Replacing $x + y$ in the first equation by $7z$ gives us $7z + z = 48{,}000$, or $z = 6000$. If $z = 6000$, then $x + y = 42{,}000$. Add $x + y = 42000$ and the second equation.

$$\begin{aligned} x + y &= 42{,}000 \\ x - y &= 6000 \\ \hline 2x \quad &= 48{,}000 \\ x &= 24{,}000 \end{aligned}$$

Use $x = 24{,}000$ in $x - y = 6000$ to get $y = 18{,}000$. So she made \$24,000 teaching, \$18,000 house painting, and \$6,000 from royalties.

53. Let $x =$ Edwin's age, $y =$ his father's age and $z =$ his grandfather's age. Since their average age is 53, we can write

$$\frac{x + y + z}{3} = 53$$

or $\quad x + y + z = 159.$

We can also write

$$\frac{1}{2}z + \frac{1}{3}y + \frac{1}{4}x = 65$$

or $\quad 12\left(\frac{1}{2}z + \frac{1}{3}y + \frac{1}{4}x\right) = 12(65)$

or $\quad 6z + 4y + 3x = 780.$

Four years ago Edwin was $x - 4$ years old and his grandfather was $z - 4$ years old. Since 4 years ago the grandfather was 4 times as old as Edwin, we can write

$$z - 4 = 4(x - 4)$$
$$z - 4 = 4x - 16$$
$$z = 4x - 12.$$

Substitute $z = 4x - 12$ into the two equations above.

$$x + y + 4x - 12 = 159$$
$$6(4x - 12) + 4y + 3x = 780$$

$$5x + y = 171$$
$$27x + 4y = 852$$

Substitute $y = 171 - 5x$ into $27x + 4y = 852$.

$$27x + 4(171 - 5x) = 852$$
$$7x + 684 = 852$$
$$7x = 168$$
$$x = 24$$

Since $z = 4x - 12$ we have $z = 84$. Since $x + y + z = 159$, we have $24 + y + 84 = 159$, or $y = 51$. So Edwin is 24, his father is 51, and his grandfather is 84 years old.

7.4 WARM-UPS

1. True
2. True.
3. True. Replace R_2 of (a) by $R_1 + R_2$ to get matrix (b).
4. False, because matrix (c) corresponds to an inconsistent system and (d) corresponds to a dependent system.
5. True, because the last row represents the equation $0 = 7$.
6. False. Replace R_2 by $2R_1 + R_2$ to get $0 = -3$ which is inconsistent.
7. False, because the system corresponding to (d) consists of two equations equivalent to $x + 3y = 5$.
8. False, because the augmented matrix is a 2×3 matrix.
9. True.
10. False, because it means to interchange R_1 and R_2.

7.4 EXERCISES

1. A matrix is a rectangular array of numbers.
3. The size of a matrix is the number of rows and columns.
5. An augmented matrix is a matrix where the entries in the first column are the coefficients of x, the entries in the second column are the coefficients of y, and the entries in the third column are the constants from a system of two linear equations in two unknowns.
7. The matrix is a 2×2 matrix because it has 2 rows and 2 columns.
9. The matrix is a 3×2 matrix because it has 3 rows and 2 columns.
11. The matrix is a 3×1 matrix because it has 3 rows and 1 column.
13. Use the coefficients 2 and -3, and the constant 9 as the first row. Use the coefficients -3 and 1, and the constant -1 as the second row.

$$\begin{bmatrix} 2 & -3 & 9 \\ -3 & 1 & -1 \end{bmatrix}$$

15. Use the coefficients 1, -1, and 1, and the constant 1 as the first row. Use the coefficients 1, 1, and -2, and the constant 3 as the second row. Use the coefficients 0, 1, and -3, and the constant 4 as the third row.

$$\begin{bmatrix} 1 & -1 & 1 & 1 \\ 1 & 1 & -2 & 3 \\ 0 & 1 & -3 & 4 \end{bmatrix}$$

17. The entries in the first row $(5, 1, -1)$ represent the equation $5x + y = -1$. The entries in the second row $(2, -3, 0)$ represent the equation $2x - 3y = 0$. So the matrix represents the following system.

$$5x + y = -1$$
$$2x - 3y = 0$$

19. The entries in the first row $(1, 0, 0, 6)$ represent the equation $x = 6$. The entries in the second row $(-1, 0, 1, -3)$ represent the equation $-x + z = -3$. The entries in the third row $(1, 1, 0, 1)$ represent the equation $x + y = 1$. So the matrix represents the following system.

$$x = 6$$
$$-x + z = -3$$
$$x + y = 1$$

21. $R_1 \leftrightarrow R_2$ means that R_1 and R_2 are interchanged. So the new matrix is

$$\begin{bmatrix} 1 & 0 & 6 \\ 0 & 2 & 4 \end{bmatrix}.$$

23. $\frac{1}{4}R_1 \rightarrow R_1$ means that $\frac{1}{4}R_1$ replaces R_1. So multiply R_1 by $\frac{1}{4}$ and write the result in place of R_1:

$$\begin{bmatrix} 1 & 3 & 4 \\ 2 & -4 & 3 \end{bmatrix}.$$

25. $R_1 + R_2 \rightarrow R_2$ means that the sum of rows 1 and 2 is used in place of R_2. So the new matrix is

$$\begin{bmatrix} 1 & 0 & -3 \\ 0 & 2 & 1 \end{bmatrix}.$$

27. $2R_1 + R_2 \rightarrow R_2$ means that the sum of twice row 1 and row 2 is used in place of R_2. So the new matrix is

$$\begin{bmatrix} 1 & 2 & 3 \\ 0 & 7 & 11 \end{bmatrix}.$$

29. Row 1 and row 2 are interchanged. This operation is written in symbols as $R_1 \leftrightarrow R_2$.

31. The second row is multiplied by $\frac{1}{5}$. In symbols $\frac{1}{5}R_2 \rightarrow R_2$

33. $\begin{bmatrix} 1 & -1 & -3 \\ 0 & 1 & 4 \end{bmatrix}$

$\begin{bmatrix} 1 & 0 & 1 \\ 0 & 1 & 4 \end{bmatrix}$ $R_2 + R_1 \rightarrow R_1$

The solution set is $\{(1, 4)\}$.

35. $\begin{bmatrix} 1 & 1 & -6 \\ 0 & 3 & 6 \end{bmatrix}$

$\begin{bmatrix} 1 & 1 & -6 \\ 0 & 1 & 2 \end{bmatrix}$ $\frac{1}{3}R_2 \rightarrow R_2$

$\begin{bmatrix} 1 & 0 & -8 \\ 0 & 1 & 2 \end{bmatrix}$ $-R_2 + R_1 \rightarrow R_1$

The solution set is $\{(-8, 2)\}$.

37. $\begin{bmatrix} 1 & -1 & 7 \\ -1 & -1 & -3 \end{bmatrix}$

$\begin{bmatrix} 1 & -1 & 7 \\ 0 & -2 & 4 \end{bmatrix}$ $R_1 + R_2 \rightarrow R_2$

$\begin{bmatrix} 1 & -1 & 7 \\ 0 & 1 & -2 \end{bmatrix}$ $-\frac{1}{2}R_2 \rightarrow R_2$

$\begin{bmatrix} 1 & 0 & 5 \\ 0 & 1 & -2 \end{bmatrix}$ $R_2 + R_1 \rightarrow R_1$

The solution set is $\{(5, -2)\}$.

39. $\begin{bmatrix} 1 & 1 & 3 \\ -3 & 1 & -1 \end{bmatrix}$

$\begin{bmatrix} 1 & 1 & 3 \\ 0 & 4 & 8 \end{bmatrix}$ $3R_1 + R_2 \rightarrow R_2$

$\begin{bmatrix} 1 & 1 & 3 \\ 0 & 1 & 2 \end{bmatrix}$ $\frac{1}{4}R_2 \rightarrow R_2$

$\begin{bmatrix} 1 & 0 & 1 \\ 0 & 1 & 2 \end{bmatrix}$ $-R_2 + R_1 \rightarrow R_1$

The solution set is $\{(1, 2)\}$.

41. $\begin{bmatrix} 2 & -1 & 3 \\ 1 & 1 & 9 \end{bmatrix}$

$\begin{bmatrix} 1 & 1 & 9 \\ 2 & -1 & 3 \end{bmatrix}$ $R_1 \leftrightarrow R_2$

$\begin{bmatrix} 1 & 1 & 9 \\ 0 & -3 & -15 \end{bmatrix}$ $-2R_1 + R_2 \rightarrow R_2$

$\begin{bmatrix} 1 & 1 & 9 \\ 0 & 1 & 5 \end{bmatrix}$ $R_2 \div (-3) \rightarrow R_2$

$\begin{bmatrix} 1 & 0 & 4 \\ 0 & 1 & 5 \end{bmatrix}$ $-R_2 + R_1 \rightarrow R_1$

The solution set is $\{(4, 5)\}$.

43. $\begin{bmatrix} 3 & -1 & 4 \\ 2 & 1 & 1 \end{bmatrix}$

$\begin{bmatrix} 1 & -2 & 3 \\ 2 & 1 & 1 \end{bmatrix}$ $-R_2 + R_1 \rightarrow R_1$

$\begin{bmatrix} 1 & -2 & 3 \\ 0 & 5 & -5 \end{bmatrix}$ $-2R_1 + R_2 \rightarrow R_2$

$\begin{bmatrix} 1 & -2 & 3 \\ 0 & 1 & -1 \end{bmatrix}$ $R_2 \div 5 \rightarrow R_2$

$\begin{bmatrix} 1 & 0 & 1 \\ 0 & 1 & -1 \end{bmatrix}$ $2R_2 + R_1 \rightarrow R_1$

The solution set is $\{(1, -1)\}$.

45. $\begin{bmatrix} 6 & -7 & 0 \\ 2 & 1 & 20 \end{bmatrix}$

$\begin{bmatrix} 0 & -10 & -60 \\ 2 & 1 & 20 \end{bmatrix}$ $-3R_2 + R_1 \rightarrow R_1$

$\begin{bmatrix} 0 & 1 & 6 \\ 2 & 1 & 20 \end{bmatrix}$ $R_1 \div (-10) \rightarrow R_1$

$\begin{bmatrix} 2 & 1 & 20 \\ 0 & 1 & 6 \end{bmatrix}$ $R_1 \leftrightarrow R_2$

$\begin{bmatrix} 2 & 0 & 14 \\ 0 & 1 & 6 \end{bmatrix}$ $-R_2 + R_1 \rightarrow R_1$

$$\begin{bmatrix} 1 & 0 & 7 \\ 0 & 1 & 6 \end{bmatrix} \quad R_1 \div 2 \rightarrow R_1$$

The solution set is $\{(7, 6)\}$.

47. $$\begin{bmatrix} 2 & -3 & 4 \\ -2 & 3 & 5 \end{bmatrix}$$

$$\begin{bmatrix} 2 & -3 & 4 \\ 0 & 0 & 9 \end{bmatrix} \quad R_1 + R_2 \rightarrow R_2$$

Since the second row represents the equation $0 = 9$, the solution set is the empty set, $\emptyset$.

49. $$\begin{bmatrix} 1 & 2 & 1 \\ 3 & 6 & 3 \end{bmatrix}$$

$$\begin{bmatrix} 1 & 2 & 1 \\ 0 & 0 & 0 \end{bmatrix} \quad -3R_1 + R_2 \rightarrow R_2$$

Since the system is equivalent to the single equation $x + 2y = 1$, the equations are dependent and the solution set is $\{(x, y) \mid x + 2y = 1\}$.

51. $$\begin{bmatrix} 1 & 1 & -1 & 4 \\ 0 & 1 & 1 & 6 \\ 0 & 0 & 1 & 2 \end{bmatrix}$$

$$\begin{bmatrix} 1 & 0 & -2 & -2 \\ 0 & 1 & 1 & 6 \\ 0 & 0 & 1 & 2 \end{bmatrix} \quad -R_2 + R_1 \rightarrow R_1$$

$$\begin{bmatrix} 1 & 0 & 0 & 2 \\ 0 & 1 & 1 & 6 \\ 0 & 0 & 1 & 2 \end{bmatrix} \quad 2R_3 + R_1 \rightarrow R_1$$

$$\begin{bmatrix} 1 & 0 & 0 & 2 \\ 0 & 1 & 0 & 4 \\ 0 & 0 & 1 & 2 \end{bmatrix} \quad -R_3 + R_2 \rightarrow R_2$$

The solution set is $\{(2, 4, 2)\}$.

53. $$\begin{bmatrix} 1 & 1 & 1 & 6 \\ 1 & -1 & 1 & 2 \\ 0 & 2 & -1 & 1 \end{bmatrix}$$

$$\begin{bmatrix} 1 & 1 & 1 & 6 \\ 0 & -2 & 0 & -4 \\ 0 & 2 & -1 & 1 \end{bmatrix} \quad -R_1 + R_2 \rightarrow R_2$$

$$\begin{bmatrix} 1 & 1 & 1 & 6 \\ 0 & 1 & 0 & 2 \\ 0 & 2 & -1 & 1 \end{bmatrix} \quad R_2 \div (-2) \rightarrow R_2$$

$-R_2 + R_1 \rightarrow R_1$

$$\begin{bmatrix} 1 & 0 & 1 & 4 \\ 0 & 1 & 0 & 2 \\ 0 & 0 & -1 & -3 \end{bmatrix}$$

$-2R_2 + R_3 \rightarrow R_3$

$$\begin{bmatrix} 1 & 0 & 1 & 4 \\ 0 & 1 & 0 & 2 \\ 0 & 0 & 1 & 3 \end{bmatrix}$$

$R_3 \div (-1) \rightarrow R_3$

$$\begin{bmatrix} 1 & 0 & 0 & 1 \\ 0 & 1 & 0 & 2 \\ 0 & 0 & 1 & 3 \end{bmatrix} \quad -R_3 + R_1 \rightarrow R_1$$

The solution set is $\{(1, 2, 3)\}$.

55. $$\begin{bmatrix} 2 & 1 & 1 & 4 \\ 1 & 1 & -1 & 1 \\ 1 & -1 & 2 & 2 \end{bmatrix}$$

$$\begin{bmatrix} 1 & 1 & -1 & 1 \\ 2 & 1 & 1 & 4 \\ 1 & -1 & 2 & 2 \end{bmatrix} \quad R_1 \leftrightarrow R_2$$

$$\begin{bmatrix} 1 & 1 & -1 & 1 \\ 0 & -1 & 3 & 2 \\ 0 & -2 & 3 & 1 \end{bmatrix} \quad -2R_1 + R_2 \rightarrow R_2$$

$-R_1 + R_3 \rightarrow R_3$

$$\begin{bmatrix} 1 & 1 & -1 & 1 \\ 0 & 1 & -3 & -2 \\ 0 & -2 & 3 & 1 \end{bmatrix} \quad -1 \cdot R_2 \rightarrow R_2$$

$$\begin{bmatrix} 1 & 0 & 2 & 3 \\ 0 & 1 & -3 & -2 \\ 0 & 0 & -3 & -3 \end{bmatrix} \quad -1 \cdot R_2 + R_1 \rightarrow R_1$$

$2R_2 + R_3 \rightarrow R_3$

$$\begin{bmatrix} 1 & 0 & 2 & 3 \\ 0 & 1 & -3 & -2 \\ 0 & 0 & 1 & 1 \end{bmatrix} \quad -\tfrac{1}{3} \cdot R_3 \rightarrow R_3$$

$$\begin{bmatrix} 1 & 0 & 0 & 1 \\ 0 & 1 & 0 & 1 \\ 0 & 0 & 1 & 1 \end{bmatrix} \quad \begin{array}{l} -2R_3 + R_1 \to R_1 \\ 3R_3 + R_2 \to R_2 \end{array}$$

The solution set is $\{(1, 1, 1)\}$.

57. $$\begin{bmatrix} 1 & 1 & -3 & 3 \\ 2 & -1 & 1 & 0 \\ 1 & -1 & 1 & -1 \end{bmatrix}$$

$$\begin{bmatrix} 1 & 1 & -3 & 3 \\ 0 & -3 & 7 & -6 \\ 0 & -2 & 4 & -4 \end{bmatrix} \quad \begin{array}{l} -2R_1 + R_2 \to R_2 \\ -R_1 + R_3 \to R_3 \end{array}$$

$$\begin{bmatrix} 1 & 1 & -3 & 3 \\ 0 & 1 & -1 & 2 \\ 0 & -2 & 4 & -4 \end{bmatrix} \quad -2R_3 + R_2 \to R_2$$

$$\begin{bmatrix} 1 & 0 & -2 & 1 \\ 0 & 1 & -1 & 2 \\ 0 & 0 & 2 & 0 \end{bmatrix} \quad \begin{array}{l} -R_2 + R_1 \to R_1 \\ 2R_2 + R_3 \to R_3 \end{array}$$

$$\begin{bmatrix} 1 & 0 & -2 & 1 \\ 0 & 1 & -1 & 2 \\ 0 & 0 & 1 & 0 \end{bmatrix} \quad R_3 \div 2 \to R_3$$

$$\begin{bmatrix} 1 & 0 & 0 & 1 \\ 0 & 1 & 0 & 2 \\ 0 & 0 & 1 & 0 \end{bmatrix} \quad \begin{array}{l} 2R_3 + R_1 \to R_1 \\ R_3 + R_2 \to R_2 \end{array}$$

The solution set is $\{(1, 2, 0)\}$.

59. $$\begin{bmatrix} 1 & -1 & -4 & -3 \\ -1 & 3 & 1 & 0 \\ 1 & 1 & 2 & 3 \end{bmatrix}$$

$$\begin{bmatrix} 1 & -1 & -4 & -3 \\ 0 & 2 & -3 & -3 \\ 0 & 2 & 6 & 6 \end{bmatrix} \quad \begin{array}{l} R_1 + R_2 \to R_2 \\ -R_1 + R_3 \to R_3 \end{array}$$

$$\begin{bmatrix} 1 & -1 & -4 & -3 \\ 0 & 2 & -3 & -3 \\ 0 & 0 & 9 & 9 \end{bmatrix} \quad -R_2 + R_3 \to R_3$$

$$\begin{bmatrix} 1 & -1 & -4 & -3 \\ 0 & 2 & -3 & -3 \\ 0 & 0 & 1 & 1 \end{bmatrix} \quad R_3 \div 9 \to R_3$$

$$\begin{bmatrix} 1 & -1 & 0 & 1 \\ 0 & 2 & 0 & 0 \\ 0 & 0 & 1 & 1 \end{bmatrix} \quad \begin{array}{l} 4R_3 + R_1 \to R_1 \\ 3R_3 + R_2 \to R_2 \end{array}$$

$$\begin{bmatrix} 1 & -1 & 0 & 1 \\ 0 & 1 & 0 & 0 \\ 0 & 0 & 1 & 1 \end{bmatrix} \quad R_2 \div 2 \to R_2$$

$$\begin{bmatrix} 1 & 0 & 0 & 1 \\ 0 & 1 & 0 & 0 \\ 0 & 0 & 1 & 1 \end{bmatrix} \quad R_2 + R_1 \to R_1$$

The solution set is $\{(1, 0, 1)\}$.

61. $$\begin{bmatrix} 1 & -1 & 1 & 1 \\ 2 & -2 & 2 & 2 \\ -3 & 3 & -3 & -3 \end{bmatrix}$$

$$\begin{bmatrix} 1 & -1 & 1 & 1 \\ 0 & 0 & 0 & 0 \\ 0 & 0 & 0 & 0 \end{bmatrix} \quad \begin{array}{l} -2R_1 + R_2 \to R_2 \\ 3R_1 + R_3 \to R_3 \end{array}$$

Since the system of equations is equivalent to the first equation, the solution set is $\{(x, y, z) \mid x - y + z = 1\}$.

63. $$\begin{bmatrix} 1 & 1 & -1 & 2 \\ 2 & -1 & 1 & 1 \\ 3 & 3 & -3 & 8 \end{bmatrix}$$

$$\begin{bmatrix} 1 & 1 & -1 & 2 \\ 0 & -3 & 3 & -3 \\ 0 & 0 & 0 & 2 \end{bmatrix} \quad \begin{array}{l} -2R_1 + R_2 \to R_2 \\ -3R_1 + R_3 \to R_3 \end{array}$$

Since the third equation is $0 = 2$, there is no solution to the system. The solution set is $\emptyset$.

65. Let x and y be the numbers.

$x + y = 12$
$x - y = 2$

$$\begin{bmatrix} 1 & 1 & 12 \\ 1 & -1 & 2 \end{bmatrix}$$

$$\begin{bmatrix} 1 & 1 & 12 \\ 0 & -2 & -10 \end{bmatrix} \quad -R_1 + R_2 \to R_2$$

$$\begin{bmatrix} 1 & 1 & 12 \\ 0 & 1 & 5 \end{bmatrix} \quad -\tfrac{1}{2}R_2 \to R_2$$

$$\begin{bmatrix} 1 & 0 & 7 \\ 0 & 1 & 5 \end{bmatrix} \quad -R_2 + R_1 \to R_1$$

So the numbers are 7 and 5.

67. Let $x =$ the length and $y =$ the width.

$x - y = 2.5$
$2x + 2y = 39$

$$\begin{bmatrix} 1 & -1 & 2.5 \\ 2 & 2 & 39 \end{bmatrix}$$

$$\begin{bmatrix} 1 & -1 & 2.5 \\ 0 & 4 & 34 \end{bmatrix} \quad -2R_1 + R_2 \to R_2$$

$$\begin{bmatrix} 1 & -1 & 2.5 \\ 0 & 1 & 8.5 \end{bmatrix} \quad \tfrac{1}{4}R_2 \to R_2$$

$$\begin{bmatrix} 1 & 0 & 11 \\ 0 & 1 & 8.5 \end{bmatrix} \quad R_2 + R_1 \to R_1$$

So the length is 11 in. and the width is 8.5 in.

69. Let $x =$ the selling price and $y =$ the buying price..

$x - y = 2$
$49x - 56y = 0$

$$\begin{bmatrix} 1 & -1 & 2 \\ 49 & -56 & 0 \end{bmatrix}$$

$$\begin{bmatrix} 1 & -1 & 2 \\ 0 & -7 & -98 \end{bmatrix} \quad -49R_1 + R_2 \to R_2$$

$$\begin{bmatrix} 1 & -1 & 2 \\ 0 & 1 & 14 \end{bmatrix} \quad -\tfrac{1}{7}R_2 \to R_2$$

$$\begin{bmatrix} 1 & 0 & 16 \\ 0 & 1 & 14 \end{bmatrix} \quad R_2 + R_1 \to R_1$$

So he buys for \$14 and sells for \$16.

71. Let $x =$ the number of four-wheeled cars, $y =$ the number of two-wheeled cars, and $z =$ the number of two-wheeled motorcycles.

$x + y + z = 50$
$4x + 3y + 2z = 192$
$x = 9(y + z)$

Write $x = 9(y + z)$ as $x - 9y - 9z = 0$:

$$\begin{bmatrix} 1 & 1 & 1 & 50 \\ 4 & 3 & 2 & 192 \\ 1 & -9 & -9 & 0 \end{bmatrix}$$

$$\begin{bmatrix} 1 & 1 & 1 & 50 \\ 0 & -1 & -2 & -8 \\ 0 & -10 & -10 & -50 \end{bmatrix}$$

$-4R_1 + R_2 \to R_2$
$-4_1 + R_3 \to R_3$

$$\begin{bmatrix} 1 & 1 & 1 & 50 \\ 0 & 1 & 2 & 8 \\ 0 & -10 & -10 & -50 \end{bmatrix}$$

$-R_2 \to R_2$

$$\begin{bmatrix} 1 & 0 & -1 & 42 \\ 0 & 1 & 2 & 8 \\ 0 & 0 & 10 & 30 \end{bmatrix}$$

$-R_2 + R_1 \to R_1$
$10R_2 + R_3 \to R_3$

$$\begin{bmatrix} 1 & 0 & -1 & 42 \\ 0 & 1 & 2 & 8 \\ 0 & 0 & 1 & 3 \end{bmatrix}$$

$\tfrac{1}{10}R_3 \to R_3$

$$\begin{bmatrix} 1 & 0 & 0 & 45 \\ 0 & 1 & 0 & 2 \\ 0 & 0 & 1 & 3 \end{bmatrix}$$

$R_3 + R_1 \to R_1$
$-2R_3 + R_2 \to R_2$

There were 45 four-wheeled cars, 2 three-wheeled cars, and 3 two-wheeled motorcycles.

7.5 WARM-UPS

1. True. because the determinant is $(-1)(-5)-(2)(3)=-1$.
2. False, because the determinant is $2\cdot 8-(-4)(4)=32$.
3. False, because if $D=0$, then Cramer's rule fails to give us the solution.
4. True, because the determinant is the value of $ad-bc$.
5. True, because this is the case where Cramer's rule fails to give the precise solution.
6. True.
7. True, because this is precisely when Cramer's rule works.
8. True, because of the definition of determinant of a 3×3 matrix.
9. True.
10. True.

7.5 EXERCISES

1. A determinant is a real number associated with a square matrix.
3. Cramer's rule works on systems that have exactly one solution.
5. A minor for an element is obtained by deleting the row and column of the element and then finding the determinant of the 2×2 matrix that remains.

7. $\begin{vmatrix} 2 & 5 \\ 3 & 7 \end{vmatrix}=2\cdot 7-3\cdot 5=-1$

9. $\begin{vmatrix} 0 & 3 \\ 1 & 5 \end{vmatrix}=0\cdot 5-1\cdot 3=-3$

11. $\begin{vmatrix} -3 & -2 \\ -4 & 2 \end{vmatrix}=-3\cdot 2-(-4)(-2)=-14$

13. $\begin{vmatrix} 0.05 & 0.06 \\ 10 & 20 \end{vmatrix}=0.05(20)-0.06(10)$
$=0.4$

15. $D=\begin{vmatrix} 1 & -1 \\ 0 & 2 \end{vmatrix}=2$

$D_x=\begin{vmatrix} -4 & -1 \\ 12 & 2 \end{vmatrix}=4$

$D_y=\begin{vmatrix} 1 & -4 \\ 0 & 12 \end{vmatrix}=12$

$x=\frac{D_x}{D}=\frac{4}{2}=2 \qquad y=\frac{D_y}{D}=\frac{12}{2}=6$
The solution set is $\{(2,6)\}$.

17. $D=\begin{vmatrix} 1 & 1 \\ 2 & 0 \end{vmatrix}=-2$

$D_x=\begin{vmatrix} 0 & 1 \\ -16 & 0 \end{vmatrix}=16$

$D_y=\begin{vmatrix} 1 & 0 \\ 2 & -16 \end{vmatrix}=-16$

$x=\frac{D_x}{D}=\frac{16}{-2}=-8 \quad y=\frac{D_y}{D}=\frac{-16}{-2}=8$
The solution set is $\{(-8,8)\}$.

19. $D=\begin{vmatrix} 2 & -1 \\ 3 & 2 \end{vmatrix}=7$

$D_x=\begin{vmatrix} 5 & -1 \\ -3 & 2 \end{vmatrix}=7$

$D_y=\begin{vmatrix} 2 & 5 \\ 3 & -3 \end{vmatrix}=-21$

$x=\frac{D_x}{D}=\frac{7}{7}=1 \qquad y=\frac{D_y}{D}=\frac{-21}{7}=-3$
The solution set is $\{(1,-3)\}$.

21. $D=\begin{vmatrix} 3 & -5 \\ 2 & 3 \end{vmatrix}=19$

$D_x=\begin{vmatrix} -2 & -5 \\ 5 & 3 \end{vmatrix}=19$

$D_y=\begin{vmatrix} 3 & -2 \\ 2 & 5 \end{vmatrix}=19$

$x=\frac{D_x}{D}=\frac{19}{19}=1 \qquad y=\frac{D_y}{D}=\frac{19}{19}=1$
The solution set is $\{(1,1)\}$.

23. $D=\begin{vmatrix} 4 & -3 \\ 2 & 5 \end{vmatrix}=26$

$$D_x = \begin{vmatrix} 5 & -3 \\ 7 & 5 \end{vmatrix} = 46 \qquad D_y = \begin{vmatrix} 4 & 5 \\ 2 & 7 \end{vmatrix} = 18$$

$x = \frac{D_x}{D} = \frac{46}{26} = \frac{23}{13}$ $y = \frac{D_y}{D} = \frac{18}{26} = \frac{9}{13}$

The solution set is $\left\{\left(\frac{23}{13}, \frac{9}{13}\right)\right\}$.

25. $D = \begin{vmatrix} 0.5 & 0.2 \\ 0.4 & -0.6 \end{vmatrix} = -0.38$

$$D_x = \begin{vmatrix} 8 & 0.2 \\ -5 & -0.6 \end{vmatrix} = -3.8$$

$$D_y = \begin{vmatrix} 0.5 & 8 \\ 0.4 & -5 \end{vmatrix} = -5.7$$

$x = \frac{D_x}{D} = \frac{-3.8}{-0.38} = 10$

$y = \frac{D_y}{D} = \frac{-5.7}{-0.38} = 15$

The solution set is $\{(10, 15)\}$.

27. Multiply the first equation by 4 and the second by 6 to eliminate the fractions.

$2x + y = 20$
$2x - 3y = -6$

$$D = \begin{vmatrix} 2 & 1 \\ 2 & -3 \end{vmatrix} = -8$$

$$D_x = \begin{vmatrix} 20 & 1 \\ -6 & -3 \end{vmatrix} = -54$$

$$D_y = \begin{vmatrix} 2 & 20 \\ 2 & -6 \end{vmatrix} = -52$$

$x = \frac{D_x}{D} = \frac{-54}{-8} = \frac{27}{4}$ $y = \frac{D_y}{D} = \frac{-52}{-8} = \frac{13}{2}$

The solution set is $\left\{\left(\frac{27}{4}, \frac{13}{2}\right)\right\}$.

29. The minor for 3 is the determinant of the matrix obtained by deleting the row and column containing 3, the first row and first column.

$$\begin{vmatrix} -3 & 7 \\ 1 & -6 \end{vmatrix} = 11$$

31. The minor for 5 is the determinant of the matrix obtained by deleting the first row and third column.

$$\begin{vmatrix} 4 & -3 \\ 0 & 1 \end{vmatrix} = 4$$

33. The minor for 7 is the determinant of the matrix obtained by deleting the second row and third column.

$$\begin{vmatrix} 3 & -2 \\ 0 & 1 \end{vmatrix} = 3$$

35. The minor for 1 is the determinant of the matrix obtained by deleting the third row and second column.

$$\begin{vmatrix} 3 & 5 \\ 4 & 7 \end{vmatrix} = 1$$

37. $1\begin{vmatrix} 3 & 1 \\ 1 & 5 \end{vmatrix} - 2\begin{vmatrix} 1 & 2 \\ 1 & 5 \end{vmatrix} + 3\begin{vmatrix} 1 & 2 \\ 3 & 1 \end{vmatrix}$

$= 1(14) - 2(3) + 3(-5) = -7$

39. $2\begin{vmatrix} 0 & 1 \\ 1 & 2 \end{vmatrix} - 1\begin{vmatrix} 1 & 0 \\ 1 & 2 \end{vmatrix} + 3\begin{vmatrix} 1 & 0 \\ 0 & 1 \end{vmatrix}$

$= 2(-1) - 1(2) + 3(1) = -1$

41. $-2\begin{vmatrix} 3 & 1 \\ 4 & 0 \end{vmatrix} + 3\begin{vmatrix} 1 & 2 \\ 4 & 0 \end{vmatrix} - 5\begin{vmatrix} 1 & 2 \\ 3 & 1 \end{vmatrix}$

$= -2(-4) + 3(-8) - 5(-5) = 9$

43. $1\begin{vmatrix} 3 & 2 \\ 2 & 3 \end{vmatrix} - 0\begin{vmatrix} 1 & 5 \\ 2 & 3 \end{vmatrix} + 0\begin{vmatrix} 1 & 5 \\ 3 & 2 \end{vmatrix}$

$= 1(5) - 0(-7) + 0(-13) = 5$

45. Expand by minors about the second column because it has two zeros in it.

$-1\begin{vmatrix} 2 & 6 \\ 4 & 1 \end{vmatrix} + 0\begin{vmatrix} 3 & 5 \\ 4 & 1 \end{vmatrix} - 0\begin{vmatrix} 3 & 5 \\ 2 & 6 \end{vmatrix}$

$= -1(-22) = 22$

47. Expand by minors about the first column because it has one zero in it.

$-2\begin{vmatrix} 1 & -1 \\ -4 & -3 \end{vmatrix} - 0\begin{vmatrix} 1 & 3 \\ -4 & -3 \end{vmatrix} + 2\begin{vmatrix} 1 & 3 \\ 1 & -1 \end{vmatrix}$

$= -2(-7) + 2(-4) = 6$

49. Expand by minors about the third column because it has two zeros in it.

$0\begin{vmatrix} 4 & -1 \\ 0 & 3 \end{vmatrix} - 0\begin{vmatrix} -2 & -3 \\ 0 & 3 \end{vmatrix} + 5\begin{vmatrix} -2 & -3 \\ 4 & -1 \end{vmatrix}$

$= 5(14) = 70$

51. Expand by minors about the second column.

$$-1\begin{vmatrix}0 & 5\\5 & 4\end{vmatrix}+0\begin{vmatrix}2 & 1\\5 & 4\end{vmatrix}-0\begin{vmatrix}2 & 1\\0 & 5\end{vmatrix}$$

$$=-1(-25)=25$$

53. $D=\begin{vmatrix}1 & 1 & 1\\1 & -1 & 1\\2 & 1 & 1\end{vmatrix}=2$

$$D_x=\begin{vmatrix}6 & 1 & 1\\2 & -1 & 1\\7 & 1 & 1\end{vmatrix}=2$$

$$D_y=\begin{vmatrix}1 & 6 & 1\\1 & 2 & 1\\2 & 7 & 1\end{vmatrix}=4$$

$$D_z=\begin{vmatrix}1 & 1 & 6\\1 & -1 & 2\\2 & 1 & 7\end{vmatrix}=6$$

$$x=\frac{D_x}{D}=\frac{2}{2}=1,\ y=\frac{D_y}{D}=\frac{4}{2}=2,$$

$$z=\frac{D_z}{D}=\frac{6}{2}=3$$

The solution set is $\{(1, 2, 3)\}$.

55. $D=\begin{vmatrix}1 & -3 & 2\\1 & 1 & 1\\1 & -1 & 1\end{vmatrix}=-2$

$$D_x=\begin{vmatrix}0 & -3 & 2\\2 & 1 & 1\\0 & -1 & 1\end{vmatrix}=2$$

$$D_y=\begin{vmatrix}1 & 0 & 2\\1 & 2 & 1\\1 & 0 & 1\end{vmatrix}=-2$$

$$D_z=\begin{vmatrix}1 & -3 & 0\\1 & 1 & 2\\1 & -1 & 0\end{vmatrix}=-4$$

$$x=\frac{D_x}{D}=\frac{2}{-2}=-1,\quad y=\frac{D_y}{D}=\frac{-2}{-2}=1,$$

$$z=\frac{D_z}{D}=\frac{-4}{-2}=2$$

The solution set is $\{(-1, 1, 2)\}$.

57. $D=\begin{vmatrix}1 & 1 & 0\\0 & 2 & -1\\1 & 1 & 1\end{vmatrix}=2$

$$D_x=\begin{vmatrix}-1 & 1 & 0\\3 & 2 & -1\\0 & 1 & 1\end{vmatrix}=-6$$

$$D_y=\begin{vmatrix}1 & -1 & 0\\0 & 3 & -1\\1 & 0 & 1\end{vmatrix}=4$$

$$D_z=\begin{vmatrix}1 & 1 & -1\\0 & 2 & 3\\1 & 1 & 0\end{vmatrix}=2$$

$$x=\frac{D_x}{D}=\frac{-6}{2}=-3,\quad y=\frac{D_y}{D}=\frac{4}{2}=2,$$

$$z=\frac{D_z}{D}=\frac{2}{2}=1$$

The solution set is $\{(-3, 2, 1)\}$.

59. $D=\begin{vmatrix}1 & 1 & -1\\2 & 2 & 1\\1 & -3 & 0\end{vmatrix}=12$

$$D_x=\begin{vmatrix}0 & 1 & -1\\6 & 2 & 1\\0 & -3 & 0\end{vmatrix}=18$$

$$D_y=\begin{vmatrix}1 & 0 & -1\\2 & 6 & 1\\1 & 0 & 0\end{vmatrix}=6$$

$$D_z=\begin{vmatrix}1 & 1 & 0\\2 & 2 & 6\\1 & -3 & 0\end{vmatrix}=24$$

$$x=\frac{D_x}{D}=\frac{18}{12}=\frac{3}{2},\quad y=\frac{D_y}{D}=\frac{6}{12}=\frac{1}{2},$$

$$z=\frac{D_z}{D}=\frac{24}{12}=2$$

The solution set is $\left\{\left(\frac{3}{2},\frac{1}{2},2\right)\right\}$.

61. $D=\begin{vmatrix}1 & 1 & 1\\0 & 2 & 2\\3 & -1 & 0\end{vmatrix}=2$

$$D_x=\begin{vmatrix}0 & 1 & 1\\0 & 2 & 2\\-1 & -1 & 0\end{vmatrix}=0$$

$$D_y = \begin{vmatrix} 1 & 0 & 1 \\ 0 & 0 & 2 \\ 3 & -1 & 0 \end{vmatrix} = 2$$

$$D_z = \begin{vmatrix} 1 & 1 & 0 \\ 0 & 2 & 0 \\ 3 & -1 & -1 \end{vmatrix} = -2$$

$$x = \frac{D_x}{D} = \frac{0}{2} = 0, \quad y = \frac{D_y}{D} = \frac{2}{2} = 1,$$

$$z = \frac{D_z}{D} = \frac{-2}{2} = -1$$

The solution set is $\{(0, 1, -1)\}$.

63. a) From the graph it appears that there should be approximately 9 servings of peas and 11 servings of beets.
b) Let $x =$ the number of servings of canned peas and $y =$ the number of servings of canned beets. In the first equation we find the total grams of protein and in the second the total grams of carbohydrates.

$$3x + y = 38$$
$$11x + 8y = 187$$

$$\text{D} = \begin{vmatrix} 3 & 1 \\ 11 & 8 \end{vmatrix} = 13 \quad D_x = \begin{vmatrix} 38 & 1 \\ 187 & 8 \end{vmatrix} = 117$$

$$D_y = \begin{vmatrix} 3 & 38 \\ 11 & 187 \end{vmatrix} = 143$$

$$x = \frac{D_x}{D} = \frac{117}{13} = 9 \quad y = \frac{D_y}{D} = \frac{143}{13} = 11$$

To get the required grams of protein and carbohydrates we need 9 servings of peas and 11 servings of beets.
65. Let $x =$ the price of a gallon of milk and $y =$ the price of the magazine. Note that she paid \$0.30 tax. The first equation expresses the total price of the goods and the second expresses the total tax.

$$x + y = 4.65$$
$$0.05x + 0.08y = 0.30$$

$$D = \begin{vmatrix} 1 & 1 \\ 0.05 & 0.08 \end{vmatrix} = 0.03$$

$$D_x = \begin{vmatrix} 4.65 & 1 \\ 0.3 & 0.08 \end{vmatrix} = 0.072$$

$$D_y = \begin{vmatrix} 1 & 4.65 \\ 0.05 & 0.3 \end{vmatrix} = 0.0675$$

$$x = \frac{D_x}{D} = \frac{0.072}{0.03} = 2.4$$

$$y = \frac{D_y}{D} = \frac{0.0675}{0.03} = 2.25$$

The price of the milk was \$2.40 and the price of the magazine was \$2.25.
67. Let $x =$ the number of singles and $y =$ the number of doubles. Write one equation for the patties and one equation for the tomato slices.

$$x + 2y = 32$$
$$2x + y = 34$$

$$D = \begin{vmatrix} 1 & 2 \\ 2 & 1 \end{vmatrix} = -3 \quad D_x = \begin{vmatrix} 32 & 2 \\ 34 & 1 \end{vmatrix} = -36$$

$$D_y = \begin{vmatrix} 1 & 32 \\ 2 & 34 \end{vmatrix} = -30$$

$$x = \frac{D_x}{D} = \frac{-36}{-3} = 12 \quad y = \frac{D_y}{D} = \frac{-30}{-3} = 10$$

He must sell 12 singles and 10 doubles.
69. Let $x =$ Gary's age and $y =$ Harry's age. Since Gary is 5 years older than Harry, $x = y + 5$. Twenty-nine years ago Gary was $x - 29$ and Harry was $y - 29$. Gary was twice as old as Harry (29 years ago) is expressed as $x - 29 = 2(y - 29)$. These equations can be rewritten as follows.

$$x - y = 5$$
$$x - 2y = -29$$

$$D = \begin{vmatrix} 1 & -1 \\ 1 & -2 \end{vmatrix} = -1$$

$$D_x = \begin{vmatrix} 5 & -1 \\ -29 & -2 \end{vmatrix} = -39$$

$$D_y = \begin{vmatrix} 1 & 5 \\ 1 & -29 \end{vmatrix} = -34$$

$$x = \frac{D_x}{D} = \frac{-39}{-1} = 39 \quad y = \frac{D_y}{D} = \frac{-34}{-1} = 34$$

So Gary is 39 and Harry is 34.

71. Let $x =$ the length of a side of the square and $y =$ the length of a side of the equilateral triangle. Since the perimeters are to be equal, $4x = 3y$. Since the total of the two perimeters is 80, $4x + 3y = 80$. Rewrite the equations as follows.

$$4x - 3y = 0$$
$$4x + 3y = 80$$

$$D = \begin{vmatrix} 4 & -3 \\ 4 & 3 \end{vmatrix} = 24$$

$$D_x = \begin{vmatrix} 0 & -3 \\ 80 & 3 \end{vmatrix} = 240$$

$$D_y = \begin{vmatrix} 4 & 0 \\ 4 & 80 \end{vmatrix} = 320$$

$$x = \frac{D_x}{D} = \frac{240}{24} = 10 \quad y = \frac{D_y}{D} = \frac{320}{24} = \frac{40}{3}$$

The length of the side of the square should be 10 feet and the length of the side of the triangle should be $40/3$ feet.

73. Let $x =$ the number of gallons of 10% solution and $y =$ the number of gallons of 25% solution. Since the total mixture is to be 30 gallons, $x + y = 30$. The next equation comes from the fact that the chlorine in the two parts is equal to the chlorine in the 30 gallons, $0.10x + 0.25y = 0.20(30)$. Rewrite the equations as follows.

$$x + \quad y = 30$$
$$0.10x + 0.25y = 6$$

$$D = \begin{vmatrix} 1 & 1 \\ 0.10 & 0.25 \end{vmatrix} = 0.15$$

$$D_x = \begin{vmatrix} 30 & 1 \\ 6 & 0.25 \end{vmatrix} = 1.5$$

$$D_y = \begin{vmatrix} 1 & 30 \\ 0.10 & 6 \end{vmatrix} = 3$$

$$x = \frac{D_x}{D} = \frac{1.5}{0.15} = 10$$

$$y = \frac{D_y}{D} = \frac{3}{0.15} = 20$$

Use 10 gallons of 10% solution and 20 gallons of 25% solution.

75. Let $x =$ Mimi's weight, $y =$ Mitzi's weight, and $z =$ Cassandra's weight. We can write 3 equations.

$$x + y + z = 175$$
$$x \quad + z = 143$$
$$y + z = 139$$

$$D = \begin{vmatrix} 1 & 1 & 1 \\ 1 & 0 & 1 \\ 0 & 1 & 1 \end{vmatrix} = -1$$

$$D_x = \begin{vmatrix} 175 & 1 & 1 \\ 143 & 0 & 1 \\ 139 & 1 & 1 \end{vmatrix} = -36$$

$$D_y = \begin{vmatrix} 1 & 175 & 1 \\ 1 & 143 & 1 \\ 0 & 139 & 1 \end{vmatrix} = -32$$

$$D_z = \begin{vmatrix} 1 & 1 & 175 \\ 1 & 0 & 143 \\ 0 & 1 & 139 \end{vmatrix} = -107$$

$$x = \frac{D_x}{D} = \frac{-36}{-1} = 36,$$

$$y = \frac{D_y}{D} = \frac{-32}{-1} = 32,$$

$$z = \frac{D_z}{D} = \frac{-107}{-1} = 107$$

So Mimi weights 36 pounds, Mitzi weighs 32 pounds, and Cassandra weighs 107 pounds.

77. Let $x =$ the number of degrees in the larger of the two acute angles, $y =$ the number of degrees in the smaller acute angle, and $z =$ the number of degrees in the right angle. We can write 3 equations about the sizes of the angles.

$$x + y + z = 180$$
$$x - y \quad = 12$$
$$z = 90$$

$$D = \begin{vmatrix} 1 & 1 & 1 \\ 1 & -1 & 0 \\ 0 & 0 & 1 \end{vmatrix} = -2$$

$$D_x = \begin{vmatrix} 180 & 1 & 1 \\ 12 & -1 & 0 \\ 90 & 0 & 1 \end{vmatrix} = -102$$

$$D_y = \begin{vmatrix} 1 & 180 & 1 \\ 1 & 12 & 0 \\ 0 & 90 & 1 \end{vmatrix} = -78$$

$$D_z = \begin{vmatrix} 1 & 1 & 180 \\ 1 & -1 & 12 \\ 0 & 0 & 90 \end{vmatrix} = -180$$

$$x = \frac{D_x}{D} = \frac{-102}{-2} = 51,$$
$$y = \frac{D_y}{D} = \frac{-78}{-2} = 39,$$
$$z = \frac{D_z}{D} = \frac{-180}{-2} = 90$$

The measures of the three angles of the triangle are 39°, 51°, and 90°.

79. If $D = 0$, then use one of the other methods.

81. No, because Cramer's rule only works on linear systems and the given system is not linear.

Enriching Your Mathematical Word Power

1. c **2.** a **3.** a **4.** d **5.** b **6.** c
7. a **8.** c **9.** d **10.** b **11.** a **12.** d

CHAPTER 7 REVIEW

1. The graph of $y = 2x - 1$ has y-intercept $(0, -1)$ and slope 2. the graph of $y = -x + 2$ has y-intercept $(0, 2)$ and slope -1. The graphs appear to intersect at $(1, 1)$. Check that $(1, 1)$ satisfies both equations. The solution set is $\{(1, 1)\}$ and the system is independent.

3. Solve $x + 2y = 4$ for y to get $y = -\frac{1}{2}x + 2$, which is the same as the second equation. So the two equations have the same graph and all points on that graph satisfy the system. The solution set to this dependent system is $\{(x, y) | x + 2y = 4\}$.

5. The graph of $y = -x$ is a line with slope -1 and y-intercept $(0, 0)$. The graph of $y = -x + 3$ is a line with slope -1 and y-intercept $(0, 3)$. Since these lines are parallel, there is no solution to the system. The system is inconsistent and the solution set is $\emptyset$.

7. Substitute $y = 3x + 11$ into $2x + 3y = 0$.

$$2x + 3(3x + 11) = 0$$
$$11x + 33 = 0$$
$$11x = -33$$
$$x = -3$$
$$y = 3(-3) + 11 = 2$$

The solution set is $\{(-3, 2)\}$ and the system is independent.

9. Substitute $x = y + 5$ into $2x - 2y = 12$.

$$2(y + 5) - 2y = 12$$
$$2y + 10 - 2y = 12$$
$$10 = 12$$

The solution set is $\emptyset$ and the system is inconsistent.

11. Write $2x - y = 3$ as $y = 2x - 3$. Substitute into $6x - 9 = 3y$.

$$6x - 9 = 3(2x - 3)$$
$$6x - 9 = 6x - 9$$

Since the last equation is an identity, the solution set is $\{(x, y) \mid 2x - y = 3\}$. The system is dependent.

13. Substitute $y = \frac{1}{2}x - 3$ into $y = \frac{1}{3}x + 2$.

$$\frac{1}{3}x + 2 = \frac{1}{2}x - 3$$
$$2x + 12 = 3x - 18$$
$$30 = x$$
$$y = \frac{1}{3}x + 2$$
$$y = \frac{1}{3} \cdot 30 + 2 = 12$$

The solution set to this independent system is $\{(30, 12)\}$.

15. Substitute $x = 1 - 2y$ into $8x + 6y = 4$:

$$8(1 - 2y) + 6y = 4$$
$$8 - 16y + 6y = 4$$
$$-10y = -4$$
$$y = \frac{2}{5}$$
$$x = 1 - 2y$$
$$x = 1 - 2 \cdot \frac{2}{5} = \frac{1}{5}$$

The solution set to this independent system is $\left\{\left(\frac{1}{5}, \frac{2}{5}\right)\right\}$

17. Multiply the first equation by 2 and the second by 3, and then add the resulting equations.

$$\begin{aligned} 10x - 6y &= -40 \\ 9x + 6y &= 21 \\ \hline 19x \quad &= -19 \\ x &= -1 \end{aligned}$$

Use $x = -1$ in $3x + 2y = 7$.

$$\begin{aligned} 3(-1) + 2y &= 7 \\ 2y &= 10 \\ y &= 5 \end{aligned}$$

The solution set is $\{(-1, 5)\}$ and the system is independent.

19. Rewrite the first equation.

$$\begin{aligned} 2(y - 5) + 4 &= 3(x - 6) \\ 2y - 10 + 4 &= 3x - 18 \\ -3x + 2y &= -12 \end{aligned}$$

Add this last equation to the original second equation.

$$\begin{aligned} -3x + 2y &= -12 \\ 3x - 2y &= 12 \\ \hline 0 &= 0 \end{aligned}$$

The two equations are just different forms of an equation for the same straight line. The solution set is $\{(x, y) \mid 3x - 2y = 12\}$ and the system is dependent.

21. Rewrite the first equation in the same form as the last so that they can be added.

$$\begin{aligned} 3x - 4(y - 5) &= x + 2 \\ 3x - 4y + 20 &= x + 2 \\ 2x - 4y &= -18 \\ x - 2y &= -9 \\ -2y + x &= -9 \end{aligned}$$

Add this last equation with the second equation.

$$\begin{aligned} -2y + x &= -9 \\ 2y - x &= 7 \\ \hline 0 &= -2 \end{aligned}$$

Since the result of the addition is a false statement, the system is inconsistent. The solution set is $\emptyset$.

23. Multiply the first equation by 8 and the second equation by 2 to clear the fractions:

$$\begin{aligned} 2x + 3y &= 3 \\ 5x - 12y &= 14 \end{aligned}$$

Multiply the first equation by 4 and add:

$$\begin{aligned} 8x + 12y &= 12 \\ 5x - 12y &= 14 \\ \hline 13x \quad &= 26 \\ x &= 2 \end{aligned}$$

Use $x = 2$ in $2x + 3y = 3$ to find y.

$$\begin{aligned} 2(2) + 3y &= 3 \\ 4 + 3y &= 3 \\ 3y &= -1 \\ y &= -\frac{1}{3} \end{aligned}$$

The solution set is $\left\{\left(2, -\frac{1}{3}\right)\right\}$. The system is independent.

25. Multiply the first equation by 100 and the second equation by 100 to clear the decimals:

$$\begin{aligned} 40x + 6y &= 1160 \\ 80x - 5y &= 1300 \end{aligned}$$

Multiply the first equation by -2 and add:

$$\begin{aligned} -80x - 12y &= -2320 \\ 80x - 5y &= 1300 \\ \hline -17y &= -1020 \\ y &= 60 \end{aligned}$$

Use $y = 60$ in $40x + 6y = 1160$ to find x.

$$\begin{aligned} 40x + 6(60) &= 1160 \\ 40x &= 800 \\ x &= 20 \end{aligned}$$

The solution set is $\{(20, 60)\}$. The system is independent.

27. Add the first and second equations.

$$\begin{aligned} x - y + z &= 4 \\ -x + 2y - z &= 0 \\ \hline y \quad &= 4 \end{aligned}$$

Add the first and third equations:

$$\begin{aligned} x - y + z &= 4 \\ -x + y - 3z &= -16 \\ \hline -2z &= -12 \\ z &= 6 \end{aligned}$$

Use $y = 4$ and $z = 6$ in $x - y + z = 4$.

$$\begin{aligned} x - 4 + 6 &= 4 \\ x &= 2 \end{aligned}$$

The solution set is $\{(2, 4, 6)\}$.

29. Add the first and second equations.

$$\begin{aligned} 2x - y - z &= 3 \\ 3x + y + 2z &= 4 \\ \hline 5x \quad + z &= 7 \qquad \text{A} \end{aligned}$$

Multiply the first equation by 2 and add the result to the last equation.

$$\begin{array}{l} 4x - 2y - 2z = 6 \\ \underline{4x + 2y - z = -4} \\ 8x \qquad - 3z = 2 \quad \text{B} \end{array}$$

Multiply equation A by 3 and add the result to equation B.

$$\begin{array}{l} 15x + 3z = 21 \\ \underline{8x - 3z = 2} \\ 23x \qquad = 23 \\ \qquad x = 1 \end{array}$$

Use $x = 1$ in $8x - 3z = 2$.

$$\begin{aligned} 8(1) - 3z &= 2 \\ -3z &= -6 \\ z &= 2 \end{aligned}$$

Use $x = 1$ and $z = 2$ in $3x + y + 2z = 4$.

$$\begin{aligned} 3(1) + y + 2(2) &= 4 \\ y &= -3 \end{aligned}$$

The solution set is $\{(1, -3, 2)\}$.

31. Add the first two equations.

$$\begin{array}{l} x + y - z = 4 \\ \underline{y + z = 6} \\ x + 2y \quad = 10 \end{array}$$

Substituting $x + 2y = 10$ into the third equation $x + 2y = 8$ yields $10 = 8$, which is false. The solution set is $\emptyset$.

33. If the first equation is multiplied by -1 the result is the second equation and if the first is multiplied by 2 the result is the third equation. So all three equations are equivalent. Any point that satisfies one of them satisfies all of them. The system is dependent. The solution set is $\{(x, y, z) \mid x - 2y + z = 8\}$.

35. $\begin{bmatrix} 1 & 1 & 7 \\ -1 & 2 & 5 \end{bmatrix}$

$\begin{bmatrix} 1 & 1 & 7 \\ 0 & 3 & 12 \end{bmatrix}$ $\quad R_1 + R_2 \to R_2$

$\begin{bmatrix} 1 & 1 & 7 \\ 0 & 1 & 4 \end{bmatrix}$ $\quad R_2 \div 3 \to R_2$

$\begin{bmatrix} 1 & 0 & 3 \\ 0 & 1 & 4 \end{bmatrix}$ $\quad -1R_2 + R_1 \to R_1$

The solution set is $\{(3, 4)\}$.

37. $\begin{bmatrix} 1 & -3 & 14 \\ 2 & 1 & 0 \end{bmatrix}$

$\begin{bmatrix} 1 & -3 & 14 \\ 0 & 7 & -28 \end{bmatrix}$ $\quad -2R_1 + R_2 \to R_2$

$\begin{bmatrix} 1 & -3 & 14 \\ 0 & 1 & -4 \end{bmatrix}$ $\quad R_2 \div 7 \to R_2$

$\begin{bmatrix} 1 & 0 & 2 \\ 0 & 1 & -4 \end{bmatrix}$ $\quad 3R_2 + R_1 \to R_1$

The solution set is $\{(2, -4)\}$.

39. $\begin{bmatrix} 1 & 1 & -1 & 0 \\ 1 & -1 & 2 & 4 \\ 2 & 1 & -1 & 1 \end{bmatrix}$

$\begin{bmatrix} 1 & 1 & -1 & 0 \\ 0 & -2 & 3 & 4 \\ 0 & -1 & 1 & 1 \end{bmatrix}$ $\quad -R_1 + R_2 \to R_2$

$-2R_1 + R_3 \to R_3$

$\begin{bmatrix} 1 & 1 & -1 & 0 \\ 0 & -1 & 1 & 1 \\ 0 & -2 & 3 & 4 \end{bmatrix}$ $\quad R_2 \leftrightarrow R_3$

$\begin{bmatrix} 1 & 1 & -1 & 0 \\ 0 & 1 & -1 & -1 \\ 0 & -2 & 3 & 4 \end{bmatrix}$ $\quad -R_2 \to R_2$

$\begin{bmatrix} 1 & 0 & 0 & 1 \\ 0 & 1 & -1 & -1 \\ 0 & 0 & 1 & 2 \end{bmatrix}$ $\quad -R_2 + R_1 \to R_1$

$2R_2 + R_3 \to R_3$

$\begin{bmatrix} 1 & 0 & 0 & 1 \\ 0 & 1 & 0 & 1 \\ 0 & 0 & 1 & 2 \end{bmatrix}$ $\quad R_3 + R_2 \to R_2$

The solution set is $\{(1, 1, 2)\}$.

41. $\begin{vmatrix} 1 & 3 \\ 0 & 2 \end{vmatrix} = 1 \cdot 2 - 0 \cdot 3 = 2$

43. $\begin{vmatrix} 0.01 & 0.02 \\ 50 & 80 \end{vmatrix} = 0.01(80) - 0.02(50)$
$= 0.8 - 1 = -0.2$

45. $D = \begin{vmatrix} 2 & -1 \\ 3 & 1 \end{vmatrix} = 5$

$$D_x = \begin{vmatrix} 0 & -1 \\ -5 & 1 \end{vmatrix} = -5$$

$$D_y = \begin{vmatrix} 2 & 0 \\ 3 & -5 \end{vmatrix} = -10$$

$$x = \frac{D_x}{D} = \frac{-5}{5} = -1$$

$$y = \frac{D_y}{D} = \frac{-10}{5} = -2$$

The solution set is $\{(-1, -2)\}$.

47. Write the system in standard form.

$$-2x + y = -3$$
$$3x - 2y = 4$$

$$D = \begin{vmatrix} -2 & 1 \\ 3 & -2 \end{vmatrix} = 1$$

$$D_x = \begin{vmatrix} -3 & 1 \\ 4 & -2 \end{vmatrix} = 2$$

$$D_y = \begin{vmatrix} -2 & -3 \\ 3 & 4 \end{vmatrix} = 1$$

$$x = \frac{D_x}{D} = \frac{2}{1} = 2 \qquad y = \frac{D_y}{D} = \frac{1}{1} = 1$$

The solution set is $\{(2, 1)\}$.

49. Expand by minors using the first column.

$$2\begin{vmatrix} 2 & 4 \\ 1 & 1 \end{vmatrix} - (-1)\begin{vmatrix} 3 & 1 \\ 1 & 1 \end{vmatrix} + 6\begin{vmatrix} 3 & 1 \\ 2 & 4 \end{vmatrix}$$
$$= 2(-2) + 2 + 6(10) = 58$$

51. Expand by minors using the second column.

$$-3\begin{vmatrix} 2 & 4 \\ -1 & 3 \end{vmatrix} + 0\begin{vmatrix} 2 & -2 \\ -1 & 3 \end{vmatrix} - 0\begin{vmatrix} 2 & -2 \\ 2 & 4 \end{vmatrix}$$
$$= -3(10) = -30$$

53. $D = \begin{vmatrix} 1 & 1 & 0 \\ 1 & 1 & 1 \\ 1 & -1 & -1 \end{vmatrix} = 2$

$$D_x = \begin{vmatrix} 3 & 1 & 0 \\ 0 & 1 & 1 \\ 2 & -1 & -1 \end{vmatrix} = 2$$

$$D_y = \begin{vmatrix} 1 & 3 & 0 \\ 1 & 0 & 1 \\ 1 & 2 & -1 \end{vmatrix} = 4$$

$$D_z = \begin{vmatrix} 1 & 1 & 3 \\ 1 & 1 & 0 \\ 1 & -1 & 2 \end{vmatrix} = -6$$

$$x = \frac{D_x}{D} = \frac{2}{2} = 1, \quad y = \frac{D_y}{D} = \frac{4}{2} = 2,$$

$$z = \frac{D_z}{D} = \frac{-6}{2} = -3$$

The solution set is $\{(1, 2, -3)\}$.

55. Let x represent the width and y represent the length, then $2x + 2y = 82$ and $y = x + 15$.

$$2x + 2(x + 15) = 82$$
$$4x + 30 = 82$$
$$4x = 52$$
$$x = 13$$
$$y = 13 + 15 = 28$$

The length is 28 feet and the width is 13 feet.

57. Let $x =$ the tens digit and $y =$ the ones digit. The value of the number is $10x + y$. If the digits are reversed the value will be $10y + x$.

We can write the following two equations.

$$x + y = 15$$
$$10x + y = 10y + x - 9$$

$$x + y = 15$$
$$9x - 9y = -9$$

$$x + y = 15$$
$$x - y = -1$$
$$2x = 14$$
$$x = 7$$
$$y = 8 \quad \text{Since } x + y = 15$$

The number is 78.

59. Let $b =$ his boat's speed in still water and $c =$ the speed of the current. Let $t =$ his time to go the same distance in the lake.

	Distance	Rate	Time
Down	$\frac{1}{2}(b + c)$	$b + c$	$\frac{1}{2}$
Back	$\frac{3}{4}(b - c)$	$b - c$	$\frac{3}{4}$
Lake	bt	b	t

Since the distance down the stream is the same as the distance going back, we can write the following equation.

$$\frac{1}{2}(b+c) = \frac{3}{4}(b-c)$$
$$\frac{1}{2}b + \frac{1}{2}c = \frac{3}{4}b - \frac{3}{4}c$$
$$2b + 2c = 3b - 3c$$
$$5c = b$$

Since $bt = \frac{1}{2}(b+c)$, we can substitute $b = 5c$ into this equation.

$$(5c)t = \frac{1}{2}(5c+c)$$
$$5ct = 3c$$
$$t = \frac{3c}{5c} = \frac{3}{5}$$

So the time in the lake is $3/5$ of an hour or 36 minutes.

61. Let $x =$ the number of liters of solution A, $y =$ the number of liters of solution B, and $z =$ the number of liters of solution C. We can write 3 equations.

$$x + y + z = 20$$
$$0.30x + 0.20y + 0.60z = 0.38(20)$$
$$z = 2x$$

Substitute $z = 2x$ into the other two equations.

$$x + y + 2x = 20$$
$$0.3x + 0.2y + 0.6(2x) = 7.6$$

$$3x + y = 20$$
$$1.5x + 0.2y = 7.6$$

Substitute $y = 20 - 3x$ into the second equation.

$$1.5x + 0.2(20 - 3x) = 7.6$$
$$0.9x + 4 = 7.6$$
$$0.9x = 3.6$$
$$x = 4$$

If $x = 4$, then $y = 20 - 3(4) = 8$ and $z = 2(4) = 8$. She should use 4 liters of 30% solution (A), 8 liters of 20% solution (B), and 8 liters of 60% solution (C).

63. Let $x =$ the number of servings of beets and $y =$ the number of servings of beans. We can write two equations.

$$x + 6y = 21$$
$$6x + 20y = 78$$

Substitute $x = 21 - 6y$ into $6x + 20y = 78$.

$$6(21 - 6y) + 20y = 78$$
$$126 - 36y + 20y = 78$$
$$-16y = -48$$
$$y = 3$$

If $y = 3$, then $x = 21 - 6(3) = 3$. It would take 3 servings of each.

Chapter 7 TEST

1. The graph of $y = -x + 4$ is a straight line with y-intercept (0, 4) and slope -1. The graph of $y = 2x + 1$ is a straight line with y-intercept (0, 1) and slope 2. The graphs appear to intersect at (1, 3). After checking that (1, 3) satisfies both equations, we can be sure that the solution set is $\{(1, 3)\}$.

2. Substitute $y = 2x - 8$ into $4x + 3y = 1$.

$$4x + 3(2x - 8) = 1$$
$$10x = 25$$
$$x = \frac{5}{2}$$

If $x = 5/2$, then $y = 2(5/2) - 8 = -3$. The solution set is $\left\{\left(\frac{5}{2}, -3\right)\right\}$.

3. Substitute $y = x - 5$ into the second equation.

$$3x - 4(y - 2) = 28 - x$$
$$3x - 4(x - 5 - 2) = 28 - x$$
$$3x - 4(x - 7) = 28 - x$$
$$28 - x = 28 - x$$

Since the last equation is an identity, the solution set is $\{(x, y) \mid y = x - 5\}$.

4. Multiply the first equation by 3 and the second by 2 and add the results.

$$\begin{aligned} 9x + 6y &= 9 \\ 8x - 6y &= -26 \\ \hline 17x \quad &= -17 \end{aligned}$$
$$x = -1$$

Use $x = -1$ in $3x + 2y = 3$ to find y.

$$3(-1) + 2y = 3$$
$$2y = 6$$
$$y = 3$$

The solution set is $\{(-1, 3)\}$.

5. Multiply the first equation by 2 and then add the result to the second equation.

$$\begin{aligned} 6x - 2y &= 10 \\ -6x + 2y &= 1 \\ \hline 0 &= 11 \end{aligned}$$

The solution set to the system is $\emptyset$.

6. The lines both have slope 3 and different y-intercepts, so they are parallel. There is no solution to the system. The system is inconsistent.

7. If we multiply the second equation by 2, the result is the same as the first equation. These are two equations for the same straight line. The system is dependent.

8. The lines have different slopes and different y-intercepts. They will intersect in exactly one point. The system is independent.

9. Add the first two equations.

$$\begin{array}{rl} x + y - z = 2 & \\ 2x - y + 3z = -5 & \\ \hline 3x \quad\ + 2z = -3 & \text{A} \end{array}$$

Multiply the first equation by 3 and add the result to the last equation.

$$\begin{array}{rl} 3x + 3y - 3z = 6 & \\ x - 3y + z = 4 & \\ \hline 4x \quad\ - 2z = 10 & \text{B} \end{array}$$

Add equations A and B.

$$\begin{array}{r} 3x + 2z = -3 \\ 4x - 2z = 10 \\ \hline 7x \quad\ = 7 \\ x = 1 \end{array}$$

Use $x = 1$ in $3x + 2z = -3$.

$$3(1) + 2z = -3$$
$$2z = -6$$
$$z = -3$$

Use $x = 1$ and $z = -3$ in $x + y - z = 2$.

$$1 + y - (-3) = 2$$
$$y = -2$$

The solution set is $\{(1, -2, -3)\}$.

10. $\begin{bmatrix} 1 & 2 & 12 \\ 3 & -1 & 1 \end{bmatrix}$

$$\begin{bmatrix} 1 & 2 & 12 \\ 0 & -7 & -35 \end{bmatrix} \quad -3R_1 + R_2 \rightarrow R_2$$

$$\begin{bmatrix} 1 & 2 & 12 \\ 0 & 1 & 5 \end{bmatrix} \quad R_2 \div (-7) \rightarrow R_2$$

$$\begin{bmatrix} 1 & 0 & 2 \\ 0 & 1 & 5 \end{bmatrix} \quad -2R_2 + R_1 \rightarrow R_1$$

The solution set is $\{(2, 5)\}$.

11. $\begin{bmatrix} 1 & -1 & -1 & 1 \\ -1 & -1 & 2 & -2 \\ -1 & -3 & 1 & -5 \end{bmatrix}$

$$\begin{bmatrix} 1 & -1 & -1 & 1 \\ 0 & -2 & 1 & -1 \\ 0 & -4 & 0 & -4 \end{bmatrix} \quad \begin{array}{l} R_1 + R_2 \rightarrow R_2 \\ R_1 + R_3 \rightarrow R_3 \end{array}$$

$$\begin{bmatrix} 1 & -1 & -1 & 1 \\ 0 & -4 & 0 & -4 \\ 0 & -2 & 1 & -1 \end{bmatrix} \quad R_2 \leftrightarrow R_3$$

$$\begin{bmatrix} 1 & -1 & -1 & 1 \\ 0 & 1 & 0 & 1 \\ 0 & -2 & 1 & -1 \end{bmatrix} \quad R_2 \div (-4) \rightarrow R_2$$

$$\begin{bmatrix} 1 & 0 & -1 & 2 \\ 0 & 1 & 0 & 1 \\ 0 & 0 & 1 & 1 \end{bmatrix} \quad \begin{array}{l} R_2 + R_1 \rightarrow R_1 \\ 2R_2 + R_3 \rightarrow R_3 \end{array}$$

$$\begin{bmatrix} 1 & 0 & 0 & 3 \\ 0 & 1 & 0 & 1 \\ 0 & 0 & 1 & 1 \end{bmatrix} \quad R_3 + R_1 \rightarrow R_1$$

The solution set is $\{(3, 1, 1)\}$.

12. $\begin{vmatrix} 2 & 3 \\ 4 & -3 \end{vmatrix} = 2(-3) - 4(3) = -18$

13. Expand by minors using the third column.

$$-1\begin{vmatrix} 2 & 3 \\ 1 & 1 \end{vmatrix} - 1\begin{vmatrix} 1 & -2 \\ 1 & 1 \end{vmatrix} + 0\begin{vmatrix} 1 & -2 \\ 2 & 3 \end{vmatrix}$$

$$= -1(-1) - 1(3) + 0(7) = -2$$

14. $D = \begin{vmatrix} 2 & -1 \\ 3 & 1 \end{vmatrix} = 5$

$D_x = \begin{vmatrix} -4 & -1 \\ -1 & 1 \end{vmatrix} = -5$

$D_y = \begin{vmatrix} 2 & -4 \\ 3 & -1 \end{vmatrix} = 10$

$x = \frac{D_x}{D} = \frac{-5}{5} = -1 \qquad y = \frac{D_y}{D} = \frac{10}{5} = 2$

The solution set is $\{(-1, 2)\}$.

15. $D = \begin{vmatrix} 1 & 1 & 0 \\ 1 & -1 & 2 \\ 2 & 1 & -1 \end{vmatrix} = 4$

$D_x = \begin{vmatrix} 0 & 1 & 0 \\ 6 & -1 & 2 \\ 1 & 1 & -1 \end{vmatrix} = 8$

$D_y = \begin{vmatrix} 1 & 0 & 0 \\ 1 & 6 & 2 \\ 2 & 1 & -1 \end{vmatrix} = -8$

$D_z = \begin{vmatrix} 1 & 1 & 0 \\ 1 & -1 & 6 \\ 2 & 1 & 1 \end{vmatrix} = 4$

$x = \frac{D_x}{D} = \frac{8}{4} = 2, \qquad y = \frac{D_y}{D} = \frac{-8}{4} = -2,$

$z = \frac{D_z}{D} = \frac{4}{4} = 1$ The solution set is $\{(2, -2, 1)\}$.

16. Let $x =$ the price of a single and $y =$ the price of a double. We can write an equation for each night.

$$5x + 12y = 390$$
$$9x + 10y = 412$$

Multiply the first equation by -9 and the second by 5, then add the results.

$$-45x - 108y = -3510$$
$$45x + 50y = 2060$$
$$-58y = -1450$$
$$y = 25$$

Use $y = 25$ in $5x + 12y = 390$ to find x.

$$5x + 12(25) = 390$$
$$5x + 300 = 390$$
$$5x = 90$$
$$x = 18$$

So singles rent for $18 per night and doubles rent for $25 per night.

17. Let $x =$ Jill's study time, $y =$ Karen's study time, and $z =$ Betsy's study time. We can write three equations.

$$x + y + z = 93$$
$$x + y = \frac{1}{2}z$$
$$x = y + 3$$

Rewrite the equations as follows.

$$x + y + z = 93$$
$$2x + 2y - z = 0$$
$$x - y \quad = 3$$

Adding the first two equations to eliminate z gives us $3x + 3y = 93$, or $x + y = 31$. Add this result to the third equation.

$$x + y = 31$$
$$x - y = 3$$
$$2x \quad = 34$$
$$x = 17$$

Since $x + y = 31$, we get $y = 14$. Use $x = 17$ and $y = 14$ in the first equation.

$$17 + 14 + z = 93$$
$$z = 62$$

So Jill studied 17 hours, Karen studied 14 hours, and Betsy studied 62 hours.

Making Connections

Chapters 1 - 7

1. $-3^4 = -(3^4) = -81$

2. $\frac{1}{3}(3) + 6 = 1 + 6 = 7$

3. $(-5)^2 - 4(-2)(6) = 25 - (-48) = 73$

4. $6 - (0.2)(0.3) = 6 - 0.06 = 5.94$

5. $5(t - 3) - 6(t - 2) = 5t - 15 - 6t + 12$
$= -t - 3$

6. $0.1(x - 1) - (x - 1) = 0.1x - 0.1 - x + 1$
$= -0.9x + 0.9$

7. $\dfrac{-9x^2 - 6x + 3}{-3} = 3x^2 + 2x - 1$

8. $\dfrac{4y - 6}{2} - \dfrac{3y - 9}{3} = 2y - 3 - (y - 3)$
$= y$

9. $3x - 5y = 7$
$$-5y = -3x + 7$$
$$y = \frac{3}{5}x - \frac{7}{5}$$
10. $Cx - Dy = W$
$$-Dy = -Cx + W$$
$$y = \frac{C}{D}x - \frac{W}{D}$$
11. $Cy = Wy - K$
$$K = Wy - Cy$$
$$K = y(W - C)$$
$$y = \frac{K}{W - C}$$
12. $A = \frac{1}{2}b(w - y)$
$$2A = bw - by$$
$$by = bw - 2A$$
$$y = \frac{bw - 2A}{b}$$
13. $2x + 3(x - 5) = 5$
$$5x - 15 = 5$$
$$5x = 20$$
$$x = 4$$
$$y = x - 5 = 4 - 5 = -1$$
The solution set is $\{(4, -1)\}$.
14. $y = 1200 - x$
$$0.05x + 0.06(1200 - x) = 67$$
$$-0.01x + 72 = 67$$
$$-0.01x = -5$$
$$x = 500$$
$$y = 1200 - x = 1200 - 500 = 700$$
The solution set is $\{(500, 700)\}$.
15. $x = 5y - 17$
$$3(5y - 17) - 15y = -51$$
$$15y - 51 - 15y = -51$$
$$-51 = -51$$
The equations are equivalent and the solution set is $\{(x, y) \mid x + 17 = 5y\}$.
16. Multiply the first equation by 100 to get $7a + 30b = 670$. This equation is inconsistent with the equation $7a + 30b = 67$. So the solution set is the empty set $\emptyset$.
17. The slope of the line is $55/99$ or $5/9$. Since the y-intercept is $(0, 55)$, the equation is $y = \frac{5}{9}x + 55$.
18. The slope of the line through $(2, -3)$ and $(-4, 8)$ is $11/-6$ or $-11/6$. Use the point-slope formula:
$$y - 8 = -\frac{11}{6}(x - (-4))$$
$$y - 8 = -\frac{11}{6}x - \frac{44}{6}$$
$$y = -\frac{11}{6}x - \frac{22}{3} + \frac{24}{3}$$
$$y = -\frac{11}{6}x + \frac{2}{3}$$

19. Any line parallel to $y = 5x$ has slope 5.
$$y - 6 = 5(x - (-4))$$
$$y - 6 = 5x + 20$$
$$y = 5x + 26$$

20. Any line perpendicular to $y = -2x + 1$ has slope $1/2$.
$$y - 7 = \frac{1}{2}(x - 4)$$
$$y - 7 = \frac{1}{2}x - 2$$
$$y = \frac{1}{2}x + 5$$

21. Any line parallel to the x-axis has slope 0 and is of the form $y = k$. Since it must go through $(3, 5)$ it is $y = 5$.

22. Any line perpendicular to the x-axis is of the form $x = k$. So the equation is $x = -7$.

23. a) From the y-intercept we see that the purchase price for A is $4000 and for B is $2000. So A has the larger purchase price.

b) Machine B makes 300,000 copies for $12,000, or $0.04 per copy. Machine A makes 300,000 copies for $9,000 or $0.03 per copy.

c) The slope of the line for machine B is 0.04 and the slope of the line for machine A is 0.03, which is the per copy cost for each machine.

d) Machine B: $y = 0.04x + 2000$
Machine A: $y = 0.03x + 4000$

e) $0.04x + 2000 = 0.03x + 4000$
$$0.01x = 2000$$
$$x = 200{,}000$$
The total costs are equal at 200,000 copies.

8.1 WARM-UPS

1. True, because both inequalities are true.
2. True, because both inequalities are correct.
3. False, because $3 > 5$ is incorrect.
4. True, because $3 \leq 10$ is correct.
5. True, because both inequalities are correct.
6. True, because both are correct.
7. False, because $0 < -2$ is incorrect.
8. True, because only numbers larger than 8 are larger than 3 and larger than 8.
9. False, because $(3, \infty) \cup [8, \infty) = (3, \infty)$.
10. True, because the numbers greater than -2 and less than 9 are between -2 and 9.

8.1 EXERCISES

1. A compound inequality consists of two inequalities joined with the words "and" or "or."
3. A compound inequality using or is true when either one or the other or both inequalities is true.
5. The inequality $a < b < c$ means that $a < b$ and $b < c$.
7. No, because $-6 > -3$ is incorrect.
9. Yes, because both inequalities are correct.
11. No, because both inequalities are incorrect.
13. No, because $-4 > -3$ is incorrect.
15. Yes, because $-4 - 3 \geq -7$ is correct.
17. Yes, because $2(-4) - 1 < -7$ is correct.
19. The solution set is the set of numbers between -1 and 4:

21. Numbers that satisfy both $x \leq 3$ and $x \leq 0$ must be less than or equal to 0. Graph the intersection of the two solution sets.

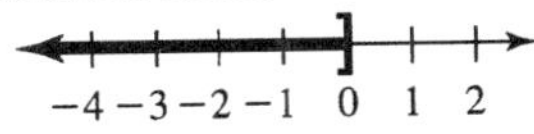

23. Numbers that satisfy $x \geq 2$ or $x \geq 5$ are greater than or equal to 2. Graph the union of the two solution sets:

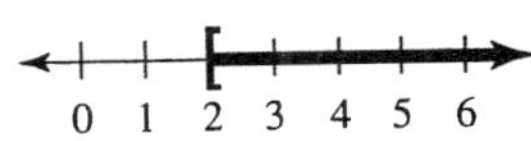

25. The union of $(-\infty, 6]$ with $(-2, \infty)$ is the interval $(-\infty, \infty)$. The union of the two solution sets consists of all real numbers:

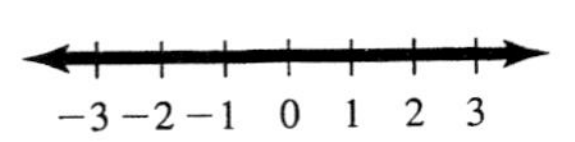

27. The solution set is $\emptyset$, because no number is greater than 9 and less than or equal to 6. There is no graph.
29. The union of the two solution sets is graphed as follows:

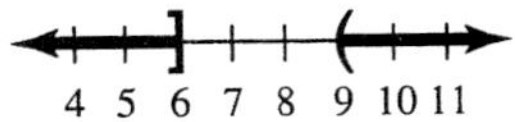

31. The solution set is $\emptyset$, because there is no intersection to the two solution sets. There is no graph.

33.
$$\begin{aligned} x - 3 > 7 &\text{ or } 3 - x > 2 \\ x > 10 &\text{ or } \quad -x > -1 \\ x > 10 &\text{ or } \quad x < 1 \end{aligned}$$
$(-\infty, 1) \cup (10, \infty)$

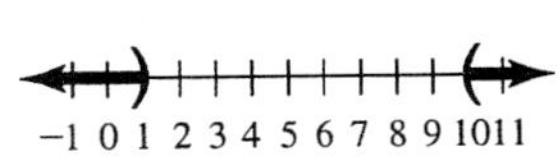

35.
$$\begin{aligned} 3 < x &\text{ and } 1 + x > 10 \\ x > 3 &\text{ and } \quad x > 9 \end{aligned}$$
$(9, \infty)$

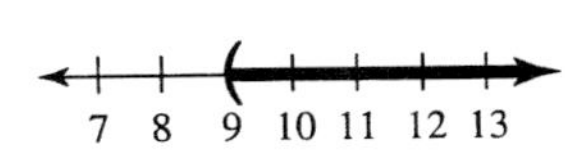

37.
$$\begin{aligned} \frac{1}{2}x > 5 &\text{ or } -\frac{1}{3}x < 2 \\ x > 10 &\text{ or } \quad x > -6 \end{aligned}$$
$(-6, \infty)$

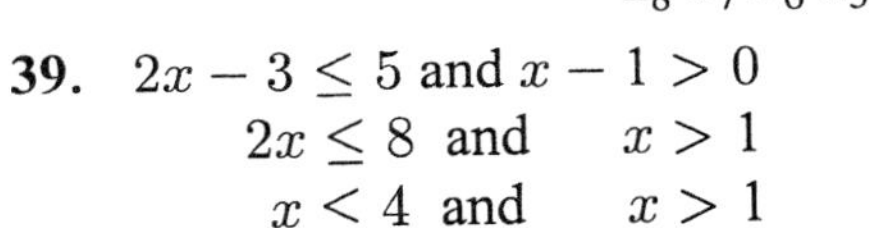

39.
$$\begin{aligned} 2x - 3 \leq 5 &\text{ and } x - 1 > 0 \\ 2x \leq 8 &\text{ and } \quad x > 1 \\ x \leq 4 &\text{ and } \quad x > 1 \end{aligned}$$
$(1, 4]$

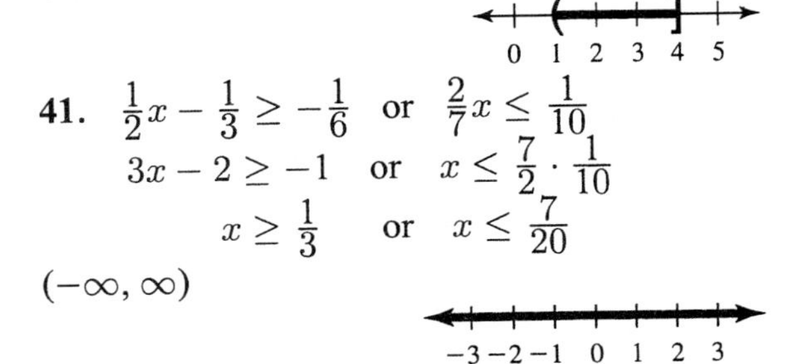

41.
$$\begin{aligned} \frac{1}{2}x - \frac{1}{3} \geq -\frac{1}{6} &\text{ or } \frac{2}{7}x \leq \frac{1}{10} \\ 3x - 2 \geq -1 &\text{ or } x \leq \frac{7}{2} \cdot \frac{1}{10} \\ x \geq \frac{1}{3} &\text{ or } x \leq \frac{7}{20} \end{aligned}$$
$(-\infty, \infty)$

43.
$$\begin{aligned} 0.5x < 2 &\text{ and } -0.6x < -3 \\ x < 4 &\text{ and } \quad x > 5 \end{aligned}$$

The solution set is $\emptyset$, because there are no numbers that are less than 4 and greater than 5. There is no graph.

45. $-3 < x + 1 < 3$
$$-3 - 1 < x + 1 - 1 < 3 - 1$$
$$-4 < x < 2$$
$(-4, 2)$

47. $5 < 2x - 3 < 11$
$$5 + 3 < 2x - 3 + 3 < 11 + 3$$
$$8 < 2x < 14$$
$$4 < x < 7$$
$(4, 7)$

49. $-1 < 5 - 3x \leq 14$
$$-6 < -3x \leq 9$$
$$\frac{-6}{-3} > \frac{-3x}{-3} \geq \frac{9}{-3}$$
$$2 > x \geq -3$$
$$-3 \leq x < 2$$
$[-3, 2)$

51. $-3 < \frac{3m+1}{2} \leq 5$
$$2(-3) < 2 \cdot \frac{3m+1}{2} \leq 2 \cdot 5$$
$$-6 < 3m + 1 \leq 10$$
$$-7 < 3m \leq 9$$
$$-\frac{7}{3} < m \leq 3$$
$(-7/3, 3]$

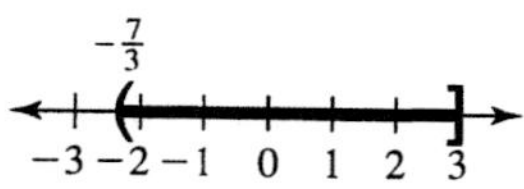

53. $-2 < \frac{1-3x}{-2} < 7$
$$-2(-2) > -2 \cdot \frac{1-3x}{-2} > -2(7)$$
$$4 > 1 - 3x > -14$$
$$3 > -3x > -15$$
$$-1 < x < 5$$
$(-1, 5)$

55. $3 \leq 3 - 5(x - 3) \leq 8$
$$3 \leq 3 - 5x + 15 \leq 8$$
$$3 \leq 18 - 5x \leq 8$$
$$-15 \leq -5x \leq -10$$
$$3 \geq x \geq 2$$
$$2 \leq x \leq 3$$
$[2, 3]$

57. $(2, \infty) \cup (4, \infty) = (2, \infty)$
59. $(-\infty, 5) \cap (-\infty, 9) = (-\infty, 5)$
61. $(-\infty, 4] \cap [2, \infty) = [2, 4]$
63. $(-\infty, 5) \cup [-3, \infty) = (-\infty, \infty)$
65. $(3, \infty) \cap (-\infty, 3] = \emptyset$
67. $(3, 5) \cap [4, 8) = [4, 5)$
69. $[1, 4) \cup (2, 6] = [1, 6]$
71. The graph shows real numbers to the right of 2: $x > 2$
73. The graph shows the real numbers to the left of 3: $x < 3$
75. This graph is the union of the numbers greater than 2 with the numbers less than or equal to -1: $x > 2$ or $x \leq -1$
77. This graph shows real numbers between -2 and 3, including -2: $-2 \leq x < 3$
79. The graph shows real numbers greater than or equal to -3: $x \geq -3$
81. The car is replaced when
$0.0004x + 20 > 40$ and
$20{,}000 - 0.2x < 12{,}000$.
Solve these inequalities to get
$x > 50{,}000$ and $x > 40{,}000$,
which is equivalent to $x > 50{,}000$. So the car is replaced when x is in the interval $(50{,}000, \infty)$.
83. The government gets involved if $20 + 0.1x < 22$ or $30 - 0.5x < 15$. Solve each inequality to get $x < 20$ or $x > 30$. So the government gets involved if x is in the union $(-\infty, 20) \cup (30, \infty)$.
85. Let $x =$ the final exam score. We write an inequality expressing the fact that 1/3 of the midterm plus 2/3 of the final must be between 70 and 79 inclusive.
$$70 \leq \frac{1}{3} \cdot 64 + \frac{2}{3}x \leq 79$$
$$210 \leq 64 + 2x \leq 237$$
$$146 \leq 2x \leq 173$$
$$73 \leq x \leq 86.5$$

87. Let $x =$ the price of the truck. The total spent will be $x + 0.08x + 84$.
$$12{,}000 \leq x + 0.08x + 84 \leq 15{,}000$$
$$12{,}000 \leq 1.08x + 84 \leq 15{,}000$$
$$11{,}916 \leq 1.08x \leq 14{,}916$$
$$\$11{,}033 \leq x \leq \$13{,}811$$

89. Let $x =$ the number of cigarettes smoked on the run, giving the equivalent of $\frac{1}{2}x$ cigarettes smoked. Thus, she smokes $3 + \frac{1}{2}x$ whole cigarettes per day and this number is between 5 and 12 inclusive:

$$5 \le 3 + \frac{1}{2}x \le 12$$
$$10 \le 6 + x \le 24$$
$$4 \le x \le 18$$

She smokes from 4 to 18 cigarettes on the run.

91. a) In 2000, we have $n = 10$.
$16.45(10) + 1062.45 = 1226.95$
In 2000, there were 1,226,950 bachelors degrees awarded.

b) $16.45n + 1062.45 = 1{,}400$
$$16.45n = 337.55$$
$$n \approx 21$$
$1990 + 21 = 2011$

c) $16.45n + 1062.45 > 1{,}400$
$$16.45n > 337.55$$
$$n > 20.52$$

$$7.79n + 326.82 > 550$$
$$7.79n > 223.18$$
$$n > 28.64$$

Both happen in the year $1990 + 29$, or 2019.

d) Either happens in the year $1990 + 21$, or 2011.

93. If $a < b$ and $a < -x < b$, we can multiply each part of this inequality by -1 to get $-a > x > -b$ or $-b < x < -a$. In words, x is between $-b$ and $-a$.

95. a) If $3 < x < 8$, then $12 < 4x < 32$.
$(12, 32)$

b) If $-2 \le x < 4$, then
$$(-5)(-2) \ge -5x > (-5)(4)$$
$$10 \ge -5x > -20$$
$$-20 < -5x \le 10.$$
$(-20, 10]$

c) If $-3 < x < 6$, then $0 < x + 3 < 9$.
$(0, 9)$

d) If $3 \le x \le 9$, then
$$\frac{3}{-3} \ge \frac{x}{-3} \ge \frac{9}{-3}$$
$$-1 \ge \frac{x}{-3} \ge -3$$
$$-3 \le \frac{x}{-3} \le -1.$$
$[-3, -1]$

8.2 WARM-UPS

1. True, because both 2 and -2 have absolute value 2.
2. False, because $|x| = 0$ has only one solution and $|x| = -1$ has no solutions.
3. False, because it is equivalent to $2x - 3 = 7$ or $2x - 3 = -7$.
4. True, because $|x| > 5$ means that x is more than 5 units away from 0 and that is true to the right of 5 or to the left of -5.
5. False, because this equation has no solution.
6. True, because only 3 satisfies the equation.
7. False, because only inequalities that express x between two numbers are written this way.
8. False, because $|x| < 7$ is equivalent to $-7 < x < 7$.
9. True, because 2 is subtracted from each side.
10. False, because if $x = 0$, then $|x| = 0$ and 0 is not positive.

8.2 EXERCISES

1. Absolute value of a number is the number's distance from 0 on the number line.
3. Since both 4 and -4 are four units from 0, $|x| = 4$ has two solutions.
5. Since the distance from 0 for every number on the number line is greater than or equal to 0, $|x| \ge 0$.
7. $|a| = 5$
$a = 5$ or $a = -5$
Solution set: $\{-5, 5\}$

9. $|x - 3| = 1$
$x - 3 = 1$ or $x - 3 = -1$
$x = 4$ or $x = 2$

Solution set: $\{2, 4\}$

11. $|3 - x| = 6$
$3 - x = 6$ or $3 - x = -6$
$-x = 3$ or $-x = -9$
$x = -3$ or $x = 9$

Solution set: $\{-3, 9\}$

13. $|3x-4|=12$
$3x-4=12$ or $3x-4=-12$
$3x=16$ or $3x=-8$
$x=\frac{16}{3}$ or $x=-\frac{8}{3}$
Solution set: $\left\{-\frac{8}{3}, \frac{16}{3}\right\}$

15. $\left|\frac{2}{3}x-8\right|=0$
$\frac{2}{3}x-8=0$
$\frac{2}{3}x=8$
$x=12$
Solution set: $\{12\}$

17. $6-0.2x=10$ or $6-0.2x=-10$
$-0.2x=4$ or $-0.2x=-16$
$x=-20$ or $x=80$
Solution set: $\{-20, 80\}$

19. Since absolute value is nonnegative, the solution set is $\emptyset$.

21. $|2(x-4)+3|=5$
$2(x-4)+3=5$ or $2(x-4)+3=-5$
$2x-5=5$ or $2x-5=-5$
$2x=10$ or $2x=0$
$x=5$ or $x=0$
Solution set: $\{0, 5\}$

23. $|7.3x-5.26|=4.215$
$7.3x-5.26=4.215$
$7.3x=9.475$
$x\approx 1.298$
or $7.3x-5.26=-4.215$
$7.3x=1.045$
$x\approx 0.143$
Solution set: $\{0.143, 1.298\}$

25. $3+|x|=5$
$|x|=2$
$x=2$ or $x=-2$
Solution set: $\{-2, 2\}$

27. $2-|x+3|=-6$
$-|x+3|=-8$
$|x+3|=8$
$x+3=8$ or $x+3=-8$
$x=5$ or $x=-11$
Solution set: $\{-11, 5\}$

29. $5-\frac{|3-2x|}{3}=4$
$15-|3-2x|=12$
$-|3-2x|=-3$
$|3-2x|=3$
$3-2x=3$ or $3-2x=-3$
$-2x=0$ or $-2x=-6$
$x=0$ or $x=3$
The solution set is $\{0, 3\}$.

31. $x-5=2x+1$ or $x-5=-(2x+1)$
$-6=x$ or $x-5=-2x-1$
$3x=4$
$x=4/3$
Solution set: $\left\{-6, \frac{4}{3}\right\}$

33. $\left|\frac{5}{2}-x\right|=\left|2-\frac{x}{2}\right|$
$\frac{5}{2}-x=2-\frac{x}{2}$ or $\frac{5}{2}-x=-\left(2-\frac{x}{2}\right)$
$5-2x=4-x$ or $\frac{5}{2}-x=-2+\frac{x}{2}$
$1=x$ or $5-2x=-4+x$
$-3x=-9$
$x=3$
Solution set: $\{1, 3\}$

35. $|x-3|=|3-x|$
$x-3=3-x$ or $x-3=-(3-x)$
$2x=6$ or $x-3=-3+x$
$x=3$ or $x-3=x-3$
The second equation is an identity. So the solution set is the set of real numbers, which is written as R or $(-\infty, \infty)$.

37. The graph shows the real numbers between -2 and 2: $|x|<2$

39. The graph shows numbers greater than 3 or less than -3: $|x|>3$

41. The graph shows numbers between -1 and 1 inclusive: $|x|\leq 1$

43. The graph shows numbers two or more units away from 0: $|x|\geq 2$

45. No, because $|x|<3$ is equivalent to $-3<x<3$.

47. Yes.

49. No, because $|x-3|\geq 1$ is equivalent to $x-3\geq 1$ or $x-3\leq -1$.

51. Yes, because the following compound inequalities are equivalent.
$|4-x|<1$
$4-x<1$ and $4-x>-1$
$4-x<1$ and $-(4-x)<1$

53. $|x|>6$
$x>6$ or $x<-6$
$(-\infty,-6)\cup(6,\infty)$

−8 −6 −4 −2 0 2 4 6 8

55. $|t| \leq 2$

$-2 \leq t \leq 2$

$[-2, 2]$

57. $|2a| < 6$

$-6 < 2a < 6$

$-3 < a < 3$

$(-3, 3)$

59. $x - 2 \geq 3$ or $x - 2 \leq -3$

$x \geq 5$ or $x \leq -1$

$(-\infty, -1] \cup [5, \infty)$

61. $\frac{1}{5}|2x - 4| < 1$

$|2x - 4| < 5$

$-5 < 2x - 4 < 5$

$-1 < 2x < 9$

$-\frac{1}{2} < x < \frac{9}{2}$

$\left(-\frac{1}{2}, \frac{9}{2}\right)$

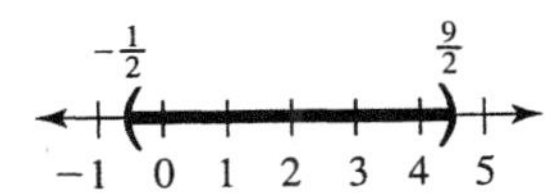

63. $-2|5 - x| \geq -14$

$|5 - x| \leq 7$

$-7 \leq 5 - x \leq 7$

$-12 \leq -x \leq 2$

$12 \geq x \geq -2$

$[-2, 12]$

65. $2|3 - 2x| - 6 \geq 18$

$2|3 - 2x| \geq 24$

$|3 - 2x| \geq 12$

$3 - 2x \geq 12$ or $3 - 2x \leq -12$

$-2x \geq 9$ or $-2x \leq -15$

$x \leq -\frac{9}{2}$ or $x \geq \frac{15}{2}$

$(-\infty, -\frac{9}{2}] \cup [\frac{15}{2}, \infty)$

67. $|x| > 0$ is true except when $x = 0$ so the solution set is the set of all real numbers except 0, which is written in interval notation as $(-\infty, 0) \cup (0, \infty)$ and graphed as follows:

69. $|x| \leq 0$ is true only when $x = 0$ because the absolute value of any nonzero real number is positive. The solution set is $\{0\}$ and it is a single point on the number line:

71. The inequality $|x - 5| \geq 0$ is satisfied by every real number because the absolute value of any real number is nonnegative. The solution set is R, or $(-\infty, \infty)$.

73. $-2|3x - 7| > 6$ is equivalent to

$|3x - 7| < -3.$

Absolute value of an expression cannot be less than a negative number. Solution set is $\emptyset$.

75. $|2x + 3| + 6 > 0$

$|2x + 3| > -6$

Since the absolute value of any expression is greater than or equal to zero, it is greater than any negative number. The solution set is R or $(-\infty, \infty)$.

77. $1 < |x + 2|$

$|x + 2| > 1$

$x + 2 > 1$ or $x + 2 < -1$

$x > -1$ or $x < -3$

$(-\infty, -3) \cup (-1, \infty)$

79. $5 > |x + 1|$

$|x| + 1 < 5$

$|x| < 4$

$-4 < x < 4$

$(-4, 4)$

81. $3 - 5|x| > -2$

$-5|x| > -5$

$|x| < 1$

$-1 < x < 1$

$(-1, 1)$

83. $|5.67x - 3.124| < 1.68$

$-1.68 < 5.67x - 3.124 < 1.68$

$1.444 < 5.67x < 4.804$

$0.255 < x < 0.847$

$(0.255, 0.847)$

85. Let $x =$ the year of the battle of Orleans. Since the difference between x and 1415 is 14 years, we have $|\ x - 1415\ | = 14$.

$$x - 1415 = 14 \quad \text{or} \quad x - 1415 = -14$$
$$x = 1429 \quad \text{or} \quad x = 1401$$

The battle Agincourt was either in 1401 or 1429.

87. Let $x =$ the weight of Kathy. The difference between their weights is less than 6 pounds is expressed by the absolute value inequality

$$|\ x - 127\ | < 6$$
$$-6 < x - 127 < 6$$
$$121 < x < 133$$

Kathy weighs between 121 and 133 pounds.

89. a) From the graph it appears that the balls are at the same height when $t = 1$ second.

b) Height of first ball is $S = -16t^2 + 50t$ and height of second ball is $S = -16t^2 + 40t + 10$. When the balls are at the same height we have

$$-16t^2 + 50t = -16t^2 + 40t + 10$$
$$50t = 40t + 10$$
$$10t = 10$$
$$t = 1$$

The balls are at the same height when $t = 1$ sec.

c) The difference between the heights is less than 5 feet when

$$|\ -16t^2 + 50t - (-16t^2 + 40t + 10)\ | < 5$$
$$|\ 10t - 10\ | < 5$$
$$-5 < 10t - 10 < 5$$
$$5 < 10t < 15$$
$$0.5 < t < 1.5$$

91. a) The equation $|m - n| = |n - m|$ is satisfied for all real numbers, because $m - n$ and $n - m$ are opposites of each other and opposites always have the same absolute value. So both m and n can be in the interval $(-\infty, \infty)$.

b) $|mn| = |m| \cdot |n|$ is satisfied for all real numbers, because of the rules for multiplying real numbers. So both m and n can be in the interval $(-\infty, \infty)$.

c) Since you cannot have 0 in a denominator, the equation is satisfied by all real numbers except if $n = 0$.

8.3 WARM-UPS

1. True, because $-3 > -3(2) + 2$.

2. False, because $3x - y > 2$ is equivalent to $y < 3x - 2$, which is below the line $3x - y = 2$.

3. True, because $3x + y < 5$ is equivalent to $y < -3x + 5$.

4. False, the region $x < -3$ is to the left of the vertical line $x = -3$.

5. True, because the word "and" is used.

6. True, because the word "or" is used.

7. False, because $(2, -5)$ does not satisfy the inequality $y > -3x + 5$.
Note that $-5 > -3(2) + 5$ is incorrect.

8. True, because $(-3, 2)$ satisfies $y \le x + 5$.

9. False, it is equivalent to the compound inequality $2x - y \le 4$ and $2x - y \ge -4$.

10. True, because in general $|\ x\ | > k$ (for a positive k) is equivalent to $x > k$ or $x < -k$.

8.3 EXERCISES

1. A compound inequality in two variables is formed by connecting two simple inequalities with "and" or "or."

3. A point satisfies an "and" inequality only if it satisfies both inequalities.

5. A test point is used to check whether all points in the region of the test point satisfy the compound inequality.

7. Check each ordered pair in $y > 5x$ and $y < -x$:

$(1, 3)$	$3 > 5(1)$ and $3 < -1$	
	$3 > 5$ and $3 < -1$	False
$(-2, 5)$	$5 > 5(-2)$ and $5 < -(-2)$	
	$5 > -10$ and $5 < 2$	False
$(-6, -4)$	$-4 > 5(-6)$ and $-4 < -(-6)$	
	$-4 > -30$ and $-4 < 6$	True
$(7, -8)$	$-8 > 5(7)$ and $-8 < -(7)$	
	$-8 > 35$ and $-8 < -7$	False

Only $(-6, -4)$ satisfies the compound inequality.

9. Check each ordered pair in

$y > -x + 1$ or $y > 4x$:

$(1, 3)$	$3 > -1 + 1$ or $3 > 4(1)$	
	$3 > 0$ or $3 > 4$	True

$(-2,5)$ $\quad 5 > -(-2)+1$ or $5 > 4(-2)$

$5 > 3$ or $5 > -8$ True

$(-6,-4)$ $\quad -4 > -(-6)+1$ or $-4 > 4(-6)$

$-4 > 7$ or $-4 > -24$ True

$(7,-8)$ $\quad -8 > -(7)+1$ or $-8 > 4(7)$

$-8 > -6$ or $-8 > 28$ False

So $(1,3)$, $(-2,5)$, and $(-6,-4)$ satisfy the compound inequality.

11. Check each ordered pair in

$|x+y| < 3$:

$(1,3)$ $\quad |1+3| < 3$

$4 < 3$ False

$(-2,5)$ $\quad |-2+5| < 3$

$3 < 3$ False

$(-6,-4)$ $\quad |-6+(-4)| < 3$

$10 < 3$ False

$(7,-8)$ $\quad |7+(-8)| < 3$

$1 < 3$ True

So $(7,-8)$ satisfies the absolute value inequality.

13. To graph $y > x$ and $y > -2x+3$, we first draw dashed lines for $y = x$ and for $y = -2x+3$. Test one point in each of the four regions to see if it satisfies the compound inequality. Test $(5, 0)$, $(0, 5)$, $(-5, 0)$ and $(0,-5)$. Only $(0, 5)$ satisfies both inequalities. So shade the region containing $(0, 5)$.

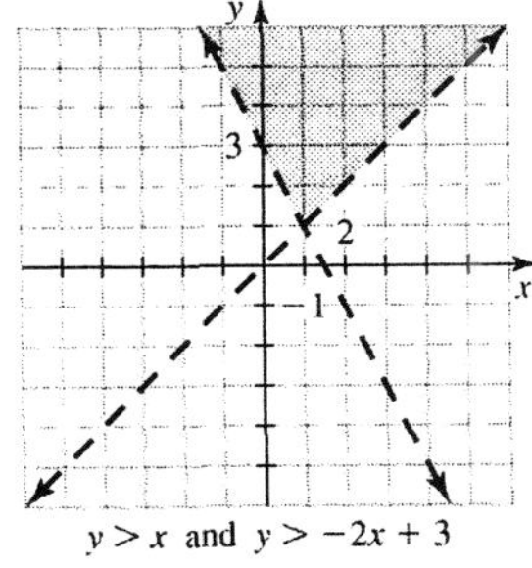

15. First graph the equations $y = x+3$ and $y = -x+2$. Test the points $(0, 5)$, $(5, 0)$, $(0, -5)$, and $(-5, 0)$ in the compound inequality. Only $(-5, 0)$ fails to satisfy the compound inequality. So shade all regions except the one containing $(-5, 0)$.

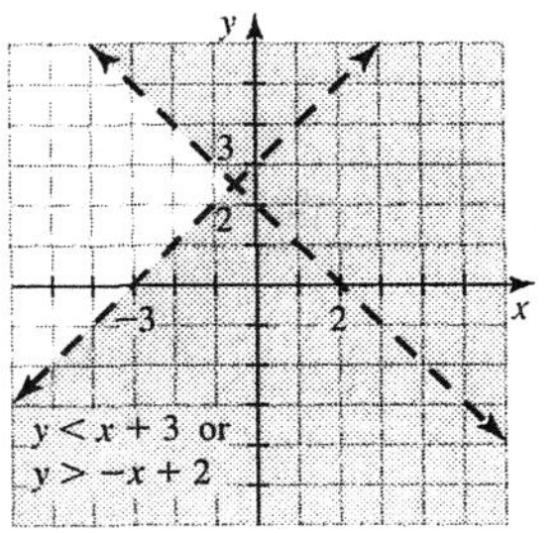

17. First graph the lines $x - 4y = 0$ and $3x + 2y = 6$. Use a dashed line for the first and a solid line for the second. Select a point in each region as a test point. Use $(0, 5)$, $(0, 1)$, $(0, -5)$, and $(6, 0)$. Only $(0, 5)$ satisfies the compound sentence $x - 4y < 0$ and $3x + 2y \geq 6$. So shade the region that contains $(0, 5)$.

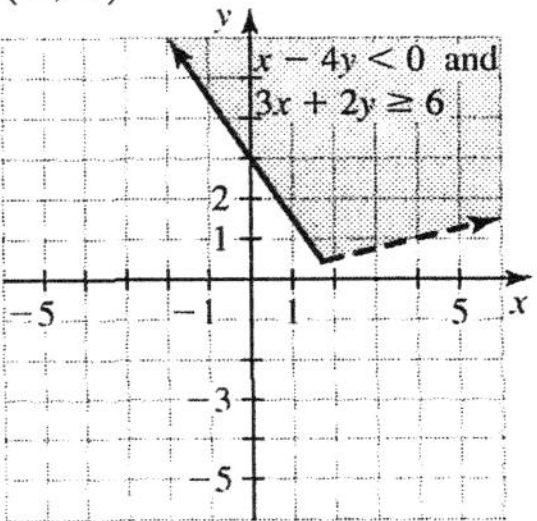

19. First graph the equations $x + y = 5$ and $x - y = 3$. Test one point in each of the 4 regions. Only points in the region containing $(0, 0)$ satisfy both inequalities. Shade that region including the boundary lines.

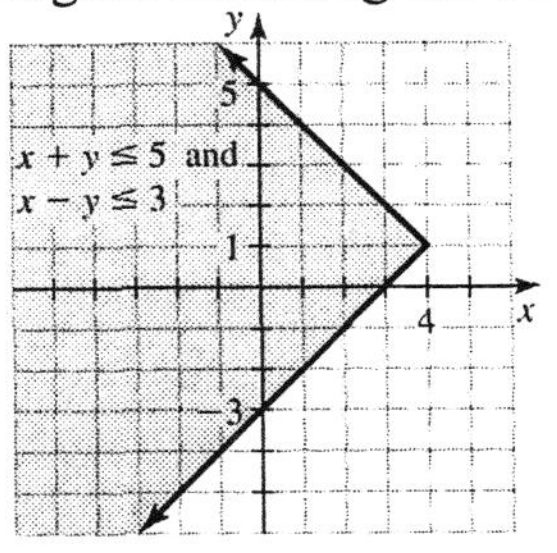

21. Graph the equations $x - 2y = 4$ and $2x - 3y = 6$. Only the region containing the point $(0, -5)$ fails to satisfy the compound inequality. Shade the other three regions including the boundary lines.

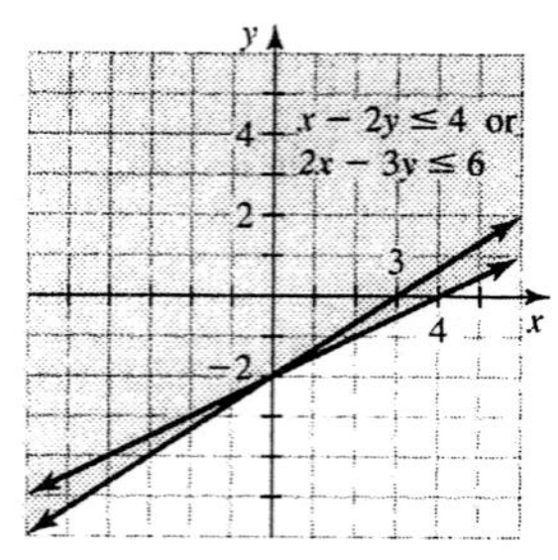

23. Graph the horizontal line $y = 2$ and the vertical line $x = 3$. Only points in the region containing $(0, 5)$ satisfy both inequalities. Shade that region with dashed boundary lines.

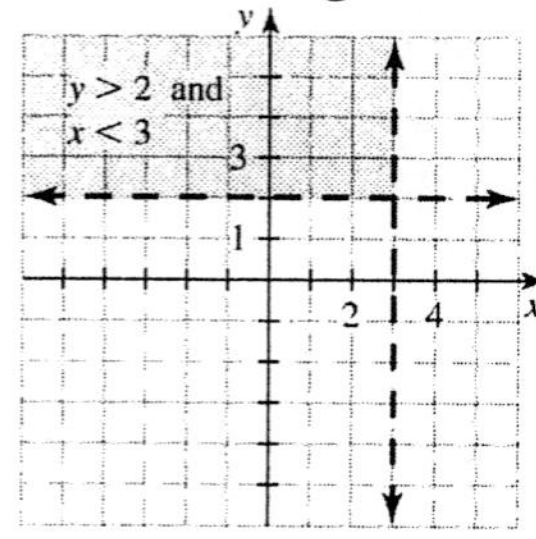

25. Graph $y = x$ and $x = 2$. Only points in the region containing $(0, 5)$ satisfy both inequalities. Shade that region and include the boundary lines.

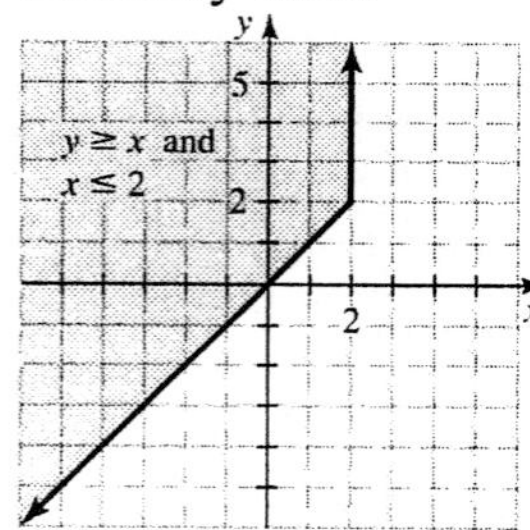

27. Graph $2x = y + 3$ and $y = 2 - x$. Only points in the region containing $(0, -5)$ fail to satisfy the compound inequality. Shade all regions except the one containing $(0, -5)$. Use dashed lines for the boundaries because of the inequality symbols.

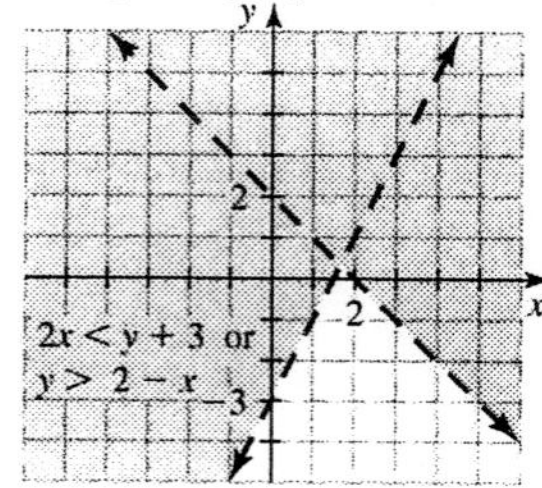

29. Graph the lines $y = x - 1$ and $y = x + 3$. Only points in the region containing $(0, 0)$ satisfy the compound inequality. Shade that region and use dashed boundary lines.

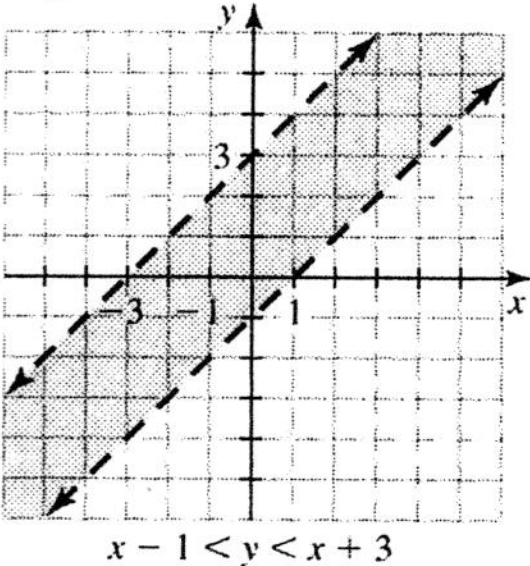

$x - 1 < y < x + 3$

31. Graph the lines $y = 0$, $y = x$, and $x = 1$. Only points inside the triangular region bounded by the lines satisfy the compound inequality. Shade that region and use solid boundary lines.

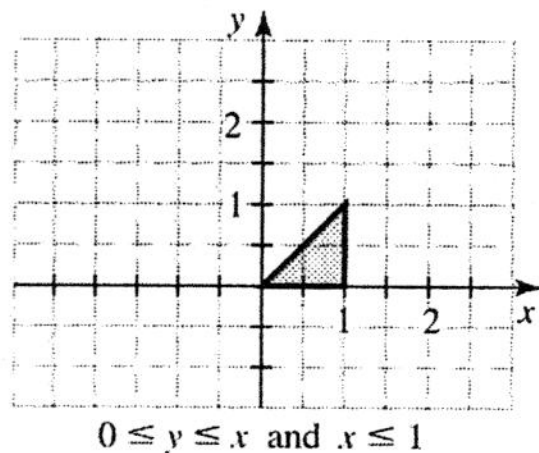

$0 \le y \le x$ and $x \le 1$

33. Graph $x = 1$, $x = 3$, $y = 2$, and $y = 5$.

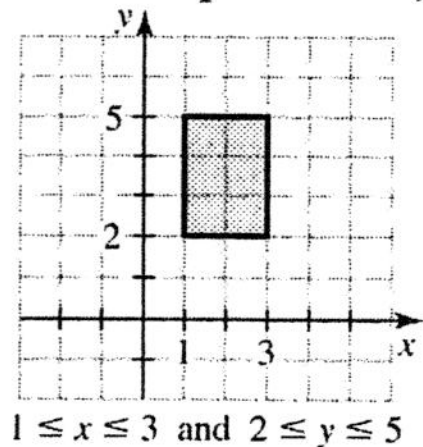

$1 \le x \le 3$ and $2 \le y \le 5$

Only points inside the rectangular region satisfy the compound inequality. Shade that region and include the boundary lines.

35. Graph the equations $x + y = 2$ and $x + y = -2$. Only points between these two parallel lines satisfy the absolute value inequality.

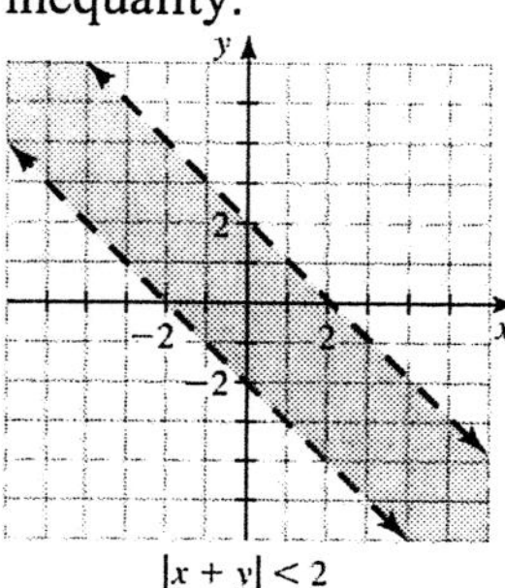

$|x + y| < 2$

37. Graph the parallel lines $2x + y = 1$ and $2x + y = -1$. Points between the lines do not satisfy the absolute value inequality. So shade the other two regions.

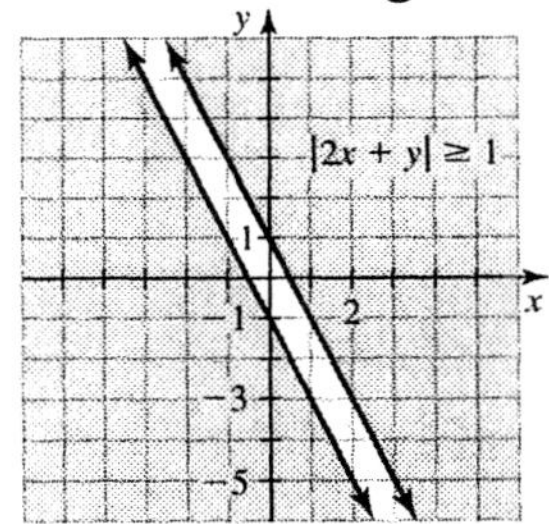

39. The inequality $|y - x| > 2$ is equivalent to

$$y - x > 2 \quad \text{or} \quad y - x < -2$$
$$y > x + 2 \quad \text{or} \quad y < x - 2$$

Graph the lines $y = x + 2$ and $y = x - 2$. Points above $y = x + 2$ together with points below $y = x - 2$ satisfy the absolute value inequality.

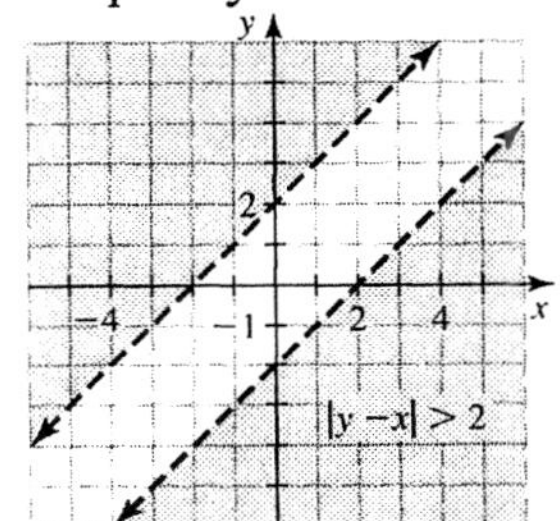

41. Graph the parallel lines $x - 2y = 4$ and $x - 2y = -4$. Points in the region between the lines satisfy the inequality.

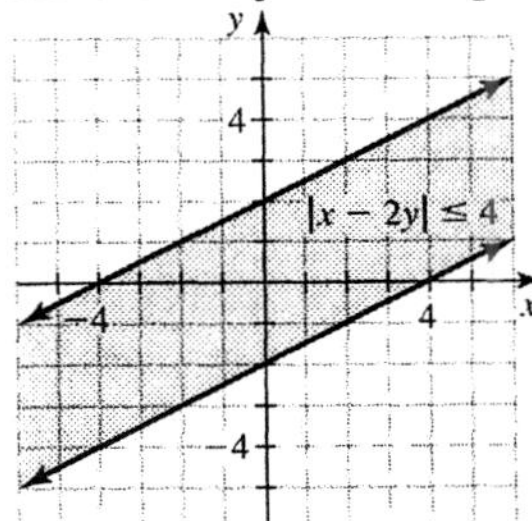

43. Graph the vertical lines $x = 2$ and $x = -2$. Points in the region between the lines do not satisfy the inequality but points in the other two regions do.

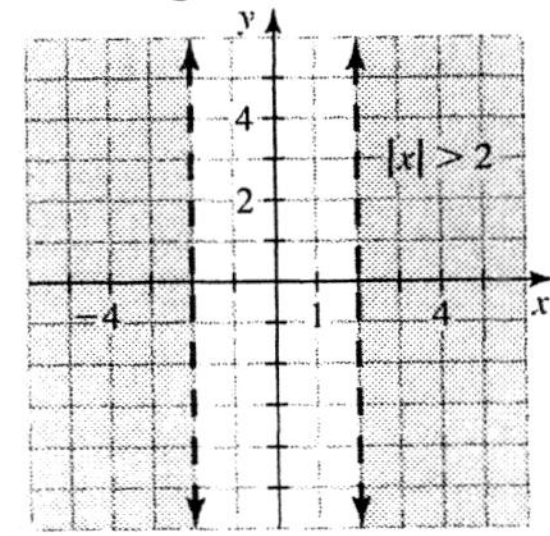

45. Graph the horizontal lines $y = 1$ and $y = -1$. Only points in the region between the lines satisfy $|y| < 1$.

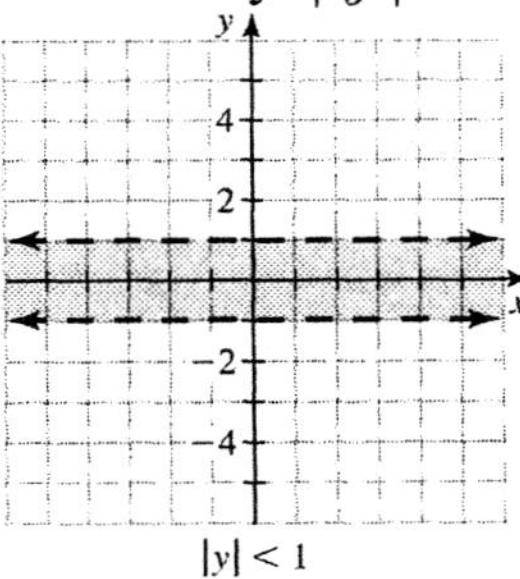

$|y| < 1$

47. Graph the lines $x = 2$, $x = -2$, $y = 3$, and $y = -3$. Only points inside the rectangular region bounded by the lines satisfy the compound inequality $|x| < 2$ and $|y| < 3$.

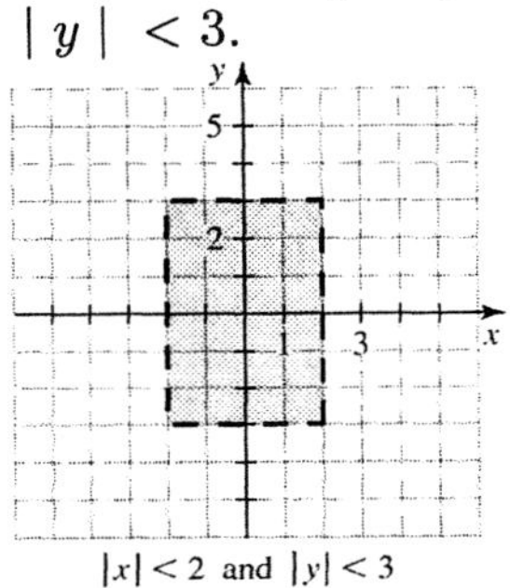

$|x| < 2$ and $|y| < 3$

49. The inequality $|x - 3| < 1$ is equivalent to $-1 < x - 3 < 1$, or $2 < x < 4$. The inequality $|y - 2| < 1$ is equivalent to $-1 < y - 2 < 1$, or $1 < y < 3$. Graph the lines $x = 2$, $x = 4$, $y = 1$, and $y = 3$. Only points inside the square region bounded by these lines satisfy the inequalities $|x - 3| < 1$ and $|y - 2| < 1$.

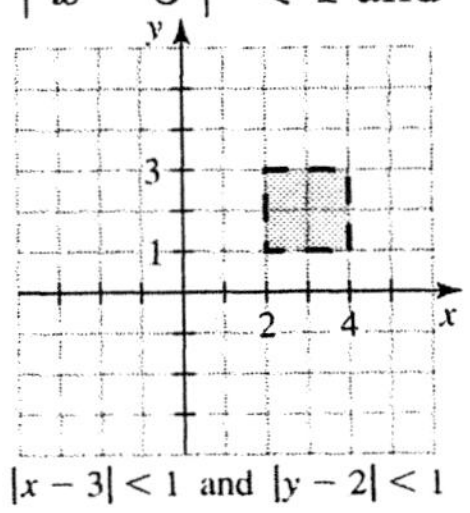

y
3
1
2
4
x

$|x - 3| < 1$ and $|y - 2| < 1$

51. Since $(0, 5)$ satisfies $y > x$ and $x < 1$ the solution set is not the empty set.

53. Points that satisfy $y < 2x - 5$ and $y > 2x + 5$ would be above the line $y = 2x + 5$ and below the line $y = 2x - 5$. Since these are parallel lines, there are no such points. The solution set is the empty set.

55. Since $(10, 0)$ satisfies $y < 2x - 5$ and the connecting word is "or", the solution set is not the empty set.
57. Points that satisfy $y < 2x$ lie below the line $y = 2x$ and points that satisfy $y > 3x$ lie above the line $y = 3x$. Since these lines are not parallel there are points that satisfy both inequalities. For example $(-5, -12)$ satisfies both. So the solution set is not the empty set.
59. This compound inequality indicates that $y < x$ and $y > x$. Points that satisfy both would be above the line $y = x$ and below the line $y = x$. There are no such points. The solution set is the empty set, $\emptyset$.
61. Since no real number has an absolute value that is less than 0, there are no points that satisfy $|y + 2x| < 0$. The solution set is the empty set, $\emptyset$.
63. Since no real number has an absolute value that is less than 0, there are no points that satisfy $|3x + 2y| \leq -4$. The solution set is the empty set, $\emptyset$.
65. Since the absolute value of any real number is nonnegative, all ordered pairs of real numbers satisfy $|x + y| > -4$. The solution set is not the empty set.
67. Let $x =$ the number of compact cars and $y =$ the number of full-size cars. We have

$$15{,}000x + 20{,}000y \leq 120{,}000$$

$$3x + 4y \leq 24$$

We also have $x \geq 0$ and $y \geq 0$ because they cannot purchase a negative number of cars. Graph $3x + 4y \leq 24$, $x \geq 0$, and $y \geq 0$.

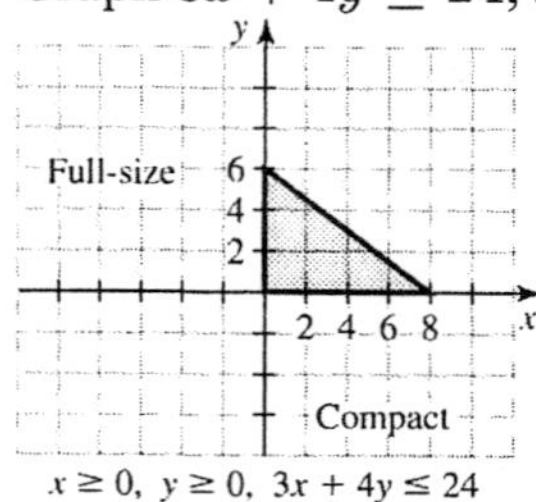

$x \geq 0,\ y \geq 0,\ 3x + 4y \leq 24$

69. Graph $3x + 4y \leq 24$, $x \geq 0$, $y \geq 0$, and $y > x$:

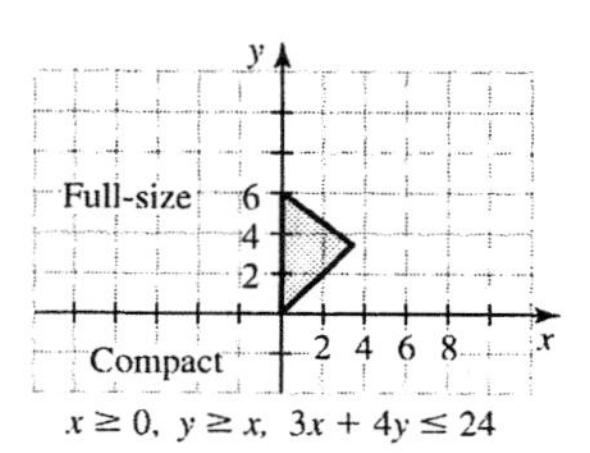

$x \geq 0,\ y \geq x,\ 3x + 4y \leq 24$

71. Graph $h \leq 187 - 0.85a$ and $h \geq 154 - 0.70a$ for $20 \leq a \leq 75$.

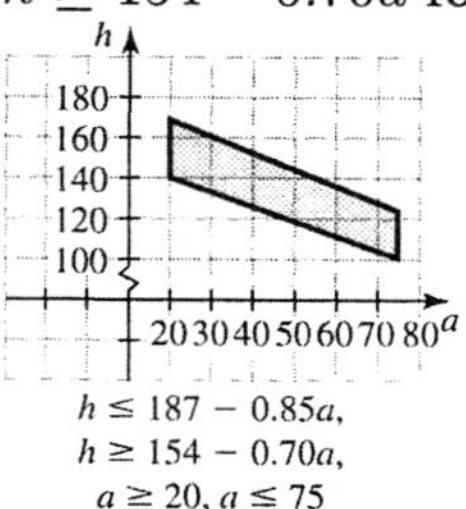

$h \leq 187 - 0.85a,$
$h \geq 154 - 0.70a,$
$a \geq 20,\ a \leq 75$

73. Let $d =$ the number of days of the *Daily Chronicle* advertising and $t =$ the number times an ad is aired on TV.

$$300d + 1000t \leq 9000$$

$$3d + 10t \leq 90$$

Graph $3d + 10t \leq 90$, $d \geq 0$, and $t \geq 0$:

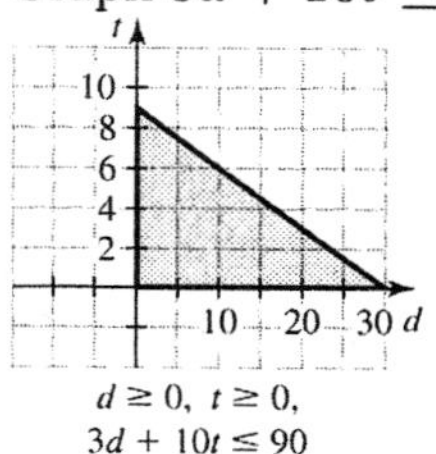

$d \geq 0,\ t \geq 0,$
$3d + 10t \leq 90$

75. The solution set to a compound inequality using "and" is the intersection of the two solution sets. The solution set to a compound inequality using "or" is the union of the two solution sets.

8.4 WARM-UPS

1. False, because $x \geq 0$ consists of the points on or to the right of the y-axis.
2. False, because $y \geq 0$ consists of the points on or above the x-axis.
3. False, because $x + y \leq 6$ consists of the points on or below the line $x + y = 6$.
4. False, because the x-intercept is (15, 0) and the y-intercept is (0, 10).
5. False, because the solution set to a system is the intersection of the solution sets.

6. True, because that is the definition of constraint.
7. False, because a linear function does not have an x^2 in it.
8. True, because $R(2,4) = 3(2) + 5(4) = 26$.
9. False, because $C(0,5) = 12(0) + 10(5) = 50$.
10. True, because the maximum or minimum of a linear function occurs at a vertex.

8.4 EXERCISES

1. A constraint is an inequality that restricts the values of the variables.
3. Constraints may be limitations on the amount of available supplies, money, or other resources.
5. The maximum or minimum of a linear function subject to linear constraints occurs at a vertex of the region determined by the constraints.
7. The graph of $x \geq 0$ is the region on or to the right of the y-axis. The graph of $y \geq 0$ is the region on or above the x-axis. The graph of $x + y \leq 5$ is the region on or below the line $x + y = 5$. The intersection of these 3 regions is shaded here.

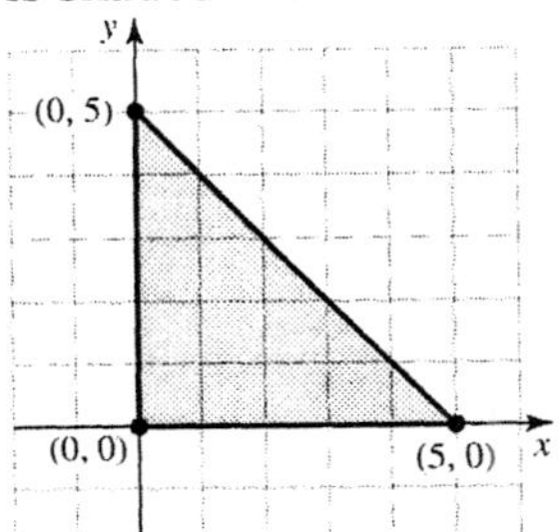

9. The graph of $x \geq 0$ is the region on or to the right of the y-axis. The graph of $y \geq 0$ is the region on or above the x-axis. The graph of $2x + y \leq 4$ is the region on or below the line $2x + y = 4$. The graph of $x + y \leq 3$ is the region on or below the line $x + y = 3$. The intersection of these regions is shaded below.

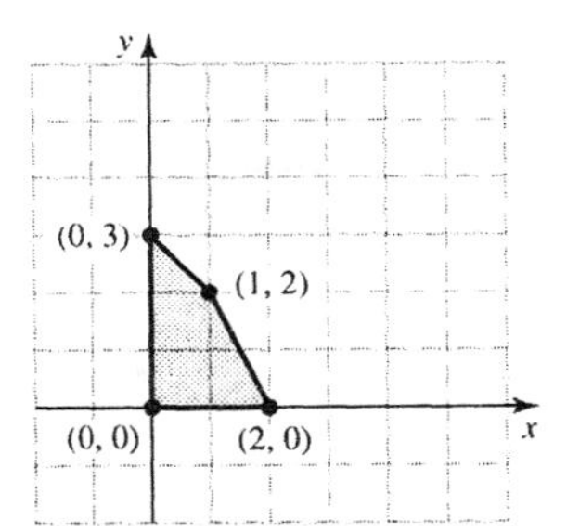

11. The graph of $x \geq 0$ is the region on or to the right of the y-axis. The graph of $y \geq 0$ is the region on or above the x-axis. The graph of $2x + y \geq 3$ is the region on or above the line $2x + y = 3$. The graph of $x + y \geq 2$ is the region on or above the line $x + y = 2$. The intersection of these regions is shaded below.

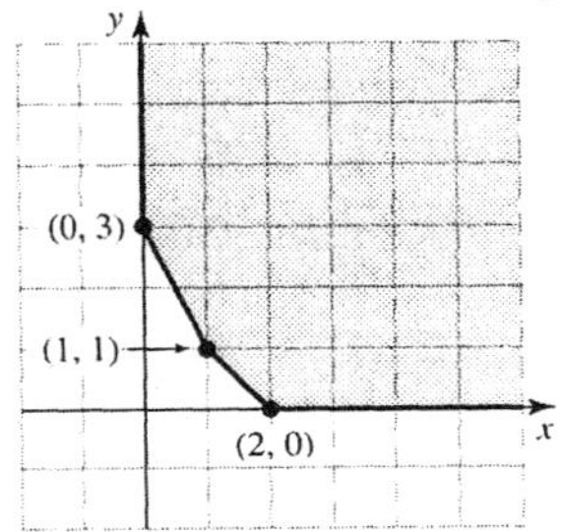

13. The graph of $x \geq 0$ is the region on or to the right of the y-axis. The graph of $y \geq 0$ is the region on or above the x-axis. The graph of $x + 3y \leq 15$ is the region on or below the line $x + 3y = 15$. The graph of $2x + y \leq 10$ is the region on or below the line $2x + y = 10$. The intersection of these regions is shaded below.

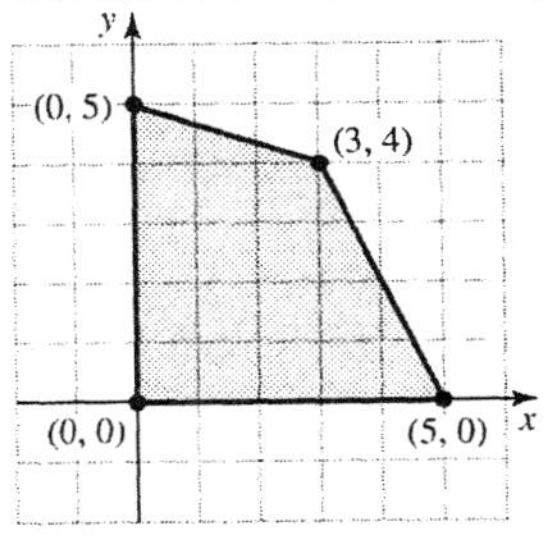

15. The graph of $x \geq 0$ is the region on or to the right of the y-axis. The graph of $y \geq 0$ is the region on or above the x-axis. The graph of $x + y \geq 4$ is the region on or above the line $x + y = 4$. The graph of $3x + y \geq 6$ is the region on or above the line $3x + y = 6$. The intersection of these regions is shaded below.

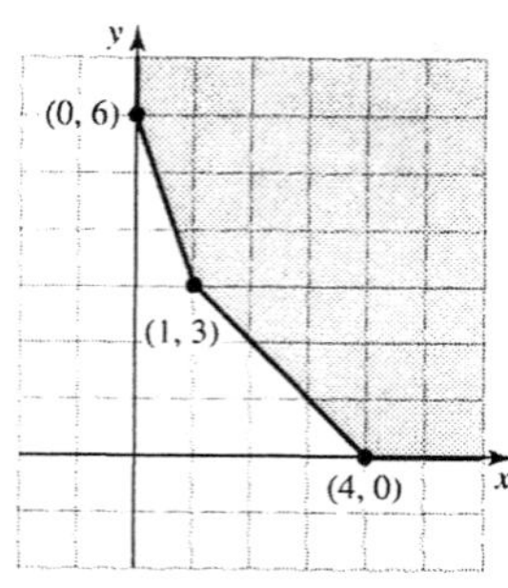

17. Since $P(x, y) = 6x + 8y$,
$P(1, 5) = 6(1) + 8(5) = 46$.
19. Since $R(x, y) = 11x + 20y$,
$R(8, 0) = 11(8) + 20(0) = 88$.
21. Since $C(x, y) = 5x + 12y$,
$C(4, 9) = 5(4) + 12(9) = 128$.
23. a) $A(0, 0) = 0$
$A(0, 80) = 9000(0) + 4000(80) = 320{,}000$
$A(50, 0) = 9000(50) + 4000(0) = 450{,}000$
$A(30, 60) = 9000(30) + 4000(60) = 510{,}000$
b) To maximize the audience they should use 30 TV ads and 60 radio ads.
25. Let $x =$ the number of doubles and $y =$ the number of triples. We can write four inequalities.
$x \geq 0, \quad y \geq 0$
$2x + 3y \leq 24$
$6x + 3y \leq 48$
Graph the system of inequalities. The vertices of the region are (0, 8), (6, 4), (8, 0), and (0, 0). Since Doubles are \$1.20 each and Triples are \$1.50 each, the revenue in dollars from x Doubles and y Triples is given by
$R(x, y) = 1.2x + 1.5y$.
Find the revenue at each vertex.
$R(0, 0) = 0$
$R(0, 8) = 1.2(0) + 1.5(8) = \$12.00$
$R(6, 4) = 1.2(6) + 1.5(4) = \$13.20$
$R(8, 0) = 1.2(8) + 1.5(0) = \$9.60$
The maximum revenue occurs at (6, 4). So they should make 6 Doubles and 4 Triples to maximize their revenue.
27. Let $x =$ the number of Doubles and $y =$ the number of Triples. The constraints and the graph are the same as in Exercise 19. Since the Doubles now sell for \$1.00 each and the Triples now sell for \$2.00 each, the revenue in dollars is given by the function is
$R(x, y) = x + 2y$. Find the revenue at each vertex.
$R(0, 0) = 0$
$R(0, 8) = 0 + 2(8) = \$16.00$
$R(6, 4) = 6 + 2(4) = \$14.00$
$R(8, 0) = 8 + 2(0) = \$8.00$
The maximum revenue occurs at (0, 8). So to maximize the revenue they should make no Doubles and 8 Triples.
29. Let $x =$ the number of cups of Doggie Dinner, and $y =$ the number of cups of Puppie Power. We can write four inequalities.
$x \geq 0, \quad y \geq 0$
$20x + 30y \geq 200$
$40x + 20y \geq 180$
Graph the region that satisfies the constraints. Since one cup of Doggie Dinner costs 16 cents and one cup of Puppie Power costs 20 cents, the total cost in dollars for x cups of DD and y cups of PP is $C(x, y) = 0.16x + 0.20y$.

Find the cost at each vertex.
$C(0, 9) = 0.16(0) + 0.20(9) = \$1.80$
$C(1.75, 5.5) = 0.16(1.75) + 0.20(5.5) = \$1.38$
$C(10, 0) = 0.16(10) + 0.20(0) = \$1.60$

The minimum cost occurs at (1.75, 5.5). To minimize the cost and satisfy the constraints she should use 1.75 cups of DD and 5.5 cups of PP.

31. If the cost of one cup of DD is 4 cents and one cup of PP is 10 cents then the total cost in dollars of x cups of DD and y cups of PP is $C(x, y) = 0.04x + 0.10y$. The inequalities and the region are the same as in Exercise 23. Find the cost at each vertex of the region.

$C(0, 9) = 0.04(0) + 0.10(9) = \$0.90$
$C(1.75, 5.5) = 0.04(1.75) + 0.10(5.5) = \$0.62$
$C(10, 0) = 0.04(10) + 0.10(0) = \$0.40$

The minimum cost occurs at (10, 0). She should use 10 cups of DD and 0 cups of PP to satisfy the constraints and minimize the cost.
33. Let $x =$ the amount invested in the laundromat and $y =$ the amount invested in the car wash. We can write four inequalities.

$x \geq 0,\ \ y \geq 0$
$x + y \leq 24{,}000$
$2x \leq y \leq 3x$

Graph the region that satisfies the system of inequalities. Since she makes 18% on the money invested in the laundry and 12% on the money invested in the car wash, the income in dollars is given by the function
$I(x, y) = 0.18x + 0.12y$.
Find the income at each vertex.
$I(0, 0) = 0$
$I(8000, 16000) = 0.18(8000) + 0.12(16000)$
$= \$3{,}360$
$I(6000, 18000) = 0.18(6000) + 0.12(18000)$
$= \$3{,}240$

The maximum income occurs at $(8000, 16000)$. So she should invest \$8000 in the laundromat and \$16,000 in the car wash.

Enriching Your Mathematical Word Power

1. c **2.** d **3.** a **4.** b
5. a **6.** d **7.** c **8.** a
9. c **10.** a **11.** b

CHAPTER 8 REVIEW

1. $x + 2 > 3$ or $x - 6 < -10$
$x > 1$ or $x < -4$
$(-\infty, -4) \cup (1, \infty)$

3. $x > 0$ and $x - 6 < 3$
$x > 0$ and $x < 9$
$(0, 9)$

5. $6 - x < 3$ or $-x < 0$
$-x < -3$ or $-x < 0$
$x > 3$ or $x > 0$
$(0, \infty)$

7. $2x < 8$ and $2(x - 3) < 6$
$2x < 8$ and $2x - 6 < 6$
$x < 4$ and $2x < 12$
$x < 4$ and $x < 6$
$(-\infty, 4)$

9. $x - 6 > 2$ and $6 - x > 0$
$x > 8$ and $-x > -6$
$x > 8$ and $x < 6$
No number is greater than 8 and less than 6.
$\emptyset$

11. $0.5x > 10$ or $0.1x < 3$
$x > 20$ or $x < 30$
Every number is either greater than 20 or less than 30. Solution set is R or $(-\infty, \infty)$

13. $-2 \leq \frac{2x - 3}{10} \leq 1$
$10(-2) \leq 10 \cdot \frac{2x - 3}{10} \leq 10(1)$
$-20 \leq 2x - 3 \leq 10$
$-17 \leq 2x \leq 13$
$-\frac{17}{2} \leq x \leq \frac{13}{2}$
$\left[-\frac{17}{2}, \frac{13}{2}\right]$

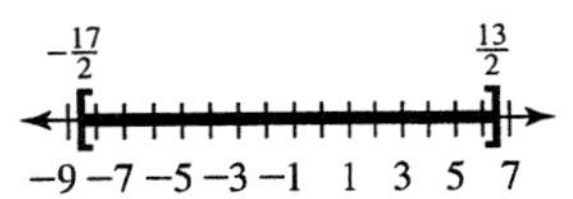

15. $[1, 4) \cup (2, \infty) = [1, \infty)$
17. $(3, 6) \cap [2, 8] = (3, 6)$
19. $(-\infty, 5) \cup [5, \infty) = (-\infty, \infty)$
21. $(-3, -1] \cap [-2, 5] = [-2, -1]$

23. $|x| + 2 = 16$
$|x| = 14$
$x = 14$ or $x = -14$
$\{-14, 14\}$

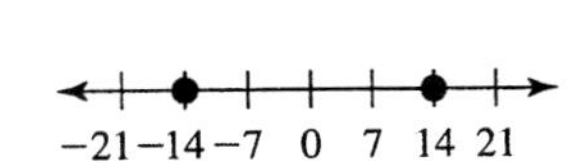

25. $|4x - 12| = 0$
$4x - 12 = 0$
$4x = 12$
$x = 3$
$\{3\}$

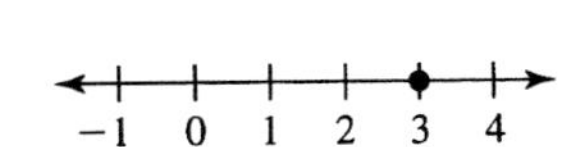

27. Since $|x| \geq 0$ for any real number, the solution set is $\emptyset$.

29. $|2x - 1| - 3 = 0$
$|2x - 1| = 3$
$2x - 1 = 3$ or $2x - 1 = -3$
$2x = 4$ or $2x = -2$
$x = 2$ or $x = -1$
$\{-1, 2\}$

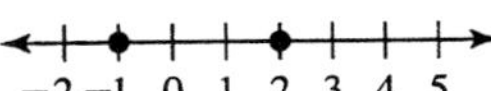

31. $$|2x| \geq 8$$
$$2x \geq 8 \quad \text{or} \quad 2x \leq -8$$
$$x \geq 4 \quad \text{or} \quad x \leq -4$$
$(-\infty, -4] \cup [4, \infty)$

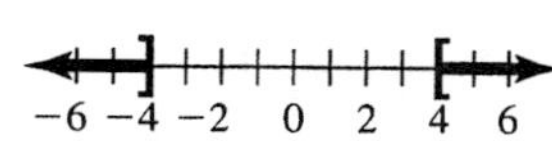

33. $\left|1 - \frac{x}{5}\right| > \frac{9}{5}$

$$1 - \frac{x}{5} > \frac{9}{5} \quad \text{or} \quad 1 - \frac{x}{5} < -\frac{9}{5}$$
$$5 - x > 9 \quad \text{or} \quad 5 - x < -9$$
$$-x > 4 \quad \text{or} \quad -x < -14$$
$$x < -4 \quad \text{or} \quad x > 14$$
$(-\infty, -4) \cup (14, \infty)$

−8 −4 0 4 8 12 16

35. Since $|x - 3| \geq 0$ for any value of x, the solution set is $\emptyset$.

37. Since $|x + 4| \geq 0$ for any value of x, $|x + 4| \geq -1$ for any x. Solution set: R

−3 −2 −1 0 1 2 3

39. $$1 - \frac{3}{2}|x - 2| < -\frac{1}{2}$$
$$2 - 3|x - 2| < -1$$
$$-3|x - 2| < -3$$
$$|x - 2| > 1$$
$$x - 2 > 1 \quad \text{or} \quad x - 2 < -1$$
$$x > 3 \quad \text{or} \quad x < 1$$
$(-\infty, 1) \cup (3, \infty)$

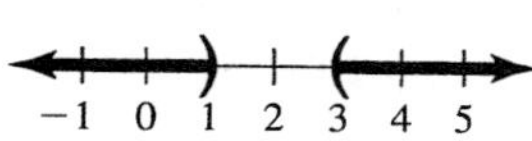

41. First graph the lines $y = 3$ and $y - x = 5$.

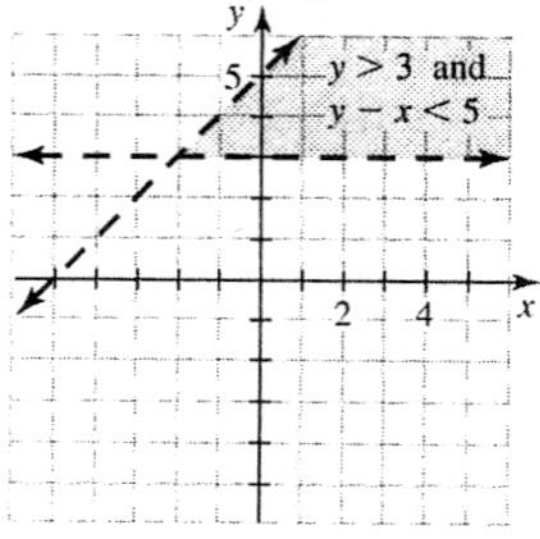

Test the points $(0, 0)$, $(0, 4)$, $(0, 6)$, and $(-6, 0)$. Only $(0, 4)$ satisfies $y > 3$ and $y - x < 5$. Shade the region containing $(0, 4)$.

43. First graph the lines $3x + 2y = 8$ and $3x - 2y = 6$. Test the points $(0, 0)$, $(0, 5)$, $(0, -6)$, and $(5, 0)$. The points $(0, 0)$, $(0, 5)$, and $(5, 0)$ satisfy the compound inequality $3x + 2y \geq 8$ or $3x - 2y \leq 6$. Shade the regions containing those points.

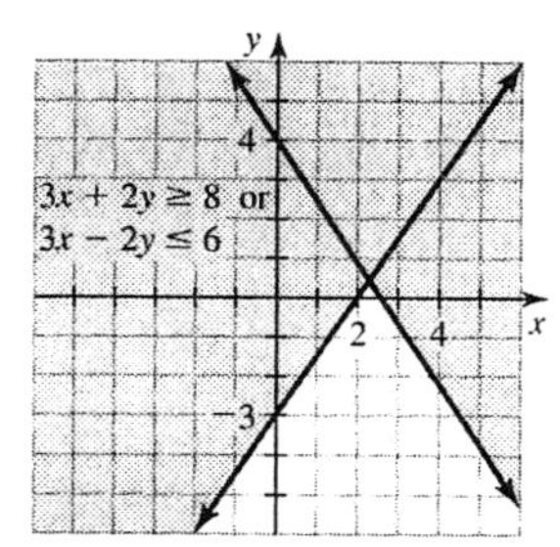

45. First graph the equations $x + 2y = 10$ and $x + 2y = -10$. Test the points $(0, 0)$, $(0, 8)$, and $(0, -8)$ in the inequality $|x + 2y| < 10$. Only $(0, 0)$ satisfies the inequality. So shade the region containing $(0, 0)$.

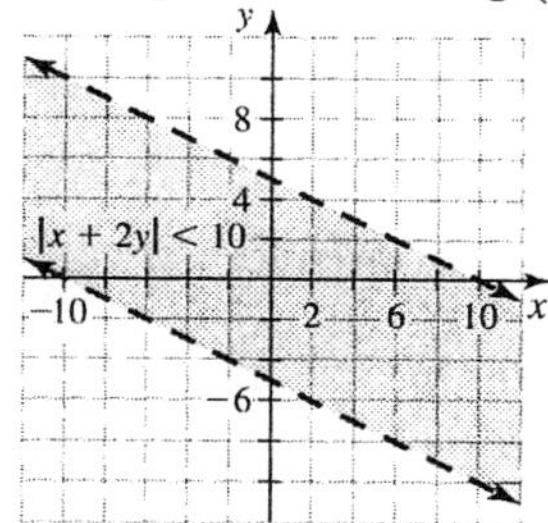

47. First graph the two vertical lines $x = 5$ and $x = -5$.

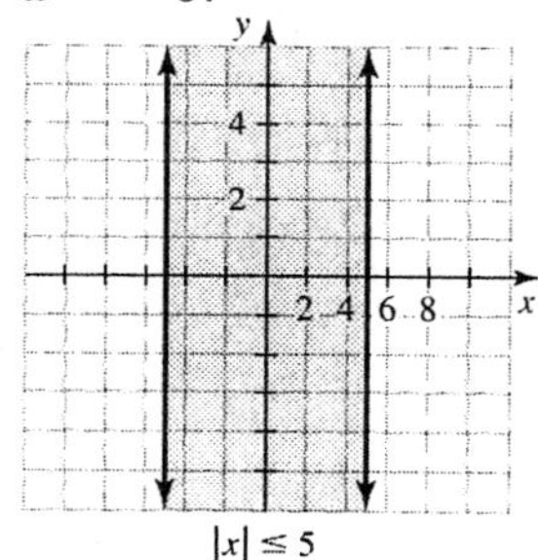

Test the points $(0, 0)$, $(8, 0)$, $(-8, 0)$ in the inequality $|x| \leq 5$. Only $(0, 0)$ satisfies the inequality. So shade the region containing $(0, 0)$, the region between the parallel lines.

49. First graph the lines $y - x = 2$ and $y - x = -2$. Test the points $(0, 0)$, $(0, 4)$, and $(0, -4)$ in the inequality $|y - x| > 2$. The points $(0, 4)$ and $(0, -4)$ satisfy the inequality. Shade the regions containing those points.

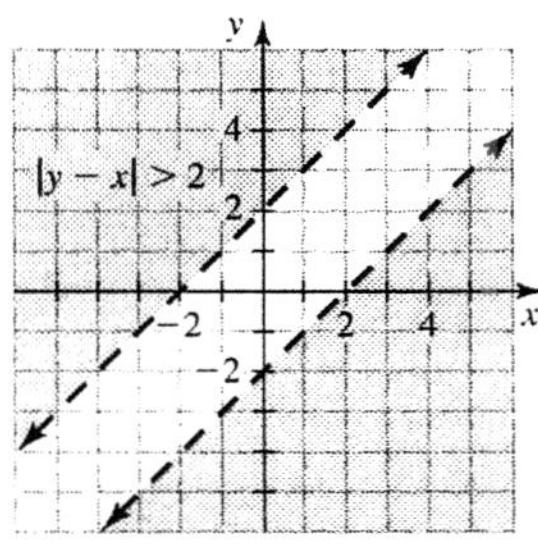

51. The graph of $x \geq 0$ consists of points on and to the right of the y-axis. The graph of $y \geq 0$ consists of points on and above the x-axis. The graph of $x + 2y \leq 6$ consists of points on and below the line $x + 2y = 6$. The graph of $x + y \leq 5$ consists of points on and below the line $x + y = 5$. Points that satisfy all 4 inequalities are shown in the following graph.

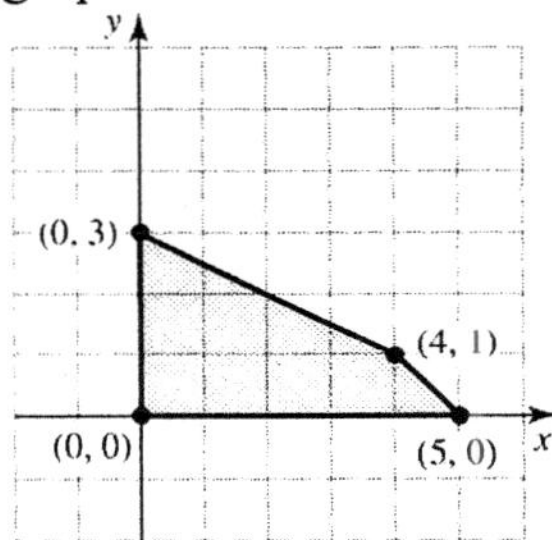

The vertices of the region are (0, 3), (4, 1), (5, 0), and (0, 0).

53. First graph the region that satisfies the four inequalities. The vertices of the region are $(0,0)$, $(0,3)$, $(2,2)$, and $(3,0)$.
Evaluate $R(x,y) = 6x + 9y$ at each vertex.
$R(0,0) = 0$
$R(0,3) = 6(0) + 9(3) = 27$
$R(2,2) = 6(2) + 9(2) = 30$
$R(3,0) = 6(3) + 9(0) = 18$
The maximum value of $R(x,y)$ is 30.

55. Let x = the rental price, $0.45x$ = the overhead per tape, and $x - 0.45x$ or $0.55x$ = the profit per tape. The rental price must be less than or equal to \$5 and satisfy the inequality

$$0.55x \geq 1.65$$
$$x \geq 3$$

The range of the rental price is $\$3 \leq x \leq \$5$.

57. Since $150 < h < 180$, we have

$$150 < 60.089 + 2.238F < 180$$
$$89.911 < 2.238F < 119.911$$
$$40.2 < F < 53.6$$

The length of the femur is in (40.2, 53.6).

59. Let x = the number on the mile marker where Dane was picked up. We can write the absolute value equation

$$|x - 86| = 5$$
$$x - 86 = 5 \quad \text{or} \quad x - 86 = -5$$
$$x = 91 \quad \text{or} \quad x = 81$$

He was either at 81 or 91.

61. The numbers to the right of 1 are described by the inequality $x > 1$.

63. The number 2 satisfies the equation $|x - 2| = 0$.

65. The numbers 3 and -3 both satisfy the equation $|x| = 3$.

67. The numbers to the left of and including -1 satisfy $x \leq -1$.

69. The numbers between -2 and 2 including the endpoints satisfy $|x| \leq 2$.

71. $x \leq 2$ or $x \geq 7$

73. The numbers greater than 3 or less than -3 satisfy $|x| > 3$.

75. $5 < x < 7$ or $|x - 6| < 1$

77. Every number except 0 has a positive absolute value and satisfies $|x| > 0$.

CHAPTER 8 TEST

1. The graph shows the numbers greater than -3 and less than or equal to 2. The inequality is $-3 < x \leq 2$.

2. The graph shows the numbers greater than 1. The inequality is $x > 1$.

3. The solution set to $x \geq 3$ is $[3, \infty)$.

4. The solution set to $x > 1$ and $x \leq 6$ is $(1, 6]$.

5. The solution set to $x < 5$ or $x > 9$ is $(-\infty, 5) \cup (9, \infty)$.

6.
$$|x| < 3$$
$$-3 < x < 3$$

The solution set is $(-3, 3)$.

7. $|x| \geq 2$
$x \geq 2$ or $x \leq -2$

The solution set is $(-\infty, -2] \cup [2, \infty)$.

8. $2x + 3 > 1$

$2x > -2$

$x > -1$

The solution set is $(-1, \infty)$.

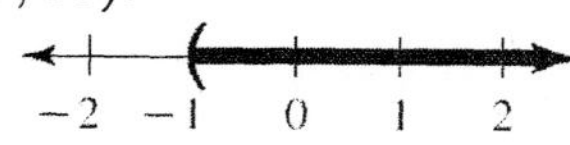

9. $|m - 6| \leq 2$

$-2 \leq m - 6 \leq 2$

$4 \leq m \leq 8$

$[4, 8]$

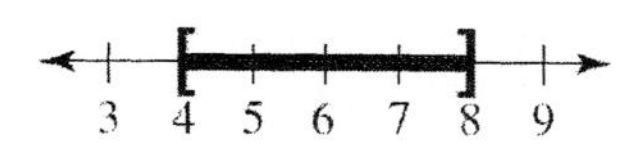

10. $2\,|\,x - 3\,| > 20$

$|\,x - 3\,| > 10$

$x - 3 > 10$ or $x - 3 < -10$

$x > 13$ or $x < -7$

$(-\infty, -7) \cup (13, \infty)$

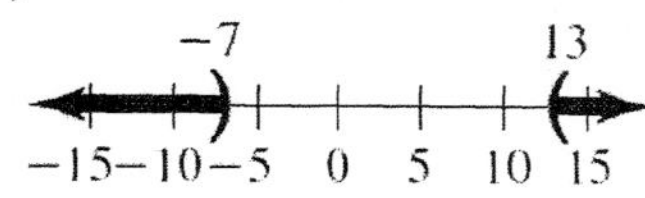

11. $2 - 3(w - 1) < -2w$

$2 - 3w + 3 < -2w$

$-w < -5$

$w > 5$

$(5, \infty)$

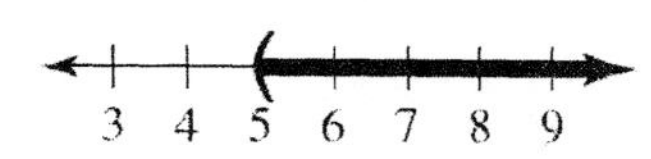

12. $3x - 2 < 7$ and $-3x \leq 15$

$3x < 9$ and $x \geq -5$

$x < 3$ and $x \geq -5$

$[-5, 3)$

13. $\frac{2}{3}y < 4$ or $y - 3 < 12$

$\frac{3}{2}\left(\frac{2}{3}y\right) < \frac{3}{2}(4)$ or $y - 3 < 12$

$y < 6$ or $y < 15$

$(-\infty, 15)$

14. The equation $|\,2x - 7\,| = -3$ has no solution because the absolute value of any real number is greater than or equal to zero. The solution set is $\emptyset$.

15. $x - 4 > 1$ or $x < 12$

$x > 5$ or $x < 12$

$(-\infty, \infty)$

16. $3x < 0$ and $x - 5 > 2$

$x < 0$ and $x > 7$

No real number is both less than 0 and greater than 7. Solution set: $\emptyset$

17. Since no real number satisfies $|2x - 5| < 0$, we need only solve $|2x - 5| = 0$, which is equivalent to $2x - 5 = 0$, or $x = 2.5$.

Solution set: $\{\,2.5\}$

18. Since no real number satisfies $|x - 3| < 0$, the solution set is $\emptyset$.

19. Since $|x - 6| \geq 0$ for any real number x, $|\,x - 6\,| \geq -6$ for any real number x. The solution set is R or $(-\infty, \infty)$.

20. Graph the vertical line $x = 2$ using a dashed line. Graph $x + y = 0$ using a dashed line. Test the points $(-2, 0)$, $(1, 0)$, $(5, 0)$, and $(3, -5)$. Only $(5, 0)$ satisfies $x > 2$ and $x + y > 0$. So shade the region containing $(5, 0)$.

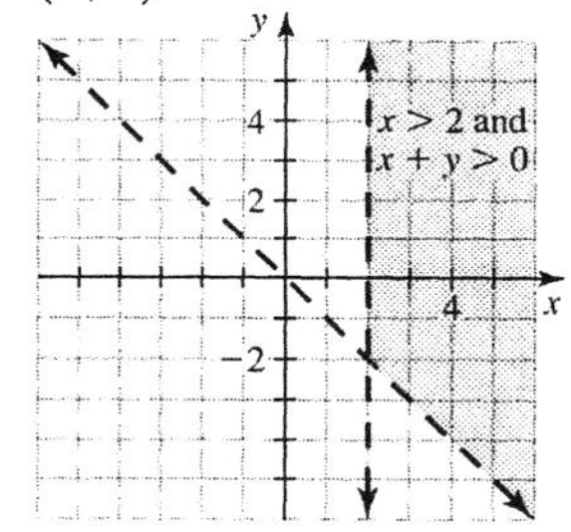

21. First graph the parallel lines $2x + y = 3$ and $2x + y = -3$, using solid lines. Test the points $(-5, 0)$, $(0, 0)$ and $(5, 0)$ in $|2x + y| \geq 3$. Both $(-5, 0)$ and $(5, 0)$ satisfy the inequality, so shade the regions containing those points.

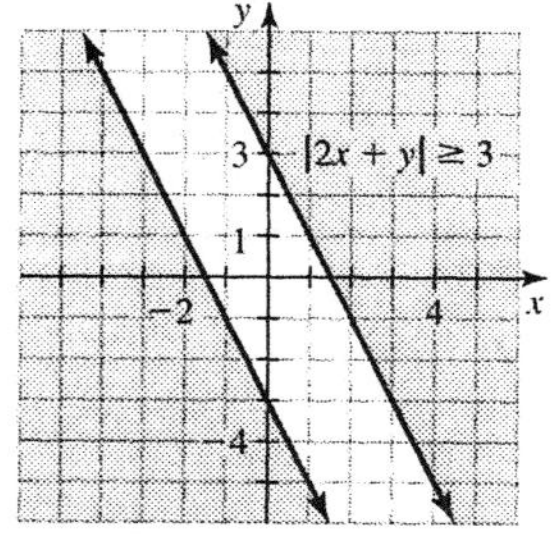

22. Graph the line $x + y = 1$ through $(0, 1)$ and $(1, 0)$. Graph the line $x - y = 2$ through the points $(0, -2)$ and $(2, 0)$. Use dashed lines. Test the points $(0, 0)$, $(0, 3)$, $(5, 0)$, and

$(0,-5)$. Only $(0,-5)$ fails to satisfy either of the inequalities. So shade other three regions.

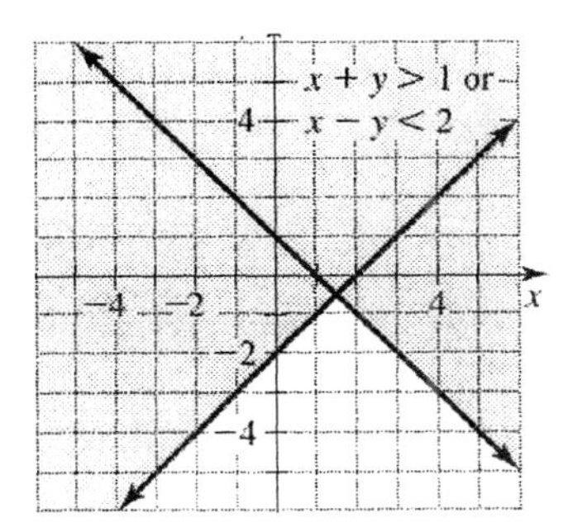

23. Let b = Brenda's salary.

$|b - 28{,}000| > 3000$

$b - 28{,}000 > 3000$ or $b - 28{,}000 < -3000$

$b > 31{,}000$ or $b < 25{,}000$

Brenda's salary is either greater than \$31,000 or less than \$25,000.

24. First graph the system of inequalities. The vertices of the region are $(0,0)$, $(0,4)$, $(3,2)$, and $(5,0)$. Evaluate the function $P(x,y) = 8x + 10y$ at each vertex.

$P(0,0) = 0$

$P(0,4) = 8(0) + 10(4) = 40$

$P(3,2) = 8(3) + 10(2) = 44$

$P(5,0) = 8(5) + 10(0) = 40$

The maximum value of the function $P(x,y)$ is 44.

Making Connections

Chapters 1 - 8

1. $5x + 6x = 11x$

2. $5x \cdot 6x = 30x^2$

3. $\frac{6x+2}{2} = \frac{6x}{2} + \frac{2}{2} = 3x + 1$

4. $5 - 4(2 - x) = 5 - 8 + 4x = 4x - 3$

5. $(30 - 1)(30 + 1) = 29 \cdot 31 = 899$

6. $(30 + 1)^2 = 31^2 = 961$

7. $(30 - 1)^2 = 29^2 = 841$

8. $(2 + 3)^2 = 5^2 = 25$

9. $2^2 + 3^2 = 4 + 9 = 13$

10. $(8 - 3)(3 - 8) = 5(-5) = -25$

11. $(-1)(3 - 8) = -1(-5) = 5$

12. $-2^2 = -(2^2) = -4$

13. $3x + 8 - 5(x - 1) = 3x + 8 - 5x + 5$
$= -2x + 13$

14. $(-6)^2 - 4(-3)2 = 36 + 24 = 60$

15. $3^2 \cdot 2^3 = 9 \cdot 8 = 72$

16. $4(-6) - (-5)(3) = -24 + 15 = -9$

17. $5x + 6x = 8x$

$11x = 8x$

$3x = 0$

$x = 0$

Solution set: $\{0\}$

18. $5x + 6x = 11x$

$11x = 11x$

This equation is an identity. Solution set is R or $(-\infty, \infty)$.

19. $5x + 6x = 0$

$11x = 0$

$x = 0$

Solution set: $\{0\}$

20. $5x + 6 = 11x$

$-6x = -6$

$x = 1$

Solution set: $\{1\}$

21. $3x + 1 = 0$

$3x = -1$

$x = -\frac{1}{3}$

Solution set: $\left\{-\frac{1}{3}\right\}$

22. $5 - 4(2 - x) = 1$

$5 - 8 + 4x = 1$

$4x = 4$

$x = 1$

Solution set: $\{1\}$

23. $x - 0.01x = 990$

$0.99x = 990$

$x = 1000$

Solution set: $\{1000\}$

24. $|5x + 6| = 11$

$5x + 6 = 11$ or $5x + 6 = -11$

$5x = 5$ or $5x = -17$

$x = 1$ or $x = -17/5$

Solution set: $\{-17/5, 1\}$

25.
$$\begin{array}{r} 2x + y = 5 \\ x - y = 7 \\ \hline 3x \quad = 12 \\ x \quad = 4 \end{array}$$

If $x = 4$, $2(4) + y = 5$ or $y = -3$.

The solution set is $\{(4, -3)\}$.

26.
$$\begin{array}{r} 3x - y = 5 \\ -3x + y = -5 \\ \hline 0 \quad = 0 \end{array}$$

The equations are equivalent. So the solution set is $\{x \mid 3x - y = 5\}$.

27. $2x + 5y = 16$
$3x - 4y = -22$

Multiply the first equation by 4 and the second by 5.

$$\begin{aligned} 8x + 20y &= 64 \\ 15x - 20y &= -110 \\ \hline 23x \quad &= -46 \\ x \quad &= -2 \end{aligned}$$

If $x = -2$, then

$$\begin{aligned} 8(-2) + 20y &= 64 \\ 20y &= 80 \\ y &= 4 \end{aligned}$$

The solution set is $\{(-2, 4)\}$.

28. $\frac{1}{2}x - \frac{2}{3}y = -6$
$\frac{3}{4}x + \frac{2}{5}y = 12$

Multiply the first equation by 12 and the second by 20.

$$\begin{aligned} 6x - 8y &= -72 \\ 15x + 8y &= 240 \\ \hline 21x \quad &= 168 \\ x \quad &= 8 \end{aligned}$$

If $x = 8$, then

$$\begin{aligned} 6(8) - 8y &= -72 \\ -8y &= -120 \\ y &= 15 \end{aligned}$$

The solution set is $\{(8, 15)\}$.

29. $2 - x < 5$
$-x < 3$
$x > -3$

So $2 - x < 5$ is equivalent to (E).

30. $x + 1 > x - 2$
$2 > -2$

All real numbers satisfy the given inequality. So it is equivalent to (F), whose solution set is also the set of all real numbers.

31. The compound inequality $x > 2$ and $x > 5$ is equivalent to $x > 5$. So the given inequality is equivalent to (G).

32. The compound inequality $x < -5$ or $x < -3$ is equivalent to $x < -3$. So the given inequality is equivalent to (D).

33. The compound inequality $x < -9$ and $x > -3$ has no solutions. So the given inequality is equivalent to (H), which also has no solutions.

34. The compound inequality $x < -3$ or $x > 3$ is equivalent to the absolute value inequality $|x| > 3$, which is (C).

35. The compound inequality $x > -3$ and $x < 3$ is equivalent to $|x| < 3$, which is (B).

36. The absolute value inequality is satisfied by every real number except -3. So it is equivalent to (I).

37. The compound inequality $y < x + 3$ and $y < x$ is equivalent to $y < x$ because $y = x + 3$ and $y = x$ are parallel lines with $y = x$ below $y = x + 3$. So the given inequality is equivalent to (A).

38. The compound inequality $y > x - 3$ or $y > x$ is equivalent to $y > x - 3$, which is (J).

39. a) From the graph it appears that the cost of renting and buying are equal at around 87,500 copies.

b) If $x =$ the number of copies made in 5 years, then the cost for renting is
$R = 60(75) + 0.06x$ dollars
or $R = 4500 + 0.06x$ dollars.
The cost if the copier is purchased is
$P = 8000 + 0.02x$ dollars.

c)
$$\begin{aligned} 60(75) + 0.06x &= 8000 + 0.02x \\ 0.04x &= 3500 \\ x &= 87{,}500 \end{aligned}$$

Five-year cost is same for 87,500 copies.

d) If 120,000 copies are made, then renting cost is \$11,700 and buying cost is \$10,400. So buying is \$1300 cheaper.

e)
$$\begin{aligned} |60(75) + 0.06x - (8000 + 0.02x)| &< 500 \\ |-3500 + 0.04x| &< 500 \\ -500 < -3500 + 0.04x &< 500 \\ 3000 < 0.04x &< 4000 \\ 75{,}000 < x &< 100{,}000 \end{aligned}$$

If the number of copies is between 75,000 and 100,000, then the plans differ by less than \$500.

9.1 WARM-UPS

1. True, because of the definition of square root.
2. False, because $\sqrt[3]{2}\cdot\sqrt[3]{2}\cdot\sqrt[3]{2}=2$.
3. True, because $(-3)^3=-27$.
4. False, because $(-5)^2=25$.
5. True, because $2^4=16$.
6. False, because $\sqrt{9}=3$.
7. False, because $(2^3)^2=2^6$.
8. False, because $\frac{\sqrt{10}}{\sqrt{2}}=\sqrt{5}$.
9. True, because $\left(\frac{1}{2}\right)^2=\frac{1}{4}$.
10. True, because $\frac{\sqrt{6}}{\sqrt{3}}=\sqrt{\frac{6}{3}}=\sqrt{2}$

9.1 EXERCISES

1. If $b^n=a$ then b is an nth root of a.
3. If $b^n=a$ then b is an even root of a provided n is even or an odd root provided n is odd.
5. The product rule for radicals says that $\sqrt[n]{a}\cdot\sqrt[n]{b}=\sqrt[n]{ab}$ provided all of these roots are real.
7. Because $6^2=36$, $\sqrt{36}=6$.
9. Because $10^2=100$, $\sqrt{100}=10$.
11. Because $3^2=9$, $\sqrt{9}=3$ and $-\sqrt{9}=-3$.
13. Because $2^3=8$, $\sqrt[3]{8}=2$.
15. Because $(-2)^3=-8$, $\sqrt[3]{-8}=-2$.
17. Because $2^5=32$, $\sqrt[5]{32}=2$.
19. Because $10^3=1000$, $\sqrt[3]{1000}=10$.
21. The expression $\sqrt[4]{-16}$ is not a real number since it is an even root of a negative number.
23. Since $(-2)^5=-32$, $\sqrt[5]{-32}=-2$.
25. Since $m\cdot m=m^2$, $\sqrt{m^2}=m$.
27. Since $(x^8)^2=x^{16}$, $\sqrt{x^{16}}=x^8$.
29. Since $(y^3)^5=y^{15}$, $\sqrt[5]{y^{15}}=y^3$.
31. Since $(y^5)^3=y^{15}$, $\sqrt[3]{y^{15}}=y^5$.
33. Since $(m)^3=m^3$, $\sqrt[3]{m^3}=m$.
35. Since $(w^3)^4=w^{12}$, $\sqrt[4]{w^{12}}=w^3$.
37. $\sqrt{9y}=\sqrt{9}\sqrt{y}=3\sqrt{y}$
39. $\sqrt{4a^2}=\sqrt{4}\sqrt{a^2}=2a$
41. $\sqrt{x^4y^2}=\sqrt{x^4}\sqrt{y^2}=x^2y$
43. $\sqrt{5m^{12}}=\sqrt{m^{12}}\sqrt{5}=m^6\sqrt{5}$
45. $\sqrt[3]{8y}=\sqrt[3]{8}\sqrt[3]{y}=2\sqrt[3]{y}$
47. $\sqrt[3]{3a^6}=\sqrt[3]{a^6}\sqrt[3]{3}=a^2\sqrt[3]{3}$
49. $\sqrt{20}=\sqrt{4}\cdot\sqrt{5}=2\sqrt{5}$
51. $\sqrt{50}=\sqrt{25}\cdot\sqrt{2}=5\sqrt{2}$
53. $\sqrt{72}=\sqrt{36}\cdot\sqrt{2}=6\sqrt{2}$
55. $\sqrt[3]{40}=\sqrt[3]{8}\cdot\sqrt[3]{5}=2\sqrt[3]{5}$
57. $\sqrt[3]{81}=\sqrt[3]{27}\cdot\sqrt[3]{3}=3\sqrt[3]{3}$
59. $\sqrt[4]{48}=\sqrt[4]{16}\cdot\sqrt[4]{3}=2\sqrt[4]{3}$
61. $\sqrt[5]{96}=\sqrt[5]{32}\cdot\sqrt[5]{3}=2\sqrt[5]{3}$
63. $\sqrt{a^3}=\sqrt{a^2}\sqrt{a}=a\sqrt{a}$
65. $\sqrt{18a^6}=\sqrt{9a^6}\sqrt{2}=3a^3\sqrt{2}$
67. $\sqrt{20x^5y}=\sqrt{4x^4}\sqrt{5xy}=2x^2\sqrt{5xy}$
69. $\sqrt[3]{24m^4}=\sqrt[3]{8m^3}\cdot\sqrt[3]{3m}=2m\sqrt[3]{3m}$
71. $\sqrt[4]{32a^5}=\sqrt[4]{16a^4}\cdot\sqrt[4]{2a}=2a\sqrt[4]{2a}$
73. $\sqrt[5]{64x^6}=\sqrt[5]{32x^5}\cdot\sqrt[5]{2x}=2x\sqrt[5]{2x}$
75. $\sqrt{48x^3y^8z^7}=\sqrt{16x^2y^8z^6}\cdot\sqrt{3xz}$
$=4xy^4z^3\sqrt{3xz}$
77. $\sqrt{\frac{t}{4}}=\frac{\sqrt{t}}{\sqrt{4}}=\frac{\sqrt{t}}{2}$
79. $\sqrt{\frac{625}{16}}=\frac{\sqrt{625}}{\sqrt{16}}=\frac{25}{4}$
81. $\frac{\sqrt{30}}{\sqrt{3}}=\sqrt{\frac{30}{3}}=\sqrt{10}$
83. $\sqrt[3]{\frac{t}{8}}=\frac{\sqrt[3]{t}}{\sqrt[3]{8}}=\frac{\sqrt[3]{t}}{2}$
85. $\sqrt[3]{\frac{-8x^6}{y^3}}=\frac{\sqrt[3]{-8x^6}}{\sqrt[3]{y^3}}=\frac{-2x^2}{y}$
87. $\sqrt{\frac{4a^6}{9}}=\frac{\sqrt{4a^6}}{\sqrt{9}}=\frac{2a^3}{3}$
89. $\sqrt{\frac{12}{25}}=\frac{\sqrt{12}}{\sqrt{25}}=\frac{\sqrt{4}\sqrt{3}}{5}=\frac{2\sqrt{3}}{5}$
91. $\sqrt{\frac{27}{16}}=\frac{\sqrt{27}}{\sqrt{16}}=\frac{\sqrt{9}\sqrt{3}}{4}=\frac{3\sqrt{3}}{4}$
93. $\sqrt[3]{\frac{a^4}{125}}=\frac{\sqrt[3]{a^4}}{\sqrt[3]{125}}=\frac{\sqrt[3]{a^3}\cdot\sqrt[3]{a}}{5}$
$=\frac{a\cdot\sqrt[3]{a}}{5}$
95. $\sqrt[3]{\frac{81}{8b^3}}=\frac{\sqrt[3]{81}}{\sqrt[3]{8b^3}}=\frac{\sqrt[3]{27}\cdot\sqrt[3]{3}}{2b}=\frac{3\sqrt[3]{3}}{2b}$
97. $\sqrt[4]{\frac{x^7}{y^8}}=\frac{\sqrt[4]{x^7}}{\sqrt[4]{y^8}}=\frac{\sqrt[4]{x^4}\cdot\sqrt[4]{x^3}}{y^2}=\frac{x\sqrt[4]{x^3}}{y^2}$

99. $\sqrt[4]{\frac{a^5}{16b^{12}}} = \frac{\sqrt[4]{a^5}}{\sqrt[4]{16b^{12}}} = \frac{\sqrt[4]{a^4}\cdot\sqrt[4]{a}}{2b^3}$
$= \frac{a\sqrt[4]{a}}{2b^3}$

101. In $\sqrt{x-2}$ we must have $x-2 \geq 0$ or $x \geq 2$. So the domain is the interval $[2,\infty)$.

103. In $\sqrt[3]{3x-7}$, $3x-7$ can be any real number. So any real number can be used for x. So the domain is the interval $(-\infty,\infty)$.

105. In $\sqrt[4]{9-3x}$ we must have $9-3x \geq 0$, $-3x \geq -9$, or $x \leq 3$. So the domain is the interval $(-\infty, 3]$.

107. In $\sqrt{2x+1}$ we must have $2x+1 \geq 0$, $2x \geq -1$, or $x \geq -1/2$. So the domain is the interval $[-1/2,\infty)$.

109. $w = 91.4 - \frac{(10.5+6.7\sqrt{20}-0.45\cdot 20)(457-5\cdot 25)}{110}$
$w \approx -4°\text{F}$
If the air temp is 25°F and wind is 30 mph, then from the graph w is approximately -10°F.

111. a) $t = \sqrt{\frac{h}{16}} = \frac{\sqrt{h}}{4}$

b) $t = \frac{\sqrt{40}}{4} = \frac{2\sqrt{10}}{4} = \frac{\sqrt{10}}{2}$ sec

c) From the graph it appears that if a diver takes 2.5 seconds then the height is 100 feet.

113. $M = 1.3\sqrt{20} \approx 5.8$ knots

115. $V = \sqrt{\frac{841\cdot 8700}{2.81\cdot 200}} \approx 114.1$ ft/sec
$114.1\ \frac{\text{ft}}{\text{sec}}\cdot\frac{1\text{ mi}}{5280\text{ ft}}\cdot\frac{3600\text{ sec}}{1\text{ hr}} \approx 77.8$ mph

117. The equations in (a), (c), and (d) are identities because the radical symbol always indicates a positive root. In (b), $\sqrt[3]{x^3}$ is negative if x is negative, but $|x|$ is positive if x is negative. So (b) is not an identity.

119. The arithmetic mean of 80 and 100 is 90. Solve $80/h = h/100$ to get $h^2 = 8000$ or $h = \sqrt{8000} \approx 89.4$. So you are better of with the arithmetic mean.

9.2 WARM-UPS

1. True, by definition of exponent $1/3$.

2. False, because $8^{5/3} = \sqrt[3]{8^5}$.

3. False, because $-16^{1/2} = -4$, while $(-16)^{1/2}$ is not a real number.

4. True, $9^{-3/2} = \frac{1}{(\sqrt{9})^3} = \frac{1}{3^3} = \frac{1}{27}$.

5. True, $\frac{\sqrt{6}}{6} = \frac{6^{1/2}}{6^1} = 6^{1/2-1} = 6^{-1/2}$.

6. True, because $\frac{2^1}{2^{1/2}} = 2^{1-1/2} = 2^{1/2}$.

7. True, because $2^{1/2}\cdot 2^{1/2} = 2^1 = 2$ and $4^{1/2} = 2$.

8. False, because $16^{-1/4} = \frac{1}{16^{1/4}} = \frac{1}{2}$.

9. True, $6^{1/6}\cdot 6^{1/6} = 6^{1/6+1/6} = 6^{2/6} = 6^{1/3}$.

10. True, $(2^8)^{3/4} = 2^{24/4} = 2^6$.

9.2 EXERCISES

1. The nth root of a is $a^{1/n}$.

3. The expression $a^{-m/n}$ means $\frac{1}{a^{m/n}}$.

5. The operations can be performed in any order, but the easiest is usually root, power, and then reciprocal.

7. $\sqrt[4]{7} = 7^{1/4}$

9. $\sqrt{5x} = (5x)^{1/2}$

11. $9^{1/5} = \sqrt[5]{9}$

13. $a^{1/2} = \sqrt{a}$

15. $25^{1/2} = \sqrt{25} = 5$

17. $(-125)^{1/3} = \sqrt[3]{-125} = -5$

19. $16^{1/4} = \sqrt[4]{16} = 2$

21. $(-4)^{1/2}$ is not a real number because it is $\sqrt{-4}$.

23. $\sqrt[3]{w^7} = w^{7/3}$

25. $\frac{1}{\sqrt[3]{2^{10}}} = 2^{-10/3}$

27. $w^{-3/4} = \sqrt[4]{\frac{1}{w^3}}$

29. $(ab)^{3/2} = \sqrt{(ab)^3}$

31. $125^{2/3} = (\sqrt[3]{125})^2 = 5^2 = 25$

33. $25^{3/2} = (\sqrt{25})^3 = 5^3 = 125$

35. $27^{-4/3} = \frac{1}{(\sqrt[3]{27})^4} = \frac{1}{3^4} = \frac{1}{81}$

37. $16^{-3/2} = \frac{1}{(\sqrt{16})^3} = \frac{1}{4^3} = \frac{1}{64}$

39. $(-27)^{-1/3} = \frac{1}{(-27)^{1/3}} = \frac{1}{-3} = -\frac{1}{3}$

41. $(-16)^{-1/4}$ is not a real number because it is a fourth root of a negative number.

43. $3^{1/3}\cdot 3^{1/4} = 3^{\frac{1}{3}+\frac{1}{4}} = 3^{7/12}$

45. $3^{1/3}\cdot 3^{-1/3} = 3^0 = 1$

47. $\frac{8^{1/3}}{8^{2/3}} = 8^{1/3-2/3} = 8^{-1/3} = \frac{1}{8^{1/3}} = \frac{1}{2}$

49. $4^{3/4} \div 4^{1/4} = 4^{3/4-1/4} = 4^{1/2} = 2$

51. $18^{1/2} \cdot 2^{1/2} = 36^{1/2} = 6$
53. $(2^6)^{1/3} = 2^{6/3} = 2^2 = 4$
55. $(3^8)^{1/2} = 3^{8/2} = 3^4 = 81$
57. $(2^{-4})^{1/2} = 2^{-2} = \frac{1}{4}$
59. $\left(\frac{3^4}{2^6}\right)^{1/2} = \frac{3^2}{2^3} = \frac{9}{8}$
61. $(x^4)^{1/4} = |x|$
63. $(a^8)^{1/2} = a^4$
65. $(y^3)^{1/3} = y$
67. $(9x^6y^2)^{1/2} = |3x^3y|$
69. $\left(\frac{81x^{12}}{y^{20}}\right)^{1/4} = \left|\frac{3x^3}{y^5}\right|$

71. $x^{1/2}x^{1/4} = x^{2/4+1/4} = x^{3/4}$
73. $(x^{1/2}y)(x^{-3/4}y^{1/2}) = x^{-1/4}y^{3/2} = \frac{y^{3/2}}{x^{1/4}}$
75. $\frac{w^{1/3}}{w^3} = w^{\frac{1}{3}-3} = w^{-8/3} = \frac{1}{w^{8/3}}$
77. $(144x^{16})^{1/2} = 12x^8$
79. $\left(\frac{a^{-1/2}}{b^{-1/4}}\right)^{-4} = \frac{a^{4/2}}{b^{4/4}} = \frac{a^2}{b}$
81. $\left(\frac{2w^{1/3}}{w^{-3/4}}\right)^3 = \frac{8w}{w^{-9/4}} = 8w^{1-(-9/4)}$
$= 8w^{13/4}$
83. $\frac{9^{1/4}h^{1/2}k^{3/2}}{9^{3/4}h^{1/3}k^2} = \frac{h^{1/6}}{9^{1/2}k^{1/2}} = \frac{h^{1/6}}{3k^{1/2}}$

85. $(9^2)^{1/2} = 9$
87. $-16^{-3/4} = -\frac{1}{2^3} = -\frac{1}{8}$
89. $125^{-4/3} = \frac{1}{5^4} = \frac{1}{625}$
91. $2^{1/2} \cdot 2^{-1/4} = 2^{2/4-1/4} = 2^{1/4}$
93. $3^{0.26}3^{0.74} = 3^{0.26+0.74} = 3^1 = 3$
95. $3^{1/4} \cdot 27^{1/4} = (3 \cdot 27)^{1/4} = 81^{1/4} = 3$
97. $\left(-\frac{8}{27}\right)^{2/3} = \frac{(-8)^{2/3}}{27^{2/3}} = \frac{4}{9}$
99. Not a real number, because the fourth root of $-1/16$ is not real.
101. $\left(\frac{9}{16}\right)^{-1/2} = \left(\frac{16}{9}\right)^{1/2} = \frac{4}{3}$
103. $-\left(\frac{25}{36}\right)^{-3/2} = -\left(\frac{36}{25}\right)^{3/2} = -\frac{216}{125}$
105. $(9x^9)^{1/2} = 9^{1/2}x^{9/2} = 3x^{9/2}$
107. $(3a^{-2/3})^{-3} = 3^{-3}a^2 = \frac{a^2}{27}$
109. $(a^{1/2}b)^{1/2}(ab^{1/2}) = a^{1/4}b^{1/2}a^1b^{1/2}$
$= a^{5/4}b$
111. $(km^{1/2})^3(k^3m^5)^{1/2} = k^3m^{3/2}k^{3/2}m^{5/2}$
$= k^{9/2}m^4$
113. $2^{1/3} \approx 2^{0.33333333} \approx 1.2599$
115. $-2^{1/2} = -(2^{0.5}) \approx -1.4142$
117. $1024^{1/10} = 2$
119. $\left(\frac{64}{15{,}625}\right)^{-1/6} = 2.5$
121. $a^{m/2} \cdot a^{m/4} = a^{m/2+m/4} = a^{3m/4}$
123. $\frac{a^{-m/5}}{a^{-m/3}} = a^{-m/5+m/3} = a^{2m/15}$
125. $(a^{-1/m}b^{-1/n})^{-mn} = a^nb^m$
127. $\left(\frac{a^{-3m}b^{-6n}}{a^{9m}}\right)^{-1/3} = \frac{a^mb^{2n}}{a^{-3m}} = a^{4m}b^{2n}$
129. $D = (12^2 + 4^2 + 3^2)^{1/2}$
$= (144 + 16 + 9)^{1/2}$
$= 169^{1/2} =$ 13 inches
131. $S = (13.0368 + 7.84(18.42)^{1/3} - 0.8(21.45))^2$
$S \approx 274.96 \text{ m}^2$
133. $r = \left(\frac{20{,}130}{10{,}000}\right)^{1/3} - 1 \approx 0.263$
$= 26.26\%$
135. $r = \left(\frac{141{,}600{,}000{,}000}{450{,}000}\right)^{1/213} - 1$
$\approx 0.061 = 6.12\%$
137. The expression $(-1)^{1/2}$ is not a real number. The rules of exponents hold only for expressions that represent real numbers. So we should not be using either rule of exponents to perform these computations. Using complex numbers from Section 7.6 we can compute $(-1)^{1/2} \cdot (-1)^{1/2} = i \cdot i = -1$.

9.3 WARM-UPS

1. False, because $\sqrt{3} + \sqrt{3} = 2\sqrt{3}$.
2. True, because $\sqrt{8} + \sqrt{2} = 2\sqrt{2} + \sqrt{2} = 3\sqrt{2}$.
3. False, because $2\sqrt{3} \cdot 3\sqrt{3} = 6 \cdot 3 = 18$.
4. False, because $\sqrt[3]{2} \cdot \sqrt[3]{2} = \sqrt[3]{4}$.
5. True, because $\sqrt{5} \cdot \sqrt{2} = \sqrt{10}$.
6. False, because $2\sqrt{5} + 3\sqrt{5} = 5\sqrt{5}$.
7. True, because $\sqrt{2}\sqrt{3} = \sqrt{6}$ and $\sqrt{2}\sqrt{2} = 2$.
8. False, because $\sqrt{12} = \sqrt{4}\sqrt{3} = 2\sqrt{3}$.
9. False, because $(\sqrt{2} + \sqrt{3})^2$
$= 2 + 2\sqrt{2}\sqrt{3} + 3 = 5 + 2\sqrt{6}$.
10. True, $(\sqrt{3} - \sqrt{2})(\sqrt{3} + \sqrt{2}) = 3 - 2 = 1$.

9.3 EXERCISES

1. Like radicals are radicals with the same index and same radicand.

3. In the product rule the radicals must have the same index, but do not have to have the same radicand.

5. $\sqrt{3}-2\sqrt{3}=1\sqrt{3}-2\sqrt{3}$
$= -1\sqrt{3}=-\sqrt{3}$

7. $5\sqrt{7x}+4\sqrt{7x}=(5+4)\sqrt{7x}=\ 9\sqrt{7x}$

9. $2\cdot\sqrt[3]{2}+3\cdot\sqrt[3]{2}=(2+3)\sqrt[3]{2}=\ 5\sqrt[3]{2}$

11. $\sqrt{3}-\sqrt{5}+3\sqrt{3}-\sqrt{5}$
$=\sqrt{3}+3\sqrt{3}-\sqrt{5}-\sqrt{5}=4\sqrt{3}-2\sqrt{5}$

13. $\sqrt[3]{2}+\sqrt[3]{x}-\sqrt[3]{2}+4\sqrt[3]{x}$
$=\sqrt[3]{2}-\sqrt[3]{2}+\sqrt[3]{x}+4\sqrt[3]{x}=\ 5\sqrt[3]{x}$

15. $\sqrt[3]{x}-\sqrt{2x}+\sqrt[3]{x}=\sqrt[3]{x}+\sqrt[3]{x}-\sqrt{2x}$
$=\ 2\sqrt[3]{x}-\sqrt{2x}$

17. $\sqrt{8}+\sqrt{28}=\sqrt{4}\sqrt{2}+\sqrt{4}\sqrt{7}$
$=\ 2\sqrt{2}+2\sqrt{7}$

19. $\sqrt{8}+\sqrt{18}=\sqrt{4}\sqrt{2}+\sqrt{9}\sqrt{2}$
$=2\sqrt{2}+3\sqrt{2}=5\sqrt{2}$

21. $2\sqrt{45}-3\sqrt{20}=2\sqrt{9}\sqrt{5}-3\sqrt{4}\sqrt{5}$
$=6\sqrt{5}-6\sqrt{5}=0$

23. $\sqrt{2}-\sqrt{8}=\sqrt{2}-2\sqrt{2}=-\sqrt{2}$

25. $\sqrt{45x^3}-\sqrt{18x^2}+\sqrt{50x^2}-\sqrt{20x^3}$
$=3x\sqrt{5x}-3x\sqrt{2}+5x\sqrt{2}-2x\sqrt{5x}$
$=\ x\sqrt{5x}+2x\sqrt{2}$

27. $2\sqrt[3]{24}+\sqrt[3]{81}=2\sqrt[3]{8}\cdot\sqrt[3]{3}+\sqrt[3]{27}\cdot\sqrt[3]{3}$
$=4\sqrt[3]{3}+3\cdot\sqrt[3]{3}=\ 7\sqrt[3]{3}$

29. $\sqrt[4]{48}-2\sqrt[4]{243}=\sqrt[4]{16\cdot 3}-2\sqrt[4]{81\cdot 3}$
$=2\sqrt[4]{3}-6\sqrt[4]{3}$
$=-4\sqrt[4]{3}$

31. $\sqrt[3]{54t^4y^3}-\sqrt[3]{16t^4y^3}$
$=\sqrt[3]{27t^3y^3}\cdot\sqrt[3]{2t}-\sqrt[3]{8t^3y^3}\cdot\sqrt[3]{2t}$
$=3ty\cdot\sqrt[3]{2t}-2ty\cdot\sqrt[3]{2t}=ty\sqrt[3]{2t}$

33. $\sqrt{3}\sqrt{5}=\sqrt{3\cdot 5}=\sqrt{15}$

35. $(2\sqrt{5})(3\sqrt{10})=6\sqrt{50}=6\sqrt{25}\sqrt{2}$
$=6\cdot 5\sqrt{2}=\ 30\sqrt{2}$

37. $(2\sqrt{7a})(3\sqrt{2a})=6\sqrt{14a^2}=6\sqrt{a^2}\sqrt{14}$
$=6a\sqrt{14}$

39. $(\sqrt[4]{9})(\sqrt[4]{27})\ =\sqrt[4]{243}=\sqrt[4]{81}\cdot\sqrt[4]{3}$
$=3\sqrt[4]{3}$

41. $(2\sqrt{3})^2=4\cdot 3=12$

43. $2\sqrt{3}(\sqrt{6}+3\sqrt{3})=2\sqrt{18}+18$
$=2\sqrt{9}\sqrt{2}+18=6\sqrt{2}+18$

45. $\sqrt{5}(\sqrt{10}-2)=\sqrt{50}-2\sqrt{5}$
$=\sqrt{25}\sqrt{2}-2\sqrt{5}=\ 5\sqrt{2}-2\sqrt{5}$

47. $\sqrt[3]{3t}\left(\sqrt[3]{9t}-\sqrt[3]{t^2}\right)=\sqrt[3]{27t^2}-\sqrt[3]{3t^3}$
$=3\sqrt[3]{t^2}-t\sqrt[3]{3}$

49. $\left(\sqrt{3}+2\right)\left(\sqrt{3}-5\right)$
$=3+2\sqrt{3}-5\sqrt{3}-10=-7-3\sqrt{3}$

51. $(\sqrt{11}-3)(\sqrt{11}+3)=11-9=2$

53. $(2\sqrt{5}-7)(2\sqrt{5}+4)$
$=20-14\sqrt{5}+8\sqrt{5}-28$
$=-8-6\sqrt{5}$

55. $(2\sqrt{3}-\sqrt{6})(\sqrt{3}+2\sqrt{6})$
$=6-\sqrt{18}+4\sqrt{18}-12$
$=-6-3\sqrt{2}+4\cdot 3\sqrt{2}=-6+9\sqrt{2}$

57. $\sqrt[3]{3}\cdot\sqrt{3}=3^{1/3}3^{1/2}=3^{5/6}=\sqrt[6]{3^5}$
$=\sqrt[6]{243}$

59. $\sqrt[3]{5}\cdot\sqrt[4]{5}=5^{1/3}5^{1/4}=5^{7/12}=\sqrt[12]{5^7}$

61. $\sqrt[3]{2}\cdot\sqrt{5}=2^{1/3}5^{1/2}=2^{2/6}5^{3/6}$
$=\sqrt[6]{2^2 5^3}=\sqrt[6]{500}$

63. $\sqrt[3]{2}\cdot\sqrt[4]{3}=2^{1/3}3^{1/4}=2^{4/12}3^{3/12}$
$=\sqrt[12]{2^4 3^3}=\sqrt[12]{432}$

65. $(\sqrt{3}-2)(\sqrt{3}+2)=3-4=\ -1$

67. $(\sqrt{5}+\sqrt{2})(\sqrt{5}-\sqrt{2})=5-2=3$

69. $(2\sqrt{5}+1)(2\sqrt{5}-1)=4\cdot 5-1=\ 19$

71. $(3\sqrt{2}+\sqrt{5})(3\sqrt{2}-\sqrt{5})=9\cdot 2-5$
$=13$

73. $(5-3\sqrt{x})(5+3\sqrt{x})=25-9x$

75. $\sqrt{300}+\sqrt{3}=10\sqrt{3}+\sqrt{3}=\ 11\sqrt{3}$

77. $2\sqrt{5}\cdot 5\sqrt{6}=10\sqrt{30}$

79. $(3+2\sqrt{7})(\sqrt{7}-2)$
$=3\sqrt{7}-6+2\cdot 7-4\sqrt{7}$
$=\ 8-\sqrt{7}$

81. $4\sqrt{w}\cdot 4\sqrt{w}=16(\sqrt{w})^2=16w$

83. $\sqrt{3x^3}\cdot\sqrt{6x^2}=\sqrt{18x^5}=\sqrt{9x^4}\cdot\sqrt{2x}$
$=3x^2\sqrt{2x}$

85. $(2\sqrt{5}+\sqrt{2})(3\sqrt{5}-\sqrt{2})$
$=30+3\sqrt{10}-2\sqrt{10}-2=\ 28+\sqrt{10}$

87. $\dfrac{\sqrt{2}}{3}+\dfrac{\sqrt{2}}{5}=\dfrac{5\sqrt{2}}{5\cdot 3}+\dfrac{3\sqrt{2}}{3\cdot 5}=\dfrac{8\sqrt{2}}{15}$

89. $(5+2\sqrt{2})(5-2\sqrt{2})=25-4\cdot 2=17$

91. $(3+\sqrt{x})^2=9+2\cdot 3\sqrt{x}+x$
$=9+6\sqrt{x}+x$

93. $(5\sqrt{x}-3)^2=25x-2\cdot 3\cdot 5\sqrt{x}+9$
$=25x-30\sqrt{x}+9$

95. $(1+\sqrt{x+2})^2=1+2\sqrt{x+2}+x+2$
$=\ x+3+2\sqrt{x+\ 2}$

97. $\sqrt{4w} - \sqrt{9w} = 2\sqrt{w} - 3\sqrt{w}$
$= -\sqrt{w}$

99. $2\sqrt{a^3} + 3\sqrt{a^3} - 2a\sqrt{4a}$
$= 2a\sqrt{a} + 3a\sqrt{a} - 4a\sqrt{a} = a\sqrt{a}$

101. $\sqrt{x^5} + 2x\sqrt{x^3}$
$= \sqrt{x^4}\sqrt{x} + 2x\sqrt{x^2}\sqrt{x}$
$= x^2\sqrt{x} + 2x^2\sqrt{x} = 3x^2\sqrt{x}$

103. $\sqrt[3]{-16x^4} + 5x\sqrt[3]{54x}$
$= -2x\sqrt[3]{2x} + 5x \cdot 3 \cdot \sqrt[3]{2x} = 13x\sqrt[3]{2x}$

105. $\sqrt[3]{2x} \cdot \sqrt{2x} = (2x)^{1/3}(2x)^{1/2} = (2x)^{5/6}$
$= \sqrt[6]{32x^5}$

107. $A = LW = \sqrt{6} \cdot \sqrt{3} = \sqrt{18}$
$= 3\sqrt{2}$ ft^2

109. $A = \frac{1}{2}h(b_1 + b_2)$
$= \frac{1}{2}\sqrt{6}\left(\sqrt{3} + \sqrt{12}\right)$
$= \frac{1}{2}\sqrt{6}\left(\sqrt{3} + 2\sqrt{3}\right)$
$= \frac{1}{2}\sqrt{6} \cdot 3\sqrt{3} = \frac{3\sqrt{18}}{2}$
$= \frac{9\sqrt{2}}{2}$ ft^2

111. No because $\sqrt{9} + \sqrt{16} = 3 + 4 = 7$ and $\sqrt{9 + 16} = \sqrt{25} = 5$.

113. a) $y^2 - 3 = \left(y + \sqrt{3}\right)\left(y - \sqrt{3}\right)$
$2a^2 - 7 = \left(\sqrt{2}a + \sqrt{7}\right)\left(\sqrt{2}a - \sqrt{7}\right)$

b) $x^2 - 8 = 0$
$\left(x + \sqrt{8}\right)\left(x - \sqrt{8}\right) = 0$
$x + \sqrt{8} = 0$ or $x - \sqrt{8} = 0$
$x = -\sqrt{8}$ or $x = \sqrt{8}$
$x = -2\sqrt{2}$ or $x = 2\sqrt{2}$
The solution set is $\left\{\pm 2\sqrt{2}\right\}$.

c) $x^2 - a = 0$
$(x + \sqrt{a})(x - \sqrt{a}) = 0$
$x + \sqrt{a} = 0$ or $x - \sqrt{a} = 0$
$x = -\sqrt{a}$ or $x = \sqrt{a}$
The solution set is $\{\pm\sqrt{a}\}$.

9.4 WARM-UPS

1. True, because $\sqrt{3}\sqrt{2} = \sqrt{6}$.

2. True, because
$\frac{2}{\sqrt{2}} = \frac{2\sqrt{2}}{\sqrt{2}\sqrt{2}} = \frac{2\sqrt{2}}{2} = \sqrt{2}$.

3. False, because $\frac{4 - \sqrt{10}}{2} = 2 - \frac{\sqrt{10}}{2}$.

4. True, because $\frac{1}{\sqrt{3}} = \frac{1 \cdot \sqrt{3}}{\sqrt{3}\sqrt{3}} = \frac{\sqrt{3}}{3}$.

5. False, because $\frac{8\sqrt{7}}{2\sqrt{7}} = 4$.

6. True, because
$(2 - \sqrt{3})(2 + \sqrt{3}) = 4 - 3 = 1$.

7. False, because $\frac{\sqrt{12}}{3} = \frac{2\sqrt{3}}{3}$.

8. True, because
$\frac{\sqrt{20}}{\sqrt{5}} = \frac{\sqrt{4}\sqrt{5}}{\sqrt{5}} = \sqrt{4} = 2$.

9. True, because $(2\sqrt{4})^2 = 4 \cdot 4 = 16$.

10. True, because
$(3\sqrt{5})^3 = 3^3 \cdot \sqrt{5^3} = 27\sqrt{125}$.

9.4 EXERCISES

1. $\frac{2}{\sqrt{5}} = \frac{2\sqrt{5}}{\sqrt{5}\sqrt{5}} = \frac{2\sqrt{5}}{5}$

3. $\frac{\sqrt{3}}{\sqrt{7}} = \frac{\sqrt{3}\sqrt{7}}{\sqrt{7}\sqrt{7}} = \frac{\sqrt{21}}{\sqrt{49}} = \frac{\sqrt{21}}{7}$

5. $\frac{1}{\sqrt[3]{4}} = \frac{1 \cdot \sqrt[3]{2}}{\sqrt[3]{4} \cdot \sqrt[3]{2}} = \frac{\sqrt[3]{2}}{\sqrt[3]{8}} = \frac{\sqrt[3]{2}}{2}$

7. $\frac{\sqrt[3]{6}}{\sqrt[3]{5}} = \frac{\sqrt[3]{6} \cdot \sqrt[3]{25}}{\sqrt[3]{5} \cdot \sqrt[3]{25}} = \frac{\sqrt[3]{150}}{\sqrt[3]{125}} = \frac{\sqrt[3]{150}}{5}$

9. $\frac{\sqrt{5}}{\sqrt{12}} = \frac{\sqrt{5}\sqrt{3}}{\sqrt{12}\sqrt{3}} = \frac{\sqrt{15}}{\sqrt{36}} = \frac{\sqrt{15}}{6}$

11. $\frac{\sqrt{3}}{\sqrt{12}} = \frac{\sqrt{3}}{\sqrt{4}\sqrt{3}} = \frac{1}{\sqrt{4}} = \frac{1}{2}$

13. $\sqrt{\frac{1}{2}} = \frac{1}{\sqrt{2}} = \frac{1 \cdot \sqrt{2}}{\sqrt{2}\sqrt{2}} = \frac{\sqrt{2}}{2}$

15. $\sqrt[3]{\frac{2}{3}} = \frac{\sqrt[3]{2} \cdot \sqrt[3]{9}}{\sqrt[3]{3} \cdot \sqrt[3]{9}} = \frac{\sqrt[3]{18}}{3}$

17. $\sqrt[3]{\frac{7}{4}} = \frac{\sqrt[3]{7} \cdot \sqrt[3]{2}}{\sqrt[3]{4} \cdot \sqrt[3]{2}} = \frac{\sqrt[3]{14}}{2}$

19. $\sqrt{\frac{x}{y}} = \frac{\sqrt{x}\sqrt{y}}{\sqrt{y}\sqrt{y}} = \frac{\sqrt{xy}}{y}$

21. $\frac{\sqrt{a^3}}{\sqrt{b^7}} = \frac{\sqrt{a^2}\sqrt{a}}{\sqrt{b^6}\sqrt{b}} = \frac{a\sqrt{a}\sqrt{b}}{b^3\sqrt{b}\sqrt{b}} = \frac{a\sqrt{ab}}{b^4}$

23. $\sqrt{\frac{a}{3b}} = \frac{\sqrt{a}\sqrt{3b}}{\sqrt{3b}\sqrt{3b}} = \frac{\sqrt{3ab}}{3b}$

25. $\sqrt[3]{\frac{a}{b}} = \frac{\sqrt[3]{a} \cdot \sqrt[3]{b^2}}{\sqrt[3]{b} \cdot \sqrt[3]{b^2}} = \frac{\sqrt[3]{ab^2}}{b}$

27. $\sqrt[3]{\frac{5}{2b^2}} = \frac{\sqrt[3]{5}\cdot\sqrt[3]{4b}}{\sqrt[3]{2b^2}\cdot\sqrt[3]{4b}} = \frac{\sqrt[3]{20b}}{2b}$

29. $\sqrt{15} \div \sqrt{5} = \sqrt{\frac{15}{5}} = \sqrt{3}$

31. $\sqrt{3} \div \sqrt{5} = \frac{\sqrt{3}}{\sqrt{5}} = \frac{\sqrt{3}\sqrt{5}}{\sqrt{5}\sqrt{5}} = \frac{\sqrt{15}}{5}$

33. $(3\sqrt{3}) \div (5\sqrt{6}) = \frac{3\sqrt{3}}{5\sqrt{6}} = \frac{3\sqrt{3}}{5\sqrt{2}\sqrt{3}}$

$= \frac{3}{5\sqrt{2}} = \frac{3\sqrt{2}}{5\sqrt{2}\sqrt{2}} = \frac{3\sqrt{2}}{10}$

35. $(2\sqrt{3}) \div (3\sqrt{6}) = \frac{2\sqrt{3}}{3\sqrt{6}} = \frac{2}{3\sqrt{2}}$

$= \frac{2\sqrt{2}}{3\sqrt{2}\sqrt{2}} = \frac{2\sqrt{2}}{3\cdot 2} = \frac{\sqrt{2}}{3}$

37. $(\sqrt{24a^2}) \div (\sqrt{72a}) = \frac{\sqrt{24a^2}}{\sqrt{72a}} = \frac{2a\sqrt{6}}{6\sqrt{2a}}$

$= \frac{a\sqrt{3}}{3\sqrt{a}} = \frac{a\sqrt{3}\sqrt{a}}{3\sqrt{a}\sqrt{a}} = = \frac{a\sqrt{3a}}{3a} = \frac{\sqrt{3a}}{3}$

39. $\sqrt[3]{20} \div \sqrt[3]{2} = \frac{\sqrt[3]{20}}{\sqrt[3]{2}} = \sqrt[3]{\frac{20}{2}} = \sqrt[3]{10}$

41. $\sqrt[4]{48} \div \sqrt[4]{3} = \frac{\sqrt[4]{48}}{\sqrt[4]{3}} = \frac{\sqrt[4]{16}\cdot\sqrt[4]{3}}{\sqrt[4]{3}} = 2$

43. $\sqrt[4]{16w} \div \sqrt[4]{w^5} = \frac{\sqrt[4]{16w}}{\sqrt[4]{w^5}} = \frac{2\sqrt[4]{w}}{w\sqrt[4]{w}}$

$= \frac{2}{w}$

45. $\frac{6+\sqrt{45}}{3} = \frac{6+3\sqrt{5}}{3} = 2+\sqrt{5}$

47. $\frac{-2+\sqrt{12}}{-2} = \frac{-2+2\sqrt{3}}{-2} = 1-\sqrt{3}$

49. $\frac{4}{2+\sqrt{8}} = \frac{4(2-\sqrt{8})}{(2+\sqrt{8})(2-\sqrt{8})}$

$= \frac{8-4\sqrt{8}}{4-8} = \frac{8-8\sqrt{2}}{-4} = -2+2\sqrt{2}$

$= 2\sqrt{2}-2$

51. $\frac{3}{\sqrt{11}-\sqrt{5}}$

$= \frac{3(\sqrt{11}+\sqrt{5})}{(\sqrt{11}-\sqrt{5})(\sqrt{11}+\sqrt{5})}$

$= \frac{3\sqrt{11}+3\sqrt{5}}{11-5} = \frac{3\sqrt{11}+3\sqrt{5}}{6}$

$= \frac{\sqrt{11}+\sqrt{5}}{2}$

53. $\frac{1+\sqrt{2}}{\sqrt{3}-1} = \frac{(1+\sqrt{2})(\sqrt{3}+1)}{(\sqrt{3}-1)(\sqrt{3}+1)}$

$= \frac{1+\sqrt{6}+\sqrt{2}+\sqrt{3}}{2}$

55. $\frac{\sqrt{2}}{\sqrt{6}+\sqrt{3}} = \frac{\sqrt{2}(\sqrt{6}-\sqrt{3})}{(\sqrt{6}+\sqrt{3})(\sqrt{6}-\sqrt{3})}$

$= \frac{\sqrt{12}-\sqrt{6}}{6-3} = \frac{2\sqrt{3}-\sqrt{6}}{3}$

57. $\frac{2\sqrt{3}}{3\sqrt{2}-\sqrt{5}}$

$= \frac{2\sqrt{3}(3\sqrt{2}+\sqrt{5})}{(3\sqrt{2}-\sqrt{5})(3\sqrt{2}+\sqrt{5})}$

$= \frac{6\sqrt{6}+2\sqrt{15}}{18-5} = \frac{6\sqrt{6}+2\sqrt{15}}{13}$

59. $(2\sqrt{2})^5 = 2^5\cdot\sqrt{2^5} = 32\sqrt{16}\sqrt{2}$

$= 128\sqrt{2}$

61. $(\sqrt{x})^5 = \sqrt{x^5} = \sqrt{x^4}\sqrt{x} = x^2\sqrt{x}$

63. $(-3\sqrt{x^3})^3 = -27\sqrt{x^9} = -27\sqrt{x^8}\sqrt{x}$

$= -27x^4\sqrt{x}$

65. $(2x\sqrt[3]{x^2})^3 = 8x^3\sqrt[3]{x^6} = 8x^3\cdot x^2 = 8x^5$

67. $(-2\sqrt[3]{5})^2 = (-2)^2\cdot\sqrt[3]{5^2} = 4\sqrt[3]{25}$

69. $(\sqrt[3]{x^2})^6 = (x^2)^{6/3} = (x^2)^2 = x^4$

71. $\frac{\sqrt{3}}{\sqrt{2}} + \frac{2}{\sqrt{2}} = \frac{\sqrt{3}\sqrt{2}}{\sqrt{2}\sqrt{2}} + \frac{2\sqrt{2}}{\sqrt{2}\sqrt{2}}$

$= \frac{\sqrt{6}}{2} + \frac{2\sqrt{2}}{2} = \frac{\sqrt{6}+2\sqrt{2}}{2}$

73. $\frac{\sqrt{3}}{\sqrt{2}} + \frac{3\sqrt{6}}{2} = \frac{\sqrt{3}\sqrt{2}}{\sqrt{2}\sqrt{2}} + \frac{3\sqrt{6}}{2}$

$= \frac{\sqrt{6}}{2} + \frac{3\sqrt{6}}{2} = \frac{4\sqrt{6}}{2} = 2\sqrt{6}$

75. $\frac{\sqrt{6}}{2}\cdot\frac{1}{\sqrt{3}} = \frac{\sqrt{2}\sqrt{3}}{2\sqrt{3}} = \frac{\sqrt{2}}{2}$

77. $(2\sqrt{w}) \div (3\sqrt{w}) = \frac{2\sqrt{w}}{3\sqrt{w}} = \frac{2}{3}$

79. $\frac{8-\sqrt{32}}{20} = \frac{8-4\sqrt{2}}{20} = \frac{4(2-\sqrt{2})}{4\cdot 5}$

$= \frac{2-\sqrt{2}}{5}$

81. $\frac{5+\sqrt{75}}{10} = \frac{5+5\sqrt{3}}{10} = \frac{5(1+\sqrt{3})}{5\cdot 2}$

$= \frac{1+\sqrt{3}}{2}$

83. $\sqrt{a}(\sqrt{a}-3) = a-3\sqrt{a}$

85. $4\sqrt{a}(a+\sqrt{a}) = 4a\sqrt{a}+4a$

87. $(2\sqrt{3m})^2 = 4\cdot 3m = 12m$

89. $\left(-2\sqrt{xy^2z}\right)^2 = (-2)^2xy^2z = 4xy^2z$

91. $\sqrt[3]{m}\left(\sqrt[3]{m^2} - \sqrt[3]{m^5}\right) = \sqrt[3]{m^3} - \sqrt[3]{m^6}$
$= m - m^2$

93. $\sqrt[3]{8x^4} + \sqrt[3]{27x^4}$
$= \sqrt[3]{8x^3} \cdot \sqrt[3]{x} + \sqrt[3]{27x^3} \cdot \sqrt[3]{x}$
$= 2x \cdot \sqrt[3]{x} + 3x \cdot \sqrt[3]{x} = 5x\sqrt[3]{x}$

95. $\left(2m\sqrt[4]{2m^2}\right)^3 = 8m^3\sqrt[4]{8m^6}$
$= 8m^3\sqrt[4]{m^4}\sqrt[4]{8m^2}$
$= 8m^3 \cdot m \cdot \sqrt[4]{8m^2} = 8m^4\sqrt[4]{8m^2}$

97. $\dfrac{x-9}{\sqrt{x}-3} = \dfrac{(x-9)(\sqrt{x}+3)}{(\sqrt{x}-3)(\sqrt{x}+3)}$
$= \dfrac{(x-9)(\sqrt{x}+3)}{x-9} = \sqrt{x}+3$

99. $\dfrac{3\sqrt{k}}{\sqrt{k}+\sqrt{7}} = \dfrac{3\sqrt{k}(\sqrt{k}-\sqrt{7})}{(\sqrt{k}+\sqrt{7})(\sqrt{k}-\sqrt{7})}$
$= \dfrac{3k - 3\sqrt{7k}}{k-7}$

101. $\dfrac{5}{\sqrt{2}-1} + \dfrac{3}{\sqrt{2}+1}$
$= \dfrac{5(\sqrt{2}+1)}{(\sqrt{2}-1)(\sqrt{2}+1)} + \dfrac{3(\sqrt{2}-1)}{(\sqrt{2}+1)(\sqrt{2}-1)}$
$= \dfrac{5\sqrt{2}+5}{2-1} + \dfrac{3\sqrt{2}-3}{2-1} = 2 + 8\sqrt{2}$

103. $\dfrac{1}{\sqrt{2}} + \dfrac{1}{\sqrt{3}} = \dfrac{\sqrt{2}}{2} + \dfrac{\sqrt{3}}{3}$
$= \dfrac{3\sqrt{2}}{3 \cdot 2} + \dfrac{2\sqrt{3}}{2 \cdot 3} = \dfrac{3\sqrt{2}+2\sqrt{3}}{6}$

105. $\dfrac{3}{\sqrt{2}-1} + \dfrac{4}{\sqrt{2}+1}$
$= \dfrac{3(\sqrt{2}+1)}{(\sqrt{2}-1)(\sqrt{2}+1)} + \dfrac{4(\sqrt{2}-1)}{(\sqrt{2}+1)(\sqrt{2}-1)}$
$= \dfrac{3\sqrt{2}+3}{1} + \dfrac{4\sqrt{2}-4}{1} = 7\sqrt{2}-1$

107. $\dfrac{\sqrt{x}}{\sqrt{x}+2} + \dfrac{3\sqrt{x}}{\sqrt{x}-2}$
$= \dfrac{\sqrt{x}(\sqrt{x}-2)}{(\sqrt{x}+2)(\sqrt{x}-2)}$
$+ \dfrac{3\sqrt{x}(\sqrt{x}+2)}{(\sqrt{x}-2)(\sqrt{x}+2)}$
$= \dfrac{x-2\sqrt{x}}{x-4} + \dfrac{3x+6\sqrt{x}}{x-4} = \dfrac{4x+4\sqrt{x}}{x-4}$

109. $\dfrac{1}{\sqrt{x}} + \dfrac{1}{1-\sqrt{x}}$
$= \dfrac{1(1-\sqrt{x})}{\sqrt{x}(1-\sqrt{x})} + \dfrac{1\sqrt{x}}{(1-\sqrt{x})\sqrt{x}}$
$= \dfrac{1-\sqrt{x}}{\sqrt{x}-x} + \dfrac{\sqrt{x}}{\sqrt{x}-x} = \dfrac{1}{\sqrt{x}-x}$
$= \dfrac{1(\sqrt{x}+x)}{(\sqrt{x}-x)(\sqrt{x}+x)} = \dfrac{x+\sqrt{x}}{x-x^2}$
$= \dfrac{x+\sqrt{x}}{x(1-x)}$

111. a) Use $(a-b)(a^2+ab+b^2) = a^3 - b^3$ with $a = x$ and $b = \sqrt[3]{2}$ to get $x^3 - 2$.
b) Use $a^3 + b^3 = (a+b)(a^2 - ab + b^2)$ to get
$x^3 + 5 = x^3 + \left(\sqrt[3]{5}\right)^3$
$= (x + \sqrt[3]{5})(x^2 - \sqrt[3]{5}\,x + \sqrt[3]{25})$
c) Use $(a-b)(a^2+ab+b^2) = a^3 - b^3$ with $a = \sqrt[3]{5}$ and $b = \sqrt[3]{2}$ to get
$\left(\sqrt[3]{5} - \sqrt[3]{2}\right)\left(\sqrt[3]{25} + \sqrt[3]{10} + \sqrt[3]{4}\right)$
$= 5 - 2 = 3$
d)
$a + b = \left(\sqrt[3]{a} + \sqrt[3]{b}\right)\left(\sqrt[3]{a^2} - \sqrt[3]{ab} + \sqrt[3]{b^2}\right)$
$a - b = \left(\sqrt[3]{a} - \sqrt[3]{b}\right)\left(\sqrt[3]{a^2} + \sqrt[3]{ab} + \sqrt[3]{b^2}\right)$

9.5 WARM-UPS

1. False, because $x^2 = 4$ is equivalent to the compound equation $x = 2$ or $x = -2$.
2. True, because the square of any real number is nonnegative.
3. False, 0 is a solution.
4. False, because -2 is not a solution to $x^3 = 8$.
5. True, because if x is positive the left side of the equation is negative and the right side is positive.
6. False, we should square each side.
7. False, extraneous roots are found but they do not satisfy the equation.
8. True, because we get $x = 49$, and 49 does not satisfy $\sqrt{x} = -7$.
9. True, because both square roots of 6 satisfy $x^2 - 6 = 0$.
10. True, we get extraneous roots only by raising each side to an even power.

9.5 EXERCISES

1. The odd-root property says that if n is an odd positive integer then $x^n = k$ is equivalent to $x = \sqrt[n]{k}$ for any real number k.

3. An extraneous solution is a solution that appears when solving an equation, but does not satisfy the original equation.

5. $x^3 = -1000$

$x = \sqrt[3]{-1000} = -10$

Solution set: $\{-10\}$

7. $32m^5 - 1 = 0$

$32m^5 = 1$

$m^5 = \frac{1}{32}$

$m = \sqrt[5]{\frac{1}{32}} = \frac{1}{2}$

Solution set: $\left\{\frac{1}{2}\right\}$

9. $(y-3)^3 = -8$

$y - 3 = \sqrt[3]{-8}$

$y - 3 = -2$

$y = 1$

Solution set: $\{1\}$

11. $\frac{1}{2}x^3 + 4 = 0$

$\frac{1}{2}x^3 = -4$

$x^3 = -8$

$x = \sqrt[3]{-8} = -2$

Solution set: $\{-2\}$

13. $x^2 = 25$

$x = \pm 5$

Solution set: $\{-5, 5\}$

15. $x^2 - 20 = 0$

$x^2 = 20$

$x = \pm\sqrt{20} = \pm 2\sqrt{5}$

Solution set: $\{-2\sqrt{5}, 2\sqrt{5}\}$

17. $x^2 = -9$

Since an even root of a negative number is not a real number, there is no real solution.

19. $(x-3)^2 = 16$

$x - 3 = \pm 4$

$x = 3 \pm 4$

$x = 3 + 4$ or $x = 3 - 4$

$x = 7$ or $x = -1$

Solution set: $\{-1, 7\}$

21. $(x+1)^2 - 8 = 0$

$(x+1)^2 = 8$

$x + 1 = \pm\sqrt{8}$

$x = -1 \pm 2\sqrt{2}$

Solution set: $\left\{-1 - 2\sqrt{2}, -1 + 2\sqrt{2}\right\}$

23. $\frac{1}{2}x^2 = 5$

$x^2 = 10$

$x = \pm\sqrt{10}$

Solution set: $\left\{-\sqrt{10}, \sqrt{10}\right\}$

25. $(y-3)^4 = 0$

$y - 3 = \sqrt[4]{0} = 0$

$y = 3$

Solution set: $\{3\}$

27. $2x^6 = 128$

$x^6 = 64$

$x = \pm\sqrt[6]{64}$

$x = \pm 2$

Solution set: $\{-2, 2\}$

29. $\sqrt{x-3} - 3 = 4$

$\sqrt{x-3} = 7$

$(\sqrt{x-3})^2 = 7^2$

$x - 3 = 49$

$x = 52$

Check 52 in the original equation.

Solution set: $\{52\}$

31. $2\sqrt{w+4} = 5$

$(2\sqrt{w+4})^2 = 5^2$

$4(w+4) = 25$

$4w + 16 = 25$

$4w = 9$

$w = \frac{9}{4}$

Check $\frac{9}{4}$ in the original equation.

Solution set: $\left\{\frac{9}{4}\right\}$

33. $\sqrt[3]{2x+3} = \sqrt[3]{x+12}$

$\left(\sqrt[3]{2x+3}\right)^3 = \left(\sqrt[3]{x+12}\right)^3$

$2x + 3 = x + 12$

$x = 9$

Check 9 in the original equation.

Solution set: $\{9\}$

35. $\sqrt{2t-4} = \sqrt{t-1}$

$(\sqrt{2t-4})^2 = (\sqrt{t-1})^2$

$2t - 4 = t - 1$

$t = 3$

Check: $\sqrt{2(3) - 4} = \sqrt{3 - 1}$

Since each side of the equation is $\sqrt{2}$, the solution set is $\{3\}$.

37. $\sqrt{4x^2+x-3}=2x$

$\left(\sqrt{4x^2+x-3}\right)^2=(2x)^2$

$4x^2+x-3=4x^2$

$x-3=0$

$x=3$

Check 3 in the original equation.
Solution set: $\{3\}$

39. $\sqrt{x^2+2x-6}=3$

$\left(\sqrt{x^2+2x-6}\right)^2=3^2$

$x^2+2x-6=9$

$x^2+2x-15=0$

$(x+5)(x-3)=0$

$x+5=0$ or $x-3=0$

$x=-5$ or $x=3$

Check 3 and -5 in the original equation.
Solution set: $\{-5, 3\}$

41. $\sqrt{2x^2-1}=x$

$\left(\sqrt{2x^2-1}\right)^2=x^2$

$2x^2-1=x^2$

$x^2=1$

$x=\pm 1$

Checking in the original we find that if $x=-1$ we get $\sqrt{1}=-1$, which is incorrect. So the solution set is $\{1\}$.

43. $\sqrt{2x^2+5x+6}=x$

$\left(\sqrt{2x^2+5x+6}\right)^2=x^2$

$2x^2+5x+6=x^2$

$x^2+5x+6=0$

$(x+2)(x+3)=0$

$x=-2$ or $x=-3$

If we use $x=-2$ in the original equation we get $\sqrt{4}=-2$, and if $x=-3$ in the original we get $\sqrt{9}=-3$. Since both of these equations are incorrect, the solution set is $\emptyset$.

45. $\sqrt{x}+\sqrt{x-3}=3$

$\sqrt{x}=3-\sqrt{x-3}$

$\left(\sqrt{x}\right)^2=\left(3-\sqrt{x-3}\right)^2$

$x=9-6\sqrt{x-3}+x-3$

$-6=-6\sqrt{x-3}$

$1=\sqrt{x-3}$

$1^2=\left(\sqrt{x-3}\right)^2$

$1=x-3$

$4=x$

Check 4 in the original equation. The solution set is $\{4\}$.

47. $\sqrt{x+2}+\sqrt{x-1}=3$

$\sqrt{x+2}=3-\sqrt{x-1}$

$\left(\sqrt{x+2}\right)^2=\left(3-\sqrt{x-1}\right)^2$

$x+2=9-6\sqrt{x-1}+x-1$

$-6=-6\sqrt{x-1}$

$1=\sqrt{x-1}$

$1^2=\left(\sqrt{x-1}\right)^2$

$1=x-1$

$2=x$

Check 2 in the original equation. The solution set is $\{2\}$.

49. $\sqrt{x+3}-\sqrt{x-2}=1$

$\sqrt{x+3}=1+\sqrt{x-2}$

$\left(\sqrt{x+3}\right)^2=\left(1+\sqrt{x-2}\right)^2$

$x+3=1+2\sqrt{x-2}+x-2$

$4=2\sqrt{x-2}$

$2=\sqrt{x-2}$

$2^2=\left(\sqrt{x-2}\right)^2$

$4=x-2$

$6=x$

Check 6 in the original equation. The solution set is $\{6\}$.

51. $\sqrt{2x+2}-\sqrt{x-3}=2$

$\sqrt{2x+2}=2+\sqrt{x-3}$

$\left(\sqrt{2x+2}\right)^2=\left(2+\sqrt{x-3}\right)^2$

$2x+2=4+4\sqrt{x-3}+x-3$

$x+1=4\sqrt{x-3}$

$(x+1)^2=\left(4\sqrt{x-3}\right)^2$

$x^2+2x+1=16(x-3)$

$x^2-14x+49=0$

$(x-7)^2=0$

$x=7$

Check 7 in the original equation.
Solution set: $\{7\}$

53. $\sqrt{4-x}-\sqrt{x+6}=2$

$\sqrt{4-x}=2+\sqrt{x+6}$

$\left(\sqrt{4-x}\right)^2=\left(2+\sqrt{x+6}\right)^2$

$4-x=4+4\sqrt{x+6}+x+6$

$-6-2x=4\sqrt{x+6}$

$$-3 - x = 2\sqrt{x+6}$$
$$(-3-x)^2 = \left(2\sqrt{x+6}\right)^2$$
$$9 + 6x + x^2 = 4(x+6)$$
$$x^2 + 2x - 15 = 0$$
$$(x+5)(x-3) = 0$$
$$x = -5 \quad \text{or} \quad x = 3$$

If $x = 3$ in the original equation we get $1 - 3 = 2$, which is incorrect. If $x = -5$ we get $3 - 1 = 2$. So the solution set is $\{-5\}$.

55.
$$\sqrt{x-5} - \sqrt{x} = 3$$
$$\sqrt{x-5} = 3 + \sqrt{x}$$
$$\left(\sqrt{x-5}\right)^2 = (3+\sqrt{x})^2$$
$$x - 5 = 9 + 6\sqrt{x} + x$$
$$-14 = 6\sqrt{x}$$
$$-\frac{7}{3} = \sqrt{x}$$
$$x = \left(-\frac{7}{3}\right)^2 = \frac{49}{9}$$

Since $49/9$ does not satisfy the original equation, the solution set is $\emptyset$.

57. $\sqrt{3x+1} + \sqrt{2x+4} = 3$
$$\sqrt{3x+1} = 3 - \sqrt{2x+4}$$
$$\left(\sqrt{3x+1}\right)^2 = (3 - \sqrt{2x+4})^2$$
$$3x + 1 = 9 - 6\sqrt{2x+4} + 2x + 4$$
$$x - 12 = -6\sqrt{2x+4}$$
$$(x-12)^2 = (-6\sqrt{2x+4})^2$$
$$x^2 - 24x + 144 = 36(2x+4)$$
$$x^2 - 96x = 0$$
$$(x)(x-96) = 0$$
$$x = 0 \quad \text{or} \quad x - 96 = 0$$
$$x = 6 \quad \text{or} \quad x = 96$$

Since 96 does not satisfy the original equation, the solution set is $\{0\}$.

59.
$$x^{2/3} = 3$$
$$\left(x^{2/3}\right)^3 = 3^3$$
$$x^2 = 27$$
$$x = \pm\sqrt{27} = \pm 3\sqrt{3}$$

Solution set: $\left\{-3\sqrt{3}, 3\sqrt{3}\right\}$

61.
$$y^{-2/3} = 9$$
$$\left(y^{-2/3}\right)^{-3} = (9)^{-3}$$
$$y^2 = \frac{1}{729}$$
$$y = \pm\sqrt{\frac{1}{729}} = \pm\frac{1}{27}$$

Solution set: $\left\{-\frac{1}{27}, \frac{1}{27}\right\}$

63.
$$w^{1/3} = 8$$
$$\left(w^{1/3}\right)^3 = 8^3$$
$$w = 512$$

Solution set: $\{512\}$

65.
$$t^{-1/2} = 9$$
$$\left(t^{-1/2}\right)^{-2} = (9)^{-2}$$
$$t = \frac{1}{81}$$

Solution set: $\left\{\frac{1}{81}\right\}$

67.
$$\left((3a-1)^{-2/5}\right)^{-5} = 1^{-5}$$
$$(3a-1)^2 = 1$$
$$3a - 1 = \pm 1$$
$$3a - 1 = 1 \quad \text{or} \quad 3a - 1 = -1$$
$$3a = 2 \quad \text{or} \quad 3a = 0$$
$$a = \frac{2}{3} \quad \text{or} \quad a = 0$$

Solution set: $\left\{0, \frac{2}{3}\right\}$

69. $(t-1)^{-2/3} = 2$
$$\left((t-1)^{-2/3}\right)^{-3} = 2^{-3}$$
$$(t-1)^2 = \frac{1}{8}$$
$$t - 1 = \pm\sqrt{\frac{1}{8}} = \pm\frac{\sqrt{2}}{4}$$
$$t = 1 \pm \frac{\sqrt{2}}{4} = \frac{4}{4} \pm \frac{\sqrt{2}}{4} = \frac{4 \pm \sqrt{2}}{4}$$

Solution set: $\left\{\frac{4-\sqrt{2}}{4}, \frac{4+\sqrt{2}}{4}\right\}$

71.
$$(x-3)^{2/3} = -4$$
$$\left((x-3)^{2/3}\right)^3 = (-4)^3$$
$$(x-3)^2 = -64$$

Because the square of any real number is nonnegative, the solution set is $\emptyset$. There is no real solution.

73.
$$2x^2 + 3 = 7$$
$$2x^2 = 4$$
$$x^2 = 2$$
$$x = \pm\sqrt{2}$$

Solution set: $\left\{-\sqrt{2}, \sqrt{2}\right\}$

75.
$$\sqrt[3]{2w+3} = \sqrt[3]{w-2}$$
$$\left(\sqrt[3]{2w+3}\right)^3 = \left(\sqrt[3]{w-2}\right)^3$$
$$2w + 3 = w - 2$$
$$w = -5$$

Solution set: $\{-5\}$

77. $(w+1)^{2/3} = -3$

$$\left((w+1)^{2/3}\right)^3 = (-3)^3$$

$$(w+1)^2 = -27$$

This equation has no real solution by the even root property. The solution set is $\emptyset$.

79. $(a+1)^{1/3} = -2$

$$\left((a+1)^{1/3}\right)^3 = (-2)^3$$

$$a+1 = -8$$

$$a = -9$$

Solution set: $\{-9\}$

81. $(4y-5)^7 = 0$

$$4y-5 = \sqrt[7]{0} = 0$$

$$4y = 5$$

$$y = \frac{5}{4}$$

Solution set: $\left\{\frac{5}{4}\right\}$

83. $\sqrt{5x^2+4x+1} - x = 0$

$$\sqrt{5x^2+4x+1} = x$$

$$5x^2+4x+1 = x^2$$

$$4x^2+4x+1 = 0$$

$$(2x+1)^2 = 0$$

$$2x+1 = 0$$

$$x = -\frac{1}{2}$$

Since $-1/2$ does not satisfy the original equation, the solution set is $\emptyset$.

85. $\sqrt{4x^2} = x+2$

$$\left(\sqrt{4x^2}\right)^2 = (x+2)^2$$

$$4x^2 = x^2+4x+4$$

$$3x^2-4x-4 = 0$$

$$(3x+2)(x-2) = 0$$

$$3x+2 = 0 \quad \text{or} \quad x-2 = 0$$

$$x = -\frac{2}{3} \quad \text{or} \quad x = 2$$

Solution set: $\left\{-\frac{2}{3}, 2\right\}$

87. $(t+2)^4 = 32$

$$t+2 = \pm\sqrt[4]{32} = \pm 2\cdot\sqrt[4]{2}$$

$$t = -2 \pm 2\cdot\sqrt[4]{2}$$

Solution set: $\left\{-2-2\sqrt[4]{2}, -2+2\sqrt[4]{2}\right\}$

89. $\sqrt{x^2-3x} = x$

$$\left(\sqrt{x^2-3x}\right)^2 = x^2$$

$$x^2-3x = x^2$$

$$-3x = 0$$

$$x = 0$$

Solution set: $\{0\}$

91. $x^{-3} = 8$

$$\left(x^{-3}\right)^{-1} = 8^{-1}$$

$$x^3 = \frac{1}{8}$$

$$x = \sqrt[3]{\frac{1}{8}} = \frac{1}{2}$$

Solution set: $\left\{\frac{1}{2}\right\}$

93. Let $x =$ the length of a side. Two sides and the diagonal of a square form a right triangle. By the Pythagorean theorem we can write the equation

$$x^2+x^2 = 8^2$$

$$2x^2 = 64$$

$$x^2 = 32$$

$$x = \pm\sqrt{32} = \pm 4\sqrt{2}$$

The length of the side is not a negative number. So the side is $4\sqrt{2}$ feet in length.

95. Let $s =$ the length of the side of the square. Since $A = s^2$ for a square, we can write the equation

$$s^2 = 50$$

$$s = \pm\sqrt{50} = \pm 5\sqrt{2}$$

Since the side of a square is not negative, the length of the side is $5\sqrt{2}$ feet.

97. Let $d =$ the length of a diagonal of the rectangle whose sides are 30 and 40 feet. By the Pythagorean theorem we can write the equation

$$d^2 = 30^2+40^2$$

$$d^2 = 2500$$

$$d = \pm\sqrt{2500} = \pm 50$$

Since the diagonal is not negative, the length of the diagonal is 50 feet.

99. a) $C = 4(23{,}245)^{-1/3}(13.5) \approx 1.89$

b)

$$C = 4d^{-1/3}b$$

$$d^{-1/3} = \frac{C}{4b}$$

$$(d^{-1/3})^{-3} = \left(\frac{C}{4b}\right)^{-3}$$

$$d = \frac{64b^3}{C^3}$$

c) The capsize screening value is less than 2 when $d > 19{,}683$ pounds.

101. If the volume of the cube is 2, each side of the cube has length $\sqrt[3]{2}$, because $V = s^3$. Let $d =$ the length of the diagonal of a side. The diagonal of a side is the diagonal of a square with sides of length $\sqrt[3]{2}$. By the Pythagorean theorem we can write

$$d^2 = \left(\sqrt[3]{2}\right)^2 + \left(\sqrt[3]{2}\right)^2$$
$$d^2 = \sqrt[3]{4} + \sqrt[3]{4}$$
$$d^2 = 2 \cdot \sqrt[3]{4} = \sqrt[3]{8} \cdot \sqrt[3]{4} = \sqrt[3]{32}$$
$$d = \left(\sqrt[3]{32}\right)^{1/2} = \left(32^{1/3}\right)^{1/2} = 32^{1/6}$$

The length of the diagonal is $\sqrt[6]{32}$ meters.

103. Let $x =$ the third side to the triangle whose given sides are 3 and 5. By the Pythagorean theorem we can find x:

$$x^2 + 3^2 = 5^2$$
$$x^2 = 16$$
$$x = 4$$

Since $x = 4$, the base of length 12 is divided into 2 parts, one of length 4 and the other of length 8. The side marked a is the hypotenuse of a right triangle with legs 3 and 8:

$$a^2 = 3^2 + 8^2$$
$$a^2 = 73$$
$$a = \sqrt{73} \text{ km}$$

105. $r = \left(\frac{S}{P}\right)^{1/n} - 1$

$$1 + r = \left(\frac{S}{P}\right)^{1/n}$$
$$\frac{S}{P} = (1 + r)^n$$
$$S = P(1 + r)^n$$

Solve for P:

$$P = \frac{S}{(1 + r)^n}$$
$$P = S(1 + r)^{-n}$$

107. $\frac{11.86^2}{5.2^3} = \frac{29.46^2}{R^3}$

$$11.86^2 R^3 = 5.2^3 \cdot 29.46^2$$
$$R^3 = \frac{5.2^3 \cdot 29.46^2}{11.86^2}$$
$$R = \sqrt[3]{\frac{5.2^3 \cdot 29.46^2}{11.86^2}} \approx 9.5 \text{ AU}$$

109. $x^2 = 3.24$

$$x = \pm\sqrt{3.24} = \pm 1.8$$

Solution set: $\{-1.8, 1.8\}$

111. $\sqrt{x - 2} = 1.73$

$$x - 2 = (1.73)^2$$
$$x = 2 + (1.73)^2 \approx 4.993$$

Solution set: $\{4.993\}$

113. $x^{2/3} = 8.86$

$$\left(x^{2/3}\right)^3 = (8.86)^3$$
$$x^2 = 695.506$$
$$x = \pm\sqrt{695.506} \approx \pm 26.372$$

Solution set: $\{-26.372, 26.372\}$

115. Since $V = s^3$, $s = \sqrt[3]{V}$.

9.6 WARM-UPS

1. True, because every real number is a complex number.
2. False, because $2 - \sqrt{-6} = 2 - i\sqrt{6}$.
3. False, because $\sqrt{-9} = 3i$.
4. True, because $(\pm 3i)^2 = -9$.
5. True, because we subtract complex numbers just like we subtract binomials.
6. True, because $i^4 = i^2 \cdot i^2 = (-1)(-1) = 1$.
7. True, because $(2 - i)(2 + i) = 4 - i^2 = 4 - (-1) = 5$.
8. False, $i^3 = i^2 \cdot i = -1 \cdot i = -i$.
9. True, $i^{48} = (i^4)^{12} = 1^{12} = 1$.
10. False, $x^2 = 0$ has only one solution.

9.6 EXERCISES

1. A complex number is a number of the form $a + bi$ where a and b are real numbers.
3. The union of the real numbers and the imaginary numbers is the set of complex numbers.
5. The conjugate of $a + bi$ is $a - bi$.
7. $(2 + 3i) + (-4 + 5i) = -2 + 8i$
9. $(2 - 3i) - (6 - 7i) = 2 - 3i - 6 + 7i$
$= -4 + 4i$
11. $(-1 + i) + (-1 - i) = -2$
13. $(-2 - 3i) - (6 - i) = -2 - 3i - 6 + i$
$= -8 - 2i$
15. $3(2 + 5i) = 3 \cdot 2 + 3 \cdot 5i = 6 + 15i$
17. $2i(i - 5) = 2i^2 - 10i = 2(-1) - 10i$
$= -2 - 10i$

19. $-4i(3 - i) = -12i + 4i^2$
$= -12i + 4(-1) = -4 - 12i$
21. $(2 + 3i)(4 + 6i) = 8 + 24i + 18i^2$
$= 8 + 24i + 18(-1) = -10 + 24i$
23. $(-1 + i)(2 - i) = -2 + 3i - i^2$
$= -2 + 3i - (-1) = -1 + 3i$
25. $(-1 - 2i)(2 + i) = -2 - 5i - 2i^2$
$= -2 - 5i - 2(-1) = -5i$
27. $(5 - 2i)(5 + 2i) = 25 - 4i^2$
$= 25 - 4(-1) = 29$

29. $(1-i)(1+i)=1-i^2=1-(-1)=2$

31. $(4+2i)(4-2i)=16-4i^2$
$=16-4(-1)=20$

33. $(3i)^2=9i^2=9(-1)=-9$

35. $(-5i)^2=(-5)^2i^2=25(-1)=-25$

37. $(2i)^4=2^4i^4=16(1)=16$

39. $i^9=(i^4)^2\cdot i=1^2\cdot i=i$

41. $(3+5i)(3-5i)=9-25i^2=9+25$
$=34$

43. $(1-2i)(1+2i)=1-4i^2=1-4(-1)$
$=5$

45. $(-2+i)(-2-i)=4-i^2=4-(-1)$
$=5$

47. $(2-i\sqrt{3})(2+i\sqrt{3})=4-3i^2$
$=4-3(-1)=7$

49. $\dfrac{3}{4+i}=\dfrac{3(4-i)}{(4+i)(4-i)}=\dfrac{12-3i}{16-i^2}$
$=\dfrac{12-3i}{17}=\dfrac{12}{17}-\dfrac{3}{17}i$

51. $\dfrac{2+i}{3-2i}=\dfrac{(2+i)(3+2i)}{(3-2i)(3+2i)}$
$=\dfrac{6+7i+2i^2}{9-4i^2}=\dfrac{4+7i}{13}=\dfrac{4}{13}+\dfrac{7}{13}i$

53. $\dfrac{4+3i}{i}=\dfrac{(4+3i)(-i)}{(i)(-i)}=\dfrac{-4i-3i^2}{-i^2}$
$=\dfrac{-4i+3}{1}=3-4i$

55. $\dfrac{2+6i}{2}=\dfrac{2}{2}+\dfrac{6i}{2}=1+3i$

57. $\dfrac{1+i}{3i-2}=\dfrac{(1+i)(3i+2)}{(3i-2)(3i+2)}$
$=\dfrac{2+5i+3i^2}{9i^2-4}=\dfrac{-1+5i}{-13}=\dfrac{1}{13}-\dfrac{5}{13}i$

59. $\dfrac{6}{3i}=\dfrac{6\cdot(-3i)}{(3i)(-3i)}=\dfrac{-18i}{-9i^2}=\dfrac{-18i}{9}=-2i$

61. $2+\sqrt{-4}=2+i\sqrt{4}=2+2i$

63. $2\sqrt{-9}+5=2i\sqrt{9}+5=2i\cdot 3+5$
$=5+6i$

65. $7-\sqrt{-6}=7-i\sqrt{6}$

67. $\sqrt{-8}+\sqrt{-18}=i\sqrt{4}\sqrt{2}+i\sqrt{9}\sqrt{2}$
$=2i\sqrt{2}+3i\sqrt{2}=5i\sqrt{2}$

69. $\dfrac{2+\sqrt{-12}}{2}=\dfrac{2+i\sqrt{4}\sqrt{3}}{2}=1+i\sqrt{3}$

71. $\dfrac{-4-\sqrt{-24}}{4}=\dfrac{-4-i\sqrt{4}\sqrt{6}}{4}$
$=\dfrac{-4}{4}-\dfrac{2i\sqrt{6}}{4}=-1-\frac{1}{2}i\sqrt{6}$

73. $\sqrt{-2}\cdot\sqrt{-6}=i\sqrt{2}\cdot i\sqrt{6}=i^2\sqrt{12}$
$=-1\cdot 2\sqrt{3}=-2\sqrt{3}$

75. $\sqrt{-3}\cdot\sqrt{-27}=i\sqrt{3}\cdot i\sqrt{27}=i^2\sqrt{81}$
$=-1\cdot 9=-9$

77. $\dfrac{\sqrt{8}}{\sqrt{-4}}=\dfrac{\sqrt{8}}{i\sqrt{4}}=\dfrac{\sqrt{2}}{i}=\dfrac{\sqrt{2}\cdot(-i)}{i(-i)}$
$=\dfrac{-i\sqrt{2}}{-i^2}=\dfrac{-i\sqrt{2}}{1}=-i\sqrt{2}$

79. $x^2=-36$
$x=\pm\sqrt{-36}=\pm 6i$
Solution set: $\{\pm 6i\}$

81. $x^2=-12$
$x=\pm\sqrt{-12}=\pm i\sqrt{4}\sqrt{3}=\pm 2i\sqrt{3}$
Solution set: $\{\pm 2i\sqrt{3}\}$

83. $2x^2+5=0$
$x^2=-\frac{5}{2}$
$x=\pm\sqrt{-\frac{5}{2}}=\pm\dfrac{i\sqrt{5}}{\sqrt{2}}=\pm\dfrac{i\sqrt{10}}{2}$
Solution set: $\left\{\pm\dfrac{i\sqrt{10}}{2}\right\}$

85. $3x^2+6=0$
$x^2=-2$
$x=\pm\sqrt{-2}=\pm i\sqrt{2}$
Solution set: $\{\pm i\sqrt{2}\}$

87. $(2-3i)(3+4i)=6-i-12i^2$
$=6-i-12(-1)=18-i$

89. $(2-3i)+(3+4i)=5+i$

91. $\dfrac{2-3i}{3+4i}=\dfrac{(2-3i)(3-4i)}{(3+4i)(3-4i)}$
$=\dfrac{6-17i+12i^2}{9-16i^2}=\dfrac{6-17i+12(-1)}{9-16(-1)}$
$=\dfrac{-6-17i}{25}=-\dfrac{6}{25}-\dfrac{17}{25}i$

93. $i(2-3i)=2i-3i^2=2i+3=3+2i$

95. $(-3i)^2=9i^2=-9$

97. $\sqrt{-12}+\sqrt{-3}=2i\sqrt{3}+i\sqrt{3}=3i\sqrt{3}$

99. $(2-3i)^2=4-12i+9i^2=-5-12i$

101. $\dfrac{-4+\sqrt{-32}}{2}=\dfrac{-4+4i\sqrt{2}}{2}$
$=-2+2i\sqrt{2}$

103. If $x=2-i$ then
$x^2-4x+5=(2-i)^2-4(2-i)+5$
$=4-4i+i^2-8+4i+5$
$=1+i^2=0$
So $2-i$ is a solution to $x^2-4x+5=0$.

105. The product rule for radicals is valid only for radicals that represent real numbers. So the product of 4 is not correct. The correct product is found as follows:
$\sqrt{-4}\cdot\sqrt{-4}=2i\cdot 2i=4i^2=-4$

Enriching Your Mathematical Word Power

1. d	**2.** b	**3.** b	**4.** b	**5.** d
6. b	**7.** c	**8.** a	**9.** a	**10.** d
11. c	**12.** a	**13.** c	**14.** d	**15.** b

CHAPTER 9 REVIEW

1. $\sqrt[5]{32} = 2$ because $2^5 = 32$.

3. $\sqrt[3]{1000} = 10$ because $10^3 = 1000$.

5. $\sqrt{72} = \sqrt{36}\sqrt{2} = 6\sqrt{2}$

7. $\sqrt{x^{12}} = x^6$ because $(x^6)^2 = x^{12}$.

9. $\sqrt[3]{x^6} = x^2$ because $(x^2)^3 = x^6$.

11. $\sqrt{2x^9} = \sqrt{x^8}\sqrt{2x} = x^4\sqrt{2x}$

13. $\sqrt{8w^5} = \sqrt{4w^4}\sqrt{2w} = 2w^2\sqrt{2w}$

15. $\sqrt[3]{16x^4} = \sqrt[3]{8x^3}\sqrt[3]{2x} = 2x\sqrt[3]{2x}$

17. $\sqrt[4]{a^9b^5} = \sqrt[4]{a^8b^4}\sqrt[4]{ab} = a^2b\sqrt[4]{ab}$

19. $\sqrt{\dfrac{x^3}{16}} = \dfrac{\sqrt{x^2}\sqrt{x}}{\sqrt{16}} = \dfrac{x\sqrt{x}}{4}$

21. In $\sqrt{2x-5}$ we must have $2x - 5 \geq 0$ or $x \geq 2.5$. So the domain is the interval $[2.5, \infty)$.

23. In $\sqrt[3]{7x-1}$, $7x - 1$ can be any real number. So any real number can be used for x. So the domain is the interval $(-\infty, \infty)$.

25. In $\sqrt[4]{-3x+1}$ we must have $-3x + 1 \geq 0$, $-3x \geq -1$, or $x \leq 1/3$. So the domain is the interval $(-\infty, 1/3]$.

27. In $\sqrt{\frac{1}{2}x+1}$ we must have $\frac{1}{2}x + 1 \geq 0$, $\frac{1}{2}x \geq -1$, or $x \geq -2$. So the domain is the interval $[-2, \infty)$.

29. $(-27)^{-2/3} = \dfrac{1}{(-3)^2} = \dfrac{1}{9}$

31. $(2^6)^{1/3} = 2^{6/3} = 2^2 = 4$

33. $100^{-3/2} = \dfrac{1}{10^3} = \dfrac{1}{1000}$

35. $\dfrac{3x^{-1/2}}{3^{-2}x^{-1}} = 3^3x^{-1/2+1} = 27x^{1/2}$

37. $(a^{1/2}b)^3(ab^{1/4})^2 = a^{3/2}b^3a^2b^{2/4}$
$= a^{7/2}b^{7/2}$

39. $(x^{1/2}y^{1/4})(x^{1/4}y) = x^{1/2+1/4}y^{1/4+1}$
$= x^{3/4}y^{5/4}$

41. $\sqrt{13}\sqrt{13} = 13$

43. $\sqrt{27} + \sqrt{45} - \sqrt{75}$
$= 3\sqrt{3} + 3\sqrt{5} - 5\sqrt{3} = 3\sqrt{5} - 2\sqrt{3}$

45. $3\sqrt{2}(5\sqrt{2} - 7\sqrt{3}) = 15 \cdot 2 - 21\sqrt{6}$
$= 30 - 21\sqrt{6}$

47. $(2 - \sqrt{3})(3 + \sqrt{2})$
$= 6 - 3\sqrt{3} + 2\sqrt{2} - \sqrt{6}$

49. $5 \div \sqrt{2} = \dfrac{5}{\sqrt{2}} = \dfrac{5\sqrt{2}}{\sqrt{2}\sqrt{2}} = \dfrac{5\sqrt{2}}{2}$

51. $\sqrt{\dfrac{2}{5}} = \dfrac{\sqrt{2}\sqrt{5}}{\sqrt{5}\sqrt{5}} = \dfrac{\sqrt{10}}{5}$

53. $\sqrt[3]{\dfrac{2}{3}} = \dfrac{\sqrt[3]{2} \cdot \sqrt[3]{9}}{\sqrt[3]{3} \cdot \sqrt[3]{9}} = \dfrac{\sqrt[3]{18}}{\sqrt[3]{27}} = \dfrac{\sqrt[3]{18}}{3}$

55. $\dfrac{2}{\sqrt{3x}} = \dfrac{2\sqrt{3x}}{\sqrt{3x}\sqrt{3x}} = \dfrac{2\sqrt{3x}}{3x}$

57. $\dfrac{\sqrt{10y^3}}{\sqrt{6}} = \dfrac{\sqrt{2}\sqrt{5y}\sqrt{y^2}}{\sqrt{2}\sqrt{3}} = \dfrac{y\sqrt{5y}}{\sqrt{3}}$
$= \dfrac{y\sqrt{5y}\sqrt{3}}{\sqrt{3}\sqrt{3}} = \dfrac{y\sqrt{15y}}{3}$

59. $\dfrac{3}{\sqrt[3]{2a}} = \dfrac{3\sqrt[3]{4a^2}}{\sqrt[3]{2a}\sqrt[3]{4a^2}} = \dfrac{3\sqrt[3]{4a^2}}{2a}$

61. $\dfrac{5}{\sqrt[4]{3x^2}} = \dfrac{5\sqrt[4]{27x^2}}{\sqrt[4]{3x^2} \cdot \sqrt[4]{27x^2}}$
$= \dfrac{5\sqrt[4]{27x^2}}{\sqrt[4]{81x^4}} = \dfrac{5\sqrt[4]{27x^2}}{3x}$

63. $(\sqrt{3})^4 = 3^{4/2} = 3^2 = 9$

65. $\dfrac{2 - \sqrt{8}}{2} = \dfrac{2 - 2\sqrt{2}}{2} = \dfrac{2(1 - \sqrt{2})}{2}$
$= 1 - \sqrt{2}$

67. $\dfrac{\sqrt{6}}{1 - \sqrt{3}} = \dfrac{\sqrt{6}(1 + \sqrt{3})}{(1 - \sqrt{3})(1 + \sqrt{3})}$

$= \dfrac{\sqrt{6} + \sqrt{18}}{1 - 3} = \dfrac{\sqrt{6} + 3\sqrt{2}}{-2} = \dfrac{-\sqrt{6} - 3\sqrt{2}}{2}$

69. $\dfrac{2\sqrt{3}}{3\sqrt{6} - \sqrt{12}} = \dfrac{2\sqrt{3}}{3\sqrt{6} - 2\sqrt{3}}$

$= \dfrac{2\sqrt{3}(3\sqrt{6} + 2\sqrt{3})}{(3\sqrt{6} - 2\sqrt{3})(3\sqrt{6} + 2\sqrt{3})}$

$= \dfrac{6\sqrt{18} + 12}{54 - 12} = \dfrac{6 \cdot 3\sqrt{2} + 12}{42} = \dfrac{6 \cdot 3\sqrt{2} + 6 \cdot 2}{6 \cdot 7}$

$= \dfrac{3\sqrt{2} + 2}{7}$

71. $\left(2w\sqrt[3]{2w^2}\right)^6 = 2^6w^6\sqrt[3]{2^6w^{12}}$
$= 2^62^2w^6w^4 = 256w^{10}$

73. $x^2 = 16$

$x = \pm 4$

Solution set: $\{-4, 4\}$

75. $(a-5)^2 = 4$

$a - 5 = \pm 2$

$a = 5 \pm 2$

$a = 5+2$ or $a = 5-2$

$a = 7$ or $a = 3$

Solution set: $\{3, 7\}$

77. $(a+1)^2 = 5$

$a + 1 = \pm\sqrt{5}$

$a = -1 \pm \sqrt{5}$

Solution set: $\{-1-\sqrt{5}, -1+\sqrt{5}\}$

79. $(m+1)^2 = -8$

Since the square root of -8 is not a real number, there is no real solution. The solution set is $\emptyset$.

81. $\sqrt{m-1} = 3$

$(\sqrt{m-1})^2 = 3^2$

$m - 1 = 9$

$m = 10$

Solution set: $\{10\}$

83. $\sqrt[3]{2x+9} = 3$

$(\sqrt[3]{2x+9})^3 = 3^3$

$2x + 9 = 27$

$2x = 18$

$x = 9$

Solution set: $\{9\}$

85. $w^{2/3} = 4$

$(w^{2/3})^3 = 4^3$

$w^2 = 64$

$w = \pm 8$

Solution set: $\{-8, 8\}$

87. $(m+1)^{1/3} = 5$

$((m+1)^{1/3})^3 = 5^3$

$m + 1 = 125$

$m = 124$

Solution set: $\{124\}$

89. $\sqrt{x-3} = \sqrt{x+2} - 1$

$\left(\sqrt{x-3}\right)^2 = \left(\sqrt{x+2} - 1\right)^2$

$x - 3 = x + 2 - 2\sqrt{x+2} + 1$

$-6 = -2\sqrt{x+2}$

$3 = \sqrt{x+2}$

$3^2 = (\sqrt{x+2})^2$

$9 = x + 2$

$7 = x$

Solution set: $\{7\}$

91. $\sqrt{5x - x^2} = \sqrt{6}$

$\left(\sqrt{5x-x^2}\right)^2 = \left(\sqrt{6}\right)^2$

$5x - x^2 = 6$

$-x^2 + 5x - 6 = 0$

$x^2 - 5x + 6 = 0$

$(x-2)(x-3) = 0$

$x = 2$ or $x = 3$

Solution set: $\{2, 3\}$

93. $\sqrt{x+7} - 2\sqrt{x} = -2$

$\sqrt{x+7} = 2\sqrt{x} - 2$

$\left(\sqrt{x+7}\right)^2 = \left(2\sqrt{x} - 2\right)^2$

$x + 7 = 4x - 8\sqrt{x} + 4$

$8\sqrt{x} = 3x - 3$

$\left(8\sqrt{x}\right)^2 = \left(3x-3\right)^2$

$64x = 9x^2 - 18x + 9$

$0 = 9x^2 - 82x + 9$

$0 = (9x-1)(x-9)$

$x = \frac{1}{9}$ or $x = 9$

Since $\sqrt{\frac{1}{9} + 7} - 2\sqrt{\frac{1}{9}} = \frac{8}{3} - \frac{2}{3} = 2$, $1/9$ is an extraneous root. The solution set is $\{9\}$.

95. $2\sqrt{x} - \sqrt{x-3} = 3$

$2\sqrt{x} = \sqrt{x-3} + 3$

$\left(2\sqrt{x}\right)^2 = \left(\sqrt{x-3} + 3\right)^2$

$4x = x - 3 + 6\sqrt{x-3} + 9$

$3x - 6 = 6\sqrt{x-3}$

$x - 2 = 2\sqrt{x-3}$

$(x-2)^2 = (2\sqrt{x-3})^2$

$x^2 - 4x + 4 = 4(x-3)$

$x^2 - 8x + 16 = 0$

$(x-4)^2 = 0$

$x = 4$

Solution set: $\{4\}$

97. $(2-3i)(-5+5i) = -10 + 25i - 15i^2$

$= -10 + 25i + 15 = 5 + 25i$

99. $(2+i) + (5-4i) = 7 - 3i$

101. $(1-i) - (2-3i) = 1 - i - 2 + 3i$

$= -1 + 2i$

103. $\frac{6+3i}{3} = \frac{6}{3} + \frac{3i}{3} = 2 + i$

105. $\frac{4-\sqrt{-12}}{2} = \frac{4 - 2i\sqrt{3}}{2} = 2 - i\sqrt{3}$

107. $\frac{2-3i}{4+i} = \frac{(2-3i)(4-i)}{(4+i)(4-i)}$

$= \frac{5-14i}{17} = \frac{5}{17} - \frac{14}{17}i$

109. $x^2 + 100 = 0$

$$x^2 = -100$$
$$x = \pm\sqrt{-100} = \pm 10i$$

The solution set is $\{\pm 10i\}$.

111. $2b^2 + 9 = 0$

$$2b^2 = -9$$
$$b^2 = -\frac{9}{2}$$
$$b = \sqrt{-\frac{9}{2}} = \pm\frac{3i}{\sqrt{2}} = \pm\frac{3i\sqrt{2}}{2}$$

The solution set is $\left\{\pm\frac{3i\sqrt{2}}{2}\right\}$.

113. False, because $2^3 \cdot 3^2 = 8 \cdot 9 = 72$.

115. True, because $(\sqrt{2})^3 = \sqrt{8} = 2\sqrt{2}$.

117. True, because $8^{200}8^{200} = (8 \cdot 8)^{200} = 64^{200}$.

119. False, because $4^{1/2} = \sqrt{4} = 2$.

121. False, because $5^2 \cdot 5^2 = 5^4 = 625$.

123. False, because $\sqrt{w^{10}} = |w^5|$.

125. False, because $\sqrt{x^6} = |x^3|$.

127. True, $\sqrt{x^8} = x^{8/2} = x^4$.

129. False, $\sqrt{16} = 4$.

131. True, $2^{600} = (2^2)^{300} = 4^{300}$.

133. False, $\frac{2+\sqrt{6}}{2} = 1 + \frac{\sqrt{6}}{2}$.

135. False, $\sqrt{\frac{4}{6}} = \sqrt{\frac{2}{3}} = \frac{\sqrt{2}\sqrt{3}}{\sqrt{3}\sqrt{3}} = \frac{\sqrt{6}}{3}$.

137. True, $81^{2/4} = 3^2 = 9 = \sqrt{81}$.

139. True, because $(a^4b^2)^{1/2}$ is nonnegative and a^2b could be negative, absolute value symbols are necessary.

141. To find the time for which $s = 12{,}000$, solve the equation

$$16t^2 = 12{,}000$$
$$t^2 = 750$$
$$t = \sqrt{750} = 5\sqrt{30}$$

The time is $5\sqrt{30}$ seconds.

143. The guy wire of length 40 is the hypotenuse of a right triangle where one leg is length 30 and the other is length x. By the Pythagorean theorem we can write

$$x^2 + 30^2 = 40^2$$
$$x^2 = 700$$
$$x = \sqrt{700} = 10\sqrt{7}$$

The wire is attached to the ground $10\sqrt{7}$ feet from the base of the antenna.

145. Let $x =$ the length of the guy wire. The height of the antenna is 200 feet and the distance from the base of the antenna to the point on the ground where the guy wire is attached is 200 feet. The guy wire is the hypotenuse of a right triangle whose legs each have length 200. By the Pythagorean theorem we can write

$$x^2 = 200^2 + 200^2$$
$$x^2 = 80{,}000$$
$$x = \sqrt{80{,}000} = \sqrt{40{,}000}\sqrt{2} = 200\sqrt{2}$$

The length of the guy wires should be $200\sqrt{2}$ feet.

147. If the volume is 40 ft^3, then each side is $\sqrt[3]{40}$ ft in length. The surface area of the six square sides is $6(\sqrt[3]{40})^2$ ft^2. Multiply by 1.1 to get $1.1 \cdot 6(\sqrt[3]{40})^2$ ft^2 as the amount of cardboard needed to make the box. Simplify:

$$6.6\left(2\sqrt[3]{5}\right)^2 = 26.4\sqrt[3]{25} \text{ ft}^2$$

This is the exact answer. You can find an approximate answer with a calculator.

149. a) $1630.3 = 993.3(1+r)^9$

$$(1+r)^9 = \frac{1630.3}{993.3}$$
$$1 + r = \left(\frac{1630.3}{993.3}\right)^{1/9}$$
$$r = \left(\frac{1630.3}{993.3}\right)^{1/9} - 1$$
$$r \approx 0.057 = 5.7\%$$

b) From the graph it appears that the annual cost of health care in 2010 will about $2300 billion.

151. $V = \sqrt{\frac{841L}{CS}} = \frac{\sqrt{841} \cdot \sqrt{L} \cdot \sqrt{CS}}{\sqrt{CS} \cdot \sqrt{CS}}$

$$V = \frac{29\sqrt{LCS}}{CS}$$

CHAPTER 9 TEST

1. $8^{2/3} = 2^2 = 4$

2. $4^{-3/2} = \frac{1}{2^3} = \frac{1}{8}$

3. $\sqrt{21} \div \sqrt{7} = \frac{\sqrt{21}}{\sqrt{7}} = \sqrt{\frac{21}{7}} = \sqrt{3}$

4. $2\sqrt{5} \cdot 3\sqrt{5} = 6 \cdot 5 = 30$

5. $\sqrt{20} + \sqrt{5} = 2\sqrt{5} + \sqrt{5} = 3\sqrt{5}$

6. $\sqrt{5} + \frac{1\sqrt{5}}{\sqrt{5}\sqrt{5}} = \sqrt{5} + \frac{\sqrt{5}}{5}$

$$= \frac{5\sqrt{5}}{5} + \frac{\sqrt{5}}{5} = \frac{6\sqrt{5}}{5}$$

7. $2^{1/2} \cdot 2^{1/2} = 2^1 = 2$

8. $\sqrt{72} = \sqrt{36}\sqrt{2} = 6\sqrt{2}$

9. $\sqrt{\frac{5}{12}} = \frac{\sqrt{5}\sqrt{3}}{\sqrt{12}\sqrt{3}} = \frac{\sqrt{15}}{\sqrt{36}} = \frac{\sqrt{15}}{6}$

10. $\frac{6+\sqrt{18}}{6} = \frac{6+3\sqrt{2}}{6} = \frac{3(2+\sqrt{2})}{3 \cdot 2}$
$= \frac{2+\sqrt{2}}{2}$

11. $(2\sqrt{3}+1)(\sqrt{3}-2)$
$= 6+\sqrt{3}-4\sqrt{3}-2 = 4-3\sqrt{3}$

12. $\sqrt[4]{32a^5y^8} = \sqrt[4]{16a^4y^8} \cdot \sqrt[4]{2a}$
$= 2ay^2\sqrt[4]{2a}$

13. $\frac{1}{\sqrt[3]{2x^2}} = \frac{1 \cdot \sqrt[3]{4x}}{\sqrt[3]{2x^2} \cdot \sqrt[3]{4x}} = \frac{\sqrt[3]{4x}}{\sqrt[3]{8x^3}}$
$= \frac{\sqrt[3]{4x}}{2x}$

14. $\sqrt{\frac{8a^9}{b^3}} = \frac{2a^4\sqrt{2a}}{b\sqrt{b}} = \frac{2a^4\sqrt{2a}\sqrt{b}}{b\sqrt{b}\sqrt{b}}$
$= \frac{2a^4\sqrt{2ab}}{b^2}$

15. $\sqrt[3]{-27x^9} = -3x^{9/3} = -3x^3$

16. $\sqrt{20m^3} = \sqrt{4m^2}\sqrt{5m} = 2m\sqrt{5m}$

17. $x^{1/2}x^{1/4} = x^{1/2+1/4} = x^{3/4}$

18. $(9y^4x^{1/2})^{1/2} = 3y^2x^{1/4}$

19. $\sqrt[3]{40x^7} = \sqrt[3]{8x^6} \cdot \sqrt[3]{5x} = 2x^2\sqrt[3]{5x}$

20. $(4+\sqrt{3})^2 = 16+8\sqrt{3}+3 = 19+8\sqrt{3}$

21. In $\sqrt{4-x}$ we must have $4-x \geq 0, -x \geq -4$, or $x \leq 4$. So the domain is the interval $(-\infty, 4]$.

22. In $\sqrt[3]{5x-3}$, $5x-3$ can be any real number. So any real number can be used for x. So the domain is the interval $(-\infty, \infty)$.

23. $\frac{2}{5-\sqrt{3}} = \frac{2(5+\sqrt{3})}{(5-\sqrt{3})(5+\sqrt{3})}$
$= \frac{2(5+\sqrt{3})}{25-3} = \frac{2(5+\sqrt{3})}{22} = \frac{5+\sqrt{3}}{11}$

24. $\frac{\sqrt{6}}{4\sqrt{3}+\sqrt{2}} = \frac{\sqrt{6}(4\sqrt{3}-\sqrt{2})}{(4\sqrt{3}+\sqrt{2})(4\sqrt{3}-\sqrt{2})}$
$= \frac{4\sqrt{18}-\sqrt{12}}{48-2} = \frac{12\sqrt{2}-2\sqrt{3}}{46}$
$= \frac{6\sqrt{2}-\sqrt{3}}{23}$

25. $(3-2i)(4+5i) = 12+7i-10i^2$
$= 22+7i$

26. $i^4 - i^5 = 1 - i^4 \cdot i = 1 - i$

27. $\frac{3-i}{1+2i} = \frac{(3-i)(1-2i)}{(1+2i)(1-2i)}$
$= \frac{3-7i+2i^2}{1-4i^2} = \frac{1-7i}{5} = \frac{1}{5} - \frac{7}{5}i$

28. $\frac{-6+\sqrt{-12}}{8} = \frac{-6+2i\sqrt{3}}{8}$
$= \frac{-3+i\sqrt{3}}{4} = -\frac{3}{4} + \frac{1}{4}i\sqrt{3}$

29.
$$(x-2)^2 = 49$$
$$x-2 = \pm 7$$
$$x = 2 \pm 7$$
$$x = 2+7 \quad \text{or} \quad x = 2-7$$
$$x = 9 \quad \text{or} \quad x = -5$$
Solution set: $\{-5, 9\}$

30.
$$2\sqrt{x+4} = 3$$
$$\left(2\sqrt{x+4}\right)^2 = (3)^2$$
$$4(x+4) = 9$$
$$4x = -7$$
$$x = -\frac{7}{4}$$
Solution set: $\left\{-\frac{7}{4}\right\}$

31.
$$w^{2/3} = 4$$
$$(w^{2/3})^3 = 4^3$$
$$w^2 = 64$$
$$w = \pm 8$$
Solution set: $\{-8, 8\}$

32.
$$9y^2 + 16 = 0$$
$$9y^2 = -16$$
$$y^2 = -\frac{16}{9}$$
$$y = \pm\sqrt{-\frac{16}{9}} = \pm\frac{4}{3}i$$
The solution set is $\left\{\pm\frac{4}{3}i\right\}$.

33.
$$\sqrt{2x^2+x-12} = x$$
$$\left(\sqrt{2x^2+x-12}\right)^2 = x^2$$
$$2x^2+x-12 = x^2$$
$$x^2+x-12 = 0$$
$$(x+4)(x-3) = 0$$
$$x+4 = 0 \quad \text{or} \quad x-3 = 0$$
$$x = -4 \quad \text{or} \quad x = 3$$
Since -4 does not check in the original equation, the solution set is $\{3\}$.

34.
$$\sqrt{x-1} + \sqrt{x+4} = 5$$
$$\sqrt{x-1} = 5 - \sqrt{x+4}$$
$$\left(\sqrt{x-1}\right)^2 = \left(5-\sqrt{x+4}\right)^2$$
$$x-1 = 25-10\sqrt{x+4}+x+4$$
$$10\sqrt{x+4} = 30$$

$$\left(\sqrt{x+4}\right)^2 = (3)^2$$
$$x + 4 = 9$$
$$x = 5$$

Solution set: $\{5\}$

35. Let $x =$ the length of the side. Since the diagonal is the hypotenuse of a right triangle, we can write the equation

$$x^2 + x^2 = 3^2$$
$$2x^2 = 9$$
$$x^2 = \frac{9}{2}$$
$$x = \sqrt{\frac{9}{2}} = \frac{3\sqrt{2}}{\sqrt{2}\sqrt{2}} = \frac{3\sqrt{2}}{2}$$

The length of each side is $\frac{3\sqrt{2}}{2}$ feet.

36. Let $x =$ one number and $x + 11 =$ the other. Since their square roots differ by 1, we can write

$$\sqrt{x+11} - \sqrt{x} = 1$$
$$\sqrt{x+11} = \sqrt{x} + 1$$
$$\left(\sqrt{x+11}\right)^2 = \left(\sqrt{x}+1\right)^2$$
$$x + 11 = x + 2\sqrt{x} + 1$$
$$10 = 2\sqrt{x}$$
$$10^2 = (2\sqrt{x})^2$$
$$4x = 100$$
$$x = 25$$
$$x + 11 = 36$$

The numbers are 25 and 36.

37. If the perimeter is 20, the sum of the length and width is 10. Let $x =$ the length and $10 - x =$ the width. Use the Pythagorean theorem to write the equation

$$x^2 + (10 - x)^2 = \left(2\sqrt{13}\right)^2$$
$$x^2 + 100 - 20x + x^2 = 52$$
$$2x^2 - 20x + 48 = 0$$
$$x^2 - 10x + 24 = 0$$
$$(x - 4)(x - 6) = 0$$
$$x = 4 \quad \text{or} \quad x = 6$$
$$10 - x = 6 \quad \text{or} \quad 10 - x = 4$$

The length and width are 4 feet and 6 feet.

38. $R = (248.530)^{2/3} \approx 39.53$ AU

$$30.08 = T^{2/3}$$
$$T = 30.08^{3/2} = 164.97 \text{ years}$$

Making Connections

Chapters 1 - 9

1. $3(x - 2) + 5 = 7 - 4(x + 3)$

$$3x - 6 + 5 = 7 - 4x - 12$$
$$3x - 1 = -4x - 5$$
$$7x = -4$$
$$x = -\frac{4}{7}$$

Solution set: $\left\{-\frac{4}{7}\right\}$

2. $\sqrt{6x+7} = 4$

$$\left(\sqrt{6x+7}\right)^2 = (4)^2$$
$$6x + 7 = 16$$
$$6x = 9$$
$$x = \frac{3}{2}$$

Solution set: $\left\{\frac{3}{2}\right\}$

3. $|\,2x + 5\,| > 1$

$$2x + 5 > 1 \quad \text{or} \quad 2x + 5 < -1$$
$$2x > -4 \quad \text{or} \quad 2x < -6$$
$$x > -2 \quad \text{or} \quad x < -3$$

$(-\infty, -3) \cup (-2, \infty)$

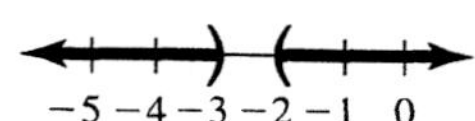

4. $8x^3 - 27 = 0$

$$x^3 = \frac{27}{8}$$
$$x = \sqrt[3]{\frac{27}{8}} = \frac{3}{2}$$

Solution set: $\left\{\frac{3}{2}\right\}$

5. $2x - 3 > 3x - 4$

$$1 > x$$

$(-\infty, 1)$

−3 −2 −1 0 1 2 3

6. $\sqrt{2x-3} - \sqrt{3x+4} = 0$

$$\left(\sqrt{2x-3}\right)^2 = \left(\sqrt{3x+4}\right)^2$$
$$2x - 3 = 3x + 4$$
$$-7 = x$$

Checking -7 gives us a square root of a negative number. So the solution set is $\emptyset$.

7. $\frac{w}{3} + \frac{w-4}{2} = \frac{11}{2}$

$$6\left(\frac{w}{3} + \frac{w-4}{2}\right) = 6\left(\frac{11}{2}\right)$$
$$2w + 3w - 12 = 33$$
$$5w = 45$$
$$w = 9$$

Solution set: $\{9\}$

8. $2(x+7)-4 = x-(10-x)$
$2x+14-4 = x-10+x$
$2x+10 = 2x-10$
$10 = -10$
Solution set: $\emptyset$

9. $(x+7)^2 = 25$
$x+7 = \pm 5$
$x = -7 \pm 5$
Solution set: $\{-12, -2\}$

10. $a^{-1/2} = 4$
$\left(a^{-1/2}\right)^{-2} = (4)^{-2}$
$a = \frac{1}{16}$
Solution set: $\left\{\frac{1}{16}\right\}$

11. $x-3 > 2$ or $x < 2x+6$
$x > 5$ or $-x < 6$
$x > 5$ or $x > -6$
$(-6, \infty)$

−8 −7 −6 −5 −4 −3 −2

12. $a^{-2/3} = 16$
$\left(a^{-2/3}\right)^{-3} = (16)^{-3}$
$a^2 = 2^{-12}$
$a = \pm\sqrt{2^{-12}} = \pm 2^{-6} = \pm\frac{1}{64}$
Solution set: $\left\{-\frac{1}{64}, \frac{1}{64}\right\}$

13. $3x^2-1 = 0$
$x^2 = \frac{1}{3}$
$x = \pm\sqrt{\frac{1}{3}} = \pm\frac{\sqrt{3}}{3}$
Solution set: $\left\{-\frac{\sqrt{3}}{3}, \frac{\sqrt{3}}{3}\right\}$

14. $5-2(x-2) = 3x-5(x-2)-1$
$5-2x+4 = 3x-5x+10-1$
$-2x+9 = -2x+9$
Solution set is all real numbers R.

15. $|\,3x-4\,| < 5$
$-5 < 3x-4 < 5$
$-1 < 3x < 9$
$-\frac{1}{3} < x < 3$
$\left(-\frac{1}{3}, 3\right)$

$-\frac{1}{3}$

−2 −1 0 1 2 3 4

16. $3x-1 = 0$
$x = \frac{1}{3}$
Solution set: $\left\{\frac{1}{3}\right\}$

17. $\sqrt{y-1} = 9$
$\left(\sqrt{y-1}\right)^2 = 9^2$
$y-1 = 81$
$y = 82$
Solution set: $\{82\}$

18. $|\,5(x-2)+1\,| = 3$
$|\,5x-10+1\,| = 3$
$|\,5x-9\,| = 3$
$5x-9 = 3$ or $5x-9 = -3$
$5x = 12$ $\quad 5x = 6$
$x = \frac{12}{5}$ or $x = \frac{6}{5}$
Solution set: $\left\{\frac{6}{5}, \frac{12}{5}\right\}$

19. $0.06x - 0.04(x-20) = 2.8$
$0.06x - 0.04x + 0.8 = 2.8$
$0.02x = 2.0$
$x = 100$
Solution set: $\{100\}$

20. $|\,3x-1\,| > -2$
Since absolute value of any quantity is greater than or equal to zero, any real number satisfies this inequality. Solution set: R

21. $\frac{3\sqrt{2}}{x} = \frac{\sqrt{3}}{4\sqrt{5}}$
$\sqrt{3}x = 12\sqrt{10}$
$x = \frac{12\sqrt{10}}{\sqrt{3}} = \frac{12\sqrt{10}\sqrt{3}}{\sqrt{3}\sqrt{3}} = 4\sqrt{30}$
Solution set: $\left\{4\sqrt{30}\right\}$

22. $\frac{\sqrt{x}-4}{x} = \frac{1}{\sqrt{x}+5}$
$x = (\sqrt{x}-4)(\sqrt{x}+5)$
$x = x + \sqrt{x} - 20$
$20 = \sqrt{x}$
$400 = x$

Solution set: $\{400\}$

23. $\dfrac{3\sqrt{2}+4}{\sqrt{2}} = \dfrac{x\sqrt{18}}{3\sqrt{2}+2}$

$6x = (3\sqrt{2}+4)(3\sqrt{2}+2)$

$6x = 18 + 18\sqrt{2} + 8$

$x = \dfrac{26+18\sqrt{2}}{6} = \dfrac{13+9\sqrt{2}}{3}$

Solution set: $\left\{\dfrac{13+9\sqrt{2}}{3}\right\}$

24. $\dfrac{x}{2\sqrt{5}-\sqrt{2}} = \dfrac{2\sqrt{5}+\sqrt{2}}{x}$

$x^2 = 20 - 2$

$x = \pm\sqrt{18} = \pm 3\sqrt{2}$

Solution set: $\{-3\sqrt{2}, 3\sqrt{2}\}$

25. $\dfrac{\sqrt{2x-5}}{x} = \dfrac{-3}{\sqrt{2x+5}}$

$-3x = 2x - 25$

$-5x = -25$

$x = 5$

Solution set: $\{5\}$

26. $\dfrac{\sqrt{6}+2}{x} = \dfrac{2}{\sqrt{6}+4}$

$2x = (\sqrt{6}+2)(\sqrt{6}+4)$

$2x = 14 + 6\sqrt{6}$

$x = 7 + 3\sqrt{6}$

Solution set: $\{7+3\sqrt{6}\}$

27. $\dfrac{x-1}{\sqrt{6}} = \dfrac{\sqrt{6}}{x}$

$x^2 - x = 6$

$x^2 - x - 6 = 0$

$(x-3)(x+2) = 0$

$x = 3$ or $x = -2$

Solution set: $\{-2, 3\}$

28. $\dfrac{x+3}{\sqrt{10}} = \dfrac{\sqrt{10}}{x}$

$x^2 + 3x = 10$

$x^2 + 3x - 10 = 0$

$(x+5)(x-2) = 0$

$x = -5$ or $x = 2$

Solution set: $\{-5, 2\}$

29. $\dfrac{1}{x} - \dfrac{1}{x-1} = -\dfrac{1}{6}$

$6x(x-1)(\frac{1}{x} - \frac{1}{x-1}) = 6x(x-1)(-\frac{1}{6})$

$6x - 6 - 6x = -x^2 + x$

$x^2 - x - 6 = 0$

$(x-3)(x+2) = 0$

$x - 3 = 0$ or $x + 2 = 0$

$x = 3$ or $x = -2$

Solution set: $\{-2, 3\}$

30. $\dfrac{1}{x^2-2x} + \dfrac{1}{x} = \dfrac{2}{3}$

$3x(x-2)(\frac{1}{x^2-2x} + \frac{1}{x}) = 3x(x-2)\frac{2}{3}$

$3 + 3x - 6 = 2x^2 - 4x$

$-2x^2 + 7x - 3 = 0$

$2x^2 - 7x + 3 = 0$

$(2x-1)(x-3) = 0$

$2x - 1 = 0$ or $x - 3 = 0$

$x = \frac{1}{2}$ or $x = 3$

Solution set: $\left\{\frac{1}{2}, 3\right\}$

31. $\dfrac{-2+\sqrt{2^2-4(1)(-15)}}{2(1)}$

$= \dfrac{-2+\sqrt{64}}{2} = 3$

32. $\dfrac{-8+\sqrt{8^2-4(1)(12)}}{2(1)}$

$= \dfrac{-8+\sqrt{16}}{2} = -2$

33. $\dfrac{-5+\sqrt{5^2-4(2)(-3)}}{2(2)}$

$= \dfrac{-5+\sqrt{49}}{4} = \dfrac{1}{2}$

34. $\dfrac{-7+\sqrt{7^2-4(6)(-3)}}{2(6)}$

$= \dfrac{-7+\sqrt{121}}{12} = \dfrac{1}{3}$

35. a) $v = -94.8 + 21.4x - 0.761x^2$

$v = -94.8 + 21.4(11) - 0.761(11)^2$

$\approx 48.5 \text{ cm}^3$

b) From the graph it appears that a moisture content of 14% will produce the maximum volume of popped corn.

c) From the graph it appears that the maximum volume is about 56 cm^3.

10.1 WARM-UPS

1. False, completing the square involves perfect square trinomials.
2. False, it is equivalent to $x - 3 = \pm 2\sqrt{3}$.
3. False, because some quadratic polynomials cannot be factored.
4. False, because one-half of $4/3$ is $2/3$ and $2/3$ squared is $4/9$.
5. True.
6. False, we must first divide each side of the equation by 2.
7. False, $x = 3/2$ or $x = -5/3$.
8. True, one-half of 3 is $3/2$ and $3/2$ squared is $9/4$.
9. False, $x = \pm 2i\sqrt{2}$.
10. False, $(x-3)^2 = 0$ is a quadratic equation with only one solution.

10.1 EXERCISES

1. In this section, quadratic equations are solved by factoring, the even root property, and completing the square.
3. The last term is the square of one-half the coefficient of the middle term.
5.
$$\begin{aligned} x^2 - x - 6 &= 0 \\ (x-3)(x+2) &= 0 \\ x - 3 = 0 \quad &\text{or} \quad x + 2 = 0 \\ x = 3 \quad &\text{or} \quad x = -2 \end{aligned}$$
Solution set: $\{-2, 3\}$
7.
$$\begin{aligned} a^2 + 2a &= 15 \\ a^2 + 2a - 15 &= 0 \\ (a+5)(a-3) &= 0 \\ a + 5 = 0 \quad &\text{or} \quad a - 3 = 0 \\ a = -5 \quad &\text{or} \quad a = 3 \end{aligned}$$
Solution set: $\{-5, 3\}$
9.
$$\begin{aligned} 2x^2 - x - 3 &= 0 \\ (2x-3)(x+1) &= 0 \\ 2x - 3 = 0 \quad &\text{or} \quad x + 1 = 0 \\ x = \frac{3}{2} \quad &\text{or} \quad x = -1 \end{aligned}$$
Solution set: $\left\{-1, \frac{3}{2}\right\}$
11.
$$\begin{aligned} y^2 + 14y + 49 &= 0 \\ (y+7)^2 &= 0 \\ y + 7 &= 0 \\ y &= -7 \end{aligned}$$
Solution set: $\{-7\}$
13.
$$\begin{aligned} a^2 - 16 &= 0 \\ (a-4)(a+4) &= 0 \\ a - 4 = 0 \quad &\text{or} \quad a + 4 = 0 \\ a = 4 \quad &\text{or} \quad a = -4 \end{aligned}$$
Solution set: $\{-4, 4\}$
15.
$$\begin{aligned} x^2 &= 81 \\ x &= \pm\sqrt{81} = \pm 9 \end{aligned}$$
Solution set: $\{-9, 9\}$
17.
$$\begin{aligned} x^2 &= \frac{16}{9} \\ x &= \pm\sqrt{\frac{16}{9}} = \pm\frac{4}{3} \end{aligned}$$
Solution set: $\left\{-\frac{4}{3}, \frac{4}{3}\right\}$
19.
$$\begin{aligned} (x-3)^2 &= 16 \\ x - 3 &= \pm\sqrt{16} \\ x &= 3 \pm 4 \\ x = 3 + 4 \quad &\text{or} \quad x = 3 - 4 \\ x = 7 \quad &\text{or} \quad x = -1 \end{aligned}$$
Solution set: $\{-1, 7\}$
21.
$$\begin{aligned} (z+1)^2 &= 5 \\ z + 1 &= \pm\sqrt{5} \\ z &= -1 \pm \sqrt{5} \end{aligned}$$
Solution set: $\{-1 - \sqrt{5}, -1 + \sqrt{5}\}$
23.
$$\begin{aligned} \left(w - \frac{3}{2}\right)^2 &= \frac{7}{4} \\ w - \frac{3}{2} &= \pm\sqrt{\frac{7}{4}} \\ w &= \frac{3}{2} \pm \frac{\sqrt{7}}{2} = \frac{3 \pm \sqrt{7}}{2} \end{aligned}$$
Solution set: $\left\{\frac{3-\sqrt{7}}{2}, \frac{3+\sqrt{7}}{2}\right\}$
25. One-half of 2 is 1, and 1 squared is 1: $x^2 + 2x + 1$
27. One-half of -3 is $-3/2$, and $-3/2$ squared is $9/4$: $x^2 - 3x + \frac{9}{4}$
29. One-half of $1/4$ is $1/8$, and $1/8$ squared is $1/64$: $y^2 + \frac{1}{4}y + \frac{1}{64}$
31. One-half of $2/3$ is $1/3$, and $1/3$ squared is $1/9$: $x^2 + \frac{2}{3}x + \frac{1}{9}$
33. $x^2 + 8x + 16 = (x+4)^2$
35. $y^2 - 5y + \frac{25}{4} = \left(y - \frac{5}{2}\right)^2$
37. $z^2 - \frac{4}{7}z + \frac{4}{49} = \left(z - \frac{2}{7}\right)^2$
39. $t^2 + \frac{3}{5}t + \frac{9}{100} = \left(t + \frac{3}{10}\right)^2$
41.
$$\begin{aligned} x^2 - 2x - 15 &= 0 \\ x^2 - 2x \quad &= 15 \\ x^2 - 2x + 1 &= 15 + 1 \end{aligned}$$

$$(x-1)^2 = 16$$
$$x-1 = \pm 4$$
$$x = 1 \pm 4$$
$$x = 5 \text{ or } x = -3$$
Solution set: $\{-3, 5\}$

43.
$$2x^2 - 4x = 70$$
$$x^2 - 2x = 35$$
$$x^2 - 2x + 1 = 35 + 1$$
$$(x-1)^2 = 36$$
$$x - 1 = \pm 6$$
$$x = 1 \pm 6$$
$$x = 7 \text{ or } x = -5$$
Solution set: $\{-5, 7\}$

45. $w^2 - w - 20 = 0$
$$w^2 - w = 20$$
$$w^2 - w + \frac{1}{4} = 20 + \frac{1}{4}$$
$$\left(w - \frac{1}{2}\right)^2 = \frac{81}{4}$$
$$w - \frac{1}{2} = \pm\frac{9}{2}$$
$$w = \frac{1}{2} \pm \frac{9}{2}$$
$$x = \frac{10}{2} = 5 \quad \text{or} \quad x = \frac{-8}{2} = -4$$
Solution set: $\{-4, 5\}$

47.
$$q^2 + 5q = 14$$
$$q^2 + 5q + \frac{25}{4} = 14 + \frac{25}{4}$$
$$\left(q + \frac{5}{2}\right)^2 = \frac{81}{4}$$
$$q + \frac{5}{2} = \pm\frac{9}{2}$$
$$q = -\frac{5}{2} \pm \frac{9}{2}$$
$$q = \frac{4}{2} = 2 \quad \text{or} \quad q = \frac{-14}{2} = -7$$
Solution set: $\{-7, 2\}$

49. $2h^2 - h - 3 = 0$
$$2h^2 - h = 3$$
$$h^2 - \frac{1}{2}h = \frac{3}{2}$$
$$h^2 - \frac{1}{2}h + \frac{1}{16} = \frac{3}{2} + \frac{1}{16}$$
$$\left(h - \frac{1}{4}\right)^2 = \frac{25}{16}$$
$$h - \frac{1}{4} = \pm\frac{5}{4}$$
$$h = \frac{1}{4} \pm \frac{5}{4}$$
Solution set: $\left\{-1, \frac{3}{2}\right\}$

51.
$$x^2 + 4x = 6$$
$$x^2 + 4x + 4 = 6 + 4$$
$$(x+2)^2 = 10$$
$$x + 2 = \pm\sqrt{10}$$
$$x = -2 \pm \sqrt{10}$$
Solution set: $\{-2 - \sqrt{10}, -2 + \sqrt{10}\}$

53. $x^2 + 8x - 4 = 0$
$$x^2 + 8x = 4$$
$$x^2 + 8x + 16 = 4 + 16$$
$$(x+4)^2 = 20$$
$$x + 4 = \pm\sqrt{20}$$
$$x = -4 \pm 2\sqrt{5}\tfrac{1}{2}$$
Solution set: $\{-4 - 2\sqrt{5}, -4 + 2\sqrt{5}\}$

55.
$$4x^2 - 4x - 1 = 0$$
$$x^2 - x - \frac{1}{4} = 0$$
$$x^2 - x = \frac{1}{4}$$
$$x^2 - x + \frac{1}{4} = \frac{1}{4} + \frac{1}{4}$$
$$\left(x - \frac{1}{2}\right)^2 = \frac{1}{2}$$
$$x - \frac{1}{2} = \pm\sqrt{\frac{1}{2}}$$
$$x = \frac{1}{2} \pm \frac{\sqrt{2}}{2}$$
$$x = \frac{1 \pm \sqrt{2}}{2}$$
Solution set: $\left\{\frac{1 \pm \sqrt{2}}{2}\right\}$

57. $2x^2 + 3x - 4 = 0$
$$2x^2 + 3x = 4$$
$$x^2 + \frac{3}{2}x = 2$$
$$x^2 + \frac{3}{2}x + \frac{9}{16} = 2 + \frac{9}{16}$$
$$\left(x + \frac{3}{4}\right)^2 = \frac{41}{16}$$
$$x + \frac{3}{4} = \pm\frac{\sqrt{41}}{4}$$
$$x = -\frac{3}{4} \pm \frac{\sqrt{41}}{4}$$
Solution set: $\left\{\frac{-3 - \sqrt{41}}{4}, \frac{-3 + \sqrt{41}}{4}\right\}$

59.
$$\sqrt{2x+1} = x - 1$$
$$(\sqrt{2x+1})^2 = (x-1)^2$$
$$2x + 1 = x^2 - 2x + 1$$
$$0 = x^2 - 4x$$
$$x(x-4) = 0$$
$$x = 0 \text{ or } x = 4$$
Since 0 is an extraneous root, the solution set is $\{4\}$.

61.
$$w = \frac{\sqrt{w+1}}{2}$$
$$2w = \sqrt{w+1}$$
$$(2w)^2 = (\sqrt{w+1})^2$$
$$4w^2 = w + 1$$
$$4w^2 - w = 1$$

$$w^2 - \frac{1}{4}w = \frac{1}{4}$$
$$w^2 - \frac{1}{4}w + \frac{1}{64} = \frac{1}{4} + \frac{1}{64}$$
$$\left(w - \frac{1}{8}\right)^2 = \frac{17}{64}$$
$$w - \frac{1}{8} = \pm\frac{\sqrt{17}}{8}$$
$$w = \frac{1}{8} \pm \frac{\sqrt{17}}{8}$$

The number $\frac{1-\sqrt{17}}{8}$ is a negative number. No negative number can be a solution to the original equation because the left side would be negative and the right side is a principal square root. The solution set is $\left\{\frac{1+\sqrt{17}}{8}\right\}$.

63. $\frac{t}{t-2} = \frac{2t-3}{t}$
$$t^2 = (t-2)(2t-3)$$
$$t^2 = 2t^2 - 7t + 6$$
$$-t^2 + 7t - 6 = 0$$
$$t^2 - 7t + 6 = 0$$
$$(t-6)(t-1) = 0$$
$$t = 6 \quad \text{or} \quad t = 1$$

Solution set: $\{1, 6\}$

65. $\frac{2}{x^2} + \frac{4}{x} + 1 = 0$
$$x^2\left(\frac{2}{x^2} + \frac{4}{x} + 1\right) = x^2(0)$$
$$2 + 4x + x^2 = 0$$
$$x^2 + 4x = -2$$
$$x^2 + 4x + 4 = -2 + 4$$
$$(x+2)^2 = 2$$
$$x + 2 = \pm\sqrt{2}$$
$$x = -2 \pm \sqrt{2}$$

Solution set: $\{-2-\sqrt{2}, -2+\sqrt{2}\}$

67. $x^2 + 2x + 5 = 0$
$$x^2 + 2x + 1 = -5 + 1$$
$$(x+1)^2 = -4$$
$$x + 1 = \pm\sqrt{-4}$$
$$x = -1 \pm 2i$$

The solution set is $\{-1-2i, -1+2i\}$.

69. $x^2 - 6x + 11 = 0$
$$x^2 - 6x + 9 = -11 + 9$$
$$(x-3)^2 = -2$$
$$x - 3 = \pm\sqrt{-2}$$
$$x = 3 \pm i\sqrt{2}$$

The solution set is $\left\{3 - i\sqrt{2}, 3 + i\sqrt{2}\right\}$.

71. $x^2 = -\frac{1}{2}$
$$x = \pm\sqrt{-\frac{1}{2}} = \pm i\frac{\sqrt{2}}{2}$$

Solution set: $\left\{\pm i\frac{\sqrt{2}}{2}\right\}$

73. $x^2 + 12 = 0$
$$x^2 = -12$$
$$x = \pm\sqrt{-12}$$
$$x = \pm 2i\sqrt{3}$$

The solution set is $\{-2i\sqrt{3}, 2i\sqrt{3}\}$

75. $5z^2 - 4z + 1 = 0$
$$z^2 - \frac{4}{5}z + \frac{4}{25} = -\frac{1}{5} + \frac{4}{25}$$
$$\left(z - \frac{2}{5}\right)^2 = \frac{-1}{25}$$
$$z - \frac{2}{5} = \pm\sqrt{\frac{-1}{25}}$$
$$z = \frac{2}{5} \pm \sqrt{\frac{-1}{25}}$$
$$z = \frac{2}{5} \pm \frac{i}{5} = \frac{2 \pm i}{5}$$

The solution set is $\left\{\frac{2 \pm i}{5}\right\}$.

77. $x^2 = -121$
$$x = \pm\sqrt{-121} = \pm 11i$$

Solution set: $\{\pm 11i\}$

79. $4x^2 + 25 = 0$
$$4x^2 = -25$$
$$x^2 = -\frac{25}{4}$$
$$x = \pm\sqrt{-\frac{25}{4}} = \pm\frac{5}{2}i$$

Solution set: $\left\{-\frac{5}{2}i, \frac{5}{2}i\right\}$

81. $\left(p + \frac{1}{2}\right)^2 = \frac{9}{4}$
$$p + \frac{1}{2} = \pm\frac{3}{2}$$
$$p = -\frac{1}{2} \pm \frac{3}{2}$$

The solution set is $\{-2, 1\}$

83. $5t^2 + 4t - 3 = 0$
$$5t^2 + 4t = 3$$
$$t^2 + \frac{4}{5}t = \frac{3}{5}$$
$$t^2 + \frac{4}{5}t + \frac{4}{25} = \frac{3}{5} + \frac{4}{25}$$
$$\left(t + \frac{2}{5}\right)^2 = \frac{19}{25}$$
$$t + \frac{2}{5} = \pm\frac{\sqrt{19}}{5}$$
$$t = -\frac{2}{5} \pm \frac{\sqrt{19}}{5}$$

Solution set: $\left\{\frac{-2-\sqrt{19}}{5}, \frac{-2+\sqrt{19}}{5}\right\}$

85. $m^2 + 2m - 24 = 0$
$$m^2 + 2m = 24$$
$$m^2 + 2m + 1 = 24 + 1$$
$$(m+1)^2 = 25$$
$$m + 1 = \pm 5$$

$$m = -1 \pm 5$$
$$m = -1 - 5 = -6 \text{ or } m = -1 + 5 = 4$$
The solution set is $\{-6, 4\}$.

87. $\left(x - 2\right)^2 = -9$
$$x - 2 = \pm\sqrt{-9}$$
$$x = 2 \pm 3i$$
Solution set: $\{2 \pm 3i\}$

89. $-x^2 + x + 6 = 0$
$$-x^2 + x = -6$$
$$x^2 - x = 6$$
$$x^2 - x + \frac{1}{4} = 6 + \frac{1}{4}$$
$$\left(x - \frac{1}{2}\right)^2 = \frac{25}{4}$$
$$x - \frac{1}{2} = \pm\frac{5}{2}$$
$$x = \frac{1}{2} \pm \frac{5}{2}$$
Solution set: $\{-2, 3\}$

91. $x^2 - 6x + 10 = 0$
$$x^2 - 6x + 9 = -10 + 9$$
$$(x - 3)^2 = -1$$
$$x - 3 = \pm\sqrt{-1} = \pm i$$
$$x = 3 \pm i$$
The solution set is $\{3 - i, 3 + i\}$.

93. $2x - 5 = \sqrt{7x + 7}$
$$(2x - 5)^2 = (\sqrt{7x + 7})^2$$
$$4x^2 - 20x + 25 = 7x + 7$$
$$4x^2 - 27x + 18 = 0$$
$$(4x - 3)(x - 6) = 0$$
$$4x - 3 = 0 \text{ or } x - 6 = 0$$
$$x = \frac{3}{4} \text{ or } x = 6$$
Since $3/4$ does not check, the solution set is $\{6\}$.

95. $\frac{1}{x} + \frac{1}{x - 1} = \frac{1}{4}$
$$4x(x - 1)\left(\frac{1}{x} + \frac{1}{x - 1}\right) = 4x(x - 1)\left(\frac{1}{4}\right)$$
$$4x - 4 + 4x = x^2 - x$$
$$-x^2 + 9x - 4 = 0$$
$$x^2 - 9x + 4 = 0$$
$$x^2 - 9x + \frac{81}{4} = -4 + \frac{81}{4}$$
$$\left(x - \frac{9}{2}\right)^2 = \frac{65}{4}$$
$$x - \frac{9}{2} = \pm\frac{\sqrt{65}}{2}$$
$$x = \frac{9}{2} \pm \frac{\sqrt{65}}{2}$$
Solution set: $\left\{\frac{9 - \sqrt{65}}{2}, \frac{9 + \sqrt{65}}{2}\right\}$

97. $(2 + \sqrt{3})^2 - 4(2 + \sqrt{3}) + 1$
$= 4 + 4\sqrt{3} + 3 - 8 - 4\sqrt{3} + 1 = 0$

$(2 - \sqrt{3})^2 - 4(2 - \sqrt{3}) + 1$
$= 4 - 4\sqrt{3} + 3 - 8 + 4\sqrt{3} + 1 = 0$

99. $(i + 1)^2 - 2(i + 1) + 2$
$= -1 + 2i + 1 - 2i - 2 + 2 = 0$

$(1 - i)^2 - 2(1 - i) + 2$
$= 1 - 2i - 1 - 2 + 2i + 2 = 0$

101. $1211.1 \cdot 8700 = 2.81A^2 \cdot 200$
$$A^2 = \frac{1211.1 \cdot 8700}{2.81 \cdot 200}$$
$$A = \sqrt{\frac{1211.1 \cdot 8700}{2.81 \cdot 200}} \approx 136.9 \text{ ft/sec}$$

103. $17{,}568 = 1500x - 3x^2$
$$3x^2 - 1500x = -17{,}568$$
$$x^2 - 500x = -5856$$
$$x^2 - 500x + 62500 = -5856 + 62500$$
$$(x - 250)^2 = 56644$$
$$x - 250 = \pm 238$$
$$x = 250 \pm 238$$
$$x = 12 \quad \text{or} \quad x = 488$$
Since x is less than 25, the answer is 12.

105. Equation (c) is not quadratic because there is no x^2 term.

109. $\{4.56, 2.74\}$

111. $\{3.53\}$

10.2 WARM-UPS

1. True.

2. False, before identifying a, b, and c, we must write the equation as $3x^2 - 4x + 7 = 0$.

3. True, this is just the quadratic formula with $a = d$, $b = e$, and $c = f$.

4. False, the quadratic formula works to solve any quadratic equation.

5. True, because $(-3)^2 - 4(2)(-4) = 9 + 32 = 41$.

6. True, if the discriminant is 0, then the quadratic equation has one real solution.

7. True.

8. True, because we can write the equation in the form $-1x^2 + 2x + 0 = 0$.

9. False, x and $6 - x$ have a sum of 6.

10. False, there can be one real solution, 2 real solutions, or 2 imaginary solutions.

10.2 EXERCISES

1. The quadratic formula can be used to solve any quadratic equation.

3. Factoring is used when the quadratic polynomial is simple enough to factor.

5. The discriminant is $b^2 - 4ac$.

7. $x^2 + 5x + 6 = 0$

$a = 1, b = 5, \ c = 6$

$$x = \frac{-5 \ \pm \ \sqrt{5^2 - 4(1)(6)}}{2(1)} = \frac{-5 \pm \sqrt{1}}{2}$$

$$= \frac{-5 \pm 1}{2} = \frac{-4}{2}, \frac{-6}{2}$$

Solution set: $\{-3, -2\}$

9. $y^2 + y - 6 = 0$

$a = 1, b = 1, c = -6$

$$y = \frac{-1 \pm \sqrt{1^2 - 4(1)(-6)}}{2(1)} = \frac{-1 \pm \sqrt{25}}{2}$$

$$= \frac{-1 \pm 5}{2} = \frac{4}{2}, \frac{-6}{2}$$

Solution set: $\{-3, 2\}$

11. $-6z^2 + 7z + 3 = 0$

$a = -6, b = 7, c = 3$

$$z = \frac{-7 \pm \sqrt{(7)^2 - 4(-6)(3)}}{2(-6)} = \frac{-7 \pm \sqrt{121}}{-12}$$

$$= \frac{-7 \pm 11}{-12} = \frac{-18}{-12}, \frac{4}{-12}$$

Solution set: $\left\{-\frac{1}{3}, \frac{3}{2}\right\}$

13. $4x^2 - 4x + 1 = 0$

$a = 4, \ b = -4, \ c = 1$

$$x = \frac{4 \pm \sqrt{16 - 4(4)(1)}}{2(4)} = \frac{4 \pm 0}{8} = \frac{1}{2}$$

Solution set: $\left\{\frac{1}{2}\right\}$

15. $-9x^2 + 6x - 1 = 0$

$a = -9, b = 6, c = -1$

$$x = \frac{-6 \pm \sqrt{36 - 4(-9)(-1)}}{2(-9)} = \frac{-6 \pm \sqrt{0}}{-18}$$

$$= \frac{1}{3}$$

Solution set: $\left\{\frac{1}{3}\right\}$

17. $16x^2 + 24x + 9 = 0$

$a = 16, b = 24, c = 9$

$$x = \frac{-24 \pm \sqrt{576 - 4(16)(9)}}{2(16)} = \frac{-24 \pm \sqrt{0}}{32}$$

$$= -\frac{3}{4}$$

Solution set: $\left\{-\frac{3}{4}\right\}$

19. $v^2 + 8v + 6 = 0$

$a = 1, b = 8, c = 6$

$$v = \frac{-8 \pm \sqrt{8^2 - 4(1)(6)}}{2(1)} = \frac{-8 \pm \sqrt{40}}{2}$$

$$= \frac{-8 \pm 2\sqrt{10}}{2} = -4 \pm \sqrt{10}$$

Solution set: $\{-4 \pm \sqrt{10}\}$

21. $x^2 + 5x - 1 = 0$: $a = 1, b = 5, c = -1$

$$x = \frac{-5 \pm \sqrt{5^2 - 4(1)(-1)}}{2(1)} = \frac{-5 \pm \sqrt{29}}{2}$$

Solution set: $\left\{\frac{-5 \pm \sqrt{29}}{2}\right\}$

23. $2t^2 - 6t + 1 = 0$

$a = 2, b = -6, c = 1$

$$t = \frac{6 \pm \sqrt{36 - 4(2)(1)}}{2(2)} = \frac{6 \pm \sqrt{28}}{4} = \frac{6 \pm 2\sqrt{7}}{4}$$

$$= \frac{2(3 \pm \sqrt{7})}{2(2)} = \frac{3 \pm \sqrt{7}}{2}$$

Solution set: $\left\{\frac{3 \pm \sqrt{7}}{2}\right\}$

25. $2t^2 - 6t + 5 = 0$

$$t = \frac{6 \pm \sqrt{36 - 4(2)(5)}}{2(2)} = \frac{6 \pm \sqrt{-4}}{4} = \frac{6 \pm 2i}{4}$$

$= \frac{3 \pm i}{2}$ Solution set: $\left\{\frac{3 \pm i}{2}\right\}$

27. $-2x^2 + 3x - 6 = 0$

$a = -2, b = 3, c = -6$

$$x = \frac{-3 \pm \sqrt{(3)^2 - 4(-2)(-6)}}{2(-2)} = \frac{-3 \pm \sqrt{-39}}{-4}$$

$$= \frac{-3 \pm i\sqrt{39}}{-4} = \frac{3 \pm i\sqrt{39}}{4}$$

Solution set: $\left\{\frac{3 \pm i\sqrt{39}}{4}\right\}$

29. $\frac{1}{2}x^2 + 13 = 5x, \quad x^2 - 10x + 26 = 0$

$$x = \frac{10 \pm \sqrt{100 - 4(1)(26)}}{2(1)} = \frac{10 \pm \sqrt{-4}}{2}$$

$= \frac{10 \pm 2i}{2} = 5 \pm i$ Solution set: $\{5 \pm i\}$

31. $x^2 - 6x + 2 = 0$

$a = 1, b = -6, c = 2$

$b^2 - 4ac = 36 - 4(1)(2) = 28$

So there are two real solutions to the equation.

33. $-2x^2 + 5x - 6 = 0$

$a = -2, b = 5, c = -6$

$b^2 - 4ac = 25 - 4(-2)(-6) = -23$

So there are no real solutions to the equation.

35. $4m^2 - 20m + 25 = 0$

$a = 4, b = -20, c = 25$

$b^2 - 4ac = 400 - 4(4)(25) = 0$

So there is one real solution to the equation.

37. $y^2 - \frac{1}{2}y + \frac{1}{4} = 0$

$a = 1, b = -\frac{1}{2}, c = \frac{1}{4}$

$b^2 - 4ac = \frac{1}{4} - 4(1)(\frac{1}{4}) = -\frac{3}{4}$

So there are no real solutions to the equation.

39. $-3t^2 + 5t + 6 = 0$

$a = -3, \ b = 5, c = 6$

$b^2 - 4ac = 25 - 4(-3)(6) = 97$

So there are two real solutions to the equation.

41. $16z^2 - 24z + 9 = 0$

$a = 16, b = -24, c = 9$

$b^2 - 4ac = (-24)^2 - 4(16)(9) = 0$

So there is one real solution to the equation.

43. $5x^2 - 7 = 0$

$a = 5, b = 0, c = -7$

$b^2 - 4ac = 0 - 4(5)(-7) = 140$

So there are two real solutions to the equation.

45. $x^2 - x = 0$

$a = 1, b = -1, c = 0$

$b^2 - 4ac = 1 - 4(1)(0) = 1$

So there are two real solutions to the equation.

47. $\frac{1}{4}y^2 + y = 1$

$4\left(\frac{1}{4}y^2 + y\right) = 4 \cdot 1$

$y^2 + 4y = 4$

$y^2 + 4y - 4 = 0$

$a = 1, b = 4, c = -4$

$$y = \frac{-4 \pm \sqrt{16 - 4(1)(-4)}}{2(1)} = \frac{-4 \pm \sqrt{32}}{2}$$

$$= \frac{-4 \pm 4\sqrt{2}}{2} = -2 \pm 2\sqrt{2}$$

Solution set: $\left\{-2 \pm 2\sqrt{2}\right\}$

49. $\frac{1}{3}x^2 + \frac{1}{2}x = \frac{1}{3}$

$6\left(\frac{1}{3}x^2 + \frac{1}{2}x\right) = 6\left(\frac{1}{3}\right)$

$2x^2 + 3x = 2$

$2x^2 + 3x - 2 = 0$

$a = 2, b = 3, c = -2$

$$x = \frac{-3 \pm \sqrt{9 - 4(2)(-2)}}{2(2)} = \frac{-3 \pm 5}{4}$$

$$= \frac{2}{4}, \frac{-8}{4}$$

Solution set: $\left\{-2, \frac{1}{2}\right\}$

51. $3y^2 + 2y - 4 = 0$

$a = 3, b = 2, c = -4$

$$y = \frac{-2 \pm \sqrt{4 - 4(3)(-4)}}{2(3)} = \frac{-2 \pm \sqrt{52}}{6}$$

$$= \frac{-2 \pm 2\sqrt{13}}{6} = \frac{-1 \pm \sqrt{13}}{3}$$

Solution set: $\left\{\frac{-1 \pm \sqrt{13}}{3}\right\}$

53. $\frac{w}{w-2} = \frac{w}{w-3}$

$w^2 - 3w = w^2 - 2w$

$0 = w$

Solution set: $\{0\}$

55. $\frac{9(3x-5)^2}{4} = 1$

$(3x - 5)^2 = \frac{4}{9}$

$3x - 5 = \pm\frac{2}{3}$

$3x = 5 + \frac{2}{3}$ or $3x = 5 - \frac{2}{3}$

$3x = \frac{17}{3}$ or $3x = \frac{13}{3}$

$x = \frac{17}{9}$ or $x = \frac{13}{9}$

Solution set: $\left\{\frac{13}{9}, \frac{17}{9}\right\}$

57. $25 - \frac{1}{3}x^2 = 0$

$-\frac{1}{3}x^2 = -25$

$x^2 = 75$

$x = \pm\sqrt{75} = \pm 5\sqrt{3}$

Solution set: $\left\{\pm 5\sqrt{3}\right\}$

59. $1 + \frac{20}{x^2} = \frac{8}{x}$

$x^2(1 + \frac{20}{x^2}) = x^2 \cdot \frac{8}{x}$

$x^2 - 8x + 20 = 0$

$x^2 - 8x + 16 = -20 + 16$

$(x - 4)^2 = -4$

$x - 4 = \pm 2i$

$x = 4 \pm 2i$

Solution set: $\{4 \pm 2i\}$

61. $(x - 8)(x + 4) = -42$

$x^2 - 4x + 10 = 0$

$x^2 - 4x + 4 = -10 + 4$

$(x - 2)^2 = -6$

$x - 2 = \pm i\sqrt{6}$

$x = 2 \pm i\sqrt{6}$

Solution set: $\{2 \pm i\sqrt{6}\}$

63. $y = \frac{3(2y + 5)}{8(y - 1)}$

$8y^2 - 8y = 6y + 15$

$8y^2 - 14y - 15 = 0$

$(4y+3)(2y-5)=0$
$4y+3=0$ or $2y-5=0$
$y=-\frac{3}{4}$ or $y=\frac{5}{2}$

Solution set: $\left\{-\frac{3}{4}, \frac{5}{2}\right\}$

65. $x=\dfrac{-3.2 \pm \sqrt{(3.2)^2-4(1)(-5.7)}}{2(1)}$

$=\dfrac{-3.2 \pm \sqrt{33.04}}{2}$ $\{-4.474, 1.274\}$

67. $x=\dfrac{7.4 \pm \sqrt{(-7.4)^2-4(1)(13.69)}}{2(1)}$

$=\dfrac{7.4 \pm \sqrt{0}}{2(1)}=3.7$ $\{3.7\}$

69. $x=\dfrac{-6.72 \pm \sqrt{(6.72)^2-4(1.85)(3.6)}}{2(1.85)}$

$=\dfrac{-6.72 \pm \sqrt{18.5184}}{3.7}$ $\{-2.979, -0.653\}$

71. $x=\dfrac{-14379 \pm \sqrt{14379^2-4(3)(243)}}{2(3)}$

$=\dfrac{-14379 \pm 14378.8986}{6}$

Solution set: $\{-4792.983, -0.017\}$

73. $x=\dfrac{-0.00075 \pm \sqrt{0.00075^2-4 \cdot 1(-0.0062)}}{2(1)}$

$=\dfrac{-0.00075 \pm 0.1574819}{2}$ $\{-0.079, 0.078\}$

75. Let $x=$ one number and $x+1=$ the other. Since their product is 16, we can write

$x(x+1)=16$
$x^2+x=16$
$x^2+x-16=0$

$x=\dfrac{-1 \pm \sqrt{1^2-4(1)(-16)}}{2(1)}=\dfrac{-1 \pm \sqrt{65}}{2}$

Since the numbers are positive,

$x=\dfrac{-1+\sqrt{65}}{2}$ and $x+1=\dfrac{-1+\sqrt{65}}{2}+\dfrac{2}{2}$

$=\dfrac{1+\sqrt{65}}{2}$. The numbers are $\dfrac{1+\sqrt{65}}{2}$ and $\dfrac{-1+\sqrt{65}}{2}$ or approximately 4.5 and 3.5.

77. Let $x=$ one of the numbers. If the numbers are to have a sum of 6, then $6-x=$ the other number. Since their product is 4, we can write the equation

$x(6-x)=4$
$-x^2+6x-4=0$
$x^2-6x+4=0$

$x=\dfrac{6 \pm \sqrt{36-4(1)(4)}}{2(1)}=\dfrac{6 \pm 2\sqrt{5}}{2}$
$=3 \pm \sqrt{5}$

If $x=3+\sqrt{5}$, then
$6-x=6-(3+\sqrt{5})=3-\sqrt{5}$.
If $x=3-\sqrt{5}$, then
$6-x=6-(3-\sqrt{5})=3+\sqrt{5}$.
So the numbers are $3+\sqrt{5}$ and $3-\sqrt{5}$ or approximately 5.2 and 0.8

79. Let $x=$ the width and $x+1=$ the length. Since the diagonal is the hypotenuse of a right triangle, we can write

$x^2+(x+1)^2=(\sqrt{3})^2$
$x^2+x^2+2x+1=3$
$2x^2+2x-2=0$
$x^2+x-1=0$

$x=\dfrac{-1 \pm \sqrt{1-4(1)(-1)}}{2(1)}=\dfrac{-1 \pm \sqrt{5}}{2}$

Since the width of a rectangle is a positive number, we have width $=\dfrac{-1+\sqrt{5}}{2}$ feet and length $=\dfrac{-1+\sqrt{5}}{2}+\dfrac{2}{2}=\dfrac{1+\sqrt{5}}{2}$ feet or width approximately 0.6 ft and length approximately 1.6 ft.

81. Let $x=$ the width, and $x+4=$ the length. Since the area is 10 square feet, we can write

$x(x+4)=10$
$x^2+4x-10=0$

$x=\dfrac{-4 \pm \sqrt{16-4(1)(-10)}}{2(1)}=\dfrac{-4 \pm 2\sqrt{14}}{2}$
$=-2 \pm \sqrt{14}$

Since $x>0$, the width is $-2+\sqrt{14}$ feet, and the length is $-2+\sqrt{14}+4=2+\sqrt{14}$ feet or approximately 1.7 ft and 5.7 ft.

83. The time it takes to reach the ground is found by solving the equation

$-16t^2+16t+96=0$
$t^2-t-6=0$
$(t-3)(t+2)=0$
$t=3$ or $t=-2$

The pine cone reaches the earth 3 seconds after it is tossed.

85. The time it takes to reach the ground is found by solving the equation

$-16t^2+10t+5=0$
$16t^2-10t-5=0$

$$t = \frac{-(-10) \pm \sqrt{(-10)^2 - 4(16)(-5)}}{2(16)}$$
$$= \frac{10 \pm \sqrt{420}}{32} = \frac{10 \pm 2\sqrt{105}}{32}$$
$$= \frac{5 \pm \sqrt{105}}{16}$$

Since $5 - \sqrt{105}$ is a negative number, the only possibility for t is $\frac{5 + \sqrt{105}}{16}$ or approximately 1.0 seconds.

87. The time it takes to reach the river is found by solving the equation

$$-16t^2 - 30t + 1000 = 0$$
$$8t^2 + 15t - 500 = 0$$
$$t = \frac{-15 \pm \sqrt{15^2 - 4(8)(-500)}}{2(8)}$$
$$= \frac{-15 \pm \sqrt{16225}}{16} = \frac{-15 \pm 127.377}{16}$$

Since the time is positive, the time is $\frac{-15 + 127.377}{16}$, or 7.0 seconds.

89. If x is the width of the border, then

$$(30 - 2x)(40 - 2x) = 704$$
$$1200 - 140x + 4x^2 = 704$$
$$4x^2 - 140x + 496 = 0$$
$$x^2 - 35x + 124 = 0$$
$$x = \frac{35 \pm \sqrt{35^2 - 4 \cdot 1 \cdot 124}}{2} = \frac{35 \pm \sqrt{729}}{2}$$
$$= \frac{35 \pm 27}{2} = 31 \text{ or } 4$$

The width of the border is 4 inches.

91. Let $x =$ the number presently sharing the cost. The cost is $\frac{100}{x}$ dollars each. If 6 more join, the cost is $\frac{100}{x} - 15$. We have

$$(x + 6)(\frac{100}{x} - 15) = 100$$
$$100 + \frac{600}{x} - 15x - 90 = 100$$
$$\frac{600}{x} - 15x - 90 = 0$$
$$600 - 15x^2 - 90x = 0$$
$$x^2 + 6x - 40 = 0$$
$$(x - 4)(x + 10) = 0$$
$$x - 4 = 0 \quad \text{or } x + 10 = 0$$
$$x = 4 \quad \text{or} \quad x = -10$$

There are 4 workers in the original group.

93. Let $x =$ the original number of melons purchased. The cost is $\frac{750}{x}$ dollars each. He sold them for $\frac{750}{x} + 2$ dollars each. When he sold $x - 100$ of them he broke even. So

$$\left(\frac{750}{x} + 2\right)(x - 100) = 750$$
$$750 + \frac{75000}{x} + 2x - 200 = 750$$
$$2x - 200 + \frac{75000}{x} = 0$$
$$2x^2 - 200x + 75000 = 0$$

$$x^2 - 100x + 37500 = 0$$
$$(x - 250)(x + 150) = 0$$
$$x - 250 = 0 \quad \text{or } x + 150 = 0$$
$$x = 250 \quad \text{or} \quad x = -150$$

So he bought 250 melons originally.

95.
$$6x^2 + 5x - 4 = 0$$
$$(3x + 4)(2x - 1) = 0$$
$$x = -\frac{4}{3} \text{ or } x = \frac{1}{2}$$
$$-\frac{4}{3} + \frac{1}{2} = -\frac{8}{6} + \frac{3}{6} = -\frac{5}{6}$$
$$-\frac{b}{a} = -\frac{5}{6}$$

The sum of the solutions is equal to $-\frac{b}{a}$.

97. The solutions to $6x^2 + 5x - 4 = 0$ are $-\frac{4}{3}$ and $\frac{1}{2}$. Their product is $-\frac{2}{3}$, which is the value of $\frac{c}{a}$

99. Since $y = x^2 - 6.33x + 3.7$ crosses the x-axis twice there are 2 real solutions.

101. Since $y = 4x^2 - 67.1x + 344$ does not cross the x-axis there are no real solutions.

103. Since the graph does not cross the x-axis there are no real solutions.

10.3 WARM-UPS

1. True, because $-1 = (-2)^2 - 5$ is correct.
2. False, the y-intercept is (0, 9).
3. True, because the solution to $x^2 - 5 = 0$ is $x = \pm\sqrt{5}$.
4. True, because $a > 0$.
5. False, because in $y = x^2 + 4$, $a > 0$.
6. True, because $x = \frac{-2}{2(1)} = -1$ and $(-1)^2 + 2(-1) = -1$.
7. True, because $x^2 + 1 = 0$ has no real solution.
8. True, because if $x = 0$, then $y = c$.
9. True, because the parabola opens downward from the vertex (0, 9)
10. False, the minimum value of y occurs when $x = 7/6$.

10.3 EXERCISES

1. The graph of $y = ax^2 + bx + c$ with $a \neq 0$ is a parabola.

3. To find the x-intercepts solve $ax^2 + bx + c = 0$.

5. The x-coordinate of the vertex is $-b/(2a)$.

7. If $x = 3$, then $y = 3^2 - 3 - 12 = -6$.
If $y = 0$, then $x^2 - x - 12 = 0$.

$$(x - 4)(x + 3) = 0$$
$$x - 4 = 0 \text{ or } x + 3 = 0$$
$$x = 4 \text{ or } \quad x = -3$$

So the ordered pairs are $(3, -6)$, $(4, 0)$, and $(-3, 0)$.

9. If $x = 4$, then
$y = -16 \cdot 4^2 + 32(4) = -128$.
It $y = 0$, then $-16x^2 + 32x = 0$.

$$-16x(x - 2) = 0$$
$$-16x = 0 \text{ or } x - 2 = 0$$
$$x = 0 \text{ or } \quad x = 2$$

The ordered pairs are $(4, -128)$, $(0, 0)$ and $(2, 0)$.

11. Because $a = 1$ in $y = x^2 + 5$ the graph opens upward.

13. Because $a = -3$ in $y = -3x^2 + 4x + 2$ the graph opens downward.

15. $y = (-2x + 3)^2 = 4x^2 - 12x + 9$
Because $a = 4$ in $y = 4x^2 - 12x + 9$ the graph opens upward.

17. The ordered pairs $(-2, 6)$, $(-1, 3)$, $(0, 2)$, $(1, 3)$, and $(2, 6)$ satisfy $y = x^2 + 2$.

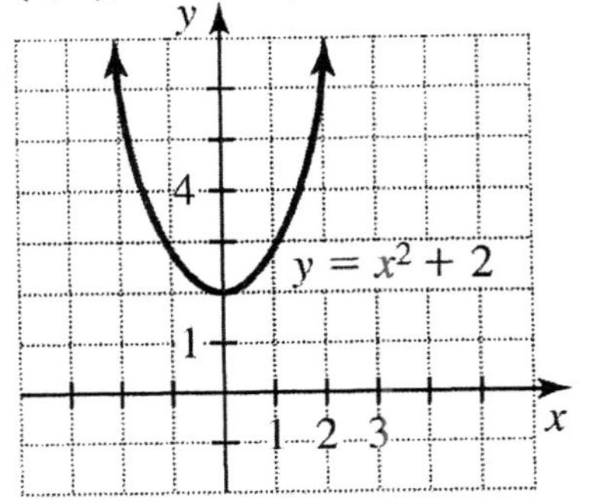

19. The ordered pairs $(-4, 4)$, $(-2, -2)$, $(0, -4)$, $(2, -2)$, and $(4, 4)$ satisfy $y = \frac{1}{2}x^2 - 4$.

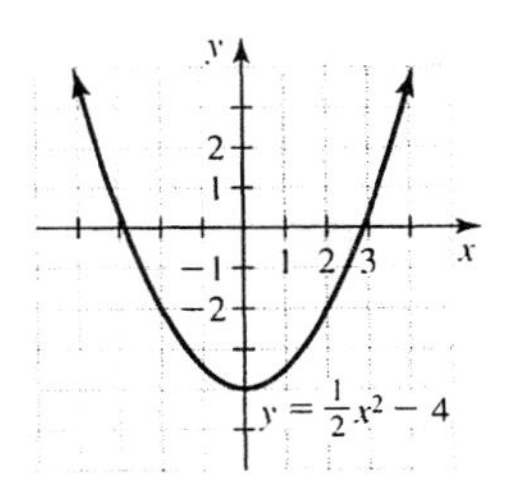

21. The ordered pairs $(-2, -3)$, $(-1, 3)$, $(0, 5)$, $(1, 3)$, and $(2, -3)$ satisfy $y = -2x^2 + 5$.

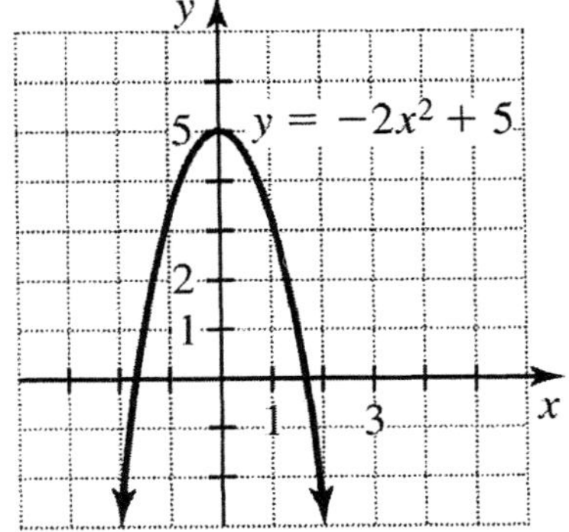

23. The ordered pairs $(-6, -7)$, $(-3, 2)$, $(0, 5)$, $(3, 2)$, and $(6, -7)$ satisfy $y = -\frac{1}{3}x^2 + 5$.

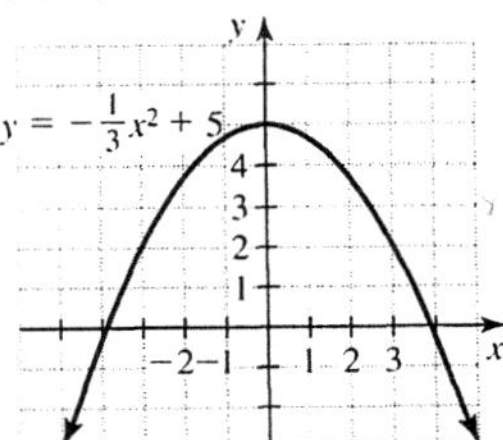

25. The ordered pairs $(0, 4)$, $(1, 1)$, $(2, 0)$, $(3, 1)$, and $(4, 4)$ satisfy $y = (x - 2)^2$.

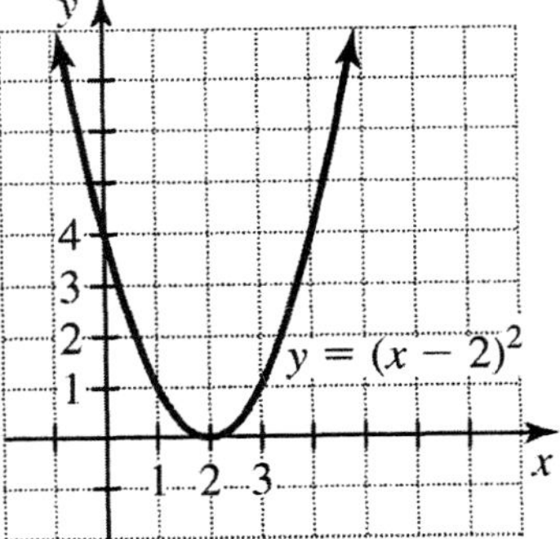

27. For $f(x) = x^2 - 9$ we have $a = 1$ and $b = 0$.

$$x = \frac{-b}{2a} = \frac{-0}{2(1)} = 0$$
$$y = 0^2 - 9 = -9$$

So the vertex is $(0, -9)$.

29. For $y = x^2 - 4x + 1$ we have $a = 1$ and $b = -4$.

$$x = \frac{-b}{2a} = \frac{4}{2(1)} = 2$$
$$y = 2^2 - 4(2) + 1 = -3$$
So the vertex is $(2, -3)$.

31. For $f(x) = -2x^2 + 20x + 1$ we have $a = -2$ and $b = 20$.
$$x = \frac{-b}{2a} = \frac{-20}{2(-2)} = 5$$
$$y = -2(5)^2 + 20(5) + 1 = 51$$
So the vertex is $(5, 51)$.

33. For $y = x^2 - x + 1$ we have $a = 1$ and $b = -1$.
$$x = \frac{-b}{2a} = \frac{1}{2(1)} = \frac{1}{2}$$
$$y = \left(\frac{1}{2}\right)^2 - \left(\frac{1}{2}\right) + 1 = \frac{3}{4}$$
So the vertex is $\left(\frac{1}{2}, \frac{3}{4}\right)$.

35. To find the y-intercept, let $x = 0$:
$f(0) = 16 - 0^2 = 16$.
So the y-intercept is $(0, 16)$.
To find the x-intercepts let $y = 0$:
$$16 - x^2 = 0$$
$$x^2 = 16$$
$$x = \pm 4$$
So the x-intercepts are $(-4, 0)$ and $(4, 0)$.

37. To find the y-intercept, let $x = 0$:
$$y = 0^2 - 2(0) - 8 = -8.$$
So the y-intercept is $(0, -8)$.
To find the x-intercepts let $y = 0$:
$$x^2 - 2x - 8 = 0$$
$$(x - 4)(x + 2) = 0$$
$$x = 4 \quad \text{or} \quad x = -2$$
So the x-intercepts are $(-2, 0)$ and $(4, 0)$.

39. To find the y-intercept, let $x = 0$:
$$f(0) = -4(0)^2 + 12(0) - 9 = -9.$$
So the y-intercept is $(0, -9)$.
To find the x-intercepts let $y = 0$:
$$-4x^2 + 12x - 9 = 0$$
$$4x^2 - 12x + 9 = 0$$
$$(2x - 3)^2 = 0$$
$$2x - 3 = 0$$
$$x = 3/2$$
So the x-intercept is $(3/2, 0)$.

41. $x = \frac{-b}{2a} = \frac{1}{2(1)} = \frac{1}{2}$.
$$f\left(\frac{1}{2}\right) = \left(\frac{1}{2}\right)^2 - \frac{1}{2} - 2 = -\frac{9}{4}$$
The vertex is $\left(\frac{1}{2}, -\frac{9}{4}\right)$. The y-intercept is $(0, -2)$.
$$x^2 - x - 2 = 0$$
$$(x - 2)(x + 1) = 0$$
$$x - 2 = 0 \quad \text{or} \quad x + 1 = 0$$
$$x = 2 \quad \text{or} \quad x = -1$$
The x-intercepts are $(-1, 0)$ and $(2, 0)$.

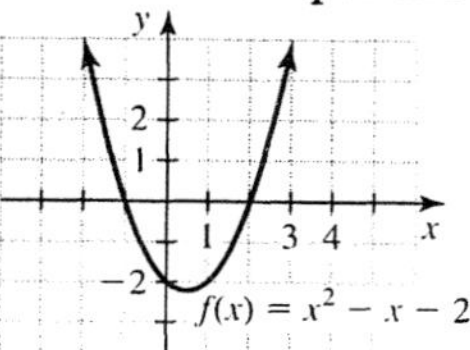

43. $x = \frac{-b}{2a} = \frac{-2}{2(1)} = -1$
$g(-1) = (-1)^2 + 2(-1) - 8 = -9$
The vertex is $(-1, -9)$. The y-intercept is $(0, -8)$.
$$x^2 + 2x - 8 = 0$$
$$(x - 2)(x + 4) = 0$$
$$x - 2 = 0 \quad \text{or} \quad x + 4 = 0$$
$$x = 2 \quad \text{or} \quad x = -4$$
The x-intercepts are $(-4, 0)$ and $(2, 0)$.

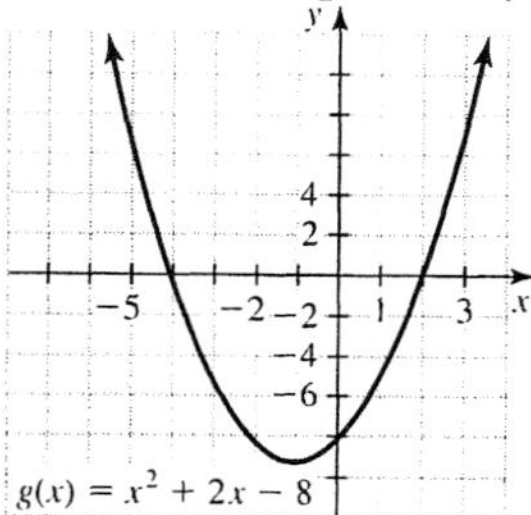

45. $y = -x^2 - 4x - 3$
$$x = \frac{-b}{2a} = \frac{4}{2(-1)} = -2$$
$$y = -(-2)^2 - 4(-2) - 3 = 1$$
Vertex $(-2, 1)$, y-intercept $(0, -3)$.
$$-x^2 - 4x - 3 = 0$$
$$x^2 + 4x + 3 = 0$$
$$(x + 1)(x + 3) = 0$$
$$x + 1 = 0 \quad \text{or} \quad x + 3 = 0$$
$$x = -1 \quad \text{or} \quad x = -3$$
The x-intercepts are $(-1, 0)$ and $(-3, 0)$.

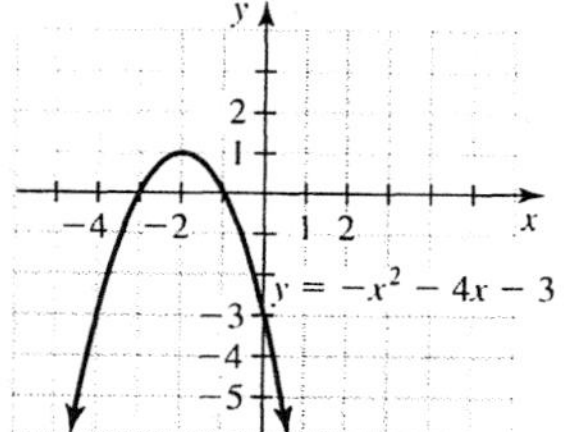

47. $h(x) = -x^2 + 3x + 4$
$$x = \frac{-b}{2a} = \frac{-3}{2(-1)} = \frac{3}{2}$$

$h\left(\frac{3}{2}\right) = -(\frac{3}{2})^2 + 3(\frac{3}{2}) + 4 = \frac{25}{4}$
Vertex $(\frac{3}{2}, \frac{25}{4})$, y-intercept $(0, 4)$.

$$-x^2 + 3x + 4 = 0$$
$$x^2 - 3x - 4 = 0$$
$$(x-4)(x+1) = 0$$
$$x - 4 = 0 \text{ or } x + 1 = 0$$
$$x = 4 \text{ or } x = -1$$

The x-intercepts are $(4, 0)$ and $(-1, 0)$.

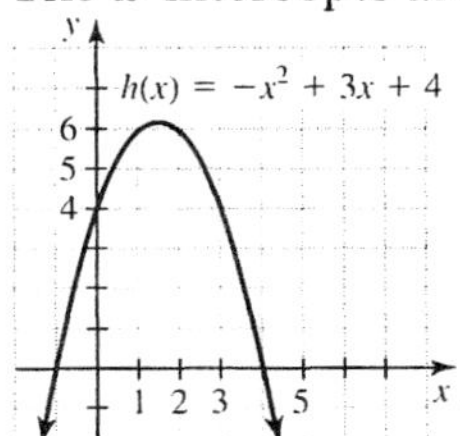

49. $a = b^2 - 6b - 16$
$$b = \frac{6}{2(1)} = 3$$
$a = 3^2 - 6(3) - 16 = -25$
Vertex $(3, -25)$, b-intercept $(0, -16)$.

$$b^2 - 6b - 16 = 0$$
$$(b-8)(b+2) = 0$$
$$b - 8 = 0 \text{ or } b + 2 = 0$$
$$b = 8 \text{ or } b = -2$$

The b-intercepts are $(8, 0)$ and $(-2, 0)$.

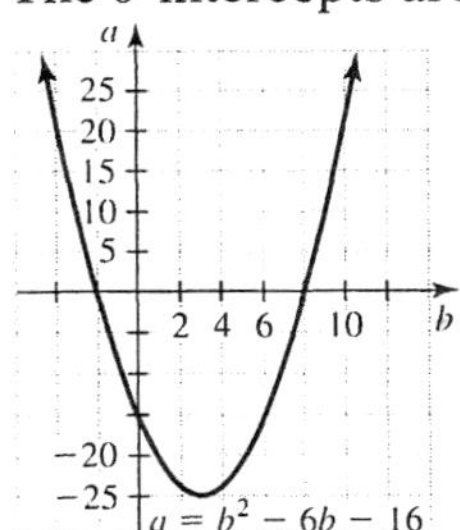

51. $y = x^2 - 8$
$$x = \frac{-b}{2a} = \frac{0}{2(1)} = 0$$
$y = 0^2 - 8 = -8$
The minimum value of y is -8.

53. $y = -3x^2 + 14$
$$x = \frac{-b}{2a} = \frac{0}{2(-3)} = 0$$
$y = -3 \cdot 0^2 + 14 = 14$
The maximum value of y is 14.

55. $y = x^2 + 2x + 3$
$$x = \frac{-b}{2a} = \frac{-2}{2(1)} = -1$$
$y = (-1)^2 + 2(-1) + 3 = 2$
The minimum value of y is 2.

57. $y = -2x^2 - 4x$
$$x = \frac{-b}{2a} = \frac{4}{2(-2)} = -1$$
$y = -2(-1)^2 - 4(-1) = 2$
The maximum value of y is 2.

59. $s(t) = -16t^2 + 64t$
$$t = \frac{-64}{2(-16)} = 2$$
$s(2) = -16 \cdot 2^2 + 64(2) = 64$
Maximum height is 64 feet.

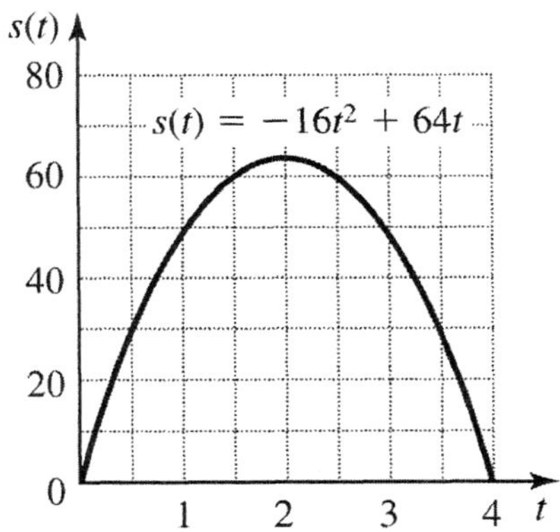

61. The minimum value of C in the formula $C = 0.009x^2 - 1.8x + 100$ is attained when
$$x = \frac{-(-1.8)}{2(0.009)} = 100.$$
The cost per hour will be at a minimum when they produce 100 balls per hour.

63. $A = -w^2 + 50w$
$$w = \frac{-50}{2(-1)} = 25$$
$A = -25^2 + 50(25) = 625$
Maximum area is 625 square meters.

65. $A(t) = -2t^2 + 32t + 12$
$$t = \frac{-32}{2(-2)} = 8$$
The nitrogen dioxide is at its maximum 8 hours after 6 A.M. or at 2 P.M.

67. Since the x-coordinate of the tower on the right is 20, we get $y = 0.0375(20)^2 = 15$. So the height of the towers is 15 meters. Use the Pythagorean theorem to find z:
$$15^2 + 20^2 = z^2$$
$$z^2 = 625$$
$$z = 25$$
So the length of the cable marked z is 25 meters.

69. The graph of $y = ax^2$ gets narrower as a gets larger.

71. The graph of $y = x^2$ has the same shape as $x = y^2$.

73. a) Vertex $(0.84, -0.68)$, x-intercepts $(1.30, 0)$, $(0.38, 0)$
b) Vertex $(5.96, 46.26)$,
x-intercepts $(12.48, 0)$, $(-0.55, 0)$

10.4 WARM-UPS

1. True, because if $w = x^2$ the equation becomes $w^2 - 5w + 6 = 0$.
2. False, the equation cannot be solved by substitution.
3. False, we can use factoring.
4. True, because $(x^{1/6})^2 = x^{2/6} = x^{1/3}$.
5. False, we should let $w = \sqrt{x}$.
6. False, because $(2^{1/2})^2 = 2^{2/2} = 2^1 = 2$.
7. False, his rate is $1/x$ of the fence per hour.
8. True, because $R = D/T$.
9. False, against a 5 mph current it will go 5 mph.
10. False, the dimensions of the bottom will be $11 - 2x$ by $14 - 2x$.

10.4 EXERCISES

1. If the coefficients are integers and the discriminant is a perfect square, then the quadratic polynomial can be factored.
3. If the solutions are a and b, then the quadratic equation $(x - a)(x - b) = 0$ has those solutions.
5. If the solutions are 3 and -7, then the factors are $x - 3$ and $x - (-7)$ or $x + 7$:
$$(x-3)(x+7) = 0$$
$$x^2 + 4x - 21 = 0$$
7. If the solutions are 4 and 1, then the factors are $x - 4$ and $x - 1$:
$$(x-4)(x-1) = 0$$
$$x^2 - 5x + 4 = 0$$
9. $(x - \sqrt{5})(x + \sqrt{5}) = 0$
$$x^2 - 5 = 0$$
11. $(x - 4i)(x + 4i) = 0$
$$x^2 + 16 = 0$$
13. $(x - i\sqrt{2})(x + i\sqrt{2}) = 0$
$$x^2 + 2 = 0$$
15. $(2x - 1)(3x - 1) = 0$
$$6x^2 - 5x + 1 = 0$$
17. $b^2 - 4ac = (-1)^2 - 4(2)4 = -31$
Polynomial is prime.
19. $b^2 - 4ac = 6^2 - 4(2)(-5) = 76$
Polynomial is prime.
21. $b^2 - 4ac = 19^2 - 4(6)(-36) = 1225$
Since $\sqrt{1225} = 35$, the polynomial is not prime.
$6x^2 + 19x - 36 = (3x - 4)(2x + 9)$
23. $b^2 - 4ac = 25 - 4(4)(-12) = 217$
Polynomial is prime.
25. $b^2 - 4ac = (-18)^2 - 4 \cdot 8(-45) = 1764$
Since $\sqrt{1764} = 42$, the polynomial is not prime.
$8x^2 - 18x - 45 = (4x - 15)(2x + 3)$
27. Let $w = x - 1$ in
$(x-1)^2 - 2(x-1) - 8 = 0$:
$$w^2 - 2w - 8 = 0$$
$$(w-4)(w+2) = 0$$
$$w - 4 = 0 \text{ or } w + 2 = 0$$
$$w = 4 \text{ or } w = -2$$
$$x - 1 = 4 \text{ or } x - 1 = -2$$
$$x = 5 \text{ or } x = -1$$
Solution set: $\{-1, 5\}$
29. Let $w = 2a - 1$ in
$(2a-1)^2 + 2(2a-1) - 8 = 0$:
$$w^2 + 2w - 8 = 0$$
$$(w+4)(w-2) = 0$$
$$w + 4 = 0 \text{ or } w - 2 = 0$$
$$w = -4 \text{ or } w = 2$$
$$2a - 1 = -4 \text{ or } 2a - 1 = 2$$
$$2a = -3 \text{ or } 2a = 3$$
$$a = -\frac{3}{2} \text{ or } a = \frac{3}{2}$$
Solution set: $\left\{-\frac{3}{2}, \frac{3}{2}\right\}$
31. Let $y = w - 1$ in
$(w-1)^2 + 5(w-1) + 5 = 0$:
$$y^2 + 5y + 5 = 0$$
$$y = \frac{-5 \pm \sqrt{25 - 4(1)(5)}}{2(1)} = \frac{-5 \pm \sqrt{5}}{2}$$
$$w - 1 = \frac{-5 \pm \sqrt{5}}{2}$$
$$w = \frac{-5 \pm \sqrt{5}}{2} + 1 = \frac{-5 \pm \sqrt{5}}{2} + \frac{2}{2} = \frac{-3 \pm \sqrt{5}}{2}$$
Solution set: $\left\{\frac{-3 \pm \sqrt{5}}{2}\right\}$
33. Let $w = x^2$ and $w^2 = x^4$ in
$x^4 - 13x^2 + 36 = 0$:
$$w^2 - 13w + 36 = 0$$

$$(w-4)(w-9)=0$$
$$w-4=0 \text{ or } w-9=0$$
$$w=4 \text{ or } w=9$$
$$x^2=4 \text{ or } x^2=9$$
$$x=\pm 2 \text{ or } x=\pm 3$$
Solution set: $\{\pm 2, \pm 3\}$

35. Let $w=x^3$ and $w^2=x^6$ in $x^6-28x^3+27=0$:
$$w^2-28w+27=0$$
$$(w-27)(w-1)=0$$
$$w-27=0 \text{ or } w-1=0$$
$$w=27 \text{ or } w=1$$
$$x^3=27 \text{ or } x^3=1$$
$$x=3 \text{ or } x=1$$
Solution set: $\{1, 3\}$

37. Let $w=x^2$ and $w^2=x^4$ in $x^4-14x^2+45=0$:
$$w^2-14w+45=0$$
$$(w-5)(w-9)=0$$
$$w-5=0 \text{ or } w-9=0$$
$$w=5 \text{ or } w=9$$
$$x^2=5 \text{ or } x^2=9$$
$$x=\pm\sqrt{5} \text{ or } x=\pm 3$$
Solution set: $\{\pm\sqrt{5}, \pm 3\}$

39. Let $w=x^3$, and $w^2=x^6$ in $x^6+7x^3=8$:
$$w^2+7w=8$$
$$w^2+7w-8=0$$
$$(w-1)(w+8)=0$$
$$w-1=0 \text{ or } w+8=0$$
$$w=1 \text{ or } w=-8$$
$$x^3=1 \text{ or } x^3=-8$$
$$x=1 \text{ or } x=-2$$
Solution set: $\{-2, 1\}$

41. Let $w=x^2+1$ in $(x^2+1)^2-11(x^2+1)=-10$:
$$w^2-11w+10=0$$
$$(w-10)(w-1)=0$$
$$w-10=0 \text{ or } w-1=0$$
$$w=10 \text{ or } w=1$$
$$x^2+1=10 \text{ or } x^2+1=1$$
$$x^2=9 \text{ or } x^2=0$$
$$x=\pm 3 \text{ or } x=0$$
Solution set: $\{0, \pm 3\}$

43. Let $w=x^2+2x$ in $(x^2+2x)^2-7(x^2+2x)+12=0$:
$$w^2-7w+12=0$$
$$(w-3)(w-4)=0$$
$$w-3=0 \text{ or } w-4=0$$
$$w=3 \text{ or } w=4$$
$$x^2+2x=3 \text{ or } x^2+2x=4$$
$$x^2+2x-3=0 \text{ or } x^2+2x-4=0$$
$$(x+3)(x-1)=0 \text{ or}$$
$$x=\frac{-2\pm\sqrt{4-4(1)(-4)}}{2}$$
$$x=-3 \text{ or } x=1 \text{ or } x=\frac{-2\pm 2\sqrt{5}}{2}$$
Solution set: $\{-1\pm\sqrt{5}, -3, 1\}$

45. Let $w=y^2+y$ in $(y^2+y)^2-8(y^2+y)+12=0$:
$$w^2-8w+12=0$$
$$(w-6)(w-2)=0$$
$$w=6 \text{ or } w=2$$
$$y^2+y=6 \text{ or } y^2+y=2$$
$$y^2+y-6=0 \text{ or } y^2+y-2=0$$
$$(y+3)(y-2)=0 \text{ or } (y+2)(y-1)=0$$
$$y=-3 \text{ or } y=2 \text{ or } y=-2 \text{ or } y=1$$
Solution set: $\{-3, -2, 1, 2\}$

47. Let $w=x^{1/2}$ and $w^2=x$ in $x-3x^{1/2}+2=0$:
$$w^2-3w+2=0$$
$$(w-2)(w-1)=0$$
$$w-2=0 \text{ or } w-1=0$$
$$w=2 \text{ or } w=1$$
$$x^{1/2}=2 \text{ or } x^{1/2}=1$$
$$(x^{1/2})^2=2^2 \text{ or } (x^{1/2})^2=1^2$$
$$x=4 \text{ or } x=1$$
Solution set: $\{1, 4\}$

49. Let $w=x^{1/3}$ and $w^2=x^{2/3}$ in $x^{2/3}+4x^{1/3}+3=0$:
$$w^2+4w+3=0$$
$$(w+3)(w+1)=0$$
$$w+3=0 \text{ or } w+1=0$$
$$w=-3 \text{ or } w=-1$$
$$x^{1/3}=-3 \text{ or } x^{1/3}=-1$$
$$(x^{1/3})^3=(-3)^3 \text{ or } (x^{1/3})^3=(-1)^3$$
$$x=-27 \text{ or } x=-1$$
Solution set: $\{-27, -1\}$

51. Let $w=x^{1/4}$ and $w^2=x^{1/2}$ in $x^{1/2}-5x^{1/4}+6=0$:
$$w^2-5w+6=0$$
$$(w-3)(w-2)=0$$
$$w-3=0 \text{ or } w-2=0$$
$$w=3 \text{ or } w=2$$
$$x^{1/4}=3 \text{ or } x^{1/4}=2$$
$$(x^{1/4})^4=3^4 \text{ or } (x^{1/4})^4=2^4$$

$x = 81$ or $x = 16$

Solution set: $\{16, 81\}$

53. Let $w = x^{1/2}$ and $w^2 = x$ in $2x - 5x^{1/2} - 3 = 0$.

$$2w^2 - 5w - 3 = 0$$
$$(2w+1)(w-3) = 0$$
$$2w + 1 = 0 \quad \text{or} \quad w - 3 = 0$$
$$w = -\frac{1}{2} \quad \text{or} \quad w = 3$$
$$x^{1/2} = -\frac{1}{2} \quad \text{or} \quad x^{1/2} = 3$$
$$x = \frac{1}{4} \quad \text{or} \quad x = 9$$

The solution $1/4$ does not check in the original equation. So the solution set is $\{9\}$.

55. Let $t = x^{-1}$ in $x^{-2} + x^{-1} - 6 = 0$:

$$t^2 + t - 6 = 0$$
$$(t+3)(t-2) = 0$$
$$t = -3 \quad \text{or} \quad t = 2$$
$$x^{-1} = -3 \quad \text{or} \quad x^{-1} = 2$$
$$x = -\frac{1}{3} \quad \text{or} \quad x = \frac{1}{2}$$

Solution set: $\left\{-\frac{1}{3}, \frac{1}{2}\right\}$

57. Let $w = x^{1/6}$ and $w^2 = x^{1/3}$ in $x^{1/6} - x^{1/3} + 2 = 0$:

$$w - w^2 + 2 = 0$$
$$w^2 - w - 2 = 0$$
$$(w-2)(w+1) = 0$$
$$w = 2 \quad \text{or} \quad w = -1$$
$$x^{1/6} = 2 \quad \text{or} \quad x^{1/6} = -1$$
$$(x^{1/6})^6 = 2^6 \quad \text{or} \quad (x^{1/6})^6 = (-1)^6$$
$$x = 64 \quad \text{or} \quad x = 1$$

The original equation is not satisfied for $x = 1$. So the solution set is $\{64\}$.

59. Let $w = \frac{1}{y-1}$ in $\left(\frac{1}{y-1}\right)^2 + \left(\frac{1}{y-1}\right) = 6$.

$$w^2 + w = 6$$
$$w^2 + w - 6 = 0$$
$$(w+3)(w-2) = 0$$
$$w = -3 \quad \text{or} \quad w = 2$$
$$\frac{1}{y-1} = -3 \quad \text{or} \quad \frac{1}{y-1} = 2$$
$$-3y + 3 = 1 \quad \text{or} \quad 2y - 2 = 1$$
$$-3y = -2 \quad \text{or} \quad 2y = 3$$
$$y = \frac{2}{3} \quad \text{or} \quad y = \frac{3}{2}$$

Solution set: $\left\{\frac{2}{3}, \frac{3}{2}\right\}$

61. Let $w = \sqrt{2x^2 - 3}$ and $w^2 = 2x^2 - 3$ in $2x^2 - 3 - 6\sqrt{2x^2 - 3} + 8 = 0$:

$$w^2 - 6w + 8 = 0$$
$$(w-4)(w-2) = 0$$
$$w = 4 \quad \text{or} \quad w = 2$$
$$\sqrt{2x^2 - 3} = 4 \quad \text{or} \quad \sqrt{2x^2 - 3} = 2$$

Square each side:

$$2x^2 - 3 = 16 \quad \text{or} \quad 2x^2 - 3 = 4$$
$$2x^2 = 19 \quad \text{or} \quad 2x^2 = 7$$
$$x^2 = \frac{19}{2} \quad \text{or} \quad x^2 = \frac{7}{2}$$
$$x = \pm\sqrt{\frac{19}{2}} \quad \text{or} \quad x = \pm\sqrt{\frac{7}{2}}$$
$$x = \pm\frac{\sqrt{38}}{2} \quad \text{or} \quad x = \pm\frac{\sqrt{14}}{2}$$

Solution set: $\left\{\pm\frac{\sqrt{14}}{2}, \pm\frac{\sqrt{38}}{2}\right\}$

63. Let $t = x^{-1}$ and $t^2 = x^{-2}$ in $x^{-2} - 2x^{-1} - 1 = 0$:

$$t^2 - 2t - 1 = 0$$

$$t = \frac{2 \pm \sqrt{4 - 4(1)(-1)}}{2} = \frac{2 \pm 2\sqrt{2}}{2}$$
$$= 1 \pm \sqrt{2}$$
$$x^{-1} = 1 \pm \sqrt{2}$$
$$x = \frac{1}{1+\sqrt{2}} \quad \text{or} \quad x = \frac{1}{1-\sqrt{2}}$$
$$x = \frac{1(1-\sqrt{2})}{(1+\sqrt{2})(1-\sqrt{2})} \quad \text{or}$$
$$x = \frac{1(1+\sqrt{2})}{(1-\sqrt{2})(1+\sqrt{2})}$$

$$x = \frac{1-\sqrt{2}}{-1} \quad \text{or} \quad x = \frac{1+\sqrt{2}}{-1}$$
$$x = -1 + \sqrt{2} \quad \text{or} \quad x = -1 - \sqrt{2}$$

Solution set: $\{-1+\sqrt{2}, -1-\sqrt{2}\}$

65. $w^2 + 4 = 0$

$$w^2 = -4$$
$$w = \pm\sqrt{-4} = \pm 2i$$

The solution set is $\{\pm 2i\}$.

67. $a^4 + 6a^2 + 8 = 0$

$$(a^2+2)(a^2+4) = 0$$
$$a^2 + 2 = 0 \quad \text{or} \quad a^2 + 4 = 0$$
$$a^2 = -2 \quad \text{or} \quad a^2 = -4$$
$$a = \pm i\sqrt{2} \quad \text{or} \quad a = \pm 2i$$

The solution set is $\left\{\pm i\sqrt{2}, \pm 2i\right\}$.

69. $m^4 - 16 = 0$

$$(m^2-4)(m^2+4) = 0$$
$$b^2 - 4 = 0 \quad \text{or} \quad m^2 + 4 = 0$$

$$b^2 = 4 \quad \text{or} \quad m^2 = -4$$
$$b = \pm 2 \quad \text{or} \quad m = \pm 2i$$
The solution set is $\{\pm 2i, \pm 2\}$.

71. $16b^4 - 1 = 0$
$$(4b^2 - 1)(4b^2 + 1) = 0$$
$$4b^2 - 1 = 0 \quad \text{or} \quad 4b^2 + 1 = 0$$
$$b^2 = \frac{1}{4} \quad \text{or} \quad b^2 = -\frac{1}{4}$$
$$b = \pm\frac{1}{2} \quad \text{or} \quad b = \pm\frac{1}{2}i$$
The solution set is $\left\{\pm\frac{1}{2}, \pm\frac{i}{2}\right\}$.

73. $x^3 + 1 = 0$
$$(x+1)(x^2 - x + 1) = 0$$
$$x + 1 = 0 \quad \text{or} \quad x^2 - x + 1 = 0$$
$$x = -1 \quad \text{or} \quad x = \frac{1 \pm \sqrt{(-1)^2 - 4(1)(1)}}{2(1)}$$
$$x = -1 \quad \text{or} \quad x = \frac{1 \pm \sqrt{-3}}{2}$$
The solution set is $\left\{\frac{1 \pm i\sqrt{3}}{2}, -1\right\}$.

75. $x^3 + 8 = 0$
$$(x+2)(x^2 - 2x + 4) = 0$$
$$x + 2 = 0 \quad \text{or} \quad x^2 - 2x + 4 = 0$$
$$x = -2 \quad \text{or} \quad x = \frac{2 \pm \sqrt{(-2)^2 - 4(1)(4)}}{2(1)}$$
$$x = -2 \quad \text{or} \quad x = \frac{2 \pm 2i\sqrt{3}}{2} = 1 \pm i\sqrt{3}$$
The solution set is $\left\{1 \pm i\sqrt{3}, -2\right\}$.

77. $a^{-2} - 2a^{-1} + 5 = 0$

Multiply each side by a^2:
$$1 - 2a + 5a^2 = 0$$
$$5a^2 - 2a + 1 = 0$$
$$a = \frac{2 \pm \sqrt{(-2)^2 - 4(5)(1)}}{2(5)}$$
$$= \frac{2 \pm 4i}{10} = \frac{1 \pm 2i}{5}$$
The solution set is $\left\{\frac{1 \pm 2i}{5}\right\}$.

79. $(2x - 1)^2 - 2(2x - 1) + 5 = 0$

Let $w = 2x - 1$:
$$w^2 - 2w + 5 = 0$$
$$w = \frac{2 \pm \sqrt{(-2)^2 - 4(1)(5)}}{2(1)}$$
$$= \frac{2 \pm 4i}{2} = 1 \pm 2i$$
Now use $w = 2x - 1$:
$$2x - 1 = 1 \pm 2i$$
$$2x = 2 \pm 2i$$
$$x = \frac{2 \pm 2i}{2} = 1 \pm i$$
The solution set is $\{1 \pm i\}$.

81. Let x = Gary's travel time and $x + 1$ = Harry's travel time. Since $R = D/T$, Gary's speed is $300/x$ and Harry's speed is $300/(x+1)$. Since Gary travels 10 mph faster, we can write the following equation.
$$\frac{300}{x} = \frac{300}{x+1} + 10$$
$$x(x+1)\left(\frac{300}{x}\right) = x(x+1)\left(\frac{300}{x+1} + 10\right)$$
$$300x + 300 = 300x + 10x^2 + 10x$$
$$0 = 10x^2 + 10x - 300$$
$$0 = x^2 + x - 30$$
$$0 = (x+6)(x-5)$$
$$x + 6 = 0 \quad \text{or} \quad x - 5 = 0$$
$$x = -6 \quad \text{or} \quad x = 5$$
Gary travels 5 hours and they arrive at 2 P.M.

83. Let x = her speed before lunch and $x - 4$ = her speed after lunch. Since $T = D/R$, her time before lunch was $60/x$ and her time after lunch was $46/(x-4)$. Since she put in one hour more after lunch, we can write the following equation.
$$\frac{46}{x-4} - 1 = \frac{60}{x}$$
$$x(x-4)\left(\frac{46}{x-4} - 1\right) = x(x-4)\left(\frac{60}{x}\right)$$
$$46x - x^2 + 4x = 60x - 240$$
$$-x^2 - 10x + 240 = 0$$
$$x^2 + 10x - 240 = 0$$
$$x = \frac{-10 \pm \sqrt{100 - 4(1)(-240)}}{2(1)}$$
$$= \frac{-10 \pm \sqrt{1060}}{2} = \frac{-10 \pm 2\sqrt{265}}{2}$$
$$= -5 \pm \sqrt{265}$$
Since $-5 - \sqrt{265}$ is a negative number, we disregard that answer. Her speed before lunch was $-5 + \sqrt{265} \approx 11.3$ mph, and her speed after lunch was 4 mph slower: $-9 + \sqrt{265} \approx 7.3$ mph.

85. Let x = Andrew's time and $x + 3$ = John's time. Andrew's rate is $1/x$ job/hr and John's rate is $1/(x+3)$ job/hr. In 8 hours Andrew does $8/x$ job and John does $8/(x+3)$ job.
$$\frac{8}{x} + \frac{8}{x+3} = 1$$

$$x(x+3)\left(\frac{8}{x}+\frac{8}{x+3}\right)=x(x+3)1$$
$$8x+24+8x=x^2+3x$$
$$0=x^2-13x-24$$

$$x=\frac{13\pm\sqrt{169-4(1)(-24)}}{2(1)}=\frac{13\pm\sqrt{265}}{2}$$

Since $\frac{13-\sqrt{265}}{2}$ is a negative number, we disregard that solution. Andrew's time is $\frac{13+\sqrt{265}}{2}\approx 14.6$ hr and John's time is 3 hours more :

$3+\frac{13+\sqrt{265}}{2}=\frac{19+\sqrt{265}}{2}\approx 17.6$ hr

87. Let $x=$ the amount of increase. The new length and width will be $30+x$ and $20+x$. Since the new area is to be 1000, we can write the following equation.

$$(20+x)(30+x)=1000$$
$$x^2+50x-400=0$$

$$x=\frac{-50\pm\sqrt{2500-4(1)(-400)}}{2(1)}$$

$$=\frac{-50\pm\sqrt{4100}}{2}=\frac{-50\pm10\sqrt{41}}{2}$$
$$=-25\pm5\sqrt{41}$$

Disregard the negative solution.
Use $x=-25+5\sqrt{41}$ to get
$30+x=5+5\sqrt{41}\approx 37.02$ feet, and
$20+x=-5+5\sqrt{41}\approx 27.02$ feet.

89. Let $x=$ the number of hours for A to empty the pool and $x+2=$ the number of hours for B to empty the pool. A's rate is $\frac{1}{x}$ pool/hr and B's rate is $\frac{1}{x+2}$ pool/hr. A works for 9 hrs and does $\frac{9}{x}$ of the pool while B works for 6 hours and does $\frac{6}{x+2}$ of the pool. Since the pool is half full we have the following equation.

$$\frac{9}{x}+\frac{6}{x+2}=\frac{1}{2}$$
$$2x(x+2)\left(\frac{9}{x}+\frac{6}{x+2}\right)=2x(x+2)\frac{1}{2}$$
$$18x+36+12x=x^2+2x$$
$$-x^2+28x+36=0$$
$$x^2-28x-36=0$$

$$x=\frac{28\pm\sqrt{28^2-4(1)(-36)}}{2}=\frac{28\pm\sqrt{928}}{2}$$
$$=\frac{28\pm4\sqrt{58}}{2}=14\pm2\sqrt{58}=-1.2 \text{ or } 29.2$$

A would take 29.2 hours working alone.

91. $$\frac{10}{W}=\frac{W}{10-W}$$
$$100-10W=W^2$$
$$W^2+10W-100=0$$

$$W=\frac{-10\pm\sqrt{10^2-4(1)(-100)}}{2}$$
$$=\frac{-10\pm10\sqrt{5}}{2}$$
$$W=-5\pm5\sqrt{5}\approx-16.2 \text{ or } 6.2$$

So the width is $-5+5\sqrt{5}$ or 6.2 meters.

93. a) $P(x)=x^4+6x^2-27$
$P(3i)=(3i)^4+6(3i)^2-27=0$
$P(-3i)=(-3i)^4+6(-3i)^2-27=0$
$P(\sqrt{3})=(\sqrt{3})^4+6(\sqrt{3})^2-27=0$
$P(-\sqrt{3})=(-\sqrt{3})^4+6(-\sqrt{3})^2-27=0$

b) All four numbers are solutions to the equation $x^4+6x^2-27=0$. The solutions occur in conjugate pairs.

95. $\{1,2\}$

97. $\{-4.25,-3.49,\ 0.49,\ 1.25\}$

10.5 WARM-UPS

1. False, solution set is $(-\infty,-2)\cup(2,\infty)$.
2. False, we do not multiply each side by a variable.
3. False, that is not how we solve a quadratic inequality.
4. False, we can solve any quadratic inequality.
5. True, that is why we make the sign graph.
6. True, because inequalities change direction when multiplied by a negative number and we do not know if the variable is positive or negative.
7. True, because the solution is based on rules for multiplying or dividing signed numbers.
8. True, multiply each side by 2.
9. True, subtract 1 from each side.
10. False, because 4 causes the denominator to be 0 and so it cannot be in the solution set.

10.5 EXERCISES

1. A quadratic inequality has the form $ax^2 + bx + c > 0$. In place of $>$ we can also use $<$, $\leq$, or $\geq$.

3. A rational inequality is an inequality involving a rational expression.

5. $x^2 + x - 6 < 0$

$(x + 3)(x - 2) < 0$

$x + 3$ − − − − 0 + + + + + + +

$x - 2$ − − − − − − − − − 0 + + + +

−3 2

The product is negative only if the factors have opposite signs. That happens when x is chosen between −3 and 2. Solution set: $(-3, 2)$

−4 −3 −2 −1 0 1 2 3

7. $x^2 - 3x - 4 \geq 0$

$(x - 4)(x + 1) \geq 0$

$x - 4$ − − − − − − 0 + + + + +

$x + 1$ − − 0 + + + + + + + + +

−1 4

The product is positive only if the factors have the same sign. That happens if x is chosen to the left of −1 or to the right of 4. The product is 0 if x is either −1 or 4.

$(-\infty, -1] \cup [4, \infty)$

−2 −1 0 1 2 3 4 5

9. $x^2 - 2x - 8 \leq 0$

$(x - 4)(x + 2) \leq 0$

$x - 4$ − − − − − − − − 0 + + +

$x + 2$ − − − 0 + + + + + + +

−2 4

The product is negative only if the factors have opposite signs. That happens if x is chosen between −2 and 4. The product is 0 if x is either −2 or 4.

$[-2, 4]$

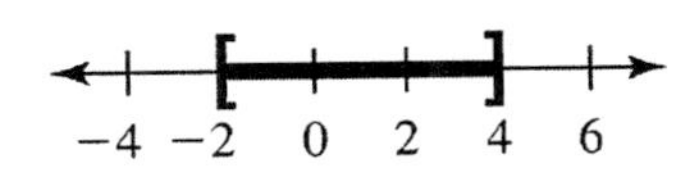

11. $2u^2 + 5u \geq 12$

$2u^2 + 5u - 12 \geq 0$

$(2u - 3)(u + 4) \geq 0$

$2u - 3$ − − − − − − − − 0 + + +

$u + 4$ − − − 0 + + + + + + +

−4 3/2

The product is positive only if the factors have the same sign. That happens if u is chosen to the left of −4 or to the right of 3/2. The product is 0 if u is either −4 or 3/2.

$(-\infty, -4] \cup \left[\frac{3}{2}, \infty\right)$

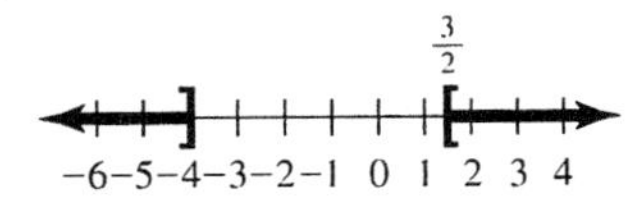

13. $4x^2 - 8x \geq 0$

$4x(x - 2) \geq 0$

$4x$ − − − 0 + + + + + + + +

$x - 2$ − − − − − − − 0 + + + +

0 2

The product is greater than zero when the signs are the same.

$(-\infty, 0] \cup [2, \infty)$

−2 −1 0 1 2 3 4

15. $5x - 10x^2 < 0$

$5x(1 - 2x) < 0$

$5x$ − − 0 + + + + + + + +

$1 - 2x$ + + + + + + 0 − − − −

0 1/2

The product is negative only if the factors have opposite signs: $(-\infty, 0) \cup \left(\frac{1}{2}, \infty\right)$

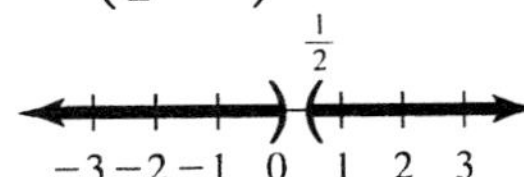

17. $x^2 + 6x + 9 \geq 0$

$(x + 3)(x + 3) \geq 0$

$x + 3$ − − − − 0 + + + +

$x + 3$ − − − − 0 + + + +

−3

The product is positive only if the factors have the same sign. That happens for every value of x except −3, in which case the product is 0. The solution set is the set of all real numbers, $(-\infty, \infty)$.

−3 −2 −1 0 1 2 3

19.
$$x^2 + 4 < 4x$$
$$x^2 - 4x + 4 < 0$$
$$(x-2)^2 < 0$$

$x - 2$ − − − − − − − 0 + + + +
$x - 2$ − − − − − − − 0 + + + +
2

The product is negative only if the factors have opposite signs. That never happens. So the solution set is the empty set $\emptyset$.

21.
$$4x^2 - 20x + 25 \leq 0$$
$$(2x-5)^2 \leq 0$$

$2x - 5$ − − − − − − − 0 + + + +
$2x - 5$ − − − − − − − 0 + + + +
5/2

The product is negative only if the factors have opposite signs. That never happens. If $x = 5/2$ then the inequality is satisfied. So the solution set consists of a single number and it is $\left\{\frac{5}{2}\right\}$.

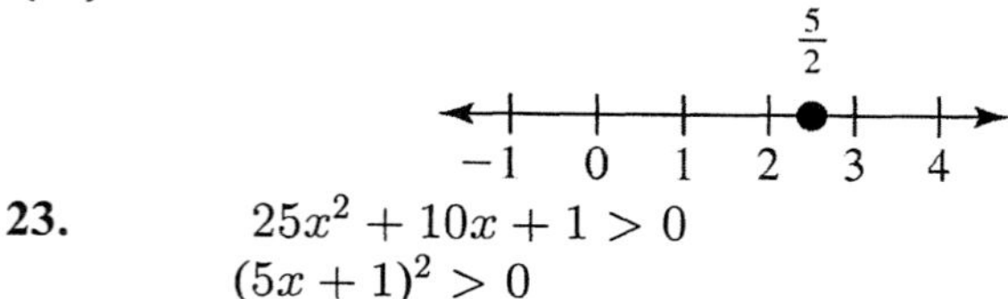

23.
$$25x^2 + 10x + 1 > 0$$
$$(5x+1)^2 > 0$$

$5x + 1$ − − − − − − − 0 + + + +
$5x + 1$ − − − − − − − 0 + + + +
−1/5

The product is positive only if the factors have the same sign. That happens everywhere except $x = -1/5$. If $x = -1/5$ then the inequality is not satisfied. So the solution set consists of all real numbers except $-1/5$: $(-\infty, -1/5) \cup (-1/5, \infty)$

$-\frac{1}{5}$
−2 −1 0 1 2

25. $\frac{1}{x} > 0$

1 + + + + + + + + + + + +
x − − − − − 0 + + + + + + +
0

The quotient is positive only if the numerator and denominator have the same sign. That happens only when x is greater than 0. So the solution set is $(0, \infty)$.

−3 −2 −1 0 1 2 3

27. $\frac{x}{x-3} > 0$

x − − − − 0 + + + + + + +
$x - 3$ − − − − − − − 0 + + + +
0 3

The quotient is positive only if the numerator and denominator have the same sign.
$(-\infty, 0) \cup (3, \infty)$

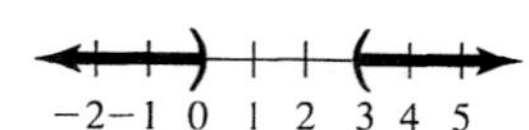

29. $\frac{x+2}{x} \leq 0$

$x + 2$ − − − − 0 + + + + + + +
x − − − − − − − − 0 + + + +
−2 0

The quotient is negative only if the numerator and denominator have opposite signs. Since the denominator must not be 0, we do not include 0 in the solution set: $[-2, 0)$

−4 −3 −2 −1 0 1 2

31. $\frac{t-3}{t+6} > 0$

$t + 6$ − − − 0 + + + + + + +
$t - 3$ − − − − − − − − 0 + + +
−6 3

The quotient is positive only if the numerator and denominator have the same sign.
$(-\infty, -6) \cup (3, \infty)$

−8 −6 −4 −2 0 2 4

33.
$$\frac{x}{x+2} > -1$$
$$\frac{x}{x+2} + 1 > 0$$
$$\frac{x}{x+2} + \frac{x+2}{x+2} > 0$$
$$\frac{2x+2}{x+2} > 0$$

$x + 2$ − − − − 0 + + + + + + +
$2x + 2$ − − − − − − − 0 + + + +
−2 −1

The quotient is positive only if the numerator and denominator have the same sign.
$(-\infty, -2) \cup (-1, \infty)$

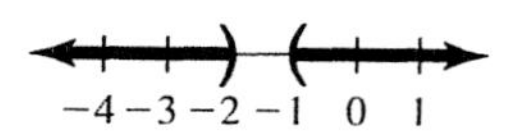

35.

$$\frac{2}{x-5} > \frac{1}{x+4}$$
$$\frac{2}{x-5} - \frac{1}{x+4} > 0$$
$$\frac{2(x+4)}{(x-5)(x+4)} - \frac{1(x-5)}{(x+4)(x-5)} > 0$$
$$\frac{x+13}{(x-5)(x+4)} > 0$$

$x+4 - - - - 0 + + + + + + +$
$x+13 - 0 + + + + + + + + + +$
$x-5 - - - - - - - - - 0 + + +$
$\quad -13 \quad -4 \quad 5$

This quotient will be positive only if an even number of the factors have negative values.
$(-13, -4) \cup (5, \infty)$

37.

$$\frac{m}{m-5} + \frac{3}{m-1} > 0$$
$$\frac{m(m-1)}{(m-5)(m-1)} + \frac{3(m-5)}{(m-1)(m-5)} > 0$$
$$\frac{m^2+2m-15}{(m-1)(m-5)} > 0$$
$$\frac{(m+5)(m-3)}{(m-1)(m-5)} > 0$$

$m-3 - - - - - - - - 0 + + + + +$
$m-1 - - - - 0 + + + + + + + + +$
$m+5 - - 0 + + + + + + + + + + +$
$m-5 - - - - - - - - - - - - 0 + +$
$\quad -5 \quad 1 \quad 3 \quad 5$

This quotient is positive only if there is an even number of factors with negative signs.
$(-\infty, -5) \cup (1, 3) \cup (5, \infty)$

39.

$$\frac{x}{x-3} \le \frac{-8}{x-6}$$
$$\frac{x}{x-3} + \frac{8}{x-6} \le 0$$
$$\frac{x(x-6)}{(x-3)(x-6)} + \frac{8(x-3)}{(x-6)(x-3)} \le 0$$
$$\frac{x^2+2x-24}{(x-3)(x-6)} \le 0$$
$$\frac{(x+6)(x-4)}{(x-3)(x-6)} \le 0$$

$x-4 - - - - - - - - 0 + + + + +$
$x-3 - - - - 0 + + + + + + + + +$
$x+6 - - 0 + + + + + + + + + + +$
$x-6 - - - - - - - - - - - - 0 + +$
$\quad -6 \quad 3 \quad 4 \quad 6$

The quotient is negative only if an odd number of factors have negative signs. Note that 3 and 6 cause the denominator to have a value of 0, and so they are excluded from the solution set.
$[-6, 3) \cup [4, 6)$

41. First solve $x^2 - 5 = 0$:

$$x^2 - 5 = 0$$
$$x^2 = 5$$
$$x = \pm\sqrt{5}$$

The two solutions to the equation divide the number line into three regions. Choose one number in each region, say -10, 0, and 10. Now evaluate $x^2 - 5$ for the numbers -10, 0, and 10.

$$(-10)^2 - 5 = 95$$
$$0^2 - 5 = -5$$
$$10^2 - 5 = 95$$

The signs of these answers indicate the solution to the inequality. The inequality is satisfied to the left of $-\sqrt{5}$ or to the right of $\sqrt{5}$. The solution set is
$(-\infty, -\sqrt{5}) \cup (\sqrt{5}, \infty)$.

43. To solve $x^2 - 2x - 5 \le 0$ we first solve $x^2 - 2x - 5 = 0$:

$$x = \frac{2 \pm \sqrt{4 - 4(1)(-5)}}{2(1)} = \frac{2 \pm 2\sqrt{6}}{2} = 1 \pm \sqrt{6}$$

The two solutions to the equation divide the number line into three regions. Choose one number in each region, say -10, 0, and 10. Now evaluate $x^2 - 2x - 5$ for the numbers -10, 0, and 10.

$$(-10)^2 - 2(-10) - 5 = 115$$
$$0^2 - 2(0) - 5 = -5$$
$$10^2 - 2(10) - 5 = 75$$

The signs of these answers indicate the solution to the inequality.
$[1 - \sqrt{6}, 1 + \sqrt{6}]$

45. To solve $2x^2 - 6x + 3 \ge 0$, we first solve $2x^2 - 6x + 3 = 0$.

$$x = \frac{6 \pm \sqrt{36 - 4(2)(3)}}{2(2)} = \frac{3 \pm \sqrt{3}}{2}$$

The two solutions to the equation divide the number line into three regions. Choose a test point in each region, say -10, 2, and 10. Evaluate the polynomial $2x^2 - 6x + 3$ at each of these points.

$2(-10)^2 - 6(-10) + 3 = 263$
$2(2)^2 - 6(2) + 3 = -1$
$2(10)^2 - 6(10) + 3 = 143$

The signs of these answers indicate the regions that satisfy the inequality.

$\left(-\infty, \frac{3-\sqrt{3}}{2}\right] \cup \left[\frac{3+\sqrt{3}}{2}, \infty\right)$

$\frac{3-\sqrt{3}}{2}$ $\frac{3+\sqrt{3}}{2}$

47. To solve $y^2 - 3y - 9 \leq 0$ we first solve $y^2 - 3y - 9 = 0$:

$$y = \frac{3 \pm \sqrt{9 - 4(1)(-9)}}{2} = \frac{3 \pm 3\sqrt{5}}{2}$$

Pick test points -10, 4, and 10, and evaluate $y^2 - 3y - 9$ for these numbers.

$(-10)^2 - 3(-10) - 9 = 121$
$4^2 - 3(4) - 9 = -5$
$10^2 - 3(10) - 9 = 61$

The signs of these answers indicate which regions satisfy the original inequality.

$\left[\frac{3 - 3\sqrt{5}}{2}, \frac{3 + 3\sqrt{5}}{2}\right]$

$\frac{3-3\sqrt{5}}{2}$ $\frac{3+3\sqrt{5}}{2}$

49. $x^2 + 5x + 12 \geq 0$

First solve $x^2 + 5x + 12 = 0$

$$x = \frac{-5 \pm \sqrt{25 - 4(1)(12)}}{2(1)} = \frac{-5 \pm \sqrt{-23}}{2}$$

Since $b^2 - 4ac$ is negative, the equation has no real solutions. So $x^2 + 5x + 12$ does not change sign. It is either always positive or always negative. To see which, test it with a real number, say 0.

$0^2 + 5(0) + 12 = 12 > 0$

So $x^2 + 5x + 12$ is positive for every real number x. The solution set is the set of all real numbers, $(-\infty, \infty)$.

51. $2x^2 + 5x + 5 < 0$

First solve $2x^2 + 5x + 5 = 0$

$$x = \frac{-5 \pm \sqrt{25 - 4(2)(5)}}{2(1)} = \frac{-5 \pm \sqrt{-15}}{2}$$

Since $b^2 - 4ac$ is negative, the equation has no real solutions. So $2x^2 + 5x + 5$ does not change sign. It is either always positive or always negative. To see which test it with a real number, say 0.

$2(0)^2 + 5(0) + 5 = 5 > 0$

So $2x^2 + 5x + 5$ is positive for every real number x. So the inequality $2x^2 + 5x + 5 < 0$ has no solution. The solution set is the empty set, $\emptyset$.

53. $-5x^2 + 2x \leq 4$

$-5x^2 + 2x - 4 \leq 0$

First solve $-5x^2 + 2x - 4 = 0$

$$x = \frac{-2 \pm \sqrt{4 - 4(-5)(-4)}}{2(-5)} = \frac{-2 \pm \sqrt{-76}}{-10}$$

Since $b^2 - 4ac$ is negative, the equation has no real solutions. So $-5x^2 + 2x - 4$ does not change sign. It is either always positive or always negative. To see which, test it with a real number, say 0.

$-5(0)^2 + 2(0) - 4 = -4 < 0$

So $-5x^2 + 2x - 4$ is negative for every real number x. The solution set is the set of all real numbers, $(-\infty, \infty)$.

55. Since x^2 is positive except when $x = 0$, all real numbers except zero satisfy $x^2 > 0$. The solution set is $(-\infty, 0) \cup (0, \infty)$.

57. Since x^2 is positive when x is nonzero and x^2 is zero, $x^2 + 4$ is actually greater than 0 for all real numbers. So all real numbers satisfy $x^2 + 4 \geq 0$. The solution set is $(-\infty, \infty)$.

59. Since 1 is positive, $1/x < 0$ is true if and only if x is negative. The solution set is $(-\infty, 0)$.

61. $x^2 \leq 9$

$x^2 - 9 \leq 0$

$(x - 3)(x + 3) \leq 0$

$x + 3$ − − − − 0 + + + + + + + +
$x - 3$ − − − − − − − − − − 0 + + +
−3 3

The product is negative only if the factors have opposite signs. That happens if x is chosen between -3 and 3. At ± 3 the product is 0. The solution set is $[-3, 3]$.

63.
$$16 - x^2 > 0$$
$$-1(16 - x^2) < -1(0)$$
$$x^2 - 16 < 0$$
$$(x - 4)(x + 4) < 0$$

$x + 4$ − − − 0 + + + + + + + + +
$x - 4$ − − − − − − − − − 0 + + + +
−4 4

The product is negative only if the value of x is between -4 and 4. The solution set is $(-4, 4)$.

65.
$$x^2 - 4x \geq 0$$
$$x(x - 4) \geq 0$$

x − − − − 0 + + + + + + + + +
$x - 4$ − − − − − − − − − 0 + + + +
0 4

The product is positive only if the factors have the same sign. The solution set is $(-\infty, 0] \cup [4, \infty)$.

67.
$$3(2w^2 - 5) < w$$
$$6w^2 - 15 < w$$
$$6w^2 - w - 15 < 0$$
$$(2w + 3)(3w - 5) < 0$$

$2w + 3$ − − − 0 + + + + + + + +
$3w - 5$ − − − − − − − − − 0 + + +
−3/2 5/3

The product is negative only if the factors have opposite signs. The solution set is $\left(-\frac{3}{2}, \frac{5}{3}\right)$.

69.
$$z^2 \geq 4(z + 3)$$
$$z^2 \geq 4z + 12$$
$$z^2 - 4z - 12 \geq 0$$
$$(z - 6)(z + 2) \geq 0$$

$z + 2$ − − − − 0 + + + + + + + +
$z - 6$ − − − − − − − − − 0 + + + +
−2 6

The product is positive only if the factors have the same sign. The solution set is $(-\infty, -2] \cup [6, \infty)$.

71.
$$(q + 4)^2 > 10q + 31$$
$$q^2 + 8q + 16 > 10q + 31$$
$$q^2 - 2q - 15 > 0$$
$$(q + 3)(q - 5) > 0$$

$q + 3$ − − − − 0 + + + + + + + +
$q - 5$ − − − − − − − − − 0 + + + +
−3 5

The product is positive only if the factors have the same sign. The solution set is $(-\infty, -3) \cup (5, \infty)$.

73.
$$\frac{1}{2}x^2 \geq 4 - x$$
$$x^2 \geq 8 - 2x$$
$$x^2 + 2x - 8 \geq 0$$
$$(x + 4)(x - 2) \geq 0$$

$x + 4$ − − − 0 + + + + + + + + +
$x - 2$ − − − − − − − − − 0 + + + +
−4 2

The product is positive only if the factors have the same sign. The solution set is $(-\infty, -4] \cup [2, \infty)$.

75. $\dfrac{x - 4}{x + 3} \leq 0$

$x + 3$ − − − − 0 + + + + + + + +
$x - 4$ − − − − − − − − − 0 + + + +
−3 4

The quotient is negative if the factors have opposite signs. The solution set is $(-3, 4]$.

77. $(x - 2)(x + 1)(x - 5) \geq 0$

$x - 2$ − − − − 0 + + + + + + + +
$x + 1$ − − 0 + + + + + + + + + +
$x - 5$ − − − − − − − − − 0 + + + +
−1 2 5

The product of these three factors is positive only if an even number of the factors have negative signs. This happens between -1 and 2, and also above 5 where no factors have negative signs. The solution set is $[-1, 2] \cup [5, \infty)$.

79.
$$x^3 + 3x^2 - x - 3 < 0$$
$$x^2(x + 3) - 1(x + 3) < 0$$
$$(x^2 - 1)(x + 3) < 0$$
$$(x - 1)(x + 1)(x + 3) < 0$$

$x + 1$ − − − − − 0 + + + + + + +
$x + 3$ − − 0 + + + + + + + + + +
$x - 1$ − − − − − − − − − − 0 + + +
−3 −1 1

The product of these three factors is negative only if an odd number of them have negative signs. This happens to the left of -3 and also between -1 and 1. The solution set is $(-\infty, -3) \cup (-1, 1)$.

81. To solve $0.23x^2 + 6.5x + 4.3 < 0$, we first solve $0.23x^2 + 6.5x + 4.3 = 0$.

$$x = \frac{-6.5 \pm \sqrt{(6.5)^2 - 4(0.23)(4.3)}}{2(0.23)}$$

$x = -27.58$ or $x = -0.68$
Test the numbers -30, -1, and 0.
$0.23(-30)^2 + 6.5(-30) + 4.3 = 16.3$
$0.23(-1)^2 + 6.5(-1) + 4.3 = -1.97$
$0.23(0)^2 + 6.5(0) + 4.3 = 4.3$

According to the signs of the values of the polynomial at the test points, the value of the polynomial is negative between the two solutions to the equation. The solution set is $(-27.58, -0.68)$.

83.
$$\frac{x}{x-2} > \frac{-1}{x+3}$$
$$\frac{x}{x-2} + \frac{1}{x+3} > 0$$
$$\frac{x(x+3)}{(x-2)(x+3)} + \frac{1(x-2)}{(x+3)(x-2)} > 0$$
$$\frac{x^2+4x-2}{(x+3)(x-2)} > 0$$

Solve $x^2 + 4x - 2 = 0$:

$$x = \frac{-4 \pm \sqrt{16 - 4(1)(-2)}}{2} = \frac{-4 \pm 2\sqrt{6}}{2}$$
$$= -2 \pm \sqrt{6} = -4.4, 0.4$$

The numbers -4.4, -3, 0.4, and 2 divide the number line into 5 regions. Pick a number in each region and test it in the original inequality. We get the solution set $(-\infty, -2 - \sqrt{6}) \cup (-3, -2 + \sqrt{6}) \cup (2, \infty)$.

85. To solve $x^2 + 5x - 50 > 0$, we first solve $x^2 + 5x - 50 = 0$:
$(x + 10)(x - 5) = 0$
$x = -10$ or $x = 5$

Test a point in each region of the number line, to find that the profit is positive if $x < -10$ or if $x > 5$. He cannot sell negative mobile homes so he must sell more than 5 to have a positive profit. He should sell 6, 7, 8, etc.

87. We must solve the inequality
$-16t^2 + 96t + 6 > 86$
$-16t^2 + 96t - 80 > 0$
$t^2 - 6t + 5 < 0$
Solve the equation $t^2 - 6t + 5 = 0$.
$(t - 5)(t - 1) = 0$
$t = 5$ or $t = 1$
Using a test point we find that the inequality is satisfied for t between 1 and 5 seconds. So the arrow is more than 86 feet high for 4 seconds.

89. **a)** From the graph it appears that the maximum height is 900 feet.
b) The projectile was above 864 feet for approximately 3 seconds.
c) $S = -16t^2 + \frac{240\sqrt{2}}{\sqrt{2}}t + 0$
or $S = -16t^2 + 240t$
Solve
$-16t^2 + 240t > 864$
$-16t^2 + 240t - 864 > 0$
$t^2 - 15t + 54 < 0$
Solve the equation
$(t - 6)(t - 9) = 0$
$t = 6$ or $t = 9$
Evaluate $t^2 - 15t + 54$ at the test points 0, 7, and 10.
$0^2 - 15(0) + 54 = 54$
$7^2 - 15(7) + 54 = -2$
$10^2 - 15(10) + 54 = 4$
So the inequality is satisfied for t in the interval $(6, 9)$. The projectile was above 864 ft for 3 sec.

91. **a)** (h, k)
b) $(-\infty, h) \cup (k, \infty)$ **c)** $(-k, -h)$
d) $(-\infty, -k] \cup [-h, \infty)$
e) $(-\infty, h] \cup (k, \infty)$ **f)** $(-k, -h]$

93. c

95. b

Enriching Your Mathematical Word Power

1. b **2.** a **3.** d **4.** c **5.** b **6.** b
7. c **8.** a **9.** c **10.** a **11.** c

CHAPTER 10 REVIEW

1. $x^2 - 2x - 15 = 0$
$(x-5)(x+3) = 0$
$x - 5 = 0$ or $x + 3 = 0$
$x = 5$ or $x = -3$
Solution set: $\{-3, 5\}$

3. $2x^2 + x = 15$
$2x^2 + x - 15 = 0$
$(2x-5)(x+3) = 0$
$2x - 5 = 0$ or $x + 3 = 0$
$x = \frac{5}{2}$ or $x = -3$
Solution set: $\left\{-3, \frac{5}{2}\right\}$

5. $w^2 - 25 = 0$
$(w-5)(w+5) = 0$
$w - 5 = 0$ or $w + 5 = 0$
$w = 5$ or $w = -5$
Solution set: $\{-5, 5\}$

7. $4x^2 - 12x + 9 = 0$
$(2x-3)^2 = 0$
$2x - 3 = 0$
$x = \frac{3}{2}$
Solution set: $\left\{\frac{3}{2}\right\}$

9. $x^2 = 12$
$x = \pm\sqrt{12} = \pm 2\sqrt{3}$
Solution set: $\left\{\pm 2\sqrt{3}\right\}$

11. $(x-1)^2 = 9$
$x - 1 = \pm 3$
$x = 1 \pm 3$
Solution set: $\{-2, 4\}$

13. $(x-2)^2 = \frac{3}{4}$
$x - 2 = \pm\sqrt{\frac{3}{4}}$
$x = 2 \pm \frac{\sqrt{3}}{2} = \frac{4}{2} \pm \frac{\sqrt{3}}{2}$
Solution set: $\left\{\frac{4 \pm \sqrt{3}}{2}\right\}$

15. $4x^2 = 9$
$x^2 = \frac{9}{4}$
$x = \pm\sqrt{\frac{9}{4}} = \pm\frac{3}{2}$
Solution set: $\left\{\pm\frac{3}{2}\right\}$

17. $x^2 - 6x + 8 = 0$
$x^2 - 6x = -8$
$x^2 - 6x + 9 = -8 + 9$
$(x-3)^2 = 1$
$x - 3 = \pm 1$
$x = 3 \pm 1$
Solution set: $\{2, 4\}$

19. $x^2 - 5x + 6 = 0$
$x^2 - 5x = -6$
$x^2 - 5x + \frac{25}{4} = -\frac{24}{4} + \frac{25}{4}$
$(x - \frac{5}{2})^2 = \frac{1}{4}$
$x - \frac{5}{2} = \pm\frac{1}{2}$
$x = \frac{5}{2} \pm \frac{1}{2}$
Solution set: $\{2, 3\}$

21. $2x^2 - 7x + 3 = 0$
$2x^2 - 7x = -3$
$x^2 - \frac{7}{2}x = -\frac{3}{2}$
$x^2 - \frac{7}{2}x + \frac{49}{16} = -\frac{24}{16} + \frac{49}{16}$
$(x - \frac{7}{4})^2 = \frac{25}{16}$
$x - \frac{7}{4} = \pm\frac{5}{4}$
$x = \frac{7}{4} \pm \frac{5}{4}$ $\left(\frac{12}{4} \text{ or } \frac{2}{4}\right)$
Solution set: $\left\{\frac{1}{2}, 3\right\}$

23. $x^2 + 4x + 1 = 0$
$x^2 + 4x = -1$
$x^2 + 4x + 4 = -1 + 4$
$(x+2)^2 = 3$
$x + 2 = \pm\sqrt{3}$
$x = -2 \pm \sqrt{3}$
Solution set: $\{-2 \pm \sqrt{3}\}$

25. $x^2 - 3x - 10 = 0$

$$x = \frac{3 \pm \sqrt{9 - 4(1)(-10)}}{2(1)} = \frac{3 \pm \sqrt{49}}{2} = \frac{3 \pm 7}{2}$$

Solution set: $\{-2, 5\}$

27. $6x^2 - 7x = 3$
$6x^2 - 7x - 3 = 0$

$$x = \frac{7 \pm \sqrt{49 - 4(6)(-3)}}{2(6)} = \frac{7 \pm \sqrt{121}}{12} = \frac{7 \pm 11}{12}$$

Solution set: $\left\{-\frac{1}{3}, \frac{3}{2}\right\}$

29. $x^2 + 4x + 2 = 0$

$$x = \frac{-4 \pm \sqrt{16 - 4(1)(2)}}{2(1)} = \frac{-4 \pm 2\sqrt{2}}{2}$$
$$= -2 \pm \sqrt{2}$$

Solution set: $\left\{-2 \pm \sqrt{2}\right\}$

31. $3x^2 + 1 = 5x$

$$3x^2 - 5x + 1 = 0$$
$$x = \frac{5 \pm \sqrt{25 - 4(3)(1)}}{2(3)} = \frac{5 \pm \sqrt{13}}{6}$$

Solution set: $\left\{\frac{5 \pm \sqrt{13}}{6}\right\}$

33. $25x^2 - 20x + 4 = 0$

$b^2 - 4ac = (-20)^2 - 4(25)(4) = 0$

One real solution.

35. $x^2 - 3x + 7 = 0$

$b^2 - 4ac = (-3)^2 - 4(1)(7) = -19$

No real solutions

37. $2x^2 - 5x + 1 = 0$

$b^2 - 4ac = (-5)^2 - 4(2)(1) = 17$

Two real solutions

39. $2x^2 - 4x + 3 = 0$

$$x = \frac{4 \pm \sqrt{16 - 4(2)(3)}}{2(2)} = \frac{4 \pm \sqrt{-8}}{4}$$
$$= \frac{4 \pm 2i\sqrt{2}}{4} = \frac{2 \pm i\sqrt{2}}{2}$$

Solution set: $\left\{\frac{2 \pm i\sqrt{2}}{2}\right\}$

41. $2x^2 - 3x + 3 = 0$

$$x = \frac{3 \pm \sqrt{9 - 4(2)(3)}}{2(2)} = \frac{3 \pm \sqrt{-15}}{4}$$
$$= \frac{3 \pm i\sqrt{15}}{4}$$

Solution set: $\left\{\frac{3 \pm i\sqrt{15}}{4}\right\}$

43. $3x^2 + 2x + 2 = 0$

$$x = \frac{-2 \pm \sqrt{4 - 4(3)(2)}}{2(3)} = \frac{-2 \pm \sqrt{-20}}{6}$$
$$= \frac{-2 \pm 2i\sqrt{5}}{6} = \frac{-1 \pm i\sqrt{5}}{3}$$

Solution set: $\left\{\frac{-1 \pm i\sqrt{5}}{3}\right\}$

45. $\frac{1}{2}x^2 + 3x + 8 = 0$

$$x^2 + 6x + 16 = 0$$
$$x = \frac{-6 \pm \sqrt{36 - 4(1)(16)}}{2(1)} = \frac{-6 \pm \sqrt{-28}}{2}$$
$$= \frac{-6 \pm 2i\sqrt{7}}{2} = -3 \pm i\sqrt{7}$$

Solution set: $\{-3 \pm i\sqrt{7}\}$

47. $f(x) = x^2 - 6x$

$x = \frac{-b}{2a} = \frac{6}{2(1)} = 3$, $f(3) = 3^2 - 6(3) = -9$

The vertex is $(3, -9)$. The y-intercept is $(0, 0)$.

$$x^2 - 6x = 0$$
$$x(x - 6) = 0$$
$$x = 0 \text{ or } x - 6 = 0$$
$$x = 0 \text{ or } x = 6$$

The x-intercepts are $(0, 0)$ and $(6, 0)$.

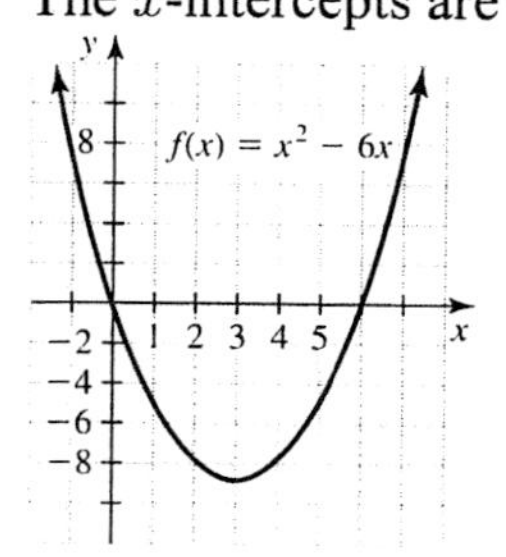

49. $g(x) = x^2 - 4x - 12$

$x = \frac{-b}{2a} = \frac{4}{2(1)} = 2$,

$g(2) = 2^2 - 4(2) - 12 = -16$

The vertex is $(2, -16)$. The y-intercept is $(0, -12)$. The solutions to $x^2 - 4x - 12 = 0$ are 6 and -2. So the x-intercepts are $(-2, 0)$ and $(6, 0)$.

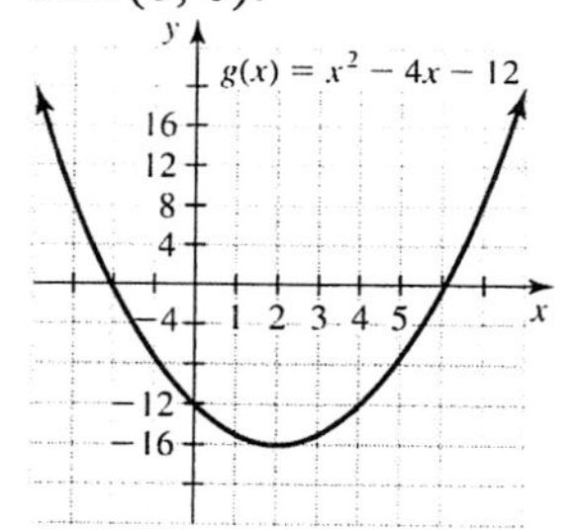

51. $h(x) = -2x^2 + 8x$

$x = \frac{-b}{2a} = \frac{-8}{2(-2)} = 2$,

$h(2) = -2 \cdot 2^2 + 8(2) = 8$

The vertex is $(2, 8)$. The y-intercept is $(0, 0)$. The solutions to $-2x^2 + 8x = 0$ are 0 and 4. So the x-intercepts are $(0, 0)$ and $(4, 0)$.

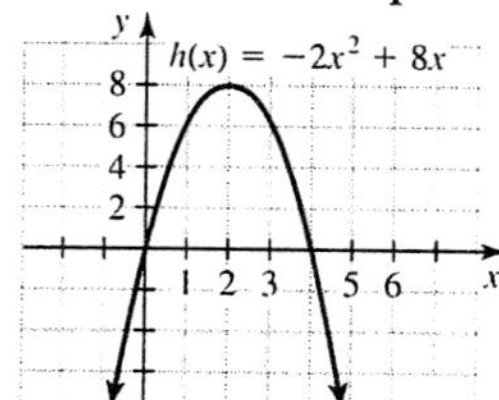

53. $y = -x^2 + 2x + 3$
$x = \frac{-b}{2a} = \frac{-2}{2(-1)} = 1,$
$y = -1^2 + 2(1) + 3 = 4$
The vertex is (1, 4). The y-intercept is (0, 3). The solutions to $-x^2 + 2x + 3 = 0$ are -1 and 3. So the x-intercepts are $(-1, 0)$ and $(3, 0)$.

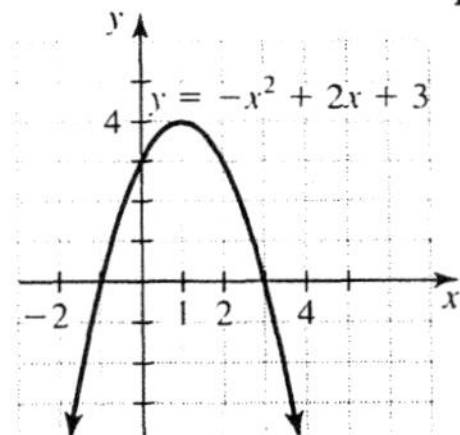

55. $f(x) = x^2 + 4x + 1$
$x = \frac{-b}{2a} = \frac{-4}{2(1)} = -2,$
$f(-2) = (-2)^2 + 4(-2) + 1 = -3$
The vertex of the graph is $(-2, -3)$ and the graph opens upward. So the minimum y-value is -3.

57. $y = -2x^2 - x + 4$
$x = \frac{-b}{2a} = \frac{1}{2(-2)} = -\frac{1}{4},$
$y = -2(-\frac{1}{4})^2 - (-\frac{1}{4}) + 4 = 4.125$

The vertex is $(-0.25, 4.125)$ and the graph opens downward. So the maximum y-value is 4.125.

59. $b^2 - 4ac = (-10)^2 - 4(8)(-3) = 196$
Since 196 is a perfect square, the polynomial is not prime.
$8x^2 - 10x - 3 = (4x + 1)(2x - 3)$

61. $b^2 - 4ac = (-5)^2 - 4(4)(2) = -7$
Since -7 is a not a perfect square, the polynomial is prime.

63. $b^2 - 4ac = (10)^2 - 4(8)(-25) = 900$
Since 900 is a perfect square, the polynomial is not prime.
$8y^2 + 10y - 25 = (4y - 5)(2y + 5)$

65. $(x + 3)(x + 6) = 0$
$x^2 + 9x + 18 = 0$

67. $(x + 5\sqrt{2})(x - 5\sqrt{2}) = 0$
$x^2 - 50 = 0$

69. $x^6 + 7x^3 - 8 = 0$
$(x^3 + 8)(x^3 - 1) = 0$
$x^3 + 8 = 0$ or $x^3 - 1 = 0$
$x^3 = -8$ or $x^3 = 1$
$x = -2$ or $x = 1$
Solution set: $\{-2, 1\}$

71. $x^4 - 13x^2 + 36 = 0$
$(x^2 - 4)(x^2 - 9) = 0$
$x^2 - 4 = 0$ or $x^2 - 9 = 0$
$x^2 = 4$ or $x^2 = 9$
$x = \pm 2$ or $x = \pm 3$
Solution set: $\{\pm 2, \pm 3\}$

73. Let $w = x^2 + 3x$ in
$(x^2 + 3x)^2 - 28(x^2 + 3x) + 180 = 0.$
$w^2 - 28w + 180 = 0$
$(w - 10)(w - 18) = 0$
$w = 10$ or $w = 18$
$x^2 + 3x = 10$ or $x^2 + 3x = 18$
$x^2 + 3x - 10 = 0$ or $x^2 + 3x - 18 = 0$
$(x + 5)(x - 2) = 0$ or $(x + 6)(x - 3) = 0$
$x = -5$ or $x = 2$ or $x = -6$ or $x = 3$
Solution set: $\{-6, -5, 2, 3\}$

75. Let $w = \sqrt{x^2 - 6x}$ and $w^2 = x^2 - 6x$ in $x^2 - 6x + 6\sqrt{x^2 - 6x} - 40 = 0.$
$w^2 + 6w - 40 = 0$
$(w + 10)(w - 4) = 0$
$w = -10$ or $w = 4$
$\sqrt{x^2 - 6x} = -10$ or $\sqrt{x^2 - 6x} = 4$
No real solution here. $x^2 - 6x = 16$
$x^2 - 6x - 16 = 0$
$(x - 8)(x + 2) = 0$
$x = 8$ or $x = -2$

Solution set: $\{-2, 8\}$

77. Let $w = t^{-1}$ and $w^2 = t^{-2}$ in $t^{-2} + 5t^{-1} - 36 = 0.$
$w^2 + 5w - 36 = 0$
$(w + 9)(w - 4) = 0$
$w = -9$ or $w = 4$
$t^{-1} = -9$ or $t^{-1} = 4$
$t = -\frac{1}{9}$ or $t = \frac{1}{4}$
Solution set: $\left\{-\frac{1}{9}, \frac{1}{4}\right\}$

79. Let $y = \sqrt{w}$ and $y^2 = w$ in $w - 13\sqrt{w} + 36 = 0.$
$y^2 - 13y + 36 = 0$
$(y - 9)(y - 4) = 0$
$y = 9$ or $y = 4$
$\sqrt{w} = 9$ or $\sqrt{w} = 4$
$w = 81$ or $w = 16$

Solution set: $\{16, 81\}$

81. $a^2 + a > 6$

$a^2 + a - 6 > 0$

$(a+3)(a-2) > 0$

$a + 3 - - - 0 + + + + + + + + +$

$a - 2 - - - - - - - - - 0 + + + +$

$-3 \qquad 2$

The inequality is satisfied if both factors are the same sign. $(-\infty, -3) \cup (2, \infty)$

83. $x^2 - x - 20 \le 0$

$(x-5)(x+4) \le 0$

$x + 4 - - - 0 + + + + + + + + +$

$x - 5 - - - - - - - - - 0 + + + +$

$-4 \qquad 5$

The inequality is satisfied if the factors have opposite signs, or if one of the factors is zero. $[-4, 5]$

85. $w^2 - w < 0$

$w(w-1) < 0$

$w - - - 0 + + + + + + + + + + +$

$w - 1 - - - - - - - - - - - 0 + + +$

$0 \qquad 1$

The inequality is satisfied if the factors have opposite signs. $(0, 1)$

87. $\dfrac{x-4}{x+2} \ge 0$

$x + 2 - - - 0 + + + + + + + + +$

$x - 4 - - - - - - - - - 0 + + + +$

$-2 \qquad 4$

The inequality is satisfied if the factors have the same sign, or if the numerator is equal to zero. $(-\infty, -2) \cup [4, \infty)$

89. $\dfrac{x-2}{x+3} < 1$

$\dfrac{x-2}{x+3} - 1 < 0$

$\dfrac{x-2}{x+3} - \dfrac{x+3}{x+3} < 0$

$\dfrac{-5}{x+3} < 0$

Since the numerator is definitely negative, the inequality is satisfied only if the denominator is positive:

$x + 3 > 0$

$x > -3$

$(-3, \infty)$

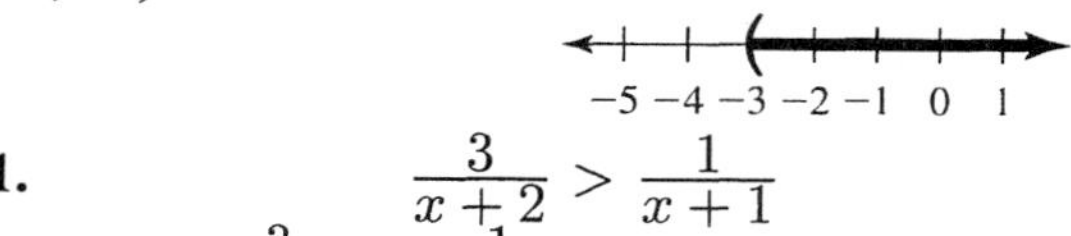

91. $\dfrac{3}{x+2} > \dfrac{1}{x+1}$

$\dfrac{3}{x+2} - \dfrac{1}{x+1} > 0$

$\dfrac{3(x+1)}{(x+2)(x+1)} - \dfrac{1(x+2)}{(x+1)(x+2)} > 0$

$\dfrac{2x+1}{(x+2)(x+1)} > 0$

$x + 1 - - - - 0 + + + + + + + +$

$x + 2 - - 0 + + + + + + + + + +$

$2x + 1 - - - - - - - - 0 + + +$

$-2 \qquad -1 \qquad -1/2$

The inequality is satisfied when an even number of the factors have negative signs. $(-2, -1) \cup \left(-\frac{1}{2}, \infty\right)$

93. $144x^2 - 120x + 25 = 0$

$(12x - 5)^2 = 0$

$12x - 5 = 0$

$x = \dfrac{5}{12}$

Solution set: $\left\{\dfrac{5}{12}\right\}$

95. $(2x+3)^2 + 7 = 12$

$(2x+3)^2 = 5$

$2x + 3 = \pm\sqrt{5}$

$x = \dfrac{-3 \pm \sqrt{5}}{2}$

Solution set: $\left\{\dfrac{-3 \pm \sqrt{5}}{2}\right\}$

97. $1 + \dfrac{20}{9x^2} = \dfrac{8}{3x}$

$9x^2\left(1 + \dfrac{20}{9x^2}\right) = 9x^2 \cdot \dfrac{8}{3x}$

$9x^2 + 20 = 24x$

$9x^2 - 24x + 20 = 0$

$x = \dfrac{24 \pm \sqrt{(-24)^2 - 4(9)(20)}}{2(9)} = \dfrac{24 \pm 12i}{18}$

$= \dfrac{4 \pm 2i}{3}$

Solution set: $\left\{\dfrac{4 \pm 2i}{3}\right\}$

99. $$\begin{aligned}\sqrt{3x^2+7x-30}&=x\\3x^2+7x-30&=x^2\\2x^2+7x-30&=0\\(2x-5)(x+6)&=0\end{aligned}$$
$$2x-5=0 \text{ or } x+6=0$$
$$x=\frac{5}{2} \text{ or } x=-6$$
Since -6 does not check, solution set is $\left\{\frac{5}{2}\right\}$.

101. $2(2x+1)^2+5(2x+1)=3$
Let $y=2x+1$:
$$\begin{aligned}2y^2+5y-3&=0\\(2y-1)(y+3)&=0\end{aligned}$$
$$2y-1=0 \text{ or } y+3=0$$
$$y=\frac{1}{2} \text{ or } y=-3$$
$$2x+1=\frac{1}{2} \text{ or } 2x+1=-3$$
$$x=-\frac{1}{4} \text{ or } x=-2$$
Solution set: $\left\{-2,-\frac{1}{4}\right\}$

103. $x^{1/2}-15x^{1/4}+50=0$ Let $y=x^{1/4}$
$$\begin{aligned}y^2-15y+50&=0\\(y-5)(y-10)&=0\end{aligned}$$
$$y=5 \text{ or } y=10$$
$$x^{1/4}=5 \text{ or } x^{1/4}=10$$
$$x=5^4 \text{ or } x=10^4$$
$$x=625 \text{ or } x=10{,}000$$
Solution set: $\{625, 10{,}000\}$

105. If $x=$ one of the numbers, then $x+4=$ the other number. Since their product is 4, we can write the equation

$$\begin{aligned}x(x+4)&=4\\x^2+4x&=4\\x^2+4x+4&=4+4\\(x+2)^2&=8\\x+2&=\pm2\sqrt{2}\\x&=-2\pm2\sqrt{2}\end{aligned}$$

Disregard $-2-2\sqrt{2}$ since it is not positive.
If $x=-2+2\sqrt{2}$ then $x+4=2+2\sqrt{2}$.
The two numbers are $-2+2\sqrt{2}\approx0.83$ and $2+2\sqrt{2}\approx4.83$.

107. Let $x=$ the width and $x-4=$ the height. The diagonal 19 is the hypotenuse of a right triangle with legs x and $x-4$. By the Pythagorean theorem we can write

$$\begin{aligned}x^2+(x-4)^2&=19^2\\x^2+x^2-8x+16&=361\\2x^2-8x-345&=0\end{aligned}$$
$$x=\frac{8\pm\sqrt{64-4(2)(-345)}}{2(2)}$$
$$=\frac{8\pm\sqrt{2824}}{4}=\frac{8\pm2\sqrt{706}}{4}$$
$$=\frac{4\pm\sqrt{706}}{2}$$
Since the width must be positive, we have
$x=\dfrac{4+\sqrt{706}}{2}$ and
$$x-4=\frac{4+\sqrt{706}}{2}-\frac{8}{2}=\frac{-4+\sqrt{706}}{2}.$$
So the width is $\dfrac{4+\sqrt{706}}{2}\approx15.3$ inches and the height is $\dfrac{-4+\sqrt{706}}{2}\approx11.3$ inches.

109. Let $x=$ the width of the border. The dimensions of the printed area will be $8-2x$ by $10-2x$. Since the printed area is to be 24 square inches, we can write the equation
$$\begin{aligned}(8-2x)(10-2x)&=24\\80-36x+4x^2&=24\\4x^2-36x+56&=0\\x^2-9x+14&=0\\(x-7)(x-2)&=0\end{aligned}$$
$$x=7 \text{ or } x=2$$

Since 7 inches is too wide for a border on an 8 by 10 piece of paper, the border must be 2 inches wide.

111. Let $x=$ width and $x+4=$ the length.
$$\begin{aligned}x(x+4)&=45\\x^2+4x-45&=0\\(x+9)(x-5)&=0\end{aligned}$$
$$x+9=0 \text{ or } x-5=0$$
$$x=-9 \text{ or } x=5$$
$$x+4=9$$
The table is 5 ft wide and 9 feet long.

113. $C(n)=0.004n^2-3.2n+660$
If $n=390$, then
$$\begin{aligned}C(390)&=0.004(390)^2-3.2(390)+660\\&=\$20.40\end{aligned}$$
$$n=\frac{3.2}{2(0.004)}=400$$
The unit cost is at a minimum for 400 starters.

115. $$12 = -16t^2 + 32t$$
$$16t^2 - 32t + 12 = 0$$
$$4t^2 - 8t + 3 = 0$$
$$t = \frac{8 \pm \sqrt{64 - 4(4)(3)}}{2(4)} = \frac{8 \pm \sqrt{16}}{8}$$
$$= 1 \pm \frac{1}{2} = 1.5 \text{ or } 0.5$$
His height was 12 ft for $t = 0.5$ sec and $t = 1.5$ sec.

CHAPTER 10 TEST

1. $2x^2 - 3x + 2 = 0$
$(-3)^2 - 4(2)(2) = 9 - 16 = -7$
The equation has no real solutions.

2. $-3x^2 + 5x - 1 = 0$
$(5)^2 - 4(-3)(-1) = 25 - 12 = 13$
The equation has 2 real solutions.

3. $4x^2 - 4x + 1 = 0$
$(-4)^2 - 4(4)(1) = 16 - 16 = 0$
The equation has 1 real solution.

4. $2x^2 + 5x - 3 = 0$
$$x = \frac{-5 \pm \sqrt{25 - 4(2)(-3)}}{2(2)} = \frac{-5 \pm \sqrt{49}}{4}$$
$$= \frac{-5 \pm 7}{4} \quad \left(\frac{2}{4} \text{ or } \frac{-12}{4}\right)$$
Solution set: $\left\{-3, \frac{1}{2}\right\}$

5. $x^2 + 6x + 6 = 0$
$$x = \frac{-6 \pm \sqrt{36 - 4(1)(6)}}{2} = \frac{-6 \pm \sqrt{12}}{2}$$
$$= \frac{-6 \pm 2\sqrt{3}}{2} = -3 \pm \sqrt{3}$$
Solution set: $\{-3 \pm \sqrt{3}\}$

6. $x^2 + 10x + 25 = 0$
$$(x + 5)^2 = 0$$
$$x + 5 = 0$$
$$x = -5$$
Solution set: $\{-5\}$

7. $2x^2 + x - 6 = 0$
$$2x^2 + x = 6$$
$$x^2 + \frac{1}{2}x = 3$$
$$x^2 + \frac{1}{2}x + \frac{1}{16} = 3 + \frac{1}{16}$$
$$(x + \tfrac{1}{4})^2 = \frac{49}{16}$$
$$x + \frac{1}{4} = \pm\frac{7}{4}$$
$$x = -\frac{1}{4} \pm \frac{7}{4} \quad \left(\frac{6}{4} \text{ or } \frac{-8}{4}\right)$$
Solution set: $\left\{-2, \frac{3}{2}\right\}$

8. $x(x + 1) = 12$
$$x^2 + x - 12 = 0$$
$$(x + 4)(x - 3) = 0$$
$$x = -4 \quad \text{or} \quad x = 3$$
Solution set: $\{-4, 3\}$

9. $a^4 - 5a^2 + 4 = 0$
$$(a^2 - 4)(a^2 - 1) = 0$$
$$a^2 - 4 = 0 \text{ or } a^2 - 1 = 0$$
$$a^2 = 4 \text{ or } a^2 = 1$$
$$a = \pm 2 \text{ or } a = \pm 1$$
Solution set: $\{\pm 1, \pm 2\}$

10. Let $w = \sqrt{x - 2}$ and $w^2 = x - 2$ in
$$x - 2 - 8\sqrt{x - 2} + 15 = 0$$
$$w^2 - 8w + 15 = 0$$
$$(w - 3)(w - 5) = 0$$
$$w = 3 \quad \text{or} \quad w = 5$$
$$\sqrt{x - 2} = 3 \quad \text{or} \quad \sqrt{x - 2} = 5$$
$$x - 2 = 9 \quad \text{or} \quad x - 2 = 25$$
$$x = 11 \quad \text{or} \quad x = 27$$
Solution set: $\{11, 27\}$

11. $x^2 + 36 = 0$
$$x^2 = -36$$
$$x = \pm\sqrt{-36} = \pm 6i$$
Solution set: $\{\pm 6i\}$

12. $x^2 + 6x + 10 = 0$
$$x = \frac{-6 \pm \sqrt{36 - 4(1)(10)}}{2(1)} = \frac{-6 \pm \sqrt{-4}}{2}$$
$$= \frac{-6 \pm 2i}{2} = -3 \pm i$$
Solution set: $\{-3 \pm i\}$

13. $3x^2 - x + 1 = 0$
$$x = \frac{1 \pm \sqrt{1 - 4(3)(1)}}{2(3)} = \frac{1 \pm \sqrt{-11}}{6}$$
Solution set: $\left\{\frac{1 \pm i\sqrt{11}}{6}\right\}$

14. The graph of $f(x) = 16 - x^2$ goes through $(-2, 12)$, $(-1, 15)$, $(0, 16)$, $(1, 15)$, and $(2, 12)$.
$$x = \frac{-b}{2a} = \frac{-0}{2(-1)} = 0$$
$$f(0) = 16$$
The vertex is $(0, 16)$. The maximum y-value is 16.
$$16 - x^2 = 0$$
$$x^2 = 16$$
$$x = \pm 4$$
The intercepts are $(0, 16)$, $(-4, 0)$, and $(4, 0)$.

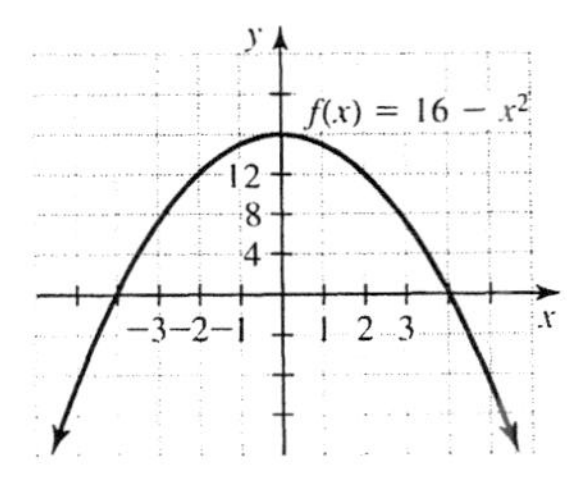

15. The graph of $g(x) = x^2 - 3x$ goes through $(0, 0)$, $(1, -2)$, $(2, -2)$, and $(3, 0)$.

$$x = \frac{-b}{2a} = \frac{-(-3)}{2(1)} = \frac{3}{2}$$

$$f(3/2) = \left(\frac{3}{2}\right)^2 - 3 \cdot \frac{3}{2} = -\frac{9}{4}$$

The vertex is $\left(\frac{3}{2}, -\frac{9}{4}\right)$. The minimum y-value is $-9/4$ or -2.25.

$$x^2 - 3x = 0$$
$$x(x - 3) = 0$$
$$x = 0 \quad \text{or} \quad x - 3 = 0$$
$$x = 0 \quad \text{or} \quad x = 3$$

The intercepts are $(0, 0)$ and $(3, 0)$.

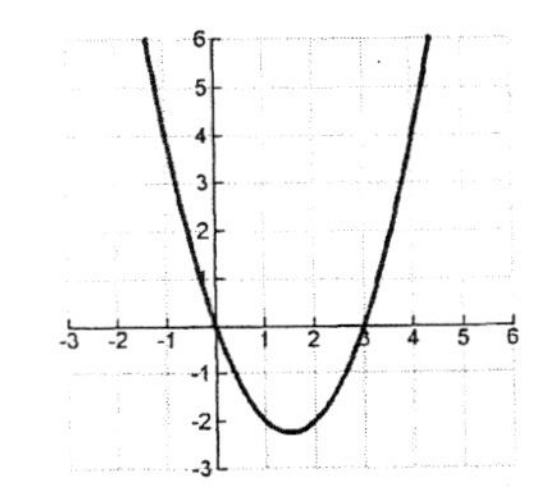

16. $(x + 4)(x - 6) = 0$

$x^2 - 2x - 24 = 0$

17. $(x - 5i)(x + 5i) = 0$

$x^2 + 25 = 0$

18.
$$w^2 + 3w < 18$$
$$w^2 + 3w - 18 < 0$$
$$(w + 6)(w - 3) < 0$$

$w + 6 \; - \; - \; - \; - \; 0 \; + \; + \; + \; + \; + \; + \; + \; +$

$w - 3 \; - \; - \; - \; - \; - \; - \; - \; - \; - \; 0 \; + \; + \; + \; +$

$\qquad -6 \qquad\qquad 3$

The inequality is satisfied when the factors have opposite signs. $(-6, 3)$

−6 −5 −4 −3 −2 −1 0 1 2 3

19.
$$\frac{2}{x - 2} < \frac{3}{x + 1}$$
$$\frac{2}{x - 2} - \frac{3}{x + 1} < 0$$
$$\frac{2(x + 1)}{(x - 2)(x + 1)} - \frac{3(x - 2)}{(x + 1)(x - 2)} < 0$$
$$\frac{-x + 8}{(x - 2)(x + 1)} < 0$$

$x - 2 \; - \; - \; - \; - \; 0 \; + \; + \; + \; + \; + \; + \; + \; +$

$x + 1 \; - \; - \; 0 \; + \; + \; + \; + \; + \; + \; + \; + \; + \; +$

$-x + 8 \; + \; + \; + \; + \; + \; + \; + \; + \; 0 \; - \; - \; -$

$\qquad -1 \qquad 2 \qquad\qquad 8$

This quotient will be negative only if an odd number of factors are negative.

$(-1, 2) \cup (8, \infty)$

−1 0 1 2 3 4 5 6 7 8 9 10

20. Let $x =$ the width and $x + 2 =$ the length. Since the area is 16 square feet, we can write the equation

$$x(x + 2) = 16$$
$$x^2 + 2x - 16 = 0$$
$$x = \frac{-2 \pm \sqrt{4 - 4(1)(-16)}}{2} = \frac{-2 \pm \sqrt{68}}{2}$$
$$= \frac{-2 \pm 2\sqrt{17}}{2} = -1 \pm \sqrt{17}$$

Since the width must be positive, we have $x = -1 + \sqrt{17}$ and $x + 2 = 1 + \sqrt{17}$. The width is $-1 + \sqrt{17}$ feet and the length is $1 + \sqrt{17}$ feet.

21. Let $x =$ time for the new computer and $x + 1 =$ time for the old computer. New computer's rate is $\frac{1}{x}$ payroll/hr and old computer's rate is $\frac{1}{x + 1}$ payroll/hr. In 3 hrs new computer does $\frac{3}{x}$ payroll and old computer does $\frac{3}{x + 1}$ payroll.

$$\frac{3}{x} + \frac{3}{x + 1} = 1$$
$$x(x + 1)\left(\frac{3}{x} + \frac{3}{x + 1}\right) = x(x + 1)1$$
$$3x + 3 + 3x = x^2 + x$$
$$0 = x^2 - 5x - 3$$
$$x = \frac{5 \pm \sqrt{5^2 - 4(1)(-3)}}{2} = \frac{5 \pm \sqrt{37}}{2}$$
$$\approx 5.5 \text{ or } -0.5$$

It takes the new computer $\frac{5 + \sqrt{37}}{2}$ or 5.5 hrs to do the payroll by itself.

22. $s(t) = -16t^2 + 48t$

$t = \frac{-48}{2(-16)} = \frac{3}{2}$, $s = -16(\frac{3}{2})^2 + 48(\frac{3}{2}) = 36$

The maximum height reached by the ball is 36 feet.

Making Connections

Chapters 1 - 10

1. $2x - 15 = 0$

$$2x = 15$$

$$x = \frac{15}{2}$$

Solution set: $\left\{\frac{15}{2}\right\}$

2. $2x^2 - 15 = 0$

$$2x^2 = 15$$

$$x^2 = \frac{15}{2}$$

$$x = \pm\sqrt{\frac{15}{2}} = \pm\frac{\sqrt{15}\sqrt{2}}{\sqrt{2}\sqrt{2}} = \pm\frac{\sqrt{30}}{2}$$

Solution set: $\left\{\pm\frac{\sqrt{30}}{2}\right\}$

3. $2x^2 + x - 15 = 0$

$$(2x - 5)(x + 3) = 0$$

$$2x - 5 = 0 \quad \text{or} \quad x + 3 = 0$$

$$x = \frac{5}{2} \quad \text{or} \quad x = -3$$

Solution set: $\left\{-3, \frac{5}{2}\right\}$

4. $2x^2 + 4x - 15 = 0$

$$x = \frac{-4 \pm \sqrt{16 - 4(2)(-15)}}{2(2)} = \frac{-4 \pm \sqrt{136}}{4}$$

$$= \frac{-4 \pm 2\sqrt{34}}{4} = \frac{-2 \pm \sqrt{34}}{2}$$

Solution set: $\left\{\frac{-2 \pm \sqrt{34}}{2}\right\}$

5. $|\,4x + 11\,| = 3$

$$4x + 11 = 3 \quad \text{or} \quad 4x + 11 = -3$$

$$4x = -8 \quad \text{or} \quad 4x = -14$$

$$x = -2 \quad \text{or} \quad x = -\frac{7}{2}$$

Solution set: $\left\{-\frac{7}{2}, -2\right\}$

6. $|\,4x^2 + 11x\,| = 3$

$$4x^2 + 11x = 3 \quad \text{or} \quad 4x^2 + 11x = -3$$

$$4x^2 + 11x - 3 = 0 \quad \text{or} \quad 4x^2 + 11x + 3 = 0$$

$$(4x - 1)(x + 3) = 0 \text{ or } \quad x = \frac{-11 \pm \sqrt{73}}{8}$$

$$4x - 1 = 0 \text{ or } x + 3 = 0$$

$$x = \frac{1}{4} \text{ or } \quad x = -3$$

Solution set: $\left\{-3, \frac{1}{4}, \frac{-11 \pm \sqrt{73}}{8}\right\}$

7. $\sqrt{x} = x - 6$

$$(\sqrt{x})^2 = (x - 6)^2$$

$$x = x^2 - 12x + 36$$

$$0 = x^2 - 13x + 36$$

$$0 = (x - 4)(x - 9)$$

$$x - 4 = 0 \quad \text{or} \quad x - 9 = 0$$

$$x = 4 \quad \text{or} \quad x = 9$$

Since $\sqrt{4} = 4 - 6$ is incorrect and $\sqrt{9} = 9 - 6$ is correct, the solution set is $\{9\}$.

8. $(2x - 5)^{2/3} = 4$

$$\left((2x - 5)^{2/3}\right)^3 = 4^3$$

$$(2x - 5)^2 = 64$$

$$2x - 5 = \pm 8$$

$$2x = 5 \pm 8$$

$$x = \frac{5 \pm 8}{2}$$

Solution set: $\left\{-\frac{3}{2}, \frac{13}{2}\right\}$

9. $1 - 2x < 5 - x$

$$-x < 4$$

$$x > -4$$

$(-4, \infty)$

10. $(1 - 2x)(5 - x) \le 0$

$$1 - 2x \; + \; + \; + 0 - - - - - - - - -$$

$$5 - x \; + + + + + + + 0 - - - -$$

$$\qquad 1/2 \qquad\qquad 5$$

The inequality is satisfied when the factors have opposite signs. The solution set is $\left[\frac{1}{2}, 5\right]$.

11. $\frac{1 - 2x}{5 - x} \le 0$ Same as last exercise, but 5 is excluded from the solution set because of the denominator $5 - x$. The solution set is $\left[\frac{1}{2}, 5\right)$.

12. $|5-x| < 3$
$-3 < 5-x < 3$
$-8 < -x < -2$
$8 > x > 2$
The solution set is (2, 8).

13. $3x-1<5$ and $-3 \le x$
$x < 2$ and $x \ge -3$
The solution set is $[-3, 2)$.

14. $x-3<1$ or $2x \ge 8$
$x<4$ or $x \ge 4$
The solution set is $(-\infty, \infty)$.

15.
$$2x-3y=9$$
$$-3y=-2x+9$$
$$y=\frac{-2x+9}{-3}$$
$$y=\frac{2}{3}x-3$$

16.
$$\frac{y-3}{x+2}=-\frac{1}{2}$$
$$y-3=-\frac{1}{2}(x+2)$$
$$y-3=-\frac{1}{2}x-1$$
$$y=-\frac{1}{2}x+2$$

17.
$$3y^2+cy+d=0$$
$$y=\frac{-c \pm \sqrt{c^2-4(3)(d)}}{2(3)}$$
$$y=\frac{-c \pm \sqrt{c^2-12d}}{6}$$

18.
$$my^2-ny-w=0$$
$$y=\frac{-(-n) \pm \sqrt{(-n)^2-4(m)(-w)}}{2m}$$
$$y=\frac{n \pm \sqrt{n^2+4mw}}{2m}$$

19.
$$\frac{1}{3}x-\frac{2}{5}y=\frac{5}{6}$$
$$30\left(\frac{1}{3}x-\frac{2}{5}y\right)=30\left(\frac{5}{6}\right)$$
$$10x-12y=25$$
$$-12y=-10x+25$$
$$y=\frac{-10}{-12}x+\frac{25}{-12}$$
$$y=\frac{5}{6}x-\frac{25}{12}$$

20.
$$y-3=-\frac{2}{3}(x-4)$$
$$y-3=-\frac{2}{3}x+\frac{8}{3}$$
$$y=-\frac{2}{3}x+\frac{8}{3}+\frac{9}{3}$$
$$y=-\frac{2}{3}x+\frac{17}{3}$$

21. $m=\frac{7-3}{5-2}=\frac{4}{3}$

22. $m=\frac{-6-5}{4-(-3)}=-\frac{11}{7}$

23. $m=\frac{0.4-0.8}{0.5-0.3}=\frac{-0.4}{0.2}=-2$

24. $m=\dfrac{-\frac{4}{3}-\frac{3}{5}}{\frac{1}{3}-\frac{1}{2}}=\dfrac{-\frac{29}{15}}{-\frac{1}{6}}=\frac{29}{15}\cdot\frac{6}{1}=\frac{58}{5}$

25. At \$20 per ticket,
$n = 48{,}000 - 400(20) = 40{,}000$.

At \$25 per ticket
$n = 48{,}000 - 400(25) = 38{,}000$

If 35,000 tickets are sold, then the price is \$32.50 per ticket.

26. If $p = \$20$, then
$R = 20(48{,}000 - 400 \cdot 20)$
$= \$800{,}000$

If $p = \$25$, then $R = 25(48{,}000 - 400 \cdot 25)$
$= \$950{,}000$

$$1{,}280{,}000 = p(48{,}000 - 400p)$$
$$1{,}280{,}000 = 48{,}000p - 400p^2$$
$$400p^2 - 48{,}000p + 1{,}280{,}000 = 0$$
$$p^2 - 120p + 3{,}200 = 0$$
$$(p-80)(p-40) = 0$$
$$p = 80 \quad \text{or} \quad p = 40$$

A revenue of \$1.28 million occurs at a price of \$40 and at a price of \$80.
The price that determines the maximum revenue is \$60 per ticket.

11.1 WARM-UPS

1. False, $\{(1, 2), (1, 3)\}$ is not a function.
2. True, because $C = \pi D$.
3. True, because no two ordered pairs have the same first coordinate and different second coordinates.
4. False, because (1, 5) and (1, 7) have the same first coordinate and different second coordinates.
5. True, because every x-coordinate in $y = x^2$ corresponds to a unique y-coordinate.
6. False, because $\{(1, 5), (3, 6), (1, 7)\}$ is a relation that is not a function.
7. True, because of the definition of domain.
8. False, because the domain of a relation or function is the set of first coordinates.
9. True, because $\sqrt{x}$ is a real number only if $x \geq 0$.
10. True, because if $h(x) = x^2 - 3$, then $h(-2) = (-2)^2 - 3 = 4 - 3 = 1$.

11.1 EXERCISES

1. To say that b is a function of a means that b is uniquely determined by a.
3. A relation is any set of ordered pairs.
5. The range of relation is the set of all possible second coordinates.
7. The number of gallons y that you get for \$10 is uniquely determined by the price per gallon x. So y is a function of x.
9. A student's test score y is not determined by the number of hours spent studying, because two students can study for the same amount of time and score differently on the test. So y is not a function of x.
11. The Fahrenheit temperature y is uniquely determined by the Celsius temperature x by a formula. So y is a function of x.
13. Two items with the same price x can have different universal product codes y. So the product code cannot be determined from the cost and y is not a function of x.
15. The total cost, C, is found by multiplying the number of toppings, t, by the cost for each topping, \$0.50, and adding on the \$5 base charge: $C = 0.50t + 5$
17. The total cost T is the price of the groceries S plus $0.09S$ or $T = S + 0.09S$. Combining like terms yields $T = 1.09S$.
19. The circumference of a circle C is found using the well-known formula $C = 2\pi r$, where r is the radius.
21. Since a square has 4 equal sides, the perimeter is 4 times the length of a side, $P = 4s$.
23. Since $A = \frac{1}{2}bh$ and $b = 10$, $A = 5h$ expresses the area as a function of the height.
25. This table has no ordered pairs with the same first coordinate and different second coordinates. So the table does give y as a function of x.
27. This table has no ordered pairs with the same first coordinate t and different second coordinates v. So the table does give v as a function of t.
29. Since the a-coordinate of 2 corresponds to both 2 and -2, P is not a function of a.
31. This table has no ordered pairs with the same first coordinate b and different second coordinates q. So the table does give q as a function of b
33. A set of ordered pairs is a function unless there are two pairs with the same first coordinate and different second coordinates. So this set is a function.
35. This set of ordered pairs is not a function because $(-2, 4)$ and $(-2, 6)$ have the same x-coordinate and different y-coordinates.
37. This set of ordered pairs is not a function because $(\pi, -1)$ and $(\pi, 1)$ have the same x-coordinate and different y-coordinates.
39. This set is a function because no two ordered pairs have the same first coordinate and different second coordinates.
41. Note that the value of y^2 is the same for a number and its opposite. So if $y = \pm 1$, $x = 2(\pm 1)^2 = 2$. So $(2, 1)$ and $(2, -1)$ both satisfy the equation.
43. Note that the value of $|2y|$ is the same for a number and its opposite. So if $y = \pm 4$, $x^2 = |2(\pm 4)| = 8$. So $(8, 4)$ and $(8, -4)$ both satisfy the equation.

45. Note that the value of x^2 or y^2 is the same for a number and its opposite. So if $y = \pm 1$, $x^2 + (\pm 1)^2 = 1$, or $x^2 = 0$. Now 0 satisfies $x^2 = 0$. So $(0, 1)$ and $(0, -1)$ both satisfy the equation.

47. Note that the value of y^4 is the same for a number and its opposite. So if $y = \pm 2$, $x = (\pm 2)^4 = 16$. So $(16, 2)$ and $(16, -2)$ both satisfy the equation.

49. Note that the value of $|y|$ is the same for a number and its opposite. So if $y = \pm 1$, $x - 2 = |\pm 1| = 1$, or $x = 3$. So $(3, 1)$ and $(3, -1)$ both satisfy the equation.

51. This relation is a function because for each value of x, $y = x^2$ determines only one value of y.

53. This relation is not a function because $(2, 1)$ and $(2, -1)$ both satisfy $x = |y| + 1$.

55. This relation is a function because for each value of x, $y = x$ determines only one value of y.

57. This relation is not a function because $(2, 1)$ and $(2, -1)$ both satisfy $x = y^4 + 1$.

59. This relation is a function because for each value of x, $y = \sqrt{x}$ determines only one value of y.

61. This relation is not a function because $(2, 1)$ and $(2, -1)$ both satisfy $|x| = |2y|$.

63. This relation is not a function because $(0, 3)$ and $(0, -3)$ both satisfy $x^2 + y^2 = 9$.

65. This relation is a function because for each value of x, $x = 2\sqrt{y}$ determines only one value of y.

67. This relation is a function because $(0, 5)$ and $(0, -5)$ both satisfy $x + 5 = |y|$.

69. Since a vertical line can be drawn to cross this graph more than once, this graph is not the graph of a function.

71. Since no vertical line can be drawn to cross this graph more than once, this graph is the graph of a function.

73. Since a vertical line can be drawn to cross this graph more than once, this graph is not the graph of a function.

75. The domain is the set of first coordinates, $\{4, 7\}$, and the range is the set of second coordinates, $\{1\}$.

77. The domain is the set of first coordinates, $\{2\}$, and the range is the set of second coordinates, $\{3, 5, 7\}$.

79. Since any number can be use in place of x in $y = x + 1$, the domain is the set of all real numbers, $(-\infty, \infty)$. Since any number can be used in place of y the range is also $(-\infty, \infty)$.

81. Since any number can be use in place of x in $y = 5 - x$, the domain is the set of all real numbers, $(-\infty, \infty)$. Since any number can be used in place of y the range is also $(-\infty, \infty)$.

83. Since $y = \sqrt{x - 2}$, we must have $x - 2 \geq 0$ or $x \geq 2$. So the domain is $[2, \infty)$. The values of y must be nonnegative. The range is $[0, \infty)$.

85. Since $y = \sqrt{2x}$, we must have $2x \geq 0$ or $x \geq 0$. So the domain is $[0, \infty)$. The values of y must be nonnegative. The range is $[0, \infty)$.

87. $f(0) = 3(0) - 2 = -2$

89. $f(4) = 3(4) - 2 = 12 - 2 = 10$

91. $g(-2) = -(-2)^2 + 3(-2) - 2$
$= -4 - 6 - 2 = -12$

93. $h(-3) = |-3 + 2| = |-1| = 1$

95. $h(-4.236) = |-4.236 + 2|$
$= |-2.236| = 2.236$

97. $f(2) = 3(2) - 2 = 4$
$g(3) = -3^2 + 3(3) - 2 = -2$
$f(2) + g(3) = 4 + (-2) = 2$

99. $g(2) = -2^2 + 3(2) - 2 = 0$
$h(-3) = |-3 + 2| = 1$
$\frac{g(2)}{h(-3)} = \frac{0}{1} = 0$

101. $f(-1) = 3(-1) - 2 = -5$
$h(-4) = |-4 + 2| = 2$
$f(-1) \cdot h(-4) = -5 \cdot 2 = -10$

103. a) $h(2) = 256 - 16(2)^2 = 192$ ft
b) $h(4) = 256 - 16(4)^2 = 0$ ft

105. The area of a square is found by squaring the length of a side: $A = s^2$ or $A(s) = s^2$.

107. The cost of the purchase, C, is the price per yard times the number of yards purchased: $C(x) = 3.98x$, $C(3) = 3.98(3) = \$11.94$.

109. The total cost, C, is found by multiplying the number of toppings, n, by the cost for each topping, \$0.50, and adding on the \$14.95 base charge: $C(n) = 0.50n + 14.95$
$C(6) = 0.50(6) + 14.95 = \$17.95$

111. The relation $y = x + 2$ is a function because there is only one y for any value of x. The relation $y > x + 2$ is not a function because $(2, 5)$ and $(2, 6)$ both satisfy $y > x + 2$.

11.2 WARM-UPS

1. True, because the graph of a function consists of all ordered pairs that are in the function.
2. True, that is why they are called linear functions.
3. True, because absolute value functions are generally v-shaped.
4. False, because any real number can be used in place of x in $f(x) = 3$.
5. True, because the graph of $y = ax^2 + bx + c$ for $a \neq 0$ is a parabola.
6. False, the domain of a quadratic function is $(-\infty, \infty)$ or R.
7. True, because $f(x)$ is just another name for the dependent variable y.
8. True, because $y^2 \geq 0$ and x must also be greater than or equal to 0.
9. False, because 1 is in the domain of the function and $(1, \infty)$ does not include 1.
10. True, because any real number can be used for x in $f(x) = ax^2 + bx + c$.

11.2 EXERCISES

1. A linear function is a function of the form $f(x) = mx + b$ where m and b are real numbers with $m \neq 0$.
3. The graph of a constant function is a horizontal line.
5. The graph of a quadratic function is a parabola.
7. The graph of $h(x) = -2$ is the same as the graph of the horizontal line $y = -2$.

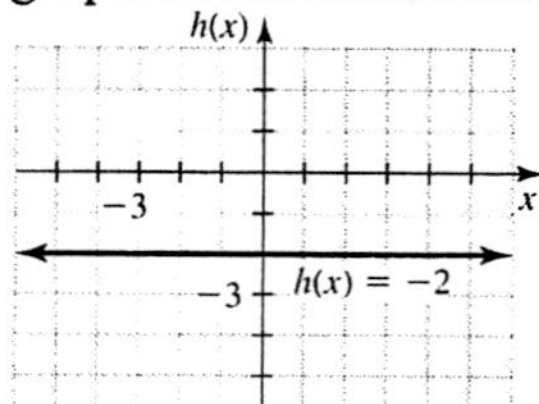

The domain of the function is R or $(-\infty, \infty)$. The only y-coordinate used is -2, and so the range is $\{-2\}$.
9. The graph of $f(x) = 2x - 1$ is the same as the graph of the linear equation $y = 2x - 1$. To draw the graph, start at the y-intercept $(0, -1)$, and use the slope of $2 = 2/1$. Rise 2 and run 1 to locate a second point on the line.

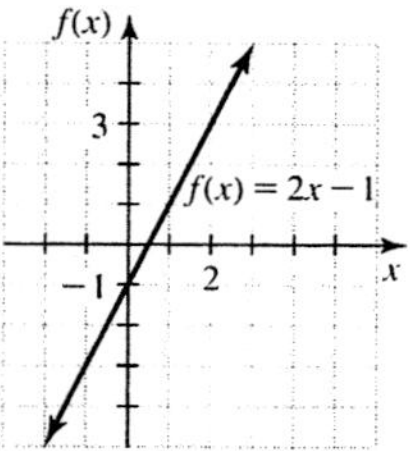

From the graph, we can see that the domain is $(-\infty, \infty)$ and the range is also $(-\infty, \infty)$.
11. Graph the line $y = (1/2)x + 2$ by locating the y-intercept (0, 2) and using a slope of $1/2$. The domain is $(-\infty, \infty)$ and the range is $(-\infty, \infty)$.

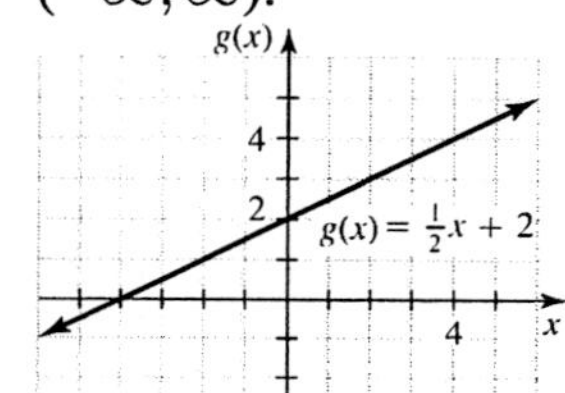

13. The graph of $y = -\frac{2}{3}x + 3$ is a straight line with y-intercept $(0, 3)$ and slope $-2/3$.

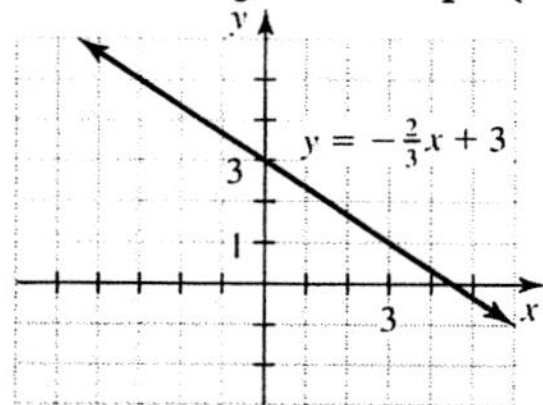

The domain is $(-\infty, \infty)$ and range is $(-\infty, \infty)$.
15. The graph of $y = -0.3x + 6.5$ is a straight line with y-intercept $(0, 6.5)$ and slope $-3/10$.

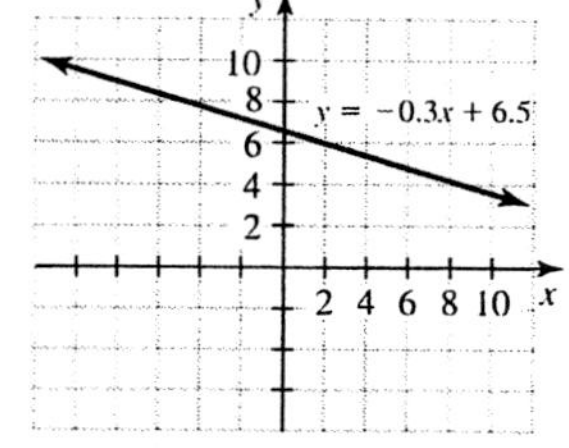

The domain is $(-\infty, \infty)$ and range is $(-\infty, \infty)$.

17. The graph of $f(x) = |x| + 1$ contains the points (0, 1), (3, 4), and (−3, 4). Plot these points and draw the v-shaped graph.

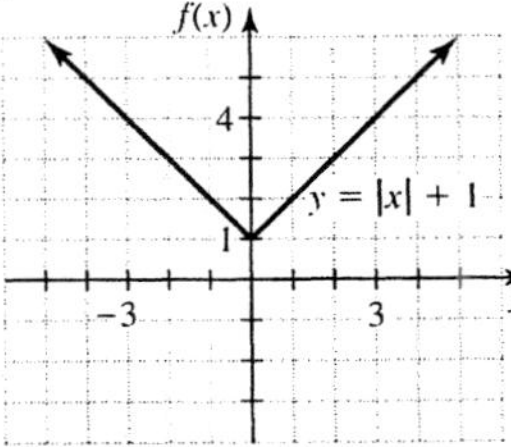

The domain is $(-\infty, \infty)$, and from the graph we can see that the range is $[1, \infty)$.

19. The graph of $h(x) = |x + 1|$ includes the points (0, 1), (1, 2), (−1, 0), and (−2, 1).

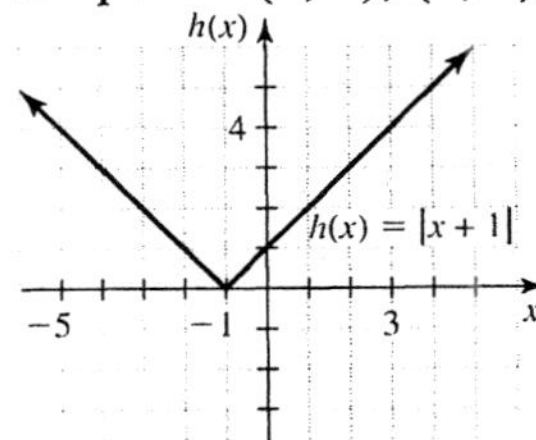

From the graph we can see that the domain is $(-\infty, \infty)$. In the vertical direction the graph is on or above the x-axis. Since all y-coordinates on the graph are greater than or equal to 0, the range is $[0, \infty)$.

21. The graph of $g(x) = |3x|$ includes the points (0, 0), (1, 3), and (−1, 3). Plot these points and draw the graph.

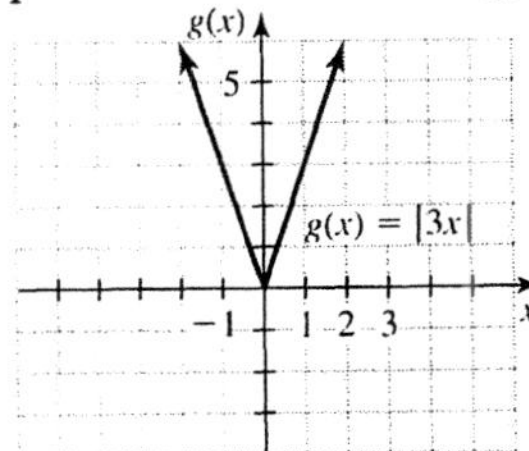

From the graph we can see that the domain is $(-\infty, \infty)$. Since all of the y-coordinates on the graph are greater than or equal to zero, the range is $[0, \infty)$.

23. The graph of $f(x) = |2x - 1|$ includes the points (0, 1), (1/2, 0), (1, 1), and (2, 3). Plot these points and draw a v-shaped graph.

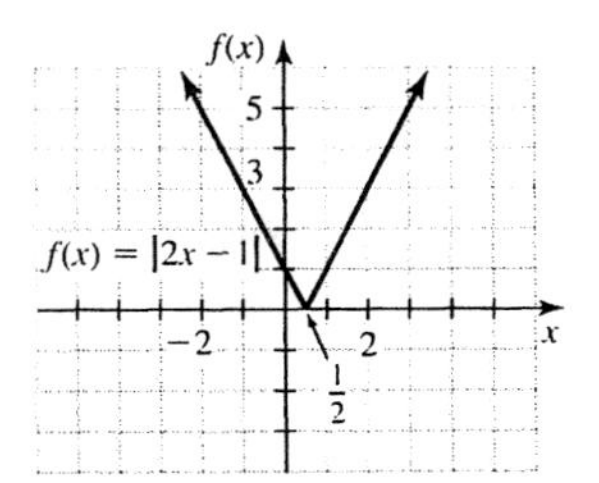

From the graph we can see that the domain is $(-\infty, \infty)$, and the range is $[0, \infty)$.

25. The graph of $f(x) = |x - 2| + 1$ includes the points (2, 1), (3, 2), (4, 3), (1, 2), and (0, 3). Plot these points and draw a v-shaped graph. Form the graph we can see that the domain is $(-\infty, \infty)$. Since all y-coordinates are greater than or equal to 1, the range is $[1, \infty)$.

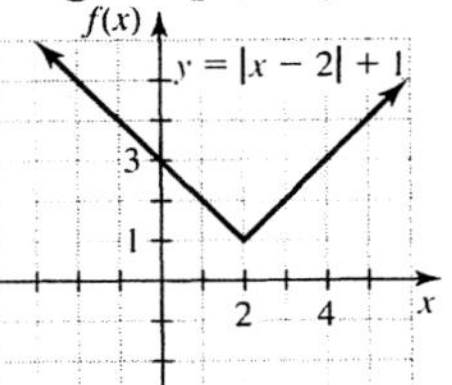

27. The graph of $y = x^2$ includes the points $(0, 0)$, $(-1, 1)$, $(1, 1)$, $(-2, 4)$, and $(2, 4)$. Plot these points and draw the parabola through them. The domain is $(-\infty, \infty)$. Since all y-coordinates are greater than or equal to zero, the range is $[0, \infty)$.

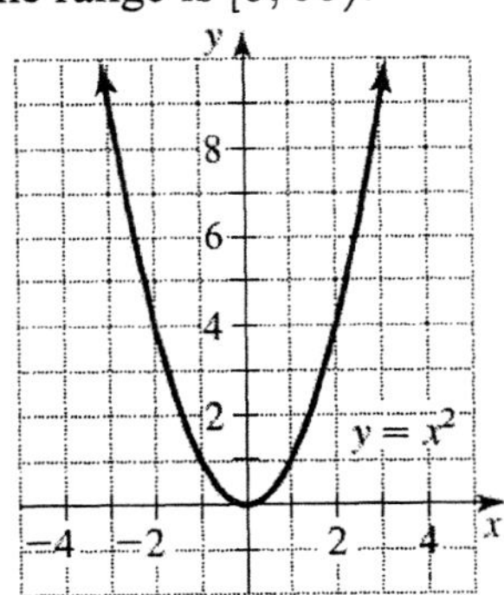

29. The graph of $g(x) = x^2 + 2$ includes the points (−2, 6), (−1, 3), (0, 2), (1, 3), and (2, 6). Plot these points and draw the parabola through them.

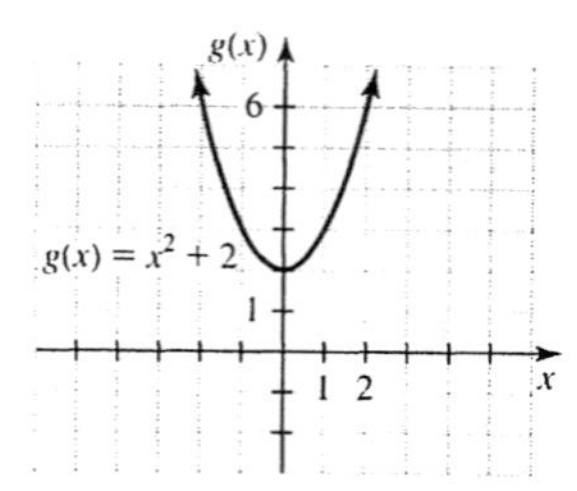

The domain is $(-\infty, \infty)$. Since no y-coordinate on the parabola is below 2, the range is $[2, \infty)$.

31. The graph of $f(x) = 2x^2$ includes the points (0, 0), (1, 2), (−1, 2), (2, 8), and (−2, 8). Plot these points and draw a parabola through them.

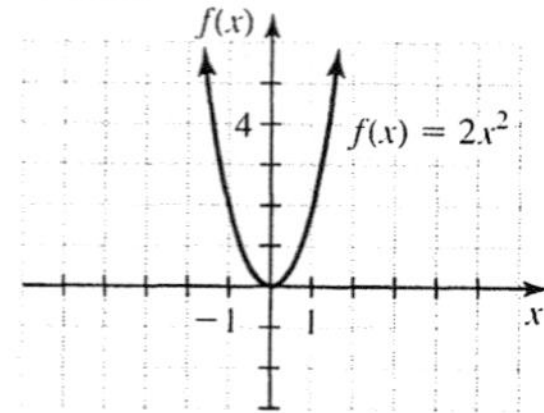

The domain is $(-\infty, \infty)$ and the range is $[0, \infty)$.

33. The graph of $y = 6 - x^2$ includes the points (−3, −3), (0, 6), (3, −3), (−2, 2), and (2, 2). Plot these points and draw a parabola through them.

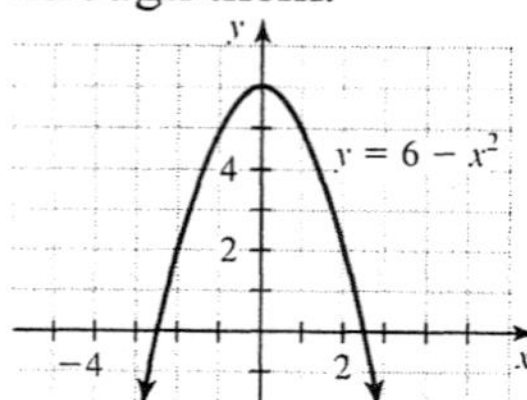

The domain is $(-\infty, \infty)$ and range is $(-\infty, 6]$.

35. The graph of $g(x) = 2\sqrt{x}$ includes the points (0, 0), (1, 2), and (4, 4). Plot these points and draw a curve through them.

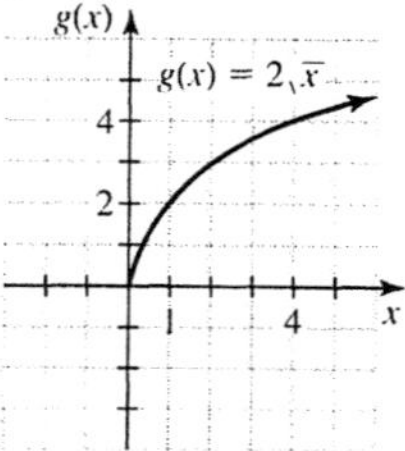

Since we must have $x \geq 0$ in $\sqrt{x}$, the domain is $[0, \infty)$. The range is $[0, \infty)$

37. The graph of $f(x) = \sqrt{x - 1}$ includes the points (1, 0), (2, 1), and (5, 2). Plot these points and draw a smooth curve through them. The curve is half of a parabola, positioned on its side.

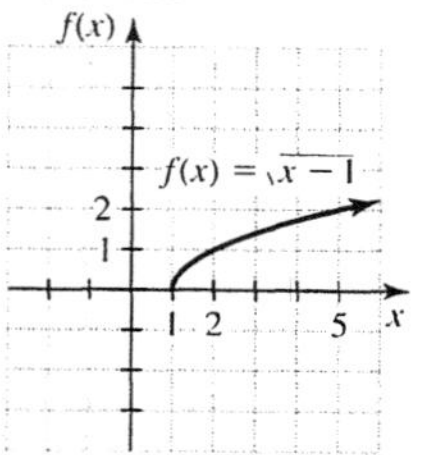

Since we must have $x - 1 \geq 0$, or $x \geq 1$ the domain is $[1, \infty)$. The range is $[0, \infty)$.

39. The graph of $h(x) = -\sqrt{x}$ includes the points (0, 0), (1, −1), and (4, −2). Plot these points and draw a curve through them.

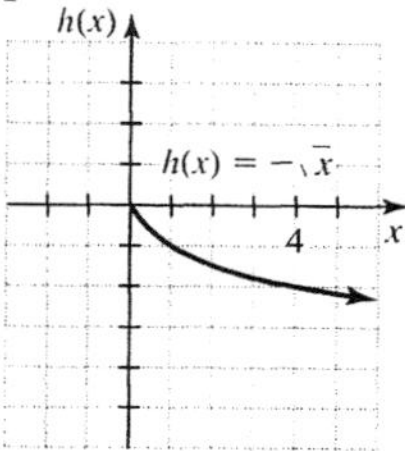

Since we must have $x \geq 0$ in $\sqrt{x}$, the domain is $[0, \infty)$. The range is $(-\infty, 0]$.

41. The graph of $y = \sqrt{x} + 2$ includes the points (0, 2), (1, 3), and (4, 4). Graph these points and draw a curve through them.

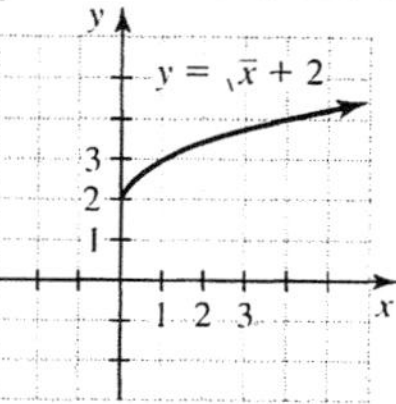

Because $\sqrt{x}$ is a real number only for nonnegative values of x, the domain is $[0, \infty)$. From the graph we see that y-coordinates go no lower than 2, so the range is $[2, \infty)$.

43. The graph of $x = |\, y \,|$ includes the points $(0, 0)$, $(1, -1)$, $(1, 1)$, $(2, -2)$, and $(2, 2)$. Note that in this case it is easier to pick the y-coordinate and then find the appropriate x-coordinate using $x = |\, y \,|$. Draw the v-shaped graph through these points.

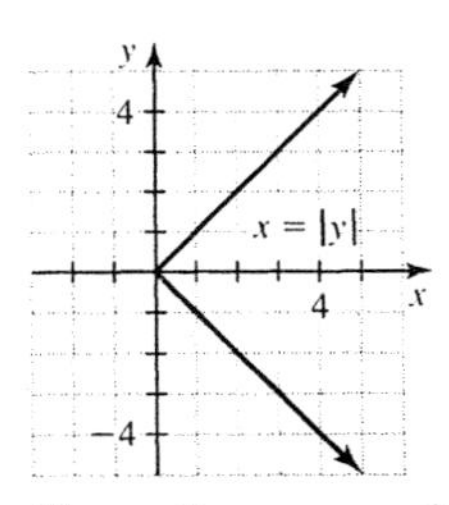

Since the x-coordinates are nonnegative, the domain is $[0, \infty)$. Since we are allowed to select any number for y, the range is $(-\infty, \infty)$.

45. To find pairs that satisfy $x = -y^2$, pick the y-coordinate first and then find the x-coordinate. The points $(0, 0)$, $(-1, 1)$, $(1, 1)$, $(-4, 2)$, and $(-4, -2)$ are on the graph.

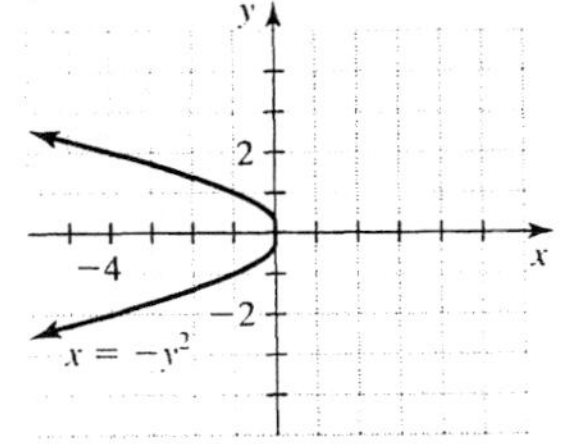

From the graph we see that only nonpositive x-coordinates are used, so the domain is $(-\infty, 0]$. Since any real number is allowable for y, the range is $(-\infty, \infty)$.

47. The equation $x = 5$ is the equation of a vertical line with an x-intercept of $(5, 0)$.

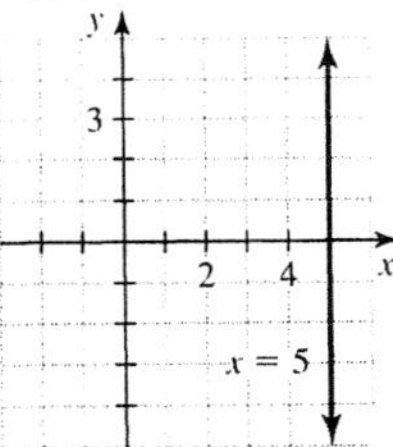

All points on this line have x-coordinate 5, so the domain is $\{5\}$. Every real number occurs as a y-coordinate, so the range is $(-\infty, \infty)$.

49. Rewrite $x + 9 = y^2$ as $x = y^2 - 9$. Select some y-coordinates and calculate the appropriate x-coordinates. The points $(-9, 0)$, $(0, 3)$, and $(0, -3)$ are on the parabola.

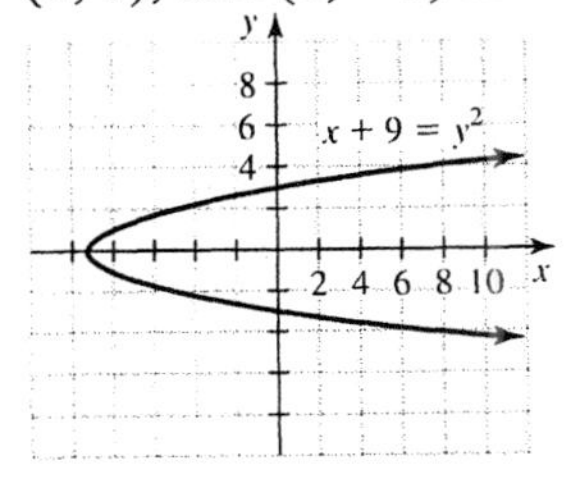

Since no x-coordinate is below -9, the domain is $[-9, \infty)$. Since we could select any value for y, the range is $(-\infty, \infty)$.

51. For the equation $x = \sqrt{y}$, we must select nonnegative values for y and find the corresponding value for x. The points $(0, 0)$, $(1, 1)$, and $(2, 4)$ are on the graph.

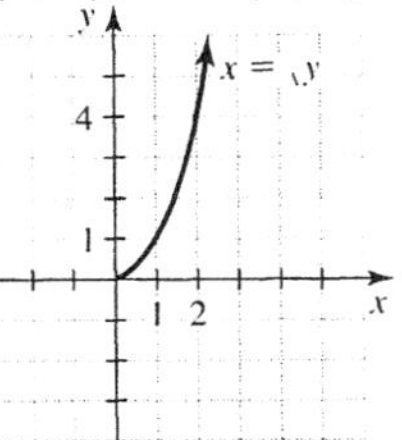

Since the x-coordinates are also nonnegative, the domain is $[0, \infty)$. Since we could only use nonnegative values for y, the range is $[0, \infty)$.

53. The graph of $x = (y - 1)^2$ includes the points $(0, 1)$, $(1, 0)$, $(1, 2)$, $(4, 3)$, and $(4, -1)$. Plot the points and draw a curve through them. From the graph we see that the domain is $[0, \infty)$ and the range is $(-\infty, \infty)$.

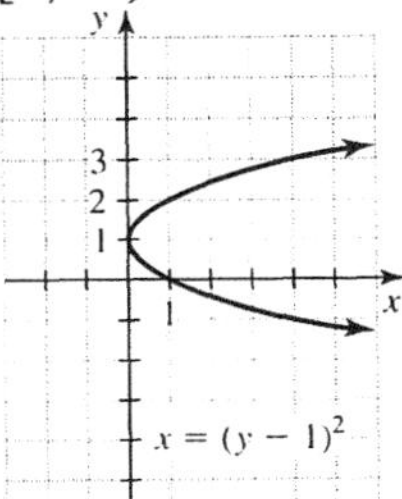

55. To graph $f(x) = 1 - |x|$, we arbitrarily select a value for x and find the corresponding value for y. The points $(-2, -1)$, $(-1, 0)$, $(0, 1)$, $(1, 0)$, and $(2, -1)$ are on the graph.

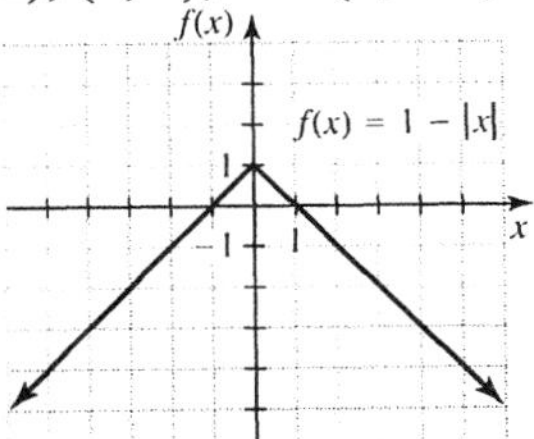

Since any real number could be used for x, the domain is $(-\infty, \infty)$. From the graph we see that the y-coordinates are not higher than 1, so the range is $(-\infty, 1]$.

57. The graph of $y = (x - 3)^2 - 1$ includes the points $(1, 3)$, $(2, 0)$, $(3, -1)$, $(4, 0)$, and $(5, 3)$.

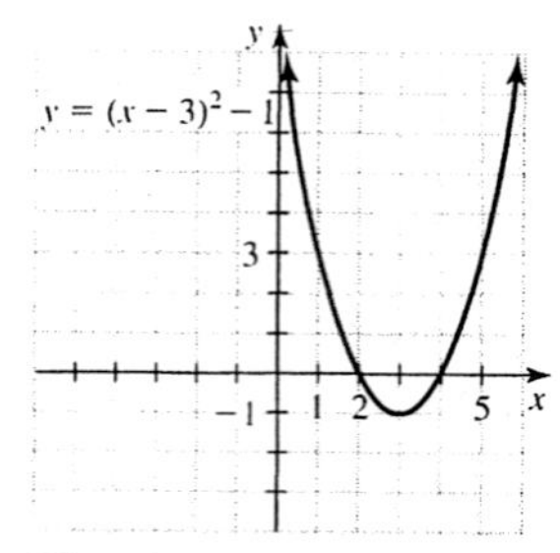

The domain is $(-\infty, \infty)$. Since the y-coordinates are greater than or equal to -1, the range is $[-1, \infty)$.

59. Graph of $y = |x + 3| + 1$ goes through $(-4, 2)$, $(-3, 1)$, $(-2, 2)$, $(-1, 3)$, $(0, 4)$.

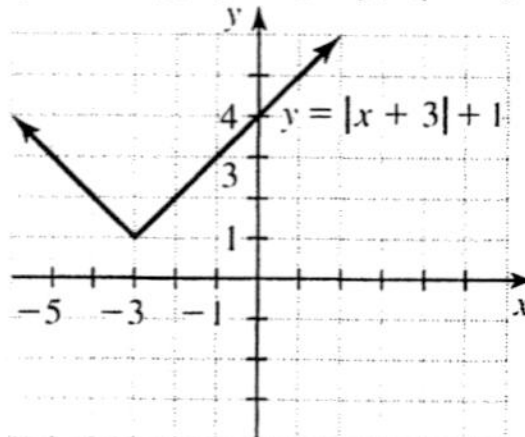

The domain is $(-\infty, \infty)$ and range is $[1, \infty)$.

61. Graph of $y = \sqrt{x} - 3$ goes through $(0, -3)$, $(1, -2)$, and $(4, -1)$. Domain is $[0, \infty)$ and range is $[-3, \infty)$.

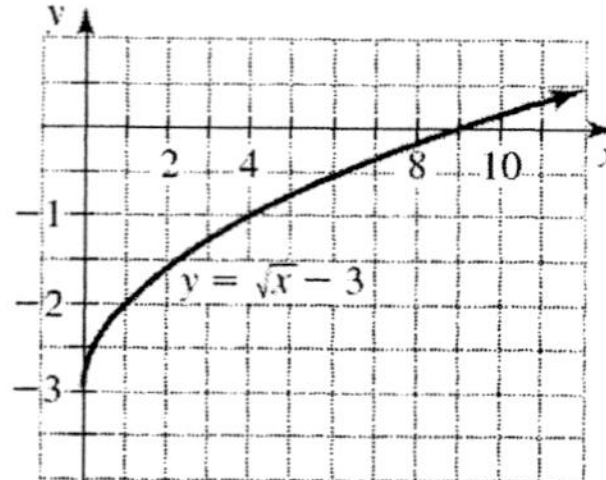

63. The graph of $y = 3x - 5$ is a straight line with y-intercept $(0, -5)$, and a slope of 3.

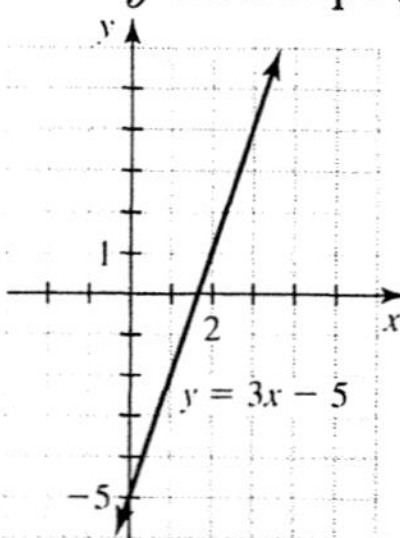

The domain is $(-\infty, \infty)$. From the graph we see that any real number could occur as a y-coordinate, so the range is $(-\infty, \infty)$.

65. The graph of $y = -x^2 + 4x - 4$ goes through $(0, -4)$, $(1, -1)$, $(2, 0)$, $(3, -1)$, and $(4, -4)$. From the graph we can see that the domain is $(-\infty, \infty)$ and range is $(-\infty, 0]$.

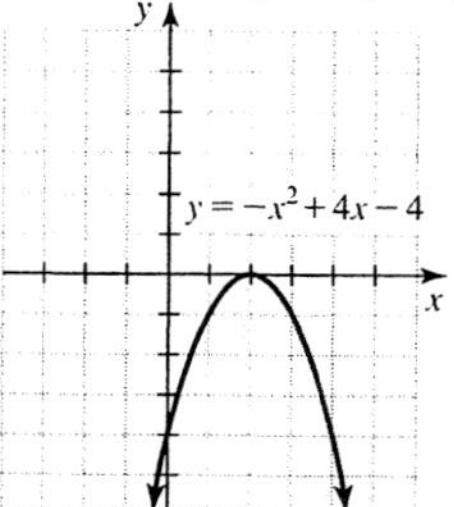

67. Graph of $f(x) = \sqrt{x^2}$ is same as graph of $y = |x|$. This graph shows us that $\sqrt{x^2} \neq x$ when $x < 0$.

69. If $a > 0$, then for large values of a the graph gets narrower and for smaller values of a the graph gets broader.

71. The graph of $y = (x - h)^2$ is a shift of the graph of $y = x^2$ to the right for $h > 0$ and to the left for $h < 0$.

73. The graph of $y = f(x - h)$ lies to the right of the graph of $y = f(x)$ when $h > 0$.

11.3 WARM-UPS

1. False, because $(-x)^2 = x^2$ and the graphs of $f(x) = x^2$ and $g(x) = (-x)^2$ are identical.

2. True, because $-2 = -(2)$.

3. True. The graph of $y = x + 3$ is 3 units above $y = x$ or 3 units to the left of $y = x$.

4. False, because $y = |x - 3|$ is 3 units to the right of $y = |x|$.

5. True, because subtracting 3 after the absolute value moves each point on $y = |x|$ down 3 units.

6. True, because the negative sign causes the reflection and the multiplication by 2 causes the stretching.

7. False, because $f(-x) = (-x - 2)^2$ $= (-1)^2(x + 2)^2 = (x + 2)^2 \neq (x - 2)^2$.

8. True, because $y = \sqrt{\frac{x}{9}} = \frac{1}{3}\sqrt{x}$. The y-coordinates on $y = \sqrt{x}$ are 3 times as large as the y-coordinates on $y = \frac{1}{3}\sqrt{x}$.

9. True, because $y = \sqrt{x - 3} + 5$ is the same as $y = \sqrt{x}$ moved 3 units to the right and 5 units upward.

10. False, because the graph of $y = x^2$ must be moved 2 units to the left, reflected in the x-

axis, and then moved 7 units downward to obtain the graph of $y = -(x + 2)^2 - 7$.

11.3 EXERCISES

1. The graph of $y = -f(x)$ is a reflection in the x-axis of the graph of $y = f(x)$.
3. The graph of $y = f(x) + k$ for $k < 0$ is a downward translation of the graph of $y = f(x)$.
5. The graph of $y = f(x - h)$ for $h < 0$ is a translation to the left of the graph of $y = f(x)$.
7. The graph of $f(x) = \sqrt{2x}$ goes through $(0, 0)$, $(2, 2)$, and $(8, 4)$. The graph of $g(x) = -\sqrt{2x}$ is a reflection in the x-axis of the graph of $f(x) = \sqrt{2x}$.

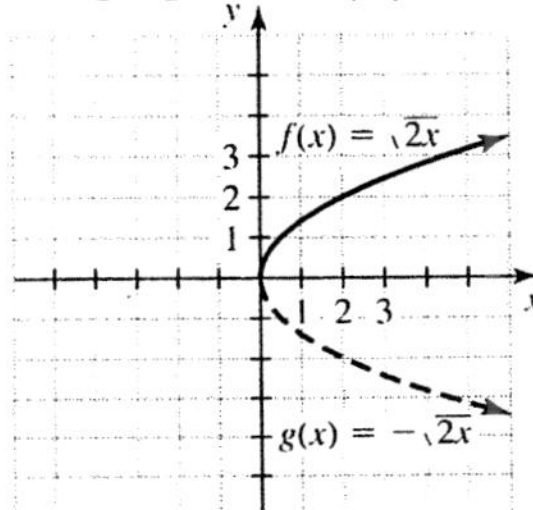

9. The graph of $f(x) = x^2 + 1$ is a parabola through $(0, 0)$, $(\pm 1, 2)$, and $(\pm 2, 5)$. The graph of $g(x) = -(x^2 + 1)$ is a reflection in the x-axis of the graph of $f(x)$.

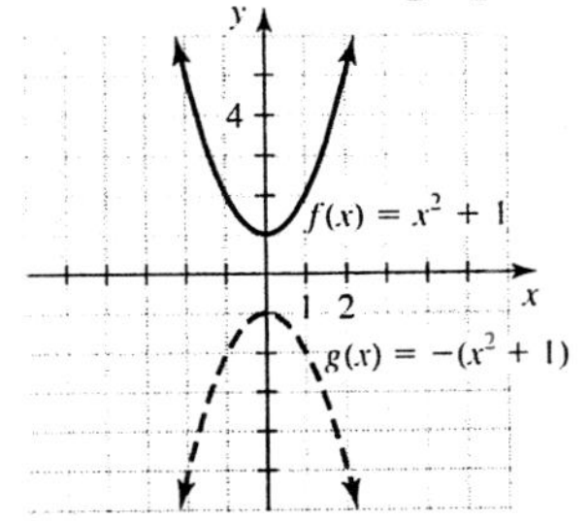

11. The graph of $y = \sqrt{x - 2}$ is half of a parabola. The graph contains the points (2, 0), (3, 1), and (6, 2). The graph of $y = -\sqrt{x - 2}$ is a reflection in the x-axis of the first graph.

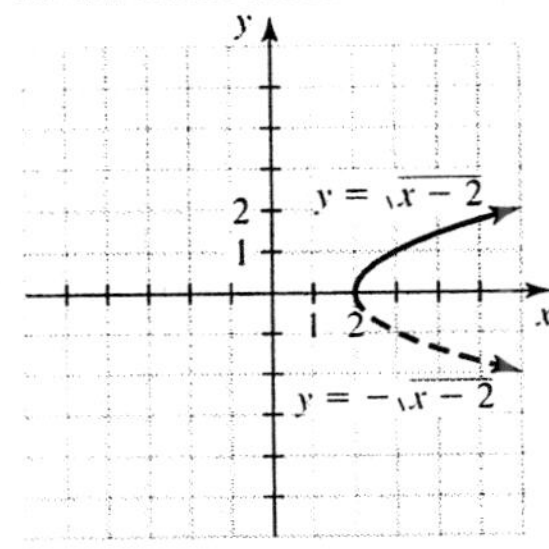

13. The graph of $f(x) = x - 3$ is a straight line through $(0, -3)$ with slope 1. The graph of $g(x) = 3 - x = -(x - 3)$ is a reflection in the x-axis of the graph of $f(x)$.

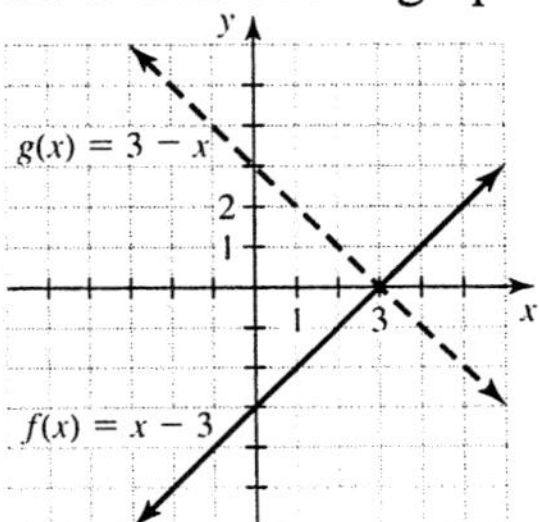

15. The graph of $g(x) = x^2$ is a parabola opening upward with vertex $(0, 0)$. The graph of $f(x) = x^2 - 4$ is a downward translation of 4 units from the graph of g. It includes the points (0,–4), (± 1, –3), and (± 2, 0).

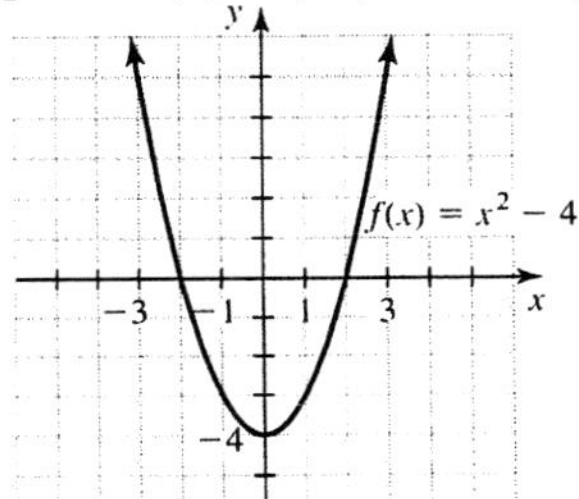

The domain is $(-\infty, \infty)$ and the range is $[-4, \infty)$.
17. The graph of $y = x$ is a straight line with slope 1 and y-intercept $(0, 0)$. The graph of $y = x + 3$ is a translation 3 units upward of the graph of $y = x$.

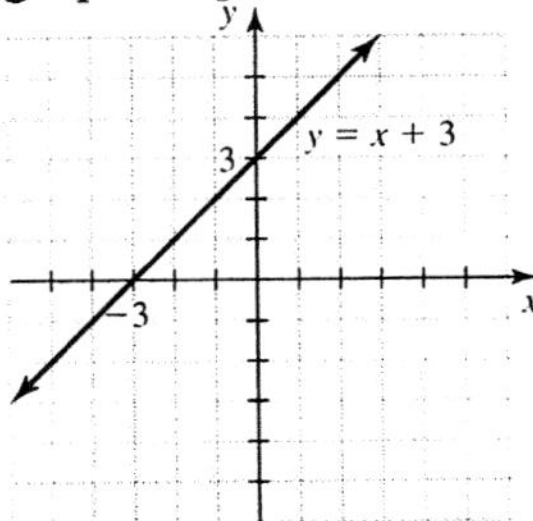

The domain is $(-\infty, \infty)$ and the range is $(-\infty, \infty)$.
19. The graph of $f(x) = (x - 3)^2$ is a translation 3 units to the right of the graph of the parabola $y = x^2$. The graph of $f(x) = (x - 3)^2$ goes through the points (2, 1), (3, 0), and (4, 1).

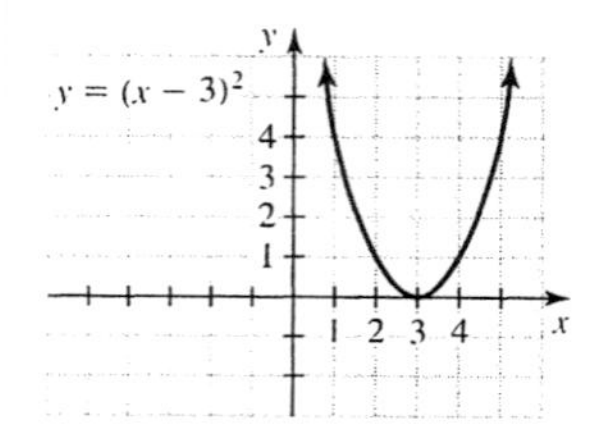

The domain is $(-\infty, \infty)$ and the range is $[0, \infty)$.

21. The graph of $y = \sqrt{x} + 1$ is a translation 1 unit upward of the graph of $y = \sqrt{x}$. The graph of $y = \sqrt{x} + 1$ goes through $(0, 1)$, $(1, 2)$, and $(4, 3)$.

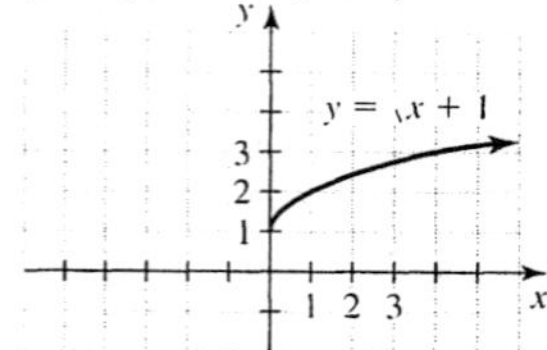

The domain is $[0, \infty)$ and the range is $[1, \infty)$.

23. The graph of $f(x) = \mid x + 2 \mid$ is a translation 2 units to the left of the graph of $y = \mid x \mid$. The graph of $f(x) = \mid x + 2 \mid$ includes the points (−3, 1), (−2, 0), and $(-1, 1)$.

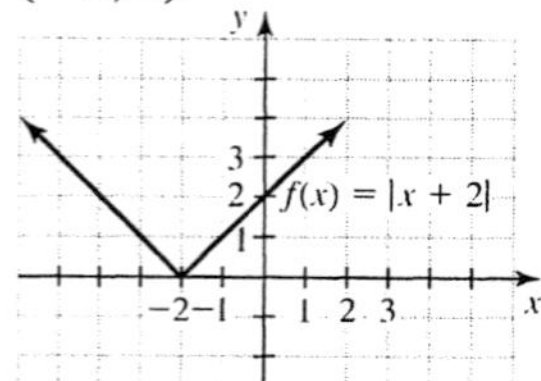

The domain is $(-\infty, \infty)$ and the range is $[0, \infty)$.

25. The graph of $y = \mid x \mid + 2$ is a translation 2 units upward of the graph of $y = \mid x \mid$. The graph of $y = \mid x \mid + 2$ goes through the points (−1, 3), (0, 2), and (1, 3).

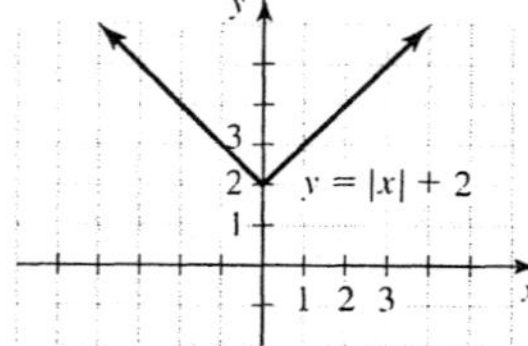

The domain is $(-\infty, \infty)$ and the range is $[2, \infty)$.

27. The graph of $f(x) = \sqrt{x - 1}$ is a translation 1 unit to the right of the graph of $y = \sqrt{x}$. The graph of $f(x) = \sqrt{x - 1}$ goes through the points (1, 0), (2, 1), and (5, 2).

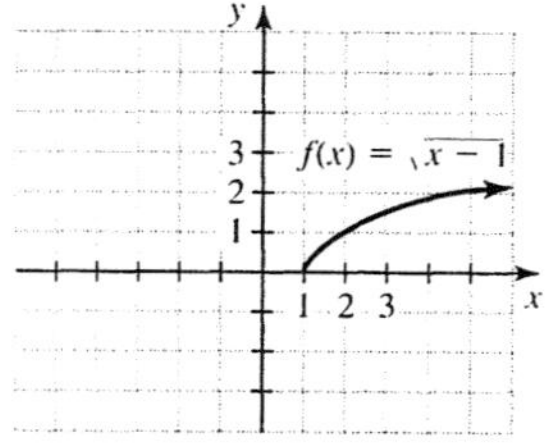

The domain is $[1, \infty)$ and the range is $[0, \infty)$.

29. The graph of $f(x) = 3x^2$ is obtained by stretching the graph of $y = x^2$. The graph of $f(x) = 3x^2$ is a parabola through (−1, 3), (0, 0), and (1, 3).

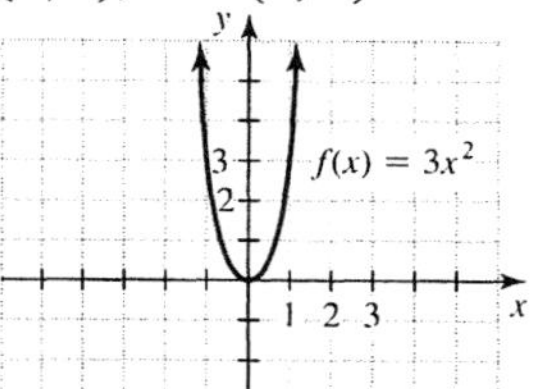

The domain is $(-\infty, \infty)$ and the range is $[0, \infty)$.

31. The graph of $y = \frac{1}{5}x$ is obtained by shrinking the graph of $y = x$. The graph of $y = \frac{1}{5}x$ is a straight line through the points (−5, −1), (0, 0), and (5, 1).

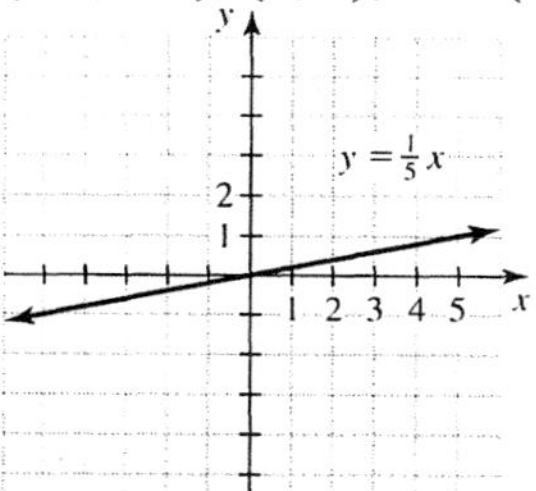

The domain is $(-\infty, \infty)$ and the range is $(-\infty, \infty)$.

33. The graph of $f(x) = 3\sqrt{x}$ is obtained by stretching the graph of $y = \sqrt{x}$. The graph of $f(x) = 3\sqrt{x}$ goes through (0, 0), (1, 3), and $(4, 6)$.

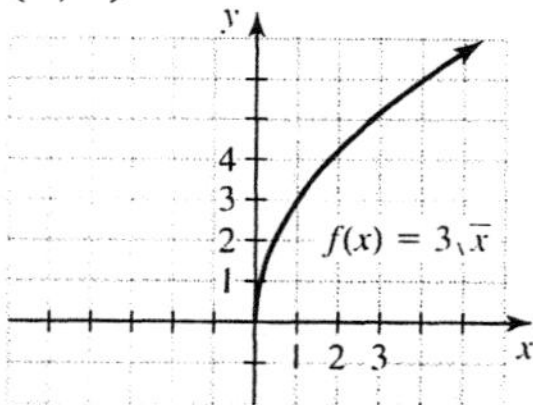

The domain is $[0, \infty)$ and the range is $[0, \infty)$.

35. The graph of $y = \frac{1}{4} \mid x \mid$ is obtained by shrinking the graph of $y = \mid x \mid$. The graph of

$y = \frac{1}{4} \mid x \mid$ goes through the points (−4, 1), (0, 0), and (4, 1).

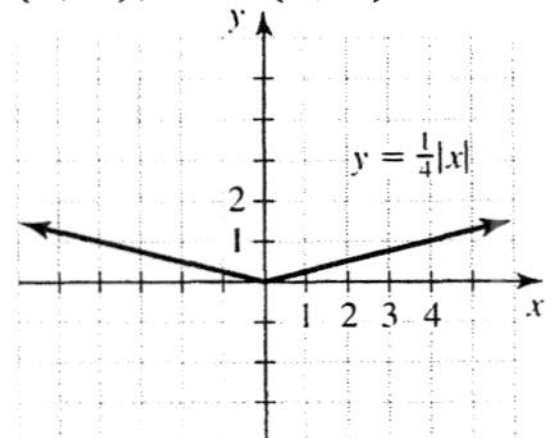

The domain is $(-\infty, \infty)$ and the range is $[0, \infty)$.

37. The graph of $y = \sqrt{x-2}+1$ is obtained by translating the graph of $y = \sqrt{x}$ two units to the right and one unit upward. The graph of $y = \sqrt{x-2}+1$ goes through the points (2, 1), (3, 2), and (6, 3).

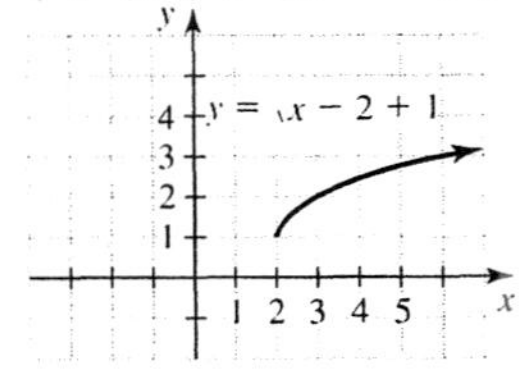

The domain is $[2, \infty)$ and the range is $[1, \infty)$.

39. The graph of $y = - \mid x+3 \mid$ is obtained from the graph of $y = \mid x \mid$ by translating it 3 units to the left and then reflecting in the x-axis. The graph of $y = - \mid x+3 \mid$ contains the points (−4, −1), (−3, 0), and (−2, −1).

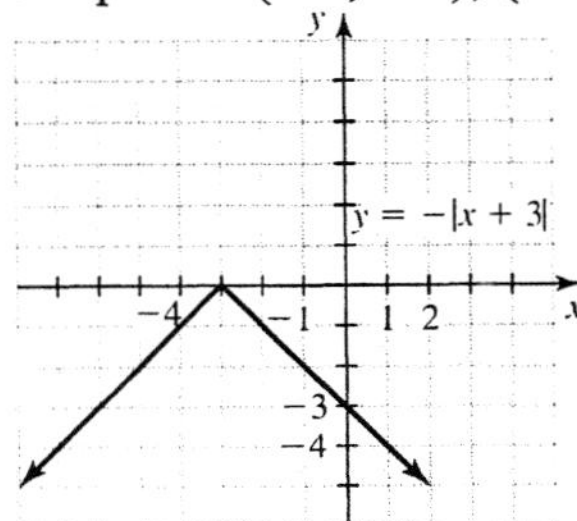

The domain is $(-\infty, \infty)$ and the range is $(-\infty, 0]$.

41. The graph of $f(x) = (x+3)^2 - 5$ is a translation of $y = x^2$ three units to the left and five units downward. The graph goes through the points (−4, −4), (−3, −5), (−2, −4), and (0, 4).

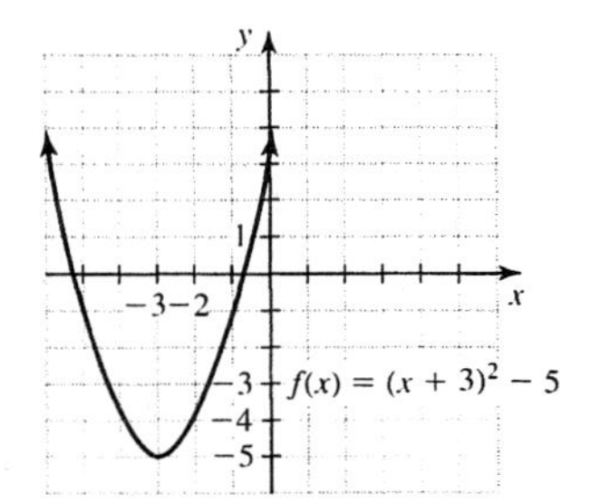

The domain is $(-\infty, \infty)$ and the range is $[-5, \infty)$.

43. The graph of $y = -\sqrt{x+1} - 2$ is obtained from the graph of $y = \sqrt{x}$ by translating it 1 unit to the left, then reflecting in the x-axis, then translating 2 units downward. The graph of $y = -\sqrt{x+1} - 2$ includes the points (−1, −2), (0, −3), and (3, −4).

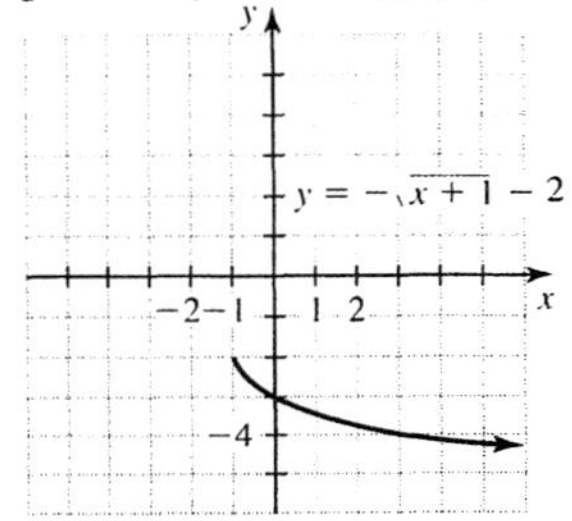

The domain is $[-1, \infty)$ and the range is $(-\infty, -2]$.

45. The graph of $f(x) = -2 \mid x-3 \mid +4$ is obtained from the graph of $y = \mid x \mid$ by translating it 3 units to the right, stretching by a factor of 2, reflecting in the x-axis, and then translating 4 units upward. The graph of $f(x) = -2 \mid x-3 \mid +4$ includes the points (2, 2), (3, 4), and (4, 2).

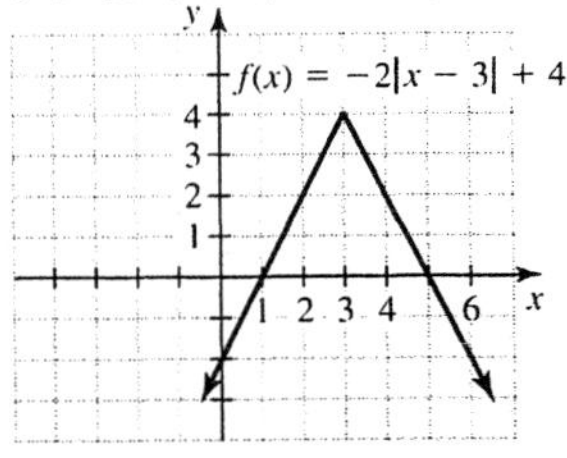

The domain is $(-\infty, \infty)$ and the range is $(-\infty, 4]$.

47. The graph of $y = -2x + 3$ is obtained from the graph of $y = x$ by stretching by a factor of 2, reflecting in the x-axis, and then translating it 3 units upward. The graph of $y = -2x + 3$ includes the points (0, 3), (1, 1), and (2, −1).

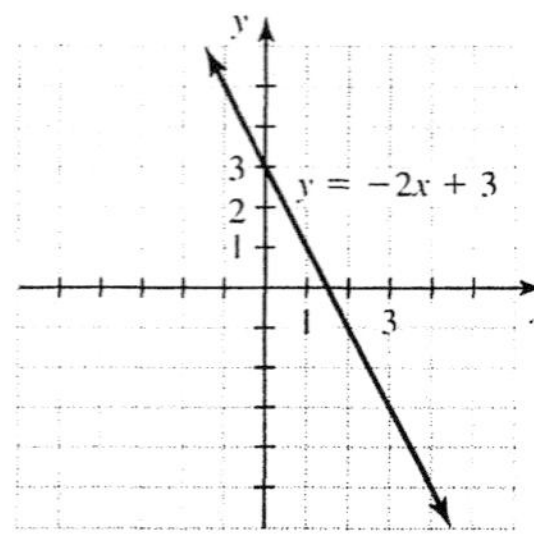

The domain is $(-\infty, \infty)$ and the range is $(-\infty, \infty)$.

49. The graph of $y = 2(x+3)^2 + 1$ is a parabola with vertex at $(-3, 1)$ and it opens upward. The graph includes the points $(-4, 3)$ and $(-2, 3)$.

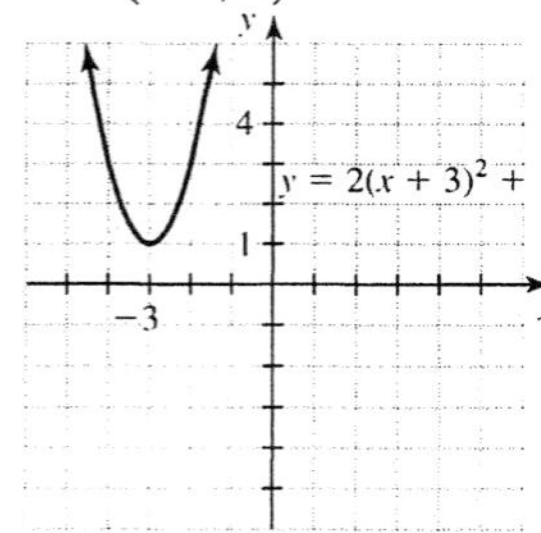

The domain is $(-\infty, \infty)$ and the range is $[1, \infty)$.

51. The graph of $y = -2(x-4)^2 + 2$ is a parabola opening downward with vertex at $(4, 2)$. The graph includes the points $(5, 0)$ and $(3, 0)$.

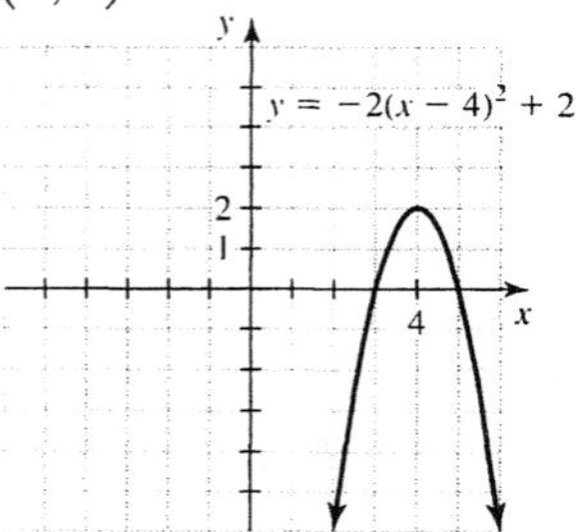

The domain is $(-\infty, \infty)$ and the range is $(-\infty, 2]$.

53. The graph of $y = -3(x-1)^2 + 6$ is a parabola opening downward with vertex at $(1, 6)$. The graph includes the points $(0, 3)$ and $(2, 3)$.

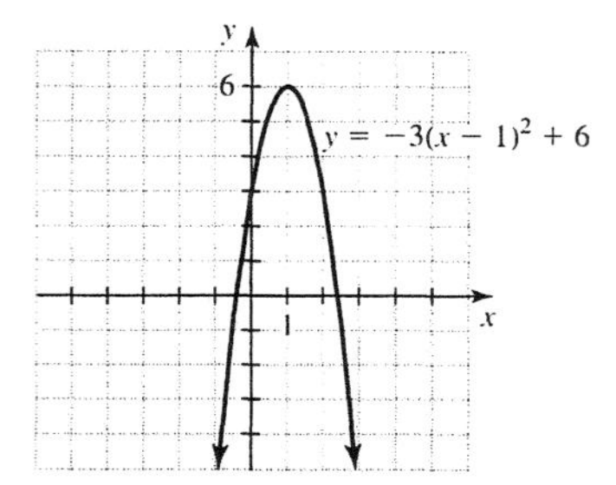

The domain is $(-\infty, \infty)$ and the range is $(-\infty, 6]$.

55. The graph of $y = 2 + \sqrt{x}$ is obtained from $y = \sqrt{x}$ by translating it 2 units upward. The graph of $y = 2 + \sqrt{x}$ includes the points (0, 2), (1, 3), and (4, 4). The graph is (d).

57. The graph of $y = 2\sqrt{x}$ is obtained by stretching the graph of $y = \sqrt{x}$. The graph of $y = 2\sqrt{x}$ goes through (0, 0), (1, 2), and (4, 4). The graph is (e).

59. The graph of $y = \frac{1}{2}\sqrt{x}$ is obtained by shrinking the graph of $y = \sqrt{x}$. The graph of $y = \frac{1}{2}\sqrt{x}$ goes through the points (0, 0), $(1, 0.5)$, and (4, 1). The graph is (h).

61. The graph of $y = -2\sqrt{x}$ is obtained by stretching the graph of $y = \sqrt{x}$ and then reflecting in the x-axis. The graph of $y = -2\sqrt{x}$ contains the points (0, 0), (1, −2), and (4, −4). The graph is (c).

63. If we translate the graph of $y = x^2$ upward 8 units, then the equation is $y = x^2 + 8$.

65. If we translate the graph of $y = \sqrt{x}$ to the left 5 units, then the equation is $y = \sqrt{x+5}$.

67. If we translate the graph of $y = |x|$ upward 5 units and the left 3 units, then the equation is $y = |x+3| + 5$.

69. Translate the graph of $f(x) = |x|$ to the right 20 units and upward 30 units to get the graph of $g(x) = |x - 20| + 30$.

11.4 WARM-UPS

1. False, because $(0, 1)$ is the y-intercept.

2. True, because if $x = 0$, then $f(0) = (0-5)^3 = -125$.

3. True, because the polynomial has three distinct factors.

4. False, the only x-intercept is $(4, 0)$.

5. False, because it touches but does not cross at $(3, 0)$.
6. True, because the exponent on $x - 7$ is even.
7. True, because the graph of any function is translated two units to the right if x is replaced by $x - 2$.
8. False, $y = x^4 + 3$ lies three units above $y = x^4$.
9. False, because if x is replace by $x + 5$ the graph is transformed 5 units to the left.
10. False, because $(2, 8)$ satisfies $y = x^3$ but does not satisfy $y = -x^3$.

11.4 EXERCISES

1. A cubic function is a third-degree polynomial function.
3. To find x-intercepts set y equal to zero and solve the resulting equation.
5. The factor corresponding to that intercept must occur with an even exponent.
7. The intercepts are $(-1, 0)$ and $(0, 1)$. The graph also goes through $(1, 2)$ and $(2, 9)$.

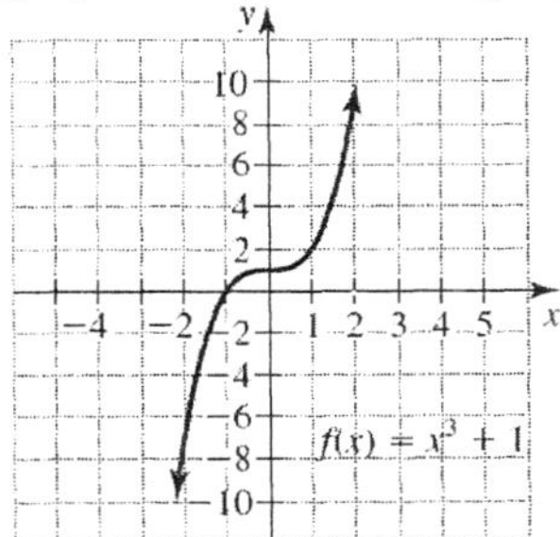

9. The intercepts are $(-3, 0)$, $(3, 0)$ and $(0, 0)$. The graph also goes through $(2, -10)$ and $(-1, 8)$.

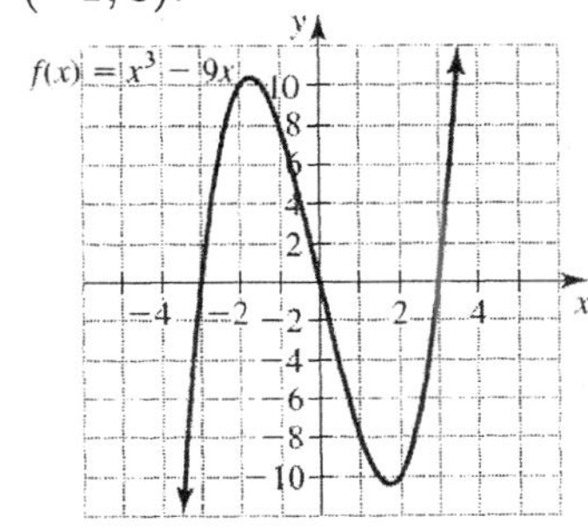

11. The intercepts are $(0, 0)$ and $(-4, 0)$. The graph also goes through $(-2, -8)$ and $(1, -5)$.

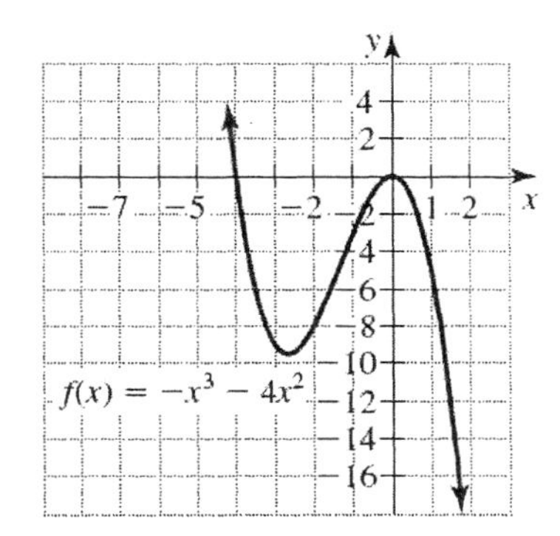

13. $f(x) = x^3 + x^2 - 4x - 4$
$= (x - 2)(x + 2)(x + 1)$
The intercepts are $(-2, 0)$, $(-1, 0)$, $(2, 0)$, and $(0, -4)$. The graph also goes through $(1, -6)$ and $(-3, -10)$.

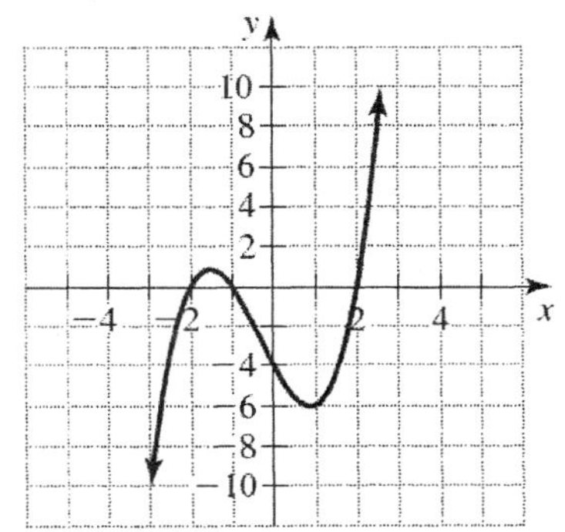

15. $f(x) = x^3 - 3x^2 - 9x + 27$
$= (x - 3)^2(x + 3)$
The intercepts are $(-3, 0)$, $(3, 0)$, and $(0, 27)$. The graph also goes through $(1, 16)$ and $(-1, 32)$.

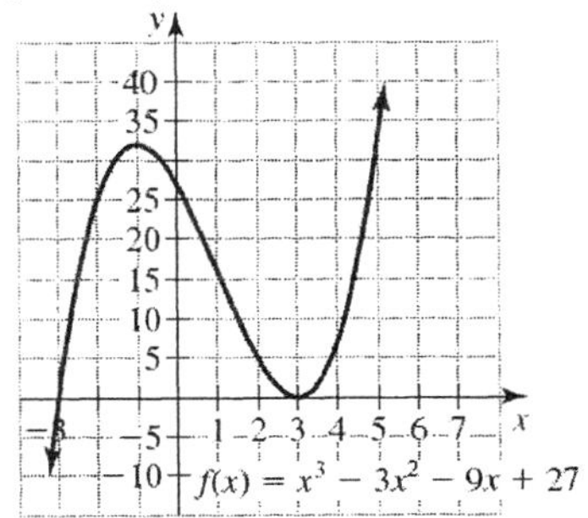

17. $f(x) = x^4 - 1$
$= (x - 1)(x + 1)(x^2 + 1)$
The intercepts are $(-1, 0)$, $(1, 0)$, and $(0, -1)$. The graph also goes through $(-2, 15)$ and $(2, 15)$.

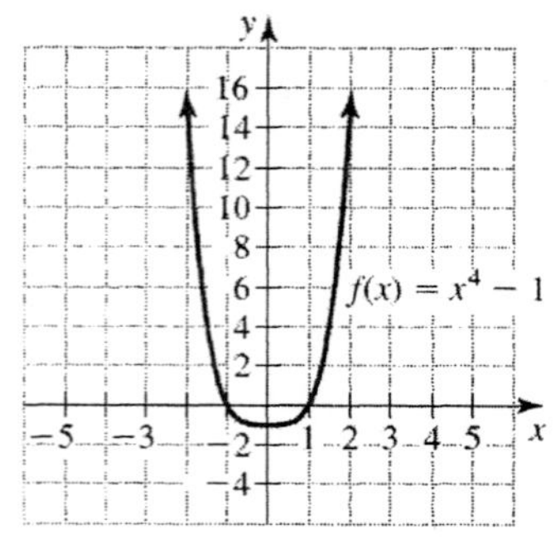

19. $f(x) = x^4 - 4x^2$
$= x^2(x-2)(x+2)$
The intercepts are $(-2,0)$, $(0,0)$, and $(2,0)$.
The graph also goes through $(-1,-3)$ and $(1,-3)$.

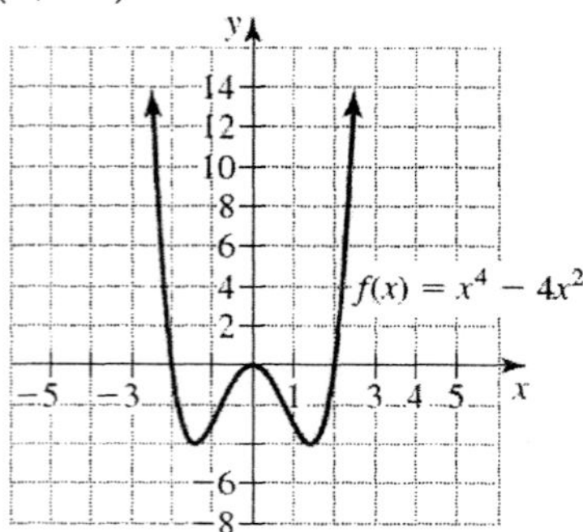

21. $f(x) = x^4 - 5x^2 + 4$
$= (x-2)(x+2)(x-1)(x+1)$
The intercepts are
$(-2,0), (-1,0), (1,0), (2,0)$, and $(0,4)$. The graph also goes through $(3,40)$ and $(-3,40)$.

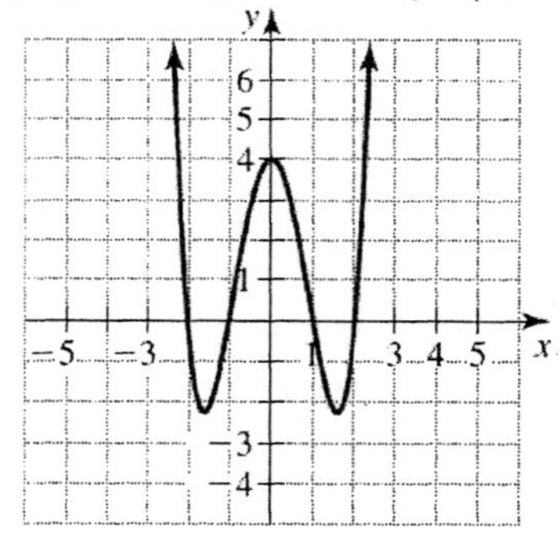

23. $f(x) = x^4 + x^3 - 4x^2 - 4x$
$= x(x+1)(x-2)(x+2)$
The intercepts are $(-2,0)$, $(-1,0)$, $(0,0)$, and $(2,0)$.
The graph also goes through $(1,-6)$ and $(3,60)$.

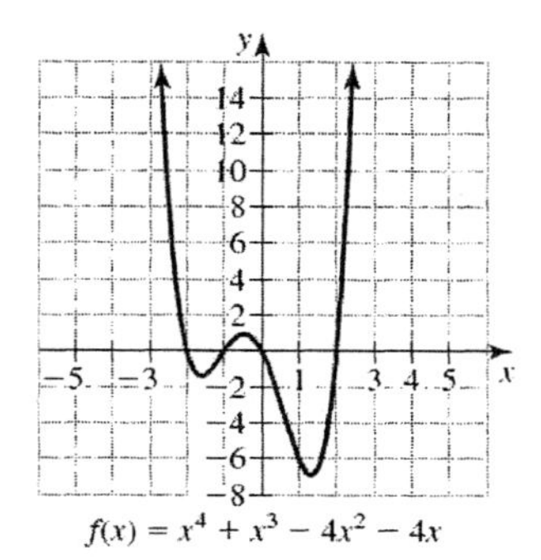

25. $f(x) = x^4 - 3x^3 - 9x^2 + 27x$
$= x(x-3)^2(x+3)$
The intercepts are $(-3,0)$, $(0,0)$, and $(3,0)$.
The graph also goes through $(2,10)$ and $(-1,-32)$.

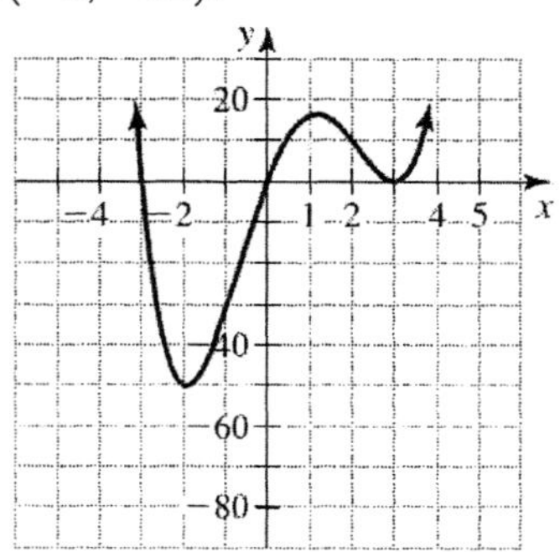

27. $f(x) = (x-2)^2(x-8)$
The x-intercepts are $(8,0)$ and $(2,0)$. The graph
crosses at $(8,0)$ because of its odd exponent; and touches the x-axis but does not cross at $(2,0)$ because of its even exponent.

29. $f(x) = (x-1)^2(x+4)^2$
The x-intercepts are $(1,0)$ and $(-4,0)$.
Does not cross at $(-4,0)$ and $(1,0)$ because both exponents are even.

31. $f(x) = (x-1)(x+4)(x-7)^2$
The x-intercepts are $(1,0)$, $(-4,0)$, and $(7,0)$.
Crosses at $(-4,0)$ and $(1,0)$ because of the odd exponents; does not cross at $(7,0)$ because of its even exponent.

33. $f(x) = x^3 + 6x^2 - x - 6$
$= (x-1)(x+1)(x+6)$
The x-intercepts are $(1,0)$, $(-1,0)$, and $(-6,0)$.
Crosses at $(-6,0)$, $(-1,0)$, and $(1,0)$ because they all have odd exponents.

35. $f(x) = -x^3 + 5x^2$
$= -x^2(x - 5)$
The x-intercepts are $(5, 0)$ and $(0, 0)$.
Crosses at $(5, 0)$; does not cross at $(0, 0)$.
37. $f(x) = x^4 - 5x^3$
$= x^3(x - 5)$
The x-intercepts are $(0, 0)$ and $(5, 0)$.
Crosses at $(0, 0)$ and $(5, 0)$ because of the odd exponent on x^3 and $(x - 5)$.
39. $f(x) = x^4 + 6x^3 + 9x^2$
$= x^2(x + 3)^2$
The x-intercepts are $(0, 0)$ and $(-3, 0)$.
Does not cross at $(-3, 0)$ and $(0, 0)$ because of the even exponents.
41. If the graph of $f(x) = x^3$ is translated 5 units to the right, x is replace by $x - 5$. If it is translated 4 units downward then 4 is subtracted as the last operation. So its equation is $f(x) = (x - 5)^3 - 4$
43. If the graph of $f(x) = x^3$ is translated 6 units to the left, then x is replace by $x + 6$. Moving it 3 units upward adds 3 as the last operation. So its equation is
$f(x) = (x + 6)^3 + 3$.
45. If the graph of $f(x) = x^3$ is reflected in the x-axis a negative sign is place so the it is the last operation performed. So the equation is $f(x) = -(x^3)$ or $f(x) = -x^3$.
47. If the graph of $f(x) = x^4$ is translated 3 units to the right, then x is replace by $x - 3$. If it is then reflected in the x-axis the negative sign must be the last operation. So the equation is $f(x) = -((x - 3)^4)$
or $f(x) = -(x - 3)^4$.
49. The graph of $f(x) = -2x + 3$ is a straight line through (0, 3) and (1, 1). The graph is (d).
51. The graph of $f(x) = -2x^3 + 3$ crosses the y-axis at (0, 3) and contains the points (1, 1) and (−1, 5). The graph is (a).
53. The graph of $f(x) = -x^4 + 3$ contains the points (0, 3), (1, 2), and (−1, 2). The graph is (c).
55. The function $f(x) = x^3 + 3x^2 - x - 3$ can be written as $f(x) = (x - 1)(x + 1)(x + 3)$. The graph crosses the x-axis at (1, 0), (−1, 0), and $(-3, 0)$. The graph is (e).

57. The graph of $f(x) = 2x - 6$ is a straight line with a y-intercept $(0, -6)$ and a slope of 2.

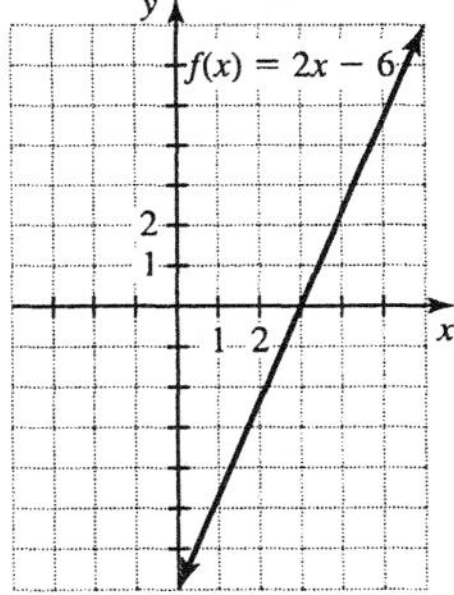

59. The graph of $f(x) = -x^2$ is a parabola opening downward. The graph includes the points (0, 0), (1, −1), (−1, −1), (2, −4), and (−2, −4).

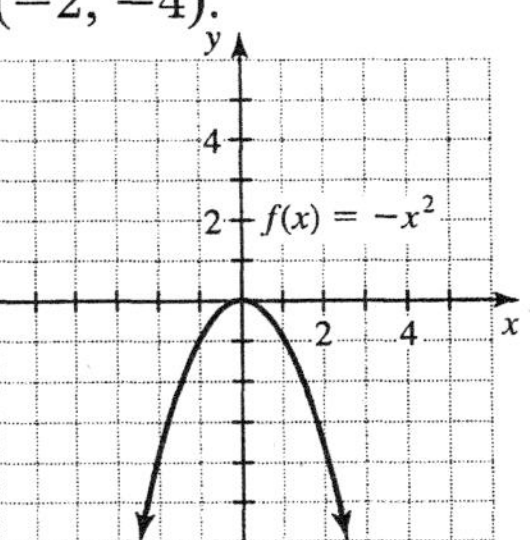

61. The function $f(x) = x^3 - 2x^2$ can be written as $f(x) = x^2(x - 2)$. The x-intercepts are (0, 0) and (2, 0). The graph crosses the x-axis at (2, 0) but does not cross at (0, 0). The graph also goes through the points (1, −1), (3, 9), and (−1, −3).

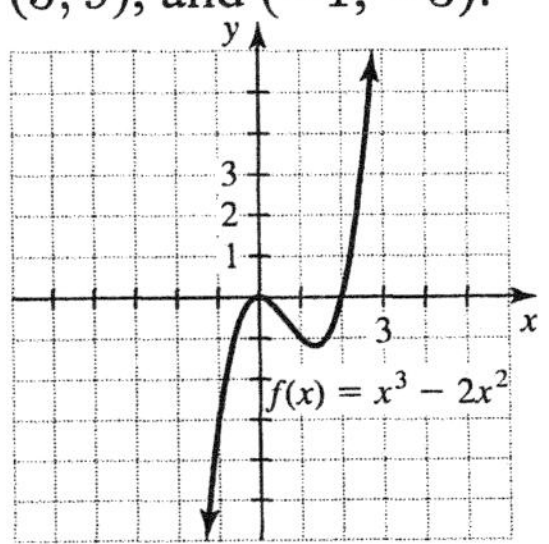

63. The graph of $f(x) = (x - 1)^2(x + 1)^2$ does not cross the x-axis at its intercepts (1, 0) and (−1, 0). The graph also includes the points (0, 1), (2, 9), (−2, 9).

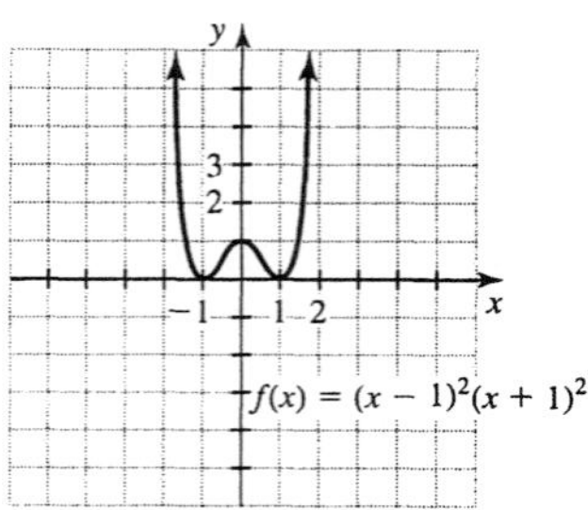

65. $f(x) = (x-1)^2(x-3)$.
The graph crosses the x-axis at the x-intercept (3, 0), but does not cross at the intercept (1, 0). The graph also includes the points (0, −3), (2, −1), and (4, 9).

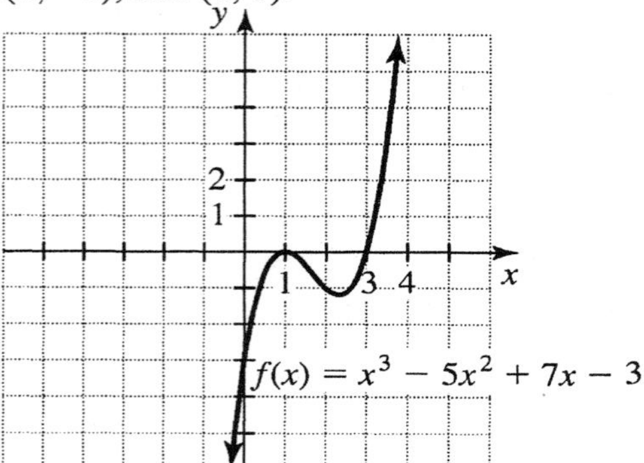

67. Write $f(x) = x^4 - 4x^3 + 4x^2$ as $f(x) = x^2(x^2 - 4x + 4) = x^2(x-2)^2$. The graph does not cross the x-axis at its x-intercepts (0, 0) and (2, 0). The graph also includes the points (1, 1), (3, 9), and (−3, 225).

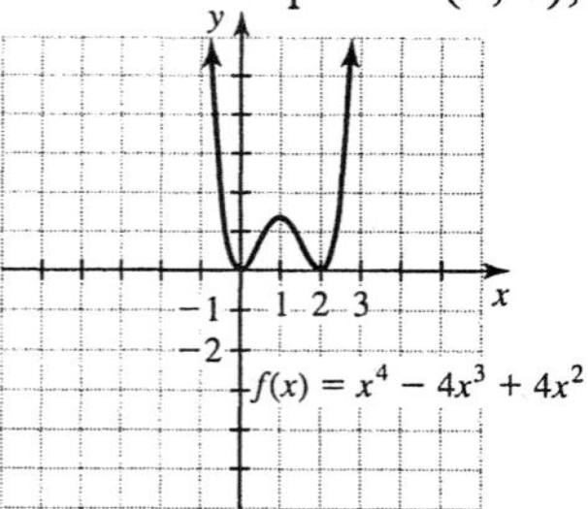

69. Get an image of the function on a graphing calculator, then graph it on paper. Adjust the window to show the x-intercepts.

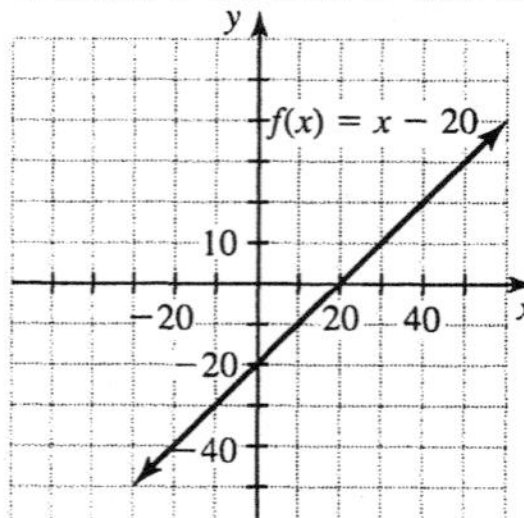

71. Get an image of the function on a graphing calculator, then graph it on paper. Adjust the window to show the x-intercepts.

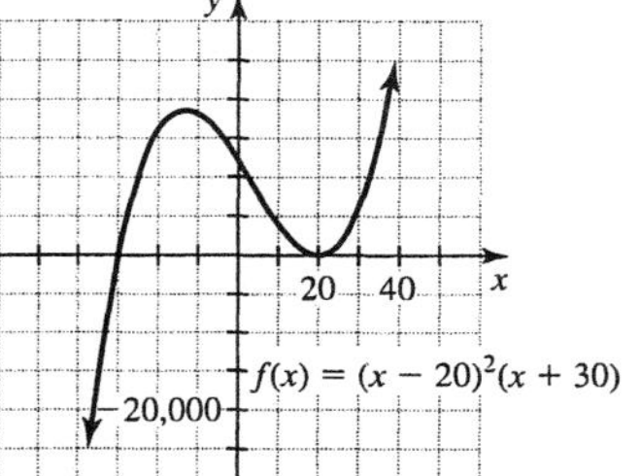

73. Get an image of the function on a graphing calculator, then graph it on paper. Adjust the window to show the x-intercepts.

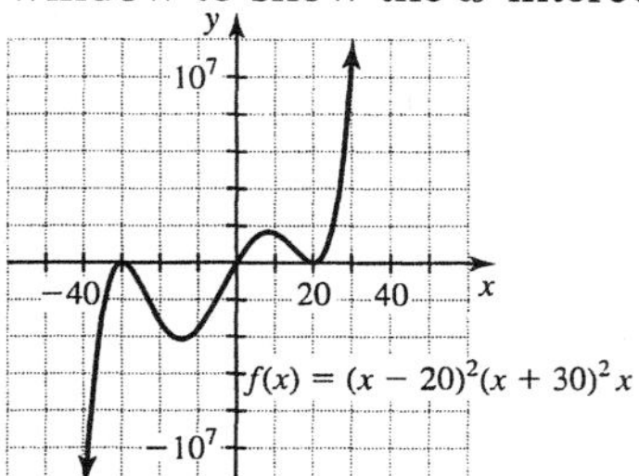

75. If the graph crosses the x-axis at $(3, 0)$ then the factor $x - 3$ should occur with an odd power. So $f(x) = x - 3$. Of course other odd powers would work also. For example, $f(x) = (x-3)^3$.

77. If the graph does not cross the x-axis at $(1, 0)$ then the factor $x - 1$ should occur with an even power. If the graph crosses the x-axis at $(-2, 0)$ then the factor $x + 2$ should occur with an odd power. So

$$f(x) = (x+2)(x-1)^2.$$

Of course other powers would work also. For example, $f(x) = (x+2)^3(x-1)^4$.

11.5 WARM-UPS

1. False, because the domain is $\{x \mid x \neq 9\}$.
2. False, because the domain is $\{x \mid x \neq -2\}$ or $(-\infty, -2) \cup (-2, \infty)$.
3. False, because the equation $x^2 + 1 = 0$ has no real solutions. The domain is the set of all real numbers, $(-\infty, \infty)$.
4. False, because the vertical asymptotes are $x = 2$ and $x = -2$.
5. True, because the degree of the denominator is larger than the degree of the numerator.

6. False, because $\frac{3x-5}{x-2} = 3 + \frac{1}{x-2}$ and $y = 3$ is the horizontal asymptote.
7. True, because the ratio of the leading coefficients is 1.
8. True, because the only solution to the equation $\frac{x-6}{x-2} = 0$ is 6.
9. True, because when x is very large, $1/x$ is approximately zero and the graph gets very close to $y = 2x - 5$.
10. False, because the graph of $f(x) = 2x - 5 + x^2$ is a parabola and it has no asymptotes.

11.5 EXERCISES

1. A rational function is of the form $f(x) = P(x)/Q(x)$ where $P(x)$ and $Q(x)$ are polynomials with no common factor and $Q(x) \neq 0$.
3. A vertical asymptote is a vertical line that is approached by the graph of a rational function.
5. An oblique asymptote is a nonhorizontal nonvertical line that is approached by the graph of a rational function.
7. First find any values of x that would make the denominator have a value of 0.

$$x - 1 = 0$$
$$x = 1$$

The domain is $\{x \mid x \neq 1\}$ or $(-\infty, 1) \cup (1, \infty)$.
9. First find any values of x that would make the denominator have a value of 0.

$$x = 0$$

The domain is $\{x \mid x \neq 0\}$ or $(-\infty, 0) \cup (0, \infty)$.
11. First find any values of x that would make the denominator have a value of 0.

$$x^2 - 16 = 0$$
$$x^2 = 16$$
$$x = \pm 4$$

The domain is $\{x \mid x \neq -4 \text{ and } x \neq 4\}$ or $(-\infty, -4) \cup (-4, 4) \cup (4, \infty)$.
13. If $x + 4 = 0$, then $x = -4$. So the vertical line $x = -4$ is an asymptote for the graph of f. The x-axis is a horizontal asymptote.
15. If $x^2 - 16 = 0$, then $x = \pm 4$. So the vertical lines $x = 4$ and $x = -4$ are asymptotes for the graph of f. The x-axis is a horizontal asymptote.
17. If $x - 7 = 0$, then $x = 7$. So the vertical line $x = 7$ is an asymptote for the graph of f. The ratio of the leading coefficients is 5. So $y = 5$ is the horizontal asymptote
19. The vertical line $x = 3$ is the vertical asymptote. Since $\frac{2x^2}{x-3} = 2x + 6 + \frac{18}{x-3}$, the line $y = 2x + 6$ is an oblique asymptote for the graph.
21. The graph of $f(x) = -\frac{2}{x}$ has the x-axis and the y-axis as asymptotes. The graph is (c).
23. The function $f(x) = \frac{x}{x-2}$ can be written as $f(x) = 1 + \frac{2}{x-2}$. The asymptotes are the lines $y = 1$ and $x = 2$. The graph is (b).
25. Write the function as $f(x) = \frac{1}{x(x-2)}$. The lines $x = 0$ and $x = 2$ are vertical asymptotes. The graph is (g).
27. Write the function as $f(x) = -\frac{1}{2}x - 2$. The graph is a straight line through (0, −2) with slope $-1/2$. The graph is (f).
29. First find any values of x that would make the denominator have a value of 0.

$$x + 4 = 0$$
$$x = -4$$

The asymptotes are $x = -4$ and the x-axis. Draw the asymptotes that we found as dashed lines. The graph goes through the points (−3, 2), (−2, 1), (−5, −2), and (−6, −1).

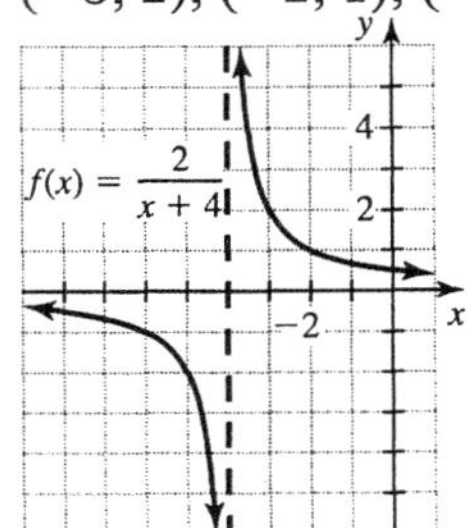

31. First find any values of x that would make the denominator have a value of 0.

$$x^2 - 9 = 0$$
$$x^2 = 9$$
$$x = \pm 3$$

The asymptotes are $x = \pm 3$ and the x-axis. Draw the asymptotes that we found as dashed lines. The graph goes through the points $(0, 0)$, $(-2, \frac{2}{5})$, $(2, -\frac{2}{5})$, $(4, \frac{4}{7})$, and $(-4, -\frac{4}{7})$.

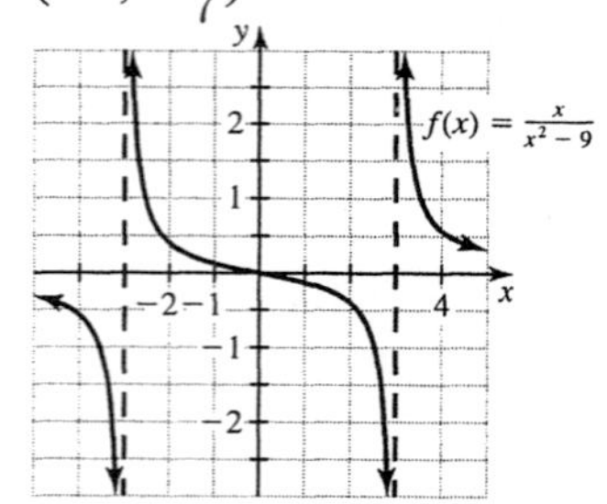

33. First find any values of x that would make the denominator have a value of 0.

$$x + 3 = 0$$
$$x = -3$$

The asymptotes are $x = -3$ and $y = 2$, from the ratio of the leading coefficients. Draw the asymptotes that we found as dashed lines. The graph goes through the points $(0, -\frac{1}{3})$, $(\frac{1}{2}, 0)$, $(-2, -5)$, and $(-4, 9)$.

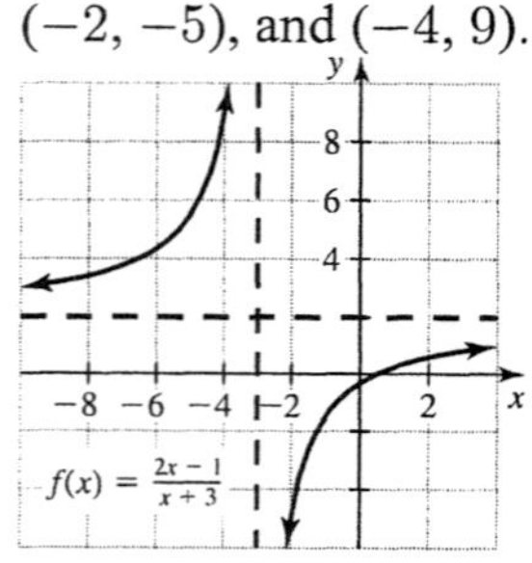

35. First find any values of x that would make the denominator have a value of 0.

$$x = 0$$

The asymptotes are $x = 0$, and $y = x - 3$, obtained from dividing the denominator into the numerator. Draw the asymptotes that we found as dashed lines. The graph goes through the points $(1, -1)$, $(-2, -\frac{11}{2})$, $(-1, -5)$, and $(3, \frac{1}{3})$.

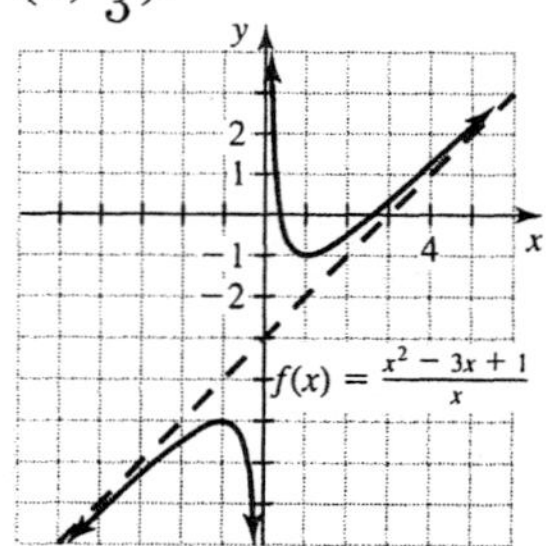

37. First find any values of x that would make the denominator have a value of 0.

$$x - 1 = 0$$
$$x = 1$$

The asymptotes are $x = 1$ and $y = 3x + 1$, obtained from dividing the denominator into the numerator. Draw the asymptotes that we found as dashed lines. The graph goes through the points $(0, 0)$, $(2, 8)$, $(\frac{2}{3}, 0)$, and $(-1, -2.5)$.

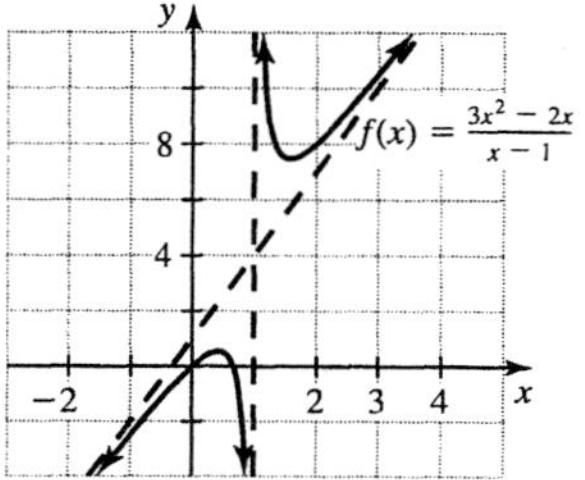

39. Since $x^2 = 0$ is equivalent to $x = 0$, the y-axis is a vertical asymptote. The x-axis is a horizontal asymptote. There are no x-intercepts or y-intercepts. The graph goes through $(1, 1)$, and $(-1, 1)$.

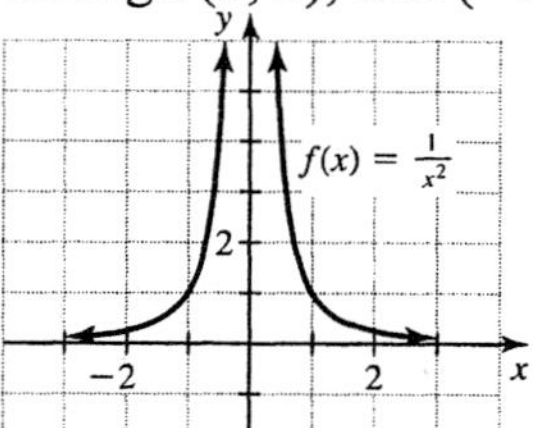

41. Write the function as $f(x) = \frac{2x - 3}{(x + 3)(x - 2)}$. The vertical asymptotes are $x = -3$ and $x = 2$. The x-axis is a horizontal asymptote. The graph goes through $(0, 0.5)$, $(1.5, 0)$, $(3, 0.5)$, and $(-4, -\frac{11}{6})$.

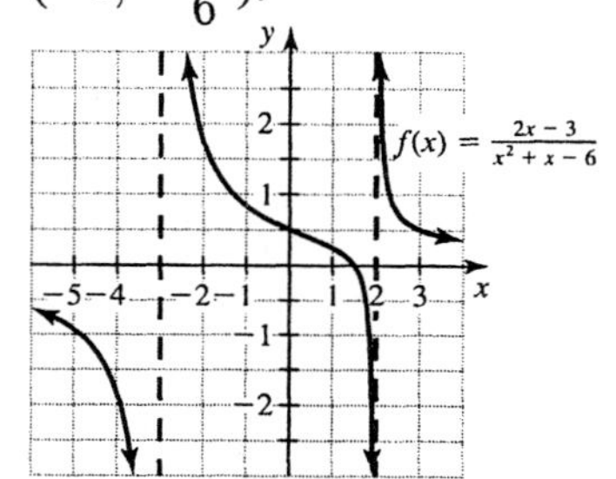

43. The graph of $f(x) = \frac{x + 1}{x^2}$ has a vertical asymptote at $x = 0$ and the x-axis is a horizontal asymptote. The graph goes through $(1, 2)$, $(2, \frac{3}{4})$, $(-1, 0)$, and $(-2, -\frac{1}{4})$.

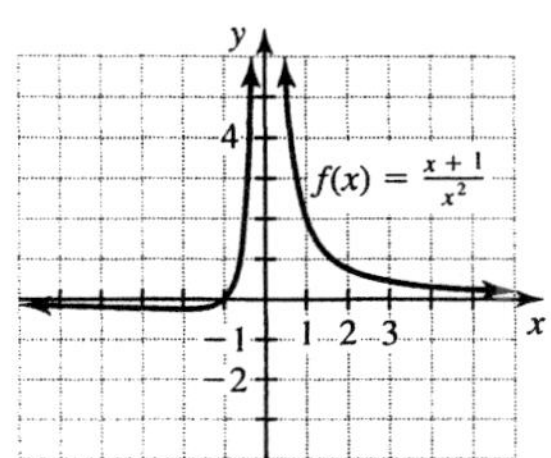

45. $f(x) = \dfrac{2x-1}{x(x-3)(x+3)}$.
The vertical asymptotes are $x = 0$, $x = 3$, and $x = -3$. The x-axis is a horizontal asymptote. The graph goes through $(\frac{1}{2}, 0)$, $(-1, -\frac{3}{8})$, $(-4, \frac{9}{28})$, and $(4, \frac{7}{28})$.

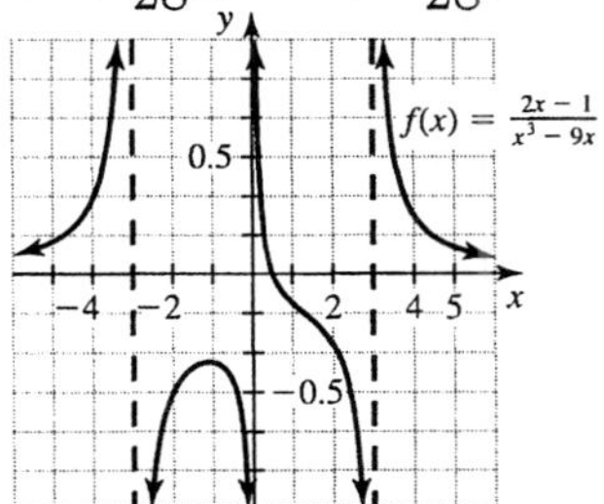

47. Write the function as $f(x) = \dfrac{x}{(x-1)(x+1)}$. The vertical asymptotes are $x = 1$ and $x = -1$. The x-axis is a horizontal asymptote. The graph goes through $(0, 0)$, $(2, \frac{2}{3})$, and $(-2, -\frac{2}{3})$.

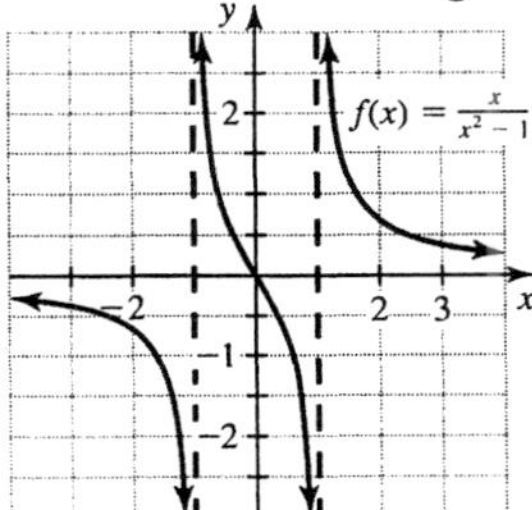

49. The function $f(x) = \dfrac{2}{x^2+1}$ has no vertical asymptotes, because the equation $x^2 + 1 = 0$ has no real solutions. The x-axis is a horizontal asymptote. The graph goes through the points $(0, 2)$, $(1, 1)$, and $(-1, 1)$.

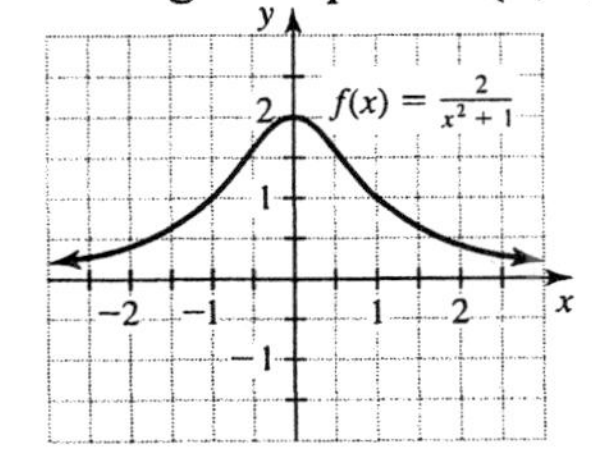

51. Use long division to write the function as $f(x) = x - 1 + \dfrac{1}{x+1}$. The line $y = x - 1$ is an oblique asymptote and the line $x = -1$ is a vertical asymptote. The graph goes through the points $(0, 0)$, $(1, 0.5)$, $(-2, -4)$, and $(-3, -4.5)$.

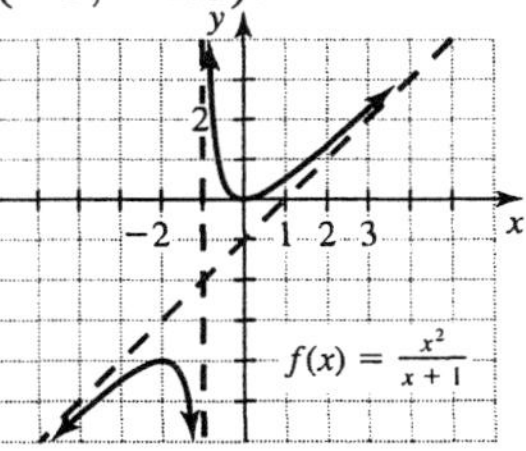

53. Write the function as $f(t) = \dfrac{36t + 500}{t(t+6)}$.
The vertical asymptotes are at $t = 0$ and $t = -6$, but neither 0 nor -6 is in the domain of the function. The t-axis is a horizontal asymptote. The average cost at time $t = 20$ is
$f(20) = \dfrac{36(20) + 500}{20(20+6)} \approx \$2.35$.
The average cost at time $t = 30$ is
$f(30) = \dfrac{36(30) + 500}{30(30+6)} \approx \$1.46$.
As t gets very large the average cost per modulator approaches 0, because the horizontal asymptote is the t-axis.

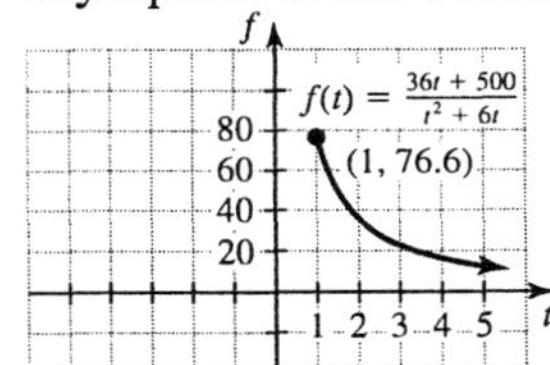

55. a) The horizontal asymptote is $y = 25{,}000$ because the ratio of the leading coefficients is 25,000.

b) $A(50{,}000)$
$$= \frac{25{,}000(50{,}000) + 700{,}000{,}000}{50{,}000}$$
$$= \$39{,}000$$

c)
$$\frac{25{,}000x + 700{,}000{,}000}{x} = 30{,}000$$
$$25{,}000x + 700{,}000{,}000 = 30{,}000x$$
$$700{,}000{,}000 = 5{,}000x$$
$$x = 140{,}000$$

d)

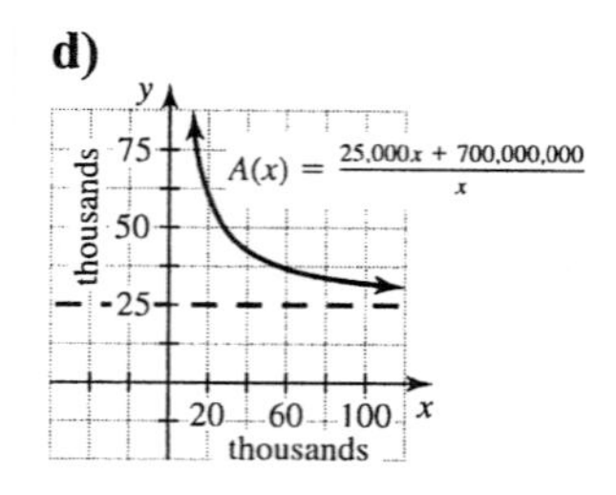

57. The graph of $f(x)$ is an asymptote for the graph of $g(x)$.
59. The graph of $f(x)$ is an asymptote for the graph of $g(x)$.
61. The graph of $f(x)$ is an asymptote for the graph of $g(x)$.
63. The graph of $f(x) = 1/x$ has the x-axis as a horizontal asymptote and the y-axis as a vertical asymptote.
65. The graph of $f(x) = \dfrac{1}{(x-3)(x+1)}$ has the x-axis as a horizontal asymptote and the vertical lines $x = 3$ and $x = -1$ as a vertical asymptotes.

11.6 WARM-UPS

1. True, because $(f - g)(x) = f(x) - g(x)$ $= x - 2 - (x + 3) = -5$.
2. True, because $(f/g)(2) = f(2)/g(2)$ $= (2 + 4)/(3 \cdot 2) = 6/6 = 1$
3. False, because if $f(x) = 3x$ and $g(x) = x + 1$, then $f(g(x)) = 3(x + 1)$ $= 3x + 3$ and $g(f(x)) = 3x + 1$.
4. False, because $(f \circ g)(x) = f(x + 2)$ $= (x + 2)^2 = x^2 + 4x + 4$.
5. False, because if $f(x) = 3x$ and $g(x) = x + 1$, then $(f \circ g)(x) = f(x + 1) = 3x + 3$ and $(f \cdot g)(x) = 3x(x + 1) = 3x^2 + 3x$.
6. False, because $g[f(x)] = \sqrt{x} - 9$ and $f[g(x)] = \sqrt{x - 9}$ which are not equal for all values of x.
7. True, because $(f \circ g)(x) = f(\frac{x}{3}) = 3 \cdot \frac{x}{3} = x$.
8. True, because b determines a and a determines c so b determines the value of c.
9. True, $F(x) = \sqrt{x - 5}$ is a composition of $g(x) = x - 5$ and $h(x) = \sqrt{x}$.
10. True, because $(g \circ h)(x) = g[h(x)]$ $= g[x - 1] = (x - 1)^2 = F(x)$.

11.6 EXERCISES

1. The basic operations of functions are addition, subtraction, multiplication, and division.
3. In the composition of functions the second function is evaluated on the result of the first function.
5.
$$\begin{aligned}(f + g)(x) &= f(x) + g(x)\\ &= 4x - 3 + x^2 - 2x\\ &= x^2 + 2x - 3\end{aligned}$$
7.
$$\begin{aligned}(f \cdot g)(x) &= f(x) \cdot g(x)\\ &= (4x - 3)(x^2 - 2x)\\ &= 4x^3 - 11x^2 + 6x\end{aligned}$$
9. Use $x = 3$ in the formula of Exercise 5:
$$(f + g)(3) = 3^2 + 2(3) - 3 = 12$$
11.
$$\begin{aligned}(f - g)(-3) &= f(-3) - g(-3)\\ &= 4(-3) - 3 - [(-3)^2 - 2(-3)]\\ &= -12 - 3 - [9 + 6]\\ &= -30\end{aligned}$$
13. Use $x = -1$ in the formula of Exercise 7:
$$\begin{aligned}(f \cdot g)(-1) &= 4(-1)^3 - 11(-1)^2 + 6(-1)\\ &= -4 - 11 - 6\\ &= -21\end{aligned}$$
15. $(f/g)(4) = \dfrac{f(4)}{g(4)} = \dfrac{4(4) - 3}{4^2 - 2(4)} = \dfrac{13}{8}$
17. Since $a = 2x - 6$, and $y = 3a - 2$ we have
$$\begin{aligned}y &= 3(2x - 6) - 2\\ y &= 6x - 18 - 2\\ y &= 6x - 20\end{aligned}$$
19. Let $d = (x + 1)/2$ in $y = 2d + 1$.
$$\begin{aligned}y &= 2\left(\frac{x + 1}{2}\right) + 1\\ y &= x + 1 + 1\\ y &= x + 2\end{aligned}$$
21. Let $m = x + 1$ in $y = m^2 - 1$.
$$\begin{aligned}y &= (x + 1)^2 - 1\\ y &= x^2 + 2x + 1 - 1\\ y &= x^2 + 2x\end{aligned}$$
23.
$$y = \frac{a - 3}{a + 2}$$
$$y = \frac{\frac{2x + 3}{1 - x} - 3}{\frac{2x + 3}{1 - x} + 2}$$

$$y=\frac{\left(\frac{2x+3}{1-x}-3\right)(1-x)}{\left(\frac{2x+3}{1-x}+2\right)(1-x)}$$
$$y=\frac{2x+3-3(1-x)}{2x+3+2(1-x)}$$
$$y=\frac{2x+3-3+3x}{2x+3+2-2x}=\frac{5x}{5}$$
$$y=x$$

25. $(g\circ f)(1)=g[f(1)]=g(-1)$
$=(-1)^2+3(-1)=-2$

27. $(f\circ g)(1)=f[g(1)]=f[4]$
$=2(4)-3=5$

29. $(f\circ f)(4)=f[f(4)]=f[5]$
$=2(5)-3=7$

31. $(h\circ f)(5)=h[f(5)]=h[7]=\frac{7+3}{2}=5$

33. $(f\circ h)(5)=f[h(5)]=f[4]=2(4)-3$
$=5$

35. $(g\circ h)(-1)=g[h(-1)]=g[1]=4$

37. $(f\circ g)(2.36)=f[g(2.36)]=f[12.6496]$
$=22.2992$

39. $(g\circ f)(x)=g[f(x)]=g[2x-3]$
$=(2x-3)^2+3(2x-3)$
$=4x^2-12x+9+6x-9$
$=4x^2-6x$

41. $(f\circ g)(x)=f[g(x)]=f[x^2+3x]$
$=2(x^2+3x)-3=2x^2+6x-3$

43. $(h\circ f)(x)=h[f(x)]=h[2x-3]$
$=\frac{2x-3+3}{2}=x$

45. $(f\circ f)(x)=f[f(x)]=f[2x-3]$
$=2(2x-3)-3=4x-9$

47. $(h\circ h)(x)=h[h(x)]=h[\frac{x+3}{2}]$
$$=\frac{\frac{x+3}{2}+3}{2}=\frac{\left(\frac{x+3}{2}+3\right)2}{2\cdot 2}$$
$$=\frac{x+3+6}{4}=\frac{x+9}{4}$$

49. $F(x)=\sqrt{x-3}=\sqrt{h(x)}=f[h(x)]$
Therefore, $F=f\circ h$.

51. $G(x)=x^2-6x+9=(x-3)^2$
$=[h(x)]^2=g[h(x)]$
Therefore $G=g\circ h$.

53. $H(x)=x^2-3=g(x)-3=h[g(x)]$
Therefore $H=h\circ g$.

55. $J(x)=x-6=x-3-3$
$=h(x)-3=h[h(x)]$
Therefore, $J=h\circ h$.

57. $K(x)=x^4=(x^2)^2=[g(x)]^2=g[g(x)]$
Therefore, $K=g\circ g$.

59. $f[g(x)]=f\left(\frac{x-5}{3}\right)=3\left(\frac{x-5}{3}\right)+5$
$=x$
$g[f(x)]=g(3x+5)=\frac{3x+5-5}{3}=x$

61. $f[g(x)]=f\left(\sqrt[3]{x+9}\right)$
$=\left(\sqrt[3]{x+9}\right)^3-9=x+9-9=x$
$g[f(x)]=g[x^3-9]=\sqrt[3]{x^3-9+9}$
$=\sqrt[3]{x^3}=x$

63. $f[g(x)]=f\left(\frac{x+1}{1-x}\right)=\frac{\frac{x+1}{1-x}-1}{\frac{x+1}{1-x}+1}$
$$=\frac{\left(\frac{x+1}{1-x}-1\right)(1-x)}{\left(\frac{x+1}{1-x}+1\right)(1-x)}=\frac{x+1-(1-x)}{x+1+1-x}$$
$$=\frac{2x}{2}=x$$
$$g[f(x)]=g\left(\frac{x-1}{x+1}\right)=\frac{\frac{x-1}{x+1}+1}{1-\frac{x-1}{x+1}}$$
$$=\frac{\left(\frac{x-1}{x+1}+1\right)(x+1)}{\left(1-\frac{x-1}{x+1}\right)(x+1)}=\frac{x-1+x+1}{x+1-(x-1)}$$
$$=\frac{2x}{2}=x$$

65. $f[g(x)]=f\left(\frac{1}{x}\right)=\frac{1}{\frac{1}{x}}=x$
$g[f(x)]=g\left(\frac{1}{x}\right)=\frac{1}{\frac{1}{x}}=x$

67. Since $f(x)=x^2$, $f(3)=3^2=9$ and the statement is true.

69. Since $(f+g)(4)=f(4)+g(4)$
$=4^2+4+5$
$=25$
the statement is false.

71. Since $(f\cdot g)(3)=f(3)\cdot g(3)$
$=3^2(3+5)=72$
the statement is true.

73. Since $(f\circ g)(2)=f(g(2))$
$=f(7)=7^2=49$
the statement is false.

75. Since $(f(g(x))=f(g(x))$
$=f(x+5)$
$=(x+5)^2$
$=x^2+10x+25$
the statement is false.

77. Since $(f \circ g)(x) = f(x+5)$
$= (x+5)^2$
$= x^2 + 10x + 25$
the statement is true.

79. a) Let $s =$ the length of a side of the square.
$10^2 = s^2 + s^2$
$2s^2 = 100$
$s^2 = 50$
$s = \sqrt{50}$
Since $A = s^2$, we have $A = 50$ ft^2

b) In general if d is the diagonal of a square and s is the length of a side, we have $d^2 = s^2 + s^2$ or $s^2 = \frac{d^2}{2}$. Since $A = s^2$, we have $A = \frac{d^2}{2}$.

81. $P(x) = R(x) - C(x)$
$= x^2 - 10x + 30 - (2x^2 - 30x + 200)$
$= -x^2 + 20x - 170$

83. Substitute $F = 0.25I$ into $J = 0.10F$.
$J = 0.10(0.25I)$
$J = 0.025I$

85. a) $x = (30.25/100)^3 \approx 0.0277$
$D = (24665/2240)/0.0277 \approx 397.8$
b) $x = (L/100)^3$
$D = (25000/2240)/x$
$D = (25000/2240)/(L/100)^3$
$D = \frac{25000}{2240} \cdot \frac{100^3}{L^3}$
$D = \frac{1.116 \times 10^7}{L^3}$
c) From the graph it appears that the displacement-length ratio decreases as the length increases.

87. The domain of $f(x) = \sqrt{x} - 4$ is $[0, \infty)$. The domain of $g(x) = \sqrt{x}$ is $[0, \infty)$.
$(g \circ f)(x) = g(\sqrt{x} - 4) = \sqrt{\sqrt{x} - 4}$
So that $\sqrt{x} - 4 \geq 0$ we must have $x \geq 16$. So the domain of $g \circ f$ is $[16, \infty)$.

89. The graph of $y = x + \sqrt{x}$ appears only for $x \geq 0$. So the domain is $[0, \infty)$. The graph starts at $(0, 0)$ and goes up from there. So the range is $[0, \infty)$.

11.7 WARM-UPS

1. False, because the inverse function is obtained by interchanging the coordinates in every ordered pair.
2. False, because the points (0, 3) and (1, 3) both satisfy $f(x) = 3$.
3. False, because the inverse of multiplication by 2 is division by 2. If $g(x) = 2x$, then $g^{-1}(x) = x/2$.
4. True, because if we interchange the coordinates in a function that is not one-to-one, we do not obtain a function.
5. True, because in an inverse function the domain and range of the function are reversed.
6. False, because it is not one-to-one. Both $(2, 16)$ and $(-2, 16)$ satisfy $f(x) = x^4$.
7. True, because taking the opposite of a number twice gives back the original number.
8. True, because the inverse function interchanges the coordinates in all of the ordered pairs of the function.
9. True, the inverse of $k(x) = 3x - 6$ is $k^{-1}(x) = (x+6)/3 = (1/3)x + 2$.
10. False, because $f^{-1}(x) = (x+4)/3$.

11.7 EXERCISES

1. The inverse of a function is a function with the same ordered pairs except that the coordinates are reversed.
3. The range of f^{-1} is the domain of f.
5. A function is one-to-one if no two ordered pairs have the same second coordinate with different first coordinates.
7. The switch and solve strategy is used to find a formula for an inverse function.
9. The function is not one-to-one because of the pairs (−2, 2) and (2, 2). Therefore, the function is not invertible.
11. The function is one-to-one. The inverse is {(4, 16), (3, 9), (0, 0)}.
It is obtained by interchanging the coordinates in each ordered pair.
13. The function is not one-to-one because of the pairs (5, 0) and (6, 0). Therefore, it is not invertible.

15. This function is one-to-one. Its inverse is obtained by interchanging the coordinates in each ordered pair: $\{(0, 0), (2, 2), (9, 9)\}$.
17. This function is not one-to-one because we can draw a horizontal line that crosses the graph twice.
19. This function is one-to-one because we cannot draw a horizontal line that crosses the graph more than once.
21. Since $(f \circ g)(x) = f(g(x)) = f(0.5x)$ $= 2(0.5x) = x$ and $(g \circ f)(x) = g(f(x))$ $= g(2x) = 0.5(2x) = x$, the functions are inverses of each other.
23. $(f \circ g)(x) = f(g(x)) = f(\frac{1}{2}x + 5)$
$= 2(\frac{1}{2}x + 5) - 10 = x + 10 - 10 = x$
$(g \circ f)(x) = g(f(x)) = g(2x - 10)$
$= \frac{1}{2}(2x - 10) + 5 = x - 5 + 5 = x$
Therefore, the functions are inverses of each other.
25. $(f \circ g)(x) = f(g(x)) = f(-x) = -(-x)$
$= x$
$(g \circ f)(x) = g(f(x)) = g(-x) = -(-x) = x$
Therefore, the functions are inverses of each other.
27. $(g \circ f)(x) = g(f(x)) = g(x^4) = (x^4)^{1/4}$
$= |\, x \,|$
For example, $g(f(-2)) = g(16) = 2$. So the functions are not inverses of each other.
29. $y = 5x,\, x = 5y,\, y = \frac{1}{5}x$
So $f^{-1}(x) = \frac{x}{5}$.
31. $y = x - 9,\, x = y - 9,\, y = x + 9$
So $g^{-1}(x) = x + 9$.
33. $y = 5x - 9,\, x = 5y - 9,\, 5y = x + 9,$
$y = \frac{x + 9}{5}$
So $k^{-1}(x) = \frac{x + 9}{5}$.
35. $y = \frac{2}{x},\, x = \frac{2}{y},\, xy = 2,\, y = \frac{2}{x}$
So $m^{-1}(x) = \frac{2}{x}$.
37. $y = \sqrt[3]{x - 4},\, x = \sqrt[3]{y - 4},\, x^3 = y - 4,$
$y = x^3 + 4$ So $f^{-1}(x) = x^3 + 4$.
39. $y = \frac{3}{x - 4},\, x = \frac{3}{y - 4},\, x(y - 4) = 3,$
$y - 4 = \frac{3}{x},\, y = \frac{3}{x} + 4$
So $f^{-1}(x) = \frac{3}{x} + 4$.
41.
$$\begin{aligned} f(x) &= \sqrt[3]{3x + 7} \\ y &= \sqrt[3]{3x + 7} \\ x &= \sqrt[3]{3y + 7} \\ x^3 &= 3y + 7 \\ x^3 - 7 &= 3y \\ \frac{x^3 - 7}{3} &= y \\ f^{-1}(x) &= \frac{x^3 - 7}{3} \end{aligned}$$
43.
$$\begin{aligned} f(x) &= \frac{x + 1}{x - 2} \\ y &= \frac{x + 1}{x - 2} \\ x &= \frac{y + 1}{y - 2} \\ x(y - 2) &= y + 1 \\ xy - 2x &= y + 1 \\ xy - y &= 2x + 1 \\ y(x - 1) &= 2x + 1 \\ y &= \frac{2x + 1}{x - 1} \\ f^{-1}(x) &= \frac{2x + 1}{x - 1} \end{aligned}$$
45.
$$\begin{aligned} f(x) &= \frac{x + 1}{3x - 4} \\ y &= \frac{x + 1}{3x - 4} \\ x &= \frac{y + 1}{3y - 4} \\ x(3y - 4) &= y + 1 \\ 3xy - 4x &= y + 1 \\ 3xy - y &= 1 + 4x \\ y(3x - 1) &= 1 + 4x \\ y &= \frac{1 + 4x}{3x - 1} \\ f^{-1}(x) &= \frac{1 + 4x}{3x - 1} \end{aligned}$$
47. The function $p(x) = \sqrt[4]{x}$ finds the fourth root. The inverse must be the fourth power. Since the domain and range of p are both the set of nonnegative real numbers, $p^{-1}(x) = x^4$ for $x \geq 0$.
49. $y = (x - 2)^2,\, x = (y - 2)^2,$
$y - 2 = \pm\sqrt{x},\, y = 2 \pm \sqrt{x}$
Since $x \geq 2$ in the function, we must have $y \geq 2$ in the inverse function. So $y = 2 - \sqrt{x}$ is not the inverse and $f^{-1}(x) = 2 + \sqrt{x}$.
51. $y = x^2 + 3,\, x = y^2 + 3,\, y^2 = x - 3,$
$y = \pm\sqrt{x - 3}$. Since $x \geq 0$ in the function, $y \geq 0$ in the inverse function. So $f^{-1}(x) = \sqrt{x - 3}$.
53.
$$\begin{aligned} f(x) &= \sqrt{x + 2} \\ y &= \sqrt{x + 2} \\ x &= \sqrt{y + 2} \\ x^2 &= y + 2 \\ x^2 - 2 &= y \end{aligned}$$

Since $y \geq 0$ in the function, $x \geq 0$ in the inverse. So $f^{-1}(x) = x^2 - 2$ for $x \geq 0$.

55. The inverse of $f(x) = 2x + 3$ is a function that subtracts 3 and then divides by 2, $f^{-1}(x) = \frac{x-3}{2} = \frac{1}{2}x - \frac{3}{2}$. Use y-intercepts and slopes to graph each straight line.

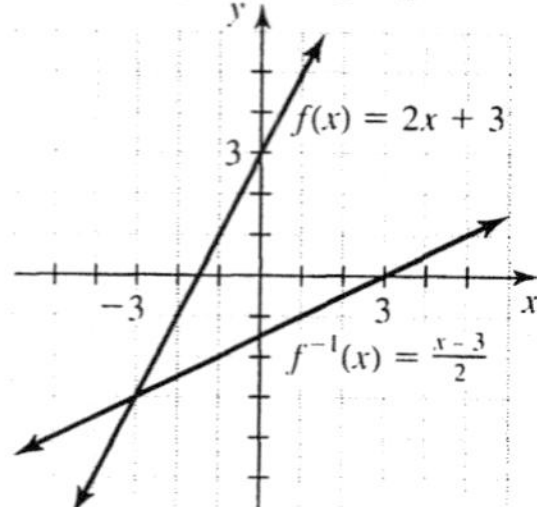

57. The graph of $f(x) = x^2 - 1$ for $x \geq 0$ contains the points $(0, -1)$, $(1, 0)$, and $(2, 3)$. For the inverse function we add 1 and then take the square root, $f^{-1}(x) = \sqrt{x+1}$. The function f^{-1} contains $(-1, 0)$, $(0, 1)$, and $(3, 2)$.

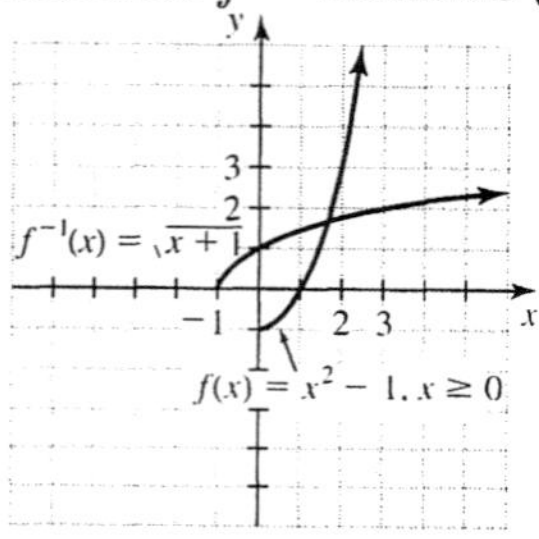

59. The graph of f is a line through $(0, 0)$ with slope 5, and the graph of $f^{-1}(x) = x/5$ is a straight line through $(0, 0)$ with slope $1/5$.

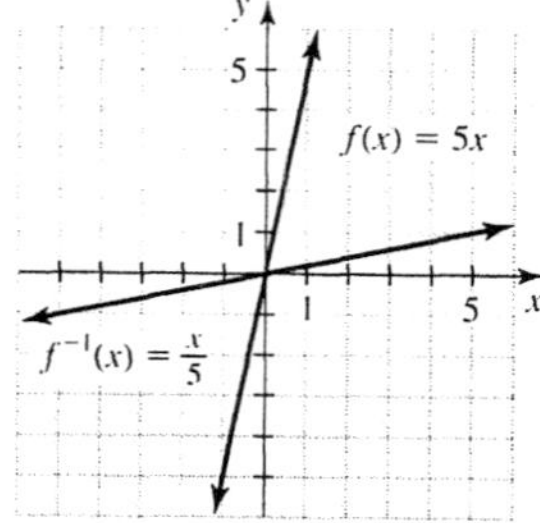

61. The inverse of cubing is the cube root. So $f^{-1}(x) = \sqrt[3]{x}$. The graph of f contains the points $(0, 0)$, $(1, 1)$, $(-1, -1)$, $(2, 8)$, and $(-2, -8)$. The graph of f^{-1} contains the points $(0, 0)$, $(1, 1)$, $(-1, -1)$, $(8, 2)$, and $(-8, -2)$.

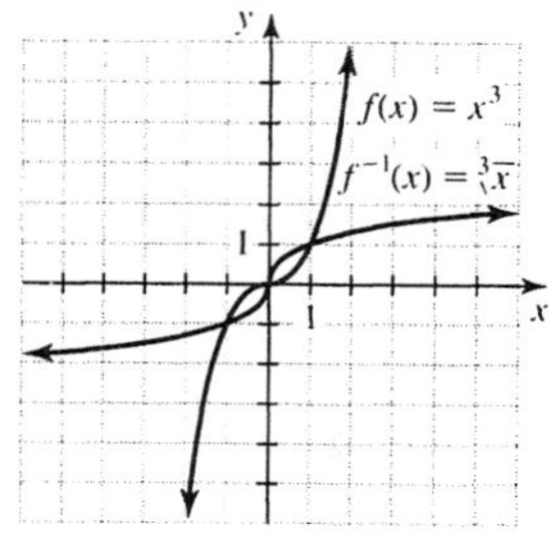

63. The inverse of subtracting 2 and then taking a square root is squaring and then adding 2. So $f^{-1}(x) = x^2 + 2$ for $x \geq 0$. The graph of f contains $(2, 0)$, $(3, 1)$, and $(6, 2)$. The graph of f^{-1} contains $(0, 2)$, $(1, 3)$, and $(2, 6)$.

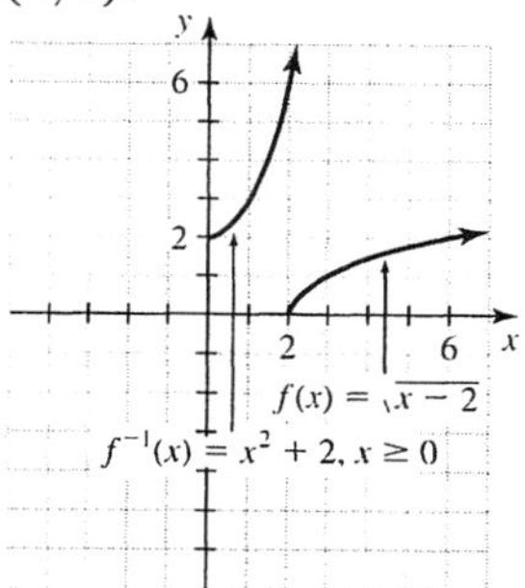

65. The inverse of multiplying by 2 is dividing by 2 which is (d)

67. The inverse of multiplying by 2 and then subtracting 1 is adding 1 and then dividing by 2 which is (i)

69. The inverse of square root is the squaring function, which is (f)

71. The inverse of subtracting 2 and then taking the cube root is cubing and then adding 2, which is (a)

73. The inverse of taking the fourth power is taking the fourth root, which is (g)

75. $(f^{-1} \circ f)(x) = f^{-1}(f(x)) = f^{-1}(x^3 - 1)$
$= \sqrt[3]{x^3 - 1 + 1} = x$

77. $(f^{-1} \circ f)(x) = f^{-1}(f(x)) = f^{-1}(\frac{1}{2}x - 3)$
$= 2(\frac{1}{2}x - 3) + 6 = x - 6 + 6 = x$

79. $(f^{-1} \circ f)(x) = f^{-1}(f(x)) = f^{-1}(\frac{1}{x} + 2)$
$= \dfrac{1}{\frac{1}{x} + 2 - 2} = \dfrac{1}{\frac{1}{x}} = x$

81. $(f^{-1} \circ f)(x) = f^{-1}(f(x)) = f^{-1}(\frac{x+1}{x-2})$

$$= \frac{2(\frac{x+1}{x-2})+1}{\frac{x+1}{x-2}-1} = \frac{\left(\frac{2x+2}{x-2}+1\right)(x-2)}{\left(\frac{x+1}{x-2}-1\right)(x-2)}$$

$$= \frac{2x+2+x-2}{x+1-x+2} = \frac{3x}{3} = x$$

83. a) $S = \sqrt{30 \cdot 50 \cdot 0.75} \approx 33.5$ mph

b) From the graph you can see that the drag factor decreases when a road gets wet because the skid marks for a given speed will be longer.

c)
$$S = \sqrt{30L \cdot 1}$$
$$S^2 = 30L$$
$$L = \frac{S^2}{30}$$

85. If x is the selling price, then the total cost is given by $T(x) = x + 0.09x + 125$ or $T(x) = 1.09x + 125$. If x is the total cost, then the function $T^{-1}(x) = \frac{x - 125}{1.09}$ gives the selling price as a function of the total cost.

87. Only the odd powers of x are one-to-one functions. So $f(x) = x^n$ is invertible if n is an odd positive integer. If n is an even positive integer, the function $f(x) = x^n$ is not one-to-one and not invertible.

89. The functions f and g are not inverses of each other because the function $y_2 = \sqrt{y_1} = \sqrt{x^2}$ is the absolute value function and not the identity function.

Enriching Your Mathematical Word Power

1. b **2.** a **3.** d **4.** b **5.** c
6. a **7.** b **8.** a **9.** a **10.** b
11. d **12.** d **13.** a **14.** b **15.** b
16. a **17.** d

CHAPTER 11 REVIEW

1. The relation is not a function because two ordered pairs $(5, 7)$ and $(5, 10)$ have the same first coordinate and different second coordinates.

3. The relation is a function because no two ordered pairs have the same x-coordinate and different y-coordinates.

5. The relation is a function because the condition that $y = x^2$ guarantees that no two ordered pairs will have the same x-coordinate and different y-coordinates.

7. This relation is not a function because the ordered pairs (16, 2) and (16, −2) are both in the relation.

9. The domain is the set of x-coordinates $\{3, 4, 5\}$, and the range is the set of y-coordinates $\{1, 5, 9\}$.

11. Since we can use any number for x in the equation $y = x + 1$, the domain is $(-\infty, \infty)$. Since y can be any real number in $y = x + 1$, the range is $(-\infty, \infty)$.

13. In the equation $y = \sqrt{x+5}$, the value of $x + 5$ must be nonnegative. So $x + 5 \geq 0$ or $x \geq -5$. The domain is $[-5, \infty)$. Since y is equal to a square root, y must be nonnegative. The range is $[0, \infty)$.

15. $f(0) = 2(0) - 5 = -5$

17. $g(0) = 0^2 + 0 - 6 = -6$

19. $g\left(\frac{1}{2}\right) = \left(\frac{1}{2}\right)^2 + \frac{1}{2} - 6$
$$= \frac{1}{4} + \frac{2}{4} - \frac{24}{4} = -\frac{21}{4}$$

21. The graph of $f(x) = 3x - 4$ is a straight line with y-intercept $(0, -4)$ and slope 3. Since any number can be used for x, the domain is $(-\infty, \infty)$. From the graph we see that any real number can occur as a y-coordinate, so the range is $(-\infty, \infty)$.

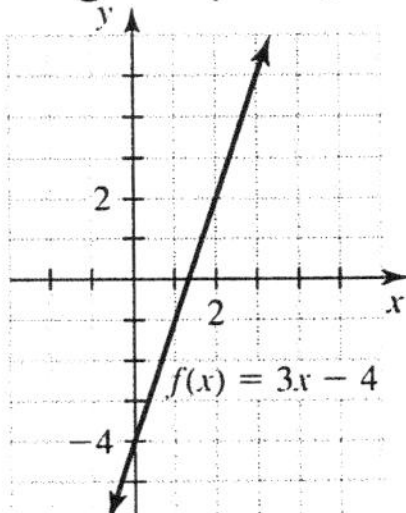

23. The graph of $h(x) = |x| - 2$ includes the points (0, −2), (1, −1), (2, 0), (−1, −1), and (−2, 0). Draw a v-shaped graph through these points.

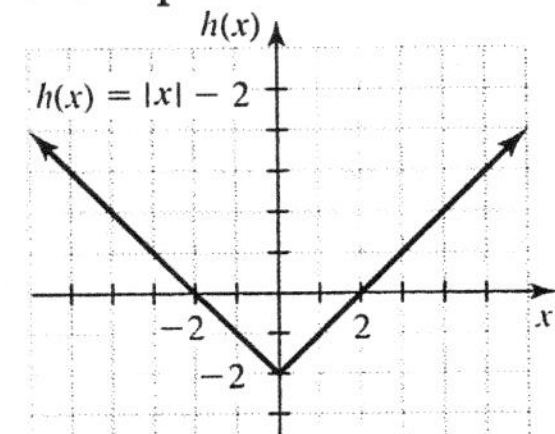

Since any real number can be used for x, the domain is $(-\infty, \infty)$. From the graph we see that all y-coordinates are greater than or equal to -2, and so the range is $[-2, \infty)$.

25. The graph of $y = x^2 - 2x + 1$ includes the points (0, 1), (1, 0), (2, 1), (3, 4), and (−1, 4). Draw a parabola through these points.

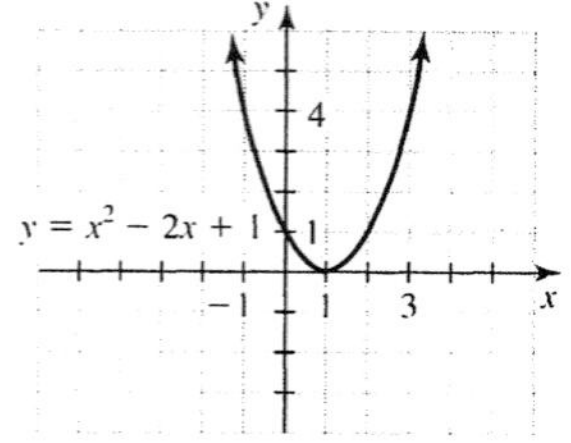

Since any number can be used for x, the domain is $(-\infty, \infty)$. From the graph we see that the y-coordinates are nonnegative, and so the range is $[0, \infty)$.

27. The graph of $k(x) = \sqrt{x} + 2$ includes the points (0, 2), (1, 3), and (4, 4).

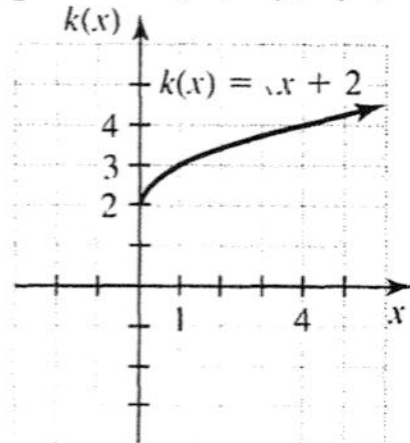

Since we can use only nonnegative numbers for x, the domain is $[0, \infty)$. From the graph we see that no y-coordinate is less than 2, and so the range is $[2, \infty)$.

29. The graph of $y = 30 - x^2$ includes the points $(-5, 5)$, $(0, 30)$, and $(5, 5)$.

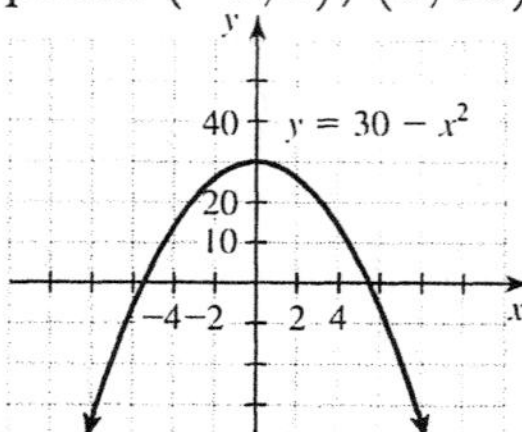

Since x can be any real number, the domain is $(-\infty, \infty)$. From the graph we see that y is at most 30. So the range is $(-\infty, 30]$.

31. The graph of $x = 2$ is the vertical line with x-intercept (2, 0).

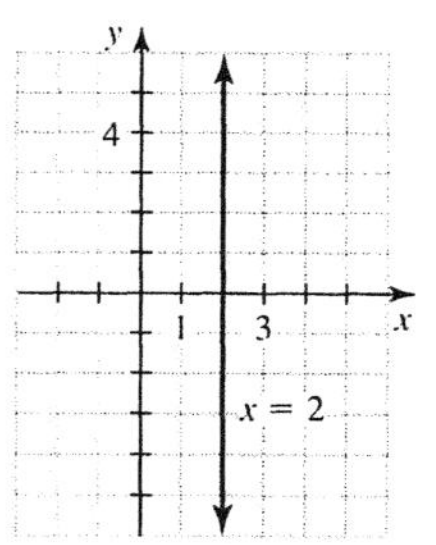

Since the only x-coordinate used on this graph is 2, the domain is $\{2\}$. Since all real numbers occur as y-coordinates, the range is $(-\infty, \infty)$.

33. The graph of $x = |y| + 1$ includes the points (1, 0), (2, 1), (2, −1), (3, 2), and (3, −2). Draw a v-shaped graph through these points.

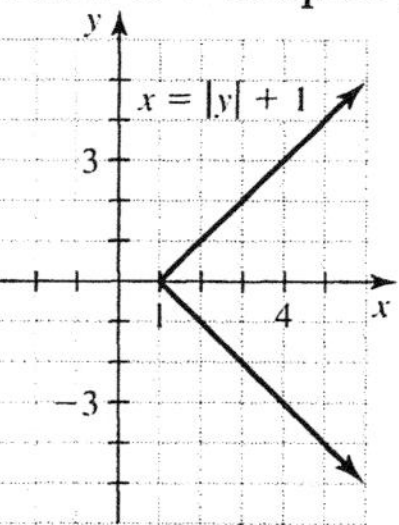

From the graph we see that the x-coordinates are not less than 1, and so the domain is $[1, \infty)$. Any number can be used for y, and so the range is $(-\infty, \infty)$.

35. The graph of $y = \sqrt{x}$ includes the points (0, 0), (1, 1), and (4, 2).

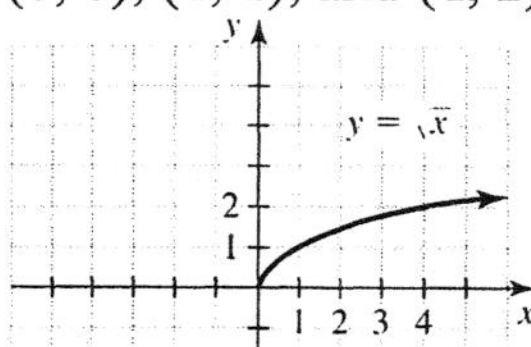

The domain is $[0, \infty)$ and the range is $[0, \infty)$.

37. The graph of $y = -2\sqrt{x}$ includes the points (0, 0), (1, −2), and (4, −4). It can be obtained from the graph of $y = \sqrt{x}$ by stretching and reflecting in the x-axis.

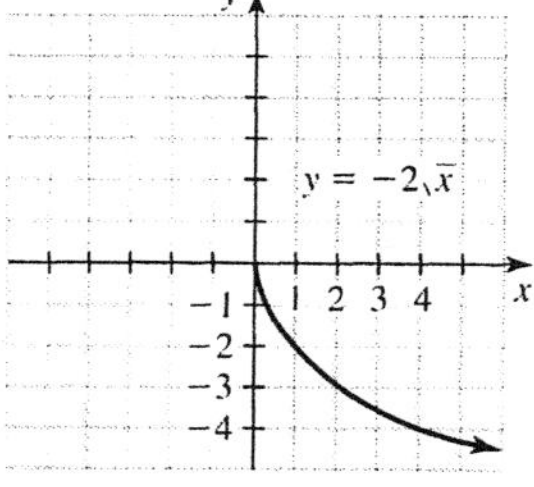

The domain is $[0, \infty)$ and the range is $(-\infty, 0]$.

39. The graph of $y = \sqrt{x-2}$ can be obtained by shifting the graph of $y = \sqrt{x}$ two spaces to the right. The graph of $y = \sqrt{x-2}$ includes the points (2, 0), (3, 1), and (6, 2).

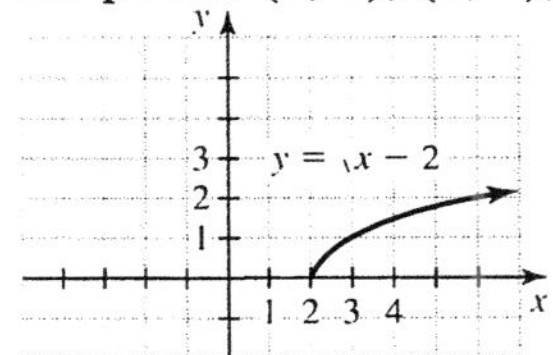

The domain is $[2, \infty)$ and the range is $[0, \infty)$.

41. The graph of $y = \frac{1}{2}\sqrt{x}$ is obtained from the graph of $y = \sqrt{x}$ by shrinking. The graph of $y = \frac{1}{2}\sqrt{x}$ includes the points (0, 0), (1, 0.5), and (4, 1).

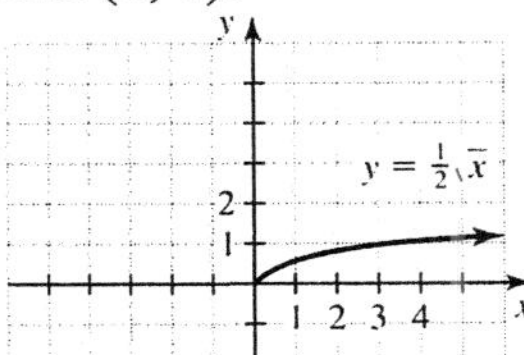

The domain is $[0, \infty)$ and the range is $[0, \infty)$.

43. The graph of $y = -\sqrt{x+1} + 3$ is obtained from the graph of $y = \sqrt{x}$ by shifting 1 unit to the left, reflecting in the x-axis, and then shifting 3 units upward. The graph of $y = -\sqrt{x+1} + 3$ includes the points $(-1, 3)$, (0, 2), and (3, 1).

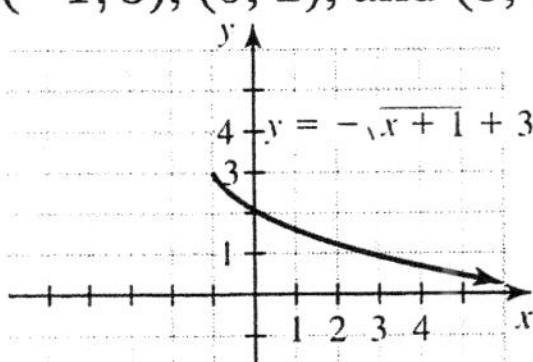

The domain is $[-1, \infty)$ and the range is $(-\infty, 3]$.

45. $f(x) = x^3 - 25x$
$= x(x-5)(x+5)$

The intercepts are $(-5, 0), (5, 0),$ and $(0, 0)$. The graph also goes through $(1, -24)$ and $(-2, 42)$.

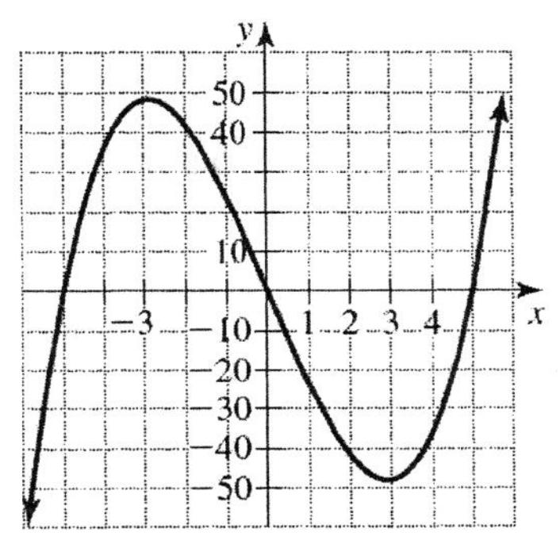

$f(x) = x^3 - 25x$

47. $f(x) = (x^2 - 4)(x - 1)$
$= (x-2)(x+2)(x-1)$

The intercepts are $(-2, 0), (1, 0), (2, 0),$ and $(0, 4)$. The graph also goes through $(3, 10)$ and $(-1, 6)$.

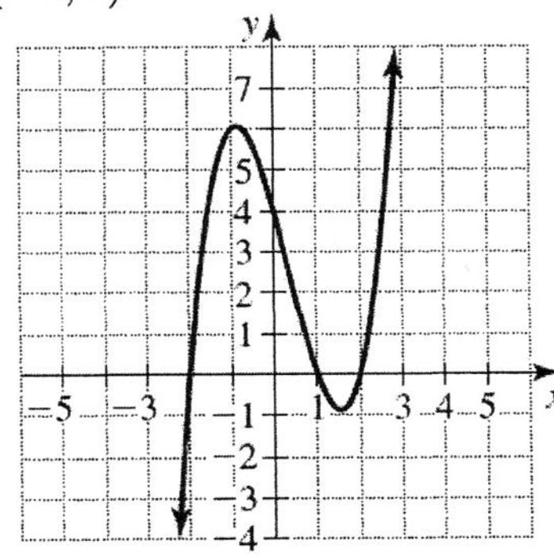

$f(x) = (x^2 - 4)(x - 1)$

49. $f(x) = x^4 - 10x^2 + 9$
$= (x-3)(x+3)(x-1)(x+1)$

The intercepts are $(-3, 0)$, $(-1, 0)$, $(1, 0)$, $(3, 0)$, and $(0, 9)$. The graph also goes through $(\pm 2, -15)$.

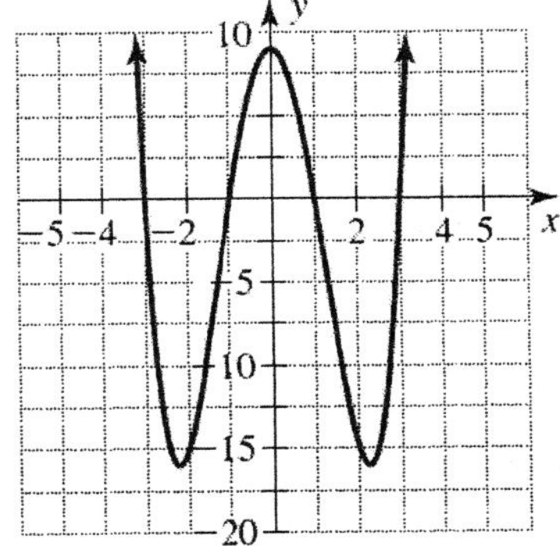

$f(x) = x^4 - 10x^2 + 9$

51. $f(x) = x^2 - 6x + 9$
$= (x-3)^2$

The graph touches but does not cross the x-axis at $(3, 0)$ because of the even exponent.

53. $f(x) = (x-3)(x+5)(x-4)^2$
The graph crosses the x-axis at $(3, 0)$ and $(-5, 0)$ because of the odd exponents, and touches but does not cross at $(4, 0)$ because of the even exponent.

55. $f(x) = x^3 - 8x^2 - 9x + 72$
$= (x-3)(x+3)(x-8)$
The graph crosses the x-axis at $(-3, 0)$, $(3, 0)$, and $(8, 0)$ because all of the exponents on the factors are 1 and 1 is odd.

57. If $2x + 3 = 0$, the $x = -\frac{3}{2}$. So the domain of the rational function is $\left\{x \mid x \neq -\frac{3}{2}\right\}$ or $(-\infty, -3/2) \cup (-3/2, \infty)$.

59. Since there is no real solution to $x^2 + 9 = 0$, the domain is the set of all real numbers or $(-\infty, \infty)$.

61. If $x - 3 = 0$, then $x = 3$. So there is a vertical asymptote at $x = 3$. If x is very large, the value of $f(x)$ is approximately 0. So the x-axis is a horizontal asymptote. The graph includes the points (0, −2/3) and (4, 2).

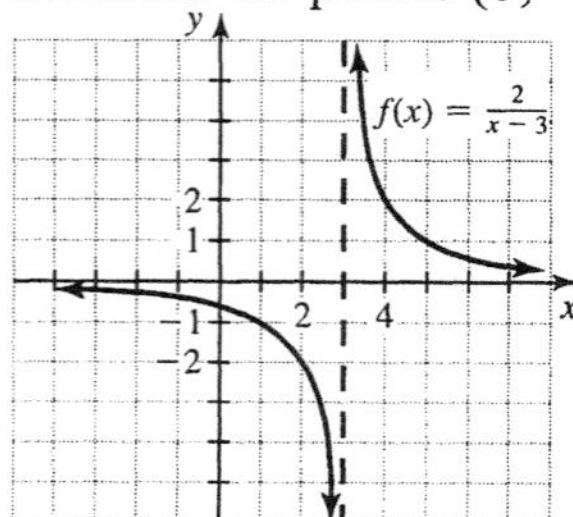

63. If $x^2 - 4 = 0$, then $x = \pm 2$. So the vertical asymptotes are $x = 2$ and $x = -2$. If x is very large, then $f(x)$ is approximately 0. So the x-axis is a horizontal asymptote. The graph goes through (0, 0), (3, $\frac{3}{5}$), and (−3, $-\frac{3}{5}$).

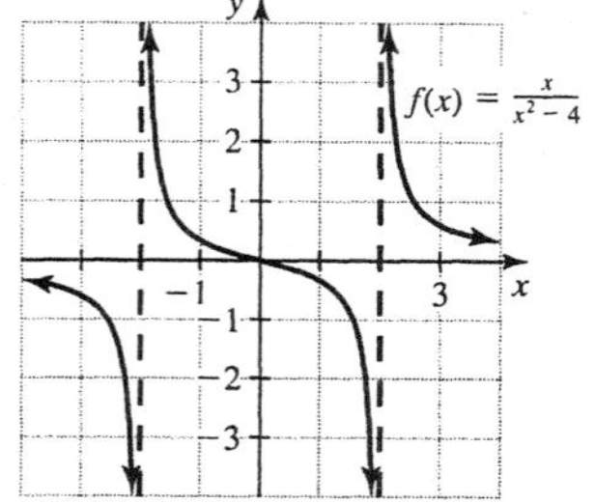

65. Use long division to rewrite the function as $f(x) = 2 + \frac{1}{x-1}$. If $x - 1 = 0$, then $x = 1$. So the graph has a vertical asymptote at $x = 1$. If x is very large, then $f(x)$ is approximately 2. So the line $y = 2$ is a horizontal asymptote. The ratio of the leading coefficients is 2. The graph includes the points (0, 1) and (2, 3).

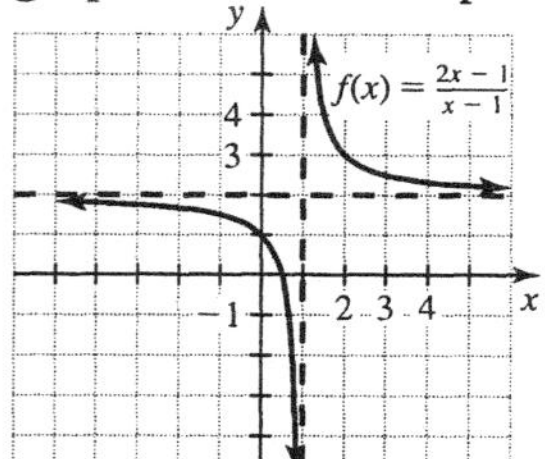

67. Use long division to rewrite the function as $f(x) = x + \frac{1}{x-2}$. If $x - 2 = 0$, then $x = 2$. So the line $x = 2$ is a vertical asymptote for the graph. If x is very large, then the value of $\frac{1}{x-2}$ is approximately 0 and the value of $f(x)$ is approximately x. So the line $y = x$ is an oblique asymptote for the graph. The graph includes the points (0, −0.5), (1, 0), (3, 4), and (4, 4.5).

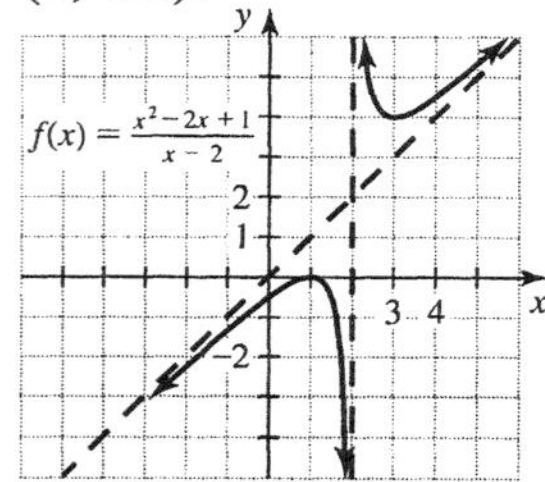

69. $f(-3) = 3(-3) + 5 = -4$

71. $(h \circ f)(\sqrt{2}) = h(f(\sqrt{2})) = h(3\sqrt{2} + 5)$
$= \frac{3\sqrt{2} + 5 - 5}{3} = \frac{3\sqrt{2}}{3} = \sqrt{2}$

73. $(g \circ f)(2) = g(f(2))$
$= g(11) = 11^2 - 2(11) = 99$

75. $(f+g)(3) = f(3) + g(3)$
$= 3 \cdot 3 + 5 + 3^2 - 2 \cdot 3 = 17$

77. $(f \cdot g)(x) = f(x) \cdot g(x)$
$= (3x + 5)(x^2 - 2x)$
$= 3x^3 + 5x^2 - 6x^2 - 10x = 3x^3 - x^2 - 10x$

79. $(f \circ f)(0) = f(f(0)) = f(5) = 3 \cdot 5 + 5$
$= 20$

81. $F(x) = |\,x + 2\,| = |\,g(x)\,| = f(g(x))$
$= (f \circ g)(x)$
Therefore, F is the same function as f composite g, and we write $F = f \circ g$.

83. $H(x) = x^2 + 2 = h(x) + 2 = g(h(x))$
$= (g \circ h)(x)$

Therefore, H is the same function as g composite h, and we write $H = g \circ h$.

85. $I(x) = x + 4 = x + 2 + 2 = g(x) + 2$
$= g(g(x)) = (g \circ g)(x)$
Therefore, I is the same function as g composite g, and we write $I = g \circ g$.

87. This function is not invertible, because it is not one-to-one. The ordered pairs have different first coordinates and the same second coordinate.

89. The function $f(x) = 8x$ is invertible. The inverse of multiplication by 8 is division by 8 and $f^{-1}(x) = x/8$.

91. The function $g(x) = 13x - 6$ is one-to-one and so it is invertible. The inverse of multiplying by 13 and subtracting 6 is adding 6 and then dividing by 13, $g^{-1}(x) = \dfrac{x+6}{13}$.

93.
$$j(x) = \frac{x+1}{x-1}$$
$$y = \frac{x+1}{x-1}$$
$$x = \frac{y+1}{y-1}$$
$$x(y-1) = y+1$$
$$xy - x = y + 1$$
$$xy - y = x + 1$$
$$y(x-1) = x+1$$
$$y = \frac{x+1}{x-1}$$
$$j^{-1}(x) = \frac{x+1}{x-1}$$

95. The function $m(x) = (x-1)^2$ is not invertible because it is not one-to-one. The ordered pairs (2, 1) and (0, 1) are both in the function.

97. The function $f(x) = 3x - 1$ is inverted by adding 1 and then dividing by 3:
$$f^{-1}(x) = \frac{x+1}{3} = \frac{1}{3}x + \frac{1}{3}$$

Use slope and y-intercept to graph both functions on the same coordinate system.

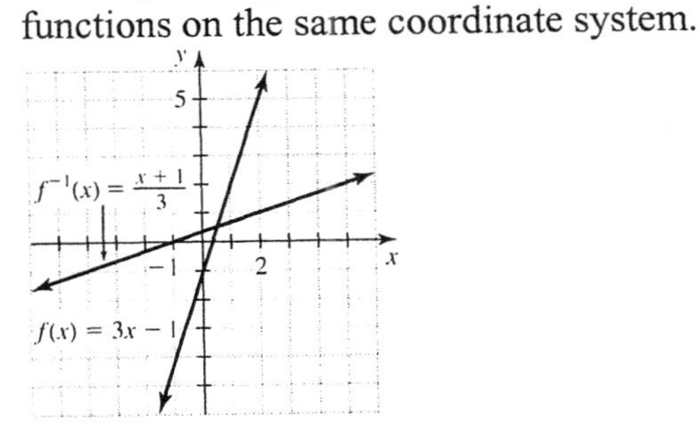

99. The function $f(x) = x^3/2$ is inverted by multiplying by 2 and then taking the cube root:
$$f^{-1}(x) = \sqrt[3]{2x}$$
The graph of f includes the points (0, 0), $(1, 1/2)$, $(-1, -1/2)$, $(2, 4)$, and $(-2, -4)$. The graph of f^{-1} includes the points (0, 0), $(1/2, 1)$, $(-1/2, -1)$, $(4, 2)$, and $(-4, -2)$. Plot these points and graph both functions on the same coordinate system.

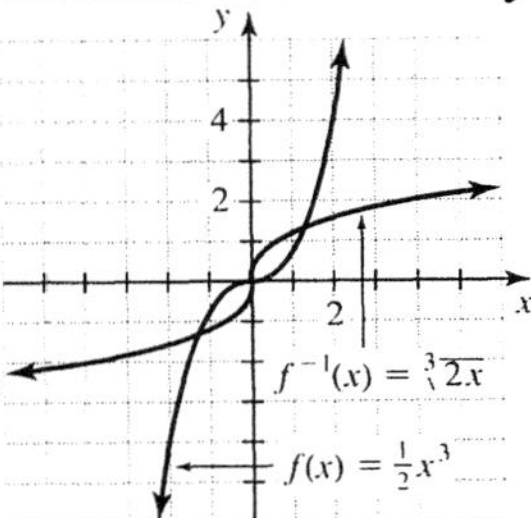

101. The graph of $f(x) = 3$ is a horizontal line with a y-intercept (0, 3).

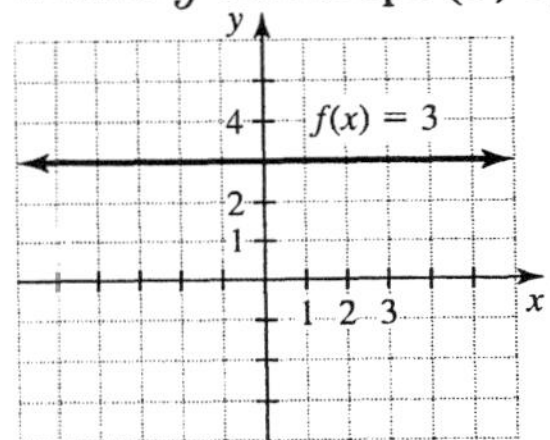

103. The graph of $f(x) = x^2 - 3$ is a parabola with vertex at $(0, -3)$. The graph also goes through $(1, -2)$ and $(-1, -2)$.

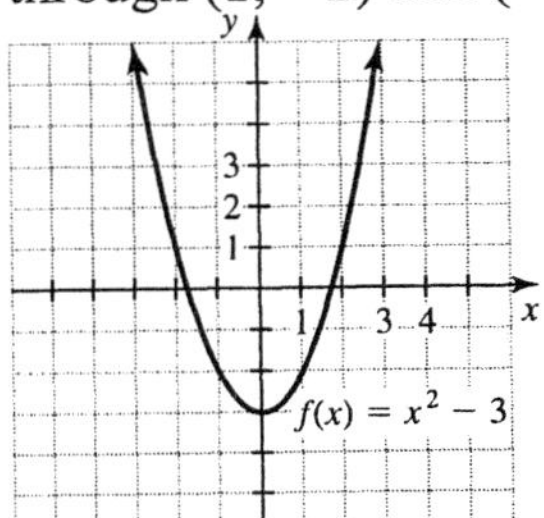

105. The graph of $f(x) = \dfrac{1}{x^2 - 3}$ has vertical asymptotes at $x = \sqrt{3}$ and $x = -\sqrt{3}$. The graph includes the points $(0, -\frac{1}{3})$, (2, 1), and $(-2, 1)$.

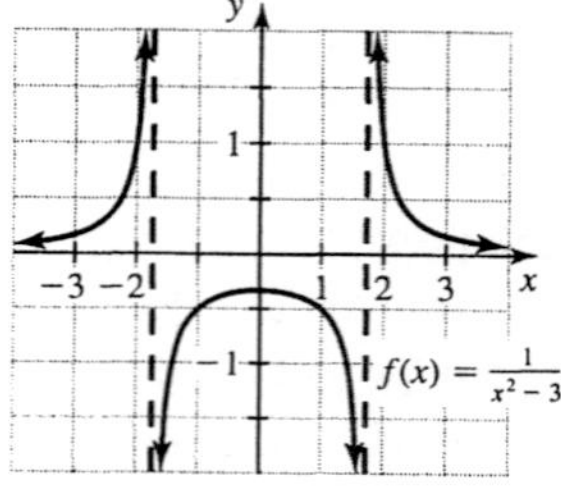

107. The graph of $f(x) = x(x-1)(x+2)$ has x-intercepts $(0, 0)$, $(1, 0)$, and $(-2, 0)$. The graph also includes the points $(-1, 2)$ and $(2, 8)$.

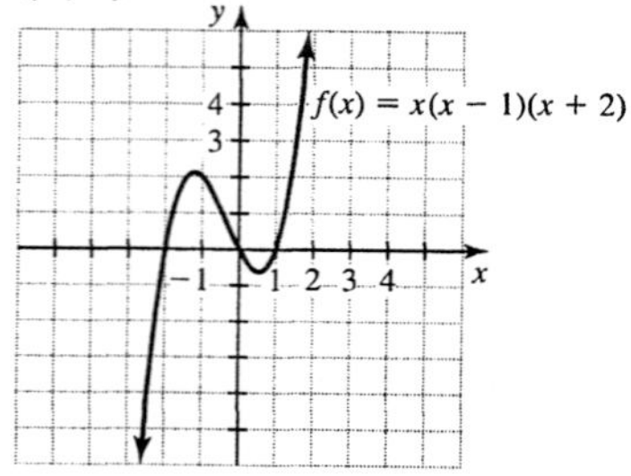

109. The area of the circle is $A = \pi r^2$. Let s = the length of the side of the square. Since the square is inscribed in the circle, the length of the diagonal of the square is the same as the diameter of the circle $2r$. Since the diagonal of the square is the hypotenuse of a right triangle whose sides are s and s, we can write the following equation.

$$\begin{aligned} s^2 + s^2 &= (2r)^2 \\ 2s^2 &= 4r^2 \\ s^2 &= 2r^2 \end{aligned}$$

Since the area of the square is s^2, we have $B = s^2$, or $B = 2r^2$. Since $r^2 = \frac{A}{\pi}$, we can write $B = 2 \cdot \frac{A}{\pi}$, or $B = \frac{2A}{\pi}$.

111. Substitute $k = 5w - 6$ into the equation $a = 3k + 2$:

$$\begin{aligned} a &= 3(5w-6) + 2 \\ a &= 15w - 18 + 2 \\ a &= 15w - 16 \end{aligned}$$

113. The area of a square as a function of the side is $A = s^2$. Solve for s to get the side as a function of the area: $s = \sqrt{A}$.

CHAPTER 11 TEST

1. This set of ordered pairs is a function because no two of the ordered pairs have the same first coordinate and different second coordinates.

2. If $f(x) = -2x + 5$, then $f(-3) = -2(-3) + 5 = 11$.

3. Since $x - 7$ must be nonnegative, we have $x - 7 \geq 0$, or $x \geq 7$. So the domain is $[7, \infty)$. Since y is equal to a square root, y must be nonnegative also. So the range is $[0, \infty)$.

4. The shipping and handling fee S is a linear function of the weight of the order n, $S = 0.50n + 3$.

5. $A(2) = -16(2)^2 + 32(2) + 6 = 6$ ft

6. The line $f(x) = -\frac{2}{3}x + 1$ has a y-intercept of $(0, 1)$ and a slope of $-2/3$. Start at $(0, 1)$ and rise -2 and go 3 units to the right to locate a second point on the line.

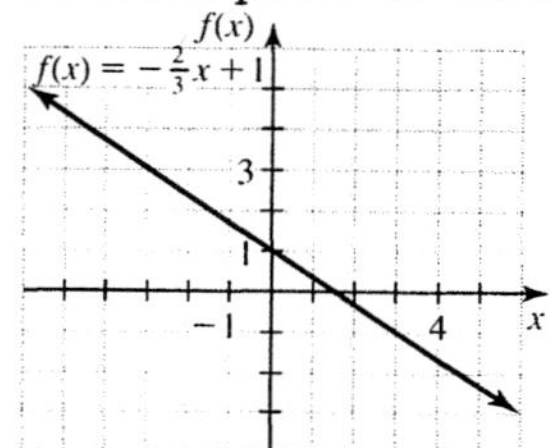

The domain is $(-\infty, \infty)$ and the range is $(-\infty, \infty)$.

7. The graph of $y = |x| - 4$ contains the points $(-2, -2)$, $(-1, -3)$, $(0, -4)$, $(1, -3)$, and $(2, -2)$. Draw a v-shaped graph through these points.

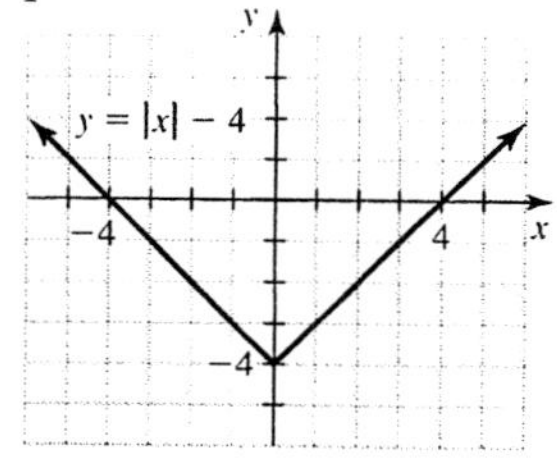

The domain is $(-\infty, \infty)$ and the range is $[-4, \infty)$.

8. The graph of $g(x) = x^2 + 2x - 8$ contains the points $(-2, -8)$, $(-1, -9)$, $(0, -8)$, $(1, -5)$, and $(2, 0)$. Draw a parabola through these points.

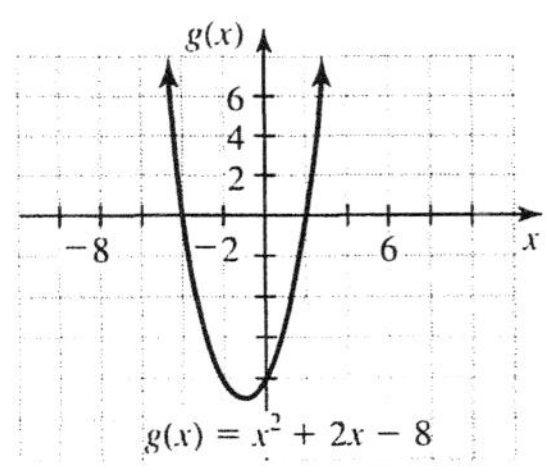

The domain is $(-\infty, \infty)$ and the range is $[-9, \infty)$.

9. The graph of $x = y^2$contains the points $(0,0)$, $(1,1)$, $(1,-1)$, $(4,-2)$, and $(4,2)$. Draw a parabola through these points.

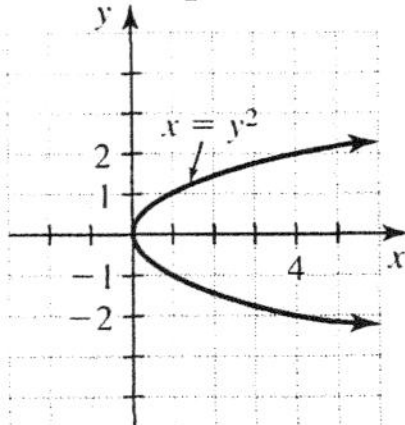

The domain is $[0, \infty)$ and the range is $(-\infty, \infty)$.

10. The graph of $y = -\mid x - 2 \mid$ is a translation and reflection of the graph of $y = \mid x \mid$. The graph contains (2, 0), (0, −2), and (4, −2).

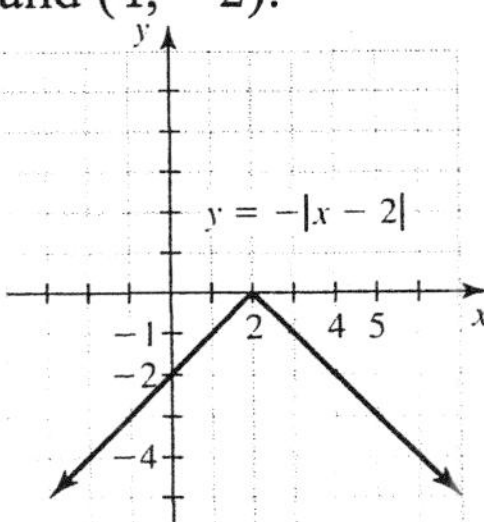

The domain is $(-\infty, \infty)$ and the range is $(-\infty, 0]$.

11. The graph of $y = \sqrt{x+5} - 2$ is obtained from the graph of $y = \sqrt{x}$ by translating 5 units to the left and 2 units downward. The graph of $y = \sqrt{x+5} - 2$ goes through (−5, −2), (−4, −1), and (−1, 0).

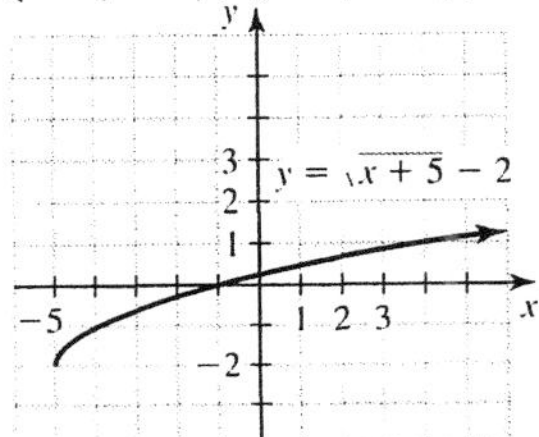

The domain is $[-5, \infty)$ and the range is $[-2, \infty)$.

12. The graph of $f(x) = (x+2)(x-2)^2$ has x-intercepts (−2, 0) and (2, 0). The graph crosses the x-axis at (−2, 0), but does not cross at (2, 0). The graph includes the points (0, 8), (1, 3), and (3, 5).

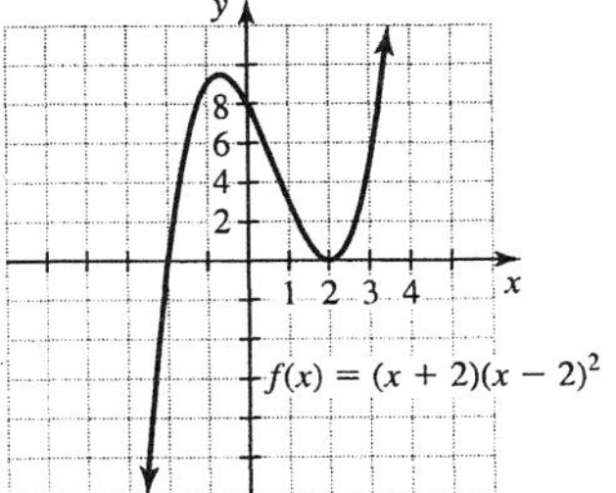

13. Write the function as $f(x) = \dfrac{1}{(x-2)^2}$.

The graph has $x = 2$ as a vertical asymptote and the x-axis as a horizontal asymptote. The graph goes through the points $(0, \frac{1}{4})$, (1, 1), and (3, 1).

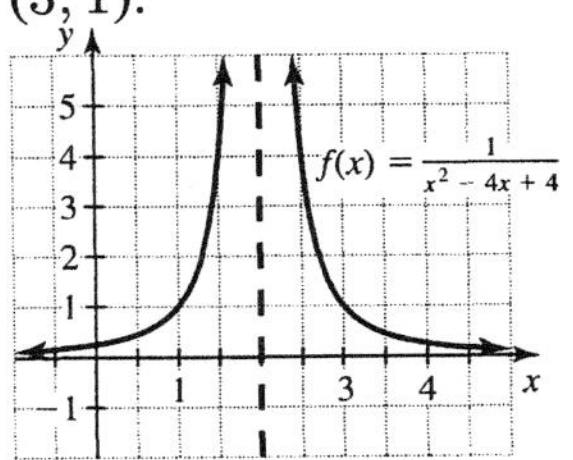

14. Use long division to write the function as $f(x) = 2 + \dfrac{1}{x-2}$. The graph has a vertical asymptote at $x = 2$ and a horizontal asymptote at $y = 2$. The graph includes the points (0, 1.5), (1, 1), and (3, 3).

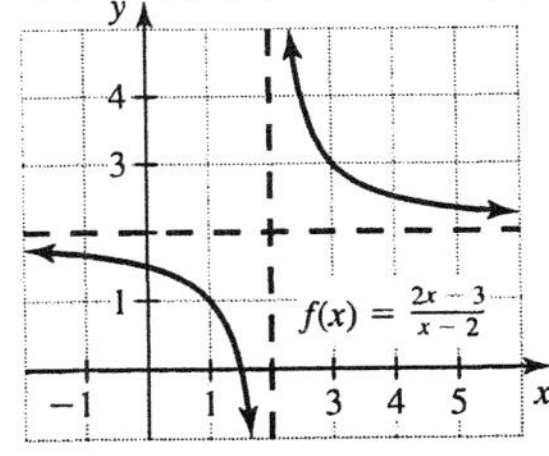

15. Write the function as $f(x) = (x-2)(x+2)(x-1)$. The x-intercepts are (2, 0), (−2, 0), and (1, 0). The y-intercept is (0, 4). The graph also goes through (−1, 6) and (3, 10).

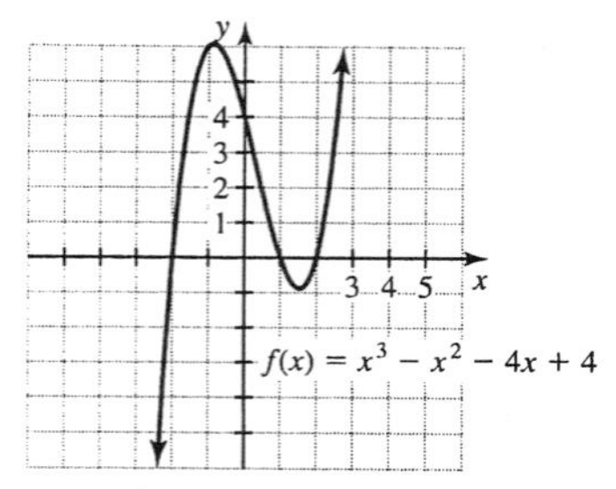

16. $f(-3) = -2(-3) + 5 = 6 + 5 = 11$

17. $(g \circ f)(-3) = g(f(-3)) = g(11)$
$= 11^2 + 4 = 125$

18. Because $f(-3) = 11$, $f^{-1}(11) = -3$.

19. Because f consists of multiplying by -2 and adding 5, f^{-1} consists of subtracting 5 and dividing by -2:
$f^{-1}(x) = \frac{x-5}{-2}$ or $f^{-1}(x) = -\frac{1}{2}x + \frac{5}{2}$

20. $(g + f)(x) = g(x) + f(x)$
$= x^2 + 4 + (-2x + 5)$
$= x^2 - 2x + 9.$

21. $(f \cdot g)(1) = f(1) \cdot g(1) = 3 \cdot 5 = 15$

22. Because f^{-1} is the inverse of f we have $(f^{-1} \circ f)(1776) = 1776$

23. $(f/g)(2) = f(2)/g(2) = 1/8$

24. $(f \circ g)(x) = f(g(x)) = f(x^2 + 4)$
$= -2(x^2 + 4) + 5 = -2x^2 - 8 + 5$
$= -2x^2 - 3$

25. $(g \circ f)(x) = g(f(x)) = g(-2x + 5)$
$= (-2x + 5)^2 + 4 = 4x^2 - 20x + 29$

26. $f(g(x)) = f(x^2) = x^2 - 7 = H(x)$
So $H = f \circ g$.

27. $g(f(x)) = g(x - 7) = (x - 7)^2$
$= x^2 - 14x + 49$

So $W = g \circ f$.

28. The function is not invertible because it is not one-to-one. The ordered pairs $(2, 3)$ and $(4, 3)$ have different first coordinates and the same second coordinate.

29. The function is one-to-one and invertible. The inverse is $\{(3, 2), (4, 3), (5, 4)\}$

30. The inverse of subtracting 5 is adding 5. So $f^{-1}(x) = x + 5$.

31. The inverse of multiplying by 3 and then subtracting 5 is adding 5 and then dividing by 3. So $f^{-1}(x) = \frac{x+5}{3}$.

32. The inverse of taking the cube root and then adding 9 is subtracting 9 and then cubing. So the inverse is $f^{-1}(x) = (x - 9)^3$.

33.
$$f(x) = \frac{2x+1}{x-1}$$
$$y = \frac{2x+1}{x-1}$$
$$x = \frac{2y+1}{y-1}$$
$$x(y-1) = 2y+1$$
$$xy - x = 2y + 1$$
$$xy - 2y = x + 1$$
$$y(x-2) = x + 1$$
$$y = \frac{x+1}{x-2}$$
$$f^{-1}(x) = \frac{x+1}{x-2}$$

Making Connections

Chapters 1 - 11

1. $125^{-2/3} = 5^{-2} = \frac{1}{25}$

2. $\left(\frac{8}{27}\right)^{-1/3} = \left(\frac{27}{8}\right)^{1/3} = \frac{3}{2}$

3. $\sqrt{18} - \sqrt{8} = 3\sqrt{2} - 2\sqrt{2} = \sqrt{2}$

4. $x^5 \cdot x^3 = x^{5+3} = x^8$

5. $16^{1/4} = 2$, because $2^4 = 16$.

6. $\frac{x^{12}}{x^3} = x^{12-3} = x^9$

7.
$$x^2 = 9$$
$$x = \pm\sqrt{9} = \pm 3$$
The solution set is $\{\pm 3\}$.

8. $x^2 = 8$, $x = \pm\sqrt{8} = \pm 2\sqrt{2}$
The solution set is $\{\pm 2\sqrt{2}\}$.

9.
$$x^2 - x = 0$$
$$x(x-1) = 0$$
$$x = 0 \text{ or } x - 1 = 0$$
The solution set is $\{0, 1\}$.

10. $x^2 - 4x - 6 = 0$

$$x = \frac{4 \pm \sqrt{(-4)^2 - 4(1)(-6)}}{2(1)}$$

$$= \frac{4 \pm \sqrt{40}}{2} = \frac{4 \pm 2\sqrt{10}}{2} = 2 \pm \sqrt{10}$$

The solution set is $\{2 \pm \sqrt{10}\}$.

11.
$$x^{1/4} = 3$$
$$(x^{1/4})^4 = 3^4$$
$$x = 81$$

Since we raised each side to an even power we must check. Since the fourth root of 81 is 3, the solution set is $\{81\}$.

12. If we raise each side to the sixth power, we will get $x = 64$. However, the sixth root of 64 is 2and not -2. So 64 does not check. The solution set is $\emptyset$.

13.
$$|x| = 8$$
$$x = 8 \text{ or } x = -8$$
The solution set is $\{\pm 8\}$.

14.
$$|5x - 4| = 21$$
$$5x - 4 = 21 \text{ or } 5x - 4 = -21$$
$$5x = 25 \quad \text{or} \quad 5x = -17$$
$$x = 5 \quad \text{or} \quad x = -\frac{17}{5}$$
The solution set is $\left\{-\frac{17}{5}, 5\right\}$.

15.
$$x^3 = 8$$
$$(x^3)^{1/3} = 8^{1/3}$$
$$x = 2$$
The solution set is $\{2\}$.

16.
$$(3x - 2)^3 = 27$$
$$3x - 2 = 3$$
$$3x = 5$$
$$x = \frac{5}{3}$$
The solution set is $\left\{\frac{5}{3}\right\}$.

17.
$$\sqrt{2x - 3} = 9$$
$$(\sqrt{2x - 3})^2 = 9^2$$
$$2x - 3 = 81$$
$$2x = 84$$
$$x = 42$$
The solution set is $\{42\}$.

18.
$$\sqrt{x - 2} = x - 8$$
$$x - 2 = (x - 8)^2$$
$$x - 2 = x^2 - 16x + 64$$
$$x^2 - 17x + 66 = 0$$
$$(x - 6)(x - 11) = 0$$
$$x = 6 \text{ or } x = 11$$
Only 11 satisfies the original equation. The solution set is $\{11\}$.

19. The graph of the set of points that satisfy the equation $y = 5$ is a horizontal straight line with y-intercept (0, 5).

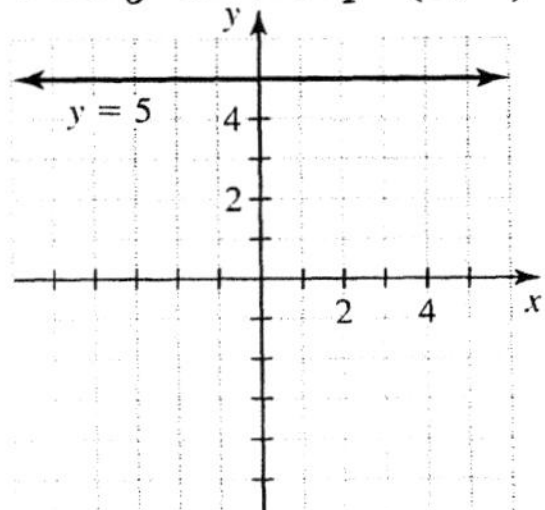

20. The graph of the set of points that satisfy $y = 2x - 5$ is a straight line with y-intercept $(0, -5)$ and slope 2.

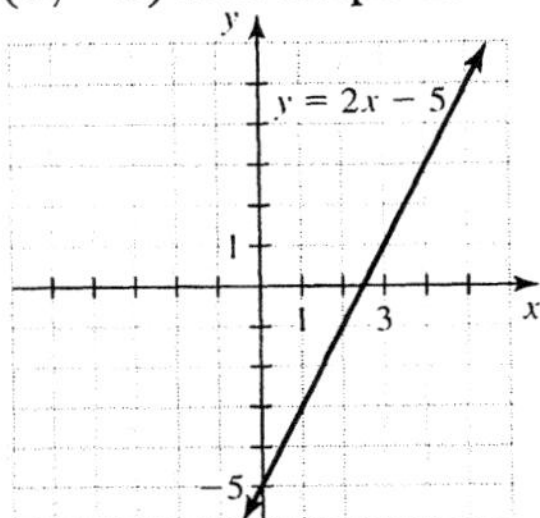

21. The graph of the set of points that satisfy $x = 5$ is a vertical line with x-intercept (5, 0).

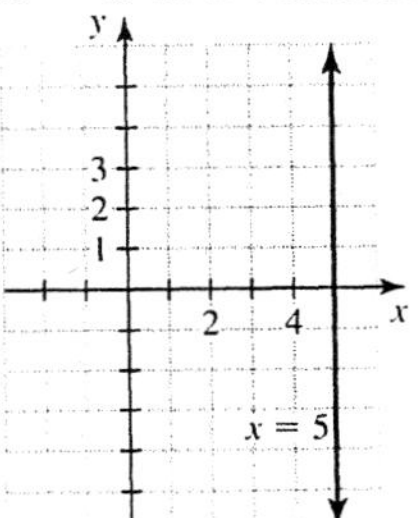

22. The graph of $3y = x$ is the same as the graph of $y = (1/3)x$, a straight line with y-intercept (0, 0) and slope $1/3$.

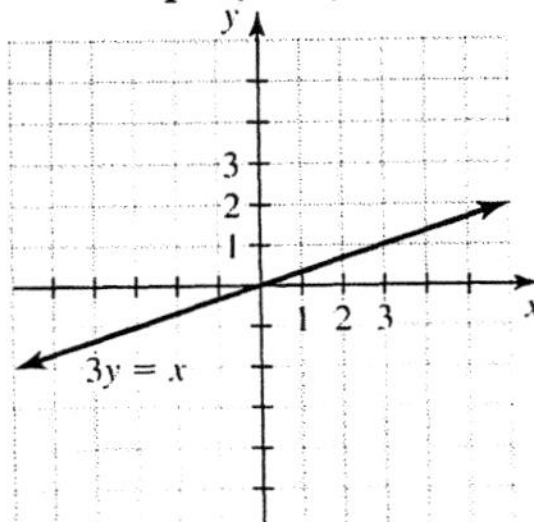

23. The graph of $y = 5x^2$ is a parabola going through (0, 0), (1, 5), and $(-1, 5)$.

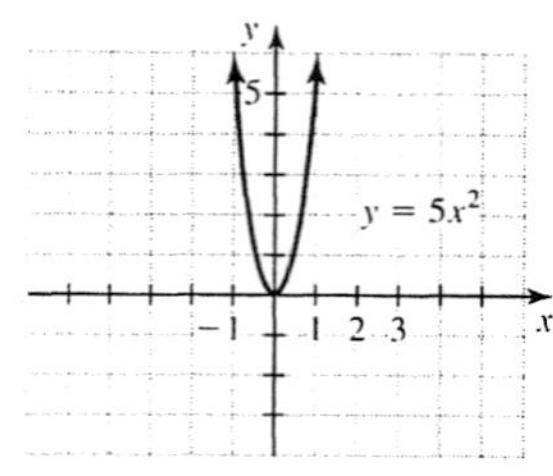

24. The graph of $y = -2x^2$ is a parabola going through $(0, 0)$, $(1, -2)$, and $(-1, -2)$.

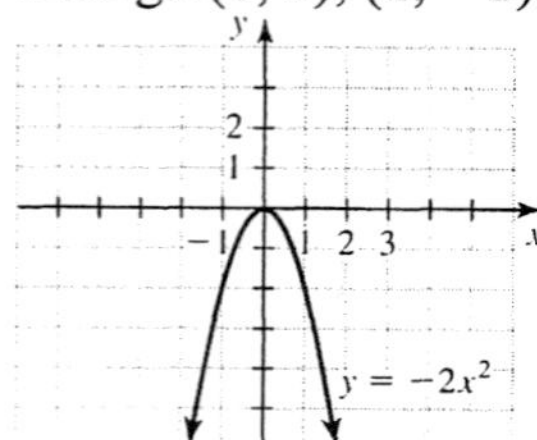

25. Because $2^2 = 4$, and $2^3 = 8$, we have $(2, 4)$ and $(3, 8)$. Because $2^1 = 2$and $2^4 = 16$, we have $(1, 2)$ and $(4, 16)$.

26. Because $4^{1/2} = 2$ and $4^{-1} = 1/4$, we have $(1/2, 2)$ and $(-1, 1/4)$. Because $4^2 = 16$ and $4^0 = 1$, we have $(2, 16)$ and $(0, 1)$.

27. In $\sqrt{x}$ we must have $x \geq 0$. So the domain is $[0, \infty)$.

28. In $\sqrt{6 - 2x}$ we must have $6 - 2x \geq 0$ or $6 \geq 2x$, or $3 \geq x$, or $x \leq 3$. So the domain is $(-\infty, 3]$.

29. Because there are no real solutions to $x^2 + 1 = 0$, we can use any real number for x in this expression. The domain is $(-\infty, \infty)$

30. Solve $x^2 - 10x + 9 = 0$ by factoring.

$$(x - 9)(x - 1) = 0$$
$$x - 9 = 0 \text{ or } x - 1 = 0$$
$$x = 9 \text{ or } \quad x = 1$$

So the domain is all real numbers except 1 and 9, which is written in interval notation as $(-\infty, 1) \cup (1, 9) \cup (9, \infty)$

31. a) $C = 0.12x + 3000$

b) Find the equation of the line through $(0, 0.15)$ and $(100{,}000, 0.25)$

$$m = \frac{0.25 - 0.15}{100{,}000 - 0} = 0.000001$$

P-intercept is $(0, 0.15)$

$$P = 0.000001x + 0.15$$
$$P = 1 \times 10^{-6}x + 0.15$$

32. a) $T = \dfrac{C}{x} + P$

$$T = \frac{3000 + 0.12x}{x} + 0.000001x + 0.15$$
$$T = \frac{3000}{x} + 0.000001x + 0.27$$

If $x = 20{,}000$, then $T = \$0.44$
If $x = 30{,}000$, then $T = \$0.40$
If $x = 90{,}000$, then $T = \$0.39$

b)

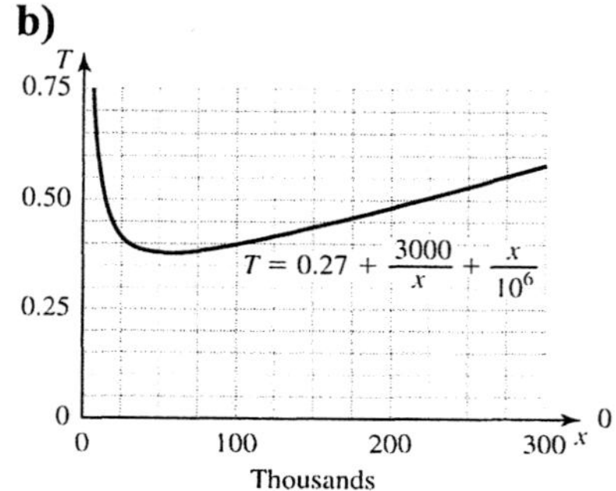

c)
$$\frac{3000}{x} + 0.000001x + 0.27 = 0.38$$
$$\frac{3000}{x} + 0.000001x = 0.11$$
$$3000 + 0.000001x^2 = 0.11x$$
$$0.000001x^2 - 0.11x + 3000 = 0$$

$$x = \frac{0.11 \pm \sqrt{0.11^2 - 4(0.000001)(3000)}}{2(0.000001)}$$
$$= 50{,}000 \text{ or } 60{,}000$$

The car will be replaced at 60,000 miles.

d) The total cost is less than or equal to \$0.38 per mile for mileage in the interval [50,000, 60,000]

12.1 WARM-UPS

1. False, because $f(-1/2) = 4^{-1/2} = 1/2$.
2. True, because $(1/3)^{-1} = 3$.
3. False, because the variable is not in the exponent.
4. True, because $(1/2)^x = (2^{-1})^x = 2^{-x}$.
5. True, because it is a one-to-one function.
6. False, because the equation $(1/3)^x = 0$ has no solution.
7. True, because $e^0 = 1$.
8. False, because the domain of the exponential function $f(x) = 2^x$ is the set of all real numbers.
9. True, because $2^{-x} = 1/(2^x)$.
10. False, at the end of 3 years the investment is worth $500(1.005)^{36}$ dollars

12.1 EXERCISES

1. An exponential function has the form $f(x) = a^x$ where $a > 0$ and $a \neq 1$.
3. The two most popular bases are e and 10.
5. The compound interest formula is $A = P(1+i)^n$.
7. $f(2) = 4^2 = 16$
9. $f(1/2) = 4^{1/2} = 2$
11. $g(-2) = (1/3)^{-2+1} = (1/3)^{-1} = 3$
13. $g(0) = (1/3)^{0+1} = (1/3)^1 = \frac{1}{3}$
15. $h(0) = -2^0 = -1$
17. $h(-2) = -2^{-2} = -1/4$
19. $h(0) = 10^0 = 1$
21. $h(2) = 10^2 = 100$
23. $j(1) = e^1 = e \approx 2.718$
25. $j(-2) = e^{-2} \approx 0.135$
27. Evaluate 4^x for $x = -2, -1, 0, 1, 2$ and place the results in the second row of the table:

$4^{-2} = \frac{1}{4^2} = \frac{1}{16}$ $\quad 4^{-1} = \frac{1}{4^1} = \frac{1}{4}$

$4^0 = 1 \quad 4^1 = 4 \quad 4^2 = 16$

x	-2	-1	0	1	2
4^x	1/16	1/4	1	4	16

29. Evaluate $\left(\frac{1}{3}\right)^x$ for $x = -2, -1, 0, 1, 2$ and place the results in the second row of the table:

$\left(\frac{1}{3}\right)^{-2} = 3^2 = 9 \quad \left(\frac{1}{3}\right)^{-1} = 3^1 = 3$

$\left(\frac{1}{3}\right)^0 = 1 \quad \left(\frac{1}{3}\right)^1 = \frac{1}{3} \quad \left(\frac{1}{3}\right)^2 = \frac{1}{9}$

x	-2	-1	0	1	2
$\left(\frac{1}{3}\right)^x$	9	3	1	1/3	1/9

31. The graph of $f(x) = 4^x$ includes the points (0, 1), (1, 4), (2, 16), and (−1, 1/4).

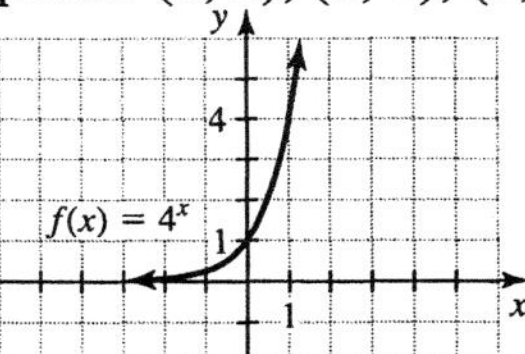

33. The graph of $h(x) = \left(\frac{1}{3}\right)^x$ includes the points (0, 1), (1, 1/3), (2, 1/9), (−1, 3), and (−2, 9).

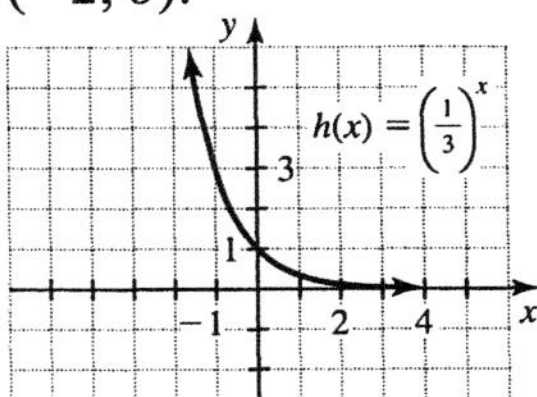

35. The graph of $y = 10^x$ includes the points (0, 1), (1, 10), and (−1, 1/10).

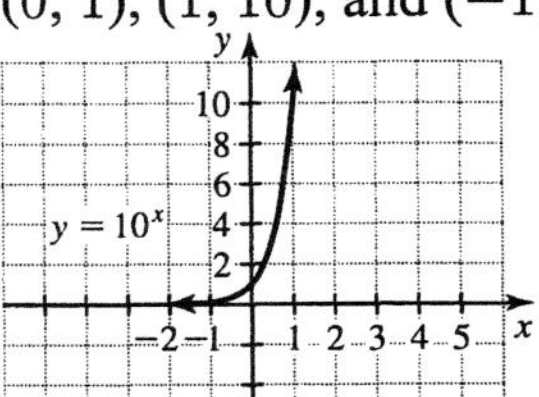

37. Evaluate 10^{x+2} for $x = -4, -3, -2, -1, 0$ and place the results in the second row of the table:

$10^{-4+2} = \frac{1}{10^2} = \frac{1}{100}$

$10^{-3+2} = \frac{1}{10^1} = \frac{1}{10}$

$10^{-2+2} = 1 \quad 10^{-1+2} = 10$

$10^{0+2} = 100$

x	-4	-3	-2	-1	0
4^x	1/100	1/10	1	10	100

39. Evaluate -2^x for $x = -2, -1, 0, 1, 2$ and place the results in the second row of the table:

$-2^{-2} = -\frac{1}{2^2} = -\frac{1}{4}$

$-2^{-1} = -\frac{1}{2^1} = -\frac{1}{2}$

$-2^0 = -1 \quad -2^1 = -2$

$-2^2 = -4$

x	-2	-1	0	1	2
-2^{2x+1}	$-1/4$	$-1/2$	-1	-2	-4

41. The graph of $y = 10^{x+2}$ includes the points (0, 100), (−1, 10), (−2, 1), and $(-3, 1/10)$.

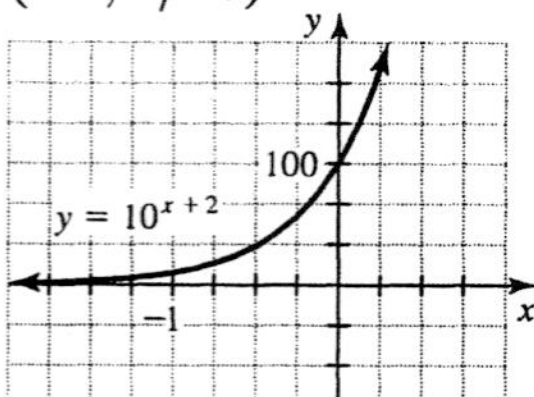

43. The graph of $f(x) = -2^x$ includes the points (0, −1), (1, −2), (2, −4), (−1, −1/2), and (−2, −1/4).

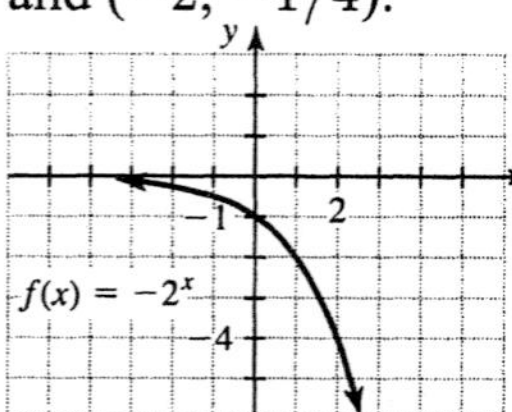

45. The graph of $g(x) = 2^{-x}$ includes the points (0, 1), (−1, 2), (−2, 4), and (1, 1/2).

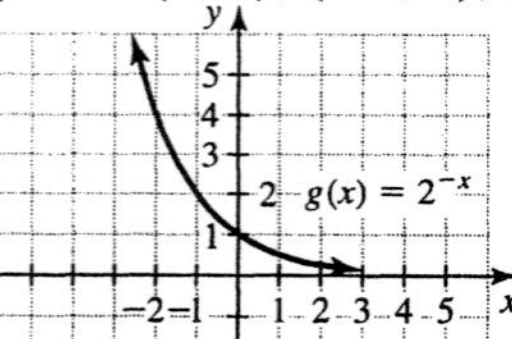

47. The graph of $f(x) = -e^x$ includes the points (0, −1), (1, −2.7), (2, −7.4), (−1, −0.4).

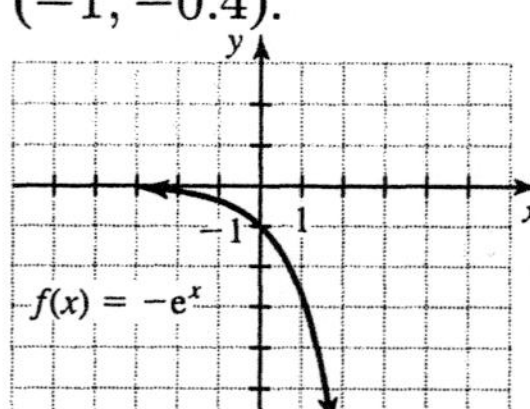

49. The graph of $H(x) = 10^{|x|}$ includes the points (0, 1), (1, 10), and (−1, 10).

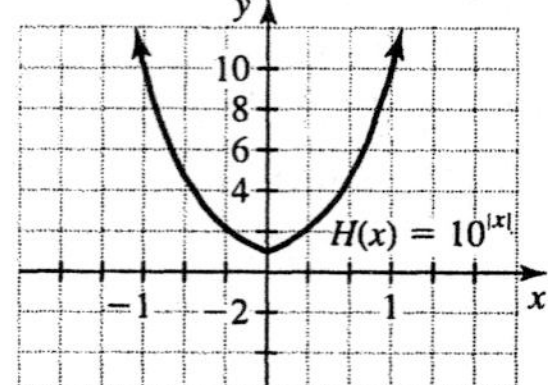

51. The graph of $P = 5000(1.05)^t$ includes the points (0, 5000), (20, 13,266), and $(-10, 3070)$.

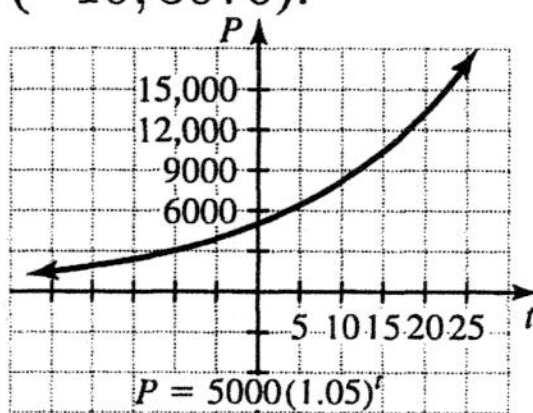

53.
$$2^x = 64$$
$$2^x = 2^6$$
$$x = 6 \quad \text{By the one-to-one property}$$
The solution set is $\{6\}$.

55.
$$10^x = 0.001$$
$$10^x = 10^{-3}$$
$$x = -3$$
The solution set is $\{-3\}$.

57.
$$2^x = \frac{1}{4}$$
$$2^x = 2^{-2}$$
$$x = -2$$
The solution set is $\{-2\}$.

59.
$$\left(\frac{2}{3}\right)^{x-1} = \frac{9}{4}$$
$$\left(\frac{2}{3}\right)^{x-1} = \left(\frac{2}{3}\right)^{-2}$$
$$x - 1 = -2$$
$$x = -1$$
The solution set is $\{-1\}$.

61.
$$5^{-x} = 25$$
$$5^{-x} = 5^2$$
$$-x = 2 \quad \text{By the one-to-one property}$$
$$x = -2$$
The solution set is $\{-2\}$.

63.
$$-2^{1-x} = -8$$
$$-2^{1-x} = -2^3$$
$$2^{1-x} = 2^3$$
$$1 - x = 3$$
$$x = -2$$
The solution set is $\{-2\}$.

65.
$$10^{|x|} = 1000$$
$$10^{|x|} = 10^3$$
$$|\,x\,| = 3$$
$$x = 3 \text{ or } x = -3$$
The solution set is $\{-3, 3\}$.

67. If $f(x) = 2^x$ and $f(x) = 4 = 2^2$, then we must have $x = 2$ by the one-to-one property of exponential functions.

69. Note that $4^{2/3} = (2^2)^{2/3} = 2^{4/3}$. So if $2^x = 2^{4/3}$, then $x = \frac{4}{3}$.
71. Note that $9 = 3^2 = (1/3)^{-2}$. So if $(1/3)^x = (1/3)^{-2}$, then $x = -2$.
73. Note that $1 = (1/3)^0$. So if $(1/3)^x = (1/3)^0$, then $x = 0$.
75. Note that $16 = 4^2$. So if $h(x) = 4^{2x-1} = 4^2$, then

$$2x - 1 = 2$$
$$2x = 3$$
$$x = \frac{3}{2}.$$

77. Since $1 = 4^0$, $h(x) = 4^{2x-1} = 4^0$ implies that $2x - 1 = 0$ or $x = \frac{1}{2}$.
79. Since $2^{-5} = \frac{1}{32}$, $2^{-3} = \frac{1}{8}$, $2^0 = 1$, $2^1 = 2$, and $2^4 = 16$, the table should read as follows.

x	-5	-3	0	1	4
2^x	$\frac{1}{32}$	$\frac{1}{8}$	1	2	16

81. Since $\left(\frac{1}{2}\right)^{-3} = 2^3 = 8$, $\left(\frac{1}{2}\right)^{-2} = 4$, $\left(\frac{1}{2}\right)^0 = 1$, $\left(\frac{1}{2}\right)^1 = \frac{1}{2}$, $\left(\frac{1}{2}\right)^5 = \frac{1}{32}$ the table should read as follows.

x	-3	-2	0	1	5
$\left(\frac{1}{2}\right)^x$	8	4	1	$\frac{1}{2}$	$\frac{1}{32}$

83. Compounded quarterly for 10 years means that interest will be paid 40 times at 1.25%.
$S = 6000\left(1 + \frac{0.05}{4}\right)^{40} = 6000(1.0125)^{40}$
$= \$9861.72$
85. a) $\$10{,}000(1 + 0.1821)^{10} \approx \$53{,}277.30$
b) From the graph it appears that the \$10,000 will be worth \$200,000approximately 18 years after 1992 or around 2010.
87. When the book is new, $t = 0$, and the value is $V = 45 \cdot 2^{-0.9(0)} = \$45$. When $t = 2$, the value is $V = 45 \cdot 2^{-0.9(2)} \approx \$12.92$.
89. A deposit of \$500 for 3 years at 7% compounded continuously amounts to $S = 500 \cdot e^{0.07(3)} \approx \$616.84$.
91. A deposit of \$80,000 at 7.5% compounded continuously for 1 year amounts to $S = 80{,}000 \cdot e^{0.075(1)} \approx \$86{,}230.73$. The interest earned is
$\$86{,}230.73 - \$80{,}000 = \$6230.73$.
93. The amount at time $t = 0$ is $A = 300 \cdot e^{-0.06(0)} = 300$ grams. The amount present after 20 years, $t = 20$, is $A = 300 \cdot e^{-0.06(20)} \approx 90.4$ grams.
One-half of the substance decays in about 12 years. The substance will never completely disappear because the t-axis is the horizontal asymptote.
95. $1 + 1 + \frac{1}{2} + \frac{1}{6} \approx 2.66666667$
$e - 2.66666667 \approx 0.0516$
$e - \left(1 + 1 + \frac{1}{2} + \frac{1}{6} + \frac{1}{24} + \frac{1}{120} + \frac{1}{720} + \frac{1}{5040}\right)$
$\approx 2.8 \times 10^{-5}$
97. The graph of $y = 3^{x-h}$ lies h units to the right of $y = 3^x$ when $h > 0$ and $|h|$ units to the left of $y = 3^x$ when $h < 0$.

12.2 WARM-UPS

1. True, because 3 is the exponent of a that produces 2.
2. False, because $b = 8^a$ is equivalent to $\log_8(b) = a$.
3. True, because the inverse of the base a exponential function is the base a logarithm function.
4. True, because the inverse of the base e exponential function is the base e logarithm function.
5. False, the domain is $(0, \infty)$.
6. False, $\log_{25}(5) = 1/2$.
7. False, $\log(-10)$ is undefined because -10 is not in the domain of the base 10 logarithm function.
8. False, $\log(0)$ is undefined.
9. True, because $\log_5(125) = 3$ and $5^3 = 125$.
10. True, because $(1/2)^{-5} = 2^5 = 32$.

12.2 EXERCISES

1. If $f(x) = 2^x$ then $f^{-1}(x) = \log_2(x)$.
3. The common logarithm uses the base 10 and the natural logarithm uses base e.
5. The one-to-one property for logarithmic functions says that if $\log_a(m) = \log_a(n)$ then $m = n$.
7. $\log_2(8) = 3$ is equivalent to $2^3 = 8$.

9. $10^2 = 100$ is equivalent to $\log(100) = 2$.
11. $y = \log_5(x)$ is equivalent to $5^y = x$.
13. $2^a = b$ is equivalent to $\log_2(b) = a$.
15. $\log_3(x) = 10$ is equivalent to $3^{10} = x$.
17. $e^3 = x$ is equivalent to $\ln(x) = 3$.
19. Because $2^2 = 4$, $\log_2(4) = 2$.
21. Because $2^4 = 16$, $\log_2(16) = 4$.
23. Because $2^6 = 64$, $\log_2(64) = 6$.
25. Because $4^3 = 64$, $\log_4(64) = 3$.
27. Because $2^{-2} = \frac{1}{4}$, $\log_2(1/4) = -2$.
29. Because $10^2 = 100$, $\log(100) = 2$.
31. Because $10^{-2} = 0.01$, $\log(0.01) = -2$.
33. Because $(1/3)^1 = 1/3$, $\log_{1/3}(1/3) = 1$.
35. Because $(1/3)^{-3} = 27$, $\log_{1/3}(27) = -3$.
37. Because $25^{1/2} = 5$, $\log_{25}(5) = 1/2$.
39. Because $e^2 = e^2$, $\ln(e^2) = 2$.
41. Use a calculator with a base 10 logarithm key to find $\log(5) \approx 0.6990$.
43. Use a calculator with a natural logarithm key to find $\ln(6.238) \approx 1.8307$.
45. Since $3^{-2} = \frac{1}{9}$, $3^{-1} = \frac{1}{3}$, $3^0 = 1$, $3^1 = 3$, and $3^2 = 9$, the table should read as follows. Note that the logarithms are the exponents.

x	$\frac{1}{9}$	$\frac{1}{3}$	1	3	9
$\log_3(x)$	-2	-1	0	1	2

47. Since $\left(\frac{1}{4}\right)^{-2} = 16$,
$\left(\frac{1}{4}\right)^{-1} = 4$, $\left(\frac{1}{4}\right)^{0} = 1$, $\left(\frac{1}{4}\right)^{1} = \frac{1}{4}$,
$\left(\frac{1}{4}\right)^{2} = \frac{1}{16}$ the table should read as follows.
Note that the logarithms are the exponents.

x	16	4	1	$\frac{1}{4}$	$\frac{1}{16}$
$\log_{1/4}(x)$	-2	-1	0	1	2

49. The graph of $f(x) = \log_3(x)$ includes the points $(3, 1)$, $(1, 0)$, and $(1/3, -1)$. All graphs of logarithm functions have similar shapes.

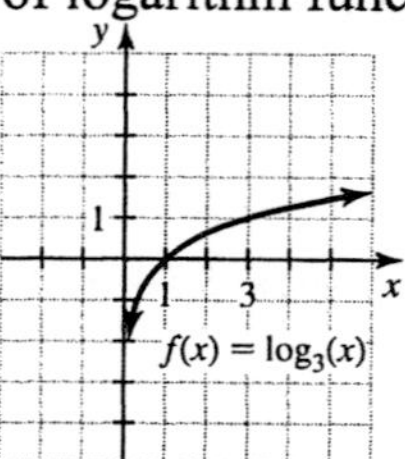

51. The graph of $y = \log_4(x)$ includes the points $(4, 1)$, $(1, 0)$, and $(1/4, -1)$.

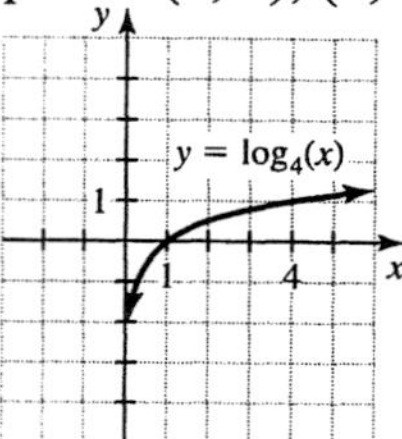

53. The graph of $y = \log_{1/4}(x)$ includes the points $(1, 0)$, $(4, -1)$, and $(1/4, 1)$.

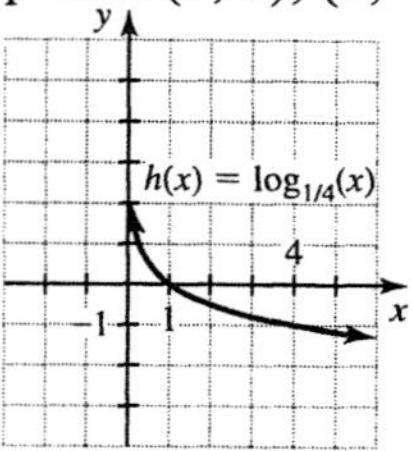

55. The graph of $h(x) = \log_{1/5}(x)$ includes the points $(1, 0)$, $(5, -1)$, and $(1/5, 1)$.

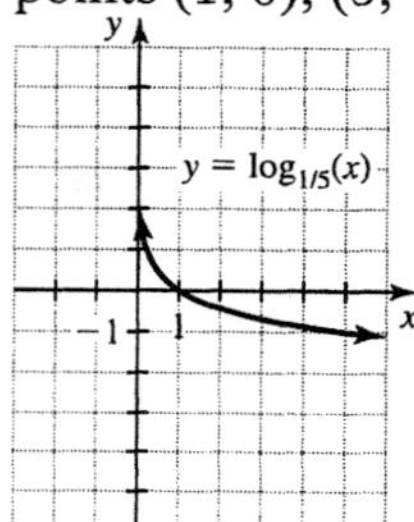

57. The inverse of $f(x) = 6^x$ is $f^{-1}(x) = \log_6(x)$.
59. The inverse of $f(x) = \ln(x)$ is $f^{-1}(x) = e^x$.
61. If $f(x) = \log_{1/2}(x)$ then $f^{-1}(x) = \left(\frac{1}{2}\right)^x$.
63. $x = (1/2)^{-2} = 2^2 = 4$
The solution set is $\{4\}$.
65. $5 = 25^x$
$x = \log_{25}(5) = 1/2$
The solution set is $\left\{\frac{1}{2}\right\}$.
67. $\log(x) = -3$
$x = 10^{-3} = 0.001$
The solution set is $\{0.001\}$.
69. $\log_x(36) = 2$
$x^2 = 36$
$x = \pm 6$

Omit -6 because the base of any logarithm function is positive. The solution set is $\{6\}$.

71. $\log_x(5) = -1$

$x^{-1} = 5$

$(x^{-1})^{-1} = 5^{-1}$

$x = \frac{1}{5}$

The solution set is $\left\{\frac{1}{5}\right\}$.

73. $\log(x^2) = \log(9)$

$x^2 = 9$

$x = \pm 3$

Both 3 and -3 check in the original equation. The solution set is $\{\pm 3\}$.

75. $3 = 10^x$

$x = \log(3) \approx 0.4771$

The solution set is $\{0.4771\}$.

77. $10^x = \frac{1}{2}$

$x = \log(1/2) = \log(0.5) \approx -0.3010$

The solution set is $\{-0.3010\}$.

79. $e^x = 7.2$

$x = \ln(7.2) \approx 1.9741$

The solution set is $\{1.9741\}$.

81. Since $2^{-2} = \frac{1}{4}$, $2^{-1} = \frac{1}{2}$, $2^0 = 1$, $2^2 = 4$, and $2^4 = 16$, the table should read as follows.

x	$\frac{1}{4}$	$\frac{1}{2}$	1	4	16
$\log_2(x)$	-2	-1	0	2	4

83. Since $\left(\frac{1}{2}\right)^{-4} = 16$, $\left(\frac{1}{2}\right)^{-2} = 4$, $\left(\frac{1}{2}\right)^{0} = 1$, $\left(\frac{1}{2}\right)^{1} = \frac{1}{2}$, $\left(\frac{1}{2}\right)^{2} = \frac{1}{4}$ the table should read as follows.

x	16	4	1	$\frac{1}{2}$	$\frac{1}{4}$
$\log_{1/2}(x)$	-4	-2	0	1	2

85. Use the continuous compounding formula.

$10{,}000 = 5000 \cdot e^{0.12t}$

$2 = e^{0.12t}$

$0.12t = \ln(2)$

$t = \frac{\ln(2)}{0.12} \approx 5.776$ years

87. To earn \$1000 in interest, the principal must increase from \$6000 to \$7000 in t years.

$7000 = 6000 \cdot e^{0.08t}$

$7/6 = e^{0.08t}$

$0.08t = \ln(7/6)$

$t = \frac{\ln(7/6)}{0.08} \approx 1.927$ years

89. a) $20{,}733 = 10{,}000 \cdot e^{r5}$

$2.0733 = e^{5r}$

$5r = \ln(2.0733)$

$r = \frac{\ln(2.0733)}{5} \approx 0.1458 = 14.58\%$

b) In 2010 the investment grows for 13 years: $A = 10{,}000e^{0.1458(13)} \approx \$66{,}576.60$

91. $\text{pH} = -\log_{10}(10^{-4.1}) = -(-4.1) = 4.1$

93. $\text{pH} = -\log(1.58 \times 10^{-7}) \approx 6.8$

95. $L = 10 \cdot \log(0.001 \times 10^{12}) = 90$ db

97. $f(x) = 5 + \log_2(x - 3)$

$y = 5 + \log_2(x - 3)$

$x = 5 + \log_2(y - 3)$

$x - 5 = \log_2(y - 3)$

$y - 3 = 2^{x-5}$

$y = 2^{x-5} + 3$

$f^{-1}(x) = 2^{x-5} + 3$

Domain of f^{-1} is $(-\infty, \infty)$ and range is $(3, \infty)$.

99. $y = \ln(e^x) = x$ for $-\infty < x < \infty$

$y = e^{\ln(x)} = x$ for $0 < x < \infty$

12.3 WARM-UPS

1. True, because $\log_2(x^2/8) = \log_2(x^2) - \log_2(8) = \log_2(x^2) - 3$.

2. False, because $\log(100) = 2$ and $\log(10) = 1$, and $2 \div 1 \neq 2 - 1$.

3. True, because $\ln(\sqrt{2}) = \ln(2^{1/2}) = \frac{1}{2} \cdot \ln(2) = \frac{\ln(2)}{2}$.

4. True, because $\log_3(17)$ is the exponent that we place on 3 to obtain 17.

5. False, because $\log_2(1/8) = -3$ and $\log_2(8) = 3$.

6. True, because $\ln(8) = \ln(2^3) = 3 \cdot \ln(2)$.

7. False, because $\ln(1) = 0$ and $e \neq 0$.

8. False, because $\log(100) = 2, \log(10) = 1$, and $2 \div 10 \neq 1$.

9. False, because $\log_2(8) = 3, \log_2(2) = 1$, $\log_2(4) = 2$, and $3 \div 1 \neq 2$.

10. True, because $\ln(2) + \ln(3) - \ln(7) = \ln(2 \cdot 3) - \ln(7) = \ln(6/7)$.

12.3 EXERCISES

1. The product rule for logarithms says that $\log_a(MN) = \log_a(M) + \log_a(N)$.

3. The power rule for logarithms says that $\log_a(M^N) = N \cdot \log_a(M)$.

5. Since $\log_a(M)$ is the exponent that you would use on a to obtain M, using $\log_a(M)$ as the exponent produces M: $a^{\log_a(M)} = M$.

7. $\log(3) + \log(7) = \log(3 \cdot 7) = \log(21)$

9. $\log_3(\sqrt{5}) + \log_3(\sqrt{x}) = \log_3(\sqrt{5x})$

11. $\log(x^2) + \log(x^3) = \log(x^2 \cdot x^3) = \log(x^5)$

13. $\ln(2) + \ln(3) + \ln(5) = \ln(2 \cdot 3 \cdot 5)$
$= \ln(30)$

15. $\log(x) + \log(x + 3) = \log(x^2 + 3x)$

17. $\log_2(x - 3) + \log_2(x + 2)$
$= \log_2(x^2 - x - 6)$

19. $\log(8) - \log(2) = \log(8/2) = \log(4)$

21. $\log_2(x^6) - \log_2(x^2) = \log_2\left(\frac{x^6}{x^2}\right)$
$= \log_2(x^4)$

23. $\log(\sqrt{10}) - \log(\sqrt{2}) = \log\left(\frac{\sqrt{10}}{\sqrt{2}}\right)$
$= \log\left(\sqrt{5}\right)$

25. $\ln(4h - 8) - \ln(4) = \ln\left(\frac{4h - 8}{4}\right)$
$= \ln(h - 2)$

27. $\log_2\left(w^2 - 4\right) - \log_2(w + 2)$
$= \log_2\left(\frac{w^2 - 4}{w + 2}\right) = \log_2\left(\frac{(w - 2)(w + 2)}{w + 2}\right)$
$= \log_2(w - 2)$

29. $\ln\left(x^2 + x - 6\right) - \ln(x + 3)$
$= \ln\left(\frac{x^2 + x - 6}{x + 3}\right) = \ln\left(\frac{(x + 3)(x - 2)}{x + 3}\right)$
$= \ln(x - 2)$

31. $\log(27) = \log(3^3) = 3\log(3)$

33. $\log(\sqrt{3}) = \log(3^{1/2}) = \frac{1}{2}\log(3)$

35. $\log(3^x) = x\log(3)$

37. $\log_2(2^{10}) = 10$

39. $5^{\log_5(19)} = 19$

41. $\log(10^8) = 8$

43. $e^{\ln(4.3)} = 4.3$

45. $\log(15) = \log(3 \cdot 5) = \log(3) + \log(5)$

47. $\log(5/3) = \log(5) - \log(3)$

49. $\log(25) = \log(5^2) = 2\log(5)$

51. $\log(75) = \log(5^2 \cdot 3) = 2\log(5) + \log(3)$

53. $\log\left(\frac{1}{3}\right) = \log(1) - \log(3)$
$= 0 - \log(3) = -\log(3)$

55. $\log(0.2) = \log(1/5) = \log(1) - \log(5)$
$= 0 - \log(5) = -\log(5)$

57. $\log(xyz) = \log(x) + \log(y) + \log(z)$

59. $\log_2(8x) = \log_2(8) + \log_2(x)$
$= 3 + \log_2(x)$

61. $\ln(x/y) = \ln(x) - \ln(y)$

63. $\log(10x^2) = \log(10) + \log(x^2)$
$= 1 + 2\log(x)$

65. $\log_5\left[\frac{(x - 3)^2}{\sqrt{w}}\right]$
$= \log_5\left[(x - 3)^2\right] - \log_5(\sqrt{w})$
$= 2\log_5(x - 3) - \frac{1}{2}\log_5(w)$

67. $\ln\left[\frac{yz\sqrt{x}}{w}\right] = \ln[yz\sqrt{x}] - \ln(w)$
$= \ln(y) + \ln(z) + \ln(\sqrt{x}) - \ln(w)$
$= \ln(y) + \ln(z) + \frac{1}{2}\ln(x) - \ln(w)$

69. $\log(x) + \log(x - 1) = \log(x^2 - x)$

71. $\ln(3x - 6) - \ln(x - 2) = \ln\left(\frac{3x - 6}{x - 2}\right)$
$= \ln(3)$

73. $\ln(x) - \ln(w) + \ln(z) = \ln(xz) - \ln(w)$
$= \ln\left(\frac{xz}{w}\right)$

75. $3 \cdot \ln(y) + 2 \cdot \ln(x) - \ln(w)$
$= \ln(y^3) + \ln(x^2) - \ln(w)$
$= \ln(x^2y^3) - \ln(w) = \ln\left(\frac{x^2y^3}{w}\right)$

77. $\frac{1}{2} \cdot \log(x - 3) - \frac{2}{3} \cdot \log(x + 1)$
$= \log\left((x - 3)^{1/2}\right) - \log\left((x + 1)^{2/3}\right)$
$= \log\left(\frac{(x - 3)^{1/2}}{(x + 1)^{2/3}}\right)$

79. $\frac{2}{3} \cdot \log_2(x - 1) - \frac{1}{4} \cdot \log_2(x + 2)$
$= \log_2\left((x - 1)^{2/3}\right) - \log_2\left((x + 2)^{1/4}\right)$
$= \log_2\left(\frac{(x - 1)^{2/3}}{(x + 2)^{1/4}}\right)$

81. False, because
$\log(56) = \log(7 \cdot 8) = \log(7) + \log(8)$.

83. True, because $\log_2(4^2) = \log_2(16) = 4$ and $(\log_2(4))^2 = (2)^2 = 4$.

85. True, because $\ln(25) = \ln(5^2) = 2 \cdot \ln(5)$.

87. False, because $\log_2(64) = 6$ and $\log_2(8) = 3$ and $6 \div 3 \neq 3$.

89. True, because $\log(1/3) = \log(1) - \log(3)$
$= 0 - \log(3) = -\log(3)$.

91. True, because
$\log_2(16^5) = 5 \cdot \log_2(16) = 5 \cdot 4 = 20.$
93. True, $\log(10^3) = 3 \cdot \log(10) = 3 \cdot 1 = 3.$
95. False, because $\log(100 + 3) = \log(103)$
and
$2 + \log(3) = \log(100) + \log(3) = \log(300).$
97. $r = \log(I) - \log(I_0)$
$r = \log\left(\frac{I}{I_0}\right)$
If $I = 100 \cdot I_0$, then
$r = \log\left(\frac{100 \cdot I_0}{I_0}\right) = \log(100) = 2.$
99. Only (b) is an identity, because it is the only one that is a correct application of a property of logarithms.
101. The graphs are the same because
$\ln(\sqrt{x}) = \ln(x^{1/2}) = \frac{1}{2}\ln(x) = 0.5 \cdot \ln(x).$
103. The graph is a straight line because
$\log(e^x) = x\log(e) \approx 0.434x$. The slope is
$\log(e)$ or approximately 0.434.

12.4 WARM-UPS

1. True, because $\log(x - 2) + \log(x + 2) = \log[(x - 2)(x + 2)] = \log(x^2 - 4).$
2. True, because of the one-to-one property of logarithms.
3. True, because of the one-to-one property of exponential functions.
4. False, because the bases are different and the one-to-one property does not apply.
5. True, because
$\log_2(x^2 - 3x + 5) = 3$ is equivalent to
$x^2 - 3x + 5 = 2^3 = 8.$
6. True, $a^x = y$ is equivalent to $\log_a(y) = x.$
7. True, if $5^x = 23$, then $\ln(5^x) = \ln(23)$, or
$x \cdot \ln(5) = \ln(23).$
8. False, because $\log_3(5) = \frac{\ln(5)}{\ln(3)}.$
9. True, because $\log_6(2) = \frac{\ln(2)}{\ln(6)}$ and
$\log_6(2) = \frac{\log(2)}{\log(6)}.$
10. False, $\log(5) \approx 0.699$ and $\ln(5) \approx 1.609.$

12.4 EXERCISES

1. Equivalent to $\log_a(x) = y$ is $a^y = x.$
3. $\log_2(x + 1) = 3$
$x + 1 = 2^3$
$x + 1 = 8$
$x = 7$
The solution set is $\{7\}$.
5. $3\log_2(x + 1) - 2 = 13$
$3\log_2(x + 1) = 15$
$\log_2(x + 1) = 5$
$x + 1 = 2^5$
$x = 31$
The solution set is $\{31\}$.
7. $12 + 2\ln(x) = 14$
$2\ln(x) = 2$
$\ln(x) = 1$
$x = e^1$
The solution set is $\{e\}$.
9. $\log(x) + \log(5) = 1$
$\log(5x) = 1$
$5x = 10^1$
$x = 2$
The solution set is $\{2\}$.
11. $\log_2(x - 1) + \log_2(x + 1) = 3$
$\log_2[(x - 1)(x + 1)] = 3$
$\log_2(x^2 - 1) = 3$
$x^2 - 1 = 2^3$
$x^2 = 9$
$x = \pm 3$
If $x = -3$ in the original equation,
$\log_2(-3 - 1)$ is undefined. The solution set is $\{3\}$.
13. $\log_2(x - 1) - \log_2(x + 2) = 2$
$\log_2\left(\frac{x-1}{x+2}\right) = 2$
$\frac{x-1}{x+2} = 2^2$
$x - 1 = 4(x + 2)$
$x - 1 = 4x + 8$
$-9 = 3x$
$-3 = x$
If $x = -3$ in the original equation, we get the logarithm of a negative number, which is undefined. The solution set is $\emptyset$.
15. $\log_2(x - 4) + \log_2(x + 2) = 4$
$\log_2(x^2 - 2x - 8) = 4$
$x^2 - 2x - 8 = 2^4$
$x^2 - 2x - 8 = 16$
$x^2 - 2x - 24 = 0$
$(x - 6)(x + 4) = 0$
$x - 6 = 0$ or $x + 4 = 0$
$x = 6$ or $x = -4$

If $x = -4$ in the original equation, we get $\log_2(-8)$, which is undefined. The solution set is $\{6\}$.

17. $$\begin{aligned}\ln(x) + \ln(x+5) &= \ln(x+1) + \ln(x+3)\\ \ln(x^2+5x) &= \ln(x^2+4x+3)\\ x^2+5x &= x^2+4x+3\\ x &= 3\end{aligned}$$

The solution set is $\{3\}$.

19. $$\begin{aligned}\log(x+3) + \log(x+4) &= \log(x^3+13x^2) - \log(x)\\ \log(x^2+7x+12) &= \log(x^2+13x)\\ x^2+7x+12 &= x^2+13x\\ 12 &= 6x\\ 2 &= x\end{aligned}$$

The solution set is $\{2\}$.

21. $$\begin{aligned}2\cdot\log(x) &= \log(20-x)\\ \log(x^2) &= \log(20-x)\\ x^2 &= 20-x\\ x^2+x-20 &= 0\\ (x-4)(x+5) &= 0\\ x-4=0 \quad &\text{or} \quad x+5=0\\ x=4 \quad &\text{or} \quad x=-5\end{aligned}$$

If $x = -5$ in the original equation we get a logarithm of a negative number, which is undefined. The solution set is $\{4\}$.

23. $$\begin{aligned}3^x &= 7\\ x &= \log_3(7)\end{aligned}$$

The solution set is $\{\log_3(7)\}$.

25. $$\begin{aligned}e^{2x} &= 7\\ 2x &= \ln(7)\\ x &= \frac{\ln(7)}{2}\end{aligned}$$

The solution set is $\left\{\frac{\ln(7)}{2}\right\}$.

27. $$\begin{aligned}2^{3x+4} &= 4^{x-1}\\ 2^{3x+4} &= (2^2)^{x-1}\\ 2^{3x+4} &= (2)^{2x-2}\\ 3x+4 &= 2x-2\\ x &= -6\end{aligned}$$

The solution set is $\{-6\}$.

29. $$\begin{aligned}(1/3)^x &= 3^{1+x}\\ (3^{-1})^x &= 3^{1+x}\\ 3^{-x} &= 3^{1+x}\\ -x &= 1+x\\ -2x &= 1\\ x &= -1/2\end{aligned}$$

The solution set is $\left\{-\frac{1}{2}\right\}$.

31. $$\begin{aligned}2^x &= 3^{x+5}\\ \ln(2^x) &= \ln(3^{x+5})\\ x\cdot\ln(2) &= (x+5)\ln(3)\\ x\cdot\ln(2) &= x\cdot\ln(3) + 5\cdot\ln(3)\\ x\cdot\ln(2) - x\cdot\ln(3) &= 5\cdot\ln(3)\\ x(\ln(2)-\ln(3)) &= 5\cdot\ln(3)\\ x &= \frac{5\cdot\ln(3)}{\ln(2)-\ln(3)}\end{aligned}$$

This is the exact solution.

Use a calculator to find an approximate solution: $\frac{5\ln(3)}{\ln(2)-\ln(3)} \approx -13.548$.

33. $$\begin{aligned}5^{x+2} &= 10^{x-4}\\ \log(5^{x+2}) &= \log(10^{x-4})\\ (x+2)\log(5) &= x-4\\ x\cdot\log(5) + 2\cdot\log(5) &= x-4\\ x\cdot\log(5) - x &= -4 - 2\cdot\log(5)\\ x[\log(5)-1] &= -4 - 2\cdot\log(5)\\ x &= \frac{-4-2\cdot\log(5)}{\log(5)-1}\\ &= \frac{4+2\log(5)}{1-\log(5)}\\ &\approx 17.932\end{aligned}$$

35. $$\begin{aligned}8^x &= 9^{x-1}\\ \ln(8^x) &= \ln(9^{x-1})\\ x\cdot\ln(8) &= (x-1)\ln(9)\\ x\cdot\ln(8) &= x\cdot\ln(9) - \ln(9)\\ x\cdot\ln(8) - x\cdot\ln(9) &= -\ln(9)\\ x[\ln(8)-\ln(9)] &= -\ln(9)\\ x = \frac{-\ln(9)}{\ln(8)-\ln(9)} &= \frac{\ln(9)}{\ln(9)-\ln(8)}\\ &\approx 18.655\end{aligned}$$

37. $\log_2(3) = \frac{\ln(3)}{\ln(2)} \approx 1.5850$

39. $\log_3(1/2) = \frac{\ln(0.5)}{\ln(3)} \approx -0.6309$

41. $\log_{1/2}(4.6) = \frac{\ln(4.6)}{\ln(0.5)} \approx -2.2016$

43. $\log_{0.1}(0.03) = \frac{\ln(0.03)}{\ln(0.1)} \approx 1.5229$

45. $$\begin{aligned}x\cdot\ln(2) &= \ln(7)\\ x &= \frac{\ln(7)}{\ln(2)}\end{aligned}$$

This is the exact solution.

Use a calculator to find the approximate value of x, $x \approx 2.807$.

47. $3x - x \cdot \ln(2) = 1$

$$x(3 - \ln(2)) = 1$$

$$x = \frac{1}{3 - \ln(2)}$$

Exact solution.

Use a calculator to find an approximate value for x: $\frac{1}{3 - \ln(2)} \approx 0.433$.

49. $3^x = 5$

$$x = \log_3(5) = \frac{\ln(5)}{\ln(3)}$$

This is the exact solution.

Use a calculator to find an approximate value for x that satisfies the equation, $\frac{\ln(5)}{\ln(3)} \approx 1.465$.

51. $2^{x-1} = 9$

$$\ln(2^{x-1}) = \ln(9)$$

$$(x-1)\ln(2) = \ln(9)$$

$$x - 1 = \frac{\ln(9)}{\ln(2)}$$

$$x = 1 + \frac{\ln(9)}{\ln(2)} \quad \text{Exact solution}$$

Use a calculator to find an approximate value for x: $1 + \frac{\ln(9)}{\ln(2)} \approx 4.170$.

53. $3^x = 20$

$$x = \log_3(20) = \frac{\ln(20)}{\ln(3)} \approx 2.727$$

55. $\log_3(x) + \log_3(5) = 1$

$$\log_3(5x) = 1$$

$$5x = 3^1$$

$$x = \frac{3}{5}$$

57. $8^x = 2^{x+1}$

$$(2^3)^x = 2^{x+1}$$

$$2^{3x} = 2^{x+1}$$

$$3x = x + 1$$

$$2x = 1$$

$$x = \frac{1}{2}$$

59. $\log_2(1 - x) = 2$

$$2^2 = 1 - x$$

$$x = 1 - 4$$

$$x = -3$$

The solutions set is $\{-3\}$.

61. $\log_3(1 - x) + \log_3(2x + 13) = 3$

$$\log_3((1 - x)(2x + 13) = 3$$

$$(1 - x)(2x + 13) = 3^3$$

$$-2x^2 - 11x + 13 = 27$$

$$-2x^2 - 11x - 14 = 0$$

$$2x^2 + 11x + 14 = 0$$

$$(2x + 7)(x + 2) = 0$$

$$2x + 7 = 0 \quad \text{or} \quad x + 2 = 0$$

$$x = -\frac{7}{2} \quad \text{or} \quad x = -2$$

The solution set is $\left\{-\frac{7}{2}, -2\right\}$.

63. $\ln(2x - 1) - \ln(x + 1) = \ln(5)$

$$\ln((2x - 1)/(x + 1)) = \ln(5)$$

$$\frac{2x - 1}{x + 1} = 5$$

$$2x - 1 = 5(x + 1)$$

$$2x - 1 = 5x + 5$$

$$-3x = 6$$

$$x = -2$$

Since $\ln(-2 + 1)$ is a logarithm of a negative number, -2 does not satisfy the original equation. So the solution set is the empty set, $\emptyset$.

65. $\log_3(x - 14) - \log_3(x - 6) = 2$

$$\log_3((x - 14)/(x - 6)) = 2$$

$$\frac{x - 14}{x - 6} = 3^2$$

$$x - 14 = 9(x - 6)$$

$$x - 14 = 9x - 54$$

$$-8x = -40$$

$$x = 5$$

Since $\log_3(5 - 14)$ is a logarithm of a negative number, 5 does not satisfy the original equation. So the solution set is the empty set, $\emptyset$

67. $\log(x + 1) + \log(x - 2) = 1$

$$\log((x + 1)(x - 2)) = 1$$

$$(x + 1)(x - 2) = 10^1$$

$$x^2 - x - 2 = 10$$

$$x^2 - x - 12 = 0$$

$$(x - 4)(x + 3) = 0$$

$$x - 4 = 0 \quad \text{or} \quad x + 3 = 0$$

$$x = 4 \quad \text{or} \quad x = -3$$

Since $\log(-3 + 1)$ is a logarithm of a negative number -3 does not satisfy the original equation. The solution set is $\{4\}$.

69. $2 \cdot \ln(x) = \ln(2) + \ln(5x - 12)$

$$\ln(x^2) = \ln(10x - 24)$$

$$x^2 = 10x - 24$$

$$x^2 - 10x + 24 = 0$$

$$(x - 4)(x - 6) = 0$$

$$x - 4 = 0 \quad \text{or} \quad x - 6 = 0$$

$$x = 4 \quad \text{or} \quad x = 6$$

The solution set is $\{4, 6\}$.

71. $\log_3(x^3+16x^2) - \log_3(x) = \log_3(36)$

$$\log_3((x^3+16x^2)/x) = \log_3(36)$$
$$\frac{x^3+16x^2}{x} = 36$$
$$x^3+16x^2 = 36x$$
$$x^3+16x^2-36x = 0$$
$$x(x^2+16x-36) = 0$$
$$x(x+18)(x-2) = 0$$
$$x = 0 \quad \text{or} \quad x = -18 \quad \text{or} \quad x = 2$$

Since $\log_3(x)$ is undefined if $x = 0$ or $x = -18$, the solution set is $\{2\}$.

73. $\log(x) + \log(x+5) = 2 \cdot \log(x+2)$

$$\log(x(x+5)) = \log((x+2)^2)$$
$$x^2+5x = x^2+4x+4$$
$$x = 4$$

The solution set is $\{4\}$.

75. $\log_7(x^2+6x+8) - \log_7(x+2) = \log_7(3)$

$$\log_7\left((x^2+6x+8)/(x+2)\right) = \log_7(3)$$
$$\frac{x^2+6x+8}{x+2} = 3$$
$$x^2+6x+8 = 3x+6$$
$$x^2+3x+2 = 0$$
$$(x+2)(x+1) = 0$$
$$x+2 = 0 \quad \text{or} \quad x+1 = 0$$
$$x = -2 \quad \text{or} \quad x = -1$$

If $x = -2$, then $\log_7(x+2)$ is undefined. So the solution set is $\{-1\}$.

77. $\ln(6) + 2 \cdot \ln(x) = \ln(38x-30) - \ln(2)$

$$\ln(6x^2) = \ln((38x-30)/2)$$
$$\ln(6x^2) = \ln(19x-15)$$
$$6x^2 = 19x-15$$
$$6x^2-19x+15 = 0$$
$$(2x-3)(3x-5) = 0$$
$$2x-3 = 0 \quad \text{or} \quad 3x-5 = 0$$
$$x = \frac{3}{2} \quad \text{or} \quad x = \frac{5}{3}$$

The solution set is $\left\{\frac{3}{2}, \frac{5}{3}\right\}$.

79. Use the formula $S = P(1+i)^n$.

$$1500 = 1000(1+0.01)^n$$
$$1.5 = (1.01)^n$$
$$n = \log_{1.01}(1.5) = \frac{\ln(1.5)}{\ln(1.01)}$$
$$\approx 40.749$$

It takes approximately 41 months.

81. Use the formula $S = P(1+i)^n$.

$$105 = 100(1+0.03/365)^n$$
$$1.05 = (1.00008)^n$$
$$n = \log_{1.00008}(1.05) = \frac{\ln(1.05)}{\ln(1.00008)} \approx 593.6$$

It takes approximately 594 days.

83. If $t = 0$, then

$$A = 10e^{-0.0001(0)} = 10 \text{ grams.}$$

If $A = 4$, then

$$4 = 10e^{-0.0001t}$$
$$0.4 = e^{-0.0001t}$$
$$-0.0001t = \ln(0.4)$$
$$t = \frac{\ln(0.4)}{-0.0001} \approx 9162.9$$

The cloth was made about 9163 years ago.

85. $y = 114.308e^{(0.265 \cdot 15.8)} \approx 7524 \text{ ft}^3/\text{sec}$

87.

$$40 = 28e^{0.05t}$$
$$\frac{10}{7} = e^{0.05t}$$
$$0.05t = \ln(10/7)$$
$$t = \frac{\ln(10/7)}{0.05} \approx 7.133$$

There will be 40 million people above the poverty level in approximately 7.1 years.

89.

$$28e^{0.05t} = 20e^{0.07t}$$
$$\frac{e^{0.05t}}{e^{0.07t}} = \frac{20}{28}$$
$$e^{-0.02t} = 5/7$$
$$-0.02t = \ln(5/7)$$
$$t = \frac{\ln(5/7)}{-0.02} \approx 16.824$$

The number of people above the poverty level will equal the number below the poverty level in approximately 16.8 years.

91.

$$\text{pH} = -\log(\text{H}^+)$$
$$3.7 = -\log(\text{H}^+)$$
$$-3.7 = \log(\text{H}^+)$$
$$\text{H}^+ = 10^{-3.7} \approx 2.0 \times 10^{-4}$$

93. $d = \log_2\left(\frac{3\sqrt[3]{2}}{2}\right) = \frac{\ln\left(\frac{3\sqrt[3]{2}}{2}\right)}{\ln(2)}$

$$\approx 0.9183$$

95. Using logarithms:

$$x^3 = 12$$
$$3\ln(x) = \ln(12)$$
$$\ln(x) = \ln(12)/3$$
$$x = e^{\ln(12)/3} \approx 2.2894$$

Using roots:

$$x^3 = 12$$
$$x = \sqrt[3]{12} = 12^{1/3} \approx 2.2894$$

97. (2.71, 6.54)

99. (1.03, 0.04), (4.74, 2.24)

Enriching Your Mathematical Word Power

1. a **2.** d **3.** b **4.** d **5.** d
6. b **7.** a **8.** b **9.** b **10.** c

CHAPTER 12 REVIEW

1. $f(-2) = 5^{-2} = \frac{1}{5^2} = \frac{1}{25}$

3. $f(3) = 5^3 = 125$

5. $g(1) = 10^{1-1} = 10^0 = 1$

7. $g(0) = 10^{0-1} = 10^{-1} = \frac{1}{10}$

9. $h(-1) = (1/4)^{-1} = 4$

11. $h(1/2) = (1/4)^{1/2} = \sqrt{\frac{1}{4}} = \frac{1}{2}$

13. If $f(x) = 25$, then $5^x = 25$, $x = 2$.

15. If $g(x) = 1000$, then $10^{x-1} = 1000$.

$$10^{x-1} = 10^3$$
$$x - 1 = 3$$
$$x = 4$$

17. If $h(x) = 32$, then $(1/4)^x = 32$.

$$(2^{-2})^x = 2^5$$
$$2^{-2x} = 2^5$$
$$-2x = 5$$
$$x = -\frac{5}{2}$$

19. If $h(x) = 1/16$, then $(1/4)^x = 1/16$, or $(1/4)^x = (1/4)^2$, $x = 2$.

21. $f(1.34) = 5^{1.34} \approx 8.6421$

23. $g(3.25) = 10^{3.25-1} = 10^{2.25} \approx 177.828$

25. $h(2.82) = (1/4)^{2.82} = (0.25)^{2.82}$
≈ 0.02005

27. $h(\sqrt{2}) = (1/4)^{\sqrt{2}} = \approx 0.1408$

29. The graph of $f(x) = 5^x$ includes the points $(0, 1)$, $(1, 5)$, and $(-1, 1/5)$.

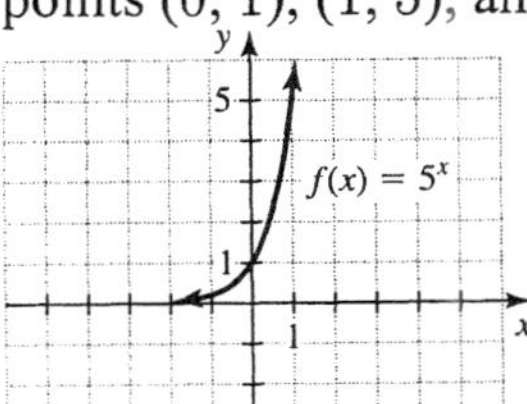

31. The graph of $y = (1/5)^x$ includes the points $(1, 1/5)$, $(0, 1)$ and $(-1, 5)$.

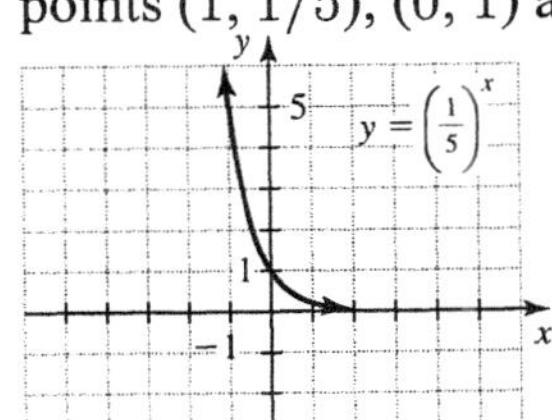

33. The graph of $y = 3^{-x}$ includes the points $(0, 1)$, $(1, 1/3)$, and $(-1, 3)$.

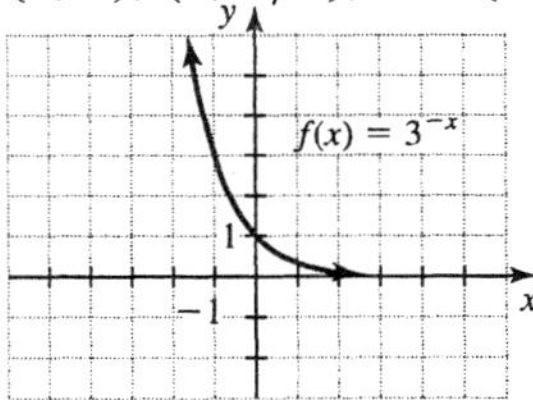

35. The graph of $y = 1 + 2^x$ includes the points $(0, 2)$, $(1, 3)$, $(2, 5)$, and $(-1, 1.5)$.

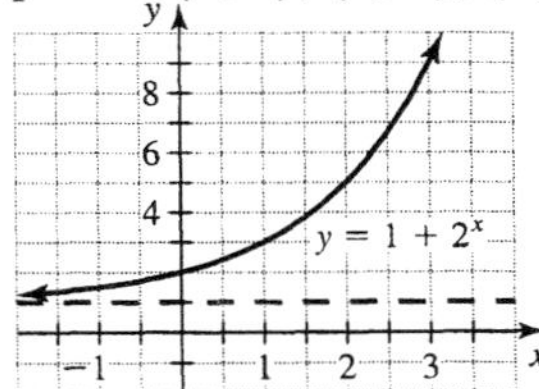

37. $\log(n) = m$

39. $k^h = t$

41. $f(1/8) = \log_2(1/8) = -3$, because $2^{-3} = 1/8$.

43. $g(0.1) = \log(0.1) = -1$, because $10^{-1} = 0.1$.

45. $g(100) = \log(100) = 2$, because $10^2 = 100$.

47. $h(1) = \log_{1/2}(1) = 0$, because $(1/2)^0 = 1$.

49. If $f(x) = 8$, then $\log_2(x) = 8$, $x = 2^8 = 256$.

51. $f(77) = \log_2(77) = \frac{\ln(77)}{\ln(2)} \approx 6.267$

53. $h(33.9) = \log_{1/2}(33.9) = \frac{\ln(33.9)}{\ln(0.5)}$
≈ -5.083

55. If $f(x) = 2.475$, then $\log_2(x) \approx 2.475$.
$x = 2^{2.475} \approx 5.560$

57. The inverse of the function $f(x) = 10^x$ is the base 10 logarithm function, $f^{-1}(x) = \log(x)$.
The graph of $f(x)$ includes the points $(1, 10)$, $(0, 1)$, and $(-1, 0.1)$. The graph of $f^{-1}(x)$ includes the points $(10, 1)$, $(1, 0)$, and $(0.1, -1)$.

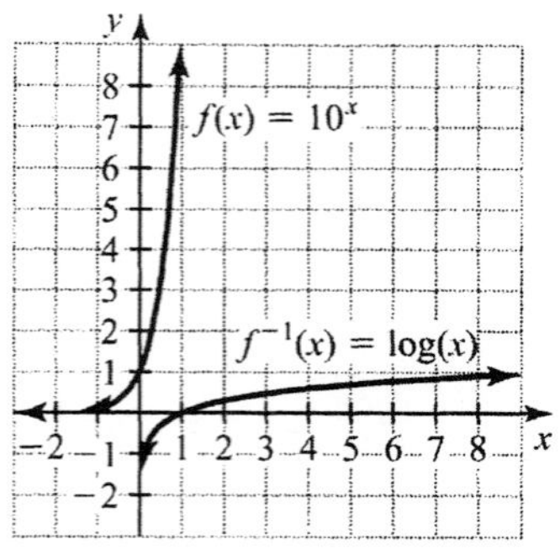

59. The inverse of the function $f(x) = e^x$ is $f^{-1}(x) = \ln(x)$. The graph of $f(x)$ includes the points (1, *e*), (0, 1), and $(-1, 1/e)$. The graph of $f^{-1}(x)$ includes the points (*e*, 1), (1, 0), and $(1/e, -1)$.

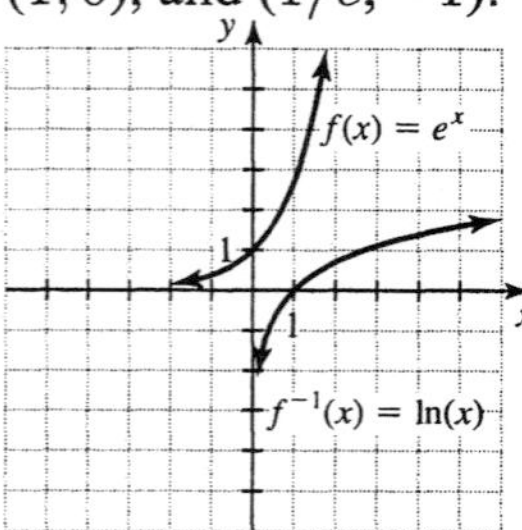

61. $\log(x^2y) = \log(x^2) + \log(y)$
$= 2\log(x) + \log(y)$

63. $\ln(16) = \ln(2^4) = 4\ln(2)$

65. $\log_5(1/x) = \log_5(1) - \log_5(x)$
$= -\log_5(x)$

67. $\frac{1}{2}\log(x+2) - 2\log(x-1)$
$= \log\left((x+2)^{1/2}\right) - \log\left((x-1)^2\right)$
$= \log\left(\frac{\sqrt{x+2}}{(x-1)^2}\right)$

69.
$$\log_2(x) = 8$$
$$x = 2^8 = 256$$
The solution set is $\{256\}$.

71.
$$\log_2(8) = x$$
$$3 = x$$
The solution set is $\{3\}$.

73.
$$x^3 = 8$$
$$x = \sqrt[3]{8} = 2$$
The solution set is $\{2\}$.

75.
$$\log_x(27) = 3$$
$$x^3 = 27$$
$$x = \sqrt[3]{27} = 3$$
The solution set is $\{3\}$.

77.
$$x \cdot \ln(3) - x = \ln(7)$$
$$x[\ln(3) - 1] = \ln(7)$$
$$x = \frac{\ln(7)}{\ln(3) - 1}$$
The solution set is $\left\{\frac{\ln(7)}{\ln(3)-1}\right\}$.

79.
$$3^x = 5^{x-1}$$
$$\ln(3^x) = \ln(5^{x-1})$$
$$x \cdot \ln(3) = (x-1)\ln(5)$$
$$x \cdot \ln(3) = x \cdot \ln(5) - \ln(5)$$
$$x \cdot \ln(3) - x \cdot \ln(5) = -\ln(5)$$
$$x[\ln(3) - \ln(5)] = -\ln(5)$$
$$x = \frac{-\ln(5)}{\ln(3) - \ln(5)} = \frac{\ln(5)}{\ln(5) - \ln(3)}$$
The solution set is $\left\{\frac{\ln(5)}{\ln(5)-\ln(3)}\right\}$.

81.
$$4^{2x} = 2^{x+1}$$
$$(2^2)^{2x} = 2^{x+1}$$
$$2^{4x} = 2^{x+1}$$
$$4x = x + 1$$
$$3x = 1$$
$$x = \frac{1}{3}$$
The solution set is $\left\{\frac{1}{3}\right\}$.

83.
$$\ln(x+2) - \ln(x-10) = \ln(2)$$
$$\ln\left(\frac{x+2}{x-10}\right) = \ln(2)$$
$$\frac{x+2}{x-10} = 2$$
$$x + 2 = 2(x - 10)$$
$$x + 2 = 2x - 20$$
$$22 = x$$
The solution set is $\{22\}$.

85.
$$\log(x) - \log(x-2) = 2$$
$$\log\left(\frac{x}{x-2}\right) = 2$$
$$\frac{x}{x-2} = 10^2$$
$$x = 100(x-2)$$
$$x = 100x - 200$$
$$-99x = -200$$
$$x = \frac{-200}{-99} = \frac{200}{99}$$
The solution set is $\left\{\frac{200}{99}\right\}$.

87.
$$6^x = 12$$
$$x = \log_6(12) = \frac{\ln(12)}{\ln(6)} \approx 1.3869$$
The solution set is $\{1.3869\}$.

89.
$$3^{x+1} = 5$$
$$x + 1 = \log_3(5)$$
$$x = -1 + \log_3(5) = -1 + \frac{\ln(5)}{\ln(3)}$$
$$\approx 0.4650$$
The solution set is $\{0.4650\}$.

91. Use the formula $S = P(1+i)^n$ with $i = 11.5\% = 0.115$, $n = 15$, and $P = \$10{,}000$.
$S = 10{,}000(1.115)^{15} \approx \$51{,}182.68$
93. Use the formula $A = A_0 e^{-0.0003t}$ with $A_0 = 218$ and $t = 1000$.
$$A = 218e^{-0.0003(1000)} = 218e^{-0.3}$$
$$\approx 161.5 \text{ grams}$$
95. The amount in Melissa's account is given by the formula $S = 1000(1.05)^t$ for any number of years t. The amount in Frank's account is given by the formula $S = 900e^{0.07t}$ for any number of years t. To find when they have the same amount, we set the two expressions equal and solve for t.
$$1000(1.05)^t = 900e^{0.07t}$$
$$\ln(1000(1.05)^t) = \ln(900e^{0.07t})$$
$$\ln(1000) + t \cdot \ln(1.05) = \ln(900) + \ln(e^{0.07t})$$
$$\ln(1000) + t \cdot \ln(1.05) = \ln(900) + 0.07t$$
$$t \cdot \ln(1.05) - 0.07t = \ln(900) - \ln(1000)$$
$$t[\ln(1.05) - 0.07] = \ln(900) - \ln(1000)$$
$$t = \frac{\ln(900) - \ln(1000)}{\ln(1.05) - 0.07} \approx 4.9675$$
The amounts will be equal in approximately 5 years.
97. $114.308e^{0.265(20.6-6.87)} \approx 4347.5 \text{ ft}^3/\text{sec}$

CHAPTER 12 TEST

1. $f(2) = 5^2 = 25$
2. $f(-1) = 5^{-1} = \frac{1}{5}$
3. $f(0) = 5^0 = 1$
4. $g(125) = \log_5(125) = 3$, because $5^3 = 125$.
5. $g(1) = \log_5(1) = 0$, because $5^0 = 1$.
6. $g(1/5) = \log_5(1/5) = -1$, because $5^{-1} = 1/5$.
7. The graph of $y = 2^x$ includes the points (0, 1), (1, 2), (2, 4), and (−1, 1/2).

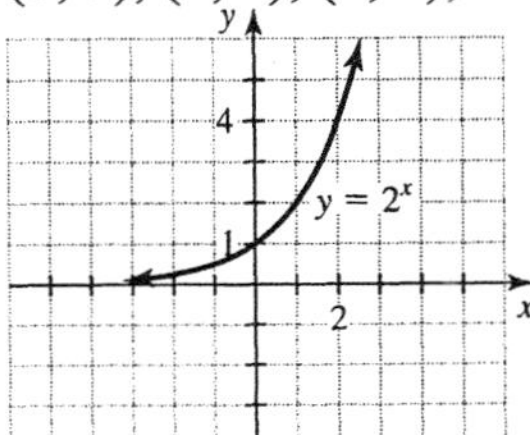

8. The graph of $f(x) = \log_2(x)$ includes the points (1, 0), (2, 1), (4, 2), and (1/2, −1).

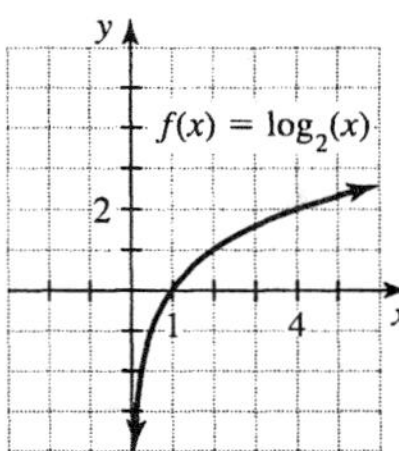

9. The graph of $y = \left(\frac{1}{3}\right)^x$ includes the points (1, 1/3), (0, 1), and (−1, 3).

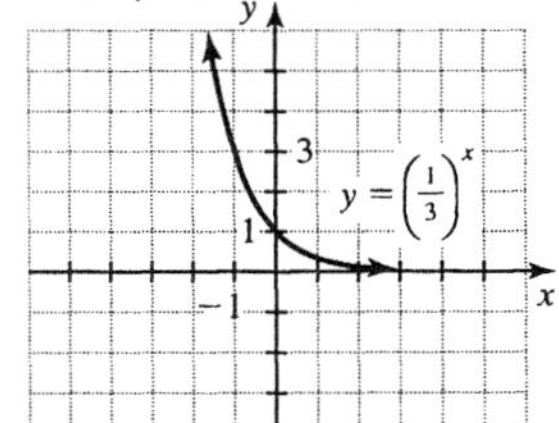

10. The graph of $g(x) = \log_{1/3}(x)$ includes the points (1, 0), (3, −1), and (1/3, 1).

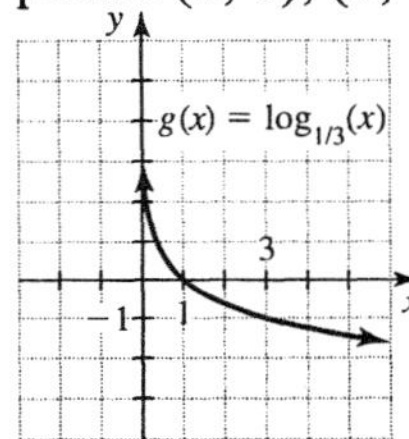

11. $\log_a(MN) = \log_a(M) + \log_a(N) = 6 + 4 = 10$
12. $\log_a(M^2/N) = 2 \cdot \log_a(M) - \log_a(N) = 2 \cdot 6 - 4 = 8$
13. $\dfrac{\log_a(M)}{\log_a(N)} = \dfrac{6}{4} = \dfrac{3}{2}$
14. $\log_a(a^3M^2) = 3 \cdot \log_a(a) + 2 \cdot \log_a(M) = 3 \cdot 1 + 2 \cdot 6 = 15$
15. $\log_a(1/N) = \log_a(1) - \log_a(N) = 0 - 4 = -4$
16. $3^x = 12$
$x = \log_3(12)$
The solution set is $\{\log_3(12)\}$ or $\{\ln(12)/\ln(3)\}$.
17. $\log_3(x) = 1/2$
$x = 3^{1/2} = \sqrt{3}$
The solution set is $\{\sqrt{3}\}$.
18.
$$5^x = 8^{x-1}$$
$$\ln(5^x) = \ln(8^{x-1})$$
$$x \cdot \ln(5) = (x-1)\ln(8)$$
$$x \cdot \ln(5) = x \cdot \ln(8) - \ln(8)$$
$$x \cdot \ln(5) - x \cdot \ln(8) = -\ln(8)$$

$$x[\ln(5) - \ln(8)] = -\ln(8)$$
$$x = \frac{-\ln(8)}{\ln(5) - \ln(8)} = \frac{\ln(8)}{-\ln(5) + \ln(8)}$$
The solution set is $\left\{ \frac{\ln(8)}{\ln(8) - \ln(5)} \right\}$.

19. $\log(x) + \log(x + 15) = 2$
$$\log(x^2 + 15x) = 2$$
$$x^2 + 15x = 10^2$$
$$x^2 + 15x - 100 = 0$$
$$(x + 20)(x - 5) = 0$$
$$x + 20 = 0 \quad \text{or} \quad x - 5 = 0$$
$$x = -20 \quad \text{or} \quad x = 5$$
If $x = -20$ in the original equation, we get a logarithm of a negative number, which is undefined. So the solution set is $\{5\}$.

20. $2 \cdot \ln(x) = \ln(3) + \ln(6 - x)$
$$\ln(x^2) = \ln(18 - 3x)$$
$$x^2 = 18 - 3x$$
$$x^2 + 3x - 18 = 0$$
$$(x + 6)(x - 3) = 0$$
$$x + 6 = 0 \quad \text{or} \quad x - 3 = 0$$
$$x = -6 \quad \text{or} \quad x = 3$$
If $x = -6$ in the original equation, we get a logarithm of a negative number, which is undefined. So the solution set is $\{3\}$.

21. $20^x = 5$
$$x = \log_{20}(5) = \frac{\ln(5)}{\ln(20)} \approx 0.5372$$
The solution set is $\{0.5372\}$.

22. $\log_3(x) = 2.75$
$$x = 3^{2.75} \approx 20.5156$$
The solution set is $\{20.5156\}$

23. To find the number present initially, let $t = 0$ in the formula:
$N = 10e^{0.4(0)} = 10e^0 = 10 \cdot 1 = 10$
To find the number present after 24 hours, let $t = 24$ in the formula:
$N = 10e^{0.4(24)} = 10e^{9.6} \approx 147{,}648$

24. To find how long it takes for the population to double, we must find the value of t for which $e^{0.4t} = 2$. Solve for t.
$$0.4t = \ln(2)$$
$$t = \frac{\ln(2)}{0.4} \approx 1.733$$
The bacteria population doubles in 1.733 hours.

Making Connections

Chapters 1 - 12

1. $(x - 3)^2 = 8$
$$x - 3 = \pm\sqrt{8}$$
$$x = 3 \pm 2\sqrt{2}$$
The solution set is $\left\{3 \pm 2\sqrt{2}\right\}$.

2. $\log_2(x - 3) = 8$
$$x - 3 = 2^8$$
$$x = 256 + 3 = 259$$
The solution set is $\{259\}$.

3. $2^{x-3} = 8$
$$2^{x-3} = 2^3$$
$$x - 3 = 3$$
$$x = 6$$
The solution set is $\{6\}$.

4. $2x - 3 = 8$
$$2x = 11$$
$$x = \frac{11}{2}$$
The solution set is $\left\{\frac{11}{2}\right\}$.

5. $|\, x - 3 \,| = 8$
$$x - 3 = 8 \quad \text{or} \quad x - 3 = -8$$
$$x = 11 \quad \text{or} \quad x = -5$$
The solution set is $\{-5, 11\}$

6. $\sqrt{x - 3} = 8$
$$(\sqrt{x - 3})^2 = 8^2$$
$$x - 3 = 64$$
$$x = 67$$
The solution set is $\{67\}$.

7. $\log_2(x - 3) + \log_2(x) = \log_2(18)$
$$\log_2(x^2 - 3x) = \log_2(18)$$
$$x^2 - 3x = 18$$
$$x^2 - 3x - 18 = 0$$
$$(x - 6)(x + 3) = 0$$
$$x - 6 = 0 \quad \text{or} \quad x + 3 = 0$$
$$x = 6 \quad \text{or} \quad x = -3$$
If $x = -3$ in the original equation, then we get a logarithm of a negative number, which is undefined. The solution set is $\{6\}$.

8. $2 \cdot \log_2(x - 3) = \log_2(5 - x)$
$$\log_2[(x - 3)^2] = \log_2(5 - x)$$
$$(x - 3)^2 = 5 - x$$
$$x^2 - 6x + 9 = 5 - x$$
$$x^2 - 5x + 4 = 0$$
$$(x - 4)(x - 1) = 0$$
$$x - 4 = 0 \quad \text{or} \quad x - 1 = 0$$
$$x = 4 \quad \text{or} \quad x = 1$$

If $x = 1$ in the original equation, we get a logarithm of a negative number. The solution set is $\{4\}$.

9.
$$\frac{1}{2}x - \frac{2}{3} = \frac{3}{4}x + \frac{1}{5}$$
$$60\left(\frac{1}{2}x - \frac{2}{3}\right) = 60\left(\frac{3}{4}x + \frac{1}{5}\right)$$
$$30x - 40 = 45x + 12$$
$$-52 = 15x$$
$$-\frac{52}{15} = x$$
The solution set is $\left\{-\frac{52}{15}\right\}$.

10. To solve $3x^2 - 6x + 2 = 0$ use the quadratic formula.
$$x = \frac{6 \pm \sqrt{(-6)^2 - 4(3)(2)}}{2(3)} = \frac{6 \pm \sqrt{12}}{6}$$
$$= \frac{6 \pm 2\sqrt{3}}{6} = \frac{3 \pm \sqrt{3}}{3}$$

The solution set is $\left\{\frac{3 \pm \sqrt{3}}{3}\right\}$.

11. The inverse of dividing by 3 is multiplying by 3. So $f^{-1}(x) = 3x$.

12. The inverse of the base 3 logarithm function is the base 3 exponential function, $1g^{-1}(x) = 3^x$.

13. The inverse of multiplying by 2 and then subtracting 4 is adding 4 and then dividing by 2, $f^{-1}(x) = \frac{x+4}{2}$.

14. The inverse of the square root function is the squaring function. To keep the domain of one function equal to the range of the inverse function we must restrict the squaring function to the nonnegative numbers:
$h^{-1}(x) = x^2$ for $x \geq 0$

15. The reciprocal function is its own inverse, $j^{-1}(x) = \frac{1}{x}$.

16. The inverse of the base 5 exponential function is the base 5 logarithm function, $k^{-1}(x) = \log_5(x)$

17. We will find the inverse for m by using the technique of interchanging x and y and then solving for y.

$$m(x) = e^{x-1}$$
$$y = e^{x-1}$$
$$x = e^{y-1}$$
$$y - 1 = \ln(x)$$
$$y = 1 + \ln(x)$$
$$m^{-1}(x) = 1 + \ln(x)$$

18. The inverse of the natural logarithm function is the base e exponential function, $n^{-1}(x) = e^x$.

19. The graph of $y = 2x$ is a straight line with slope 2 and y-intercept (0, 0). Start at the origin and rise 2 and go 1 to the right to find a second point on the line.

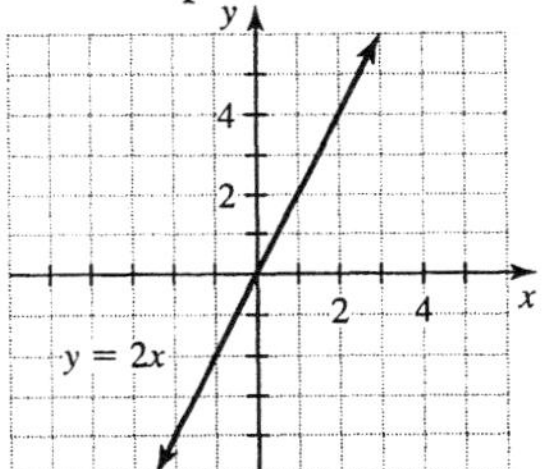

20. The graph of $y = 2^x$ includes the points (0, 1), (1, 2), (2, 2), and (−1, 1/2).

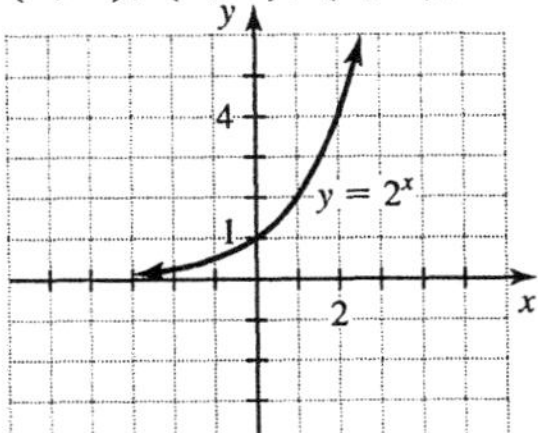

21. The graph of $y = x^2$ is a parabola through (0, 0), (1, 1), (2, 4), (−1, 1), and (−2, 4).

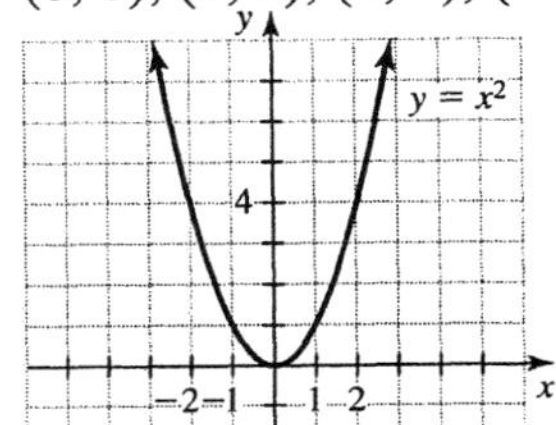

22. The graph of $y = \log_2(x)$ includes the points (1, 0), (2, 1), (1/2, −1) and (4, 2).

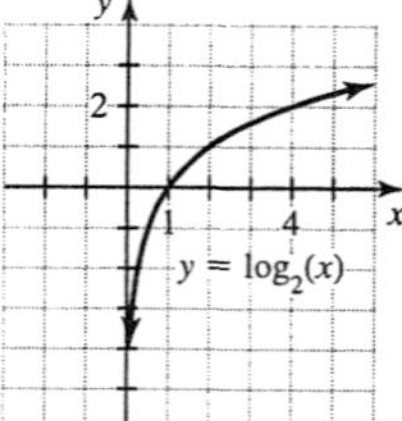

23. The graph of $y = \frac{1}{2}x - 4$ is a straight line with slope 1/2 and y-intercept (0, −4). Start at (0, −4), rise 1 and go 2 to the right to locate a second point on the line.

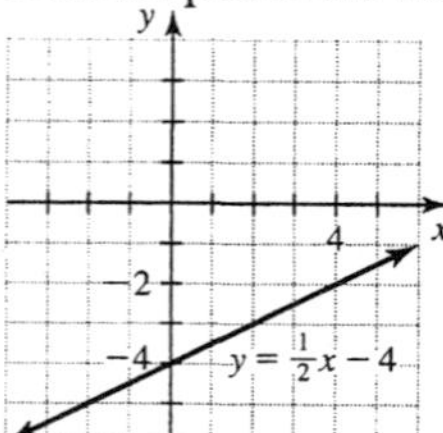

24. The graph of $y = |\,2 - x\,|$ is a v-shaped graph through (−2, 4), (−1, 3), (0, 2), (1, 1), (2, 0), (3, 1), (4, 2), and (5, 3).

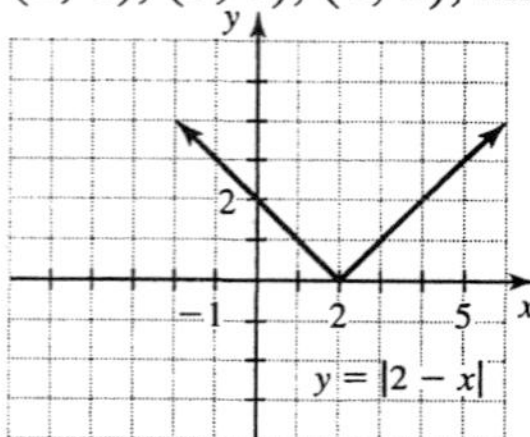

25. The graph of $y = 2 - x^2$ is a parabola with a vertex at (0, 2). It also includes the points (1, 1), (−1, 1), (2, −2), and (−2, −2).

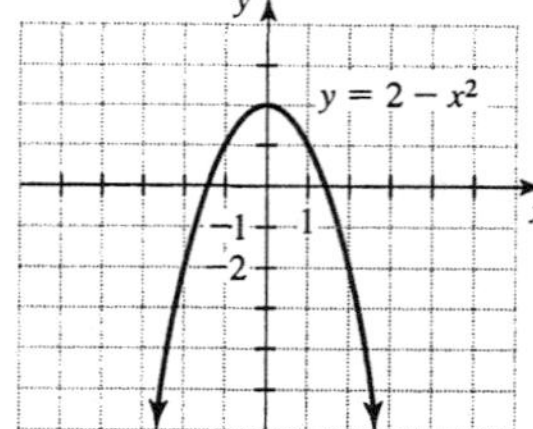

26. Note that e^2 is not a variable. The value of e^2 is approximately 7.389. The graph of $y = e^2$ is a straight line with 0 slope and y-intercept (0, 7.389).

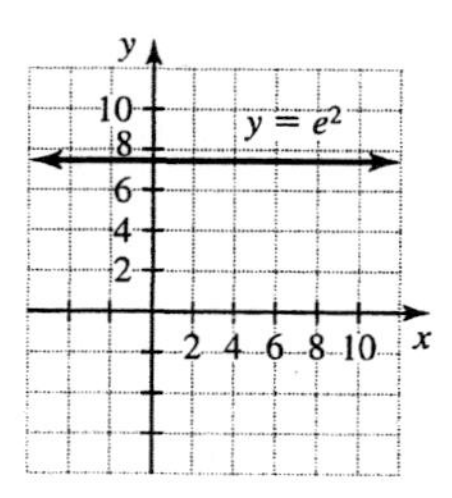

27. a) The graph of $n = 1.51t + 125.5$ is a straight line through (0, 125.5) and (10, 140.6). The graph of $n = 125.6e^{0.011t}$ is an exponential curve through (0, 125.6), (5, 132.7), (10, 140.2), and (15, 148.1).

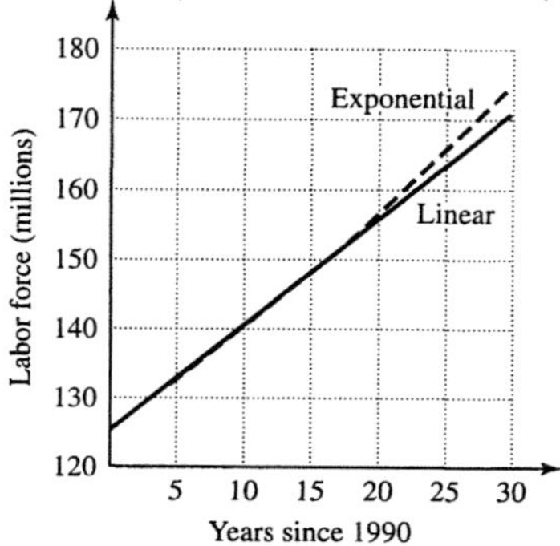

b) In 2010, $t = 20$:
$n = 1.51(20) + 125.5 \approx 155.7$ million
or $n = 125.6e^{0.011(20)} \approx 156.5$ million
c) Look up current data at www.bls.gov.

28. a) Use $D = RT$ to get $2d_1 = v(0.270)$ or $d_1 = 0.135v$.
b) Using $D = RT$ we get $2d_2 = v(0.432)$ or $d_2 = 0.216v$.
c) Using the Pythagorean theorem we get $(d_2)^2 = (d_1)^2 + 250^2$. Use $d_1 = 0.135v$ and $d_2 = 0.216v$ to get

$$\begin{aligned}(0.216v)^2 &= (0.135v)^2 + 250^2\\ 0.028431v^2 &= 250^2\\ v^2 &= 2198304.667\\ v &= 1482.668091 \text{ m/sec}\end{aligned}$$

$d_1 = 0.135v \approx 200.2$ meters

13.1 WARM-UPS

1. True, because any equation of the form $y = ax^2 + bx + c$ has a graph that is a parabola.
2. False, because absolute value has a v-shaped graph.
3. False, because $-4 \neq \sqrt{5(3)+1}$.
4. True, because $y = \sqrt{x}$ lies entirely in the first quadrant, and $y = -x - 2$ has no points in the first quadrant.
5. False, because we can also use addition.
6. True.
7. True, because a 30-60-90 triangle is half of an equilateral triangle.
8. True, because for any rectangular solid we have $V = LWH$.
9. True, because the surface area consists of 6 rectangles, of which two have area LW, two have area WH, and two have area LH.
10. True, because the area of a triangle is $(bh)/2$.

13.1 EXERCISES

1. If the graph of an equation is not a straight line then it is called nonlinear.
3. Graphing is not an accurate method for solving a system and the graphs might be difficult to draw.
5. The graph of $y = x^2$ is a parabola and the graph of $x + y = 6$ is a straight line.

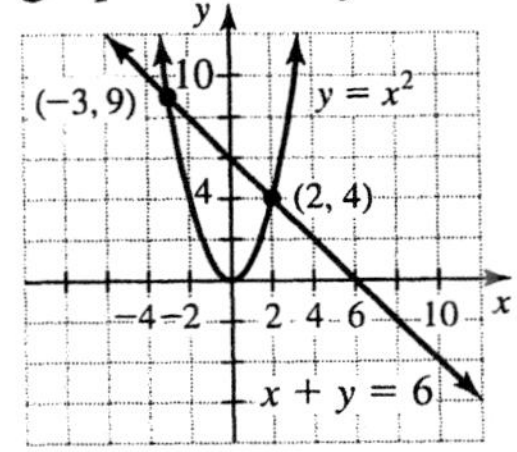

To solve the system, substitute $y = x^2$ into $x + y = 6$.

$$x + x^2 = 6$$
$$x^2 + x - 6 = 0$$
$$(x+3)(x-2) = 0$$
$$x = -3 \text{ or } x = 2$$

If $x = -3$, $y = (-3)^2 = 9$, and if $x = 2$, $y = 2^2 = 4$. The solution set to the system is $\{(2,4), (-3,9)\}$.

7. The graph of $y = |\,x\,|$ is v-shaped, and the graph of $2y - x = 6$ is a straight line. To solve the system, substitute $x = 2y - 6$ into $y = |\,x\,|$.

$$y = |\,2y - 6\,|$$
$$y = 2y - 6 \text{ or } y = -(2y - 6)$$
$$6 = y \quad \text{or} \quad y = -2y + 6$$
$$3y = 6$$
$$y = 2$$

Use $y = 6$ in $x = 2y - 6$ to get $x = 6$. Use $y = 2$ in $x = 2y - 6$ to get $x = -2$. The graphs intersect at the points $(-2, 2)$ and $(6, 6)$.

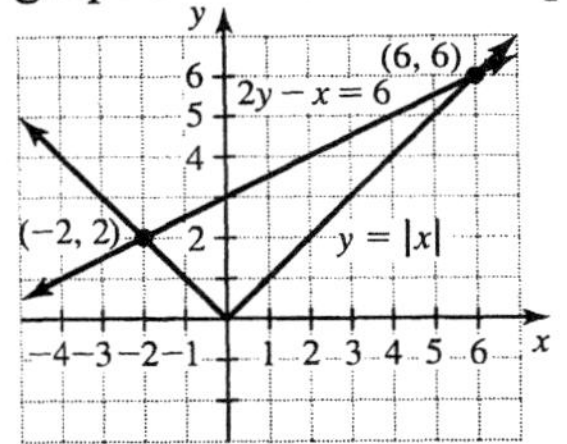

The solution set is $\{(-2,2), (6,6)\}$.

9. The graph of $y = \sqrt{2x}$ includes the points $(0,0)$, $(2,2)$, and $(4.5, 3)$. The graph of $x - y = 4$ is a straight line with slope 1 and y-intercept $(0, -4)$.

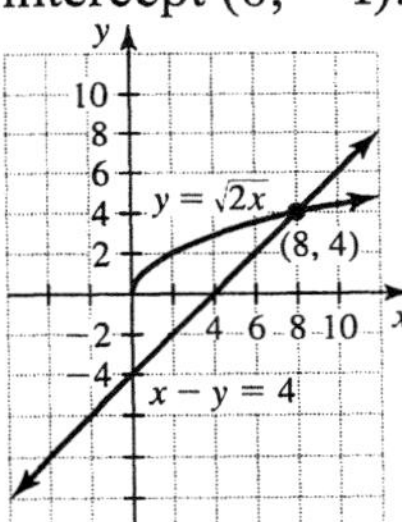

Substitute $y = \sqrt{2x}$ into $y = x - 4$.

$$\sqrt{2x} = x - 4$$
$$(\sqrt{2x})^2 = (x-4)^2$$
$$2x = x^2 - 8x + 16$$
$$0 = x^2 - 10x + 16$$
$$0 = (x-8)(x-2)$$
$$x = 8 \quad \text{or} \quad x = 2$$
$$y = \sqrt{2(8)} \text{ or } y = \sqrt{2(2)}$$
$$= 4 \quad \text{or} \quad = 2$$

Since we squared both sides, we must check. The pair $(8, 4)$ satisfies both equations, but $(2, 2)$ does not. The solution set to the system is $\{(8, 4)\}$.

11. The graph of $4x - 9y = 9$ is a straight line. The equation $xy = 1$ can be written as $y = 1/x$. Substitute $y = 1/x$ into $4x - 9y = 9$.

$$4x - 9\left(\frac{1}{x}\right) = 9$$
$$4x^2 - 9 = 9x$$
$$4x^2 - 9x - 9 = 0$$
$$(4x + 3)(x - 3) = 0$$
$$x = -\frac{3}{4} \quad \text{or} \quad x = 3$$
$$y = \frac{1}{-3/4} = -\frac{4}{3} \qquad y = \frac{1}{3}$$

The solution set is $\left\{\left(-\frac{3}{4}, -\frac{4}{3}\right), \left(3, \frac{1}{3}\right)\right\}$.

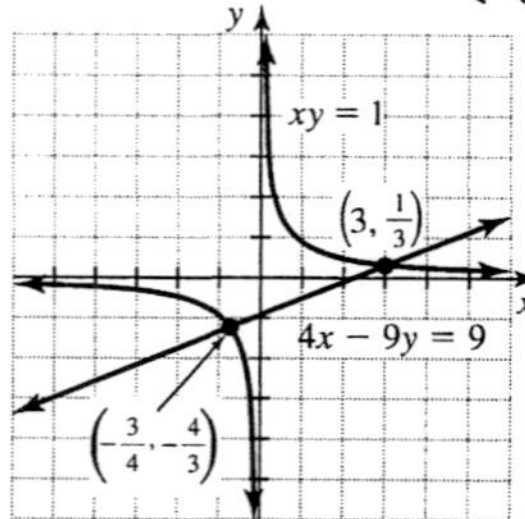

13. The graph of $y = x^2$ is a parabola opening upward, and the graph of $y = -x^2 + 1$ is a parabola opening downward.

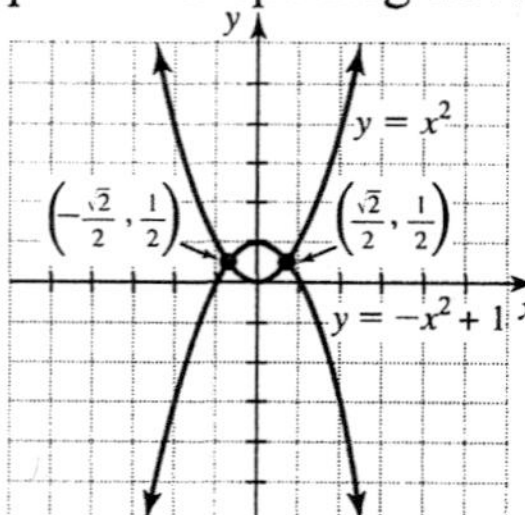

Substitute $y = x^2$ into $y = -x^2 + 1$.

$$x^2 = -x^2 + 1$$
$$2x^2 = 1$$
$$x^2 = \frac{1}{2}$$
$$x = \pm\sqrt{\frac{1}{2}} = \pm\frac{\sqrt{2}}{2}$$

Since $y = x^2$, $y = \frac{1}{2}$ for either value of x. The solution set is $\left\{\left(\frac{\sqrt{2}}{2}, \frac{1}{2}\right), \left(-\frac{\sqrt{2}}{2}, \frac{1}{2}\right)\right\}$.

15. Substitute $y = x$ into $xy = 1$:

$$xy = 1$$
$$xx = 1$$
$$x^2 = 1$$
$$x = \pm 1$$

Use $y = x$ to get $y = \pm 1$. The solution set is $\{(-1, -1), (1, 1)\}$.

17. Substitute $y = 2$ into $y = x^2$:

$$y = x^2$$
$$2 = x^2$$
$$x = \pm\sqrt{2}$$
$$x = -\sqrt{2} \quad \text{or} \quad x = \sqrt{2}$$
$$y = 2 \qquad y = 2$$

The solution set is $\left\{\left(-\sqrt{2}, 2\right), \left(\sqrt{2}, 2\right)\right\}$.

19. Write $y = x^2 - 5$ as $x^2 = y + 5$ and substitute into $x^2 + y^2 = 25$

$$y + 5 + y^2 = 25$$
$$y^2 + y - 20 = 0$$
$$(y + 5)(y - 4) = 0$$
$$y = -5 \quad \text{or} \quad y = 4$$

Use $y = -5$ in $x^2 = y + 5$ to get $x^2 = 0$ or $x = 0$. Use $y = 4$ in $x^2 = y + 5$ to get $x^2 = 9$ or $x = \pm 3$. The solution set to the system is $\{(0, -5), (3, 4), (-3, 4)\}$.

21. Substitute $y = x + 1$ into $xy - 3x = 8$.

$$x(x + 1) - 3x = 8$$
$$x^2 + x - 3x = 8$$
$$x^2 - 2x - 8 = 0$$
$$(x - 4)(x + 2) = 0$$
$$x = 4 \quad \text{or} \quad x = -2$$
$$y = 5 \qquad y = -1$$

The solution set is $\{(4, 5), (-2, -1)\}$.

23. Write $xy - x = 8$ as $xy = x + 8$, and substitute for xy in $xy + 3x = -4$.

$$x + 8 + 3x = -4$$
$$4x = -12$$
$$x = -3$$

Use $x = -3$ in $xy = x + 8$.

$$-3y = -3 + 8$$
$$-3y = 5$$
$$y = -5/3$$

The solution set is $\left\{\left(-3, -\frac{5}{3}\right)\right\}$.

25. If we add $x^2 + y^2 = 8$ and $x^2 - y^2 = 2$ we get $2x^2 = 10$.

$$x^2 = 5$$
$$x = \pm\sqrt{5}$$

Use $x = \sqrt{5}$ in $y^2 = 8 - x^2$, to get $y = \pm\sqrt{3}$. Use $x = -\sqrt{5}$ in $y^2 = 8 - x^2$, to get $y = \pm\sqrt{3}$.

The solution set contains four points, $\left\{(\sqrt{5}, \sqrt{3}), (\sqrt{5}, -\sqrt{3}), (-\sqrt{5}, \sqrt{3}), (-\sqrt{5}, -\sqrt{3})\right\}$

27. Write $x^2 + 2y^2 = 8$ as $x^2 = 8 - 2y^2$ and substitute this equation into $2x^2 - y^2 = 1$.

$$2(8 - 2y^2) - y^2 = 1$$
$$16 - 4y^2 - y^2 = 1$$
$$-5y^2 = -15$$
$$y^2 = 3$$
$$y = \pm\sqrt{3}$$

If $y = \sqrt{3}$, then $x^2 = 8 - 2(\sqrt{3})^2 = 2$, or $x = \pm\sqrt{2}$.
If $y = -\sqrt{3}$, then $x^2 = 8 - 2(-\sqrt{3})^2 = 2$, or $x = \pm\sqrt{2}$. The solution set is
$\left\{(\sqrt{2}, \sqrt{3}), (\sqrt{2}, -\sqrt{3}), (-\sqrt{2}, \sqrt{3}), (-\sqrt{2}, -\sqrt{3})\right\}$

29. If we just add the equations as they are given, then y is eliminated.

$$\frac{1}{x} - \frac{1}{y} = 5$$
$$\frac{2}{x} + \frac{1}{y} = -3$$
$$\overline{\frac{3}{x} \qquad = 2}$$
$$3 = 2x$$
$$3/2 = x$$

Use $x = 3/2$ in $\frac{1}{x} - \frac{1}{y} = 5$.

$$\frac{1}{3/2} - \frac{1}{y} = 5$$
$$\frac{2}{3} - \frac{1}{y} = 5$$
$$2y - 3 = 15y$$
$$-3 = 13y$$
$$-3/13 = y$$

The solution set is $\left\{\left(\frac{3}{2}, -\frac{3}{13}\right)\right\}$.

31. Multiply the first equation by -3 and add the result to the second equation.

$$\frac{-6}{x} + \frac{3}{y} = \frac{-15}{12}$$
$$\frac{1}{x} - \frac{3}{y} = -\frac{5}{12}$$
$$\overline{\frac{-5}{x} \qquad = \frac{-20}{12}}$$
$$-20x = -60$$
$$x = 3$$

Use $x = 3$ in the first equation to find y.

$$\frac{2}{3} - \frac{1}{y} = \frac{5}{12}$$
$$12y\left(\frac{2}{3} - \frac{1}{y}\right) = 12y \cdot \frac{5}{12}$$
$$8y - 12 = 5y$$
$$3y = 12$$
$$y = 4$$

The solution set is $\{(3, 4)\}$.

33. Substitute $y = 20/x^2$ into $xy + 2 = 6x$.

$$x \cdot \frac{20}{x^2} + 2 = 6x$$
$$\frac{20}{x} + 2 = 6x$$
$$20 + 2x = 6x^2$$
$$0 = 6x^2 - 2x - 20$$
$$3x^2 - x - 10 = 0$$
$$(3x + 5)(x - 2) = 0$$
$$x = -5/3 \quad \text{or} \quad x = 2$$

Use $y = 20/x^2$ to get

$$y = 36/5 \qquad y = 5.$$

The solution set is $\left\{\left(-\frac{5}{3}, \frac{36}{5}\right), (2, 5)\right\}$.

35. Substitute $y = 7 - x$ in $x^2 + xy - y^2 = -11$.

$$x^2 + x(7 - x) - (7 - x)^2 = -11$$
$$x^2 + 7x - x^2 - 49 + 14x - x^2 = -11$$
$$-x^2 + 21x - 38 = 0$$
$$x^2 - 21x + 38 = 0$$
$$(x - 2)(x - 19) = 0$$
$$x = 2 \quad \text{or} \quad x = 19$$

Since $y = 7 - x$

$$y = 5 \qquad y = -12$$

The solution set is $\{(2, 5), (19, -12)\}$.

37. If $x^2 = y$, then $x^4 = y^2$. Substitute for x^4 in the equation $3y - 2 = x^4$.

$$3y - 2 = y^2$$
$$0 = y^2 - 3y + 2$$
$$0 = (y - 1)(y - 2)$$
$$y = 1 \quad \text{or} \quad y = 2$$

Use $y = 1$ in $x^2 = y$ to get $x^2 = 1$, or $x = \pm 1$. Use $y = 2$ in $x^2 = y$ to get $x^2 = 2$, or $x = \pm\sqrt{2}$. The solution set is
$\left\{(\sqrt{2}, 2), (-\sqrt{2}, 2), (1, 1), (-1, 1)\right\}$.

39. Eliminate y by substitution.

$$\log_2(x - 1) = 3 - \log_2(x + 1)$$
$$\log_2(x - 1) + \log_2(x + 1) = 3$$
$$\log_2[x^2 - 1] = 3$$
$$x^2 - 1 = 8$$
$$x^2 = 9$$
$$x = \pm 3$$

If $x = -3$, we get a logarithm of a negative number. If $x = 3$, then $y = \log_2(3 - 1) = 1$. The solution set is $\{(3, 1)\}$.

41. Use substitution to eliminate y.

$$\log_2(x - 1) = 2 + \log_2(x + 2)$$
$$\log_2(x - 1) - \log_2(x + 2) = 2$$
$$\log_2\left(\frac{x - 1}{x + 2}\right) = 2$$

$$\frac{x-1}{x+2} = 4$$
$$x - 1 = 4x + 8$$
$$-9 = 3x$$
$$-3 = x$$

If $x = -3$ in either of the original equations, we get a logarithm of a negative number. So the solution set is the empty set, $\emptyset$.

43. Use substitution to eliminate y.

$$2^{3x+4} = 4^{x-1}$$
$$2^{3x+4} = (2^2)^{x-1}$$
$$2^{3x+4} = 2^{2x-2}$$
$$3x + 4 = 2x - 2$$
$$x = -6$$

If $x = -6$, then $y = 4^{-6-1} = 4^{-7}$. So the solution set is $\{(-6, 4^{-7})\}$.

45. Let $x =$ the length of one leg and $y =$ the length of the other. We write one equation for the area: $3 = \frac{1}{2}xy$. We write the other equation from the Pythagorean theorem: $x^2 + y^2 = 15$.

Substitute $y = 6/x$ into $x^2 + y^2 = 15$.

$$x^2 + (\tfrac{6}{x})^2 = 15$$
$$x^2 + \frac{36}{x^2} = 15$$
$$x^4 + 36 = 15x^2$$
$$x^4 - 15x^2 + 36 = 0$$
$$(x^2 - 3)(x^2 - 12) = 0$$
$$x = \sqrt{3} \quad \text{or} \quad x = \sqrt{12} = 2\sqrt{3}$$

If $x = \sqrt{3}$, then $y = 6/\sqrt{3} = 2\sqrt{3}$. If $x = 2\sqrt{3}$, then $y = 6/(2\sqrt{3}) = \sqrt{3}$. So the lengths of the legs are $\sqrt{3}$ feet and $2\sqrt{3}$ feet.

47. Let $h =$ the height of each triangle, and $b =$ the length of the base of each triangle. Since the 7 triangles are to have a total area of 3,500, the area of each must be 500: $\frac{1}{2}hb = 500$. The ratio of the height to base must be 1 to 4 is expressed as $\frac{h}{b} = \frac{1}{4}$. These two equations can be written as $hb = 1000$ and $b = 4h$. Use substitution.

$$h(4h) = 1000$$
$$h^2 = 250$$
$$h = \sqrt{250} = 5\sqrt{10}$$

Since $b = 4h$, $b = 20\sqrt{10}$. So the height is $5\sqrt{10}$ inches and the base is $20\sqrt{10}$ inches.

49. Let $x =$ the number of hours for pump A to fill the tank alone, and $y =$ the number of hours for pump B to fill the tank alone.

$$\frac{1}{x} + \frac{1}{y} = \frac{1}{6}$$
$$\frac{1}{y} - \frac{1}{x} = \frac{1}{12}$$

Adding these two equations eliminates x.

$$\frac{2}{y} = \frac{1}{6} + \frac{1}{12}$$
$$\frac{2}{y} = \frac{1}{4}$$
$$y = 8$$

Use $y = 8$ in the first equation.

$$\frac{1}{x} + \frac{1}{8} = \frac{1}{6}$$
$$\frac{1}{x} = \frac{1}{24}$$
$$x = 24$$

It would take pump A 24 hours to fill the tank alone and pump B 8 hours to fill the tank alone.

51. Let $x =$ the time for Jan to do the job alone. Let $y =$ the time for Beth to do the job alone. Since they do the job together in 24 minutes we have the equation

$$\frac{1}{x} + \frac{1}{y} = \frac{1}{24}.$$

In 50 minutes the job was completed, but not by working together. This implies that the total of their times alone is 100, $x + y = 100$. Substitute $y = 100 - x$ into the first equation.

$$\frac{1}{x} + \frac{1}{100 - x} = \frac{1}{24}$$
$$24(100 - x) + 24x = x(100 - x)$$
$$2400 - 24x + 24x = 100x - x^2$$
$$x^2 - 100x + 2400 = 0$$
$$(x - 40)(x - 60) = 0$$
$$x = 40 \quad \text{or} \quad x = 60$$

Since $y = 100 - x$

$$y = 60 \qquad y = 40$$

Since Jan is the faster worker, it would take her 40 minutes to complete the catfish by herself.

53. Let $x =$ the length and $y =$ the width. The area is 72 is expressed as $xy = 72$. The perimeter is 34 is expressed as $2x + 2y = 34$ or $x + y = 17$. Substitute $y = 17 - x$ into $xy = 72$.

$$x(17 - x) = 72$$
$$-x^2 + 17x - 72 = 0$$
$$x^2 - 17x + 72 = 0$$
$$(x - 9)(x - 8) = 0$$

$$x = 9 \quad \text{or} \quad x = 8$$
$$y = 8 \qquad\quad y = 9$$

The rectangular area is 8 feet by 9 feet.

55. Let x and y represent the numbers.

$$x + y = 8$$
$$xy = 20$$

Substitute $y = 8 - x$ into $xy = 20$.

$$x(8 - x) = 20$$
$$-x^2 + 8x - 20 = 0$$
$$x^2 - 8x + 20 = 0$$
$$x = \frac{8 \pm \sqrt{64 - 4(1)(20)}}{2} = \frac{8 \pm \sqrt{-16}}{2}$$
$$= 4 \pm 2i$$

If $x = 4 + 2i$, then $y = 8 - (4 + 2i) = 4 - 2i$.
If $x = 4 - 2i$, then $y = 8 - (4 - 2i) = 4 + 2i$.
So the numbers are $4 - 2i$ and $4 + 2i$.

57. Let $x =$ the length of the side of the square, and $y =$ the height of the triangle. The total height of 10 feet means that $x + y = 10$. The total area is 72 means that $x^2 + \frac{1}{2}xy = 72$. Substitute $y = 10 - x$ into the second equation.

$$x^2 + \frac{1}{2}x(10 - x) = 72$$
$$x^2 + 5x - \frac{1}{2}x^2 = 72$$
$$\frac{1}{2}x^2 + 5x - 72 = 0$$
$$x^2 + 10x - 144 = 0$$
$$(x - 8)(x + 18) = 0$$
$$x = 8 \quad \text{or} \quad x = -18$$
$$y = 2$$

The side of the square is 8 feet and the height of the triangle is 2 feet.

59. a) $(1.71, 1.55), (-2.98, -3.95)$
b) $(1, 1), (0.40, 0.16)$
c) $(1.17, 1.62), (-1.17, -1.62)$

13.2 WARM-UPS

1. False, because if the focus is (2, 3) and the directrix is $y = 1$, then the vertex is (2, 2).
2. True, because the focus is 1 unit above the vertex which is (0, 1).
3. True, because in $y = a(x - h)^2 + k$, the vertex is (h, k).
4. False, because $y = 6x + 3x + 2$ is equivalent to $y = 9x + 2$, which is a straight line.
5. False, because $y = -x^2 + 2x + 9$ opens downward.
6. True, the vertex is (0, 0) and so is the y-intercept.
7. True, because it opens upward from (2, 3) and it cannot intersect the x-axis.
8. True, because the focus is above the directrix in a parabola the opens upward.
9. True, because the vertex is $(2, k)$ and the axis of symmetry is a vertical line through the vertex.
10. True, because $1/(4 \cdot \frac{1}{4}) = 1$.

13.2 EXERCISES

1. A parabola is the set of all points in a plane that are equidistant from a given line and a fixed point not on the line.
3. A parabola can be written in the forms $y = ax^2 + bx + c$ or $y = a(x - h)^2 + k$.
5. We use completing the square to convert $y = ax^2 + bx + c$ into $y = a(x - h)^2 + k$.
7. $\sqrt{(5 - 4)^2 + (-2 - (-3))^2} = \sqrt{1 + 1}$
$= \sqrt{2}$
9. $\sqrt{(4 - 6)^2 + (2 - 5)^2} = \sqrt{4 + 9} = \sqrt{13}$
11. $\sqrt{(1 - 3)^2 + (-3 - 5)^2} = \sqrt{4 + 64}$
$= 2\sqrt{17}$
13. $\sqrt{(-3 - 4)^2 + (-6 - (-2))^2}$
$= \sqrt{49 + 16} = \sqrt{65}$
15. The vertex is (0, 0). Since $2 = 1/(4p)$, $p = 1/8$. So the focus is $(0, 1/8)$ and the directrix is $y = -1/8$.
17. The vertex is (0, 0). Since $-1/4 = 1/(4p)$, $4p = -4$, or $p = -1$. So the focus is $(0, -1)$ and the directrix is $y = 1$.
19. The vertex is (3, 2). Since $1/2 = 1/(4p)$, $p = 1/2$. So the focus is (3, 2.5) and the directrix is $y = 1.5$.
21. The vertex is $(-1, 6)$. Since $-1 = 1/(4p)$, $p = -1/4$. So the focus is $(-1, 5.75)$ and the directrix is $y = 6.25$.
23. Since the distance between the focus and directrix is 4, $p = 2$ and $a = \frac{1}{4p} = \frac{1}{4(2)} = \frac{1}{8}$. Since the vertex is half way between the focus

and directrix, the vertex is (0, 0). Use the form $y = a(x-h)^2 + k$ to get the equation.

$y = \frac{1}{8}(x-0)^2 + 0$

$y = \frac{1}{8}x^2$

25. Since the distance between the focus and directrix is 1 and the focus is below the directrix,
$p = -\frac{1}{2}$ and $a = \frac{1}{4p} = \frac{1}{4(-\frac{1}{2})} = -\frac{1}{2}$. Since the vertex is half way between the focus and directrix, the vertex is (0, 0). Use the form $y = a(x-h)^2 + k$ to get the equation.

$y = -\frac{1}{2}(x-0)^2 + 0$

$y = -\frac{1}{2}x^2$

27. Since the distance between the focus and directrix is 1 and the parabola opens upward, $p = \frac{1}{2}$ and $a = \frac{1}{4p} = \frac{1}{4(\frac{1}{2})} = \frac{1}{2}$. Since the vertex is half way between the focus and directrix, the vertex is $\left(3, \frac{3}{2}\right)$. Use the form $y = a(x-h)^2 + k$ to get the equation.

$y = \frac{1}{2}(x-3)^2 + \frac{3}{2}$

$y = \frac{1}{2}x^2 - 3x + \frac{9}{2} + \frac{3}{2}$

$y = \frac{1}{2}x^2 - 3x + 6$

29. Since the distance between the focus and directrix is 4 and the parabola opens downward, $p = -2$ and
$a = \frac{1}{4p} = \frac{1}{4(-2)} = -\frac{1}{8}$. Since the vertex is half way between the focus and directrix, the vertex is $(1, 0)$. Use the form $y = a(x-h)^2 + k$ to get the equation.

$y = -\frac{1}{8}(x-1)^2 + 0$

$y = -\frac{1}{8}x^2 + \frac{1}{4}x - \frac{1}{8}$

31. Since the distance between the focus and directrix is $\frac{1}{2}$ and the parabola opens upward, $p = \frac{1}{4}$ and $a = \frac{1}{4p} = \frac{1}{4(0.25)} = 1$. Since the vertex is half way between the focus and directrix, the vertex is $(-3, 1)$. Use the form $y = a(x-h)^2 + k$ to get the equation.

$y = 1(x+3)^2 + 1$

$y = x^2 + 6x + 10$

33. $y = x^2 - 6x + 1$

$y = x^2 - 6x + 9 - 9 + 1$

$y = (x-3)^2 - 8$

The vertex is (3, −8). Because $a = 1$, the parabola opens upward, and $p = 1/4$. The focus is (3, −7.75) and the directrix is $y = -8.25$. The axis of symmetry is the vertical line through the vertex, $x = 3$.

35. $y = 2x^2 + 12x + 5$

$y = 2(x^2 + 6x) + 5$

$y = 2(x^2 + 6x + 9) + 5 - 18$

$y = 2(x+3)^2 - 13$

The vertex is (−3, −13) and the parabola opens upward. Because $a = 2$, $p = 1/8 = 0.125$ The focus is (−3, −12.875) and the directrix is $y = -13.125$. The axis of symmetry is $x = -3$.

37. $y = -2x^2 + 16x + 1$

$y = -2(x^2 - 8x) + 1$

$y = -2(x^2 - 8x + 16) + 1 + 32$

$y = -2(x-4)^2 + 33$

The vertex is (4, 33) and the parabola opens downward. Because $a = -2$, $p = -1/8$ $= -0.125$. The focus is (4, $32\frac{7}{8}$) and the directrix is $y = 33\frac{1}{8}$. The axis of symmetry is $x = 4$.

39. $y = 5x^2 + 40x$

$y = 5(x^2 + 8x)$

$y = 5(x^2 + 8x + 16) - 80$

$y = 5(x+4)^2 - 80$

The vertex is (−4, − 80) and the parabola opens upward. Because $a = 5$, $p = 1/20$. The focus is (−4, $-79\frac{19}{20}$) and the directrix is $y = -80\frac{1}{20}$. The axis of symmetry is $x = -4$.

41. The x-coordinate of the vertex is $x = \frac{-b}{2a}$.

$x = \frac{-(-4)}{2(1)} = 2$

$y = 2^2 - 4(2) + 1 = -3$

The vertex is (2, −3) and the parabola opens upward. Because $a = 1$, $p = 1/4$. The focus is (2, $-2\frac{3}{4}$) and the directrix is $y = -3\frac{1}{4}$. Axis of symmetry is $x = 2$.

43. The x-coordinate of the vertex is $x = \frac{-b}{2a}$.

$x = \frac{-(2)}{2(-1)} = 1$

$y = -1^2 + 2(1) - 3 = -2$

The vertex is (1, −2) and the parabola opens downward. Because $a = -1$, p $= -1/4$. The focus is $(1, -2\frac{1}{4})$ and the directrix is $y = -1\frac{3}{4}$. The axis of symmetry is $x = 1$.

45. The x-coordinate of the vertex is $x = \frac{-b}{2a}$.

$$x = \frac{-(-6)}{2(3)} = 1$$
$$y = 3(1^2) - 6(1) + 1 = -2$$

The vertex is $(1, -2)$ and the parabola opens upward. Because $a = 3$, $p = 1/12$. The focus is $(1, -1\frac{11}{12})$ and the directrix is $y = -2\frac{1}{12}$. The axis of symmetry is $x = 1$.

47. The x-coordinate of the vertex is $x = \frac{-b}{2a}$.

$$x = \frac{-(-3)}{2(-1)} = -\frac{3}{2}$$
$$y = -(-\tfrac{3}{2})^2 - 3(-\tfrac{3}{2}) + 2 = \frac{17}{4}$$

The vertex is $(-\frac{3}{2}, \frac{17}{4})$ and the parabola opens downward. Because $a = -1$, $p = -1/4$. The focus is $(-\frac{3}{2}, 4)$ and the directrix is $y = \frac{9}{2}$. The axis of symmetry is $x = -\frac{3}{2}$.

49. The x-coordinate of the vertex is $x = \frac{-b}{2a}$.

$$x = \frac{-(0)}{2(3)} = 0$$
$$y = 3(0)^2 + 5 = 5$$

The vertex is $(0, 5)$ and the parabola opens upward. Because $a = 3$, $p = 1/12$. The focus is $(0, 5\frac{1}{12})$ and the directrix is $y = 4\frac{11}{12}$. The axis of symmetry is $x = 0$.

51. For $x = (y-2)^2 + 3$ the vertex is $(3, 2)$. Since $a = 1$ we have $p = 1/4$ and the parabola opens to the right. The focus is $\left(\frac{13}{4}, 2\right)$ and the directrix is $x = \frac{11}{4}$.

53. For $x = \frac{1}{4}(y-1)^2 - 2$ the vertex is $(-2, 1)$. Since $a = 1/4$ we have $p = 1$ and the parabola opens to the right. The focus is $(-1, 1)$ and the directrix is $x = -3$.

55. For $x = -\frac{1}{2}(y-2)^2 + 4$ the vertex is $(4, 2)$. Since $a = -1/2$ we have $p = -1/2$ and the parabola opens to the left. The focus is $\left(\frac{7}{2}, 2\right)$ and the directrix is $x = \frac{9}{2}$.

57. For $y = (x-2)^2 + 3$ the vertex is $(2, 3)$ and the parabola opens upward. The graph also goes through $(0, 7)$, $(1, 4)$, $(3, 4)$, and $(4, 7)$.

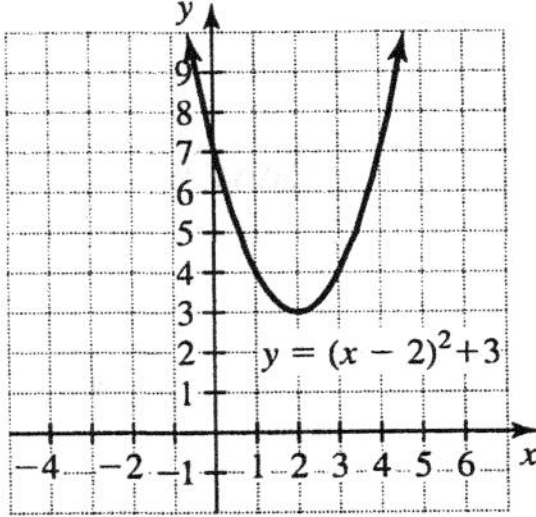

59. For $y = -2(x-1)^2 + 3$ the vertex is $(1, 3)$ and the parabola opens downward. The graph also goes through $(-1, -5)$, $(0, 1)$, $(2, 1)$, and $(3, -5)$.

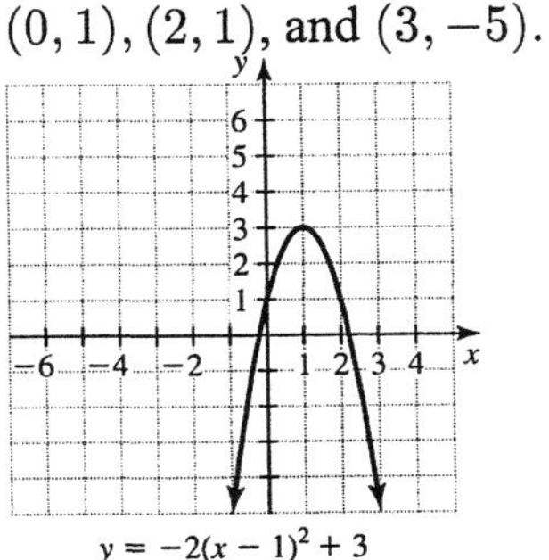

61. For $x = (y-2)^2 + 3$ the vertex is $(3, 2)$ and the parabola opens to the right. The graph also goes through $(4, 3)$, $(4, 1)$, $(7, 4)$, and $(7, 0)$.

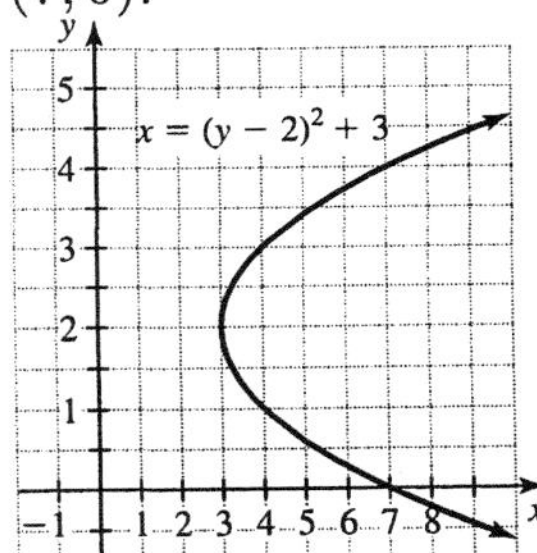

63. For $x = -2(y-1)^2 + 3$ the vertex is $(3, 1)$ and the parabola opens to the left. The graph also goes through $(1, 2)$, $(1, 0)$, $(-5, 3)$, and $(-5, -1)$.

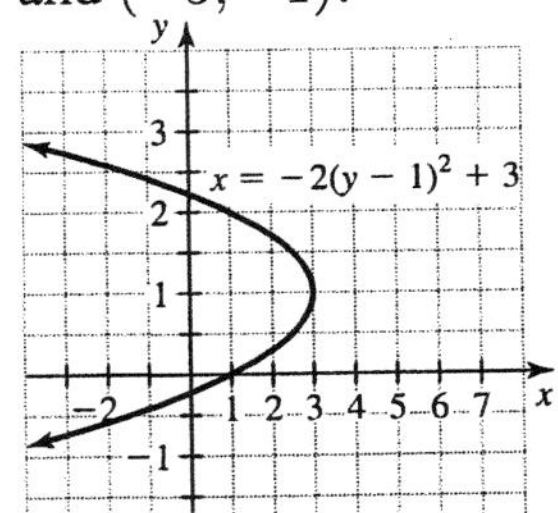

65. The distance from the vertex to the focus is 15. So $p = 15$ and $a = 1/(4 \cdot 15) = 1/60$. the equation of the parabola is $y = \frac{1}{60}x^2$.

67. The graph of $y = -x^2 + 3$ is a parabola opening downward, and the graph of $y = x^2 + 1$ is a parabola opening upward. To find the points of intersection, eliminate y by substitution.

$$x^2 + 1 = -x^2 + 3$$
$$2x^2 = 2$$
$$x^2 = 1$$
$$x = \pm 1$$

If $x = \pm 1$, then $y = 2$ (since $y = x^2 + 1$). The solution set for the system is $\{(-1, 2), (1, 2)\}$.

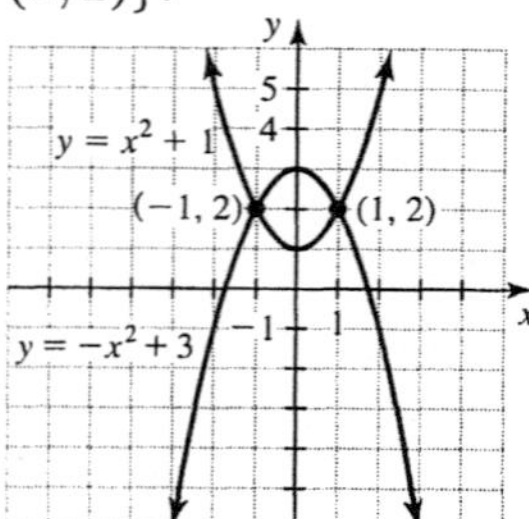

69. The graph of $y = x^2 - 2$ is a parabola opening upward, and the graph of $y = 2x - 3$ is a straight line.

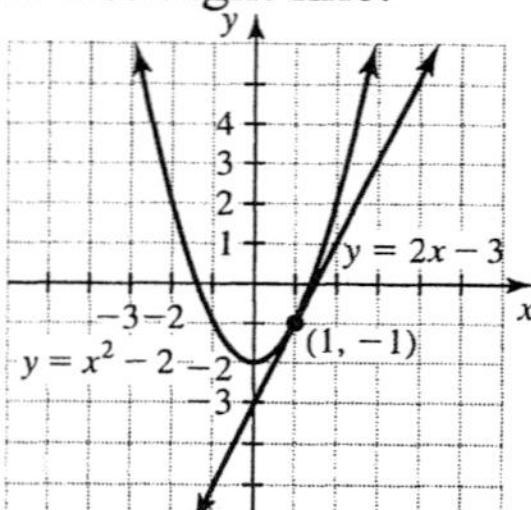

We eliminate y by substitution.

$$x^2 - 2 = 2x - 3$$
$$x^2 - 2x + 1 = 0$$
$$(x - 1)^2 = 0$$
$$x = 1$$

If $x = 1$, then $y = -1$ (because $y = 2x - 3$). The solution set for the system is $\{(1, -1)\}$.

71. The graph of $y = x^2 + 3x - 4$ is a parabola opening upward, and the graph of $y = -x^2 - 2x + 8$ is a parabola opening downward.

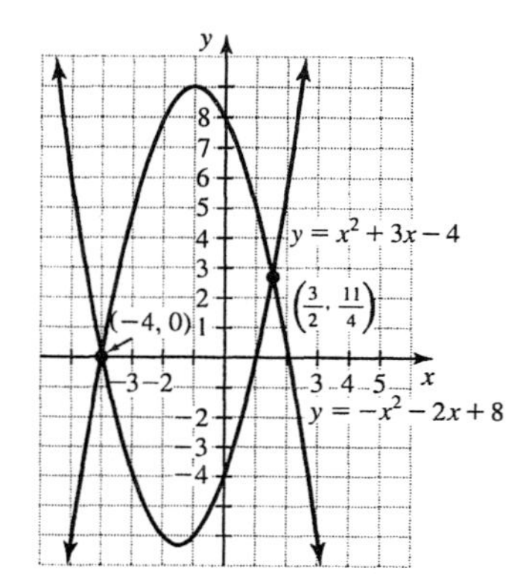

Eliminate y by substitution.

$$x^2 + 3x - 4 = -x^2 - 2x + 8$$
$$2x^2 + 5x - 12 = 0$$
$$(2x - 3)(x + 4) = 0$$
$$x = 3/2 \quad \text{or} \quad x = -4$$

If $x = 3/2$, then

$$y = (3/2)^2 + 3(3/2) - 4 = 11/4.$$

If $x = -4$, then

$$y = (-4)^2 + 3(-4) - 4 = 0.$$

The solution set for the system is $\left\{\left(\frac{3}{2}, \frac{11}{4}\right), (-4, 0)\right\}$.

73. The graph of $y = x^2 + 3x - 4$ is a parabola opening upward. The graph of $y = 2x + 2$ is a straight line with slope 2 and y-intercept (0, 2).

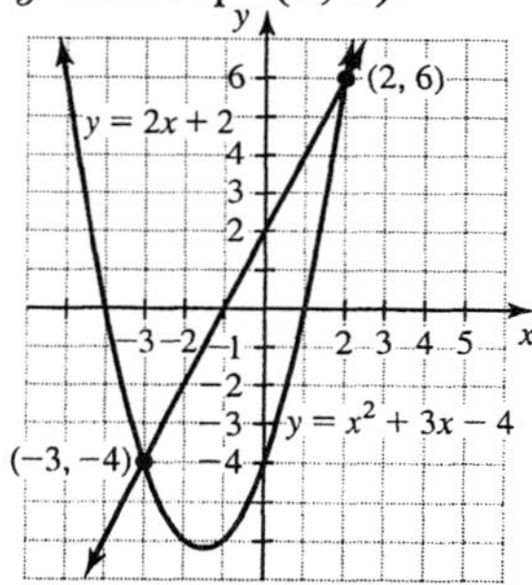

Use substitution to eliminate y.

$$x^2 + 3x - 4 = 2x + 2$$
$$x^2 + x - 6 = 0$$
$$(x + 3)(x - 2) = 0$$
$$x = -3 \quad \text{or} \quad x = 2$$

Since $y = 2x + 2$

$$y = -4 \qquad y = 6$$

The solution set for the system is $\{(-3, -4), (2, 6)\}$.

75. On the x-axis, the y-coordinate is zero. So we solve $x^2 - 2x - 3 = 0$.

$$(x - 3)(x + 1) = 0$$

$$x - 3 = 0 \text{ or } x + 1 = 0$$
$$x = 3 \text{ or } \quad x = -1$$
The points of intersection are $(3, 0)$ and $(-1, 0)$.

77. Substitute $y = 4$ into $y = 0.01x^2$:
$$4 = 0.01x^2$$
$$400 = x^2$$
$$\pm 20 = x$$
The points of intersection are $(-20, 4)$ and $(20, 4)$

79. Substitute to get $x = (x^2)^2$ or $x = x^4$.
$$x = x^4$$
$$x^4 - x = 0$$
$$x(x^3 - 1) = 0$$
$$x(x - 1)(x^2 + x + 1) = 0$$
$$x = 0 \text{ or } \quad x - 1 = 0 \quad \text{or} \quad x^2 + x + 1 = 0$$
$$x = 0 \text{ or } \quad x = 1$$
There are no real solutions to $x^2 + x + 1 = 0$. If $x = 0$, then $y = 0^2 = 0$. If $x = 1$, then $y = 1^2 = 1$. So the points of intersection are $(0, 0)$ and $(1, 1)$.

81. The distance from (x, y) to the focus $(p, 0)$ is $\sqrt{(x - p)^2 + (y - 0)^2}$. The distance from (x, y) to the vertical line $x = -p$ is the distance between the points (x, y) and $(-p, y)$, which is $\sqrt{(x - (-p))^2 + (y - y)^2}$.
Set these equal, square each side, then simplify to get
$$x^2 - 2xp + p^2 + y^2 = x^2 + 2xp + p^2.$$
Solve for x to get $x = \frac{1}{4p}y^2$ or $x = ay^2$ where $a = 1/(4p)$.

83. The graphs have identical shapes.

13.3 WARM-UPS

1. False, the radius of a circle is a positive real number.
2. False, the coordinates of the center do not satisfy the equation of the circle, only points on the circle satisfy the equation.
3. True, the center is (0, 0).
4. False, the equation of the circle centered at the origin with radius 9 is $x^2 + y^2 = 81$.
5. False, because $(x - 2)^2 + (y - 3)^2 = 4$ has radius 2.
6. False, because $(x - 3)^2 + (y + 5)^2 = 9$ is a circle of radius 3 centered at $(3, -5)$
7. True, because the distance from $(-3, -1)$ to (0, 0) is the radius.
8. False, the center is (3, 4).
9. True, because there is only an x^2-term and no x-term.
10. False, because if we complete the square for x then the right side will no longer be 4.

13.3 EXERCISES

1. A circle is the set of all points in a plane that lie at a fixed distance from a fixed point.
3. Use $h = 0$, $k = 0$, and $r = 4$ in the standard equation $(x - h)^2 + (y - k)^2 = r^2$ to get the equation $x^2 + y^2 = 16$.
5. Use $h = 0$, $k = 3$, and $r = 5$ in the standard equation $(x - h)^2 + (y - k)^2 = r^2$ to get the equation $x^2 + (y - 3)^2 = 25$.
7. Use $h = 1$, $k = -2$, and $r = 9$ in the standard equation $(x - h)^2 + (y - k)^2 = r^2$ to get the equation $(x - 1)^2 + (y + 2)^2 = 81$.
9. Use $h = 0$, $k = 0$, and $r = \sqrt{3}$ in the standard equation $(x - h)^2 + (y - k)^2 = r^2$ to get the equation $x^2 + y^2 = 3$.
11. Use $h = -6$, $k = -3$, and $r = 1/2$ in the standard equation $(x - h)^2 + (y - k)^2 = r^2$ to get the equation $(x + 6)^2 + (y + 3)^2 = \frac{1}{4}$.
13. Use $h = 1/2$, $k = 1/3$, and $r = 0.1$ in the standard equation $(x - h)^2 + (y - k)^2 = r^2$ to get the equation
$$\left(x - \frac{1}{2}\right)^2 + \left(y - \frac{1}{3}\right)^2 = 0.01.$$
15. Compare $(x - 3)^2 + (y - 5)^2 = 2$ with $(x - h)^2 + (y - k)^2 = r^2$ (*where the center is (h, k) and radius is r*), to get a center at (3, 5) and radius $\sqrt{2}$.
17. Compare $x^2 + \left(y - \frac{1}{2}\right)^2 = \frac{1}{2}$ with $(x - h)^2 + (y - k)^2 = r^2$ (*where the center is (h, k) and radius is r*), to get a center at $\left(0, \frac{1}{2}\right)$ and radius $\sqrt{\frac{1}{2}}$ or $\frac{\sqrt{2}}{2}$.
19. Divide each side by 4 to get $x^2 + y^2 = \frac{9}{4}$, which has center (0, 0) and radius $\sqrt{\frac{9}{4}}$ or $\frac{3}{2}$.

21. Rewrite the equation as $(x-2)^2+y^2=3$, which has center (2, 0) and radius $\sqrt{3}$.

23. The graph of $x^2+\ y^2=9$ is a circle with radius 3, centered at (0, 0).

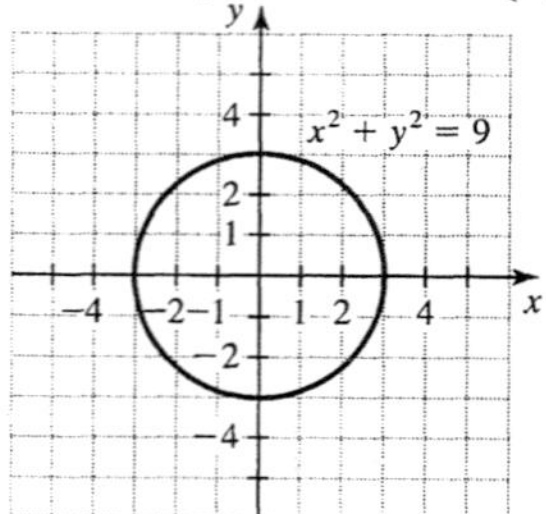

25. The graph of $x^2+(y-3)^2=9$ is a circle with radius 3, centered at (0, 3).

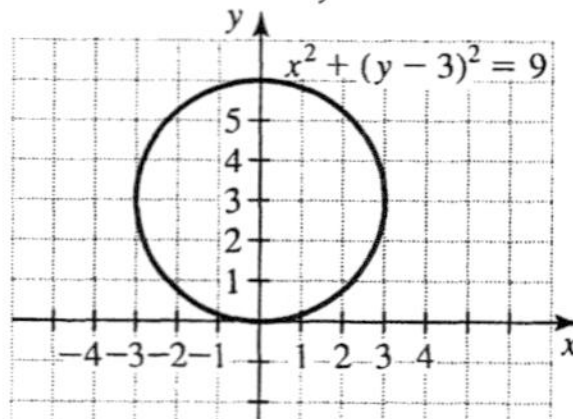

27. The graph of $(x+1)^2+(y-1)^2=2$ is a circle with radius $\sqrt{2}$, centered at $(-1, 1)$.

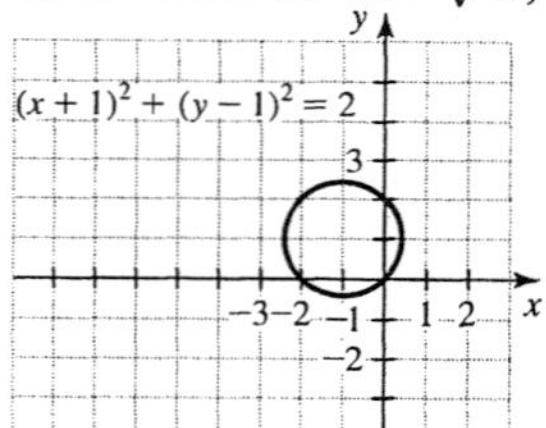

29. The graph of $(x-4)^2+(y+3)^2=16$ is a circle of radius 4, centered at $(4, -3)$.

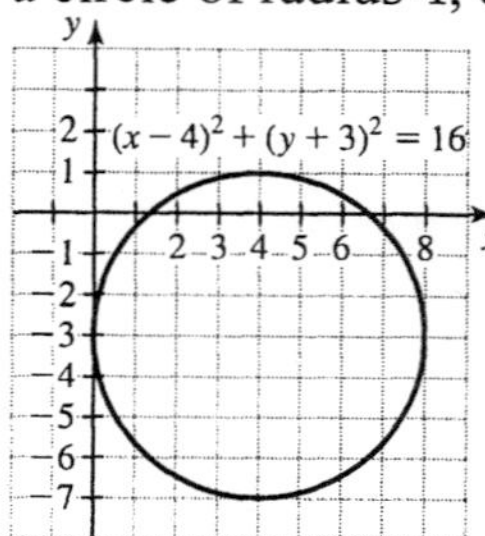

31. The graph of $\left(x-\frac{1}{2}\right)^2+\left(y+\frac{1}{2}\right)^2=\frac{1}{4}$ is a circle of radius $1/2$ and center $(1/2, -1/2)$.

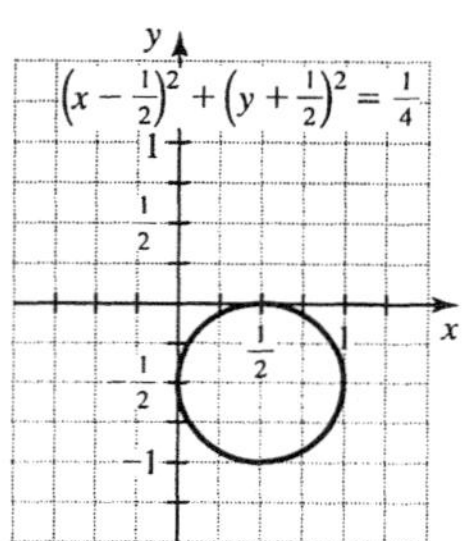

33.
$$\begin{aligned} x^2+4x+\ \ y^2+6y\ \ &=0\\ x^2+4x+4+y^2+6y+9&=0+4+9\\ (x+2)^2+(y+3)^2&=13 \end{aligned}$$
The circle has radius $\sqrt{13}$ and center $(-2, -3)$.

35.
$$\begin{aligned} x^2-2x+\ \ y^2-4y-3&=0\\ x^2-2x+1+y^2-4y+4&=3+1+4\\ (x-1)^2+(y-2)^2&=8 \end{aligned}$$
This circle has radius $\sqrt{8}=2\sqrt{2}$ and center (1, 2).

37.
$$\begin{aligned} x^2-10x+y^2-8y&=-32\\ x^2-10x+25+y^2-8y+16&\\ &=-32+25+16\\ (x-5)^2+(y-4)^2&=9 \end{aligned}$$
This circle has radius 3 and center (5, 4).

39.
$$\begin{aligned} x^2-x+\ \ y^2+y\ \ &=0\\ x^2-x+\frac{1}{4}+y^2+y+\frac{1}{4}&=\frac{1}{4}+\frac{1}{4}\\ \left(x-\frac{1}{2}\right)^2+\left(y+\frac{1}{2}\right)^2&=\frac{1}{2} \end{aligned}$$
Center is $(1/2, -1/2)$ and radius is $\sqrt{\frac{1}{2}}$ or $\frac{\sqrt{2}}{2}$.

41.
$$\begin{aligned} x^2-3x+\ \ y^2-y\ \ &=1\\ x^2-3x+\frac{9}{4}+y^2-y+\frac{1}{4}&=1+\frac{9}{4}+\frac{1}{4}\\ \left(x-\frac{3}{2}\right)^2+\left(y-\frac{1}{2}\right)^2&=\frac{7}{2} \end{aligned}$$
Center is $(3/2, 1/2)$ and radius $\sqrt{\frac{7}{2}}$ or $\frac{\sqrt{14}}{2}$.

43.
$$\begin{aligned} x^2-\frac{2}{3}x+\ \ y^2+\frac{3}{2}y\ \ &=0\\ x^2-\frac{2}{3}x+\frac{1}{9}+y^2+\frac{3}{2}y+\frac{9}{16}&=0+\frac{1}{9}+\frac{9}{16}\\ \left(x-\frac{1}{3}\right)^2+\left(y+\frac{3}{4}\right)^2&=\frac{97}{144} \end{aligned}$$
Center is $(1/3, -3/4)$ and radius is $\sqrt{\frac{97}{144}}$ or $\frac{\sqrt{97}}{12}$.

45. The graph of $x^2 + y^2 = 10$ is a circle centered at (0, 0) with radius $\sqrt{10}$. The graph of $y = 3x$ is a straight line through (0, 0) with slope 3. Use substitution to eliminate y.

$$x^2 + (3x)^2 = 10$$
$$10x^2 = 10$$
$$x^2 = 1$$
$$x = 1 \quad \text{or} \quad x = -1$$
$$y = 3 \quad \text{or} \quad y = -3 \quad \text{Since } y = 3x$$

The solution set is $\{(1, 3), (-1, -3)\}$.

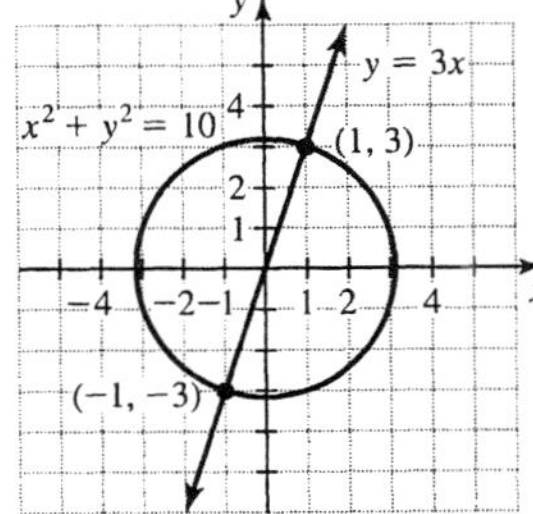

47. The graph of $x^2 + y^2 = 9$ is a circle centered at (0, 0) with radius 3. The graph of $y = x^2 - 3$ is a parabola opening upward.

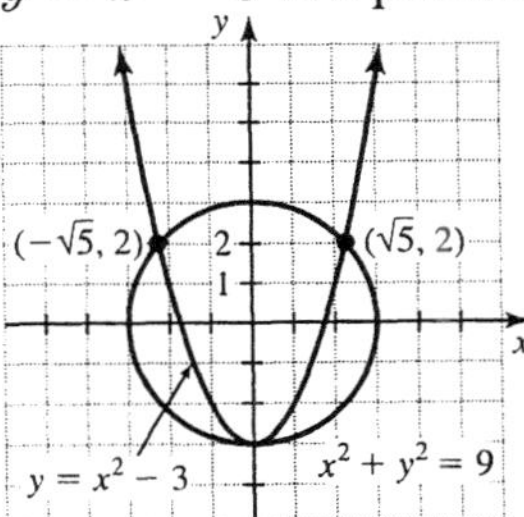

Use $x^2 = y + 3$ to eliminate x.

$$y + 3 + y^2 = 9$$
$$y^2 + y - 6 = 0$$
$$(y + 3)(y - 2) = 0$$
$$y = -3 \quad \text{or} \quad y = 2$$

If $y = -3$, then $x^2 = -3 + 3 = 0$ or $x = 0$. If $y = 2$, then $x^2 = 2 + 3 = 5$ or $x = \pm\sqrt{5}$.

The solution set is

$$\left\{(0, -3), \left(\sqrt{5}, 2\right), \left(-\sqrt{5}, 2\right)\right\}.$$

49. The graph of $(x - 2)^2 + (y + 3)^2 = 4$ is a circle centered at (2, −3) with radius 2. The graph of $y = x - 3$ is a line through (0, −3) with slope 1.

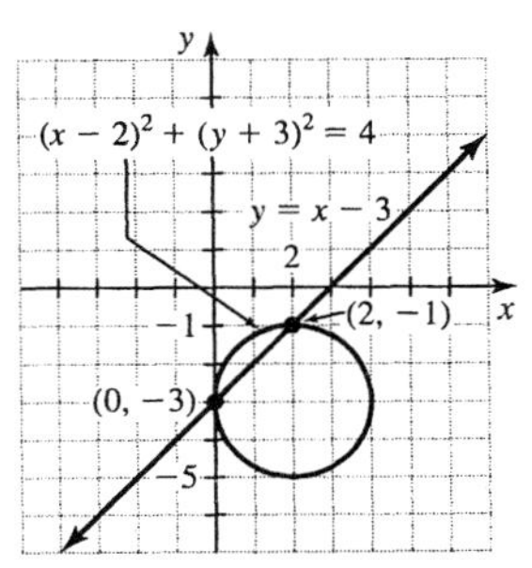

Use substitution to eliminate y.

$$(x - 2)^2 + (x - 3 + 3)^2 = 4$$
$$x^2 - 4x + 4 + x^2 = 4$$
$$2x^2 - 4x = 0$$
$$x^2 - 2x = 0$$
$$x(x - 2) = 0$$
$$x = 0 \quad \text{or} \quad x = 2$$

Since $y = x - 3$

$$y = -3 \qquad y = -1$$

The solution set is $\{(0, -3), (2, -1)\}$.

51. To find the y-intercepts, use $x = 0$ in the equation of the circle.

$$(0 - 1)^2 + (y - 2)^2 = 4$$
$$1 + y^2 - 4y + 4 = 4$$
$$y^2 - 4y + 1 = 0$$
$$y = \frac{4 \pm \sqrt{16 - 4(1)(1)}}{2(1)} = \frac{4 \pm 2\sqrt{3}}{2}$$
$$= 2 \pm \sqrt{3}$$

The y-intercepts are $(0, 2 + \sqrt{3})$ and $(0, 2 - \sqrt{3})$.

53. The radius is the distance from (2, −5) to (0, 0).

$$r = \sqrt{(2 - 0)^2 + (-5 - 0)^2} = \sqrt{4 + 25}$$
$$= \sqrt{29}$$

55. The radius is the distance from (2, 3) to (−2, −1).

$$r = \sqrt{(-2 - 2)^2 + (-1 - 3)^2} = \sqrt{16 + 16}$$
$$= \sqrt{32}$$

The equation of a circle centered at (2, 3) with radius $\sqrt{32}$ is $(x - 2)^2 + (y - 3)^2 = 32$.

57. Substitute $y^2 = 9 - x^2$ into $(x - 5)^2 + y^2 = 9$ to eliminate y.

$$(x - 5)^2 + 9 - x^2 = 9$$
$$x^2 - 10x + 25 + 9 - x^2 = 9$$
$$-10x = -25$$
$$x = 5/2$$

Use $x = 5/2$ in $y^2 = 9 - x^2$ to find y.

$$y^2 = 9 - \left(\frac{5}{2}\right)^2 = \frac{11}{4}$$

$$y = \pm\sqrt{\frac{11}{4}} = \pm\frac{\sqrt{11}}{2}$$

The circles intersect at $\left(\frac{5}{2}, -\frac{\sqrt{11}}{2}\right)$ and $\left(\frac{5}{2}, \frac{\sqrt{11}}{2}\right)$.

59. From the equation of the bore we see that the radius is $\sqrt{83.72}$. The volume is given by $V = \pi r^2 h = \pi \cdot 83.72 \cdot 2874 \approx 755{,}903 \text{ mm}^3$.

61. Since both x^2 and y^2 are nonnegative for real values of x and y, the only way their sum could be zero is to choose both x and y equal to zero. So the graph consists of (0, 0) only.

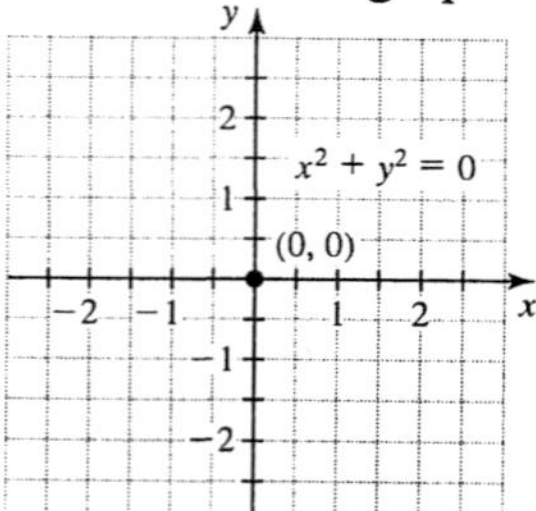

63. If we square both sides of $y = \sqrt{1 - x^2}$ we get $x^2 + y^2 = 1$, which is a circle with center (0, 0) and radius 1. In the equation $y = \sqrt{1 - x^2}$, y must be nonnegative. So the graph of $y = \sqrt{1 - x^2}$ is the top half of the graph of $x^2 + y^2 = 1$.

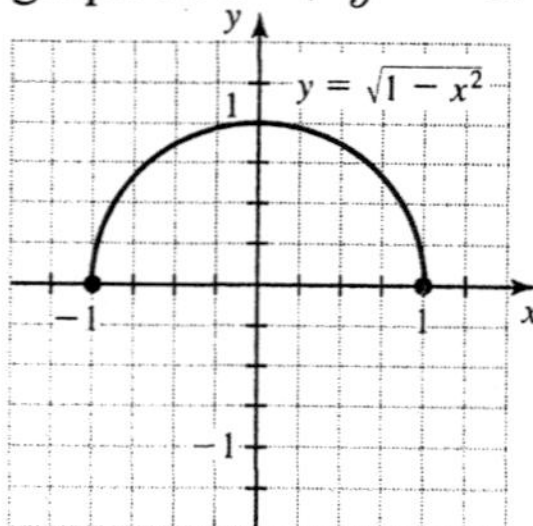

65. B and D can be any real numbers, but A must equal C. So that the radius is positive, we must also have

$\frac{E}{A} + \left(\frac{B}{2A}\right)^2 + \left(\frac{D}{2A}\right)^2 > 0$, or

$4AE + B^2 + D^2 > 0$.

No ordered pairs satisfy $x^2 + y^2 = -9$. So there is no graph.

67. Solve for y to get $y = \pm\sqrt{4 - x^2}$. Then graph $y_1 = \sqrt{4 - x^2}$ and $y_2 = -\sqrt{4 - x^2}$

69. $y = \pm\sqrt{x}$

71.
$$x = y^2 + 2y + 1$$
$$x = (y + 1)^2$$
$$y + 1 = \pm\sqrt{x}$$
$$y = -1 \pm \sqrt{x}$$

13.4 WARM-UPS

1. False, the x-intercepts are $(-6, 0)$ and $(6, 0)$.

2. False, because both x^2 and y^2 must appear in the equation for an ellipse.

3. True, because if the foci coincide then every point on the ellipse will be a fixed distance from one fixed point.

4. True, because if we divide each side of the equation by 2, then it fits the standard equation for an ellipse.

5. True, because if we use $x = 0$ in the equation, we get $y = \pm\sqrt{3}$.

6. False, because both x^2 and y^2 must appear in the equation of a hyperbola.

7. False, because in this hyperbola there are no y-intercepts.

8. True, because it has y-intercepts at $(0, -3)$ and $(0, 3)$.

9. True, because if we divide each side of the equation by 4, then it fits the standard form for the equation of a hyperbola.

10. True, the asymptotes are the extended diagonals of the fundamental rectangle.

13.4 EXERCISES

1. An ellipse is the set of all points in a plane such that the sum of their distances from two fixed points is constant.

3. The center of an ellipse is the point that is midway between the foci.

5. The equation of an ellipse centered at (h, k) is $\frac{(x-h)^2}{a^2} + \frac{(y-k)^2}{b^2} = 1$.

7. The asymptotes of a hyperbola are the extended diagonals of the fundamental rectangle.

9. The graph of $\frac{x^2}{9} + \frac{y^2}{4} = 1$ is an ellipse with x-intercepts $(-3, 0)$ and $(3, 0)$, and y-intercepts $(0, 2)$ and $(0, -2)$.

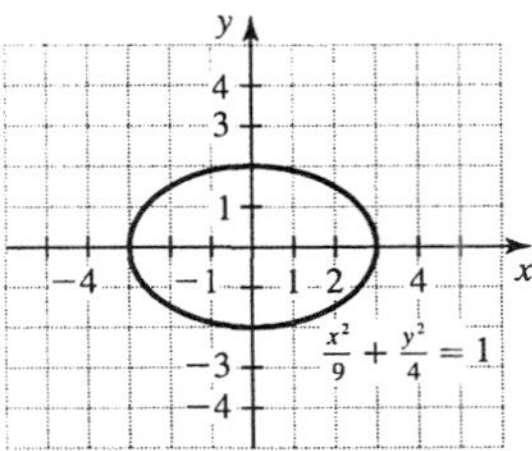

11. The graph of $\frac{x^2}{9}+y^2=1$ is an ellipse with x-intercepts $(-3,\ 0)$ and $(3,\ 0)$, and y-intercepts $(0,\ -1)$ and $(0,\ 1)$.

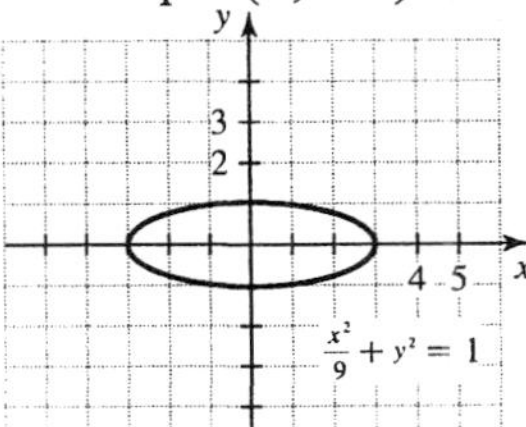

13. The graph of $\frac{x^2}{36}+\frac{y^2}{25}=1$ is an ellipse with x-intercepts $(-6,\ 0)$ and $(6,\ 0)$, and y-intercepts $(0,\ -5)$ and $(0,\ 5)$.

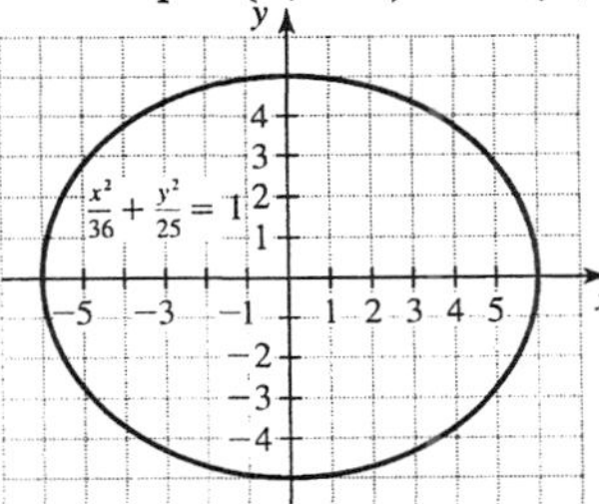

15. The graph of $\frac{x^2}{24}+\frac{y^2}{5}=1$ is an ellipse with x-intercepts $(-\sqrt{24},\ 0)$ and $(\sqrt{24},\ 0)$, and y-intercepts $(0,\ -\sqrt{5})$ and $(0,\ \sqrt{5})$.

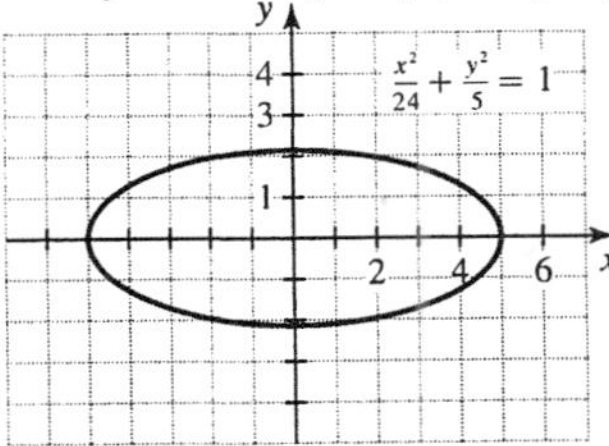

17. The graph $9x^2+16y^2=144$ is an ellipse with x-intercepts $(-4,\ 0)$ and $(4,\ 0)$, and y-intercepts $(0,\ -3)$ and $(0,\ 3)$.

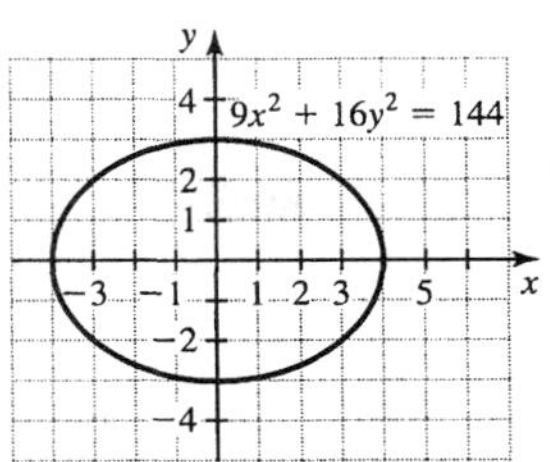

19. The graph of $25x^2+y^2=25$ is an ellipse with x-intercepts $(-1,\ 0)$ and $(1,\ 0)$, and y-intercepts $(0,\ -5)$ and $(0,\ 5)$.

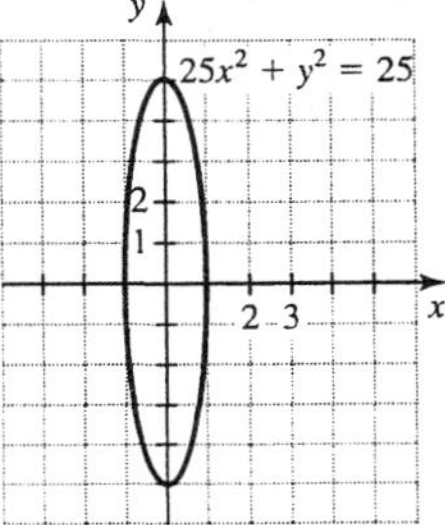

21. The graph of $4x^2+9y^2=1$ is an ellipse with x-intercepts $(1/2,\ 0)$ and $(-1/2,\ 0)$, and y-intercepts $(0,\ 1/3)$ and $(0,\ -1/3)$.

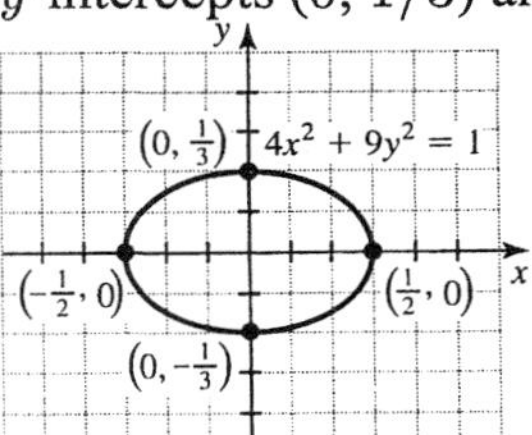

23. The graph of $\frac{(x-3)^2}{4}+\frac{(y-1)^2}{9}=1$ is an ellipse centered at $(3,\ 1)$. The 4 in the denominator indicates that the ellipse passes through points that are 2 units to the right and 2 units to the left of the center: $(5,\ 1)$ and $(1, 1)$. The 9 in the denominator indicates that the ellipse passes through points that are 3 units above and 3 units below the center: $(3,\ 4)$ and $(3,\ -2)$.

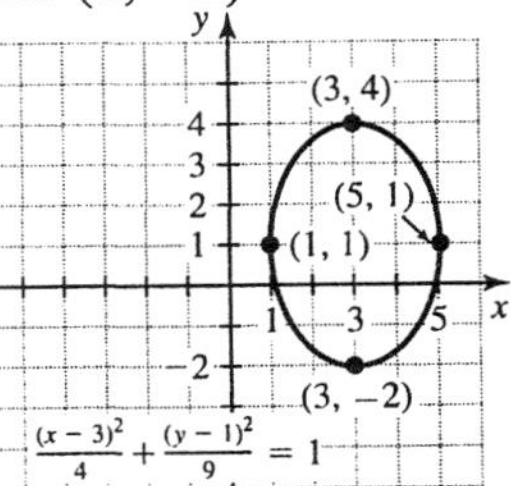

25. The graph of $\frac{(x+1)^2}{16}+\frac{(y-2)^2}{25}=1$ is an ellipse centered at $(-1,\ 2)$. The 16 in the denominator indicates that the ellipse passes through points that are 4 units to the right and 4 units to the left of the center: (3, 2) and $(-5, 2)$. The 25 in the denominator indicates that the ellipse passes through points that are 5 units above and 5 units below the center: $(-1, 7)$ and $(-1, -3)$.

27. The graph of $(x-2)^2 + \frac{(y+1)^2}{36} = 1$ is an ellipse centered at $(2, -1)$. The 1 in the denominator indicates that the ellipse passes through points that are 1 unit to the right and 1 unit to the left of the center: $(3, -1)$ and $(1, -1)$. The 36 in the denominator indicates that the ellipse passes through points that are 6 units above and 6 units below the center: (2, 5) and $(2, -7)$.

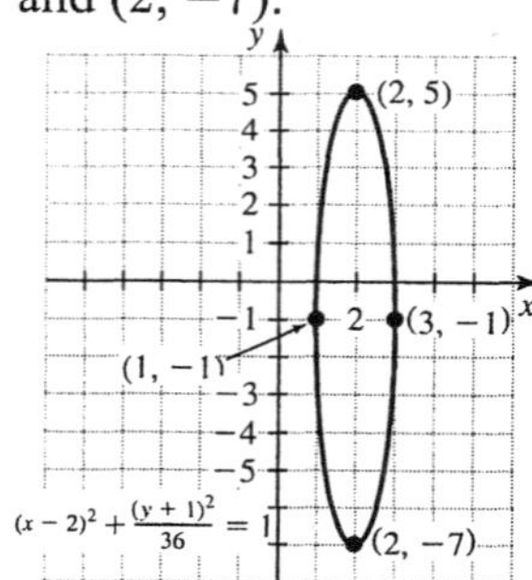

29. The graph of $\frac{x^2}{4} - \frac{y^2}{9} = 1$ is a hyperbola centered at (0, 0) with x-intercepts at $(-2, 0)$ and (2, 0). There are no y-intercepts. Use 9 to determine the size of the fundamental rectangle. The fundamental rectangle passes through (0, 3) and $(0, -3)$. extend the diagonals of the rectangle to determine the asymptotes. The hyperbola opens to the left and right. The equations of the asymptotes are $y = \pm \frac{3}{2}x$.

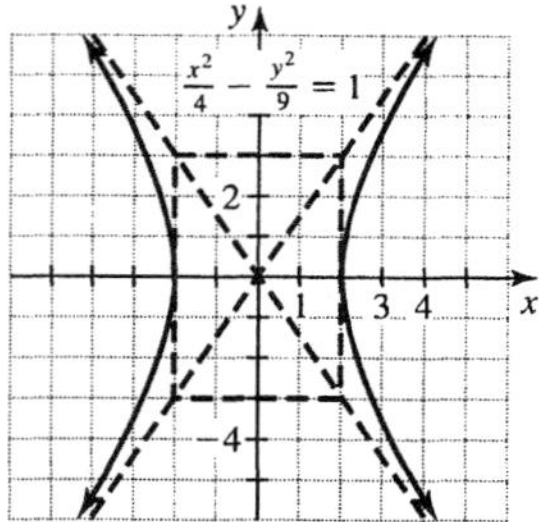

31. The graph of $\frac{y^2}{4} - \frac{x^2}{25} = 1$ is a hyperbola with y-intercepts $(0, -2)$ and (0, 2). The fundamental rectangle passes through the y-intercepts, $(-5, 0)$, and (5, 0). The extended diagonals determine the asymptotes, $y = \pm \frac{2}{5}x$.

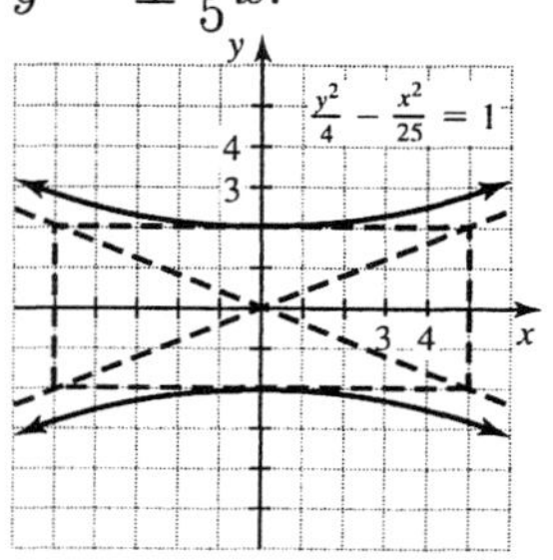

33. The graph of $\frac{x^2}{25} - y^2 = 1$ is a hyperbola with x-intercepts $(-5, 0)$ and (5, 0). The fundamental rectangle passes through the intercepts, $(0, -1)$, and (0, 1). The extended diagonals determine the asymptotes. The equations of the asymptotes are $y = \pm \frac{1}{5}x$.

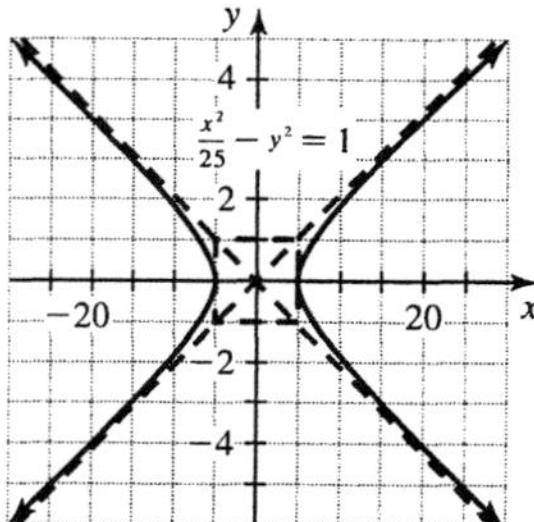

35. The graph of $x^2 - \frac{y^2}{25} = 1$ is a hyperbola with x-intercepts $(-1, 0)$ and (1, 0). Plot (0, 5) and $(0, -5)$ for the fundamental rectangle and extend the diagonals. The equations of the asymptotes are $y = \pm 5x$.

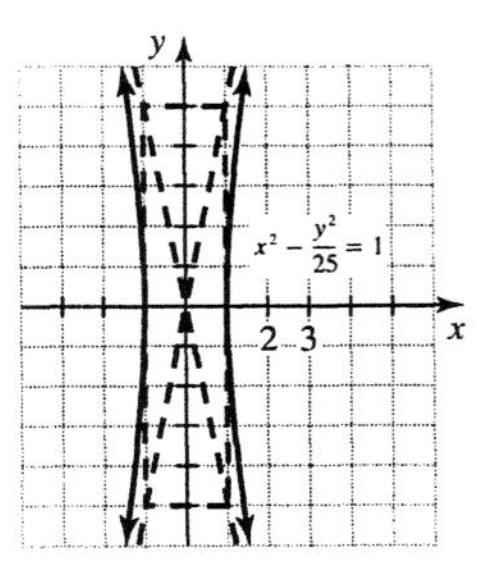

37. Divide each side of $9x^2 - 16y^2 = 144$ by 144 to get $\frac{x^2}{16} - \frac{y^2}{9} = 1$. The graph is a hyperbola with x-intercepts at $(-4, 0)$ and $(4, 0)$. The fundamental rectangle passes through the y-intercepts, $(0, -3)$, and $(0, 3)$. The extended diagonals determine the asymptotes. The equations of the asymptotes are $y = \pm\frac{3}{4}x$.

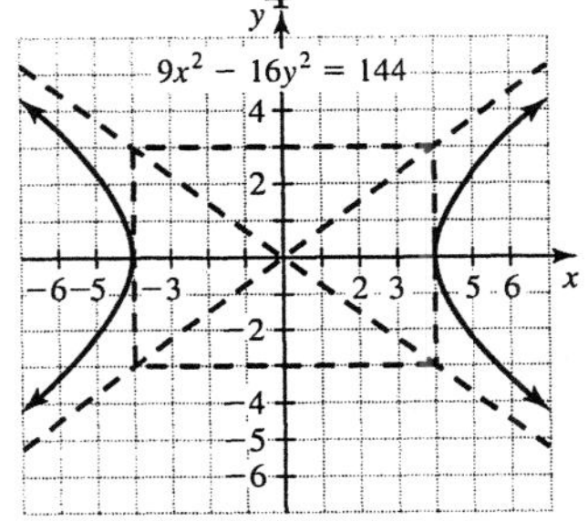

39. The graph of $x^2 - y^2 = 1$ is a hyperbola with x-intercepts $(-1, 0)$ and $(1, 0)$. The fundamental rectangle passes through the y-intercepts, $(0, -1)$, and $(0, 1)$. The extended diagonals determine the asymptotes, $y = \pm x$.

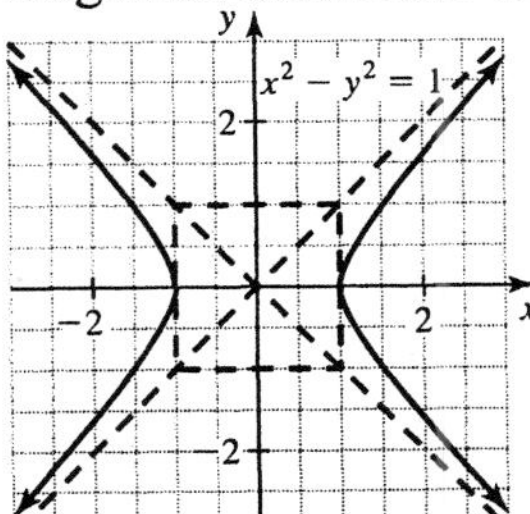

41. The graph of $\frac{(x-2)^2}{4} - (y+1)^2 = 1$ is a hyperbola centered at $(2, -1)$ and opening left and right. It is a transformation of $\frac{x^2}{4} - y^2 = 1$, which has its fundamental rectangle through $(\pm 2, 0)$ and $(0, \pm 1)$. Move the fundamental rectangle 2 to the right and down 1 unit. Then draw the asymptotes and the hyperbola.

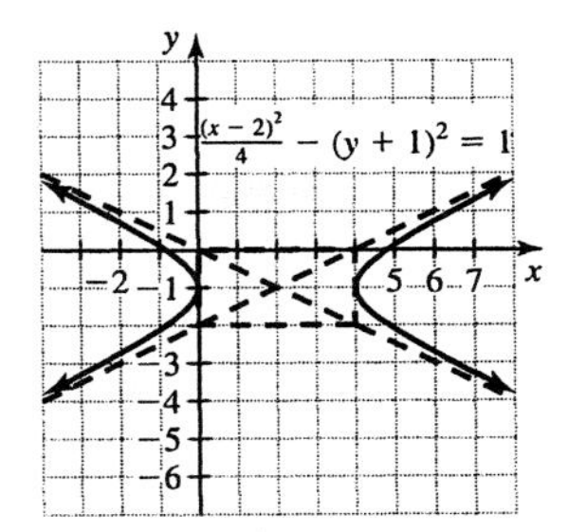

43. The graph of

$$(x+1)^2/16 - (y-1)^2/9 = 1$$

is a hyperbola centered at $(-1, 1)$ and opening left and right. It is a transformation of $x^2/16 - y^2/9 = 1$, which has its fundamental rectangle through $(\pm 4, 0)$ and $(0, \pm 3)$. Move the fundamental rectangle 1 to the left and up 1 unit. Then draw the asymptotes and the hyperbola.

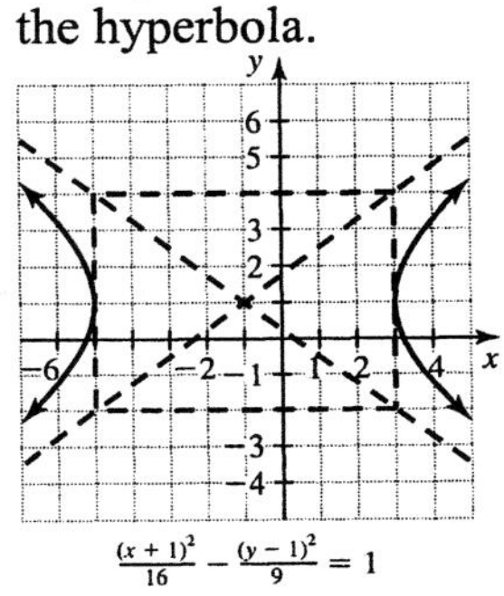

45. The graph of

$$(y-2)^2/9 - (x-4)^2/4 = 1$$

is a hyperbola centered at $(4, 2)$ and opening up and down. It is a transformation of $y^2/9 - x^2/4 = 1$, which has its fundamental rectangle through $(\pm 2, 0)$ and $(0, \pm 3)$. Move the fundamental rectangle 4 to the right and up 2 units. Then draw the asymptotes and the hyperbola.

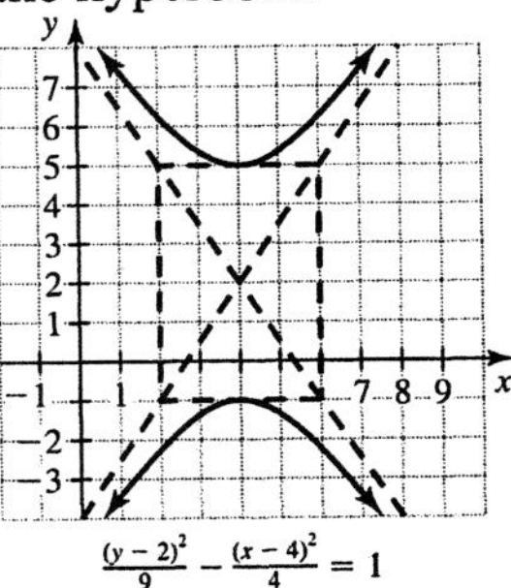

47. The graph of $y = x^2 + 1$ is a parabola because the equation has the form $y = ax^2 + bx + c$.

49. The graph of $x^2 - y^2 = 1$ is a hyperbola because the equation has the form
$\frac{x^2}{a^2} - \frac{y^2}{b^2} = 1.$

51. The graph of $\frac{x^2}{2} + y^2 = 1$ is an ellipse because the equation has the form
$\frac{x^2}{a^2} + \frac{y^2}{b^2} = 1.$

53. The graph of $(x-2)^2 + (y-4)^2 = 9$ is a circle because the equation has the form
$(x-h)^2 + (y-k)^2 = r^2.$

55. First graph the hyperbola and the ellipse.

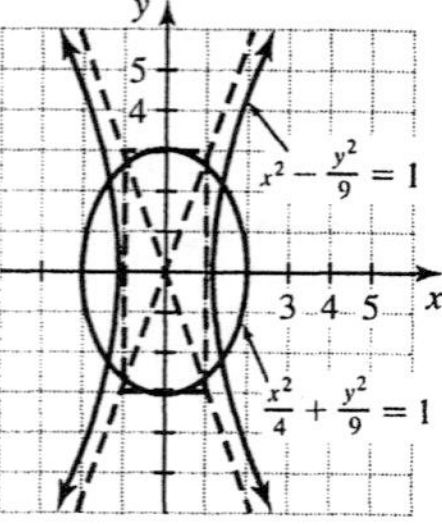

If we add the equations the variable y will be eliminated and we get the following equation.

$$\frac{x^2}{4} + x^2 = 2$$

Now multiply both sides by 4:

$$x^2 + 4x^2 = 8$$
$$5x^2 = 8$$
$$x^2 = \frac{8}{5}$$
$$x = \pm\sqrt{\frac{8}{5}} = \pm\frac{2\sqrt{10}}{5}$$

Use $x^2 = 8/5$ in the second equation.

$$\frac{8}{5} - \frac{y^2}{9} = 1$$
$$-\frac{y^2}{9} = -\frac{3}{5}$$
$$y^2 = \frac{27}{5}$$
$$y = \pm\sqrt{\frac{27}{5}} = \pm\frac{3\sqrt{15}}{5}$$

The graphs intersect at $\left(\frac{2\sqrt{10}}{5}, \frac{3\sqrt{15}}{5}\right)$, $\left(\frac{2\sqrt{10}}{5}, -\frac{3\sqrt{15}}{5}\right)$, $\left(-\frac{2\sqrt{10}}{5}, \frac{3\sqrt{15}}{5}\right)$, and $\left(-\frac{2\sqrt{10}}{5}, -\frac{3\sqrt{15}}{5}\right)$.

57. The graphs are an ellipse and a circle.

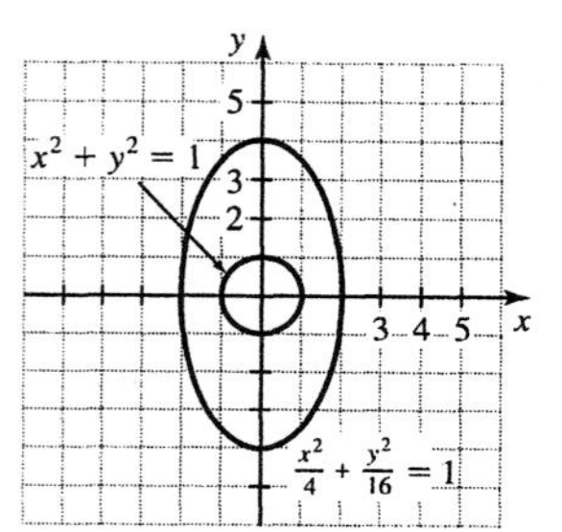

From the graph we can see that there are no points of intersection. If we eliminate y by substituting $y^2 = 1 - x^2$ into the first equation, we get the following equation.

$$\frac{x^2}{4} + \frac{1 - x^2}{16} = 1$$

Now multiply both sides by 16:

$$4x^2 + 1 - x^2 = 16$$
$$3x^2 = 15$$
$$x^2 = 5$$

From $y^2 = 1 - x^2$, we get $y^2 = -4$, which has no real solution. There are no points of intersection.

59. The graphs are a circle and a hyperbola.

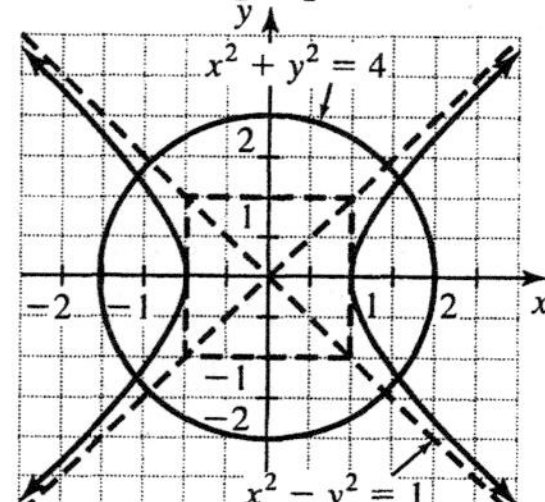

It appears that there are 4 points of intersection. To find them, add the equations to get the following equation.

$$2x^2 = 5$$
$$x^2 = \frac{5}{2}$$
$$x = \pm\sqrt{\frac{5}{2}} = \pm\frac{\sqrt{10}}{2}$$

From $y^2 = 4 - x^2$, we get $y^2 = 4 - \frac{5}{2} = \frac{3}{2}$, or
$y = \pm\sqrt{\frac{3}{2}} = \pm\frac{\sqrt{6}}{2}$.

The graphs intersect at $\left(\frac{\sqrt{10}}{2}, \frac{\sqrt{6}}{2}\right)$, $\left(\frac{\sqrt{10}}{2}, -\frac{\sqrt{6}}{2}\right)$, $\left(-\frac{\sqrt{10}}{2}, \frac{\sqrt{6}}{2}\right)$, and $\left(-\frac{\sqrt{10}}{2}, -\frac{\sqrt{6}}{2}\right)$.

61. The graphs are an ellipse and a circle.

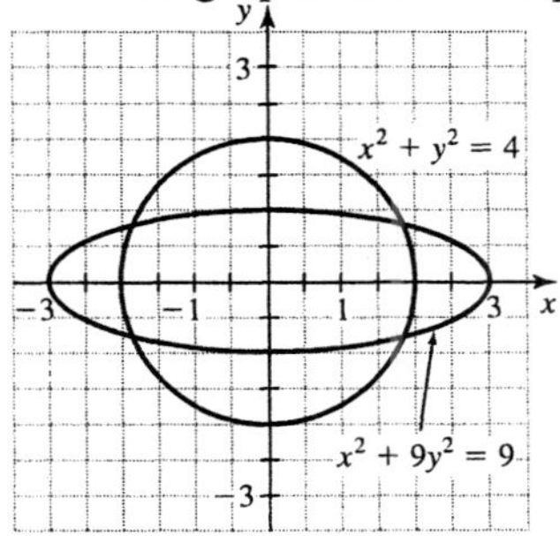

The graphs appear to intersect at 4 points. To find the points, substitute $y^2 = 4 - x^2$ into the first equation.

$$x^2 + 9(4 - x^2) = 9$$
$$-8x^2 + 36 = 9$$
$$-8x^2 = -27$$
$$x^2 = \frac{27}{8}$$
$$x = \pm\sqrt{\frac{27}{8}} = \pm\frac{3\sqrt{6}}{4}$$

Use $x^2 = 27/8$ in $y^2 = 4 - x^2$.

$$y^2 = 4 - \frac{27}{8} = \frac{5}{8}$$
$$y = \pm\sqrt{\frac{5}{8}} = \pm\frac{\sqrt{10}}{4}$$

The graphs intersect at $\left(\frac{3\sqrt{6}}{4}, \frac{\sqrt{10}}{4}\right)$, $\left(\frac{3\sqrt{6}}{4}, -\frac{\sqrt{10}}{4}\right)$, $\left(-\frac{3\sqrt{6}}{4}, \frac{\sqrt{10}}{4}\right)$, and $\left(-\frac{3\sqrt{6}}{4}, -\frac{\sqrt{10}}{4}\right)$.

63. Graph the parabola and the ellipse.

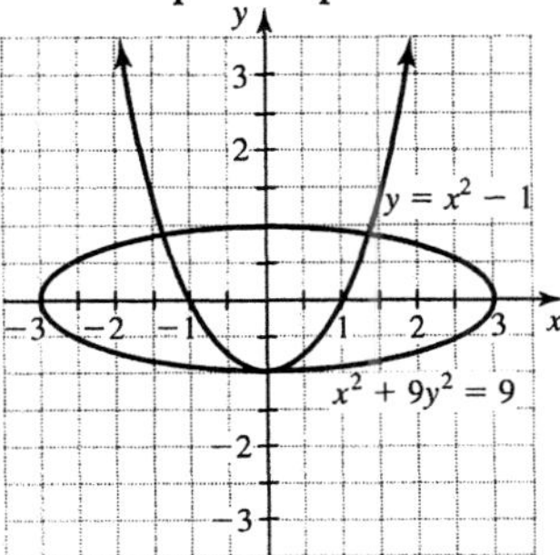

The graphs appear to intersect at 3 points. To find them, substitute $x^2 = y + 1$ into the first equation.

$$y + 1 + 9y^2 = 9$$
$$9y^2 + y - 8 = 0$$
$$(9y - 8)(y + 1) = 0$$
$$y = \frac{8}{9} \quad \text{or} \quad y = -1$$

If $y = -1$, then $x^2 = -1 + 1 = 0$ or $x = 0$. If $y = 8/9$, then

$$x^2 = \frac{8}{9} + 1 = \frac{17}{9} \quad \text{or} \quad x = \pm\frac{\sqrt{17}}{3}.$$

The three points of intersection are $\left(\frac{\sqrt{17}}{3}, \frac{8}{9}\right)$, $\left(-\frac{\sqrt{17}}{3}, \frac{8}{9}\right)$, and $(0, -1)$.

65. Graph the hyperbola and the line.

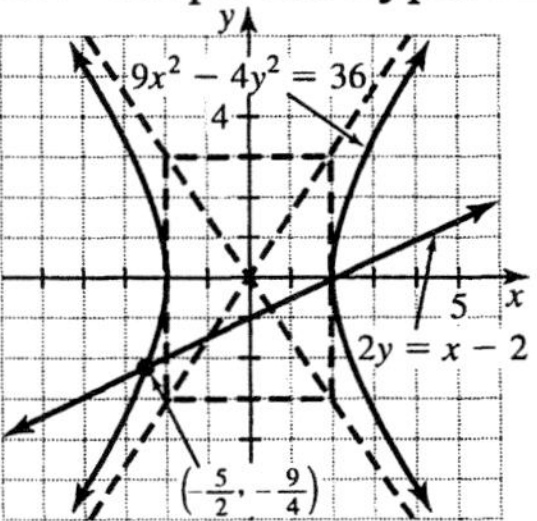

To find the points of intersection, substitute $x = 2y + 2$ into the first equation.

$$9(2y + 2)^2 - 4y^2 = 36$$
$$9(4y^2 + 8y + 4) - 4y^2 = 36$$
$$36y^2 + 72y + 36 - 4y^2 = 36$$
$$32y^2 + 72y = 0$$
$$4y^2 + 9y = 0$$
$$y(4y + 9) = 0$$
$$y = 0 \quad \text{or} \quad y = -9/4$$

If $y = 0$, then $x = 2(0) + 2 = 2$. If $y = -9/4$, then $x = 2(-9/4) + 2 = -5/2$. The two points of intersection are $(2, 0)$ and $(-5/2, -9/4)$.

67. a) From the graph it appears that the boat is approximately at (2.5, 1.5)

b) Substitute $x^2 = 1 + 3y^2$ into $4y^2 - x^2 = 1$.

$$4y^2 - (1 + 3y^2) = 1$$
$$y^2 = 2$$
$$y = \pm\sqrt{2}$$
$$x^2 = 1 + 3(\pm\sqrt{2})^2 = 1 \pm 3(2) = 1 \pm 6$$
$$= 7 \text{ or } -5$$

If $x^2 = 7$, then $x = \pm\sqrt{7}$. If $x^2 = -5$, there is no real solution. Since the boat is in the first quadrant, the boat is at $(\sqrt{7}, \sqrt{2})$.

13.5 WARM-UPS

1. False, because $x^2 + y = 4$ is the equation of a parabola.

2. True, because if we divide each side by 9 we get the standard form for an ellipse.

3. True, because it can be written as $y^2 - x^2 = 1$.

4. True, because $2(0)^2 - 0 < 3$ is correct.

5. False, because $0 > 0^2 - 3(0) + 2$ is incorrect.
6. False, because the origin is on the parabola $x^2 = y$.
7. False, because (0, 0) does not satisfy the inequality.
8. True, because $x^2 + y^2 = 4$ is a circle of radius 2 centered at (0, 0), and the points inside this circle satisfy the inequality.
9. True, because both $0^2 - 4^2 < 1$ and $4 > 0^2 - 2(0) + 3$ are correct.
10. True, because both $0^2 + 0^2 < 1$ and $0 < 0^2 + 1$ are correct.

13.5 EXERCISES

1. Graph the parabola $y = x^2$. Since (0, 4) satisfies $y > x^2$, shade the region containing $(0, 4)$.

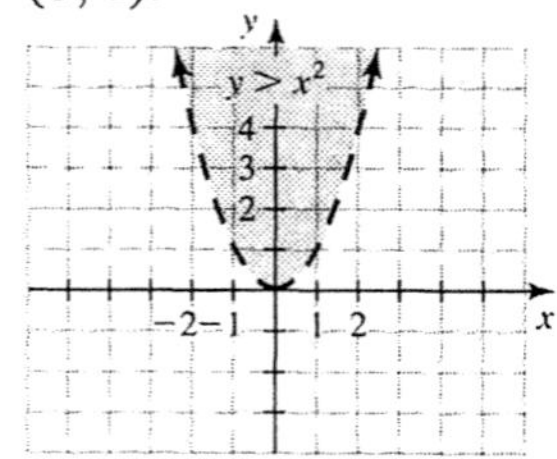

3. First graph the parabola $y = x^2 - x$. Since (5, 0) satisfies $y < x^2 - x$, we shade the region containing (5, 0).

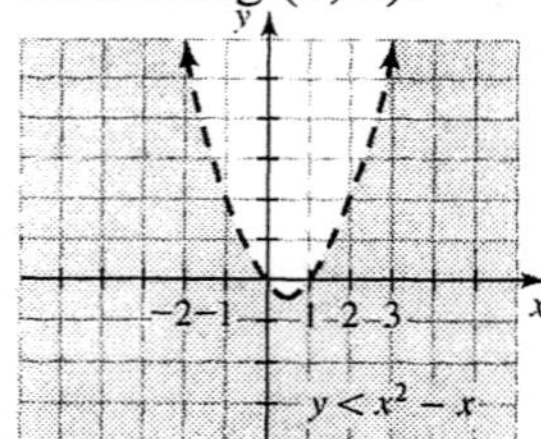

5. First graph the parabola $y = x^2 - x - 2$. Since (0, 5) satisfies $y > x^2 - x - 2$, shade the region containing (0, 5).

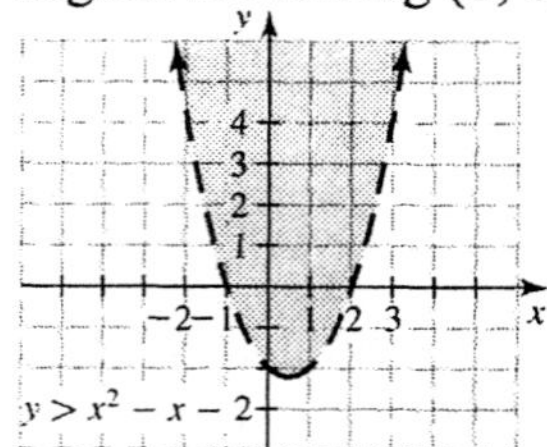

7. First graph the circle $x^2 + y^2 = 9$ using a solid curve. Since (0, 0) satisfies $x^2 + y^2 \leq 9$, shade the inside of the circle.

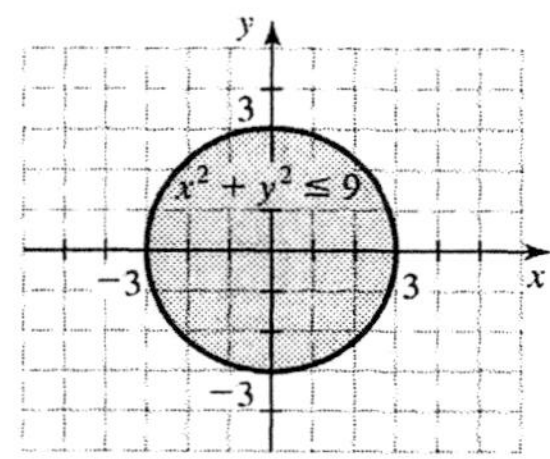

9. First graph the ellipse $x^2 + 4y^2 = 4$ using a dashed curve. Since (0, 0) does not satisfy the inequality $x^2 + 4y^2 > 4$, we shade the region outside the ellipse.

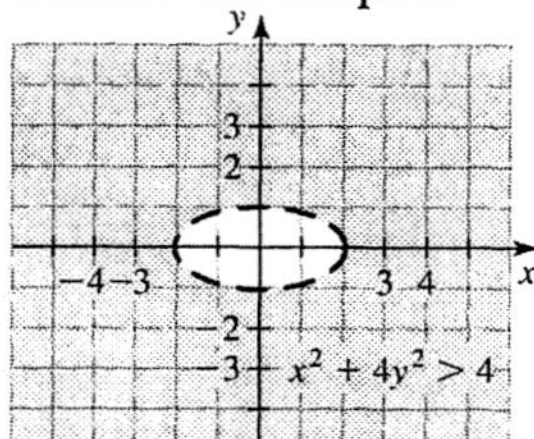

11. First divide both sides of the inequality by 36.

$$\frac{x^2}{9} - \frac{y^2}{4} < 1$$

Now graph the hyperbola $\frac{x^2}{9} - \frac{y^2}{4} = 1$. Testing a point in each of the three regions, we find that only points in the region containing the origin satisfy $4x^2 - 9y^2 < 36$.

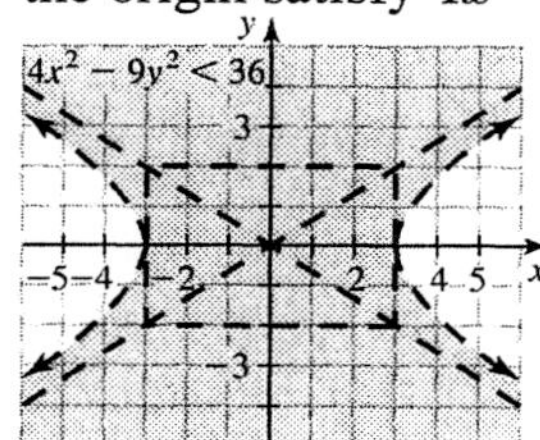

13. First graph the circle centered at (2, 3) with radius 2, using a dashed curve. Since $(2, 3)$ satisfies $(x - 2)^2 + (y - 3)^2 < 4$, shade the region inside the circle.

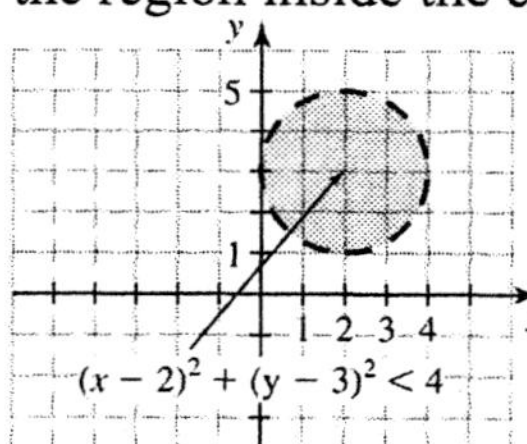

15. First graph the circle centered at (0, 0) with radius 1. Since (0, 0) does not satisfy $x^2 + y^2 > 1$, shade the region outside the circle.

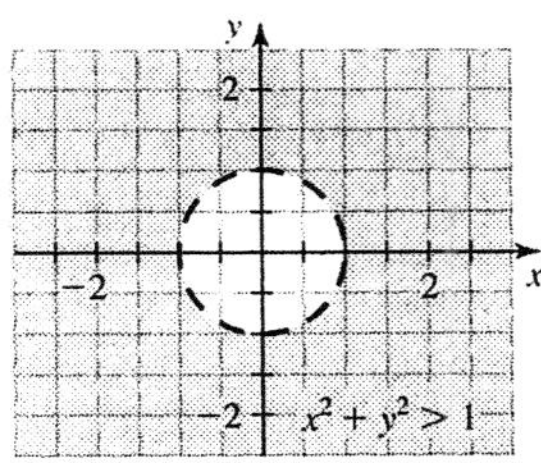

17. Graph the hyperbola $4x^2 - y^2 = 4$, using a dashed curve. After testing a point in each of the three regions, we see that only points in the region containing (0, 0) fail to satisfy $4x^2 - y^2 > 4$.

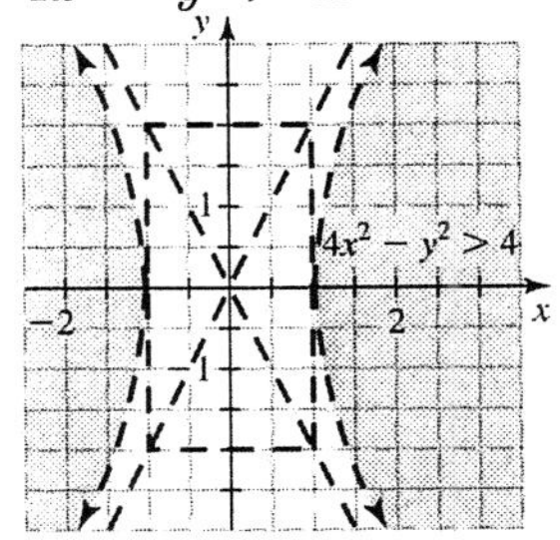

19. Graph the hyperbola $y^2 - x^2 = 1$. After testing a point in each of the three regions, we see that only points in the region containing (0, 0) satisfy $y^2 - x^2 \leq 1$.

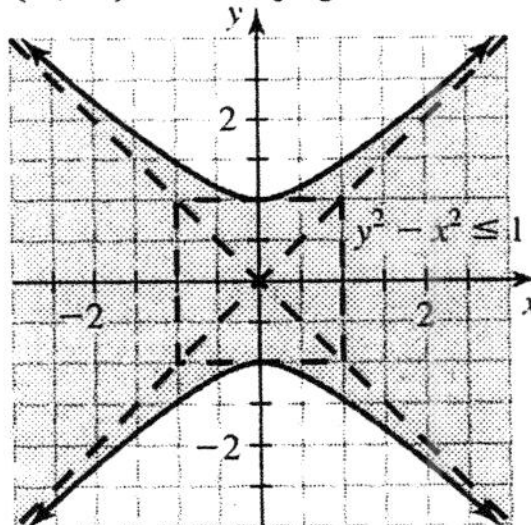

21. The graph of $y = x$ is a line with slope 1 and y-intercept (0, 0). Since $(5, -5)$ satisfies $x > y$, we shade the region below the line.

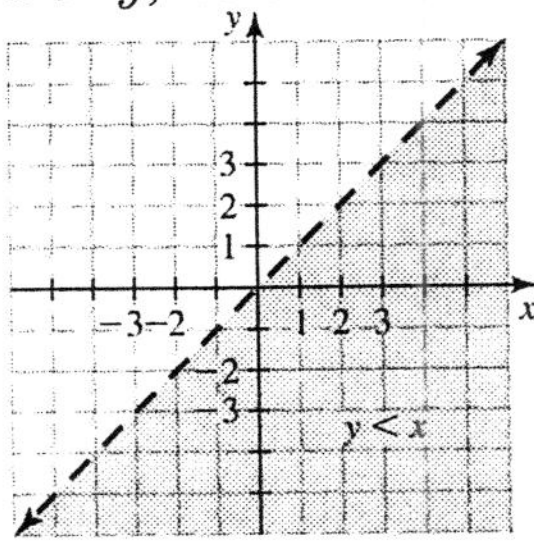

23. Check $(3, -4)$ in each inequality:

$$\begin{array}{ll} x^2 + y^2 \leq 25 & y \leq x^2 \\ 3^2 + (-4)^2 \leq 25 & -4 \leq 3^2 \\ 25 \leq 25 & -4 \leq 9 \\ \text{True} & \text{True} \end{array}$$

Since $(3, -4)$ satisfies both inequalities it is in the solution set to the system.

25. Check $(3, -4)$ in each inequality:

$$\begin{array}{ll} x - y > 1 & y > (x-2)^2 + 3 \\ 3 - (-4) > 1 & -4 > (3-2)^2 + 3 \\ 7 > 1 & -4 > 4 \\ \text{True} & \text{False} \end{array}$$

Since $(3, -4)$ does not satisfy both inequalities it is not in the solution set to the system.

27. Graph the circle $x^2 + y^2 = 9$ and the line $y = x$. The graph of $x^2 + y^2 < 9$ is the region inside the circle. the graph of $y > x$ is the region above the line. The intersection of these two regions is shown as follows.

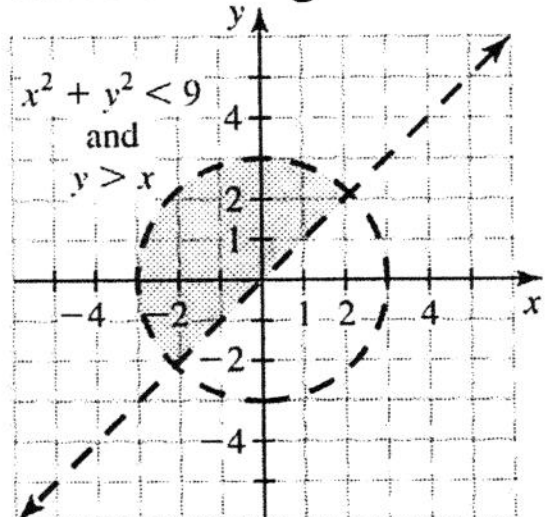

We could have used test points to determine which of the 4 regions determined by the circle and the line satisfies both inequalities.

29. The graph of $x^2 - y^2 = 1$ is a hyperbola. The graph of $x^2 + y^2 = 4$ is a circle of radius 2. Points that satisfy $x^2 + y^2 < 4$ are inside the circle. The hyperbola divides the plane into 3 regions. Points in the 2 regions not containing (0, 0) are the points that satisfy $x^2 - y^2 > 1$. The intersection of these regions is shown in the following graph.

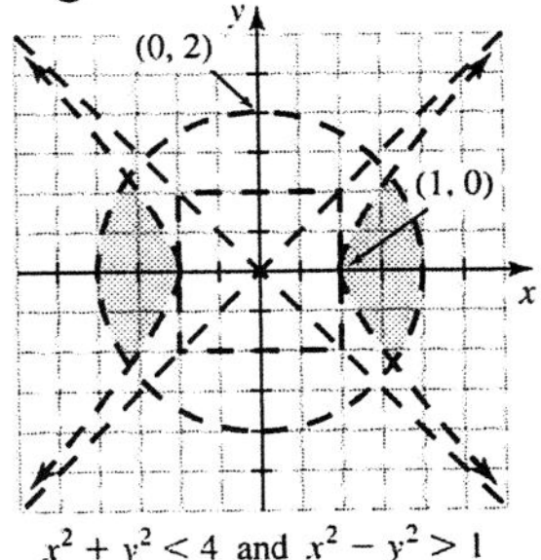

31. The graph of $y = x^2 + x$ is a parabola opening upward, with x-intercepts at (0, 0) and $(-1, 0)$. The graph of $y > x^2 + x$ is the region inside this parabola. The graph of $y = 5$ is a horizontal line through (0, 5). The graph of $y < 5$ is the region below this horizontal line.

The points that satisfy both inequalities are the points inside the parabola and below $y = 5$, as shown in the following graph.

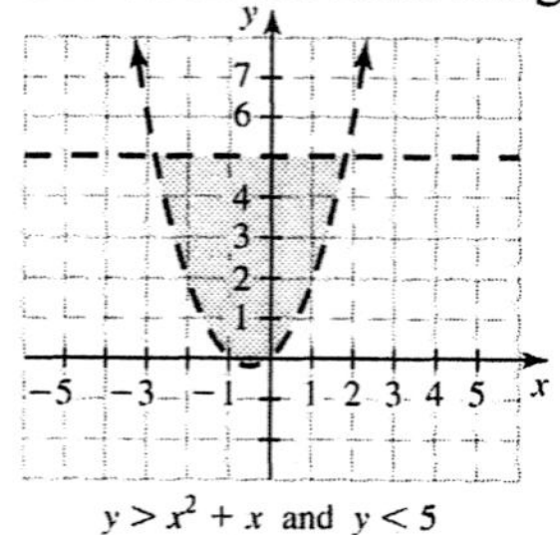

33. The graph of $y = x + 2$ is a line with y-intercept (0, 2) and slope 1. The graph of $y \geq x + 2$ is the region above and including the line. The graph of $y = 2 - x$ is a line with y-intercept (0, 2) and slope -1. The graph of $y \leq 2 - x$ is the region below and including this line. The region above the first line and below the second is shown in the following graph. We could have tested a point in each of the 4 regions determined by the two lines. We would find that only points in the region shown satisfy both inequalities.

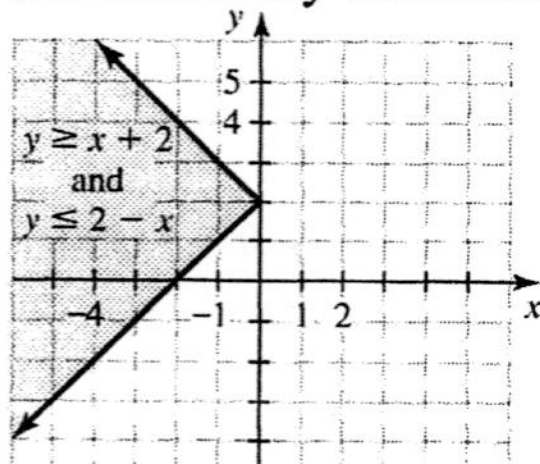

35. The graph of $4x^2 - y^2 = 4$ is a hyperbola opening to the left and right. Points that satisfy $4x^2 - y^2 < 4$ are in the region between the two branches of the hyperbola. The graph of $x^2 + 4y^2 = 4$ is an ellipse passing through $(0, 1)$, $(0, -1)$, $(2, 0)$, and $(-2, 0)$. Points that satisfy $x^2 + 4y^2 > 4$ are outside the ellipse. Points that satisfy both inequalities are outside the ellipse and between the two branches of the hyperbola.

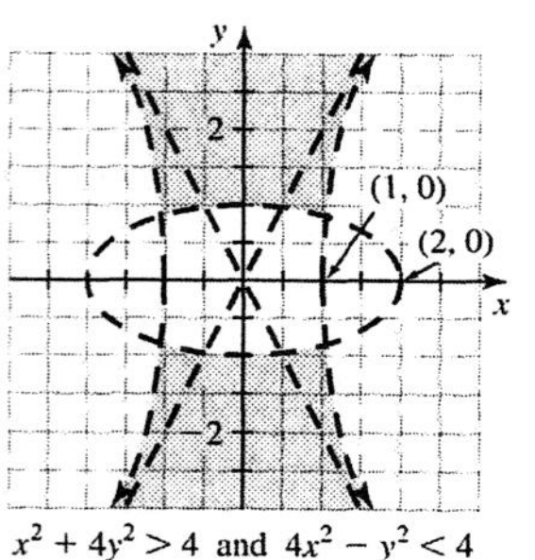

37. The graph of $x - y = 0$ is the same as the line $y = x$ (through (0, 0) with slope 1). Points that satisfy $x - y < 0$ are above the line $y = x$. The graph of $y + x^2 = 1$ is the same as the parabola $y = -x^2 + 1$, which opens downward. Points that satisfy $y + x^2 < 1$ are below the parabola. The points that satisfy both inequalities, below the parabola and above the line, are shown in the following graph.

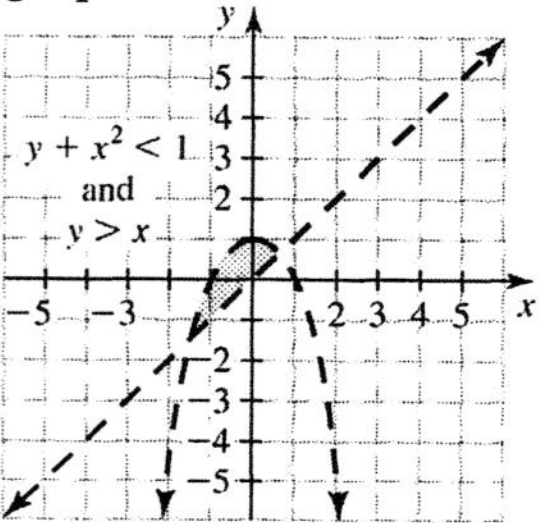

39. The graph of $y = 5x - x^2$ is a parabola opening downward. Points that satisfy $y < 5x - x^2$ are the points below the parabola. The graph of $x^2 + y^2 = 9$ is a circle of radius 3. Points that satisfy $x^2 + y^2 < 9$ are inside the circle. Points that satisfy both inequalities are the points inside the circle and below the parabola.

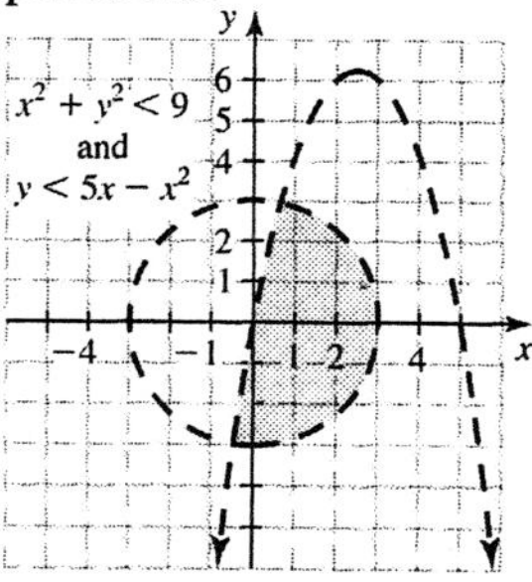

41. Points that satisfy $y \geq 3$ are above or on the horizontal line $y = 3$. Points that satisfy $x \leq 1$ are on or to the left of the vertical line $x = 1$. Points that satisfy both inequalities are shown in the following graph. The points graphed are the points with x-coordinate less than or equal to 1, and y-coordinate greater than or equal to 3.

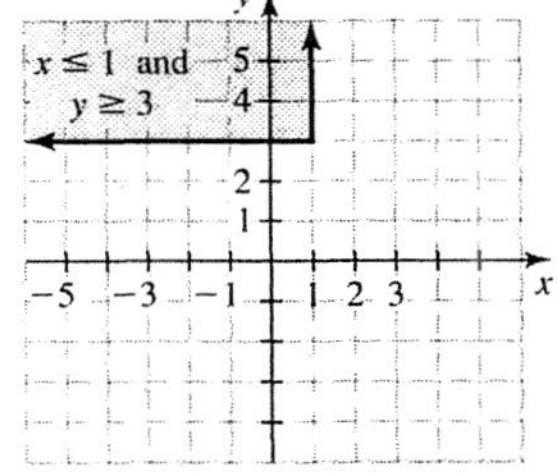

43. The graph of $4y^2 - 9x^2 = 36$ is a hyperbola opening up and down. Points that satisfy $4y^2 - 9x^2 < 36$ are the points between the two branches of the hyperbola. The graph of $x^2 + y^2 = 16$ is a circle of radius 4. Points that satisfy $x^2 + y^2 < 16$ are inside the circle. The points that satisfy both inequalities are indicated as follows.

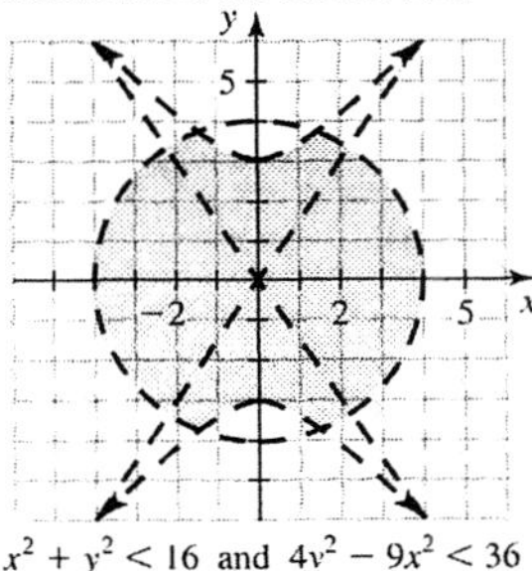
$x^2 + y^2 < 16$ and $4y^2 - 9x^2 < 36$

45. The graph of $y = x^2$ is a parabola opening upward. Points that satisfy $y < x^2$ are below this parabola. The graph of $x^2 + y^2 = 1$ is a circle of radius 1. Points that satisfy $x^2 + y^2 < 1$ are inside the circle. The graph of the system of inequalities consists of points that are below the parabola and inside the circle.

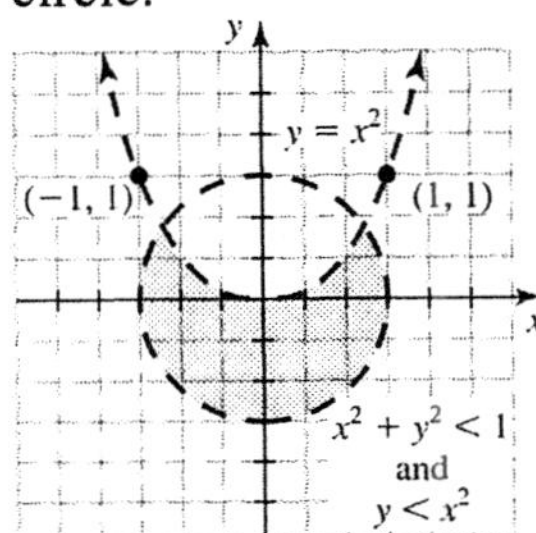

47. Let $x =$ the number of paces he walked to the east and $y =$ the number of paces he walked to the north. So $x + y \geq 50$, $x^2 + y^2 \leq 50^2$, and $y > x$. The graph of this system follows.

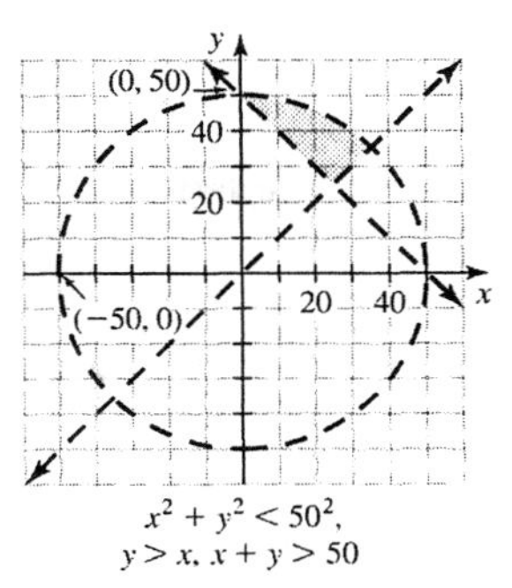

$x^2 + y^2 < 50^2$,
$y > x$, $x + y > 50$

49. From the graphs it appears that there is no intersection of the two regions and there is no solution to the system.

Enriching Your Mathematical Word Power

1. c **2.** a **3.** d **4.** a **5.** c
6. d **7.** b **8.** d **9.** c **10.** a

CHAPTER 13 REVIEW

1. The graph of $y = x^2$ is a parabola and the graph of $y = -2x + 15$ is a straight line.

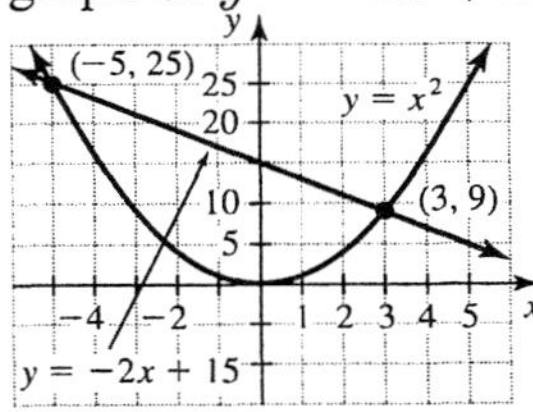

Use substitution to eliminate y.

$$x^2 = -2x + 15$$
$$x^2 + 2x - 15 = 0$$
$$(x + 5)(x - 3) = 0$$
$$x = -5 \text{ or } x = 3$$
$$y = 25 \qquad y = 9 \text{ Since } y = x^2$$

The solution set is $\{(3, 9), (-5, 25)\}$.

3. First graph $y = 3x$ and $y = 1/x$.

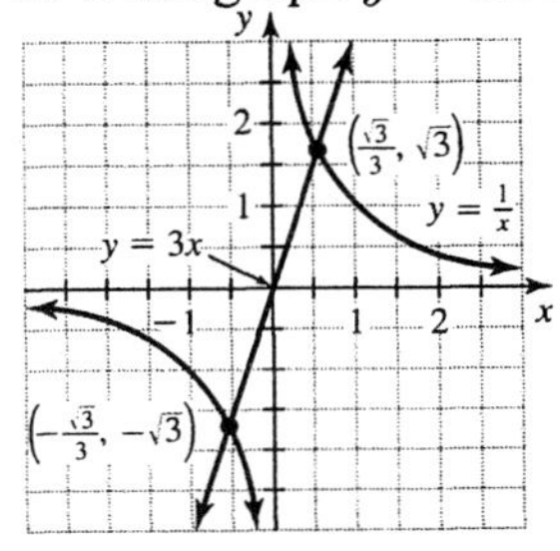

Use substitution to eliminate y.

$$3x = \frac{1}{x}$$
$$3x^2 = 1$$
$$x^2 = \frac{1}{3}$$

$$x = \pm\sqrt{\frac{1}{3}} = \pm\frac{\sqrt{3}}{3}$$

If $x = \frac{\sqrt{3}}{3}$, then $y = \sqrt{3}$. If $x = -\frac{\sqrt{3}}{3}$, then $y = -\sqrt{3}$. The solution set is $\left\{\left(\frac{\sqrt{3}}{3}, \sqrt{3}\right), \left(-\frac{\sqrt{3}}{3}, -\sqrt{3}\right)\right\}$.

5. The second equation can be written as $x^2 = 3y$. Substitute this equation into the first equation.

$$\begin{aligned} 3y + y^2 &= 4 \\ y^2 + 3y - 4 &= 0 \\ (y+4)(y-1) &= 0 \\ y = -4 \text{ or } y &= 1 \end{aligned}$$

If $y = -4$, then $x^2 = 3(-4)$ has no solution. If $y = 1$, then $x^2 = 3(1)$ gives us $x = \pm\sqrt{3}$. The solution set is $\left\{(\sqrt{3}, 1), (-\sqrt{3}, 1)\right\}$.

7. Use substitution to eliminate y.

$$\begin{aligned} x^2 + (x+2)^2 &= 34 \\ x^2 + x^2 + 4x + 4 &= 34 \\ 2x^2 + 4x - 30 &= 0 \\ x^2 + 2x - 15 &= 0 \\ (x+5)(x-3) &= 0 \\ x = -5 \text{ or } x &= 3 \\ y = -3 \quad y = 5 \quad &\text{Since } y = x + 2 \end{aligned}$$

The solution set is $\{(-5, -3), (3, 5)\}$.

9. Use substitution to eliminate y.

$$\begin{aligned} \log(x-3) &= 1 - \log(x) \\ \log(x-3) + \log(x) &= 1 \\ \log(x^2 - 3x) &= 1 \\ x^2 - 3x &= 10 \\ x^2 - 3x - 10 &= 0 \\ (x-5)(x+2) &= 0 \\ x = 5 \text{ or } x &= -2 \end{aligned}$$

If $x = 5$, then $y = \log(5-3) = \log(2)$. If $x = -2$, we get a logarithm of a negative number. So the solution set is $\{(5, \log(2))\}$.

11. Use substitution to eliminate x.

$$\begin{aligned} y^2 &= 2(12 - y) \\ y^2 &= 24 - 2y \\ y^2 + 2y - 24 &= 0 \\ (y+6)(y-4) &= 0 \\ y = -6 \text{ or } y &= 4 \end{aligned}$$

If $y = -6$, then $-6 = x^2$ has no solution. If $y = 4$, then $x^2 = 4$ and $x = \pm 2$. The solution set is $\{(2, 4), (-2, 4)\}$.

13. $\sqrt{(3-1)^2 + (3-1)^2} = \sqrt{4+4} = 2\sqrt{2}$

15. $\sqrt{(2-(-4))^2 + (-8-6)^2}$
$= \sqrt{36 + 196} = 2\sqrt{58}$

17. For $y = x^2 + 3x - 18$, use $x = -b/(2a)$ to find the x-coordinate of the vertex. The vertex is $(-\frac{3}{2}, -\frac{81}{4})$ and the axis of symmetry is $x = -3/2$. Since $a = 1$, $p = 1/4$. The focus is $(-\frac{3}{2}, -20)$ and the directrix is $y = -\frac{41}{2}$.

19. For $y = x^2 + 3x + 2$, use $x = -b/(2a)$ to find the x-coordinate of the vertex. The vertex is $(-\frac{3}{2}, -\frac{1}{4})$ and the axis of symmetry is $x = -3/2$. Since $a = 1$, $p = 1/4$. The focus is $(-\frac{3}{2}, 0)$ and the directrix is $y = -\frac{1}{2}$.

21. For $y = -\frac{1}{2}(x-2)^2 + 3$ the vertex is $(2, 3)$ and the parabola opens down. The axis of symmetry is $x = 2$. Since $a = -1/2$, $p = -1/2$. The focus is $(2, \frac{5}{2})$ and the directrix is $y = \frac{7}{2}$.

23.
$$\begin{aligned} y &= 2x^2 - 8x + 1 \\ y &= 2(x^2 - 4x) + 1 \\ y &= 2(x^2 - 4x + 4 - 4) + 1 \\ y &= 2(x^2 - 4x + 4) - 8 + 1 \\ y &= 2(x-2)^2 - 7 \end{aligned}$$

Vertex $(2, -7)$

25.
$$\begin{aligned} y &= -\tfrac{1}{2}x^2 - x + \tfrac{1}{2} \\ y &= -\tfrac{1}{2}(x^2 + 2x) + \tfrac{1}{2} \\ y &= -\tfrac{1}{2}(x^2 + 2x + 1 - 1) + \tfrac{1}{2} \\ y &= -\tfrac{1}{2}(x+1)^2 + \tfrac{1}{2} + \tfrac{1}{2} \\ y &= -\tfrac{1}{2}(x+1)^2 + 1 \end{aligned}$$

Vertex $(-1, 1)$

27. The graph of $x^2 + y^2 = 100$ is a circle centered at $(0, 0)$ with radius 10.

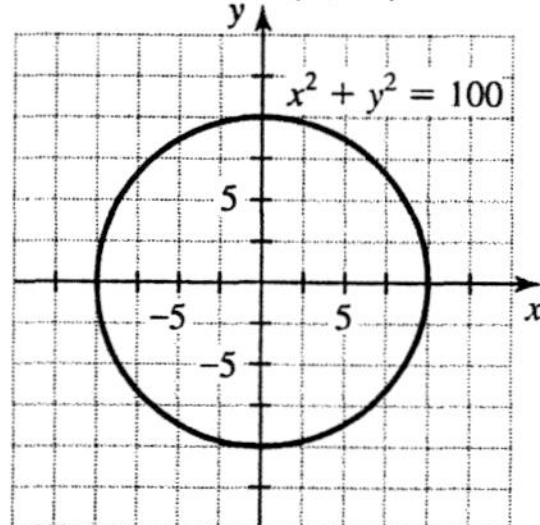

29. The graph of $(x-2)^2 + (y+3)^2 = 81$ is a circle centered at $(2, -3)$ with radius 9.

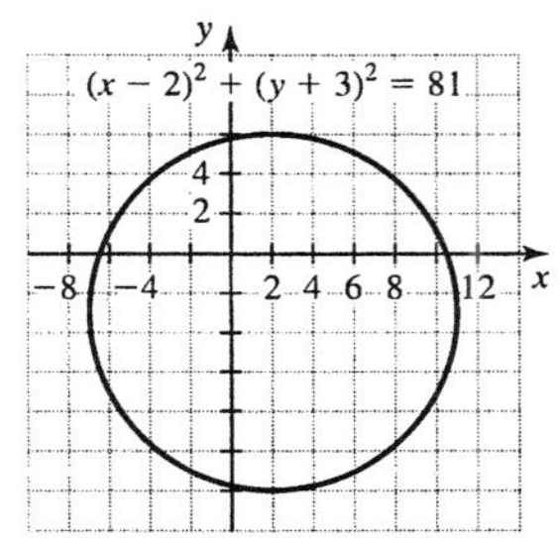

31. $9y^2 + 9x^2 = 4$

$x^2 + y^2 = \frac{4}{9}$

The graph is a circle with center (0, 0) and radius 2/3.

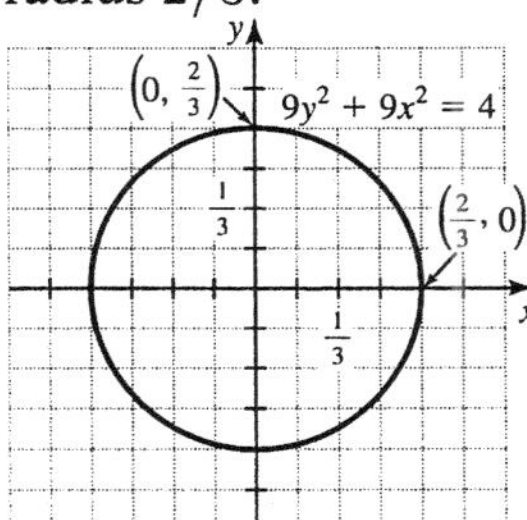

33. The equation of a circle with center (h, k) and radius r is $(x - h)^2 + (y - k)^2 = r^2$. The equation of a circle with center (0, 3) and radius 6 is $x^2 + (y - 3)^2 = 36$.

35. The equation of a circle with center $(2, -7)$ and radius 5 is $(x - 2)^2 + (y + 7)^2 = 25$.

37. The graph of $\frac{x^2}{36} + \frac{y^2}{49} = 1$ is an ellipse with x-intercepts at $(-6, 0)$ and $(6, 0)$, y-intercepts at $(0, -7)$ and $(0, 7)$, and centered at the origin.

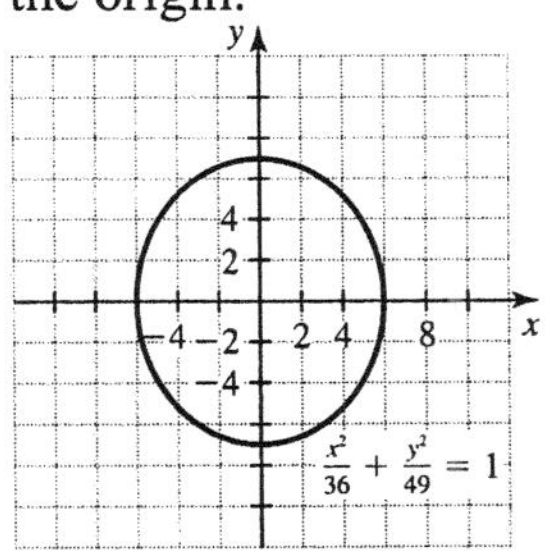

39. The graph of $25x^2 + 4y^2 = 100$ is an ellipse with x-intercepts $(-2, 0)$ and $(2, 0)$, y-intercepts $(0, -5)$ and $(0, 5)$, and centered at the origin.

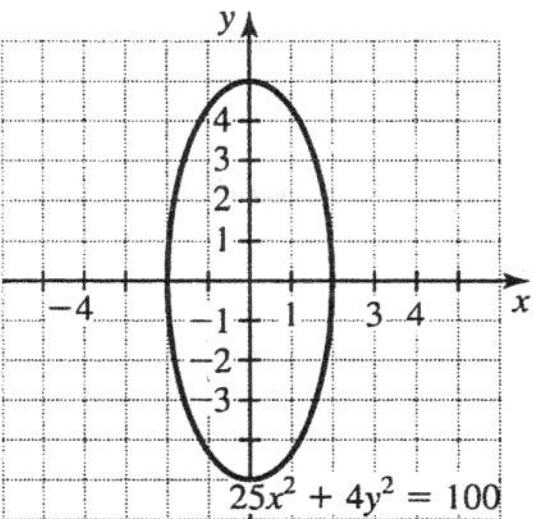

41. The graph of $\frac{x^2}{49} - \frac{y^2}{36} = 1$ is a hyperbola with x-intercepts $(-7, 0)$ and $(7, 0)$. The fundamental rectangle passes through the x-intercepts and the points $(0, -6)$ and $(0, 6)$. Extend the diagonals of the fundamental rectangle to get the asymptotes and draw a hyperbola opening to the left and right.

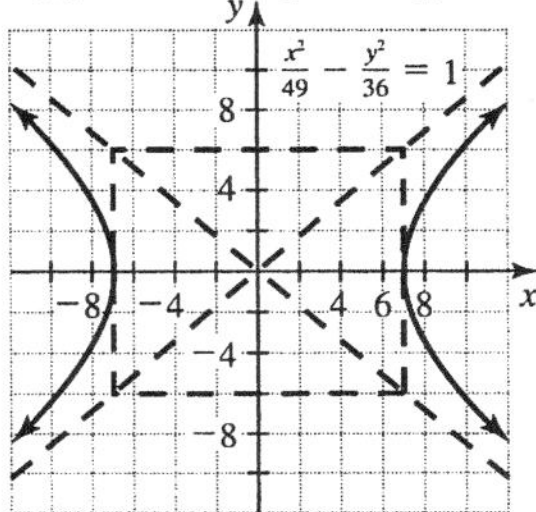

43. Write $4x^2 - 25y^2 = 100$ as $\frac{x^2}{25} - \frac{y^2}{4} = 1$. The graph has x-intercepts at $(-5, 0)$ and $(5, 0)$. The fundamental rectangle passes through the x-intercepts and the points $(0, -2)$ and $(0, 2)$. Extend the diagonals of the rectangle to get the asymptotes and draw a hyperbola opening to the left and right.

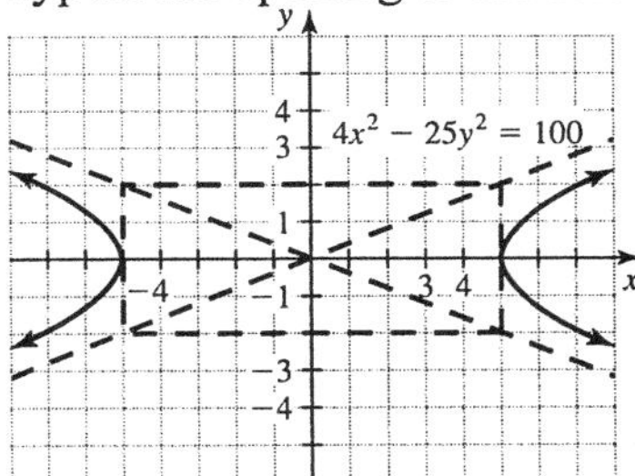

45. First graph the line $4x - 2y = 3$. It goes through $(0, -3/2)$ and $(3/4, 0)$. Since (0, 0) fails to satisfy $4x - 2y > 3$, we shade the region not containing (0, 0).

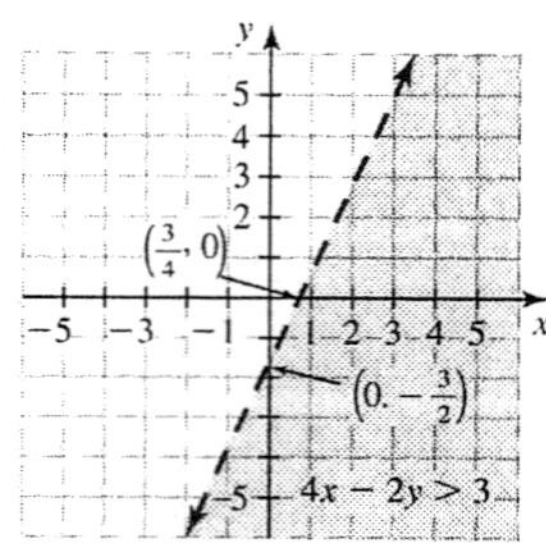

47. Write $y^2 < x^2 - 1$ as $x^2 - y^2 > 1$. Graph the hyperbola $x^2 - y^2 = 1$ and test a point in each of the three regions: $(-5, 0)$, $(0, 0)$, and $(5, 0)$. Since $(-5, 0)$ and $(5, 0)$ satisfy the inequality, we shade the regions containing those points.

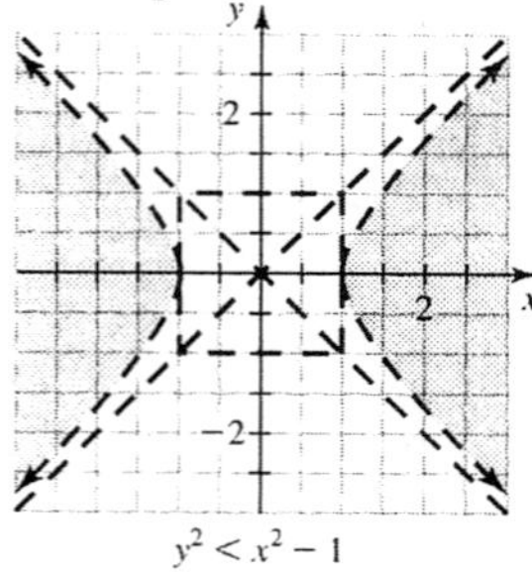

49. Write $4x^2 + 9y^2 > 36$ as $\frac{x^2}{9} + \frac{y^2}{4} > 1$. First graph the ellipse $\frac{x^2}{9} + \frac{y^2}{4} = 1$ through $(-3, 0)$, $(3, 0)$, $(0, -2)$, and $(0, 2)$. Since $(0, 0)$ fails to satisfy the inequality, we shade the region outside of the ellipse.

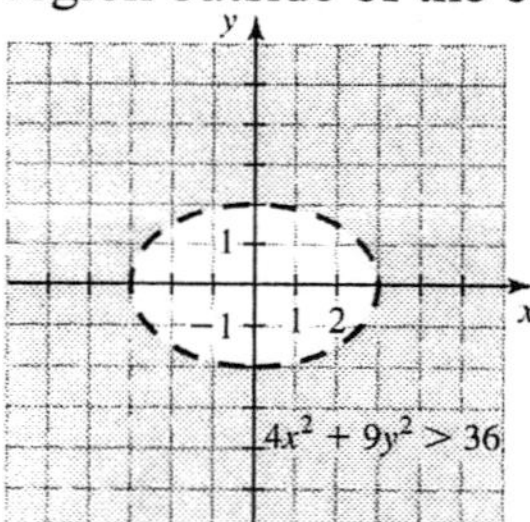

51. The graph of $y < 3x - x^2$ is the region below the parabola $y = 3x - x^2$. The graph of $x^2 + y^2 < 9$ is the region inside the circle $x^2 + y^2 = 9$. Points that are inside the circle and below the parabola are shown in the following graph.

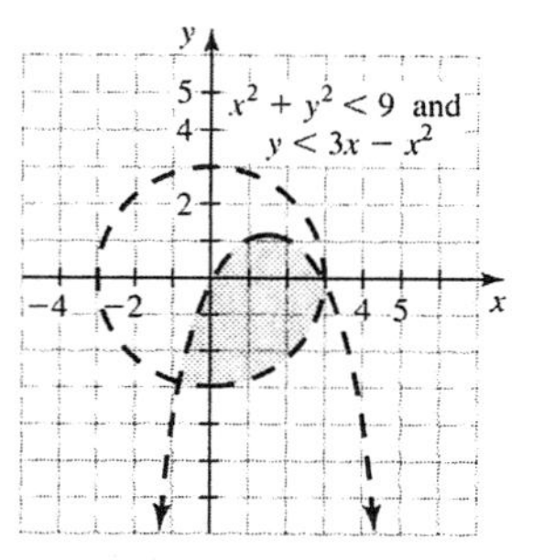

53. The set of points that satisfy $4x^2 + 9y^2 > 36$ is the set of points outside the ellipse $4x^2 + 9y^2 = 36$. The set of points that satisfy $x^2 + y^2 < 9$ is the set of points inside the circle $x^2 + y^2 = 9$. The solution set to the system consists of points that are outside the ellipse and inside the circle, as shown in the following graph.

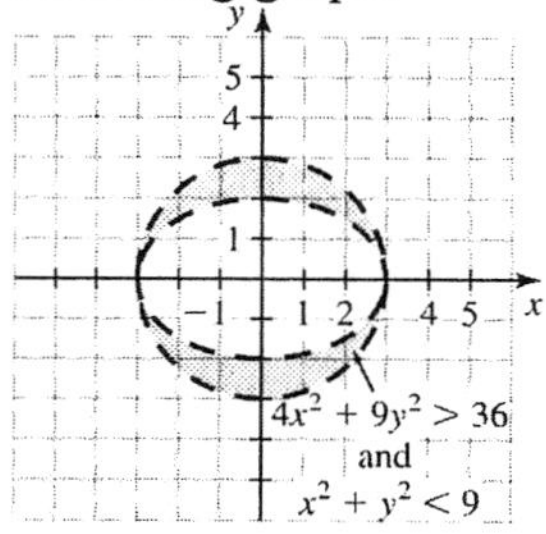

55. The equation $x^2 = y^2 + 1$ is the equation of a hyperbola because it could be written as $x^2 - y^2 = 1$.
57. The equation $x^2 = 1 - y^2$ is the equation of a circle because it could be written as $x^2 + y^2 = 1$.
59. The equation $x^2 + x = 1 - y^2$ is the equation of a circle because we could write $x^2 + x + y^2 = 1$, and then complete the square to get the standard equation for a circle.
61. The equation $x^2 + 4x = 6y - y^2$ is the equation of a circle because we could write it as $x^2 + 4x + y^2 - 6y = 0$, and then complete the squares for both x and y to get the standard equation of a circle.
63. The equation is the equation of a hyperbola in standard form.
65. The equation $4y^2 - x^2 = 8$ is the equation of a hyperbola because we could divide by 8 to get the standard equation $\frac{y^2}{2} - \frac{x^2}{8} = 1$.
67. Write $x^2 = 4 - y^2$ as $x^2 + y^2 = 4$ to see that it is the equation of a circle of radius 2 centered at the origin.

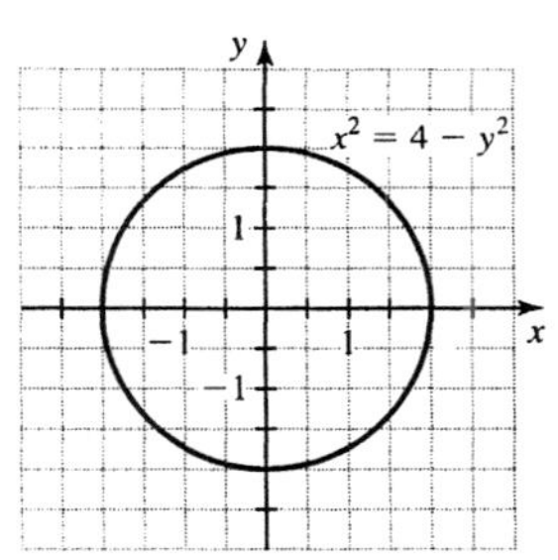

69. Write $x^2 = 4y + 4$ as $y = \frac{1}{4}x^2 - 1$ to see that it is the equation of a parabola opening upward with vertex at $(0, -1)$.

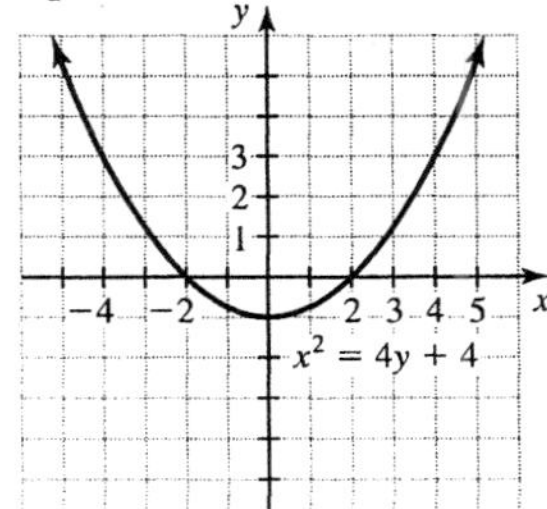

71. Write $x^2 = 4 - 4y^2$ as $\frac{x^2}{4} + y^2 = 1$ to see that it is the equation of an ellipse centered at $(0, 0)$ and passing through $(0, -1)$, $(0, 1)$, $(-2, 0)$, and $(2, 0)$.

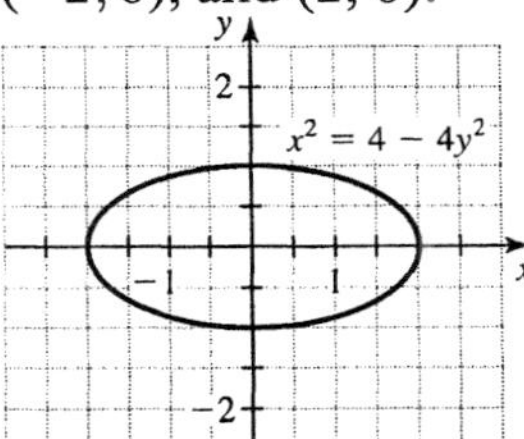

73. Write $x^2 = 4 - (y - 4)^2$ as $x^2 + (y - 4)^2 = 4$ to see that it is the equation of a circle of radius 2 centered at $(0, 4)$.

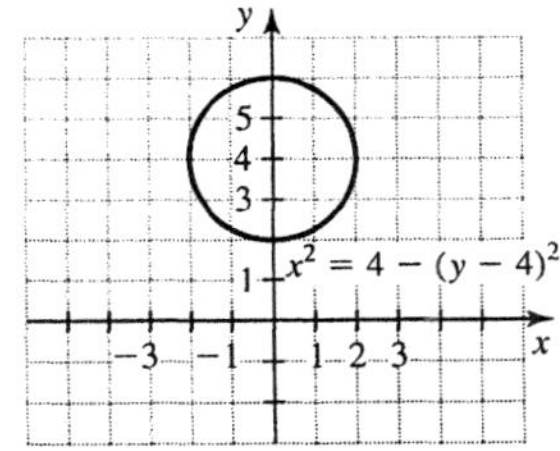

75. The radius of the circle is the distance between $(0, 0)$ and $(3, 4)$.

$$r = \sqrt{(3-0)^2 + (4-0)^2} = \sqrt{9 + 16} = \sqrt{25} = 5$$

The equation of the circle centered at $(0, 0)$ with radius 5 is $x^2 + y^2 = 25$.

77. Use the center $(-1, 5)$ and radius 6 in the form $(x - h)^2 + (y - k)^2 = r^2$ to get the equation $(x + 1)^2 + (y - 5)^2 = 36$.

79. The vertex is half way between the focus and directrix at $(1, 3)$. Since the distance from $(1, 3)$ to $(1, 4)$ is 1, we have $p = 1$, and $a = 1/4$. So the equation is

$$y = \frac{1}{4}(x - 1)^2 + 3.$$

81. Since the vertex is below the focus, $p = 1/4$ and $a = 1$. The equation is $y = 1(x - 0)^2 + 0$ or $y = x^2$.

83. Since the vertex is $(0, 0)$ the parabola has equation $y = ax^2$. Since $(3, 2)$ is on the parabola, $2 = a \cdot 3^2$, or $a = 2/9$. The equation is $y = \frac{2}{9}x^2$.

85. Substitute $y = -x + 1$ into $x^2 + y^2 = 25$ to eliminate y.

$$x^2 + (-x + 1)^2 = 25$$
$$x^2 + x^2 - 2x + 1 = 25$$
$$2x^2 - 2x - 24 = 0$$
$$x^2 - x - 12 = 0$$
$$(x - 4)(x + 3) = 0$$
$$x = 4 \quad \text{or} \quad x = -3$$
$$y = -3 \qquad y = 4 \quad \text{Since } y = -x + 1$$

The solution set is $\{(4, -3), (-3, 4)\}$

87. If we add the two equations to eliminate y, we get the equation $5x^2 = 25$. Solving this equation gives $x^2 = 5$ or $x = \pm\sqrt{5}$. Use $x^2 = 5$ in the first equation.

$$4(5) + y^2 = 4$$
$$y^2 = -16$$

There are no real numbers that satisfy the system of equations. The solution set is $\emptyset$.

89. Let x = the length and y = the width. We can write the following 2 equations.

$$2x + 2y = 16$$
$$xy = 12$$

The first equation can be written as $y = 8 - x$. Substitute $y = 8 - x$ into the second equation.

$$x(8 - x) = 12$$
$$-x^2 + 8x - 12 = 0$$
$$x^2 - 8x + 12 = 0$$
$$(x - 6)(x - 2) = 0$$
$$x = 6 \quad \text{or} \quad x = 2$$
$$y = 2 \qquad y = 6 \quad \text{Since } y = 8 - x$$

So the length is 6 feet and the width is 2 feet.

CHAPTER 13 TEST

1. The graph of $x^2 + y^2 = 25$ is a circle of radius 5 centered at (0, 0).

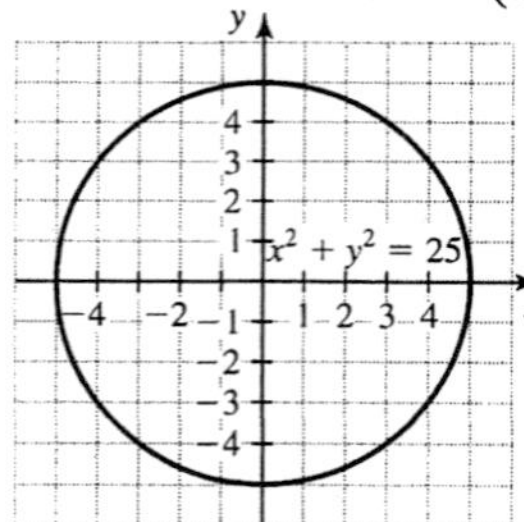

2. The graph of $\frac{x^2}{16} - \frac{y^2}{25} = 1$ is a hyperbola centered at the origin. the x-intercepts are $(-4, 0)$ and $(4, 0)$. The fundamental rectangle passes through the x-intercepts, $(0, -5)$, and $(0, 5)$. Extend the diagonals of the fundamental rectangle to obtain the asymptotes. The hyperbola opens to the left and right.

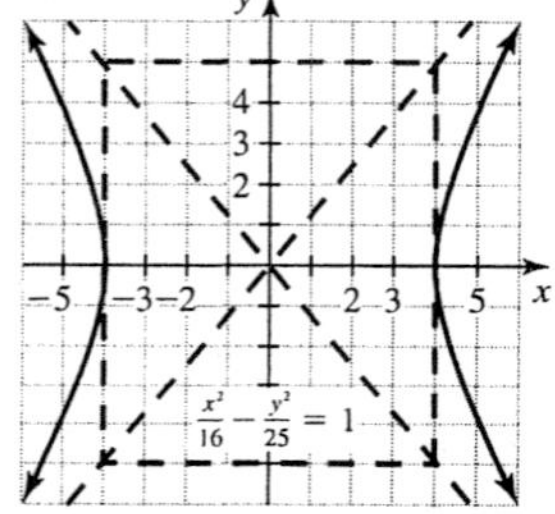

3. The graph of $y^2 + 4x^2 = 4$ is an ellipse with x-intercepts at $(-1, 0)$ and $(1, 0)$. Its y-intercepts are $(0, -2)$ and $(0, 2)$.

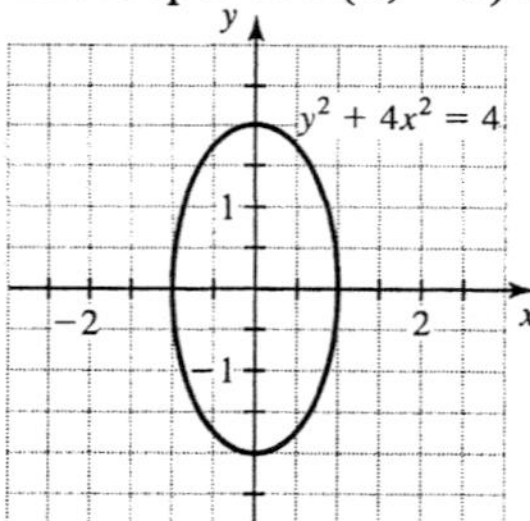

4. The graph of $y = x^2 + 4x + 4$ is a parabola opening upward with y-intercept $(0, 4)$. We can write this equation as $y = (x + 2)^2$. In this form we see that the vertex is $(-2, 0)$.

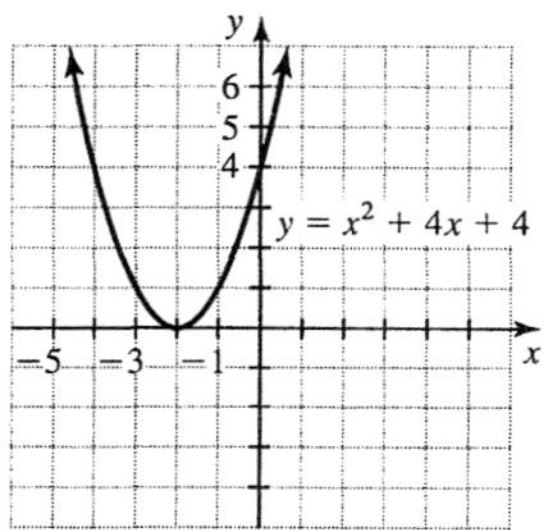

5. Write $y^2 - 4x^2 = 4$ as $\frac{y^2}{4} - x^2 = 1$. The graph is a hyperbola with y-intercepts at $(0, -2)$ and $(0, 2)$. The fundamental rectangle passes through the y-intercepts, $(-1, 0)$, and $(1, 0)$. Extend the diagonals to get the asymptotes. The hyperbola opens up and down.

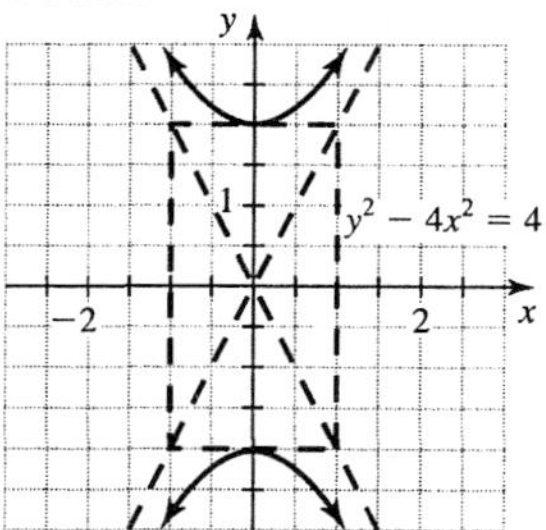

6. The graph of $y = -x^2 - 2x + 3$ is a parabola opening downward. The vertex is at $(-1, 4)$. The y-intercept is $(0, 3)$. The x-intercepts are $(-3, 0)$ and $(1, 0)$.

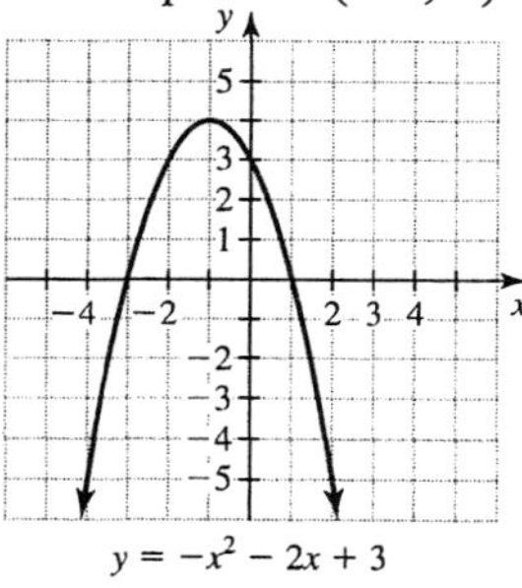

7. First graph the hyperbola $\frac{x^2}{9} - \frac{y^2}{9} = 1$. Since $(0, 0)$ satisfies the inequality, the region containing $(0, 0)$ is shaded.

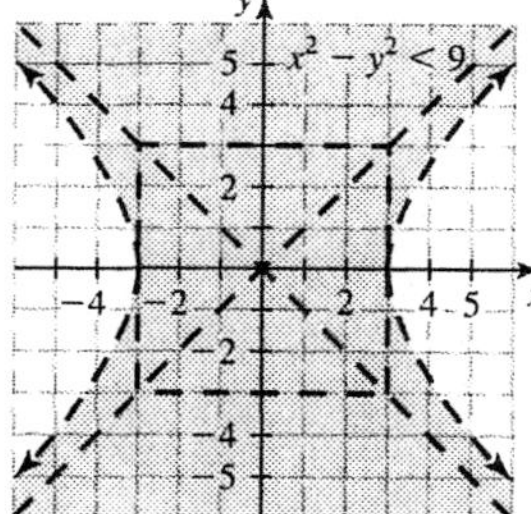

8. First graph the circle $x^2 + y^2 = 9$, centered at (0, 0) with radius 3. Since (0, 0) fails to satisfy $x^2 + y^2 > 9$, we shade the region outside the circle.

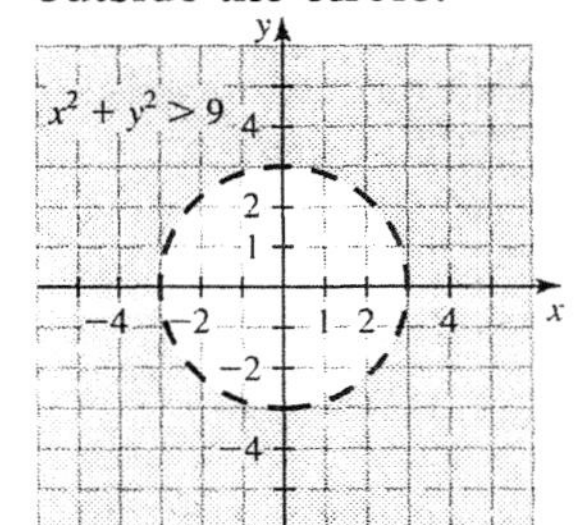

9. The graph of $y = x^2 - 9$ is a parabola opening upward. Its vertex is (0, −9). To find its x-intercepts solve $x^2 - 9 = 0$. The x-intercepts are (−3, 0) and (3, 0). The graph of $y > x^2 - 9$ is the region containing the origin because (0, 0) satisfies the inequality.

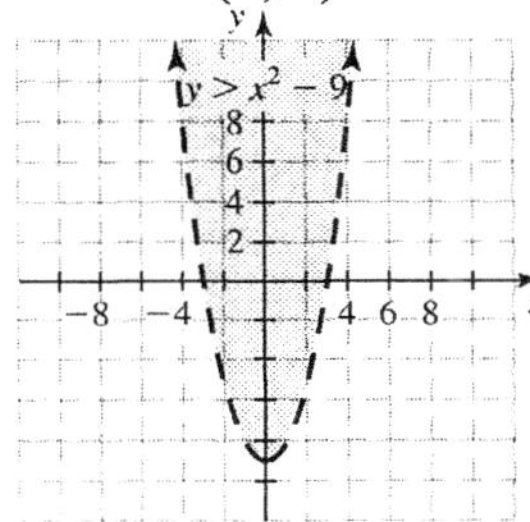

10. The graph of $x^2 + y^2 < 9$ is the region inside the circle $x^2 + y^2 = 9$. To find the graph of $x^2 - y^2 > 1$ first graph the hyperbola $x^2 - y^2 = 1$. By testing points, we can see that the two regions not containing the origin satisfy $x^2 - y^2 > 1$. The points in these two regions that are also inside the circle are the points that satisfy both inequalities of the system.

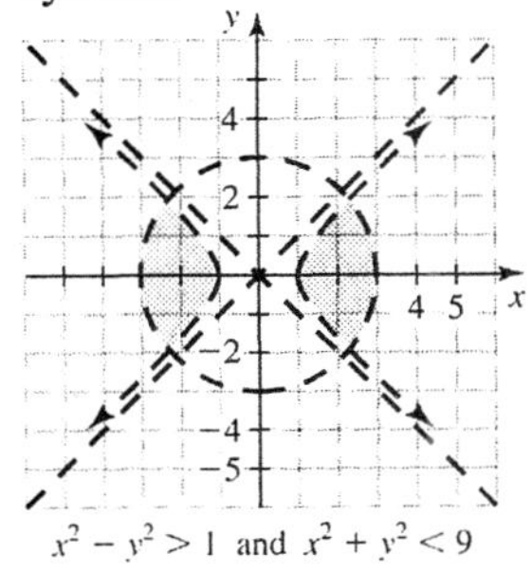

$x^2 - y^2 > 1$ and $x^2 + y^2 < 9$

11. The graph of $y = -x^2 + x$ is a parabola opening downward. Points below this parabola satisfy $y < -x^2 + x$. The graph of $y = x - 4$ is a line through (0, −4) with slope 1. Points below this line satisfy $y < x - 4$. The solution set to the system of inequalities consists of points below the parabola and below the line.

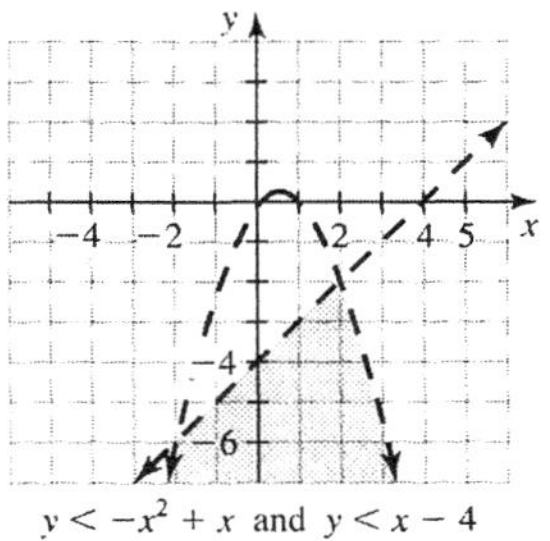

$y < -x^2 + x$ and $y < x - 4$

12. Use substitution to eliminate y.

$$x^2 - 2x - 8 = 7 - 4x$$
$$x^2 + 2x - 15 = 0$$
$$(x + 5)(x - 3) = 0$$
$$x = -5 \text{ or } x = 3$$
$$y = 27 \qquad y = -5 \text{ Since } y = 7 - 4x$$

The solution set is $\{(-5, 27), (3, -5)\}$.

13. Substitute $x^2 = y$ into $x^2 + y^2 = 12$.

$$y + y^2 = 12$$
$$y^2 + y - 12 = 0$$
$$(y + 4)(y - 3) = 0$$
$$y = -4 \text{ or } y = 3$$

If $y = -4$, the $x^2 = -4$ has no solution. If $y = 3$, then $x^2 = 3$ or $x = \pm\sqrt{3}$. The solution set is $\{(\sqrt{3}, 3), (-\sqrt{3}, 3)\}$.

14. $\sqrt{(-1 - 1)^2 + (4 - 6)^2} = \sqrt{4 + 4}$
$= \sqrt{8} = 2\sqrt{2}$

15. Complete the square to get the equation into the standard form.

$$x^2 + 2x + \quad y^2 + 10y \quad = 10$$
$$x^2 + 2x + 1 + y^2 + 10y + 25 = 10 + 1 + 25$$
$$(x + 1)^2 + (y + 5)^2 = 36$$

The center is (−1, −5) and the radius is 6.

16. For $y = x^2 + x + 3$ the x-coordinate of the vertex is

$$x = \frac{-b}{2a} = \frac{-1}{2(1)} = -\frac{1}{2}.$$

Use $x = -1/2$ in $y = x^2 + x + 3$, to get the vertex $(-\frac{1}{2}, \frac{11}{4})$. Since $a = 1$, we have $p = 1/4$. The focus is $(-\frac{1}{2}, 3)$ and the directrix is $y = \frac{5}{2}$. The axis of symmetry is $x = -\frac{1}{2}$ and the parabola opens up.

17. $y = \frac{1}{2}x^2 - 3x - \frac{1}{2}$
$y = \frac{1}{2}(x^2 - 6x) - \frac{1}{2}$

$y = \frac{1}{2}(x^2 - 6x + 9 - 9) - \frac{1}{2}$
$y = \frac{1}{2}(x^2 - 6x + 9) - \frac{9}{2} - \frac{1}{2}$
$y = \frac{1}{2}(x - 3)^2 - 5$

18. The radius of the circle is the distance between $(-1, 3)$ and $(2, 5)$.
$r = \sqrt{(-1-2)^2 + (3-5)^2} = \sqrt{9+4}$
$= \sqrt{13}$
The equation of the circle with center $(-1, 3)$ and radius $\sqrt{13}$ is $(x+1)^2 + (y-3)^2 = 13$.

19. Let $x =$ the length and $y =$ the width. We can write the following two equations.
$2x + 2y = 42$
$xy = 108$
If we solve $2x + 2y = 42$ for y, we get $y = 21 - x$.
$x(21 - x) = 108$
$-x^2 + 21x - 108 = 0$
$x^2 - 21x + 108 = 0$
$(x - 9)(x - 12) = 0$
$x = 9$ or $x = 12$
$y = 12$ $y = 9$ Since $y = 21 - x$
The length is 12 feet and the width is 9 feet.

Making Connections

Chapters 1 - 13

1. The graph of $y = 9x - x^2$ is a parabola that opens downward. Solve $9x - x^2 = 0$ to find the x-intercepts.
$x(9 - x) = 0$
$x = 0$ or $x = 9$
The x-intercepts are $(0, 0)$ and $(9, 0)$.
The x-coordinate of the vertex is
$x = \frac{-(9)}{2(-1)} = \frac{9}{2}$.
The vertex is $(9/2, 81/4)$.

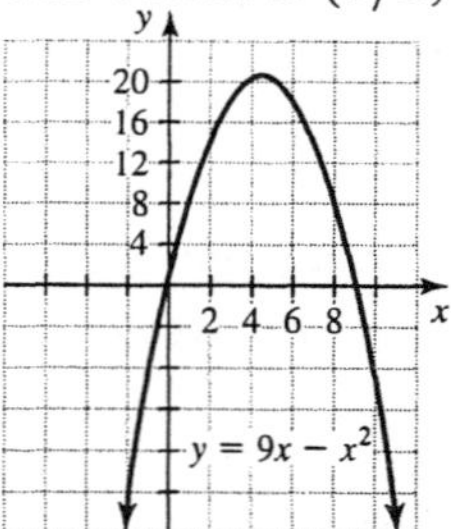

2. The graph of $y = 9x$ is a line through $(0, 0)$ with slope 9.

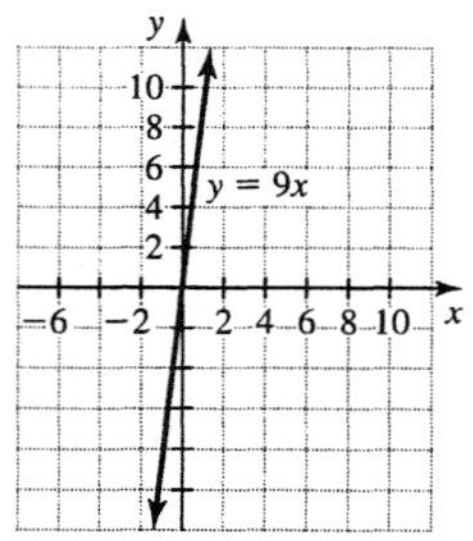

3. The graph of $y = (x - 9)^2$ is a parabola opening upward, with vertex at $(9, 0)$.

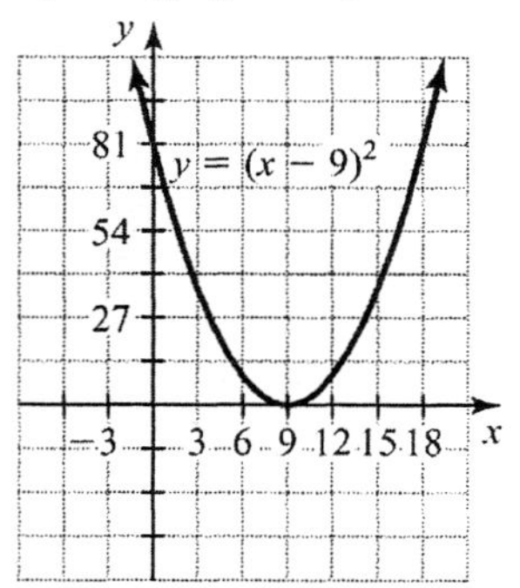

4. Write $y^2 = 9 - x^2$ as $x^2 + y^2 = 9$ to see that it is the equation of a circle of radius 3 centered at the origin.

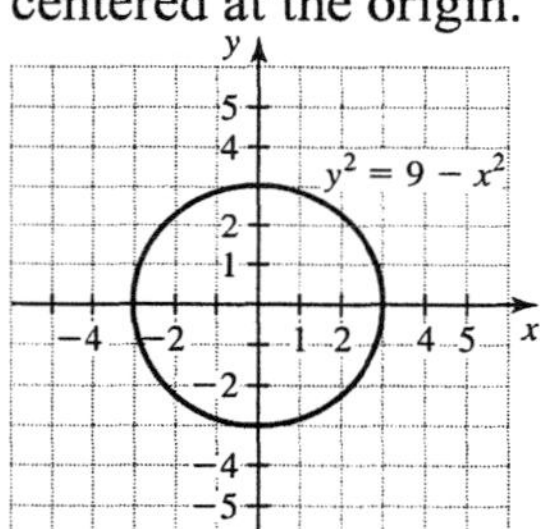

5. The graph of $y = 9x^2$ is a parabola opening upward with vertex at $(0, 0)$.

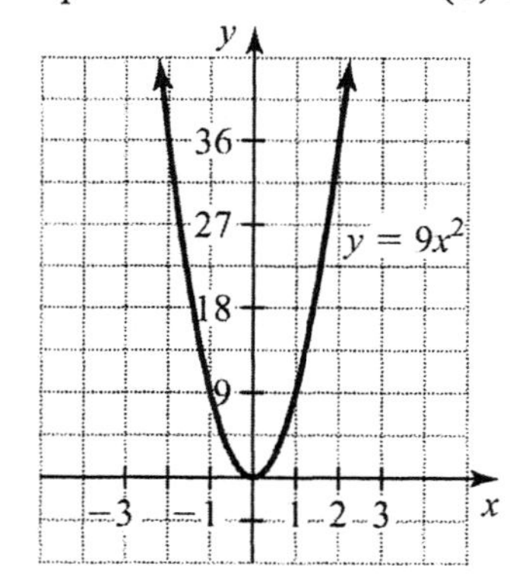

6. The graph of $y = \mid 9x \mid$ is v-shaped. It contains the points (0, 0), (−1, 9), and (1, 9).

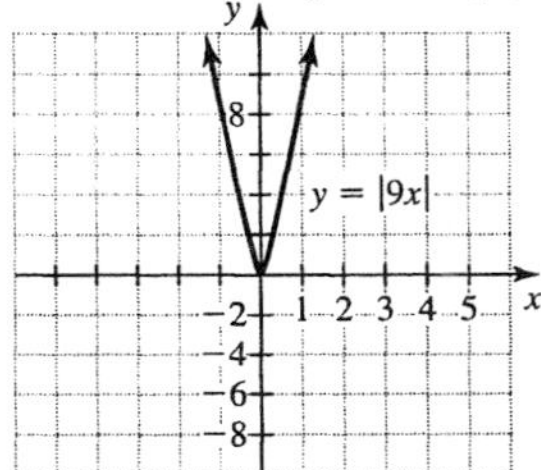

7. Write $4x^2 + 9y^2 = 36$ as $\frac{x^2}{9} + \frac{y^2}{4} = 1$ to see that it is the equation of an ellipse through (−3, 0), (3, 0), (0, 2), and (0, −2).

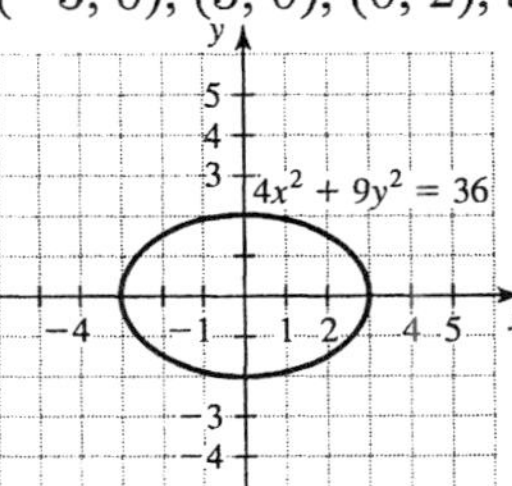

8. Write $4x^2 - 9y^2 = 36$ as $\frac{x^2}{9} - \frac{y^2}{4} = 1$ to see that it is the equation of a hyperbola with x-intercepts (−3, 0) and (3, 0). The fundamental rectangle goes through the x-intercepts, (0, −2), and (0, 2). Extend the diagonals of the rectangle to get the asymptotes.

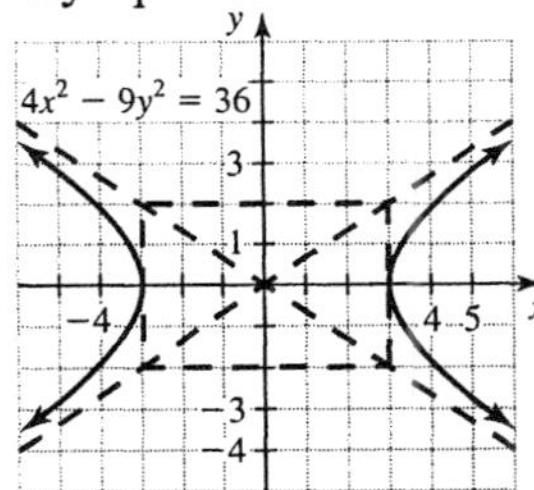

9. The graph of $y = 9 - x$ is a line through (0, 9) with slope −1.

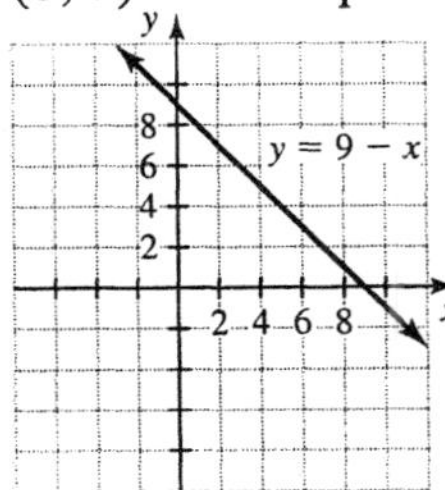

10. The graph of $y = 9^x$ goes through (0, 1), (1, 9), and (−1, 1/9).

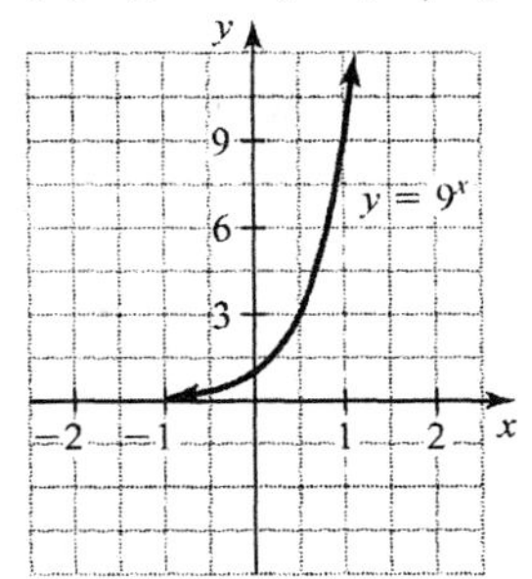

11. $(x + 2y)^2 = x^2 + 2(x)(2y) + (2y)^2$
$= x^2 + 4xy + 4y^2$

12. $(x + y)(x^2 + 2xy + y^2)$
$= x(x^2 + 2xy + y^2) + y(x^2 + 2xy + y^2)$
$= x^3 + 2x^2y + xy^2 + x^2y + 2xy^2 + y^3$
$= x^3 + 3x^2y + 3xy^2 + y^3$

13. $(a + b)^3 = (a + b)(a^2 + 2ab + b^2)$
$= a(a^2 + 2ab + b^2) + b(a^2 + 2ab + b^2)$
$= a^3 + 2a^2b = ab^2 + a^2b + 2ab^2 + b^3$
$= a^3 + 3a^2b + 3ab^2 + b^3$

14. $(a - 3b)^2 = a^2 - 2(a)(3b) + (3b)^2$
$= a^2 - 6ab + 9b^2$

15. $(2a + 1)(3a - 5) = 6a^2 + 3a - 10a - 5$
$= 6a^2 - 7a - 5$

16. $(x - y)(x^2 + xy + y^2)$
$= x(x^2 + xy + y^2) - y(x^2 + xy + y^2)$
$= x^3 + x^2y + xy^2 - x^2y - xy^2 - y^3$
$= x^3 - y^3$

17. Multiply the second equation by −2 and add the result to the first equation.

$$\begin{aligned} 2x - 3y &= -4 \\ -2x - 4y &= -10 \\ \hline -7y &= -14 \\ y &= 2 \end{aligned}$$

Use $y = 2$ in $x + 2y = 5$ to find x.
$x + 2(2) = 5$
$x = 1$
The solution set is $\{(1, 2)\}$.

18. Substitute $y = 7 - x$ into $x^2 + y^2 = 25$.

$$\begin{aligned} x^2 + (7 - x)^2 &= 25 \\ x^2 + 49 - 14x + x^2 &= 25 \\ 2x^2 - 14x + 24 &= 0 \\ x^2 - 7x + 12 &= 0 \\ (x - 3)(x - 4) &= 0 \end{aligned}$$

$x = 3$ or $x = 4$

$y = 4$ $\quad y = 3$ Since $y = 7 - x$

The solution set is $\{(3, 4), (4, 3)\}$.

19. Adding the first and second equations to eliminate z, we get $3x - 3y = 9$. Adding the second and third equations to eliminate z, we get $2x - y = 4$. Divide $3x - 3y = 9$ by -3 and add the result to $2x - y = 4$.

$$\begin{aligned} -x + y &= -3 \\ 2x - y &= 4 \\ \hline x &= 1 \end{aligned}$$

Use $x = 1$ in $-x + y = -3$.

$$\begin{aligned} -1 + y &= -3 \\ y &= -2 \end{aligned}$$

Use $x = 1$ and $y = -2$ in $x + y + z = 2$.

$$\begin{aligned} 1 + (-2) + z &= 2 \\ z &= 3 \end{aligned}$$

The solution set is $\{(1, -2, 3)\}$.

20. Substitute $y = x^2$ into $y - 2x = 3$.

$$\begin{aligned} x^2 - 2x &= 3 \\ x^2 - 2x - 3 &= 0 \\ (x - 3)(x + 1) &= 0 \end{aligned}$$

$x = 3$ or $x = -1$

$y = 9$ $\quad y = 1$ Since $y = x^2$

The solution set is $\{(-1, 1), (3, 9)\}$.

21.
$$\begin{aligned} ax + b &= 0 \\ ax &= -b \\ x &= -\frac{b}{a} \end{aligned}$$

22. Use the quadratic formula with $a = w$, $b = d$, and $c = m$.

$$x = \frac{-d \pm \sqrt{d^2 - 4wm}}{2w}$$

23.
$$\begin{aligned} A &= \tfrac{1}{2}h(B + b) \\ 2A &= h(B + b) \\ 2A &= hB + hb \\ 2A - bh &= hB \\ B &= \frac{2A - bh}{h} \end{aligned}$$

24.
$$\begin{aligned} \frac{1}{x} + \frac{1}{y} &= \frac{1}{2} \\ 2xy\left(\frac{1}{x} + \frac{1}{y}\right) &= 2xy\left(\frac{1}{2}\right) \\ 2y + 2x &= xy \\ 2y &= xy - 2x \\ 2y &= x(y - 2) \\ x &= \frac{2y}{y - 2} \end{aligned}$$

25.
$$\begin{aligned} L &= m + mxt \\ L &= m(1 + xt) \\ m &= \frac{L}{1 + xt} \end{aligned}$$

26.
$$\begin{aligned} y &= 3a\sqrt{t} \\ (y)^2 &= (3a\sqrt{t})^2 \\ y^2 &= 9a^2t \\ \frac{y^2}{9a^2} &= t \\ t &= \frac{y^2}{9a^2} \end{aligned}$$

27. First find the slope.

$$m = \frac{-3 - 1}{2 - (-4)} = \frac{-4}{6} = -\frac{2}{3}$$

Use point-slope form with $(2, -3)$.

$$\begin{aligned} y - (-3) &= -\tfrac{2}{3}(x - 2) \\ y + 3 &= -\tfrac{2}{3}x + \tfrac{4}{3} \\ y &= -\tfrac{2}{3}x - \tfrac{5}{3} \end{aligned}$$

28. Write $2x - 4y = 5$ as $y = \frac{1}{2}x - \frac{5}{4}$. The slope of any line perpendicular to this line is -2. The line through $(0, 0)$ with slope -2 is $y = -2x$.

29. The radius is the distance between (2, 5) and (−1, −1).

$$r = \sqrt{(-1-2)^2 + (-1-5)^2} = \sqrt{9+36}$$
$$= \sqrt{45}$$

The equation of circle with center (2, 5) and radius $\sqrt{45}$ is $(x-2)^2 + (y-5)^2 = 45$.

30. Use completing the square to get the equation into standard form for a circle.

$$x^2 + 3x + y^2 - 6y = 0$$
$$x^2 + 3x + \frac{9}{4} + y^2 - 6y + 9 = 0 + \frac{9}{4} + 9$$
$$(x + \tfrac{3}{2})^2 + (y-3)^2 = \frac{45}{4}$$

The center is $(-\frac{3}{2}, 3)$ and radius is $\sqrt{\frac{45}{4}}$, or $\frac{3\sqrt{5}}{2}$.

31. $2i(3+5i) = 6i + 10i^2 = -10 + 6i$

32. $i^6 = i^4 \cdot i^2 = 1(-1) = -1$

33. $(2i - 3) + (6 - 7i) = 3 - 5i$

34. $(3 + i\sqrt{2})^2 = 9 + 6i\sqrt{2} + 2i^2$
$= 7 + 6i\sqrt{2}$

35. $(2-3i)(5-6i) = 10 - 15i - 12i + 18i^2$
$= -8 - 27i$

36. $(3 - i) + (-6 + 4i) = -3 + 3i$

37. $(5-2i)(5+2i) = 25 - 4i^2 = 29$

38. $\frac{2-3i}{2i} = \frac{(2-3i)(-i)}{2i(-i)} = \frac{-2i+3i^2}{2(1)}$
$= -\frac{3}{2} - i$

39. $\frac{(4+5i)(1+i)}{(1-i)(1+i)} = \frac{4+9i+5i^2}{2}$

$= \frac{-1+9i}{2} = -\frac{1}{2} + \frac{9}{2}i$

40. $\frac{4-\sqrt{-8}}{2} = \frac{4-2i\sqrt{2}}{2} = 2 - i\sqrt{2}$

41. a) $m = \frac{250-200}{0.30-0.40} = \frac{50}{-0.1} = -500$
$q - 250 = -500(x - 0.30)$
$q - 250 = -500x + 150$
$q = -500x + 400$

b) $R = qx = (-500x + 400)x$
$R = -500x^2 + 400x$

c) The graph is a parabola through $(0,0)$, $(0.80, 0)$, $(0.40, 80)$

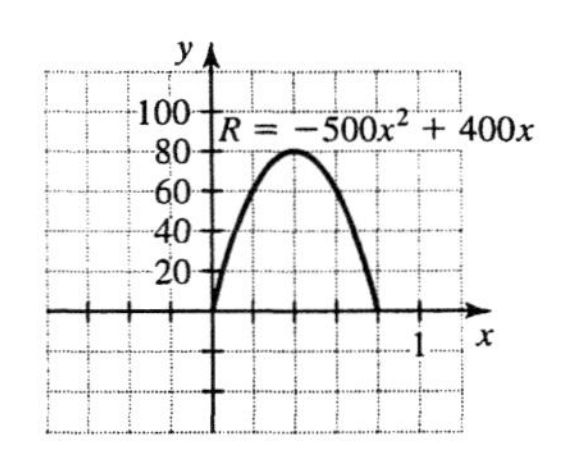

d) The maximum revenue occurs at the vertex of the parabola.

$$x = \frac{-400}{2(-500)} = 0.40$$

The maximum revenue occurs when bananas are \$0.40 per pound.

e) To find the maximum revenue use the formula for revenue with $x = 0.40$.
$R = -500(0.40)^2 + 400(0.40) = 80$
The maximum revenue is \$80 when the bananas are \$0.40 per pound.

14.1 WARM-UPS

1. True, because the formula $a_n = 2n$ will produce even numbers when n is a positive integer.
2. True, because the formula $a_n = 2n - 1$ will produce odd numbers when n is a positive integer.
3. True, by the definition of sequence.
4. False, because the domain of a finite sequence is the set of positive integers less than or equal to some fixed positive integer.
5. False, because if $n = 1$, $a_1 = (-1)^2 \cdot 1^2 = 1$.
6. False, because the independent variable is n.
7. True.
8. False, because the 6th term is $a_6 = (-1)^7 2^6 = -64$.
9. True.
10. True, because the sequence is $a_n = 2^n$ and $a_{10} = 2^{10} = 1024$.

14.1 EXERCISES

1. A sequence is a list of numbers.
3. A finite sequence is a function whose domain is the set of positive integers less than or equal to some fixed positive integer.
5. The terms of the sequence a_n are found by multiplying the integers from 1 through 5 by 2, because of the formula $a_n = 2n$. So the terms are $2, 4, 6, 8$, and 10.
7. The terms of the sequence a_n are found by squaring the integers from 1 through 8, because of the formula $a_n = n^2$. So the terms are 1, 4, 9, 16, 25, 36, 49, and 64.
9. $b_1 = \frac{(-1)^1}{1} = -1$, $b_2 = \frac{(-1)^2}{2} = \frac{1}{2}$,
$b_3 = \frac{(-1)^3}{3} = -\frac{1}{3}$, $b_4 = \frac{(-1)^4}{4} = \frac{1}{4}$,
$b_5 = \frac{(-1)^5}{5} = -\frac{1}{5}$, etc.
The 10 terms of the sequence are
$-1, \frac{1}{2}, -\frac{1}{3}, \frac{1}{4}, -\frac{1}{5}, \frac{1}{6}, -\frac{1}{7}, \frac{1}{8}, -\frac{1}{9}$, and $\frac{1}{10}$.

11. $c_1 = (-2)^{1-1} = 1$, $c_2 = (-2)^{2-1} = -2$,
$c_3 = (-2)^{3-1} = 4$, $c_4 = (-2)^{4-1} = -8$,
$c_5 = (-2)^{5-1} = 16$
The five terms are 1, -2, 4, -8, and 16.
13. $a_1 = 2^{-1} = \frac{1}{2}$, $a_2 = 2^{-2} = \frac{1}{4}$,
$a_3 = 2^{-3} = \frac{1}{8}$, $a_4 = 2^{-4} = \frac{1}{16}$, etc.
The six terms are $\frac{1}{2}, \frac{1}{4}, \frac{1}{8}, \frac{1}{16}, \frac{1}{32}$, and $\frac{1}{64}$.
15. $b_1 = 2(1) - 3 = -1$, $b_2 = 2(2) - 3 = 1$,
$b_3 = 2(3) - 3 = 3$, $b_4 = 2(4) - 3 = 5$,
$b_5 = 2(5) - 3 = 7$, $b_6 = 2(6) - 3 = 9$,
$b_7 = 2(7) - 3 = 11$
The seven terms are -1, 1, 3, 5, 7, 9, and 11.
17. $c_1 = 1^{-1/2} = 1$, $c_2 = 2^{-1/2} = \frac{1}{\sqrt{2}}$,
$c_3 = 3^{-1/2} = \frac{1}{\sqrt{3}} = \frac{\sqrt{3}}{3}$, $c_4 = 4^{-1/2} = \frac{1}{2}$,
$c_5 = 5^{-1/2} = \frac{1}{\sqrt{5}} = \frac{\sqrt{5}}{5}$
The five terms are $1, \frac{\sqrt{2}}{2}, \frac{\sqrt{3}}{3}, \frac{1}{2}$, and $\frac{\sqrt{5}}{5}$.
19. $a_1 = \frac{1}{1^2 + 1} = \frac{1}{2}$, $a_2 = \frac{1}{2^2 + 2} = \frac{1}{6}$,
$a_3 = \frac{1}{3^2 + 3} = \frac{1}{12}$, $a_4 = \frac{1}{4^2 + 4} = \frac{1}{20}$
The first four terms are $\frac{1}{2}, \frac{1}{6}, \frac{1}{12}$, and $\frac{1}{20}$.
21. $b_1 = \frac{1}{2(1) - 5} = -\frac{1}{3}$,
$b_2 = \frac{1}{2(2) - 5} = -1$,
$b_3 = \frac{1}{2(3) - 5} = 1$, $b_4 = \frac{1}{2(4) - 5} = \frac{1}{3}$
The first four terms are $-\frac{1}{3}$, -1, 1, and $\frac{1}{3}$.
23. $c_1 = (-1)^1(1 - 2)^2 = -1$,
$c_2 = (-1)^2(2 - 2)^2 = 0$,
$c_3 = (-1)^3(3 - 2)^2 = -1$,
$c_4 = (-1)^4(4 - 2)^2 = 4$
The first four terms are -1, 0, -1, and 4.

25. $a_1 = \frac{(-1)^{2(1)}}{1^2} = 1$, $a_2 = \frac{(-1)^{2(2)}}{2^2} = \frac{1}{4}$,
$a_3 = \frac{(-1)^{2(3)}}{3^2} = \frac{1}{9}$, $a_4 = \frac{(-1)^{2(4)}}{4^2} = \frac{1}{16}$
The first four terms are $1, \frac{1}{4}, \frac{1}{9}$, and $\frac{1}{16}$.

27. This sequence is a sequence of odd integers starting at 1. Even integers are represented by $2n$ and odd integers are all one less than an even integer. Try the formula $a_n = 2n - 1$. To see that it is correct, find a_1 through a_5.

29. A sequence of alternating ones and negative ones is obtained by using a power of -1. Since we want a_1 to be positive when $n = 1$, we use the $n + 1$ power on -1. A formula for the general term is $a_n = (-1)^{n+1}$.
31. This sequence is a sequence of even integers starting at 0. Even integers can be generated by $2n$, but to make the first one 0 (when $n = 1$), we use the formula $a_n = 2n - 2$.
33. This sequence is a sequence consisting of the positive integral multiples of 3. Try the formula $a_n = 3n$ to see that it generates the appropriate numbers.
35. Each term in this sequence is one larger than the corresponding term in the sequence of Exercise 31. Try the formula $a_n = 3n + 1$ to see that it generates the given sequence.
37. To get the alternating signs on the terms of the sequence, we use a power of -1. Next we observe that the numbers 1, 2, 4, 8, 16 are powers of 2. The formula $a_n = (-1)^n 2^{n-1}$ will produce the given sequence. Remember that n always starts at 1, so in this case we use $n - 1$ as the power of 2 to get $2^0 = 1$.
39. Notice that the numbers 0, 1, 4, 9, 16, . . . are the squares of the nonnegative integers: 0^2, $1^2, 2^2, 3^2, 4^2, \ldots$. We would use n^2 to generate squares, but since n starts at 1, we use $a_n = (n-1)^2$ to get the given sequence.
41. After the first penalty the ball is on the 4-yard line. After the second penalty the ball is on the 2-yard line. After the third penalty the ball is on the 1-yard line, and so on. The sequence of five terms is 4, 2, 1, $\frac{1}{2}$, $\frac{1}{4}$.
43. To find the amount of increase, we take 5% of \$35,960 to get \$1798 increase. We could just multiply \$35,960 by 1.05 to obtain the new price. In either case the price the next year is \$37,758. To find the price for the next year we multiply the last year's price by 1.05: $1.05(\$37{,}758) = \$39{,}646$. Repeating this process gives us the prices \$37,758, \$39,646, \$41,628, \$43,710, \$45,895, and \$48,190 as the price of the truck through 2010.
45. Of the \$1 million, 80% is respent in the community. So \$800,000 is respent. Of the \$800,000 we have 80% respent in the community. So \$640,000 is respent in the community, and so on. The first four terms of the sequence are \$1,000,000, \$800,000, \$640,000, \$512,000.
47. Possible vertical repeats are 27 in., 13.5 in., 9 in., 6.75 in., and 5.4 in.
49. If we put the word great in front of the word grandparents 35 times, then we have 2^{37} of this type of relative. Use a calculator to find that $2^{37} \approx 137{,}438{,}953{,}472$. This is certainly larger than the present population of the earth.
53. a) Use a calculator to find

$$a_{100} = (0.999)^{100} = 0.9048,$$
$$a_{1000} = (0.999)^{1000} = 0.3677,$$
and $$a_{10000} = (0.999)^{10000} = 0.00004517.$$

b) a_n approaches zero as n gets larger and larger.

14.2 WARM-UPS

1. True, because of the definition of series.
2. False, the sum of a series can be any real number.
3. False, there are 9 terms in 2^3 through 10^3.
4. False, because the terms in the first series are opposites of the terms in the second series.
5. False, the ninth term is

$$\frac{(-1)^9}{(9+1)(9+2)} = -\frac{1}{110}.$$

6. True, because the terms are -2 and 4 and the sum is 2.
7. True, because of the distributive property.
8. True, because the notation indicates to add the number 4 five times.
9. True, because $2i + 7i = 9i$.
10. False, because in the series on the left side the 1 is added in three times and in the series on the right side the 1 is only added in once.

14.2 EXERCISES

1. Summation notation provides a way of writing a sum without writing out all of the terms.
3. A series is the indicated sum of the terms of a sequence.

5. $\sum_{i=1}^{5} i = 1 + 2 + 3 + 4 + 5 = 15$

7. $\sum_{i=1}^{4} i^2 = 1^2 + 2^2 + 3^2 + 4^2$

$= 1 + 4 + 9 + 16 = 30$

9. $\sum_{j=0}^{5}(2j - 1) = (2 \cdot 0 - 1) + (2 \cdot 1 - 1)$

$+ (2 \cdot 2 - 1) + (2 \cdot 3 - 1)$
$+ (2 \cdot 4 - 1) + (2 \cdot 5 - 1)$
$= -1 + 1 + 3 + 5 + 7 + 9 = 24$

11. $\sum_{i=1}^{5} 2^{-i} = 2^{-1} + 2^{-2} + 2^{-3} + 2^{-4} + 2^{-5}$

$= \frac{1}{2} + \frac{1}{4} + \frac{1}{8} + \frac{1}{16} + \frac{1}{32} = \frac{31}{32}$

13. $\sum_{i=1}^{10} 5i^0 = 5(1)^0 + 5(2)^0 + 5(3)^0 + 5(4)^0$

$+ 5(5)^0 + 5(6)^0 + 5(7)^0$
$+ 5(8)^0 + 5(9)^0 + 5(10)^0$
$= 5 + 5 + 5 + 5 + 5 + 5 + 5 + 5 + 5 + 5$
$= 50$

15. $\sum_{i=1}^{3}(i - 3)(i + 1) = (1 - 3)(1 + 1)$

$+ (2 - 3)(2 + 1) + (3 - 3)(3 + 1)$
$= -4 + (-3) + 0 = -7$

17. $\sum_{j=1}^{10}(-1)^j$

$= (-1)^1 + (-1)^2 + ... + (-1)^{10}$
$= -1 + 1 - 1 + 1 - 1 + 1 - 1$
$+ 1 - 1 + 1$

$= 0$

19. The sum of the first six positive integers is written as $\sum_{i=1}^{6} i$. There are other ways to indicate this series in summation notation but this is the simplest.

21. To get the signs of the terms to alternate from positive to negative we use a power of -1. To get odd integers we use the formula $2i - 1$. So this series is written $\sum_{i=1}^{6}(-1)^i(2i - 1)$.

23. This series consists of the squares of the first six positive integers. It is written in summation notation as $\sum_{i=1}^{6} i^2$.

25. This series consists of the reciprocals of the positive integers. If the index i goes from 1 to 4, we must use $2 + i$ to get the numbers 3 through 6. So the series is written $\sum_{i=1}^{4} \frac{1}{2 + i}$.

27. The terms of this series are logarithms of positive integers. If the index i goes from 1 to 3, we must use $i + 1$ to get the numbers 2 through 4. The series is written $\sum_{i=1}^{3} \ln(i + 1)$.

29. Since the subscripts range from 1 through 4, we let i range from 1 through 4: $\sum_{i=1}^{4} a_i$.

31. The subscripts on x range from 3 through 50, so i ranges from 1 through 48: $\sum_{i=1}^{48} x_{i+2}$.

33. The subscripts on w range from 1 through n, so i ranges from 1 through n: $\sum_{i=1}^{n} w_i$.

35. If $j = 0$ when $i = 1$, then $i = j + 1$ and j ranges from 0 through 4: $\sum_{j=0}^{4} (j + 1)^2$

37. If $j = 1$ when $i = 0$, then $i = j - 1$ and j ranges from 1 through 13. If we substitute $i = j - 1$ into $2i - 1$ we get $2i - 1 = 2(j - 1) - 1 = 2j - 3$. So the series is written $\sum_{j=1}^{13} (2j - 3)$.

39. If $j = 1$ when $i = 4$, then $i = j + 3$ and j ranges from 1 through 5. Substitute $j + 3$ for i to get $\sum_{j=1}^{5} \frac{1}{j + 3}$.

41. If $j = 0$ when $i = 1$, then $i = j + 1$ and j ranges from 0 through 3. The exponent $2i + 3$ becomes $2(j + 1) + 3 = 2j + 5$. The series is written as $\sum_{j=0}^{3} x^{2j + 5}$.

43. If $j = 0$ when $i = 1$, then $i = j + 1$ and j ranges from 0 through $n - 1$. Replacing i by $j + 1$ gives us the series $\sum_{j=0}^{n-1} x^{j + 1}$.

45. $\sum_{i=1}^{6} x^i = x + x^2 + x^3 + x^4 + x^5 + x^6$

47. $\sum_{j=0}^{3}(-1)^j x_j$
$= (-1)^0 x_0 + (-1)^1 x_1 + (-1)^2 x_2 + (-1)^3 x_3$
$= x_0 - x_1 + x_2 - x_3$

49. $\sum_{i=1}^{3} ix^i = 1x^1 + 2x^2 + 3x^3$
$= x + 2x^2 + 3x^3$

51. On the first jump he has moved $\frac{1}{2}$ yard. On the second jump he moves $\frac{1}{4}$ yard. On the third jump he moves $\frac{1}{8}$ yard, and so on. To express the reciprocals of the powers of 2, we use 2^{-i}. His total movement after nine jumps is expressed as the series $\sum_{i=1}^{9} 2^{-i}$.

53. $\sum_{i=1}^{4} 1{,}000{,}000(0.8)^{i-1}$.

55. A sequence is basically a list of numbers. A series is the indicated sum of the terms of a sequence.

14.3 WARM-UPS

1. False, because the common difference is -2 $(1 - 3 = -2)$.
2. False, because the difference between consecutive terms is not constant.
3. False, because the difference between the consecutive terms is sometimes 2 and sometimes -2.
4. False, because the nth term is given by $a_n = a_1 + (n - 1)d$.
5. False, because the second must be 7.5, making $d = 2.5$ and the fourth term 12.5.
6. False, because if the first is 6 and the third is 2, the second must be 4 in order to have a common difference.
7. True, because that is the definition of arithmetic series.
8. True, because the series is $5 + 7 + 9 + 11 + 13$ and the common difference is 2.
9. True, because of the formula for the sum of an arithmetic series.
10. False, because there are 11 even integers from 8 through 28 inclusive and the sum is $\frac{11}{2}(8 + 28)$.

14.3 EXERCISES

1. An arithmetic sequence is one in which each term after the first is obtained by adding a fixed amount to the previous term.
3. An arithmetic series is an indicated sum of an arithmetic sequence.
5. The common difference is $d = 2$ and the first term is $a_1 = 2$. Use the formula $a_n = a_1 + (n - 1)d$ to find that the nth term is $a_n = 2 + (n - 1)2 = 2n$.
7. The common difference is $d = 6$ and the first term is $a_1 = 0$. Use the formula $a_n = a_1 + (n - 1)d$ to find that the nth term is $a_n = 0 + (n - 1)6 = 6n - 6$.
9. The common difference is $d = 5$ and the first term is 7. The nth term is $a_n = 7 + (n - 1)5 = 5n + 2$.
11. The common difference is $d = 2$ and the first term is $a_1 = -4$. Use the formula $a_n = a_1 + (n - 1)d$ to find that the nth term is $a_n = -4 + (n - 1)2 = 2n - 6$.
13. The common difference is $d = 1 - 5$ $= -4$ and the first term is $a_1 = 5$. The nth term is $a_n = 5 + (n - 1)(-4) = -4n + 9$.
15. The common difference is $d = -9 - (-2)$ $= -7$ and the first term is -2. The nth term is $a_n = -2 + (n - 1)(-7) = -7n + 5$.

17. The common difference is $d = -2.5 - (-3) = 0.5$ and the first term is -3. The nth term is $a_n = -3 + (n - 1)(0.5) = 0.5n - 3.5$.

19. The common difference is $d = -6.5 - (-6) = -0.5$ and the first term is -6. The nth term is $a_n = -6 + (n - 1)(-0.5)$ $= -0.5n - 5.5$.

21. $a_1 = 9 + (1 - 1)4 = 9$,
$a_2 = 9 + (2 - 1)4 = 13$,
$a_3 = 9 + (3 - 1)4 = 17$,
$a_4 = 9 + (4 - 1)4 = 21$,
$a_5 = 9 + (5 - 1)4 = 25$
The first five terms of the arithmetic sequence are 9, 13, 17, 21, and 25.

23. $a_1 = 7 + (1 - 1)(-2) = 7,$
$a_2 = 7 + (2 - 1)(-2) = 5,$
$a_3 = 7 + (3 - 1)(-2) = 3,$
$a_4 = 7 + (4 - 1)(-2) = 1,$
$a_5 = 7 + (5 - 1)(-2) = -1$
The first five terms of the arithmetic sequence are 7, 5, 3, 1, and −1.

25. $a_1 = -4 + (1 - 1)3 = -4,$
$a_2 = -4 + (2 - 1)3 = -1,$
$a_3 = -4 + (3 - 1)3 = 2,$
$a_4 = -4 + (4 - 1)3 = 5,$
$a_5 = -4 + (5 - 1)3 = 8$
The first five terms of the arithmetic sequence are −4, −1, 2, 5, and 8.

27. $a_1 = -2 + (1 - 1)(-3) = -2,$
$a_2 = -2 + (2 - 1)(-3) = -5,$
$a_3 = -2 + (3 - 1)(-3) = -8,$
$a_4 = -2 + (4 - 1)(-3) = -11,$
$a_5 = -2 + (5 - 1)(-3) = -14$
The first five terms of the arithmetic sequence are −2, −5, −8, −11, and −14.

29. $a_1 = -4(1) - 3 = -7,$
$a_2 = -4(2) - 3 = -11,$
$a_3 = -4(3) - 3 = -15,$
$a_4 = -4(4) - 3 = -19,$
$a_5 = -4(5) - 3 = -23$
The first five terms of the arithmetic sequence are −7, −11, −15, −19, and −23.

31. $a_1 = 0.5(1) + 4 = 4.5,$
$a_2 = 0.5(2) + 4 = 5, a_3 = 0.5(3) + 4 = 5.5,$
$a_4 = 0.5(4) + 4 = 6, a_5 = 0.5(5) + 4 = 6.5$
The first five terms of the arithmetic sequence are 4.5, 5, 5.5, 6, and 6.5.

33. $a_1 = 20(1) + 1000 = 1020,$
$a_2 = 20(2) + 1000 = 1040,$
$a_3 = 20(3) + 1000 = 1060,$
$a_4 = 20(4) + 1000 = 1080,$
$a_5 = 20(5) + 1000 = 1100$
The first five terms of the arithmetic sequence are 1020, 1040, 1060, 1080, and 1100.

35. Use $a_1 = 9$, $n = 8$, and $d = 6$ in the formula $a_n = a_1 + (n - 1)d$.
$$a_8 = 9 + (8 - 1)6 = 51$$

37. Use $a_1 = 6$, $a_{20} = 82$, and $n = 20$ in the formula $a_n = a_1 + (n - 1)d$.
$82 = 6 + (20 - 1)d$
$82 = 6 + 19d$
$76 = 19d$
$4 = d$

39. Use $a_7 = 14$, $d = -2$, and $n = 7$ in the formula $a_n = a_1 + (n - 1)d$.
$14 = a_1 + (7 - 1)(-2)$
$14 = a_1 - 12$
$26 = a_1$

41. From the fact that the fifth term is 13 and the first term is −3, we can find the common difference.
$13 = -3 + (5 - 1)d$
$13 = -3 + 4d$
$16 = 4d$
$4 = d$
Use $a_1 = -3$, $n = 6$, and $d = 4$ in the formula $a_n = a_1 + (n - 1)d$.
$$a_6 = -3 + (6 - 1)4 = 17$$

43. Use $a_1 = 1$, $a_{48} = 48$, and $n = 48$ in the formula $S_n = \frac{n}{2}(a_1 + a_n)$.
$$S_{48} = \frac{48}{2}(1 + 48) = 1176$$

45. To find n, use $a_1 = 8$, $d = 2$, and $a_n = 36$ in the formula $a_n = a_1 + (n - 1)d$.
$36 = 8 + (n - 1)2$
$36 = 8 + 2n - 2$
$30 = 2n$
$15 = n$
Use $a_1 = 8$, $a_{15} = 36$, and $n = 15$ in the formula $S_n = \frac{n}{2}(a_1 + a_n)$.
$$S_{15} = \frac{15}{2}(8 + 36) = 330$$

47. To find n, use $a_1 = -1$, $d = -6$, and $a_n = -73$ in the formula $a_n = a_1 + (n - 1)d$.
$-73 = -1 + (n - 1)(-6)$
$-78 = -1 - 6n + 6$
$-78 = -6n$
$13 = n$

Use $a_1 = -1$, $a_{13} = -73$, and $n = 13$ in the formula $S_n = \frac{n}{2}(a_1 + a_n)$.
$$S_{13} = \frac{13}{2}(-1 + (-73)) = -481$$

49. To find n, use $a_1 = -6$, $d = 5$, and $a_n = 64$ in the formula $a_n = a_1 + (n - 1)d$.

$64 = -6 + (n - 1)5$
$64 = -6 + 5n - 5$
$75 = 5n$
$15 = n$
Use $a_1 = -6$, $a_{15} = 64$, and $n = 15$ in the formula $S_n = \frac{n}{2}(a_1 + a_n)$.
$S_{15} = \frac{15}{2}(-6 + 64) = 435$

51. To find n, use $a_1 = 20$, $d = -8$, and $a_n = -92$ in the formula $a_n = a_1 + (n - 1)d$.
$-92 = 20 + (n - 1)(-8)$
$-92 = 20 - 8n + 8$
$-120 = -8n$
$15 = n$
Use $a_1 = 20$, $a_{15} = -92$, and $n = 15$ in the formula $S_n = \frac{n}{2}(a_1 + a_n)$.
$S_{15} = \frac{15}{2}(20 + (-92)) = -540$

53. $\sum_{i=1}^{12}(3i - 7) = -4 + (-1) + ... + 29$
Use $a_1 = -4$, $a_{12} = 29$, and $n = 12$ in the formula $S_n = \frac{n}{2}(a_1 + a_n)$.
$S_{12} = \frac{12}{2}(-4 + 29) = 150$

55. $\sum_{i=1}^{11}(-5i + 2) = -3 + (-8) + ... + (-53)$
Use $a_1 = -3$, $a_{11} = -53$, and $n = 11$ in the formula $S_n = \frac{n}{2}(a_1 + a_n)$.
$S_{11} = \frac{11}{2}(-3 + (-53)) = -308$

57. Use $a_1 = \$22{,}000$, $n = 7$, and $d = \$500$ in the formula $a_n = a_1 + (n - 1)d$.
$a_7 = 22{,}000 + (7 - 1)500 = \$25{,}000$

59. The students read 5 pages the first day, 7 pages the second day, 9 pages the third day, and so on. To find the number they read on the 31st day, let $n = 31$, $d = 2$, and $a = 5$ in the formula $a_n = a_1 + (n - 1)d$.
$a_{31} = 5 + (31 - 1)2 = 65$
To find the sum $5 + 7 + 9 + ... + 65$, use $n = 31$, $a_1 = 5$, and $a_{31} = 65$ in the formula $S_n = \frac{n}{2}(a_1 + a_n)$.
$S_{31} = \frac{31}{2}(5 + 65) = 1085$

61. The only sequence that does not have a common difference is (b) and so it is not arithmetic.

14.4 WARM-UPS

1. False, because the ratio of two consecutive terms is not constant.
2. False, there is a common ratio of 2 between adjacent terms.
3. True, because the general form for a geometric sequence is $a_n = a_1 r^{n-1}$.
4. True, because if $n = 1$, then $3(2)^{-1+3} = 12$.
5. True, because $a_1 = 12$ and $a_2 = 6$ gives $r = 1/2$.
6. True, because of the definition of geometric series.
7. False, because we have a formula for the sum of a finite geometric series.
8. False, because $a_1 = 6$.
9. True, because this is the correct formula for the sum of all of the terms of an infinite geometric series with first term 10 and ratio $1/2$.
10. False, because there is no sum for an infinite geometric series with a ratio of 2.

14.4 EXERCISES

1. A geometric sequence is one in which each term after the first is obtained by multiplying the preceding term by a constant.
3. A geometric series is an indicated sum of a geometric sequence.
5. The approximate value of r^n when n is large and $|r| < 1$ is 0.
7. Since the first term is 1 and the common ratio is 2, the nth term is $a_n = 1(2)^{n-1} = 2^{n-1}$.
9. Since the first term is $1/3$ and the common ratio is 3, the nth term is $a_n = \frac{1}{3}(3)^{n-1}$.
11. Since the first term is 64 and the common ratio is $1/8$, the nth term is $a_n = 64\left(\frac{1}{8}\right)^{n-1}$.
13. Since the first term is 8 and the common ratio is $-1/2$, the nth term is $a_n = 8\left(-\frac{1}{2}\right)^{n-1}$.
15. Since the first term is 2 and the common ratio is $-4/2 = -2$, the nth term is $a_n = 2(-2)^{n-1}$.

17. Since the first term is $-1/3$ and the common ratio is $(-1/4)/(-1/3) = 3/4$, the nth term is $a_n = -\frac{1}{3}\left(\frac{3}{4}\right)^{n-1}$.

19. $a_1 = 2(1/3)^{1-1} = 2$,
$a_2 = 2(1/3)^{2-1} = 2/3$,
$a_3 = 2(1/3)^{3-1} = 2/9$,
$a_4 = 2(1/3)^{4-1} = 2/27$,
$a_5 = 2(1/3)^{5-1} = 2/81$
The first 5 terms are 2, $\frac{2}{3}$, $\frac{2}{9}$, $\frac{2}{27}$, and $\frac{2}{81}$.

21. $a_1 = (-2)^{1-1} = 1$,
$a_2 = (-2)^{2-1} = -2$, $\quad a_3 = (-2)^{3-1} = 4$,
$a_4 = (-2)^{4-1} = -8$, $\quad a_5 = (-2)^{5-1} = 16$
The first 5 terms are 1, -2, 4, -8, and 16.

23. $a_1 = 2^{-1} = 1/2$,
$a_2 = 2^{-2} = 1/4$, $\quad a_3 = 2^{-3} = 1/8$,
$a_4 = 2^{-4} = 1/16$, $\quad a_5 = 2^{-5} = 1/32$
The first 5 terms are $\frac{1}{2}$, $\frac{1}{4}$, $\frac{1}{8}$, $\frac{1}{16}$, and $\frac{1}{32}$.

25. $a_1 = (0.78)^1 = 0.78$,
$a_2 = (0.78)^2 = 0.6084$,
$a_3 = (0.78)^3 = 0.4746$,
$a_4 = (0.78)^4 = 0.3702$, $a_5 = (0.78)^5 = 0.2887$
The first 5 terms are 0.78, 0.6084, 0.4746, 0.3702, and 0.2887.

27. Use $a_4 = 40$, $n = 4$, and $r = 2$ in the formula $a_n = a_1 r^{n-1}$.

$$40 = a_1(2)^{4-1}$$
$$40 = 8a_1$$
$$5 = a_1$$

29. Use $a_4 = 2/9$, $n = 4$, and $a_1 = 6$ in the formula $a_n = a_1 r^{n-1}$.

$$\frac{2}{9} = 6r^{4-1}$$
$$\frac{1}{27} = r^3$$
$$\frac{1}{3} = r$$

31. Use $r = 1/3$, $n = 4$, and $a_1 = -3$ in the formula $a_n = a_1 r^{n-1}$.

$$a_4 = -3\left(\frac{1}{3}\right)^{4-1} = -3\left(\frac{1}{27}\right) = -\frac{1}{9}$$

33. Use $r = 1/2$, $a_1 = 1/2$, and $a_n = 1/512$ in the formula $a_n = a_1 r^{n-1}$ to find n.

$$\frac{1}{512} = \frac{1}{2}\left(\frac{1}{2}\right)^{n-1}$$
$$\frac{1}{2^9} = \left(\frac{1}{2}\right)^n$$
$$n = 9$$

Use $n = 9$, $a_1 = 1/2$, and $r = 1/2$ in the formula $S_n = \frac{a_1(1 - r^n)}{1 - r}$.

$$S_9 = \frac{\frac{1}{2}\left(1 - \left(\frac{1}{2}\right)^9\right)}{1 - \frac{1}{2}} = 1 - \frac{1}{512} = \frac{511}{512}$$

35. Use $n = 5$, $a_1 = 1/2$, and $r = -1/2$ in the formula $S_n = \frac{a_1(1 - r^n)}{1 - r}$.

$$S_5 = \frac{\frac{1}{2}\left(1 - \left(-\frac{1}{2}\right)^5\right)}{1 - \left(-\frac{1}{2}\right)} = \frac{\frac{1}{2}\left(\frac{33}{32}\right)}{\frac{3}{2}} = \frac{11}{32}$$

37. First determine the number of terms. Since $r = 2/3$, the nth term is $30\left(\frac{2}{3}\right)^{n-1}$. Solve

$$30\left(\frac{2}{3}\right)^{n-1} = \frac{1280}{729}$$
$$\left(\frac{2}{3}\right)^{n-1} = \frac{128}{2187} = \left(\frac{2}{3}\right)^7$$
$$n - 1 = 7$$
$$n = 8$$

Use $n = 8$, $a_1 = 30$, and $r = 2/3$ in the formula $S_n = \frac{a_1(1 - r^n)}{1 - r}$.

$$S_8 = \frac{30\left(1 - \left(\frac{2}{3}\right)^8\right)}{1 - \left(\frac{2}{3}\right)} = \frac{30\left(\frac{6305}{6561}\right)}{\frac{1}{3}}$$
$$= \frac{63050}{729} \approx 86.4883$$

39. $$\sum_{i=1}^{10} 5(2)^{i-1} = S_{10} = \frac{5(1 - (2)^{10})}{1 - (2)}$$
$$= \frac{5(-1023)}{-1} = 5115$$

41. $$\sum_{i=1}^{6} (0.1)^i = S_6 = \frac{0.1(1 - (0.1)^6)}{1 - (0.1)}$$
$$= \frac{0.1(0.999999)}{0.9} = 0.111111$$

43. $$\sum_{i=1}^{6} 100(0.3)^i = S_6$$
$$= \frac{100(0.3)(1 - (0.3)^6)}{1 - (0.3)}$$
$$= \frac{100(0.3)(1 - (0.3)^6)}{0.7} = 42.8259$$

45. Use $a_1 = 1/8$ and $r = 1/2$ in the formula for the sum of an infinite geometric series $S = \frac{a_1}{1-r}$.

$$S = \frac{\frac{1}{8}}{1-\frac{1}{2}} = \frac{\frac{1}{8}}{\frac{1}{2}} = \frac{1}{4}$$

47. Use $a_1 = 3$ and $r = 2/3$ in $S = \frac{a_1}{1-r}$.

$$S = \frac{3}{1-\frac{2}{3}} = \frac{3}{\frac{1}{3}} = 9$$

49. Use $a_1 = 4$ and $r = -1/2$ in $S = \frac{a_1}{1-r}$.

$$S = \frac{4}{1-(-\frac{1}{2})} = \frac{4}{\frac{3}{2}} = \frac{8}{3}$$

51. Use $a_1 = 0.3$ and $r = 0.3$ in $S = \frac{a_1}{1-r}$.

$$S = \frac{0.3}{1-0.3} = \frac{0.3}{0.7} = \frac{3}{7}$$

53. Use $a_1 = 3$ and $r = 0.5$ in $S = \frac{a_1}{1-r}$.

$$S = \frac{3}{1-0.5} = \frac{3}{0.5} = 6$$

55. Use $a_1 = 0.3$ and $r = 0.1$ in $S = \frac{a_1}{1-r}$.

$$S = \frac{0.3}{1-0.1} = \frac{0.3}{0.9} = \frac{1}{3}$$

57. Use $a_1 = 0.12$ and $r = 0.01$ in $S = \frac{a_1}{1-r}$.

$$S = \frac{0.12}{1-0.01} = \frac{0.12}{0.99} = \frac{12}{99} = \frac{4}{33}$$

59. We want the sum of the geometric series $2000(1.12)^{45} + 2000(1.12)^{44} + ... + 2000(1.12)$.
Note that the last deposit is made at the beginning of the 45th year and earns interest for only one year. Rewrite the series as $2000(1.12) + 2000(1.12)^2 + ... + 2000(1.12)^{45}$ where the $a_1 = 2000(1.12)$, $n = 45$, and $r = 1.12$.

$$S_{45} = \frac{2000(1.12)(1-(1.12)^{45})}{1-1.12} = \$3{,}042{,}435.27$$

61. We want the sum of the finite geometric series $1 + 2 + 4 + 8 + 16 + ... + 2^{30}$, which has 31 terms and a ratio of 2.

$$S_{31} = \frac{1(1-2^{31})}{1-2} = 2^{31} - 1 = 2{,}147{,}483{,}647 \text{ cents} = \$21{,}474{,}836.47$$

63. Use $r = 0.80$, $a_1 = 1{,}000{,}000$ in $S = \frac{a_1}{1-r}$.

$$S = \frac{1{,}000{,}000}{1-0.80} = \$5{,}000{,}000$$

65. Only sequence (d) is not geometric because it is the only one that does not have a constant ratio.

67. Use $a_1 = 24/100 = 0.24$ and $r = 1/100 = 0.01$ in the formula for $S = \frac{a_1}{1-r}$.

$$S = \frac{0.24}{1-0.01} = \frac{0.24}{0.99} = \frac{8}{33}$$

14.5 WARM-UPS

1. False, because there are 13 terms in a binomial to the 12th power.
2. False, because the 7th term has variable part a^6b^6.
3. False, because if $x = 1$ the equation is incorrect.
4. True, because the signs alternate in any expansion of a difference.
5. True, because we can obtain it from the 7th line.
6. True, because $1 + 4 + 6 + 4 + 1 = 2^4$.
7. True, because of the binomial theorem.
8. True, because $2^n = (1+1)^n$
$= \sum_{i=0}^{n} \frac{n!}{(n-i)!i!} 1^{n-i} 1^i = \sum_{i=0}^{n} \frac{n!}{(n-i)!i!}$,
and the last sum is the sum of the coefficients in the nth row.
9. True, by definition of 0! and 1!.
10. True, because $\frac{7 \cdot 6 \cdot 5 \cdot 4 \cdot 3 \cdot 2 \cdot 1}{5 \cdot 4 \cdot 3 \cdot 2 \cdot 1 \cdot 2 \cdot 1} = 21$

14.5 EXERCISES

1. The sum obtained for a power of a binomial is called a binomial expansion.
3. The expression $n!$ is the product of the positive integers from 1 through n.

5. $\frac{4!}{4!0!} = \frac{4 \cdot 3 \cdot 2 \cdot 1}{4 \cdot 3 \cdot 2 \cdot 1 \cdot 1} = 1$

7. $\frac{5!}{2!3!} = \frac{5 \cdot 4}{2} = 10$

9. $\frac{8!}{5!3!} = \frac{8 \cdot 7 \cdot 6}{3 \cdot 2 \cdot 1} = 56$

11. The coefficients in the 3rd row of Pascal's triangle are 1, 3, 3, 1. Use these coefficients with the pattern for the exponents.

$$(x+1)^3 = 1x^3 1^0 + 3x^2 1^1 + 3x 1^2 + 1x^0 1^3 = x^3 + 3x^2 + 3x + 1$$

13. The coefficients in the 3rd row of Pascal's triangle are 1, 3, 3, 1. Use these coefficients with the pattern for the exponents.
$(a+2)^3 = 1a^3 2^0 + 3a^2 2^1 + 3a2^2 + 1a^0 2^3$
$= a^3 + 6a^2 + 12a + 8$

15. The coefficients in the 5th row of Pascal's triangle are 1, 5, 10, 10, 5, 1. Use these coefficients with the pattern for the exponents.
$(r+t)^5$
$= r^5 + 5r^4t + 10r^3t^2 + 10r^2t^3 + 5rt^4 + t^5$

17. The coefficients in the 3rd row are 1, 3, 3, 1. Use these coefficients with the pattern for the exponents, and alternate the signs.
$(m-n)^3 = m^3 - 3m^2n + 3mn^2 - n^3$

19. Use the coefficients 1, 3, 3, 1 and let $y = 2a$ in the binomial theorem.
$(x+2a)^3$
$= 1x^3(2a)^0 + 3x^2(2a)^1 + 3x(2a)^2 + (2a)^3$
$= x^3 + 6ax^2 + 12a^2x + 8a^3$

21. Use the coefficients 1, 4, 6, 4, 1 in the binomial theorem.
$(x^2-2)^4 = (x^2)^4 - 4(x^2)^3 2 + 6(x^2)^2 2^2 - 4x^2 2^3 + 1(x^2)^0 2^4$
$= x^8 - 8x^6 + 24x^4 - 32x^2 + 16$

23. Use the coefficients from the 7th line 1, 7, 21, 35, 35, 21, 7, 1 and alternate the signs of the terms.
$(x-1)^7 = x^7 - 7x^6 + 21x^5 - 35x^4 + 35x^3 - 21x^2 + 7x - 1$

25. Use the binomial theorem to write the first 4 terms of $(a-3b)^{12}$.
$\frac{12!}{12!0!}a^{12}b^0 - \frac{12!}{11!1!}a^{11}b^1 + \frac{12!}{10!2!}a^{10}b^2 - \frac{12!}{9!3!}a^9b^3$
$= a^{12} - 36a^{11}b + 594a^{10}b^2 - 5940a^9b^3$

27. Use the binomial theorem to write the first 4 terms of $(x^2+5)^9$.
$\frac{9!}{9!0!}(x^2)^9 5^0 + \frac{9!}{8!1!}(x^2)^8 5^1 + \frac{9!}{7!2!}(x^2)^7 5^2 + \frac{9!}{6!3!}(x^2)^6 5^3$
$= x^{18} + 45x^{16} + 900x^{14} + 10500x^{12}$

29. Use the binomial theorem to write the first 4 terms of $(x-1)^{22}$.
$\frac{22!}{22!0!}x^{22}1^0 - \frac{22!}{21!1!}x^{21}1^1 + \frac{22!}{20!2!}x^{20}1^2 - \frac{22!}{19!3!}x^{19}1^3$
$= x^{22} - 22x^{21} + 231x^{20} - 1540x^{19}$

31. Use the binomial theorem to write the first 4 terms of $\left(\frac{x}{2}+\frac{y}{3}\right)^{10}$.
$\frac{10!}{10!0!}\left(\frac{x}{2}\right)^{10}\left(\frac{y}{3}\right)^0 + \frac{10!}{9!1!}\left(\frac{x}{2}\right)^9\left(\frac{y}{3}\right)^1 + \frac{10!}{8!2!}\left(\frac{x}{2}\right)^8\left(\frac{y}{3}\right)^2 + \frac{10!}{7!3!}\left(\frac{x}{2}\right)^7\left(\frac{y}{3}\right)^3$
$= \frac{x^{10}}{1024} + \frac{5x^9y}{768} + \frac{5x^8y^2}{256} + \frac{5x^7y^3}{144}$

33. Use the formula for the kth term of $(x+y)^n$ with $k = 6$ and $n = 13$.
$\frac{13!}{(13-6+1)!(6-1)!}a^{13-6+1}w^{6-1}$
$= \frac{13!}{8!5!}a^8w^5 = 1287a^8w^5$

35. Use the formula for the kth term with $k = 8$ and $n = 16$.
$\frac{16!}{(16-8+1)!(8-1)!}m^{16-8+1}(-n)^{8-1}$
$= \frac{16!}{9!7!}m^9(-n)^7 = -11440m^9n^7$

37. Use the formula for the kth term with $k = 4$ and $n = 8$.
$\frac{8!}{(8-4+1)!(4-1)!}x^{8-4+1}(2y)^{4-1}$
$= \frac{8!}{5!3!}x^5(2y)^3 = 56x^5 8y^3 = 448x^5y^3$

39. Use the formula for the kth term with $k = 7$ and $n = 20$.
$\frac{20!}{(20-7+1)!(7-1)!}(2a^2)^{20-7+1}b^{7-1}$
$= \frac{20!}{14!6!}(2a^2)^{14}b^6$
$= 635{,}043{,}840a^{28}b^6$

41. Use $n = 8$, $x = a$, and $y = b$ in the binomial theorem with summation notation.
$$(a+m)^8 = \sum_{i=0}^{8}\frac{8!}{(8-i)!\,i!}a^{8-i}m^i$$

43. Use $n = 5$, $x = a$, and $y = -2x$ in the binomial theorem with summation notation.
$$(a+(-2x))^5 = \sum_{i=0}^{5}\frac{5!}{(5-i)!\,i!}a^{5-i}(-2x)^i$$
$$= \sum_{i=0}^{5}\frac{5!\,(-2)^i}{(5-i)!\,i!}a^{5-i}x^i$$

45. $(a+(b+c))^3$
$= a^3 + 3a^2(b+c) + 3a(b+c)^2 + (b+c)^3$
$= a^3 + 3a^2b + 3a^2c + 3ab^2 + 6abc + 3ac^2 + b^3 + 3b^2c + 3bc^2 + c^3$
$= a^3 + b^3 + c^3 + 3a^2b + 3a^2c + 3ab^2 + 3ac^2 + 3b^2c + 3bc^2 + 6abc$

Enriching Your Mathematical Word Power

1. a **2.** d **3.** c **4.** b **5.** a
6. c **7.** d **8.** b **9.** d **10.** a

CHAPTER 14 REVIEW

1. $a_1 = 1^3$, $a_2 = 2^3$, $a_3 = 3^3$, $a_4 = 4^3$, $a_5 = 5^3$
The terms of the sequence are 1, 8, 27, 64, 125.

3. $c_1 = (-1)^1(2 \cdot 1 - 3) = 1$,
$c_2 = (-1)^2(2 \cdot 2 - 3) = 1$,
$c_3 = (-1)^3(2 \cdot 3 - 3) = -3$,
$c_4 = (-1)^4(2 \cdot 4 - 3) = 5$,
$c_5 = (-1)^5(2 \cdot 5 - 3) = -7$
$c_6 = (-1)^6(2 \cdot 6 - 3) = 9$
The terms are 1, 1, −3, 5, −7, 9.

5. $a_1 = -\frac{1}{1} = -1$, $a_2 = -\frac{1}{2}$, $a_3 = -\frac{1}{3}$
The first three terms are $-1, -\frac{1}{2}, -\frac{1}{3}$.

7. $b_1 = \frac{(-1)^{2 \cdot 1}}{2 \cdot 1 + 1} = \frac{1}{3}$, $b_2 = \frac{(-1)^{2 \cdot 2}}{2 \cdot 2 + 1} = \frac{1}{5}$
$b_3 = \frac{(-1)^{2 \cdot 3}}{2 \cdot 3 + 1} = \frac{1}{7}$
The first three terms are $\frac{1}{3}, \frac{1}{5}$, and $\frac{1}{7}$.

9. $c_1 = \log_2(2^{1+3}) = \log_2(2^4) = 4$
$c_2 = \log_2(2^{2+3}) = \log_2(2^5) = 5$
$c_3 = \log_2(2^{3+3}) = \log_2(2^6) = 6$
The first three terms are 4, 5, and 6.

11. $\sum_{i=1}^{3} i^3 = 1^3 + 2^3 + 3^3 = 36$

13. $\sum_{n=1}^{5} n(n-1)$
$= 1(1-1) + 2(2-1) + 3(3-1)$
$+ 4(4-1) + 5(5-1)$
$= 0 + 2 + 6 + 12 + 20 = 40$

15. The terms in the series are reciprocals of even integers. Even integers are usually represent as $2i$, but to get 4 in the denominator when $i = 1$, we use $2(i+1)$.

$$\sum_{i=1}^{\infty} \frac{1}{2(i+1)}$$

17. The terms *in* this series are the squares of integers. Squares are usually represented as i^2, but to get the first term 0 when $i = 1$, we use $(i-1)^2$.

$$\sum_{i=1}^{\infty} (i-1)^2$$

19. To get alternating signs for the terms, we use a power of −1. If we use $(-1)^i$, then $i = 1$ makes the first term negative. So we use $(-1)^{i+1}$.

$$\sum_{i=1}^{\infty} (-1)^{i+1} x_i$$

21. $a_1 = 6 + (1-1)5 = 6$
Since the common difference is 5, the first four terms are 6, 11, 16, and 21.

23. $a_1 = -20 + (1-1)(-2) = -20$
Since the common difference is −2, the first four terms are −20, −22, −24, and −26.

25. $a_1 = 1000(1) + 2000 = 3000$
Since the common difference is 1000, the first four terms are 3000, 4000, 5000, and 6000.

27. Use $a_1 = 1/3$, $d = 1/3$, and the formula $a_n = a_1 + (n-1)d$.
$a_n = \frac{1}{3} + (n-1)\frac{1}{3} = \frac{n}{3}$

29. Use $a_1 = 2$, $d = 2$, and the formula $a_n = a_1 + (n-1)d$.
$a_n = 2 + (n-1)(2) = 2n$

31. Use $a_1 = 1$, $a_{24} = 24$, $n = 24$, and the formula $S_n = \frac{n}{2}(a_1 + a_n)$.
$S_{24} = \frac{24}{2}(1 + 24) = 300$

33. Use $a_1 = 1/6$, $d = 1/3$, $a_n = 11/2$, and the formula $a_n = a_1 + (n-1)d$ to find n.
$\frac{11}{2} = \frac{1}{6} + (n-1)\frac{1}{3}$
$33 = 1 + (n-1)2$
$32 = 2n - 2$
$34 = 2n$
$17 = n$
Now use $n = 17$, $a_1 = 1/6$, $a_{17} = 11/2$, and the formula $S_n = \frac{n}{2}(a_1 + a_n)$ to find the sum.
$S_{17} = \frac{17}{2}\left(\frac{1}{6} + \frac{11}{2}\right) = \frac{17}{2}\left(\frac{34}{6}\right) = \frac{289}{6}$

35. Use $a_1 = -1$, $a_7 = 11$, $n = 7$, and the formula $S_n = \frac{n}{2}(a_1 + a_n)$ to find the sum.
$S_7 = \frac{7}{2}(-1 + 11) = 35$

37. $a_1 = 3\left(\frac{1}{2}\right)^{1-1} = 3$, $a_2 = 3\left(\frac{1}{2}\right)^{2-1} = \frac{3}{2}$,
$a_3 = 3\left(\frac{1}{2}\right)^{3-1} = \frac{3}{4}$, $a_4 = 3\left(\frac{1}{2}\right)^{4-1} = \frac{3}{8}$
The first four terms are 3, $\frac{3}{2}, \frac{3}{4}$, and $\frac{3}{8}$.

39. $a_1 = 2^{1-1} = 1, a_2 = 2^{1-2} = \frac{1}{2}$
$a_3 = 2^{1-3} = \frac{1}{4}, \quad a_4 = 2^{1-4} = \frac{1}{8}$
The first four terms are 1, $\frac{1}{2}$, $\frac{1}{4}$, and $\frac{1}{8}$.

41. $a_1 = 23(10)^{-2(1)} = 0.23,$
$a_2 = 23(10)^{-2(2)} = 0.0023,$
$a_3 = 23(10)^{-2(3)} = 0.000023,$
$a_4 = 23(10)^{-2(4)} = 0.00000023$
The first four terms of the geometric sequence are 0.23, 0.0023, 0.000023, and 0.00000023.

43. Use $a_1 = 1/2, r = 6$, and the formula $a_n = a_1 r^{n-1}$.
$$a_n = \frac{1}{2}(6)^{n-1}$$

45. Use $a_1 = 7/10, r = 1/10$, and the formula $a_n = a_1 r^{n-1}$.
$$a_n = 0.7(0.1)^{n-1}$$

47. Use $a_1 = 1/3, r = 1/3, n = 4$, and the formula $S_n = \dfrac{a_1(1 - r^n)}{1 - r}$.
$$S_4 = \frac{\frac{1}{3}\left(1 - \left(\frac{1}{3}\right)^4\right)}{1 - \frac{1}{3}} = \frac{\frac{1}{3}\left(\frac{80}{81}\right)}{\frac{2}{3}} = \frac{40}{81}$$

49. Use $a_1 = 0.3, r = 0.1, n = 10$, and the formula $S_n = \dfrac{a_1(1 - r^n)}{1 - r}$.
$$S_{10} = \frac{0.3\left(1 - (0.1)^{10}\right)}{1 - 0.1} = \frac{0.3(0.9999999999)}{0.9}$$
$$= 0.3333333333$$
Your calculator may not give ten 3's after the decimal point, but doing this computation without a calculator does give ten 3's and this is the exact answer.

51. Use $a_1 = 1/4, r = 1/3$, and the formula for the sum of an infinite geometric series $S = \dfrac{a_1}{1 - r}$.
$$S = \frac{\frac{1}{4}}{1 - \frac{1}{3}} = \frac{\frac{1}{4}}{\frac{2}{3}} = \frac{3}{8}$$

53. Use $a_1 = 18$, r $= 2/3$, and the formula for the sum of an infinite geometric series $\text{S} = \dfrac{a_1}{1 - r}$.
$$\text{S} = \frac{18}{1 - \frac{2}{3}} = \frac{18}{\frac{1}{3}} = 54$$

55. The coefficients for the fifth power of a binomial are 1, 5, 10, 10, 5, and 1.
$$(m + n)^5 = m^5 + 5m^4n + 10m^3n^2 + 10m^2n^3 + 5mn^4 + n^5$$

57. The coefficients for the third power of a binomial are 1, 3, 3, and 1. Alternate the signs because it is a difference to a power.
$$(a^2 - 3b)^3 = 1(a^2)^3(3b)^0 - 3(a^2)^2(3b)^1 + 3(a^2)^1(3b)^2 - 1(a^2)^0(3b)^3$$
$$= a^6 - 9a^4b + 27a^2b^2 - 27b^3$$

59. Use $n = 12$ and $k = 5$ in the formula for the kth term.
$$\frac{12!}{(12 - 5 + 1)!(5 - 1)!}x^{12-5+1}y^{5-1} = \frac{12!}{8!4!}x^8y^4$$
$$= 495x^8y^4$$

61. Use $n = 14$ and $k = 3$ in the formula for the kth term.
$$\frac{14!}{(14 - 3 + 1)!(3 - 1)!}(2a)^{14-3+1}(-b)^{3-1}$$
$$= \frac{14!}{12!2!}(2a)^{12}(-b)^2 = 372{,}736a^{12}b^2$$

63. Use the binomial theorem expressed in summation notation, with $n = 7$.
$$(a + w)^7 = \sum_{i=0}^{7} \frac{7!}{(7 - i)!\, i!}\, a^{7-i}\, w^i$$

65. The sequence has neither a constant difference nor a constant ratio. So it is neither arithmetic nor geometric.

67. There is a constant difference of 3. So the sequence is an arithmetic sequence.

69. There is a constant difference of 2. So the sequence is an arithmetic sequence.

71. Use $a_1 = 6, n = 4, a_4 = 1/30$, and the formula $a_n = a_1 r^{n-1}$.
$$\frac{1}{30} = 6r^{4-1}$$
$$\frac{1}{180} = r^3$$
$$r = \sqrt[3]{\frac{1}{180}} = \frac{1}{\sqrt[3]{180}} = \frac{\sqrt[3]{150}}{30}$$

73. $$\sum_{i=1}^{5} \frac{(-1)^i}{i!} = \frac{(-1)^1}{1!} + \frac{(-1)^2}{2!} + \frac{(-1)^3}{3!} + \frac{(-1)^4}{4!} + \frac{(-1)^5}{5!}$$
$$= -1 + \frac{1}{2} - \frac{1}{6} + \frac{1}{24} - \frac{1}{120}$$

75. This is the summation notation for the binomial expansion of $(a + b)^5$.
$$\frac{5!}{5!0!}a^5b^0 + \frac{5!}{4!1!}a^4b^1 + \frac{5!}{3!2!}a^3b^2 + \frac{5!}{2!3!}a^2b^3 + \frac{5!}{1!4!}a^1b^4 + \frac{5!}{0!5!}a^0b^5$$
$$= a^5 + 5a^4b + 10a^3b^2 + 10a^2b^3 + 5ab^4 + b^5$$

77. There are 26 terms because in the expansion of $(x + y)^n$ there are $n + 1$ terms.

79. The first \$3000 earns interest for 16 years. The second \$3000 earns interest for 15 years, and so on. The last \$3000 earns interest for 1 year. The total in the account at the end of 16 years is the sum of the following series.
$3000(1.1) + 3000(1.1)^2 + ... + 3000(1.1)^{16}$
This is a geometric series with $n = 16$, $a_1 = 3000(1.1)$ and $r = 1.1$.
$$S_{16} = \frac{3000(1.1)(1 - (1.1)^{16})}{1 - 1.1} = \$118{,}634.11$$
81. We compute a new balance 16 times by multiplying by $1 + 0.10$ each time.
$3000(1.10)^{16} = \$13{,}784.92$

CHAPTER 14 TEST

1. $a_1 = -10 + (1 - 1)6 = -10$
$a_2 = -10 + (2 - 1)6 = -4$
$a_3 = -10 + (3 - 1)6 = 2$
$a_4 = -10 + (4 - 1)6 = 8$
The first four terms are -10, -4, 2, and 8.

2. $a_1 = 5(0.1)^{1-1} = 5$
$a_2 = 5(0.1)^{2-1} = 0.5$
$a_3 = 5(0.1)^{3-1} = 0.05$
$a_4 = 5(0.1)^{4-1} = 0.005$
The first four terms are 5, 0.5, 0.05, and 0.005.

3. $a_1 = \frac{(-1)^1}{1!} = -1$, $a_2 = \frac{(-1)^2}{2!} = \frac{1}{2}$,
$a_3 = \frac{(-1)^3}{3!} = -\frac{1}{6}$, $a_4 = \frac{(-1)^4}{4!} = \frac{1}{24}$
The first four terms are -1, $\frac{1}{2}$, $-\frac{1}{6}$, and $\frac{1}{24}$.

4. $a_1 = \frac{2(1) - 1}{(1)^2} = 1$, $a_2 = \frac{2(2) - 1}{(2)^2} = \frac{3}{4}$,
$a_3 = \frac{2(3) - 1}{(3)^2} = \frac{5}{9}$, $a_4 = \frac{2(4) - 1}{(4)^2} = \frac{7}{16}$
The first four terms are 1, $\frac{3}{4}$, $\frac{5}{9}$, and $\frac{7}{16}$.

5. The sequence is an arithmetic sequence with $a_1 = 7$ and $d = -3$. So the general term is $a_n = 7 + (n - 1)(-3) = 7 - 3n + 3$
$= 10 - 3n$.

6. This sequence is a geometric sequence with $a_1 = -25$, and $r = -1/5$. So the general term is $a_n = -25\left(-\frac{1}{5}\right)^{n-1}$.

7. This sequence is a sequence of even integers, which we can represent as $2n$. To get the signs to alternate, we use a power of -1. So the general term is $a_n = (-1)^{n-1}2n$.

8. This sequence is a sequence of squares of the positive integers. So the general term is $a_n = n^2$.

9. $\sum_{i=1}^{5}(2i + 3) = 2(1) + 3 + 2(2) + 3$
$+ 2(3) + 3 + 2(4) + 3 + 2(5) + 3$
$= 5 + 7 + 9 + 11 + 13$

10. $\sum_{i=1}^{6} 5(2)^{i-1} = 5(2)^{1-1} + 5(2)^{2-1}$
$+ 5(2)^{3-1} + 5(2)^{4-1} + 5(2)^{5-1} + 5(2)^{6-1}$
$= 5 + 10 + 20 + 40 + 80 + 160$

11. $\sum_{i=0}^{4} \frac{4!}{(4 - i)!i!} m^{4-i}q^i = \frac{4!}{4!0!} m^4q^0$
$+ \frac{4!}{3!1!} m^3q^1 + \frac{4!}{2!2!} m^2q^2$
$+ \frac{4!}{1!3!} m^1q^3 + \frac{4!}{0!4!} m^0q^4$
$= m^4 + 4m^3q + 6m^2q^2 + 4mq^3 + q^4$

12. Use $a_1 = 9$, $a_{20} = 66$, and $n = 20$ in the formula for the sum of an arithmetic series.
$S_{20} = \frac{20}{2}(9 + 66) = 10(75) = 750$

13. Use $a_1 = 10$, $n = 5$, and $r = 1/2$ in the formula for the sum of a finite geometric series.
$$S_5 = \frac{10\left(1 - \left(\frac{1}{2}\right)^5\right)}{1 - \frac{1}{2}} = \frac{10\left(\frac{31}{32}\right)}{\frac{1}{2}} = \frac{155}{8}$$

14. Use $a_1 = 0.35$ and $r = 0.93$ in the formula for the sum of an infinite geometric series.
$S = \frac{0.35}{1 - 0.93} = \frac{0.35}{0.07} = 5$

15. Use $a_1 = 2$, $a_{100} = 200$, and $n = 100$ in the formula for the sum of a finite arithmetic series.
$S_{100} = \frac{100}{2}(2 + 200) = 50(202) = 10{,}100$

16. Use $a_1 = 1/4$ and $r = 1/2$ in the formula for the sum of an infinite geometric series.
$S = \frac{1/4}{1 - 1/2} = \frac{1/4}{1/2} = \frac{1}{2}$

17. Use $a_1 = 2$, $r = 1/2$ and $a_n = 1/128$ to find n:
$$2\left(\frac{1}{2}\right)^{n-1} = \frac{1}{128}$$
$$\left(\frac{1}{2}\right)^{n-1} = \frac{1}{256} = \frac{1}{2^8}$$
$$n - 1 = 8$$
$$n = 9$$

Use $a_1 = 2$, $n = 9$, and $r = 1/2$ to find the sum of the 9 terms.

$$S_9 = \frac{2(1-(1/2)^9)}{1-1/2} = \frac{511}{128} \approx 3.9922$$

18. Use $a_1 = 3$, $a_5 = 48$, $n = 5$ in the formula for the general term of a geometric sequence.

$$48 = 3r^{5-1}$$
$$16 = r^4$$
$$\pm 2 = r$$

19. Use $a_1 = 1$, $a_{12} = 122$, $n = 12$, and the formula for the general term of an arithmetic sequence.

$$122 = 1 + (12-1)d$$
$$121 = 11d$$
$$11 = d$$

20. Use $n = 15$ and $k = 5$ in the formula for the kth term of a binomial expansion.

$$\frac{15!}{(15-5+1)!(5-1)!}r^{15-5+1}(-t)^{5-1}$$
$$= \frac{15!}{11!4!}r^{11}t^4 = 1365r^{11}t^4$$

21. Use $n = 8$ and $k = 4$ in the formula for the kth term of a binomial expansion.

$$\frac{8!}{(8-4+1)!(4-1)!}(a^2)^{8-4+1}(-2b)^{4-1}$$
$$= \frac{8!}{5!3!}a^{10}(-2)^3b^3 = -448a^{10}b^3$$

22. $800(1.10)^1 + 800(1.10)^2 + \ldots + 800(1.10)^{25}$

$$= \frac{800(1.10)(1-1.10^{25})}{1-1.10} = \$86{,}545.41$$

Making Connections

Chapters 1 - 14

1. $f(3) = 3^2 - 3 = 9 - 3 = 6$

2. $f(n) = n^2 - 3$

3. $f(x+h) = (x+h)^2 - 3$
$= x^2 + 2xh + h^2 - 3$

4. $f(x) - g(x) = x^2 - 3 - (2x - 1)$
$= x^2 - 2x - 2$

5. $g(f(3)) = g(6) = 2(6) - 1 = 11$

6. $(f \circ g)(2) = f(g(2)) = f(3) = 3^2 - 3 = 6$

7. $m(16) = \log_2(16) = 4$

8. $(h \circ m)(32) =$
$h(m(32)) = h(5) = 2^5 = 32$

9. $h(-1) = 2^{-1} = 1/2$

10. $h^{-1}(8) = \log_2(8) = 3$

11. $m^{-1}(0) = 2^0 = 1$

12. $(m \circ h)(x) = m(h(x)) = m(2^x)$
$= \log_2(2^x) = x$

So $(m \circ h)(x) = x$.

13. If y varies directly as x, then $y = kx$. Since $y = -6$ when $x = 4$, we have $-6 = 4k$, or $k = -3/2$. When $x = 9$, we can use the original formula with $k = -3/2$.

$$y = -\frac{3}{2}(9) = -\frac{27}{2}$$

14. If a varies inversely as b, then $a = k/b$. If $a = 2$ when $b = -4$, we can find k.

$$2 = \frac{k}{-4}$$
$$-8 = k$$

To find a, use $k = -8$ and $b = 3$.

$$a = \frac{-8}{3} = -\frac{8}{3}$$

15. If y varies directly as w and inversely as t, then $y = (kw)/t$. Use $y = 16$, $w = 3$, and $t = -4$ to find k.

$$16 = \frac{k(3)}{-4}$$
$$-64 = 3k$$
$$\frac{-64}{3} = k$$

To find y, use $k = -64/3$, $w = 2$, and $t = 3$.

$$y = -\frac{64(2)}{3(3)} = -\frac{128}{9}$$

16. If y varies jointly as h and the square of r, then $y = khr^2$. Use $y = 12$, $h = 2$, and $r = 3$ to find k.

$$12 = k(2)(3)^2$$
$$12 = 18k$$
$$\frac{2}{3} = k$$

To find y, use $k = 2/3$, $h = 6$, and $r = 2$.

$$y = \frac{2}{3}(6)(2)^2 = 16$$

17. The graph of $x > 3$ is the region to the right of the vertical line $x = 3$. The graph of $x + y < 0$ is the region below the line $y = -x$. The region to the right of $x = 3$ and below $y = -x$ is shown in the following graph.

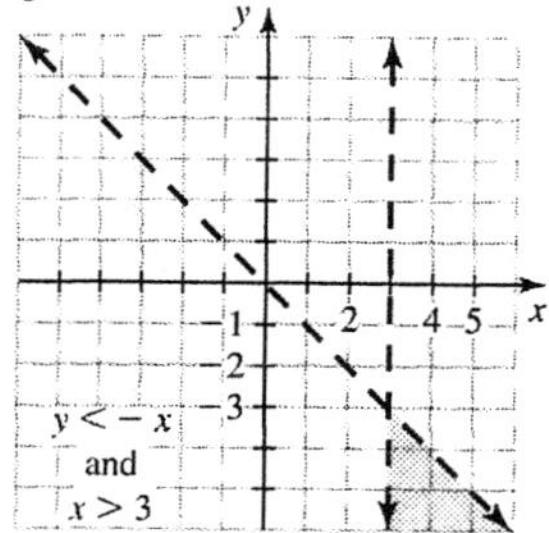

18. The inequality $|x - y| \geq 2$ is equivalent to the compound inequality $x - y \geq 2$ or $x - y \leq -2$. The graph of $x - y \geq 2$ is the region on or below the line $y = x - 2$. The graph of $x - y \leq -2$ is the region on or above the line $y = x + 2$. Since the word or is used, the graph of the compound inequality is the union of these two regions as shown in the following diagram.

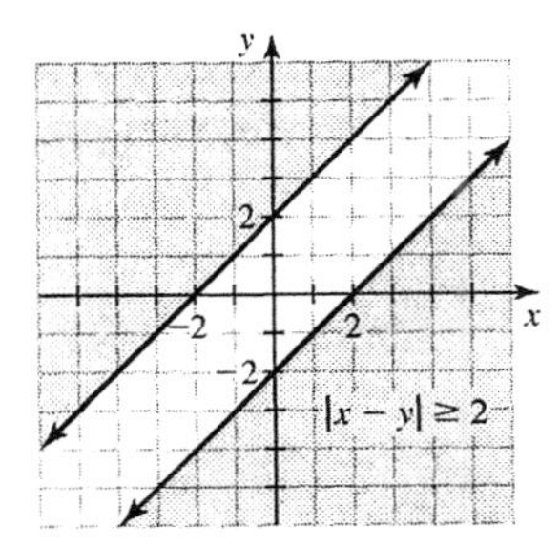

19. The graph of $y < -2x + 3$ is the region below the line $y = -2x + 3$. The graph of $y > 2^x$ is the region above the curve $y = 2^x$. Since the word and is used, the graph of the compound inequality is the intersection of these two regions, the points that lie above $y = 2^x$ and below $y = -2x + 3$. We could have used a test point in each of the four regions to see which region satisfies both inequalities.

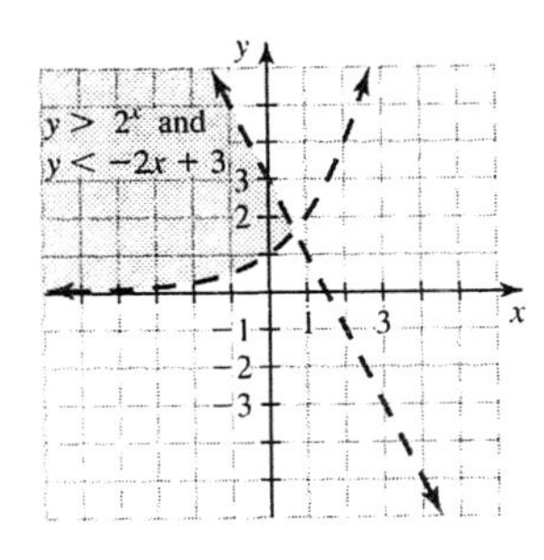

20. The inequality $|y + 2x| < 1$ is equivalent to $-1 < y + 2x < 1$. This inequality is also written as

$$y + 2x > -1 \quad \text{and} \quad y + 2x < 1$$
$$y > -2x - 1 \text{ and} \qquad y < -2x + 1$$

The graph of $y > -2x - 1$ is the region above the line $y = -2x - 1$. The graph of $y < -2x + 1$ is the region below the line $y = -2x + 1$. The points that satisfy both inequalities are the points that lie between these two parallel lines.

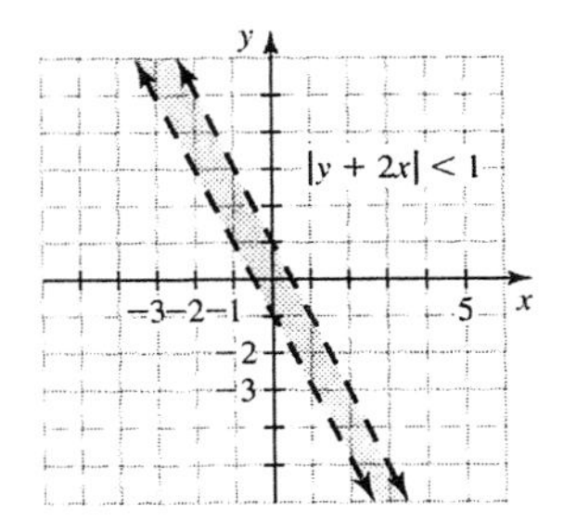

21. The graph of $x^2 + y^2 = 4$ is a circle of radius 2 centered at the origin. Since (0, 0) satisfies the inequality $x^2 + y^2 < 4$, we shade the region inside the circle.

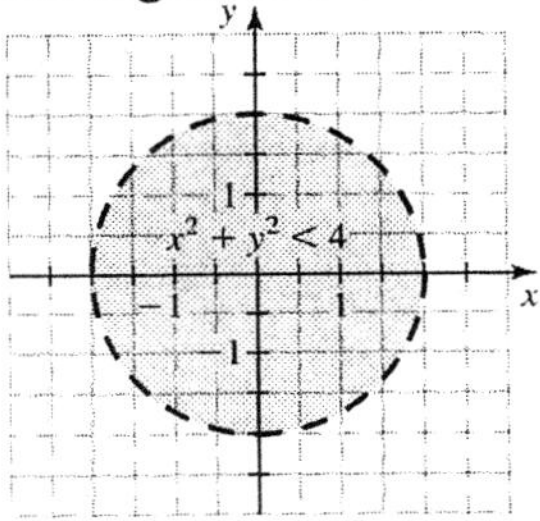

22. The graph of $x^2 - y^2 = 1$ is a hyperbola with x-intercepts $(-1, 0)$ and $(1, 0)$. The fundamental rectangle passes through the x-intercepts and $(0, 1)$ and $(0, -1)$. Extend the diagonals for the asymptotes. The hyperbola opens to the left and right. Test a point in each region to see that only points in the region containing the origin satisfy the inequality.

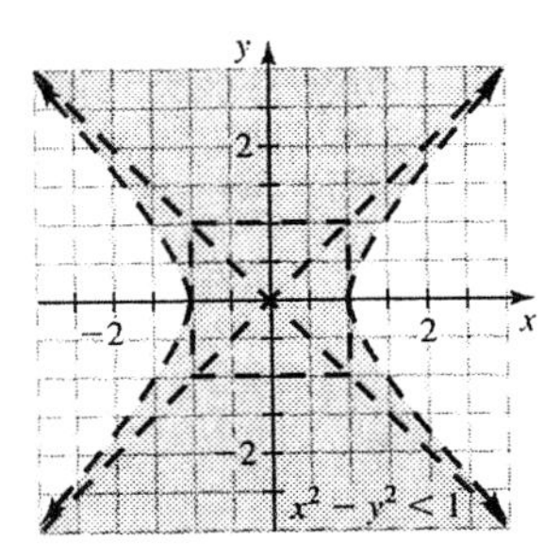

23. Graph the curve $y = \log_2(x)$ and shade the region below the curve to show the graph of $y < \log_2(x)$.

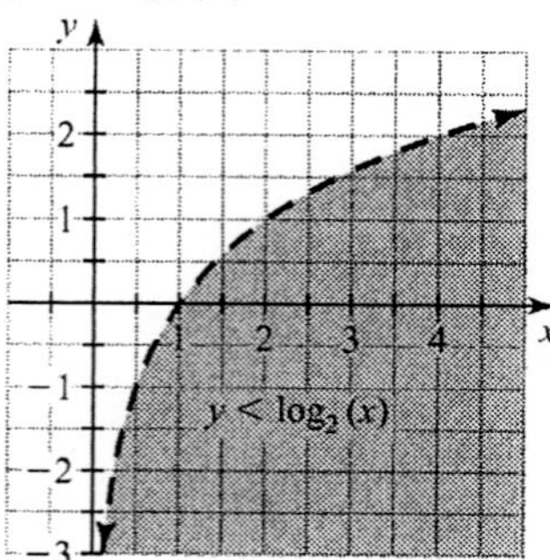

24. Write $x^2 + 2y < 4$ as $y < -\frac{1}{2}x^2 + 2$, to see that the boundary is a parabola opening downward with vertex at (0, 2).

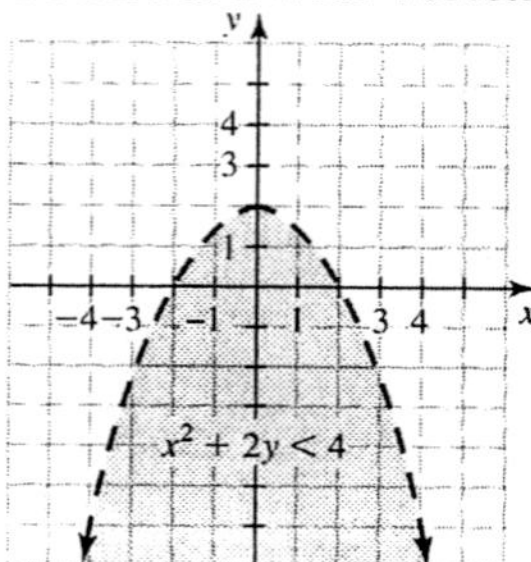

25. The graph of $\frac{x^2}{4} + \frac{y^2}{9} < 1$ is the region inside the ellipse $\frac{x^2}{4} + \frac{y^2}{9} = 1$. The graph of $y > x^2$ is the region above the parabola $y = x^2$. The points that satisfy the compound inequality are inside the ellipse and above the parabola as shown in the diagram.

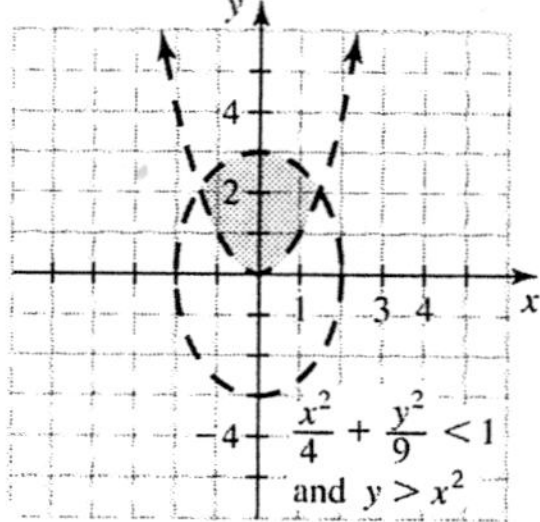

26. $\frac{a}{b} + \frac{b}{a} = \frac{a(a)}{b(a)} + \frac{b(b)}{a(b)} = \frac{a^2 + b^2}{ab}$

27. $1 - \frac{3}{y} = \frac{y}{y} - \frac{3}{y} = \frac{y-3}{y}$

28. $\frac{x-2}{x^2-9} - \frac{x-4}{x^2-2x-3}$

$$= \frac{(x-2)(x+1)}{(x-3)(x+3)(x+1)} - \frac{(x-4)(x+3)}{(x-3)(x+1)(x+3)}$$

$$= \frac{x^2-x-2}{(x-3)(x+3)(x+1)} - \frac{x^2-x-12}{(x-3)(x+3)(x+1)}$$

$$= \frac{10}{(x-3)(x+3)(x+1)}$$

29. $\frac{x^2-16}{2x+8} \cdot \frac{4x^2+16x+64}{x^3-16}$

$$\frac{(x-4)(x+4)}{2(x+4)} \cdot \frac{4(x^2+4x+16)}{x^3-16}$$

$$= \frac{2\,(x^3-64)}{x^3-16}$$

30. $\frac{(a^2b)^3}{(ab^2)^4} \cdot \frac{ab^3}{a^{-4}b^2} = \frac{a^6b^3}{a^4b^8} \cdot \frac{ab^3}{a^{-4}b^2}$

$$= \frac{a^7b^6}{b^{10}} = \frac{a^7}{b^4}$$

31. $\frac{x^2y}{(xy)^3} \div \frac{xy^2}{x^2y^4} = \frac{x^2y}{x^3y^3} \cdot \frac{x^2y^4}{xy^2}$

$$= \frac{x^4y^5}{x^4y^5} = 1$$

32. $8^{2/3} = (\sqrt[3]{8})^2 = 2^2 = 4$

33. $16^{-5/4} = \frac{1}{(\sqrt[4]{16})^5} = \frac{1}{(2)^5} = \frac{1}{32}$

34. $-4^{1/2} = -\sqrt{4} = -2$

35. $27^{-2/3} = \frac{1}{(\sqrt[3]{27})^2} = \frac{1}{3^2} = \frac{1}{9}$

36. $-2^{-3} = -\frac{1}{2^3} = -\frac{1}{8}$

37. $2^{-3/5} \cdot 2^{-7/5} = 2^{-10/5} = 2^{-2} = \frac{1}{2^2} = \frac{1}{4}$

38. $5^{-2/3} \div 5^{1/3} = 5^{-\frac{2}{3}-\frac{1}{3}} = 5^{-1} = \frac{1}{5}$

39. $(9^{1/2} + 4^{1/2})^2 = (3+2)^2 = 5^2 = 25$

40. a) Age 4 years 3 months is 4.25 years.

$h(4.25)$

$$= 79.041 + 6.39(4.25) - e^{3.261-0.993(4.25)}$$

$$= 105.8 \text{ cm}$$

$105.8 \text{ cm} \cdot \frac{1 \text{ in}}{2.54 \text{ cm}} = 41.7 \text{ in.}$

c) Using a graphing calculator we get that a child who has a height of 80 cm has an age of 1.3 years

Appendix A

1. The perimeter is $3 + 4 + 5$ or 12 in.
2. The area is $\frac{1}{2} \cdot 12 \cdot 4$ or 24 ft^2.
3. Since the sum of the 3 angles of a triangle is 180°, then third angle is $180 - 30 - 90$ or 60°.
4. Since $A = \frac{1}{2}bh$, we have $36 = \frac{1}{2} \cdot 12h$ or $36 = 6h$, or $h = 6$ ft.
5. Since the side opposite 30° is one-half of the hypotenuse, the length of the hypotenuse is 20 cm.
6.
$$A = \tfrac{1}{2}h(b_1 + b_2)$$
$$A = \tfrac{1}{2}12(4 + 20) = 6(24)$$
$$= 144 \text{ cm}^2$$
7. The area of a right triangle is one-half the product of the lengths of the legs.
$A = \frac{1}{2}(6)(8) = 24\text{ft}^2$
8. The hypotenuse of a right triangle is the longest side. So the hypotenuse is 13 ft.
9. Since $a^2 + b^2 = c^2$ for a right triangle, we have $a^2 + 40^2 = 50^2$ or $a^2 = 900$ and $a = 30$ cm.
10. Since $5^2 + 10^2 \neq 11^2$, the triangle is not a right triangle.
11. Since $7^2 + 24^2 = 15^2$, the triangle is a right triangle and its area is one-half the product of the lengths of the legs.
$A = \frac{1}{2}(7)(24) = 84 \text{ yd}^2$
12. Since the opposite side of a parallelogram are equal, the perimeter is $2(9) + 2(6)$ or 30 in.
13. The area of a parallelogram is bh. So the area is 32 ft^2.
14. Since a rhombus has four equal sides, it perimeter is 5(4) or 20 km.
15. The perimeter is $2L + 2W$ or $2(1.5) + 2(2)$ or 7 ft. The area is LW or $(1.5)(2)$ or 3 ft^2.
16.
$$P = 2L + 2W$$
$$60 = 2L + 2(8)$$
$$44 = 2L$$
$$22 = L$$
The length is 22 yd.
17. Since $A = \pi r^2$, $A = \pi(4)^2 \approx 50.3\,\text{ft}^2$.
18. Since $C = \pi d$, we have
$$C = \pi \cdot 12 \approx 37.7 \text{ ft.}$$
19. $V = \frac{1}{3}\pi r^2 h = \frac{1}{3}\pi(4)^2 9 \approx 150.80 \text{ cm}^3$
20. $S = \pi r\sqrt{r^2 + h^2} = \pi(12)\sqrt{12^2 + 20^2}$
$\approx 879.29 \text{ ft}^2$
21. $V = LWH = 12(6)(4) = 288 \text{ in.}^3$
$S = 2LW + 2LH + 2WH = 288 \text{ in.}^2$
22.
$$V = LWH$$
$$120 = 30h$$
$$4 = h$$
The height is 4 cm.
23. $A = s^2 = 10^2 = 100 \text{ mi}^2$
$P = 4s = 4(10) = 40 \text{ mi}$
24.
$$A = s^2$$
$$25 = s^2$$
$$5 = s$$
$P = 4s = 4(5) = 20$ km
25.
$$P = 4s$$
$$26 = 4s$$
$$s = 6.5$$
$A = s^2 = 6.5^2 = 42.25 \text{ cm}^2$
26. $V = \frac{4}{3}\pi r^3 = \frac{4}{3}\pi(2)^3 \approx 33.510 \text{ ft}^3$
$S = 4\pi r^2 = 4\pi(2)^2 \approx 50.265 \text{ ft}^2$
27. $V = \pi r^2 h = \pi(2)^2 6 \approx 75.4 \text{ in.}^3$
$S = 2\pi rh + 2\pi r^2 = 2\pi(2)(6) + 2\pi(2)^2$
$\approx 100.5 \text{ in.}^2$
28. Since complementary angles have a sum of 90°, the other angle is $90 - 34$ or 56°.
29. Since an isosceles triangle has two equal sides, $x + 2(12) = 29$ or $x = 5$.
So the short side is 5 cm.
30. Since the corresponding sides of similar triangles are proportional,
$$\frac{10}{25} = \frac{8}{x}$$
$$10x = 200$$
$$x = 20$$
Each side of the larger triangle is 2.5 times as large as the corresponding side in the smaller triangle. So the sides are 20 in. and $(2.5)(6)$ or 15 in.
31. Since supplementary angles have a sum of 180°, the other angle is $180 - 31$ or 149°.
32. Since all three sides of an equilateral triangle have the same length, the perimeter is 3(4) or 12 km.
33. The length of a side of an equilateral triangle is one-third the length of the perimeter. So the side has length $30/3$ or 10 yd.

Appendix B

1. A set is a collection of objects.
2. A finite set has a fixed number of elements and an infinite set does not.
3. A Venn diagram is used to illustrate relationships between sets.
4. The intersection of two sets consists of elements that are in both sets, while the union of two sets consists of elements that are in one, the other, or both sets.
5. A is a subset of B if every member of set A is also a member of set B.
6. The empty set is a subset of every set.
7. False, 6 is not odd.
8. False, because 8 is not odd.
9. True, because $1 \in A$ but $1 \notin B$.
10. False, because A is a finite set.
11. True, because $C = \{1, 2, 3, 4, 5\}$.
12. False, because 4 is a member of B.
13. False, because $5 \in A$.
14. True, since $6 \in B$ but $6 \notin C$.
15. False, because 0 is not a natural number.
16. False, because 2.5 is not a natural number.
17. False, because N is infinite and C is finite.
18. False, because N is infinite and A is finite.
19. Note that $A = \{1, 3, 5, 7, 9\}$ and $B = \{2, 4, 6, 8\}$. Since A and B have no numbers in common, $A \cap B = \emptyset$.
20. Note that $A = \{1, 3, 5, 7, 9\}$ and $B = \{2, 4, 6, 8\}$. Since $1, 2, 3, 4, 5, 6, 7, 8,$ and 9 are in either A or B, $A \cup B = \{1, 2, 3, 4, 5, 6, 7, 8, 9\}$.
21. Note that $A = \{1, 3, 5, 7, 9\}$ and $C = \{1, 2, 3, 4, 5\}$. Since $1, 3,$ and 5 are in both A and C, $A \cap C = \{1, 3, 5\}$.
22. Note that $A = \{1, 3, 5, 7, 9\}$ and $C = \{1, 2, 3, 4, 5\}$. Since $1, 2, 3, 4, 5, 7,$ and 9 are in either A or C, $A \cup C = \{1, 2, 3, 4, 5, 7, 9\}$.
23. The elements of B together with those of C give us $B \cup C = \{1, 2, 3, 4, 5, 6, 8\}$.
24. The only elements that belong to both B and C are 2 and 4. So $B \cap C = \{2, 4\}$.
25. Since the empty set has no members, forming the union with another set does not add any new members. So $A \cup \emptyset = A$.
26. Since the empty set has no members, forming the union with another set does not add any new members. So $B \cup \emptyset = B$.
27. Since $\emptyset$ has no members in common with A, $A \cap \emptyset = \emptyset$.
28. Since the empty set has no members in common with B, $B \cap \emptyset = \emptyset$.
29. Since every member of A is also a member of N, $A \cap N = A = \{1, 3, 5, 7, 9\}$.
30. Since every member of A is a member of N, $A \cup N = N = \{1, 2, 3, \ldots\}$.
31. Since the members of A are odd and the members of B are even, they have no members in common. So $A \cap B = \emptyset$.
32. Since 1 is a member of both A and C, $A \cap C \neq \emptyset$.
33. $A \cup B = \{1, 2, 3, 4, 5, 6, 7, 8, 9\}$ from Exercise 20.
34. Since A and B have no members in common, $A \cap B = \emptyset$.
35. Take the elements that B has in common with C to get $B \cap C = \{2, 4\}$.
36. Since the numbers 1, 2, 3, 4, 5, 6, and 8 are either members of B or C, we have $B \cup C = \{1, 2, 3, 4, 5, 6, 8\}$.
37. Since 3 is not in B, $3 \notin A \cap B$.
38. Since 3 is a member of A and 3 is a member of C, we have $3 \in A \cap C$.
39. Since 4 is in both sets, $4 \in B \cap C$.
40. Since 8 is a member of B, we have $8 \in B \cup C$.
41. True, since each member of A is a counting number.
42. True, since 2, 4, 6, and 8 are counting numbers.
43. True, since both 2 and 3 are members of C.
44. False, since 2 is a member of C, but 2 is not a member of A.
45. True, since $6 \in B$ but $6 \notin C$.
46. True, since 2 is a member of C, but 2 is not a member of A.
47. True, since $\emptyset$ is a subset of every set.
48. True, since the empty set is a subset of every set.

49. False, since $1 \in A$ but $1 \notin \emptyset$.
50. False, since $2 \in$ B but $2 \notin \emptyset$.
51. True, since $A \cap B = \emptyset$ and $\emptyset$ is a subset of any set.
52. True, since $B \cap C = \{2, 4\}$.
53. Using all numbers that belong to D or to E yields $D \cup E = \{2, 3, 4, 5, 6, 7, 8\}$.
54. D and E have no members in common. So $D \cap E = \emptyset$.
55. Using only numbers that belong to both D and F gives $D \cap F = \{3, 5\}$.
56. $D \cup F$ consists of numbers that are either in D or in F. So $D \cup F = \{1, 2, 3, 4, 5, 7\}$.
57. Using all numbers that belong to E or to F gives $E \cup F = \{1, 2, 3, 4, 5, 6, 8\}$.
58. The only numbers that belong to both E and F are 2 and 4. So $E \cap F = \{2, 4\}$.
59. From Exercise 53, $D \cup E = \{2, 3, 4, 5, 6, 7, 8\}$ and $F = \{1, 2, 3, 4, 5\}$. Only 2, 3, 4, and 5 are in both sets. So $(D \cup E) \cap F = \{2, 3, 4, 5\}$.
60. $D \cup F = \{1, 2, 3, 4, 5, 7\}$ and $E = \{2, 4, 6, 8\}$. The only elements that these sets have in common are 2 and 4. So $(D \cup F) \cap E = \{2, 4\}$.
61. Form the union of $E \cap F = \{2, 4\}$ with $D = \{3, 5, 7\}$ to get $D \cup (E \cap F) = \{2, 3, 4, 5, 7\}$.
62. $D = \{3, 5, 7\}$ and $F \cap E = \{2, 4\}$. Put all of the elements together in one set to form the union.
$D \cup (F \cap E) = \{2, 3, 4, 5, 7\}$
63. Form the union of $D \cap F = \{3, 5\}$ with $E \cap F = \{2, 4\}$ to get $(D \cap F) \cup (E \cap F) = \{2, 3, 4, 5\}$.
64. $D \cap E = \emptyset$ and $F \cap E = \{2, 4\}$. So $(D \cap E) \cup (F \cap E) = \{2, 4\}$.
65. Intersect $D \cup E = \{2, 3, 4, 5, 6, 7, 8\}$ with $D \cup F = \{1, 2, 3, 4, 5, 7\}$ to get $(D \cup E) \cap (D \cup F) = \{2, 3, 4, 5, 7\}$.
66. $D \cup F = \{1, 2, 3, 4, 5, 7\}$ and $D \cup E = \{2, 3, 4, 5, 6, 7, 8\}$. So $(D \cup F) \cap (D \cup E) = \{2, 3, 4, 5, 7\}$.
67. Since every element of D is an odd natural number, $D \subseteq \{x | x \text{ is an odd natural number}\}$.
68. Since E consists of exactly the even natural numbers less than 9, $E = \{x | x \text{ is an even natural number smaller than } 9\}$.
69. Since 3 is an element of D, $3 \in D$.
70. Since 3 is an element of D, $\{3\} \subseteq D$.
71. Since D and E have no elements in common, $D \cap E = \emptyset$.
72. Since $D \cap E = \emptyset$, and the empty set is a subset of every set, $D \cap E \subseteq D$.
73. Since $D \cap F = \{3, 5\}$, $D \cap F \subseteq F$.
74. Since 3 is not an element of both E and F, $3 \notin E \cap F$.
75. Since $E \cap F = \{2, 4\}$ and $E = \{2, 4, 6, 8\}$, $E \not\subseteq E \cap F$.
76. Since $E \cup F = \{1, 2, 3, 4, 5, 6, 8\}$ and $E = \{2, 4, 6, 8\}$, $E \subseteq E \cup F$.
77. Since $D \cup F = \{1, 2, 3, 4, 5, 7\}$ and $F \cup D = \{1, 2, 3, 4, 5, 7\}$, $D \cup F = F \cup D$.
78. Since $E \cap F = \{2, 4\}$ and $F \cap E = \{2, 4\}$, $E \cap F = F \cap E$.
79. The set of even natural numbers less than 20 is $\{2, 4, 6, \ldots, 18\}$.
80. The set of natural numbers greater than 6 is $\{7, 8, 9, \ldots\}$.
81. The set of odd natural numbers greater than 11 is $\{13, 15, 17, \ldots\}$.
82. The set of odd natural numbers less than 14 is $\{1, 3, 5, \ldots, 13\}$.
83. The set of even natural numbers between 4 and 79 is $\{6, 8, 10, \ldots, 78\}$. Note that 4 is not in the set because 4 is not between 4 and 79.
84. The set of odd natural numbers between 12 and 57 is $\{13, 15, 17, \ldots, 55\}$. Note that 57 is not in the set because 57 is not between 12 and 57.
85. Answers may vary. Two possible answers are $\{x \mid x \text{ is a natural number between 2 and 7}\}$ or $\{x \mid x \text{ is a natural number between 2.5 and 6.3}\}$.
86. Answers may vary. Two possible answers are $\{x \mid x \text{ is an odd natural number less than 8}\}$ or $\{x \mid x \text{ is an odd natural number between 0 and 7.5}\}$.
87. Answers may vary. $\{x \mid x \text{ is an odd natural number greater than 4}\}$

88. Answers may vary. One possibility is $\{x \mid x \text{ is a natural number greater than } 3\}$.
89. Answers may vary. One possibility is $\{x \mid x \text{ is an even natural number between 5 and 83}\}$
90. Answers may vary. One possibility is $\{x \mid x \text{ is an odd natural number between 8 and 52}\}$
91. False, since 5 is a counting number and $5 \notin A$.
92. False, since B has only 3 elements.
93. False, because the set has a specific number of members.
94. False, since $1 \notin B$.
95. True, since $3 \in A$.
96. True, because 3 and 4 are the only numbers that belong to both A and B.
97. True, because every member of C is also a member of B.
98. False, since $1 \in A$ but $1 \notin B$.
99. True, since $\emptyset$ is a subset of every set.
100. True, since $1 \in A$ and $1 \notin C$.

Appendix C

1. The set of real numbers greater than 2 is written in interval notation as $(2, \infty)$ and consists of numbers to the right of 2 on the number line.

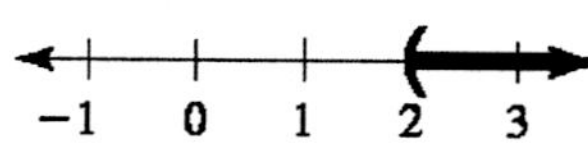

2. The set of real numbers less than or equal to -1 written in interval notation as $(-\infty, -1]$ and graphed as follows.

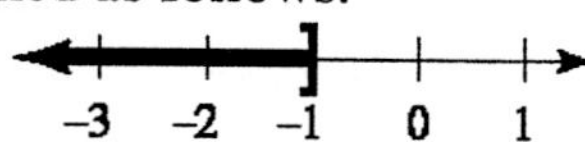

3. The set of real numbers between 0 and 1 is written in interval notation as $(0, 1)$ and graphed as follows.

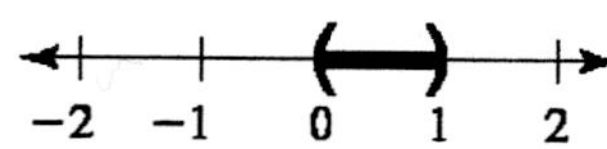

4. The set of real numbers greater than -4 and less than or equal to -2 is written in interval notation as $(-4, -2]$ and graphed as follows.

−5 −4 −3 −2 −1 0 1

5. $\frac{3}{4} \cdot \frac{7}{9} = \frac{3 \cdot 7}{4 \cdot 3 \cdot 3} = \frac{7}{12}$
6. $\frac{1}{4} + \frac{5}{6} = \frac{3}{12} + \frac{10}{12} = \frac{13}{12}$
7. $\frac{8}{9} \div 4 = \frac{8}{9} \cdot \frac{1}{4} = \frac{2}{9}$
8. $-4^2 - 3^3 = -16 - 27 = -43$
9. $|3 - 2^2| - |7 - 19| = 1 - (12) = -11$
10. $\dfrac{-3 - 5}{-2 - (-1)} = \dfrac{-8}{-1} = 8$
11. $3(x + 4) = 3x + 12$ illustrates the distributive property.
12. $x \cdot 7 = 7x$ illustrates the commutative property of multiplication.
13. $4 + (9 + y) = (4 + 9) + y$ illustrates the associative property of addition.
14. $0 + 3 = 3$ illustrates the additive identity property.
15.
$$\begin{aligned} 5x - (3 - 8x) &= 5x - 3 + 8x \\ &= 13x - 3 \end{aligned}$$
16.
$$\begin{aligned} x + 3 - 0.2(5x - 30) &= x + 3 - x + 6 \\ &= 9 \end{aligned}$$
17. $(-3x)(-5x) = 15x^2$
18. $\dfrac{3x + 12}{-3} = -x - 4$
19.
$$\begin{aligned} 11x - 2 &= 3 \\ 11x &= 5 \\ x &= 5/11 \end{aligned}$$
The solution set is $\left\{\frac{5}{11}\right\}$.
20.
$$\begin{aligned} 4x - 5 &= 12x + 11 \\ -8x &= 16 \\ x &= -1 \end{aligned}$$
The solution set is $\{-2\}$.
21.
$$\begin{aligned} 3(x - 6) &= 3x - 6 \\ 3x - 12 &= 3x - 6 \\ -12 &= 6 \end{aligned}$$
The solution set is $\emptyset$.
22.
$$\begin{aligned} x - 0.1x &= 0.9x \\ 0.9x &= 0.9x \end{aligned}$$
All real numbers satisfy this equation.
23.
$$\begin{aligned} 5x - 3y &= 9 \\ -3y &= -5x + 9 \\ y &= \frac{5}{3}x - 3 \end{aligned}$$
24.
$$\begin{aligned} ay + b &= 0 \\ ay &= -b \\ y &= -\frac{b}{a} \end{aligned}$$
25.
$$\begin{aligned} a &= t - by \\ by &= t - a \\ y &= \frac{t - a}{b} \end{aligned}$$

26. $\frac{a}{2} + \frac{y}{3} = \frac{3a}{4}$
$6a + 4y = 9a$
$4y = 3a$
$y = \frac{3}{4}a$

27. Let x, $x+1$, and $x+2$ represent the integers.
$x + x + 1 + x + 2 = 102$
$3x = 99$
$x = 33$
$x + 1 = 34$
$x + 2 = 35$
The integers are $33, 34$, and 35.

28. Let x be the width and $x+4$ the length.
$2x + 2(x+4) = 100$
$4x + 8 = 100$
$4x = 92$
$x = 23$
The width is 23 in.

29. Let h represent the height. The area of a triangle is $A = \frac{1}{2}bh$.
$\frac{1}{2}(400)h = 44{,}000$
$200h = 44{,}000$
$h = 220$
The height is 220 ft.

30. Let x represent the number of pounds of peanuts.
$x = 0.20(x + 400)$
$x = 0.20x + 80$
$0.80x = 80$
$x = 100$
He should use 100 pounds of peanuts.

31. $3x - 4 \le 11$
$3x \le 15$
$x \le 5$
$(-\infty, 5]$

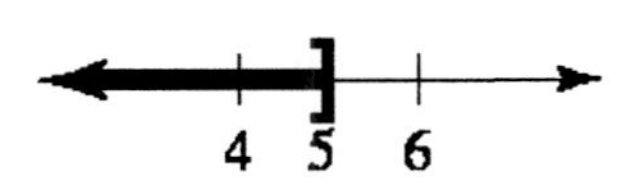

32. $5 - 7w > 26$
$-7w > 21$
$w < -3$

$(-\infty, -3)$

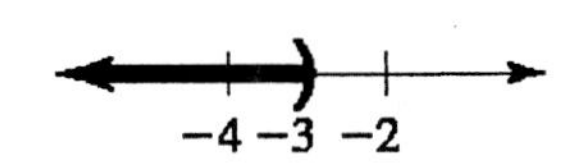

33. $-1 < 2a - 9 \le 7$
$8 < 2a \le 16$
$4 < a \le 8$
$(4, 8]$

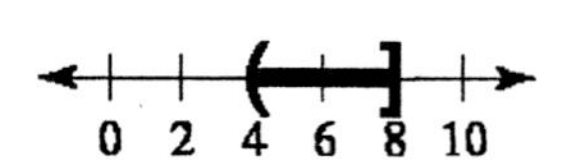

34. $5 < 6 - x < 6$
$-1 < -x < 0$
$1 > x > 0$
$0 < x < 1$
$(0, 1)$

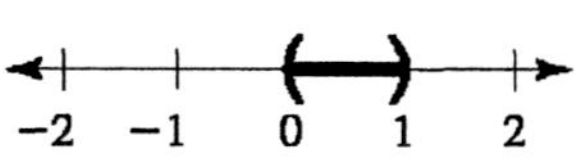

35. If $x = 0$, then $y = \frac{2}{3}(0) - 2 = -2$. If $y = 0$, then $\frac{2}{3}x - 2 = 0$ or $x = 3$. So the intercepts are $(0, -2)$ and $(3, 0)$.

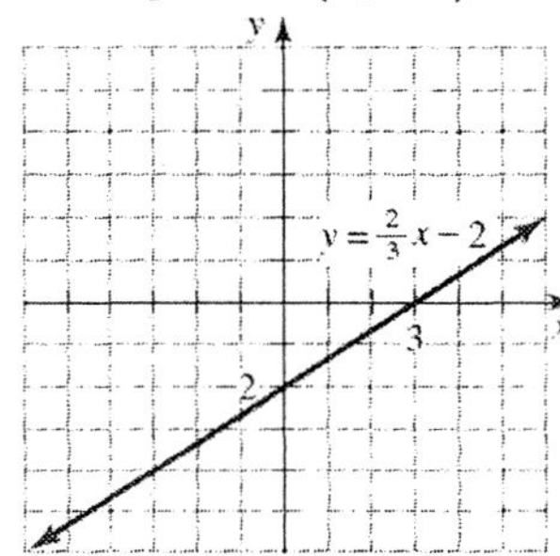

36. If $x = 0$, then $-5y = 150$ or $y = -30$. If $y = 0$, then $3x = 150$ or $x = 50$. So the intercepts are $(0, -30)$ and $(50, 0)$.

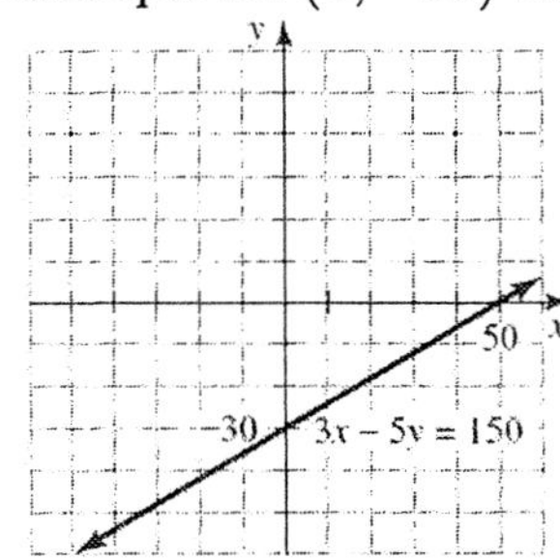

37. The graph of $y = 2$ is a horizontal line through its y-intercept $(0, 2)$.

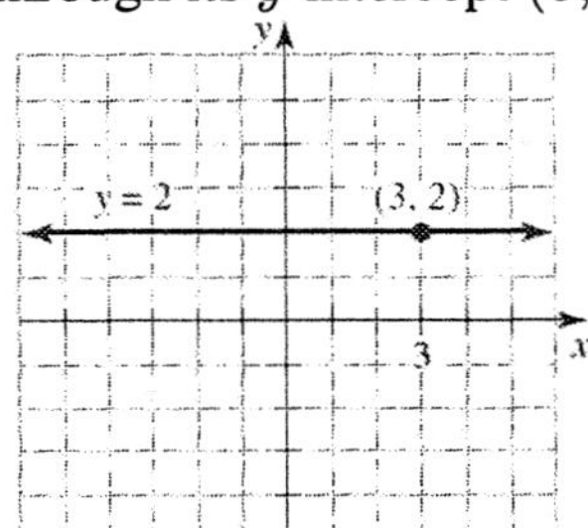

38. The graph of $x = 2$ is a vertical line through its x-intercept $(2, 0)$.

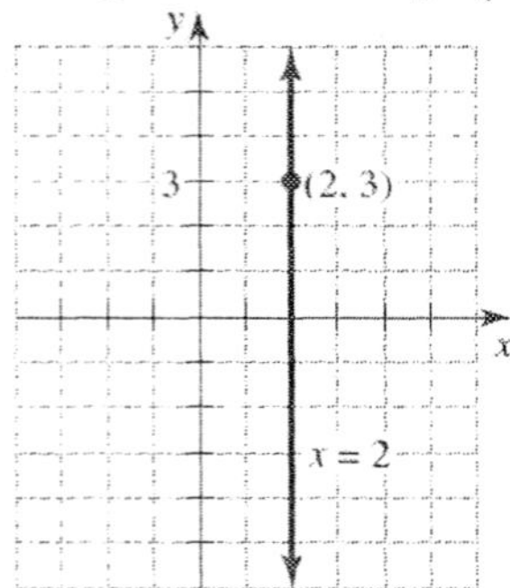

39. The line passing through the points $(1, 2)$ and $(3, 6)$ has slope $\frac{2-6}{1-3}$ or 2.

40. The line $y = \frac{1}{2}x - 4$ has slope $\frac{1}{2}$ because the slope in $y = mx + b$ form is m.

41. Write $2x + 3y = 9$ in slope intercept form:

$$\begin{aligned} 2x + 3y &= 9 \\ 3y &= -2x + 9 \\ y &= -\frac{2}{3}x + 3 \end{aligned}$$

The line parallel to $2x + 3y = 9$ has the same which is $-\frac{2}{3}$.

42. The slope of $y = -3x + 5$ is -3. The line perpendicular to has slope $\frac{1}{3}$, the opposite and reciprocal of the other slope.

43. The line passing through the points $(0, 3)$ and $(2, 11)$ has slope $\frac{11-3}{2-0}$ or 4, and y-intercept $(0, 3)$. So the equation is $y = 4x + 3$.

44. The line passing through the points $(-2, 4)$ and $(1, -2)$ has slope $\frac{4-(-2)}{-2-1}$ or -2.

$$\begin{aligned} y - 4 &= -2(x - (-2)) \\ y - 4 &= -2x - 4 \\ y &= -2x \end{aligned}$$

45. The line through $(3, 5)$ that is parallel to $x = 4$ has x-intercept $(3, 0)$. So its equation is $x = 3$.

46. The line through $(0, 8)$ that is perpendicular to $y = \frac{1}{2}x$ has slope -2 and y-intercept $(0, 8)$. So the equation is $y = -2x + 8$.

47. The time that it takes to mow a large lawn varies inversely with the number of mowers working on the job. So $t = k/m$. If it takes 30 hours with 3 mowers, then $30 = k/3$ or $k = 90$. So with 5 mowers, $t = 90/5 = 14$ hours.

48. The cost of installing ceramic floor tile in a rectangular room varies jointly with the length and the width of the room. So $C = kLW$. If the cost is \$810 for a 9 ft by 12 ft room, then $810 = k(9)(12)$ or $k = 7.5$. So the cost for a 14 ft by 18 ft room is $C = 7.5(14)(18) = \$1890$.

49. The line $3x - 4y = 12$ goes through $(0, -3)$ and $(4, 0)$. Since $(0, 0)$ does not satisfy $3x - 4y > 12$, we shade below the line.

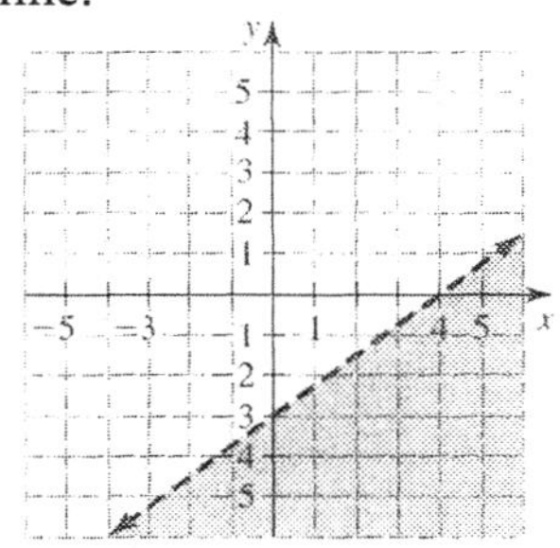

50. The line $y = 3x + 2$ goes through $(0, 2)$ and $(1, 5)$. Since $(0, 0)$ satisfies $y \leq 3x + 2$ we shade below the line.

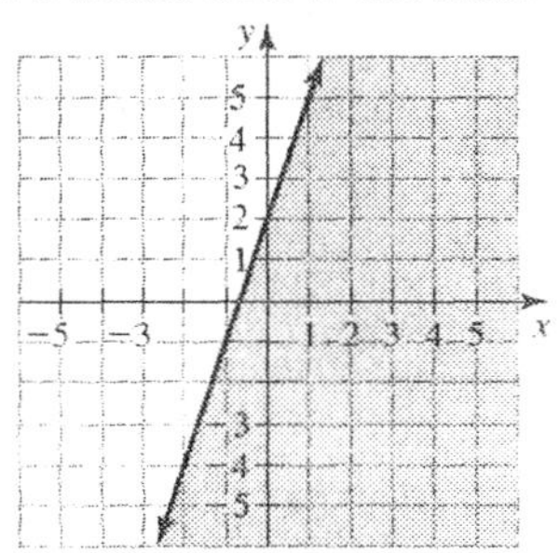

51. First graph the vertical line $x = 2$ as a dashed line and then shade to the right for the inequality $x > -2$.

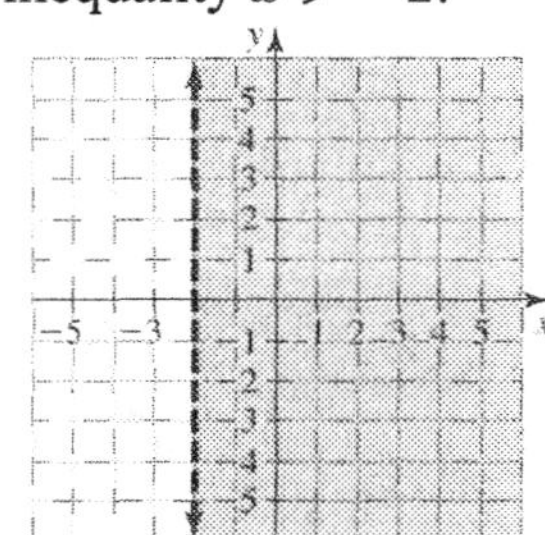

52. First graph the horizontal line $y = 4$ as a solid line and then shade below for the inequality $y \leq 4$.

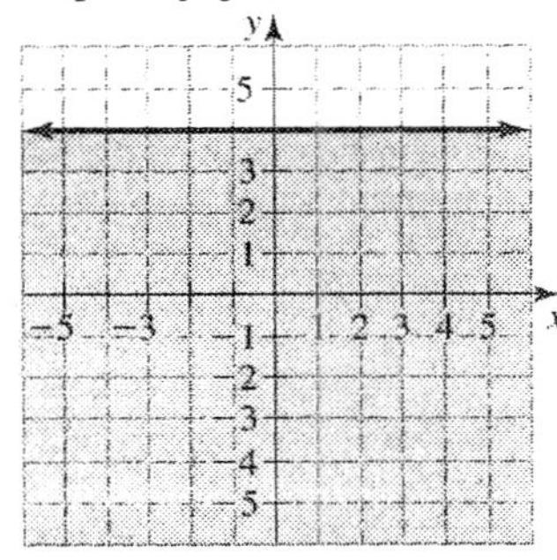

53. $(x^2 - 3x + 2) - (3x^2 + 9x - 4)$
$= x^2 - 3x + 2 - 3x^2 - 9x + 4$
$= -2x^2 - 12x + 6$

54. $-3x^2(-2x^2 - 3) = 6x^4 + 9x^3$

55. $(x + 7)(x - 9) = x^2 - 2x - 63$

56. $(x + 2)(x^2 - 2x + 4)$
$= x(x^2 - 2x + 4) + 2(x^2 - 2x + 4)$
$= x^3 - 2x^2 + 4x + 2x^2 - 4x + 8$
$= x^3 + 8$

57. $(4w^2 - 3)^2$
$= (4w^2)^2 - 2(4w^2)(3) + 3^2$
$= 16w^4 - 24w^2 + 9$

58. $(-8m^7) \div (2m^2) = -4m^5$

59. $(-9y^3 - 6y^2 + 3y) \div (3y)$
$= -3y^2 - 2y + 1$

60.
$$\begin{array}{r} x^2 + x + 2 \\ x - 3 \overline{) x^3 - 2x^2 - x - 6} \\ \underline{x^3 - 3x^2} \\ x^2 - x \\ \underline{x^2 - 3x} \\ 2x - 6 \\ \underline{2x - 6} \\ 0 \end{array}$$

$(x^3 - 2x^2 - x - 6) \div (x - 3)$
$= x^2 + x + 2$

61. $-8x^4 \cdot 4x^3 = -32x^7$

62. $3x(5x^2)^3 = 3x \cdot 125x^6 = 375x^7$

63. $\dfrac{-6x^2y^3}{-2x^{-3}y^4} = \dfrac{3x^{2-(-3)}}{y^{4-3}} = \dfrac{3x^5}{y}$

64. $\left(\dfrac{2a^2}{a^{-3}}\right)^3 = \dfrac{8a^6}{a^{-9}} = 8a^{15}$

65. $400{,}000 \cdot 600 = 4 \times 10^5 \cdot 6 \times 10^2$
$= 24 \times 10^7 = 2.4 \times 10^1 \times 10^7$
$= 2.4 \times 10^8$

66. $(9 \times 10^3)(2 \times 10^6) = 18 \times 10^9$
$= 1.8 \times 10^1 \times 10^9$
$= 1.8 \times 10^{10}$

67. $(2 \times 10^{-3})^4 = 16 \times 10^{-12}$
$= 1.6 \times 10^1 \times 10^{-12}$
$= 1.6 \times 10^{-11}$

68. $\dfrac{2 \times 10^{-9}}{2000} = \dfrac{2 \times 10^{-9}}{2 \times 10^3} = 1 \times 10^{-12}$

69. $24x^2y^3 + 18xy^5$
$= 6xy^3(4x + 3y^2)$

70. $x^2 + 2x + ax + 2a$
$= x(x + 2) + a(x + 2)$
$= (x + a)(x + 2)$

71. $4m^2 - 49$
$= (2m)^2 - 7^2$
$= (2m - 7)(2m + 7)$

72. $x^2 - 3x - 54$
$= (x - 9)(x + 6)$

73. $6t^2 - 11t - 10$
$= (2t - 5)(3t + 2)$

74. $4w^2 - 36w + 81$
$= (2w)^2 - 2(9)(2w) + 9^2$
$= (2w - 9)^2$

75. $2a^3 - 6a^2 - 108a$
$= 2a(a^2 - 3a - 54$
$= 2a(a - 9)(a + 6)$

76. $w^3 - 27 = (w - 3)(w^2 + 3w + 9)$

77. $x^2 = x$
$x^2 - x = 0$
$x(x - 1) = 0$
$x = 0$ or $x = 1$

The solution set is $\{0, 1\}$.

78. $2x^3 - 8x = 0$
$2x(x^2 - 4) = 0$
$2x(x - 2)(x + 2) = 0$
$x = 0$ or $x = 2$ or $x = -2$

The solution set is $\{-2, 0, 2\}$.

79. $a^2 + a - 6 = 0$
$(a + 3)(a - 2) = 0$
$a = -3$ or $a = 2$

The solution set is $\{-3, 2\}$.

80. $(b - 2)(b + 3) = 24$
$b^2 + b - 6 = 24$
$b^2 + b - 30 = 0$
$(x + 6)(b - 5) = 0$
$b = -6$ or $b = 5$

The solution set is $\{-6, 5\}$.

81. Let x and $10 - x$ be the numbers whose sum is 10.

$$x(10 - x) = 21$$
$$-x^2 + 10x = 21$$
$$x^2 - 10x + 21 = 0$$
$$(x - 7)(x - 3) = 0$$
$$x = 7 \text{ or } x = 3$$
$$10 - x = 3 \text{ or } 10 - x = 7$$

The numbers are 3 and 7.

82. Let x represent the width and $x + 14$ the length. Use the Pythagorean theorem.

$$x^2 + (x + 14)^2 = 26^2$$
$$x^2 + x^2 + 28x + 196 = 676$$
$$2x^2 + 28x - 480 = 0$$
$$x^2 + 14x - 240 = 0$$
$$(x + 24)(x - 10) = 0$$
$$x = -24 \text{ or } x = 10$$
$$x + 14 = 24$$

The length is 24 in. and the width is 10 in.

83. $\frac{5x}{2} + \frac{3x}{4} = \frac{10x}{4} + \frac{3x}{4} = \frac{13x}{4}$

84. $\frac{5}{x-2} - \frac{3}{2-x} = \frac{5}{x-2} + \frac{3}{x-2}$

$$= \frac{8}{x-2}$$

85. $\frac{9}{x^2 - 9} + \frac{2x}{x-3}$

$$= \frac{9}{x^2-9} + \frac{2x(x+3)}{(x-3)(x+3)}$$
$$= \frac{9}{x^2-9} + \frac{2x^2+6x}{(x-3)(x+3)}$$
$$= \frac{2x^2+6x+9}{(x-3)(x+3)}$$

86. $\frac{2}{a-5} + \frac{3}{a+4}$

$$= \frac{2(a+4)}{(a-5)(a+4)} + \frac{3(a-5)}{(a+4)(a-5)}$$
$$= \frac{2a+8+3a-15}{(a-5)(a+4)}$$
$$= \frac{5a-7}{(a-5)(a+4)}$$

87. $\frac{w^3}{2w-4} \cdot \frac{w^2-4}{w}$

$$= \frac{w^3}{2(w-2)} \cdot \frac{(w-2)(w+2)}{w}$$
$$= \frac{w^3+2w^2}{2}$$

88. $\frac{5ab^2}{6a^2b^3} \div \frac{10a}{21b^6} = \frac{5ab^2}{6a^2b^3} \cdot \frac{21b^6}{10a}$

$$= \frac{105ab^8}{60a^3b^3} = \frac{7b^5}{4a^2}$$

89. $\frac{2}{x} = \frac{3}{4}$

$$3x = 8$$
$$x = 8/3$$

The solution set is $\left\{\frac{8}{3}\right\}$.

90. $\frac{1}{w-3} = \frac{2}{w+5}$

$$w + 5 = 2(w - 3)$$
$$w + 5 = 2w - 6$$
$$11 = w$$

The solution set is $\{11\}$.

91. $\frac{1}{x} + \frac{3}{7} = \frac{1}{3x}$

$$21x\left(\frac{1}{x} + \frac{3}{7}\right) = 21x \cdot \frac{1}{3x}$$
$$21 + 9x = 7$$
$$9x = -14$$
$$x = -14/9$$

The solution set is $\left\{-\frac{14}{9}\right\}$.

92. $\frac{3}{a-1} + \frac{1}{a+2} = \frac{17}{10}$

$$10(a-1)(a+2)\left(\frac{3}{a-1} + \frac{1}{a+2}\right)$$
$$= 10(a-1)(a+2)\frac{17}{10}$$
$$30a + 60 + 10a - 10$$
$$= 17a^2 + 17a - 34$$
$$40a + 50 = 17a^2 + 17a - 34$$
$$0 = 17a^2 - 23a - 84$$
$$0 = (a - 3)(17a + 28)$$
$$a = 3 \text{ or } a = -28/17$$

The solution set is $\left\{-\frac{28}{17}, 3\right\}$.

93. $\frac{3}{y} = \frac{5}{x}$

$$5y = 3x$$
$$y = \frac{3}{5}x$$

94. $a = \frac{1}{2}y(w - c)$

$$2a = y(w - c)$$
$$y = \frac{2a}{w-c}$$

95. $\frac{y-3}{x+5} = -3$

$$y - 3 = -3(x + 5)$$
$$y - 3 = -3x - 15$$
$$y = -3x - 12$$

96. $\frac{3}{y} + \frac{1}{2} = \frac{1}{t}$

$$2yt\left(\frac{3}{y} + \frac{1}{2}\right) = 2yt \cdot \frac{1}{t}$$
$$6t + yt = 2y$$
$$yt - 2y = -6t$$
$$y(t - 2) = -6t$$
$$y = \frac{-6t}{t-2}$$
$$y = \frac{6t}{2-t}$$

Appendix D

R.1 EXERCISES

1. The set of real numbers between 0and 3 inclusive is written as $[0, 3]$.

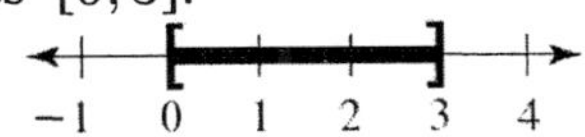

2. The set of real numbers between -2 and 5 is written as $(-2, 5)$.

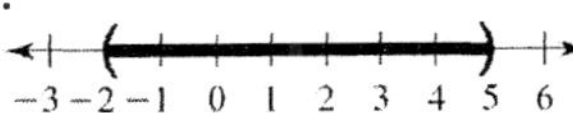

3. The set of real numbers greater than or equal to -4 and less than 0 is written as $[-4, 0)$.

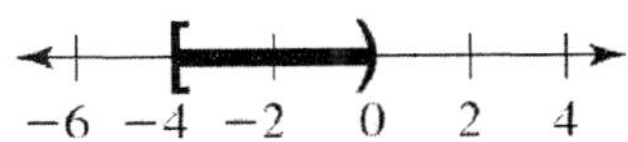

4. The set of real numbers greater than 3 and less than or equal to 8 is written as $(3, 8]$.

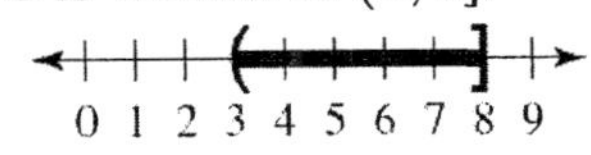

5. The set of real numbers less than -1 is written as $(-\infty, -1)$.

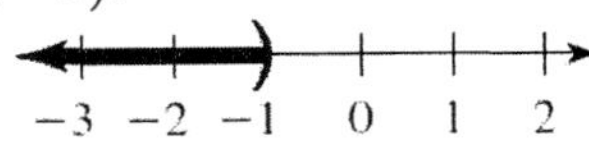

6. The set of real numbers less than or equal to 6 is written as $(-\infty, 6]$.

−6 −3 0 3 6 9

7. The set of real numbers greater than or equal to 50 is written as $[50, \infty)$.

−50 0 50 100

8. The set of real numbers greater than -10 is written as $(-10, \infty)$.

−20 −10 0 10

9. The interval $(-2, 9)$ represents the set of real numbers between -2 and 9.

10. The interval $[-4, -3]$ represents the set of real numbers between -4 and -3 inclusive.

11. The interval $[11, 13)$ represents the set of real numbers greater than or equal to 11 and less than 13.

12. The interval $(22, 26]$ represents the set of real numbers greater than 22 and less than or equal to 26.

13. The interval $(0, \infty)$ represents the set of real numbers greater than 0.

14. The interval $[99, \infty)$ represents the set of real numbers greater than or equal to 99.

15. The interval $(-\infty, -6]$ represents the set of real numbers less than or equal to -6.

16. The interval $(-\infty, 18)$ represents the set of real numbers less than 18.

17. $|-1| = -(-1) = 1$

18. $|-9.35| = -(-9.35) = 9.35$

19. $|0| = 0$

20. $|5 - 5| = |0| = 0$

21. $|50| = 50$

22. $|6.87| = 6.87$

23. $\frac{1}{2} = \frac{1 \cdot 10}{2 \cdot 10} = \frac{10}{20}$

24. $\frac{2}{3} = \frac{2 \cdot 6}{3 \cdot 6} = \frac{12}{18}$

25. $\frac{3}{4} = \frac{3 \cdot 6}{4 \cdot 6} = \frac{18}{24}$

26. $\frac{7}{8} = \frac{7 \cdot 7}{8 \cdot 7} = \frac{49}{56}$

27. $\frac{12}{20} = \frac{4 \cdot 3}{4 \cdot 5} = \frac{3}{5}$

28. $\frac{16}{24} = \frac{8 \cdot 2}{8 \cdot 3} = \frac{2}{3}$

29. $\frac{14}{48} = \frac{2 \cdot 7}{2 \cdot 24} = \frac{7}{24}$

30. $\frac{24}{84} = \frac{2 \cdot 12}{7 \cdot 12} = \frac{2}{7}$

31. $\frac{6}{10} = \frac{3 \cdot 2}{5 \cdot 2} = \frac{3}{5}$

32. $\frac{7}{14} = \frac{7 \cdot 1}{7 \cdot 2} = \frac{1}{2}$

33. $\frac{28}{49} = \frac{4 \cdot 7}{7 \cdot 7} = \frac{4}{7}$

34. $\frac{48}{72} = \frac{2 \cdot 24}{3 \cdot 24} = \frac{2}{3}$

35. $\frac{36}{108} = \frac{36 \cdot 1}{36 \cdot 3} = \frac{1}{3}$

36. $\frac{51}{68} = \frac{3 \cdot 17}{4 \cdot 17} = \frac{3}{4}$

37. $\frac{30}{100} = \frac{3 \cdot 10}{10 \cdot 10} = \frac{3}{10}$

38. $\frac{400}{1000} = \frac{2 \cdot 200}{5 \cdot 200} = \frac{2}{5}$

39. $\frac{3}{8} \cdot \frac{2}{3} = \frac{3}{2 \cdot 2 \cdot 2} \cdot \frac{2}{3} = \frac{1}{4}$

40. $\frac{2}{5} \cdot \frac{15}{26} = \frac{2}{5} \cdot \frac{3 \cdot 5}{2 \cdot 13} = \frac{3}{13}$
41. $\frac{1}{4} \div \frac{5}{2} = \frac{1}{2 \cdot 2} \cdot \frac{2}{5} = \frac{1}{10}$
42. $\frac{3}{7} \div \frac{9}{14} = \frac{3}{7} \cdot \frac{2 \cdot 7}{3 \cdot 3} = \frac{2}{3}$
43. $\frac{2}{5} \cdot 25 = \frac{2}{5} \cdot 5 \cdot 5 = 10$
44. $\frac{5}{8} \cdot 40 = \frac{5}{8} \cdot 5 \cdot 8 = 25$
45. $\frac{2}{3} \div 5 = \frac{2}{3} \cdot \frac{1}{5} = \frac{2}{15}$
46. $6 \div \frac{1}{7} = 6 \cdot 7 = 42$
47. $\frac{1}{8} + \frac{2}{3} = \frac{1 \cdot 3}{8 \cdot 3} + \frac{2 \cdot 8}{3 \cdot 8} = \frac{19}{24}$
48. $\frac{1}{5} + \frac{3}{4} = \frac{1 \cdot 4}{5 \cdot 4} + \frac{3 \cdot 5}{4 \cdot 5} = \frac{19}{20}$
49. $\frac{5}{12} - \frac{5}{18} = \frac{5}{2^2 \cdot 3} - \frac{5}{2 \cdot 3^2}$
$= \frac{5 \cdot 3}{2^2 \cdot 3^2} - \frac{5 \cdot 2}{2^2 \cdot 3^2}$
$= \frac{15}{36} - \frac{10}{36} = \frac{5}{36}$
50. $\frac{5}{16} - \frac{1}{12} = \frac{5}{2^4} - \frac{1}{2^2 \cdot 3}$
$= \frac{5 \cdot 3}{2^4 \cdot 3} - \frac{1 \cdot 2^2}{2^4 \cdot 3}$
$= \frac{15}{48} - \frac{4}{48}$
$= \frac{11}{48}$
51. $\frac{5}{8} + 2 = \frac{5}{8} + \frac{16}{8} = \frac{21}{8}$
52. $\frac{3}{7} + 1 = \frac{3}{7} + \frac{7}{7} = \frac{10}{7}$
53. $-20 + (-6) = -(20 + 6) = -26$
54. $-19 + (-8) = -(19 + 8) = -27$
55. $-30 + 7 = -(30 - 7) = -23$
56. $18 + (-9) = 18 - 9 = 9$
57. $6 + (-5) = 6 - 5 = 1$
58. $-7 + 12 = 12 - 7 = 5$
59. $-30 - 6 = -(30 + 6) = -36$
60. $-15 - 12 = -(15 + 12) = -27$
61. $20 - (-4) = 20 + 4 = 24$
62. $88 - (-12) = 88 + 12 = 100$
63. $-3 - (-5) = -3 + 5 = 2$
64. $-9 - (-6) = -9 + 6 = -3$
65. $(-3)(-60) = 3(60) = 180$
66. $(-8)(-12) = 8(12) = 96$
67. $(-7)(12) = -(7 \cdot 12) = -84$
68. $(13)(-3) = -(13 \cdot 3) = -39$
69. $(-30) \div (-2) = 30 \div 2 = 15$
70. $(-90) \div (-15) = 90 \div 15 = 6$
71. $-40 \div 5 = -(40 \div 5) = -8$
72. $100 \div (-20) = -(100 \div 20) = -5$
73. $0 \div (-7) = 0$
74. $0 \div (-2000) = 0$
75. $-3^2 - 9^2 = -9 - 81 = -90$
76. $(-4)^3 - 5^2 = -64 - 25 = -89$
77. $(4 + 2^3)(1 - 4) = 12(-3) = -36$
78. $(4 - 5)^3(3 - 6^2) = -1(-33) = 33$
79. $3 + 5 \cdot 7 = 3 + 35 = 38$
80. $10 - 6 \cdot 2 = 10 - 12 = -2$
81. $2^4 - 3 \cdot 7 = 16 - 21 = -5$
82. $3 \cdot 2^5 - 5 \cdot 2^4 = 96 - 80 = 16$
83. $|3 - 9| - |5 - 8| = 6 - 3 = 3$
84. $2|3 - 5 \cdot 4| = 2(17) = 34$
85. $|-6| - 3|2 - 2^3| = 6 - 3(6)$
$= 6 - 18 = -12$
86. $|5 \cdot 4 - 10| - |-3 \cdot 2| = 10 - 6 = 4$
87. $\frac{-4 - 2}{1 - 3} = \frac{-6}{-2} = 3$
88. $\frac{-2^2 - 3^3}{1 - (-30)} = \frac{-4 - 27}{31} = -1$
89. $\frac{-3 \cdot 5 - 2}{1 - 3 \cdot 6} = \frac{-17}{-17} = 1$
90. $\frac{4 - 2 \cdot 7}{2 - 3 \cdot 2^2} = \frac{-10}{2 - 12} = \frac{-10}{-10} = 1$
91. The sum of $5x$ and $-3y$ is $5x + (-3y)$. If $x = -2$ and $y = 5$, then the value is $5(-2) + (-3 \cdot 5) = -10 - 15 = -25$.
92. The difference of a^3 and b^3 is $a^3 - b^3$. The value for $a = -2$ and $b = 4$ is $(-2)^3 - 4^3 = -8 - 64 = -72$.
93. The product of $a + b$ and $a^2 - ab + b^2$ is $(a + b)(a^2 - ab + b^2)$. If $a = -1$ and $b = -3$ the value is
$(-1 + (-3))((-1)^2 - (-1)(-3) + (-3)^2)$
$= (-4)(7) = -28$.
94. The quotient of $x - 7$ and $7 - x$ is $\frac{x - 7}{7 - x}$. Evaluating for $x = 9$ we get $\frac{9 - 7}{7 - 9} = -1$.
95. The square of $2x - 3$ is $(2x - 3)^2$. Evaluating for $x = 5$ we get
$(2 \cdot 5 - 3)^2 = 7^2 = 49$.
96. The cube of $a - b$ is $(a - b)^3$. Evaluating for $a = 3$ and $b = -1$ we have $(3 - (-1))^3 = 4^3 = 64$.
97. The expression $a^3 - b^3$ is a difference because the last operation to perform is subtraction.
98. The expression $a^2 + b^2$ is a sum because the last operation to perform is addition.
99. The expression $5a - b$ is a difference because the last operation to perform is subtraction.

100. The expression $5(a-b)$ is a product because the last operation to perform is multiplication.

101. The expression $\frac{6-a}{6a}$ is a quotient because the last operation to perform is division.

102. The expression $(5a-b)^2$ is a square because the last operation to perform is squaring.

103. The expression $(3a)^3$ is a cube because the last operation to perform is cubing.

104. The expression $3+a^3$ is a sum because the last operation to perform is addition.

105. The equation $a(3)=3a$ is true for all real numbers a because of the commutative property of multiplication.

106. The equation $3+a=a+3$ is true for every real number a because of the commutative property of addition.

107. The equation $5(x+1)=5x+5$ is true for any real number x because of the distributive property.

108. The equation
$(w^2+8)+7=w^2+(8+7)$
is true for any real number w because of the associative property of addition.

109. The equation $5\cdot 1=5$ illustrates the multiplicative identity property.

110. The equation $3+0=3$ illustrates the additive identity property.

111. The equation $m^2\cdot 0=0$ is true for every real number m because of the multiplication property of zero.

112. The equation $6\cdot\frac{1}{6}=1$ illustrates the multiplicative inverse property.

113. The equation $3(5x)=(3\cdot 5)x$ is true for any real number x because of the associative property of multiplication.

114. The equation $a+(-a)=0$ is true for any real number a because of the additive inverse property.

115. $(2x-9)+(7-3x)=-x-2$

116. $(-3x-y)+(9y-8x)=-11x+8y$

117. $5+3(4+x)=5+12+3x$
$=3x+17$

118. $x+7(x+y)=x+7x+7y$
$=8x+7y$

119. $6+7xy-4(3-6xy)$
$=6+7xy-12+24xy$
$=31xy-6$

120. $4+3a-5(4-7a)$
$=4+3a-20+35a$
$=38a-16$

121. $(-2a)(5b)-5(4ab)$
$=-10ab-20ab$
$=-30ab$

122. $(-x)(-y)-5(-4xy)$
$=xy+20xy$
$=21xy$

123. $\frac{3(4-2x)}{6}=\frac{12-6x}{6}=2-x$

124. $\frac{2(3x-3y)}{6}=\frac{6x-6y}{6}=x-y$

125. $\frac{44-2x}{-2}=-22+x$

126. $\frac{20+8x}{-4}=-5-2x$

R.2 EXERCISES

1. $x-9=-2$
$x=-2+9$
$x=7$
The solution set is $\{7\}$.

2. $w-8=7$
$w=7+8$
$w=15$
The solution set is $\{15\}$.

3. $n+5=-3$
$n=-3-5$
$n=-8$
The solution set is $\{-8\}$.

4. $z+4=21$
$z=21-4$
$z=17$
The solution set is $\{17\}$.

5. $3a=51$
$\frac{3a}{3}=\frac{51}{3}$
$a=17$
The solution set is $\{17\}$.

6. $-5b=45$
$\frac{-5b}{-5}=\frac{45}{-5}$
$b=-9$
The solution set is $\{-9\}$.

7. $\frac{3}{4}x=-6$
$\frac{4}{3}\cdot\frac{3}{4}x=\frac{4}{3}\cdot -6$

$x = -8$
The solution set is $\{-8\}$.

8. $-\frac{5}{3}m = 15$
$-\frac{3}{5}\left(-\frac{5}{3}m\right) = -\frac{3}{5} \cdot 15$
$m = -9$
The solution set is $\{-9\}$.

9. $4a - 1 = 49$
$4a = 50$
$a = 50/4$
The solution set is $\left\{\frac{25}{2}\right\}$.

10. $3b + 2 = 0$
$3b = -2$
$b = -2/3$
The solution set is $\left\{-\frac{2}{3}\right\}$.

11. $14 - 2x = 6 - x$
$14 - x = 6$
$-x = -8$
$x = 8$
The solution set is $\{8\}$.

12. $-7 - 5x = 12 - 4x$
$-7 - 12 = -4x + 5x$
$-19 = x$
The solution set is $\{-19\}$.

13. $2x - 3 = 4x + 9$
$-3 - 9 = 4x - 2x$
$-12 = 2x$
$-6 = x$
The solution set is $\{-6\}$.

14. $5 - 4x = 3 - 2x$
$5 - 3 = -2x + 4x$
$2 = 2x$
$1 = x$
The solution set is $\{1\}$.

15. $x - 3 = 2 + 3(x + 1)$
$x - 3 = 2 + 3x + 3$
$-3 - 5 = 3x - x$
$-8 = 2x$
$-4 = x$
The solution set is $\{-4\}$.

16. $-3(x - 4) = 2x + 7$
$-3x + 12 = 2x + 7$
$12 - 7 = 2x + 3x$
$5 = 5x$
$1 = x$
The solution set is $\{1\}$.

17. $\frac{13}{15}x - \frac{4}{5}x = \frac{1}{5}x - \frac{2}{15}x$
$13x - 12x = 3x - 2x$
$x = x$
The solution set is $(-\infty, \infty)$ and the equation is an identity.

18. $x + \frac{2}{3} = \frac{1}{3}(3x + 1) + \frac{1}{3}$
$3x + 2 = 3x + 1 + 1$
$3x + 2 = 3x + 2$
The solution set is $(-\infty, \infty)$ and the equation is an identity.

19. $\frac{w}{12} - \frac{w}{4} = \frac{w}{3} - 12$
$w - 3w = 4w - 144$
$-2w = 4w - 144$
$-6w = -144$
$w = 24$
The solution set is $\{24\}$. The equation is a conditional equation.

20. $\frac{a}{6} - 5 = \frac{a}{15} - 2$
$5a - 150 = 2a - 60$
$3a = 90$
$a = 30$
The solution set is $\{30\}$. The equation is a conditional equation.

21. $0.05a - 0.7 = 0.12a + 0.7$
$5a - 70 = 12a + 70$
$-7a = 140$
$a = -20$
The solution set is $\{-20\}$. The equation is a conditional equation.

22. $0.03(z - 4) = 0.05z + 0.8$
$3(z - 4) = 5z + 80$
$3z - 12 = 5z + 80$
$-2z = 92$
$z = -46$
The solution set is $\{-46\}$. The equation is a conditional equation.

23. $\frac{1}{8}y - \frac{1}{9}y = \frac{1}{72}y + \frac{1}{2}$
$9y - 8y = y + 36$
$y = y + 36$
$0 = 36$
The solution set is $\emptyset$. The equation is an inconsistent equation.

24. $\frac{1}{6}m + \frac{1}{7}m = \frac{13}{42}m - \frac{1}{21}$
$7m + 6m = 13m - 2$
$13m = 13m - 2$
$0 = -2$

The solution set is $\emptyset$. The equation is an inconsistent equation.

25. $\frac{5}{3}t - 2\left(\frac{2}{3}t + 1\right) = 3\left(\frac{t}{3} - \frac{1}{9}\right) - \frac{7}{3}t$

$$\frac{5}{3}t - \frac{4}{3}t - 2 = t - \frac{1}{3} - \frac{7}{3}t$$
$$5t - 4t - 6 = 3t - 1 - 7t$$
$$t - 6 = -4t - 1$$
$$5t = 5$$
$$t = 1$$

The solution set is $\{1\}$. The equation is a conditional equation.

26. $\frac{3}{2}v - 4\left(\frac{v}{2} + \frac{5}{2}\right) = \frac{v}{2} - (v + 10)$

$$\frac{3}{2}v - 2v - 10 = \frac{v}{2} - v - 10$$
$$3v - 4v - 20 = v - 2v - 20$$
$$-v - 20 = -v - 20$$

The solution set is $(-\infty, \infty)$. The equation is an identity.

27. $0.001x + 0.02 = 0.2(0.1x - 0.03)$

$$0.001x + 0.02 = 0.02x - 0.006$$
$$x + 20 = 20x - 6$$
$$26 = 19x$$
$$26/19 = x$$

The solution set is $\left\{\frac{26}{19}\right\}$. The equation is a conditional equation.

28. $0.2(0.3q + 0.04) = 0.005q - 0.087$

$$0.06q + 0.008 = 0.005q - 0.087$$
$$60q + 8 = 5q - 87$$
$$55q = -95$$
$$q = -95/55 = -19/11$$

The solution set is $\left\{-\frac{19}{11}\right\}$. The equation is a conditional equation.

29. $D = RT$

$$R = \frac{D}{T}$$

30. $E = mc^2$

$$m = \frac{E}{c^2}$$

31. $K = \frac{1}{2}mv^2$

$$2K = mv^2$$
$$m = \frac{2K}{v^2}$$

32. $A = \frac{1}{2}bh$

$$2A = bh$$
$$b = \frac{2A}{h}$$

33. $P = 2L + 2W$

$$2L = P - 2W$$
$$L = \frac{P - 2W}{2}$$

34. $A = \frac{1}{2}h(b_1 + b_2)$

$$2A = hb_1 + hb_2$$
$$hb_2 = 2A - hb_1$$
$$b_2 = \frac{2A - hb_1}{h}$$

35. $A = P + Prt$

$$Prt = A - P$$
$$r = \frac{A - P}{Pt}$$

36. $2x - 3y = 6$

$$3y = 2x - 6$$
$$y = \frac{2x - 6}{3}$$

37. Use $D = RT$ with $D = 400$ and $R = 40$.

$$400 = 40T$$
$$10 = T$$

The trip took 10 hr.

38. Use $A = \frac{1}{2}bh$ with $A = 10$, $b = 4$ to find h.

$$10 = \frac{1}{2}4h$$
$$10 = 2h$$
$$5 = h$$

The other leg is 5 m.

39. Use $P = 2L + 2W$.

$$150 = 2(45) + 2W$$
$$60 = 2W$$
$$30 = W$$

The width is 30 m.

40. Use $P = 2L + 2W$.

$$53 = 2(14) + 2W$$
$$25 = 2W$$
$$12.5 = W$$

The width is 12.5 cm.

41. $K = \frac{1}{2}mv^2$

$$1800 = \frac{1}{2}m(30)^2$$
$$3600 = 900m$$
$$4 = m$$

The mass of the object is 4 kg.

42. $A = \frac{1}{2}h(b_1 + b_2)$

$$40 = \frac{1}{2} \cdot 4(b_1 + 12)$$
$$40 = 2b_1 + 24$$
$$16 = 2b_1$$
$$8 = b_1$$

The length of the upper base is 8 cm.

43. The sum of a^2 and b^2 is written as $a^2 + b^2$.
44. The number x increased by 5 is written as $x + 5$.
45. The number y decreased by 6 is written as $y - 6$.
46. The difference between a and b is expressed as $a - b$.
47. The product of a and b^2 is expressed as ab^2.
48. Ten percent of x is expressed as $0.10x$.
49. The quotient of x and y is expressed as $\frac{x}{y}$.
50. The number 14 divided by x is expressed as $\frac{14}{x}$.
51. One-half of x is expressed as $\frac{1}{2}x$.
52. Two-thirds of y is expressed as $\frac{2}{3}y$.
53. Twice the sum of a and b is expressed as $2(a + b)$.
54. The square of the sum of a and b is expressed as $(a + b)^2$.
55. Let x represent the width and $x + 6$ represent the length.

$$\begin{aligned} 2x + 2(x + 6) &= 84 \\ 4x + 12 &= 84 \\ 4x &= 72 \\ x &= 18 \\ x + 6 &= 24 \end{aligned}$$

The length is 24 in. and the width is 18 in.
56. Let x represent the width and $2x - 5$ represent the length.

$$\begin{aligned} 2x + 2(2x - 5) &= 170 \\ 6x - 10 &= 170 \\ 6x &= 180 \\ x &= 30 \\ 2x - 5 &= 55 \end{aligned}$$

The length is 55 m and the width is 30 m.
57.

	D	R	T
Mon	$8x$	x	8
Tues	$10(x - 15)$	$x - 15$	10

$$\begin{aligned} 8x + 10(x - 15) &= 840 \\ 8x + 10x - 150 &= 840 \\ 18x &= 990 \\ x &= 55 \end{aligned}$$

His average speed on the first day was 55 mph.
58.

	D	R	T
Wed	$50x$	50	x
Thurs	$64(x + 2)$	64	$x + 2$

$$\begin{aligned} 50x + 64(x + 2) &= 1040 \\ 50x + 64x + 128 &= 1040 \\ 114x &= 912 \\ x &= 8 \end{aligned}$$

His traveling time on Wednesday was 8 hr.
59. Let x represent the amount of cement to be added.

$$\begin{aligned} x + 0.17(10,000) &= 0.18(x + 10,000) \\ x + 1700 &= 0.18x + 1800 \\ 0.82x &= 100 \\ x &\approx 121.95 \end{aligned}$$

Add approximately 121.95 lbs of cement.
60. Let x represent the amount of pure water to be added.

$$\begin{aligned} 0.12(100) &= 0.08(x + 100) \\ 12 &= 0.08x + 8 \\ 4 &= 0.08x \\ 50 &= x \end{aligned}$$

Add 50 oz of water.
61. Let x represent the number of liters of 50% solution to be added.

$$\begin{aligned} 0.50x + 0.20(10) &= 0.30(x + 10) \\ 0.50x + 2 &= 0.30x + 3 \\ 0.20x &= 1 \\ x &= 5 \end{aligned}$$

Add 5 L of 50% solution.
62. Let x represent the percentage of juice in the 200 liters of punch.

$$\begin{aligned} x(200) + 0.30(100) &= 0.20(300) \\ 200x + 30 &= 60 \\ 200x &= 30 \\ x &= 0.15 = 15\% \end{aligned}$$

The percentage of fruit juice in the 200 liters of punch is 15%.
63.

	D	R	T
Police	20	100	1/5
Speeder	$7x/30$	x	7/30

$$\begin{aligned} \frac{7x}{30} &= 20 \\ 7x &= 600 \\ x &\approx 85.71 \end{aligned}$$

The speeder was traveling approximately 85.71 mph.

64.

	D	R	T
G	$80x$	80	x
A	$100(x-\frac{1}{3})$	100	$x-\frac{1}{3}$

$$\begin{aligned}100(x-\tfrac{1}{3}) &= 80x\\ 100x-\frac{100}{3} &= 80x\\ 300x-100 &= 240x\\ 60x &= 100\\ x &= 100/60 = 1\tfrac{2}{3}\end{aligned}$$

The Anderson's will catch up with the Garcia's after the Garcias have traveled one and two-thirds hours, which is one hour 40 minutes, or at 8:40 AM.

65. $3x-1 \geq 14$
$3x \geq 15$
$x \geq 5$

$[5,\infty)$

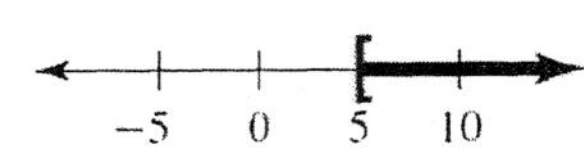

66. $2x+5 \leq 17$
$2x \leq 12$
$x \leq 6$

$(-\infty,6]$

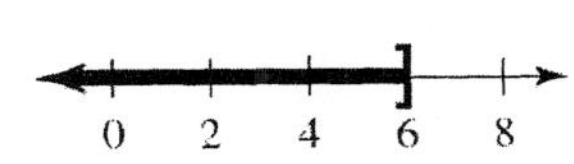

67. $4-3y<0$
$-3y<-4$
$y>4/3$

$\left(\frac{4}{3},\infty\right)$

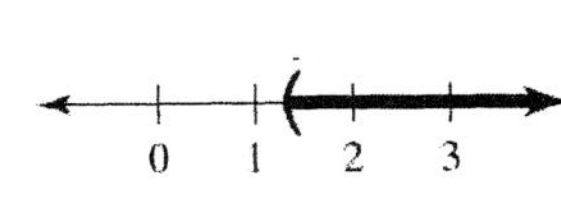

68. $5-t>0$
$-t>-5$
$t<5$

$(-\infty,5)$

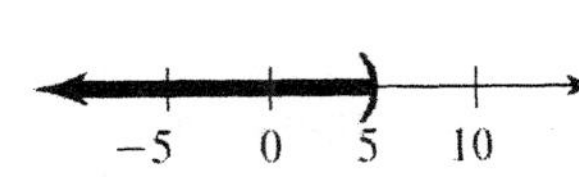

69. $-\frac{1}{2}n+6<7$
$n-12>-14$
$n>-2$

$(-2,\infty)$

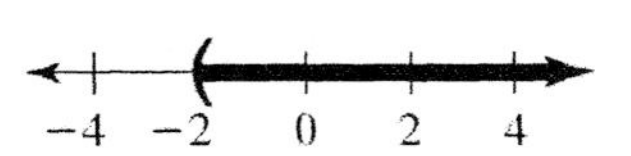

70. $-\frac{3}{4}m-1>5$
$3m+4<-20$
$3m<-24$
$m<-8$

$(-\infty,-8)$

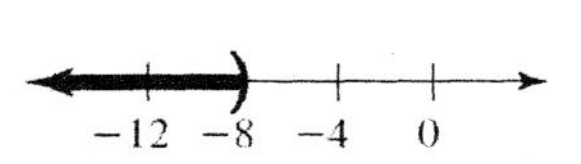

71. $5x+7<2x-8$
$3x<-15$
$x<-5$

$(-\infty,-5)$

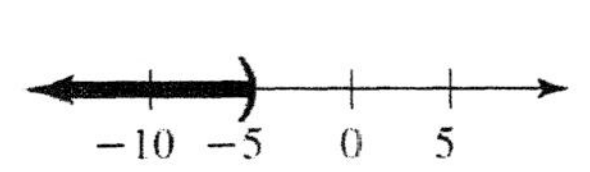

72. $6w-9>w+31$
$5w>40$
$w>8$

$(8,\infty)$

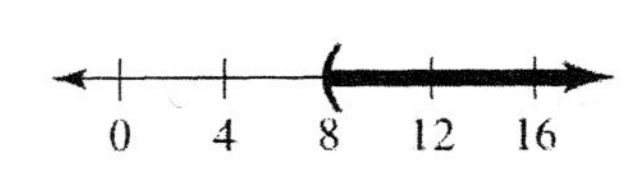

73. $-2z+3<z-6$
$-3z<-9$
$z>3$

$(3,\infty)$

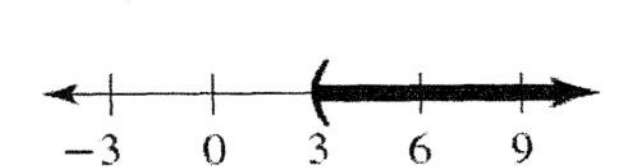

74. $-5x-8<2x+13$
$-7x<21$
$x>-3$

$(-3,\infty)$

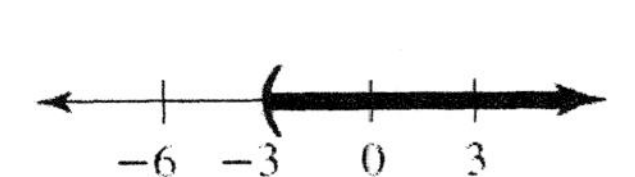

75. $-1 \leq 2b+3<19$
$-4 \leq 2b<16$
$-2 \leq b<8$

$[-2,8)$

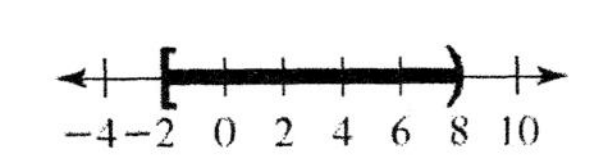

76. $1<5a-4 \leq 21$
$5<5a \leq 25$
$1<a \leq 5$

$(1,5]$

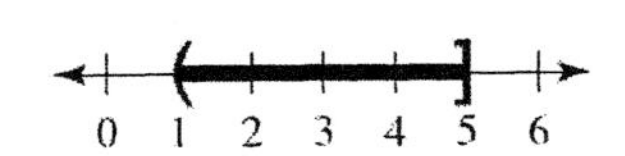

77. $-5 < 3 - 2w < 31$
$-8 < -2w < 28$
$4 > w > -14$
$-14 < w < 4$

$(-14, 4)$

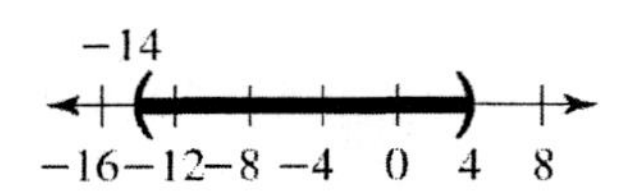

78. $4 \leq 1 - x \leq 5$
$3 < -x \leq 4$
$-3 > x \geq -4$
$-4 \leq x \leq -3$

$[-4, -3]$

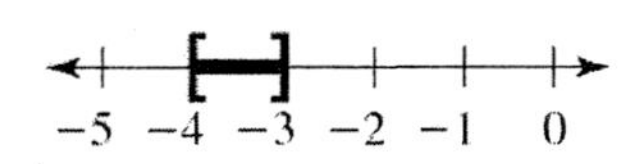

R.3 EXERCISES

1. Let $x = 0$ in $3x - 4y = 12$ to get $-4y = 12$ or $y = -3$. Let $y = 0$ to get $3x = 12$ or $x = 4$. So the intercepts are $(4, 0)$ and $(0, -3)$.

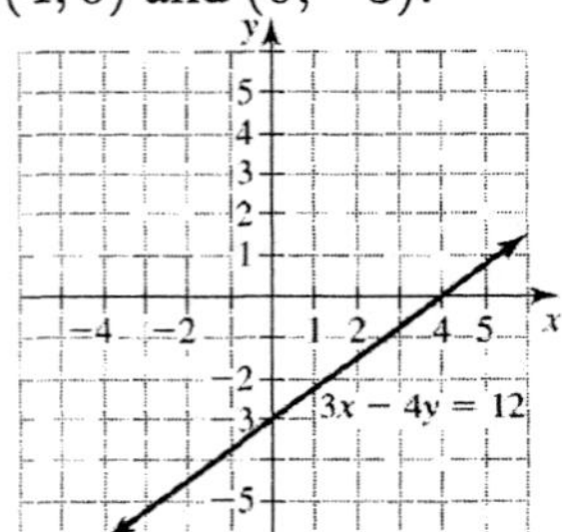

2. Let $x = 0$ in $x - 2y = 10$ to get $-2y = 10$ or $y = -5$. Let $y = 0$ to get $x = 10$. The intercepts are $(10, 0)$ and $(0, -5)$.

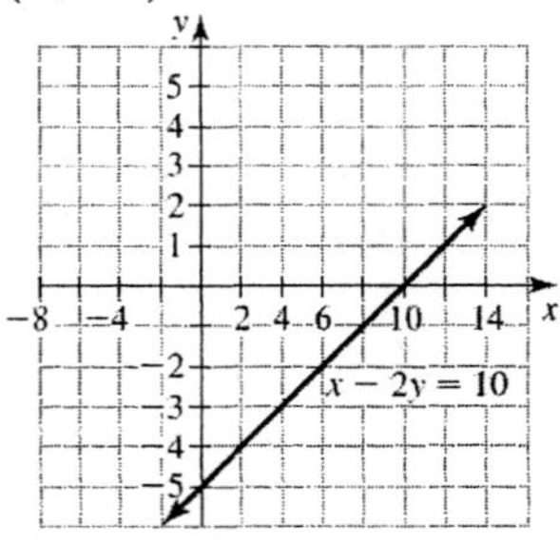

3. Let $x = 0$ in $2x + y = 6$ to get $y = 6$. Let $y = 0$ to get $2x = 6$ or $x = 3$. The intercepts are $(3, 0)$ and $(0, 6)$.

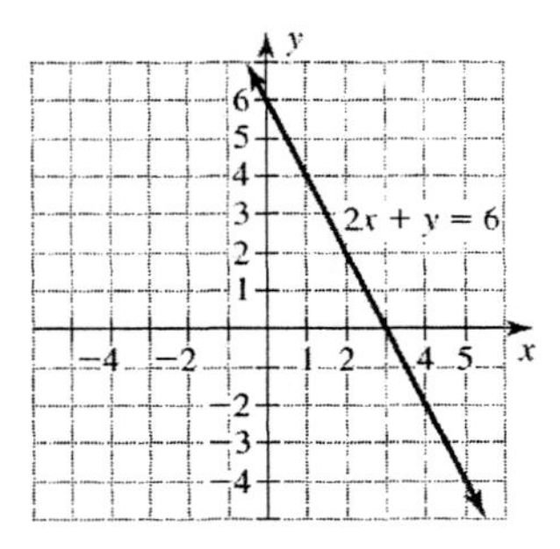

4. Let $x = 0$ in $3x + 7y = 21$ to get $7y = 21$ or $y = 3$. Let $y = 0$ to get $3x = 21$ or $x = 7$. The intercepts are $(7, 0)$ and $(0, 3)$.

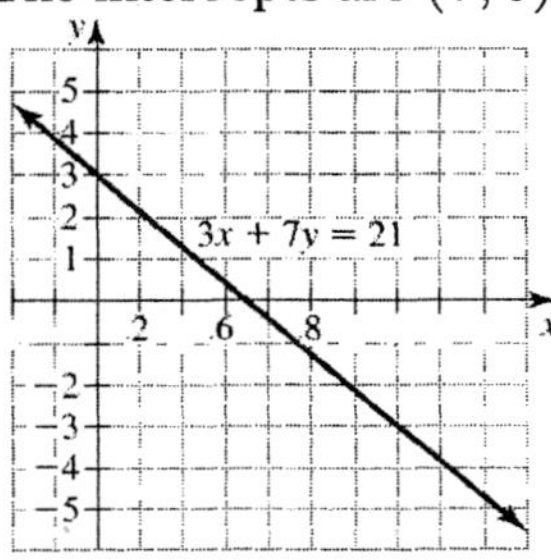

5. The graph of $x = -3$ is a vertical line with x-intercept $(-3, 0)$.

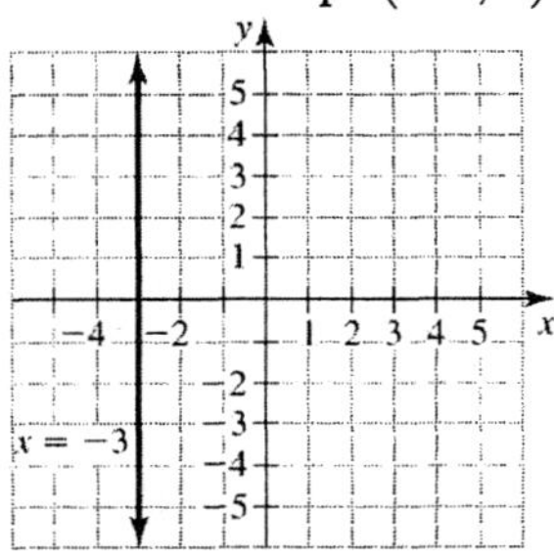

6. The graph of $x = 5$ is a vertical line with x-intercept $(5, 0)$.

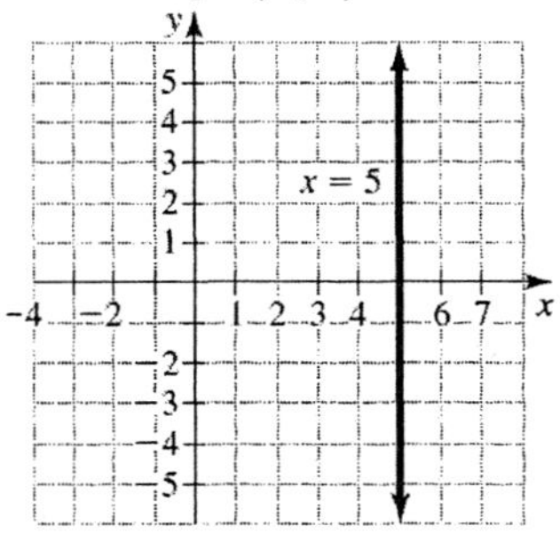

7. The graph of $y = 2$ is a horizontal line with y-intercept $(0, 2)$.

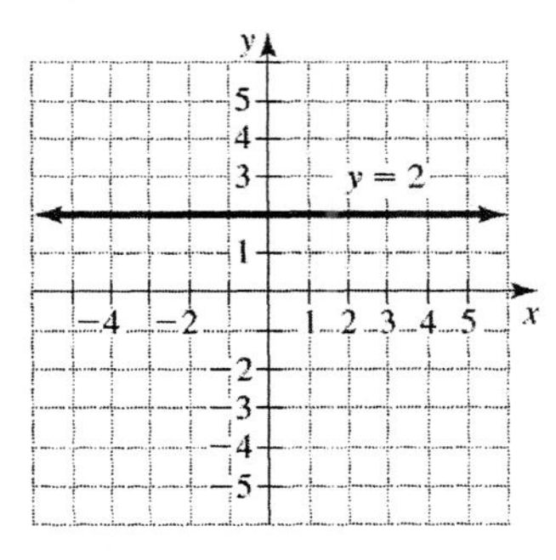

8. The graph of $y = -4$ is a horizontal line with y-intercept $(0, -4)$.

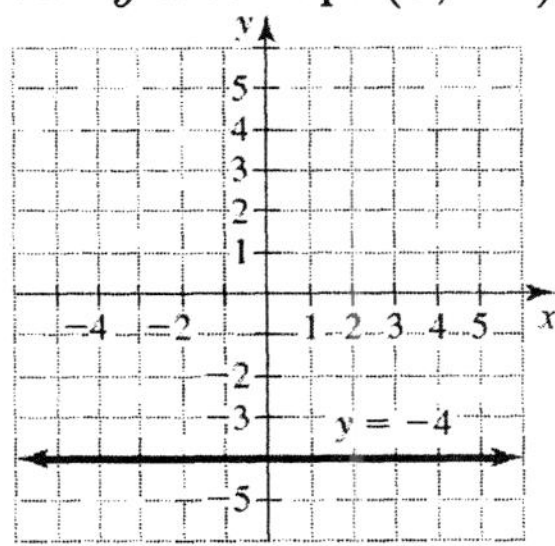

9. If $x = 0$ in $y = \frac{1}{2}x - 30$, then $y = -30$. If $y = 0$, then $\frac{1}{2}x - 30 = 0$ or $x = 60$. The intercepts are $(60, 0)$ and $(0, -30)$.

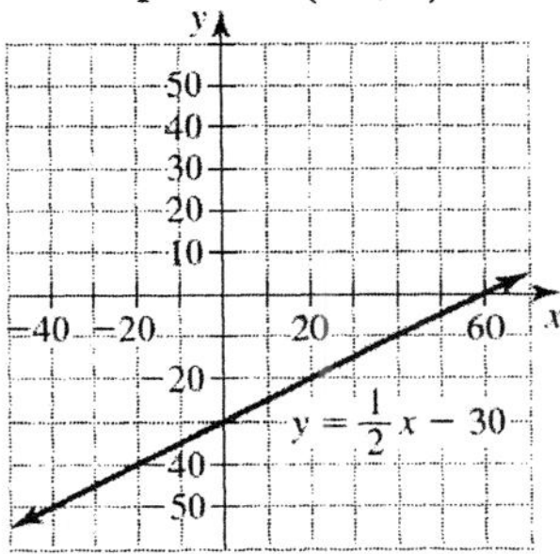

10. If $x = 0$ in $y = -\frac{2}{3}x + 20$, then $y = 20$. If $y = 0$, then $-\frac{2}{3}x + 20 = 0$ or $x = 30$. The intercepts are $(30, 0)$ and $(0, 20)$.

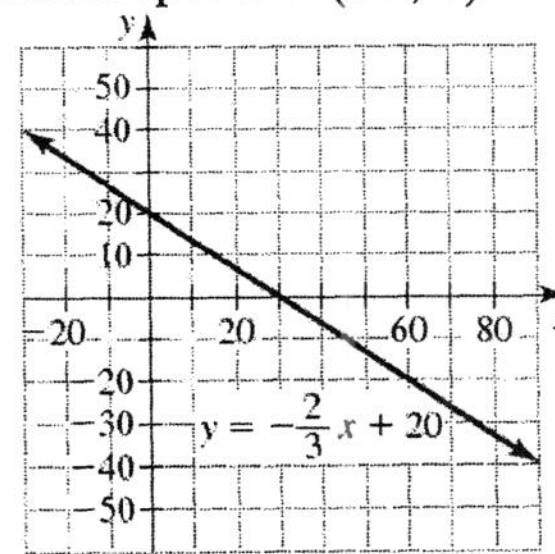

11. The line through $(-2, 1)$ and $(3, 6)$ has slope $\dfrac{6-1}{3-(-2)} = \frac{5}{5} = 1$.

12. The line through $(-1, -3)$ and $(5, 5)$ has slope $\dfrac{5-(-3)}{5-(-1)} = \frac{8}{6} = \frac{4}{3}$.

13. The line through $(-3, 3)$ and $(1, -1)$ has slope $\dfrac{-1-3}{1-(-3)} = \frac{-4}{4} = -1$.

14. The line through $(0, 0)$ and $(-5, -5)$ has slope $\dfrac{-5-0}{-5-0} = \frac{-5}{-5} = 1$.

15. The line through $(2, 1)$ and $(2, 7)$ has no slope because it is a vertical line.

16. The line through $(-3, -1)$ and $(-3, 4)$ has no slope because it is a vertical line.

17. The line through $(4, 1)$ and $(-2, 1)$ has slope $\dfrac{1-1}{-2-4} = \frac{0}{-6} = 0$.

18. The line through $(-3, 5)$ and $(3, 5)$ has slope $\dfrac{5-5}{3-(-3)} = \frac{0}{6} = 0$.

19. The line through $(1, 4)$ and $(4, 16)$ has slope $\dfrac{16-4}{4-1} = \frac{12}{3} = 4$. A line parallel to the line through $(1, 4)$ and $(4, 16)$ has the same slope 4.

20. The line through $(3, 2)$ and $(-6, 2)$ has slope $\dfrac{2-2}{-6-3} = \frac{0}{-9} = 0$. A line parallel to the line through $(3, 2)$ and $(-6, 2)$ has the same slope 0.

21. The line through $(-1, -1)$ and $(2, 3)$ has slope $\dfrac{3-(-1)}{2-(-1)} = \frac{4}{3}$. A line perpendicular to the line through $(-1, -1)$ and $(2, 3)$ has slope that is the opposite and reciprocal of $4/3$, which is $-\frac{3}{4}$.

22. The line through $(-5, 8)$ and $(5, -8)$ has slope $\dfrac{-8-8}{5-(-5)} = \frac{-16}{10} = -\frac{8}{5}$. A line perpendicular to the line through $(-5, 8)$ and $(5, -8)$ has slope that is the opposite and reciprocal of $-8/5$, which is $\frac{5}{8}$.

23. A line perpendicular to the line $x = 3$ is a horizontal line. All horizontal lines have slope 0.

24. A line parallel to the line $y = -5$ is a horizontal line. All horizontal lines have slope 0.

25. For $y = \frac{1}{3}x + 1$ the slope is $\frac{1}{3}$ and the y-intercept is $(0, 1)$. The line goes through $(0, 1)$ and $(3, 2)$.

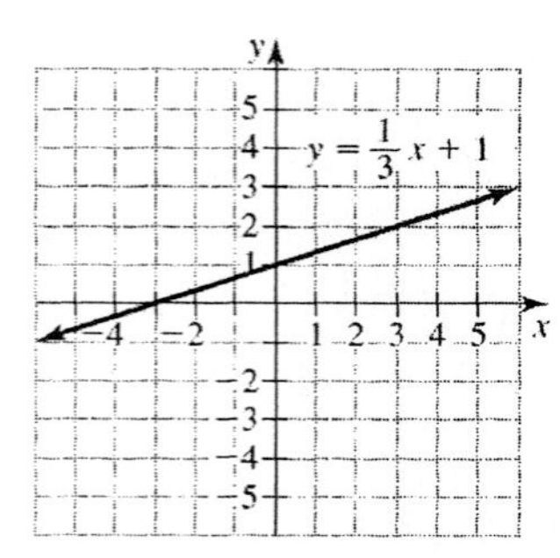

26. For $y = \frac{2}{3}x - 2$ the slope is $\frac{2}{3}$ and the y-intercept is $(0, -2)$. The line goes through $(3, 0)$.

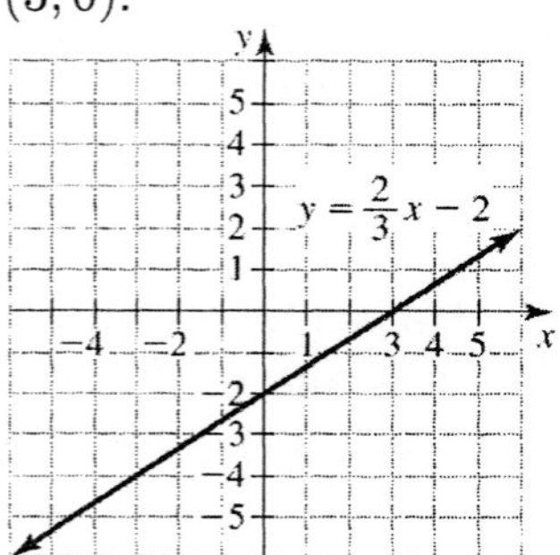

27. For $y = -3x + 4$ the slope is -3 and the y-intercept is $(0, 4)$. The line goes through $(1, 1)$.

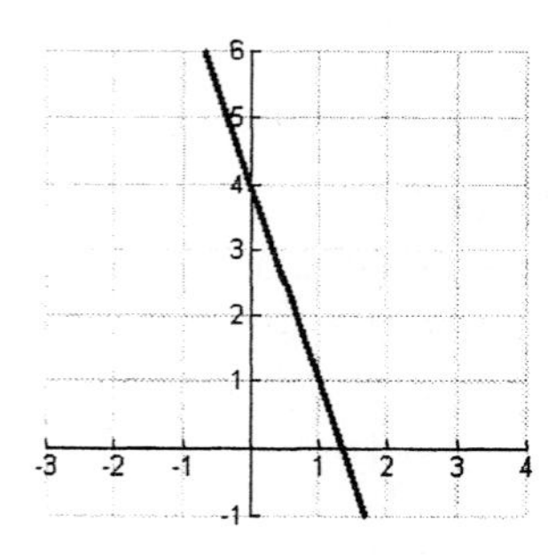

28. For $y = 2x - 5$ the slope is 2 and the y-intercept is $(0, -5)$. The line goes through $(0, -5)$ and $(1, -3)$.

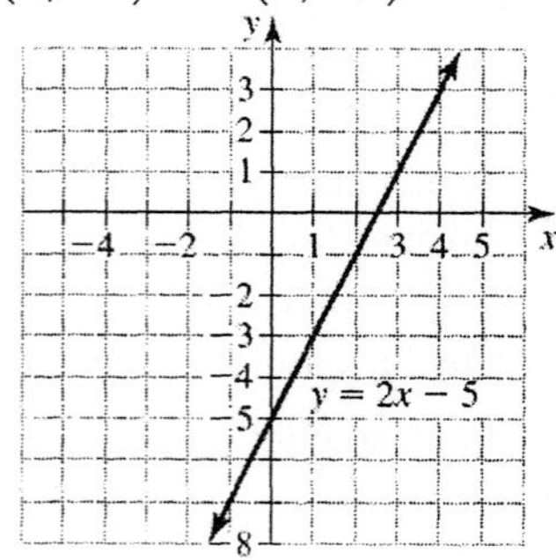

29. Write $x - y = 5$ as $y = x - 5$ to see that the slope is 1 and the y-intercept is $(0, -5)$.

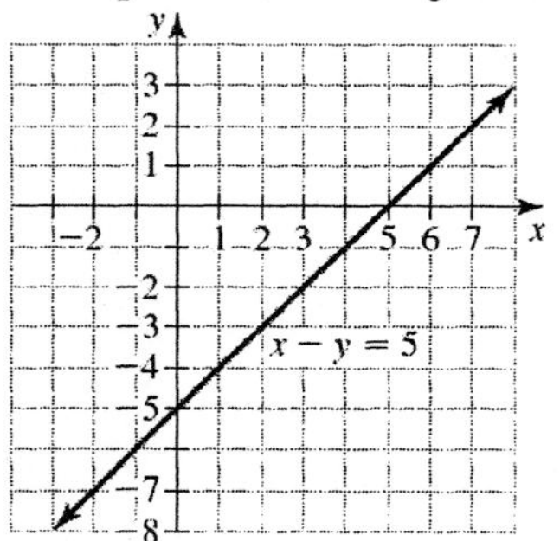

30. Write $x + 2y = 4$ as $y = -\frac{1}{2}x + 2$ to see that the y-intercept is $-\frac{1}{2}$ and the slope is $(0, 2)$. The line goes through $(2, 0)$.

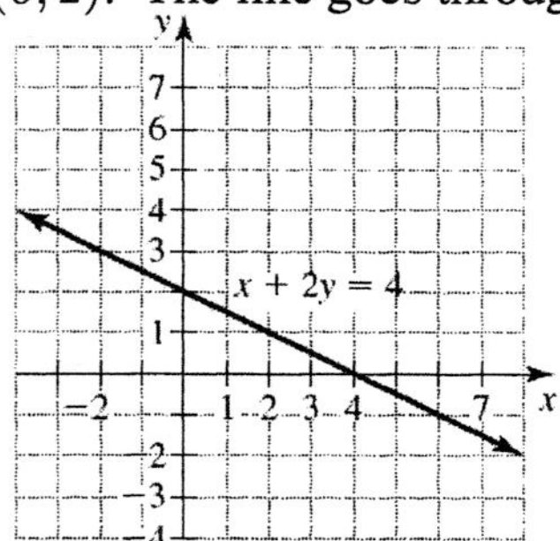

31. Write $3x - 5y = 10$ as $y = \frac{3}{5}x - 2$. The slope is $\frac{3}{5}$ and the y-intercept is $(0, -2)$. The line also goes through $(5, 1)$.

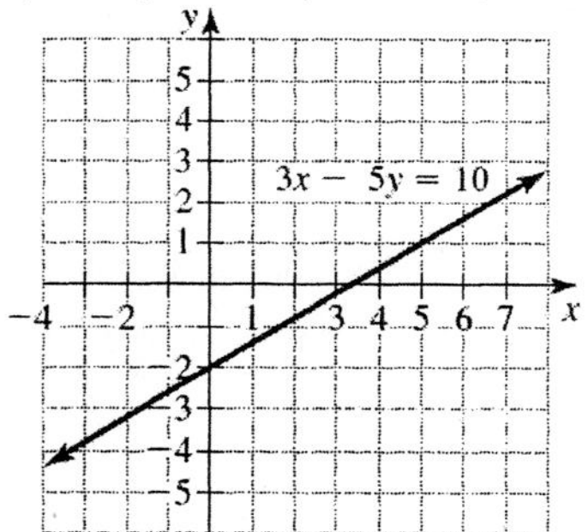

32. Write $-2x + 3y = 9$ as $y = \frac{2}{3}x + 3$. The slope is $\frac{2}{3}$ and the y-intercept is $(0, 3)$. The line also goes through $(3, 5)$.

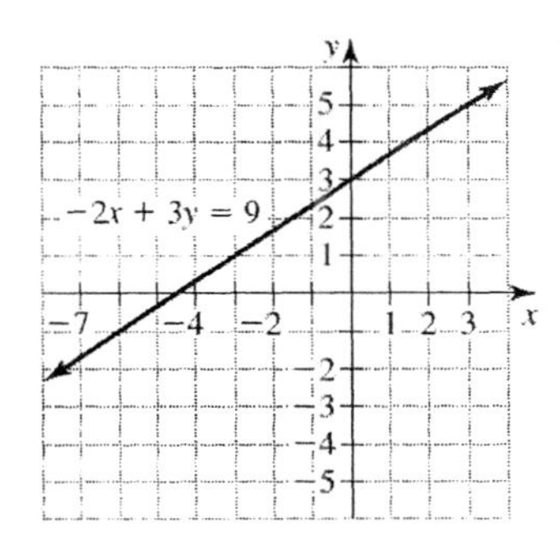

33. For $y = 4$ the slope is 0 and the y-intercept is $(0, 4)$. The graph is a horizontal line.

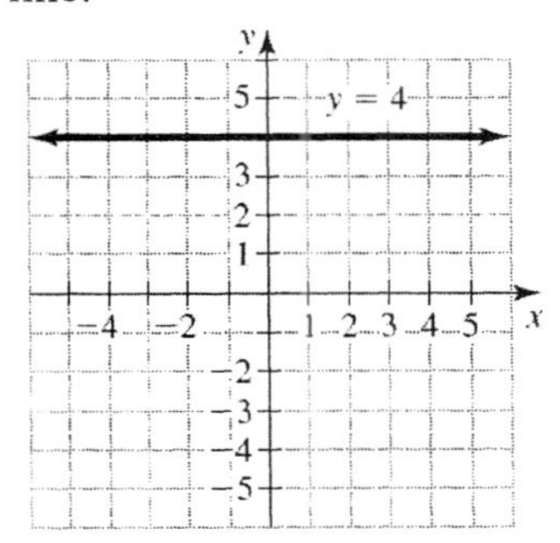

34. For $y = -5$ the slope is 0 and the y-intercept is $(0, -5)$. The graph is a horizontal line.

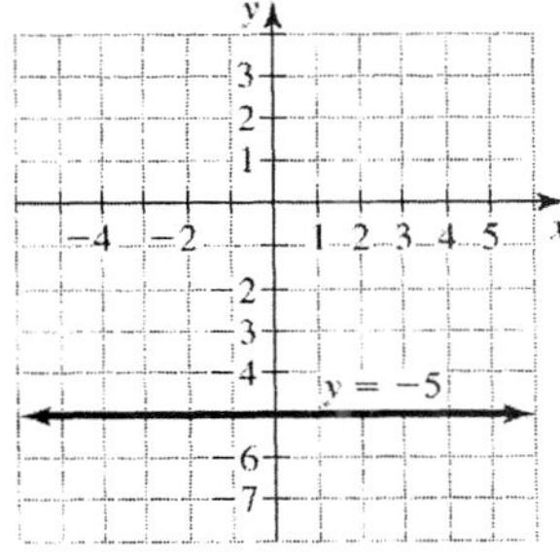

35. The line through $(0, -2)$ and $(5, 0)$ has slope $\frac{0-(-2)}{5-0} = \frac{2}{5}$. Since the y-intercept is $(0, -2)$ the equation is $y = \frac{2}{5}x - 2$.

36. The line through $(0, 5)$ and $(-3, -4)$ has slope $\frac{-4-5}{-3-0} = 3$. Since the y-intercept is $(0, 5)$ the equation is $y = 3x + 5$.

37. The line through $(0, 6)$ that is parallel to $y = \frac{2}{7}x + 3$ has slope $\frac{2}{7}$ and y-intercept $(0, 6)$. So the equation is $y = \frac{2}{7}x + 6$.

38. The line through $(0, -2)$ that is parallel to $y = -5x - 4$ has slope -5 and y-intercept $(0, -2)$. So the equation is $y = -5x - 2$.

39. Write $x - 4y = 1$ as $y = \frac{1}{4}x - \frac{1}{4}$. The line through $(0, 12)$ that is perpendicular to $x - 4y = 1$ has slope -4 and y-intercept $(0, 12)$. So the equation is $y = -4x + 12$.

40. Write $3x - y = 2$ as $y = 3x - 2$ to see that its slope is 3. The line through $(0, -14)$ that is perpendicular to $3x - y = 2$ has slope $-\frac{1}{3}$ and y-intercept $(0, -14)$. So its equation is $y = -\frac{1}{3}x - 14$.

41. The line through $(0, 3)$ that is parallel to $y = -1$ has slope 0 and y-intercept $(0, 3)$. Its equation is $y = 3$.

42. The line through $(0, 5)$ that is perpendicular to $x = 3$ is a horizontal line through $(0, 5)$. The equation is $y = 5$.

43. The line through $(-2, 4)$ and $(3, 7)$ has slope $\frac{7-4}{3-(-2)} = \frac{3}{5}$.

$$\begin{aligned} y - y_1 &= m(x - x_1) \\ y - 4 &= \tfrac{3}{5}(x - (-2)) \\ 5y - 20 &= 3x + 6 \\ -3x + 5y &= 26 \\ 3x - 5y &= -26 \end{aligned}$$

44. The line through $(2, -5)$ and $(3, 9)$ has slope $\frac{9-(-5)}{3-2} = 14$.

$$\begin{aligned} y - y_1 &= m(x - x_1) \\ y - 9 &= 14(x - 3) \\ y - 9 &= 14x - 42 \\ -14x + y &= -33 \\ 14x - y &= 33 \end{aligned}$$

45. The line through $(-1, 3)$ and $(5, 0)$ has slope $\frac{0-3}{5-(-1)} = \frac{-3}{6} = -\frac{1}{2}$.

$$\begin{aligned} y - y_1 &= m(x - x_1) \\ y - 0 &= -\tfrac{1}{2}(x - 5) \\ 2y &= -x + 5 \\ x + 2y &= 5 \end{aligned}$$

46. The line through $(2, 0)$ and $(-6, 8)$ has slope $\frac{8-0}{-6-2} = \frac{8}{-8} = -1$.

$$\begin{aligned} y - y_1 &= m(x - x_1) \\ y - 0 &= -1(x - 2) \\ y &= -x + 2 \\ x + y &= 2 \end{aligned}$$

47. The line through $(1, -4)$ that is parallel to $y = \frac{2}{3}x + 6$ has slope $\frac{2}{3}$.

$$\begin{aligned} y - y_1 &= m(x - x_1) \\ y - (-4) &= \tfrac{2}{3}(x - 1) \\ y + 4 &= \tfrac{2}{3}x - \tfrac{2}{3} \\ 3y + 12 &= 2x - 2 \\ -2x + 3y &= -14 \\ 2x - 3y &= 14 \end{aligned}$$

48. The line through $(3, -5)$ that is parallel to $y = -\frac{1}{4}x - 9$ has slope $-\frac{1}{4}$.

$$\begin{aligned} y - y_1 &= m(x - x_1) \\ y - (-5) &= -\tfrac{1}{4}(x - 3) \\ y + 5 &= -\tfrac{1}{4}x + \tfrac{3}{4} \\ 4y + 20 &= -x + 3 \\ x + 4y &= -17 \end{aligned}$$

49. Write $2x + y = 5$ as $y = -2x + 5$. The line through $(2, -5)$ that is perpendicular to $2x + y = 5$ has slope $\frac{1}{2}$.

$$\begin{aligned} y - y_1 &= m(x - x_1) \\ y - (-5) &= \tfrac{1}{2}(x - 2) \\ y + 5 &= \tfrac{1}{2}x - 1 \\ 2y + 10 &= x - 2 \\ -x + 2y &= -12 \\ x - 2y &= 12 \end{aligned}$$

50. Write $4x - y = 2$ as $y = 4x - 2$ to get a slope of 4. The line through $(3, -6)$ that is perpendicular to $4x - y = 2$ has slope $-1/4$.

$$\begin{aligned} y - y_1 &= m(x - x_1) \\ y - (-6) &= -\tfrac{1}{4}(x - 3) \\ y + 6 &= -\tfrac{1}{4}x + \tfrac{3}{4} \\ 4y + 24 &= -x + 3 \\ x + 4y &= -21 \end{aligned}$$

51. Since distance varies directly with the time, $d = kt$. If $d = 200$ when $t = 4$, then $200 = k(4)$ or $k = 50$ mph.

52. Since distance varies directly with the average speed, $d = ks$. If $d = 15$ when $s = 3$, then $15 = k(3)$ or $k = 5$. The constant of variation is 5 hr.

53. Since the time that it takes to pick the entire orange grove varies inversely with the number of pickers, $t = k/p$. If 30 pickers can pick the entire grove in 14 hours hours, then $14 = k/30$ or $k = 420$. With 40 pickers we have $t = 420/40 = 10.5$ hr.

54. Since the number of cookies each scout receives varies inversely as the number of scouts in attendance, $c = k/s$. When 4 scouts are in attendance each scout receives 12 cookies. So $12 = k/4$ or $k = 48$. If 16 scouts are in attendance, then $c = 48/16 = 3$ cookies.

55. Since the cost of wood laminate flooring for a rectangular room varies jointly as the length and width of the room, $C = kLW$. If the cost is \$1148.16 for a 12 ft by 16 ft room, then $1148.16 = k(12)(16)$ or $k = 5.98$. So the cost for a room that is 10 ft by 14 ft is $C = 5.98(10)(14) = \$837.20$.

56. Since the cost for a custom oak bookcase varies jointly with the width and height, $C = kWH$. If a bookcase that is 7 ft high and 30 in. (or 2.5 feet) wide costs \$441, then $441 = k(2.5)(7)$ or $k = 25.2$. The cost for a bookcase that is 32 in. wide and 6 ft high is $C = 24.2(32/12)(6) = \$403.20$.

57. Graph a dashed line for $3x - 2y = 6$ through the intercepts $(0, -3)$ and $(2, 0)$. Since $(0, 0)$ does not satisfy $3x - 2y > 6$, shade the side of the line that does not contain $(0, 0)$.

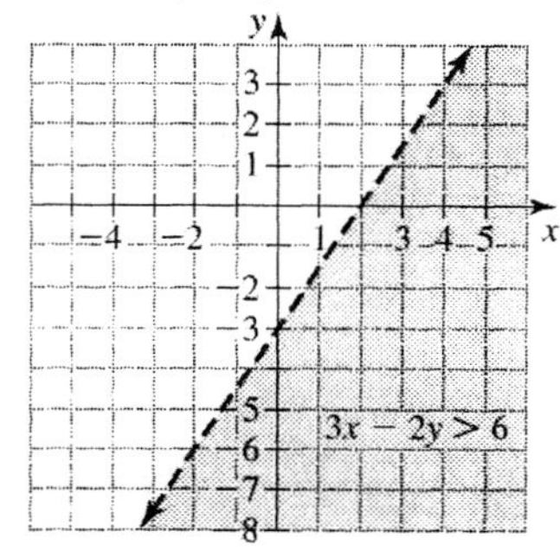

58. Graph a dashed line for $x - y = 5$ through $(0, -5)$ and $(5, 0)$. Since $(0, 0)$ satisfies $x - y < 5$ shade the side of the line that contains $(0, 0)$.

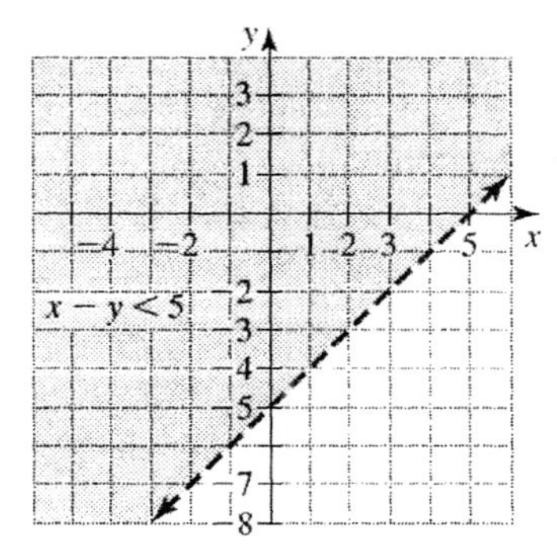

59. Graph a solid line for $x + 3y = 9$ through $(0, 3)$ and $(9, 0)$. Since $(0, 0)$ satisfies $x + 3y \leq 9$, shade the region containing $(0, 0)$.

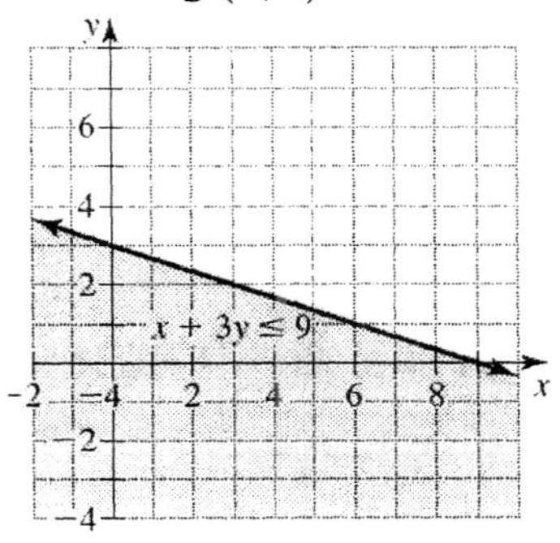

60. Graph a solid line for $6x + y = 12$ through $(0, 12)$ and $(2, 0)$. Since $(0, 0)$ does not satisfy $6x + y \geq 12$, shade the region that does not contain $(0, 0)$.

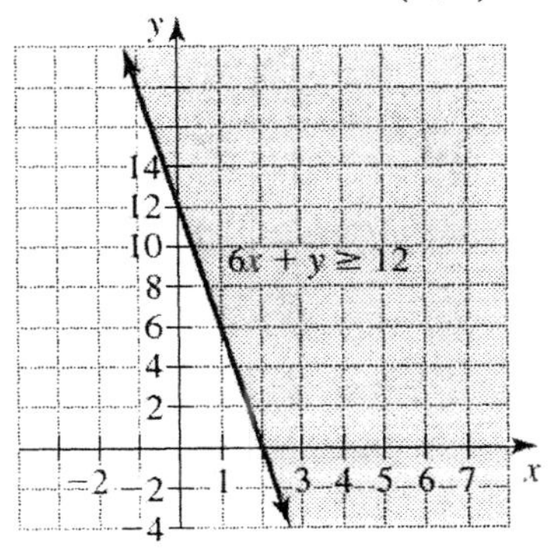

61. Graph a solid line for $y = -x + 3$ through $(0, 3)$ and $(3, 0)$. The region on or below the line satisfies $y \leq -x + 3$.

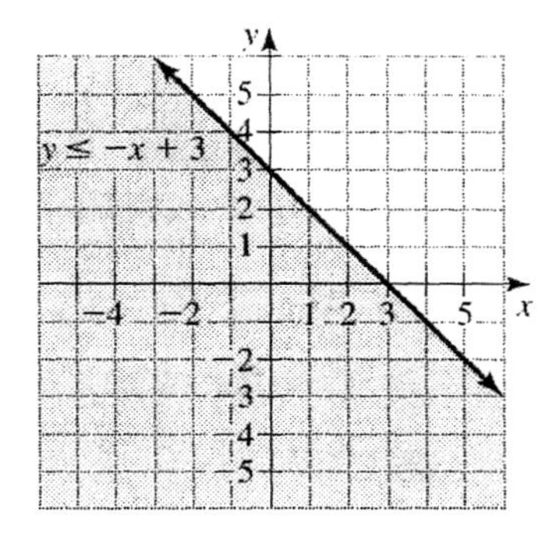

62. Graph a solid line for $y = 2x + 1$ through $(0, 1)$ and $(1, 3)$. The region on or above the line satisfies $y \geq 2x + 1$.

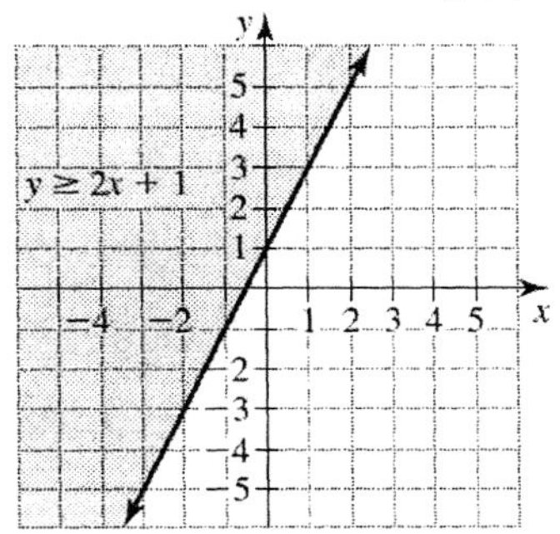

63. Graph a dashed line for $y = 3x - 4$ through $(0, -4)$ and $(1, -1)$. The region above the line satisfies $y > 3x - 4$.

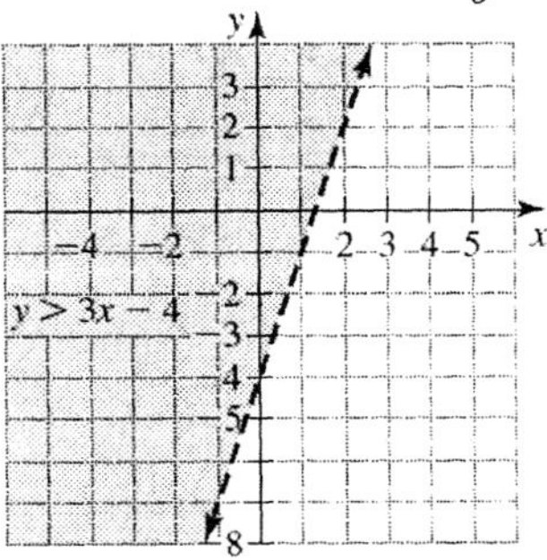

64. Graph a dashed line for $y = -2x + 2$ through $(0, 2)$ and $(1, 0)$. The region below the line satisfies $y < -2x + 2$.

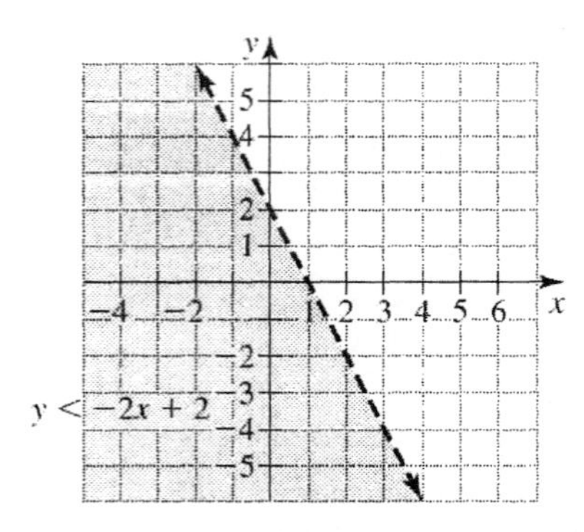

65. Graph a dashed vertical line for $x = 2$ through $(2, 0)$. The region to the left of the line satisfies $x < 2$.

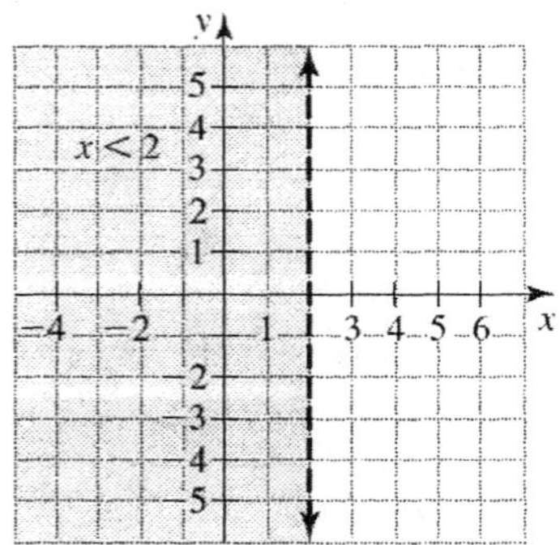

66. Graph a dashed vertical line for $x = -3$ through $(-3, 0)$. The region to the right of the line satisfies $x > -3$.

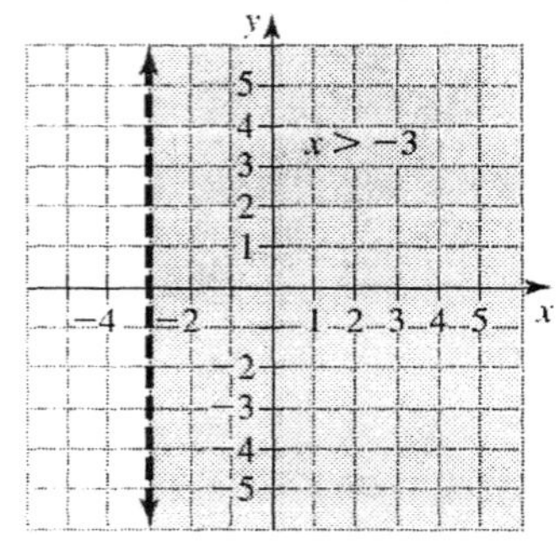

67. Graph a solid vertical line for $x = -1$ through $(-1, 0)$. The region on or to the right of this line satisfies $x \geq -1$.

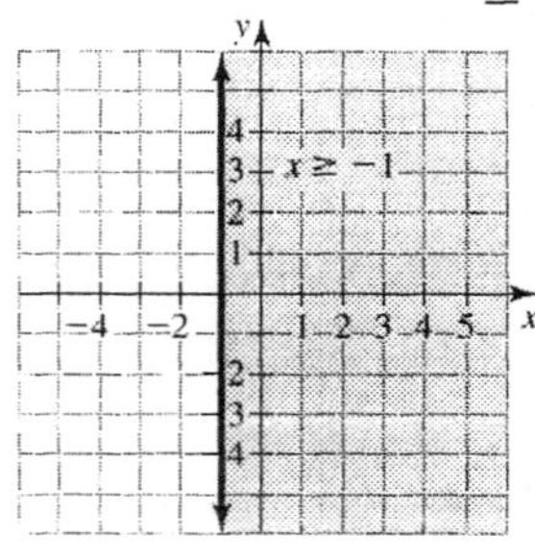

68. Graph a solid vertical line for $x = 5$ through $(5, 0)$. The region on or to the left of the vertical line satisfies $x \leq 5$.

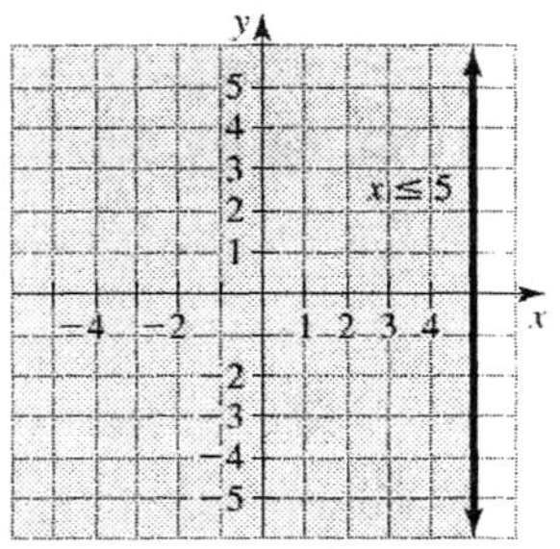

69. Graph a dashed horizontal line for $y = 4$ through $(0, 4)$. The region below this line satisfies $y < 4$.

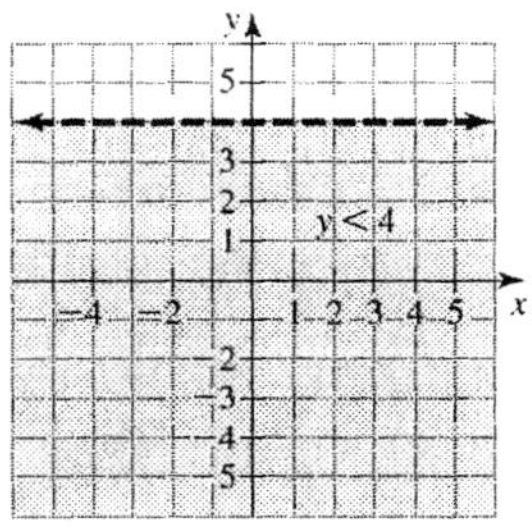

70. Graph a dashed horizontal line for $y = -2$ through $(0, -2)$. The region above this line satisfies $y > -2$.

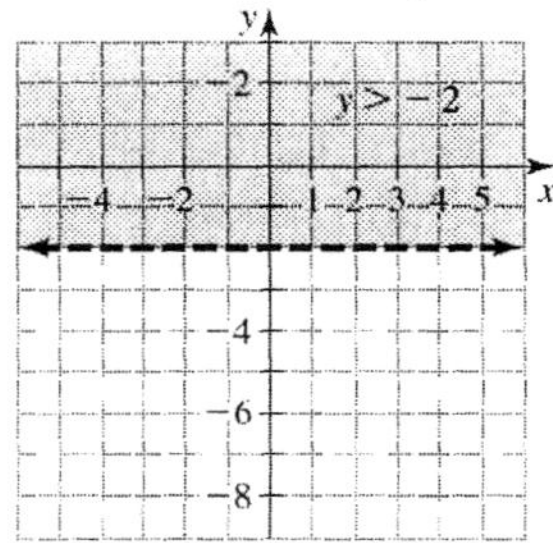

R.4 EXERCISES

1. $(x^2 + 2x) + (x^3 - 5x)$
$= x^3 + x^2 + 2x - 5x$
$= x^3 + x^2 - 3x$

2. $(x^2 - 3x) + (-2x^3 - 9x)$
$= -2x^3 + x^2 - 3x - 9x$
$= -2x^3 + x^2 - 12x$

3. $(-w^2 + 5w - 1) + (2w^2 - w - 5)$
$= 2w^2 - w^2 + 5w - w - 1 - 5$
$= w^2 + 4w - 6$

4. $(-2a^2 + 6a - 9) + (5a^2 - 3a + 8)$
$= 5a^2 - 2a^2 + 6a - 3a - 9 + 8$
$= 3a^2 + 3a - 1$

5. $(-2y^2 + 6y) - (y^2 + 5y)$
$= -2y^2 + 6y - y^2 - 5y$
$= -3y^2 + y$

6. $(-5z + 7) - (-6z - 8)$
$= -5z + 7 + 6z + 8$
$= z + 15$

7. $(-3t^2 + 5t + 1) - (-t^2 + 4t - 2)$
$= -3t^2 + 5t + 1 + t^2 - 4t + 2$
$= -2t^2 + t + 3$

8. $(-n^2 - 3n + 9) - (-4n^2 - 2n + 1)$
$= -n^2 - 3n + 9 + 4n^2 + 2n - 1$
$= 3n^2 - n + 8$

9. $2x(4x - 3) = 2x \cdot 4x - 2x \cdot 3$
$= 8x^2 - 6x$

10. $5x(-6x + 2) = 5x(-6x) + 5x(2)$
$= -30x^2 + 10x$

11. $-2a(a^2 - 4a + 9)$
$= -2a \cdot a^2 - (-2a)(4a) + (-2a)9$
$= -2a^3 + 8a^2 - 18a$

12. $-3b(2b^2 - 5b - 1)$
$= -3b \cdot 2b^2 - (-3b)(5b) - (-3b)(1)$
$= -6b^3 + 15b^2 + 3b$

13. $6w^2(w^3 - w^2 + w + 3)$
$= 6w^2w^3 - 6w^2w^2 + 6w^2w + 6w^23$
$= 6w^5 - 6w^4 + 6w^3 + 18w^2$

14. $5t^3(2t^3 + t^2 - 8t - 3)$
$= 5t^32t^3 + 5t^3t^2 - 5t^38t - 5t^33$
$= 10t^6 + 5t^5 - 40t^4 - 15t^3$

15. $(x + 2)(x + 4) = (x + 2)x + (x + 2)4$
$= x^2 + 2x + 4x + 8$
$= x^2 + 6x + 8$

16. $(a + 5)(a + 7) = (a + 5)a + (a + 5)7$
$= a^2 + 5a + 7a + 35$
$= a^2 + 12a + 35$

17. $(2s - 3)(3s + 1)$
$= (2s - 3)3s + (2s - 3)1$
$= 6s^2 - 9s + 2s - 3$
$= 6s^2 - 7s - 3$

18. $(4t - 1)(t - 2)$
$= (4t - 1)t - (4t - 1)(2)$
$= 4t^2 - t - 8t + 2$
$= 4t^2 - 9t + 2$

19. $(-2x^2 + 1)(-3x^2 - 5)$
$= (-2x^2 + 1)(-3x^2) - (-2x^2 + 1)(5)$
$= 6x^4 - 3x^2 + 10x^2 - 5$
$= 6x^4 + 7x^2 - 5$

20. $(-x^3 + 5x)(2x^3 - 3x)$
$= (-x^3 + 5x)(2x^3) - (-x^3 + 5x)(3x)$
$= -2x^6 + 10x^4 + 3x^4 - 15x^2$
$= -2x^6 + 13x^4 - 15x^2$

21. $(x - 3)(x^2 + 3x - 9)$
$= (x)(x^2 + 3x - 9) - 3(x^2 + 3x - 9)$
$= x^3 + 3x^2 - 9x - 3x^2 - 9x + 27$
$= x^3 - 18x + 27$

22. $(a - 2)(a^2 + 5a - 8)$
$= (a)(a^2 + 5a - 8) - 2(a^2 + 5a - 8)$
$= a^3 + 5a^2 - 8a - 2a^2 - 10a + 16$
$= a^3 + 3a^2 - 18a + 16$

23. $(w + 3)(3w^2 + 5w - 2)$
$= w(3w^2 + 5w - 2) + 3(3w^2 + 5w - 2)$
$= 3w^3 + 5w^2 - 2w + 9w^2 + 15w - 6$
$= 3w^3 + 14w^2 + 13w - 6$

24. $(m + 7)(-2m^2 + 4m - 9)$
$= m(-2m^2 + 4m - 9)$
$+ 7(-2m^2 + 4m - 9)$
$= -2m^3 + 4m^2 - 9m - 14m^2 + 28m - 63$
$= -2m^3 - 10m^2 + 19m - 63$

25. $(a + m)(b + n) = ab + an + mb + mn$

26. $(x + t)(y + s) = xy + sx + yt + st$

27. $(x + 2)(x - 6) = x^2 - 6x + 2x - 12$
$= x^2 - 4x - 12$

28. $(x - 5)(x + 3) = x^2 + 3x - 5x - 15$
$= x^2 - 2x - 15$

29. $(2a + 1)(3a - 4) = 6a^2 - 8a + 3a - 4$
$= 6a^2 - 5a - 4$

30. $(3b - 7)(5b - 9)$
$= 15b^2 - 27a - 35b + 63$
$= 15b^2 - 62b + 63$

31. $(-2x + 1)(-5x + 7)$
$= 10x^2 - 14x - 5x + 7$
$= 10x^2 - 19x + 7$

32. $(-3x - 2)(-x - 6)$
$= 3x^2 + 18x + 2x + 12$
$= 3x^2 + 20x + 12$

33. $(2a^3 - 6)(5a^3 + 3)$
$= 10a^6 + 6a^3 - 30a^3 - 18$
$= 10a^6 - 24a^3 - 18$

34. $(4w^3 - 5)(3w^3 - 7)$
$= 12w^6 - 28w^3 - 15w^3 + 35$
$= 12w^6 - 43w^3 + 35$

35. $(4x^4 - x)(4x^4 + x)$
$= 16x^8 + 4x^5 - 4x^5 - x^2$
$= 16x^8 - x^2$

36. $(5a^4 + x^3)(5a^4 - x^3)$
$= 25a^8 - 5a^4x^3 + 5a^4x^3 - x^6$
$= 25a^8 - x^6$

37. $(x+5)^2 = x^2 + 2(x)(5) + 5^2$
$= x^2 + 10x + 25$

38. $(y+3)^2 = y^2 + 2(y)(3) + 3^2$
$= y^2 + 6y + 9$

39. $(2t+7)^2 = (2t)^2 + 2(2t)(7) + 7^2$
$= 4t^2 + 28t + 49$

40. $(3w+4)^2 = (3w)^2 + 2(3w)(4) + 4^2$
$= 9w^2 + 24w + 16$

41. $(s-2)^2 = s^2 - 2(s)(2) + 2^2$
$= s^2 - 4s + 4$

42. $(h-3)^2 = h^2 - 2(h)(3) + 3^2$
$= h^2 - 6h + 9$

43. $(3y-5)^2 = (3y)^2 - 2(3y)(5) + 5^2$
$= 9y^2 - 30y + 25$

44. $(6x-1)^2 = (6x)^2 - 2(6x)(1) + 1^2$
$= 36x^2 - 12x + 1$

45. $(3q+4)(3q-4) = (3q)^2 - 4^2$
$= 9q^2 - 16$

46. $(5m+6)(5m-6) = (5m)^2 - 6^2$
$= 25m^2 - 36$

47. $(2x^2 - 3n)(2x^2 + 3n) = (2x^2)^2 - (3n)^2$
$= 4x^4 - 9n^2$

48. $(5t^2 - 3m)(5t^2 + 3m)$
$= (5t^2)^2 - (3m)^2 = 25t^4 - 9m^2$

49. $(6x^8) \div (3x^2) = 2x^{8-2} = 2x^6$

50. $(-12a^{14}) \div (-3a^2) = 4a^{14-2} = 4a^{12}$

51. $(-4w^5) \div (2w^4) = -2w^{5-4} = -2w$

52. $(20b^{12}) \div (-5b^{10}) = -4b^{12-10} = -4b^2$

53. $(3x^{22}) \div (6x^{20}) = \frac{1}{2}x^{22-20} = \frac{1}{2}x^2$

54. $(-4t^{16}) \div (8t^8) = -\frac{1}{2}t^{16-8} = -\frac{1}{2}t^8$

55. $(30x^3 - 20x^2 + 10x) \div (10x)$
$= \frac{30x^3}{10x} - \frac{20x^2}{10x} + \frac{10x}{10x}$
$= 3x^2 - 2x + 1$

56. $(25a^3 + 20a^2 + 5a) \div (-5a)$
$= \frac{25a^3}{-5a} + \frac{20a^2}{-5a} + \frac{5a}{-5a}$
$= -5a^2 - 4a - 1$

57. $(-3x^5 - 9x^4 + 3x^3 - 6x^2) \div (-3x^2)$
$= \frac{-3x^5}{-3x^2} - \frac{9x^4}{-3x^2} + \frac{3x^3}{-3x^2} - \frac{6x^2}{-3x^2}$
$= x^3 + 3x^2 - x + 2$

58. $(-8w^4 - 6w^3 + 4w^2) \div (-2w^2)$
$= \frac{-8w^4}{-2w^2} - \frac{6w^3}{-2w^2} + \frac{4w^2}{-2w^2}$
$= 4w^2 + 3w - 2$

59.
$$
\begin{array}{r l}
 & x^2 - 3x - 3 \\
x - 1 & \overline{)x^3 - 4x^2 + 0 \cdot x + 3} \\
 & \underline{x^3 - x^2} \\
 & -3x^2 + 0 \cdot x \\
 & \underline{-3x^2 + 3x} \\
 & -3x + 3 \\
 & \underline{-3x + 3} \\
 & 0
\end{array}
$$
$(x^3 - 4x^2 + 3) \div (x - 1) = x^2 - 3x + 3$

60.
$$
\begin{array}{r l}
 & 2x^2 - 2x + 5 \\
x + 1 & \overline{)2x^3 + 0 \cdot x^2 + 3x + 5} \\
 & \underline{2x^3 + 2x^2} \\
 & -2x^2 + 3x \\
 & \underline{-2x^2 - 2x} \\
 & 5x + 5 \\
 & \underline{5x + 5} \\
 & 0
\end{array}
$$
$(2x^3 + 3x + 5) \div (x + 1) = 2x^2 - 2x + 5$

61.
$$
\begin{array}{r l}
 & x^2 - 2x - 3 \\
x - 2 & \overline{)x^3 - 4x^2 + x + 6} \\
 & \underline{x^3 - 2x^2} \\
 & -2x^2 + x \\
 & \underline{-2x^2 + 4x} \\
 & -3x + 6 \\
 & \underline{-3x + 6} \\
 & 0
\end{array}
$$
$(x^3 - 4x^2 + x + 6) \div (x - 2)$
$= x^2 - 2x - 3$

62.
$$
\begin{array}{r l}
 & 2x^2 - 3x + 4 \\
x + 3 & \overline{)2x^3 + 3x^2 - 5x + 12} \\
 & \underline{2x^3 + 6x^2} \\
 & -3x^2 - 5x \\
 & \underline{-3x^2 - 9x} \\
 & 4x + 12 \\
 & \underline{4x + 12} \\
 & 0
\end{array}
$$
$(2x^3 + 3x^2 - 5x + 12) \div (x + 3)$
$= 2x^2 - 3x + 4$

63.

$$\begin{array}{r} x^2 - x + 2 \\ 2x+1\overline{)2x^3 - x^2 + 3x + 2} \\ \underline{2x^3 + x^2} \\ -2x^2 + 3x \\ \underline{-2x^2 - x} \\ 4x + 2 \\ \underline{4x + 2} \\ 0 \end{array}$$

$$(2x^3 - x^2 + 3x + 2) \div (2x + 1) = x^2 - x + 2$$

64.

$$\begin{array}{r} x^2 + x - 3 \\ 2x-3\overline{)2x^3 - x^2 - 9x + 9} \\ \underline{2x^3 - 3x^2} \\ 2x^2 - 9x \\ \underline{2x^2 - 3x} \\ -6x + 9 \\ \underline{-6x + 9} \\ 0 \end{array}$$

$$(2x^3 - x^2 - 9x + 9) \div (2x - 3) = x^2 + x - 3$$

65. $\dfrac{3x^5 \cdot 5x^9}{45x^{14}} = \dfrac{15x^{14}}{45x^{14}} = \dfrac{1}{3}$

66. $\dfrac{-2y^3 \cdot 4y^6}{-24y^9} = \dfrac{-8y^9}{-24y^9} = \dfrac{1}{3}$

67. $\dfrac{2a^2 \cdot 6a^4}{(-2a^2)^3} = \dfrac{12a^6}{-8a^6} = -\dfrac{3}{2}$

68. $\dfrac{(-2w^3)(8w^{15})}{(-2w^3)^6} = \dfrac{-16w^{18}}{64w^{18}} = -\dfrac{1}{4}$

69. $(-3a^2)^3(2a^4)^5 = -27a^6 \cdot 32a^{20} = -864a^{26}$

70. $(-2b)^4(2b^3)^2 = 16b^4(4b^6) = 64b^{10}$

71. $(-5x^3)^2(-2x^2)^3 = 25x^6(-8x^6) = -200x^{12}$

72. $(-5y^5)^2(-3y^3)^2 = 25y^{10}(9y^6) = 225y^{16}$

73. $\left(\dfrac{3x^4}{6y^2}\right)^3 = \dfrac{27x^{12}}{216y^6} = \dfrac{x^{12}}{8y^6}$

74. $\left(\dfrac{-2q^3}{4p^2}\right)^3 = \dfrac{-8q^9}{64p^6} = -\dfrac{q^9}{8p^6}$

75. $\left(\dfrac{2ab \cdot 3a^2b}{-4b^2}\right)^2 = \left(\dfrac{6a^3b^2}{-4b^2}\right)^2 = \dfrac{36a^6b^4}{16b^4} = \dfrac{9a^6}{4}$

76. $\left(\dfrac{3ab^2 \cdot 6a^2b^3}{(2ab)^3}\right)^2 = \left(\dfrac{18a^3b^5}{8a^3b^3}\right)^2 = \left(\dfrac{9b^2}{4}\right)^2 = \dfrac{81b^4}{16}$

77. $\dfrac{-3a^{-2} \cdot 4a^3}{2a^{-5}} = \dfrac{-12a}{2a^{-5}} = -6a^6$

78. $\dfrac{-5b^{-6} \cdot 6b^{-5}}{2b^{-20}} = \dfrac{-30b^{-11}}{2b^{-20}} = -15b^9$

79. $\dfrac{3w^{-7} \cdot 5w^4}{30w^{-9}} = \dfrac{15w^{-3}}{30w^{-9}} = \dfrac{w^6}{2}$

80. $\dfrac{4t^{-5} \cdot 8t^9}{16t^{18}} = \dfrac{32t^4}{16t^{18}} = \dfrac{2}{t^{14}}$

81. $(-3x^{-1})^{-4} = (-3)^{-4}x^4 = \dfrac{x^4}{81}$

82. $(-5y^{-6})^{-2} = (-5)^{-2}y^{12} = \dfrac{y^{12}}{25}$

83. $(-2a^2)^{-5}(a^{-3})^6 = (-2)^{-5}a^{-10}a^{-18} = -\dfrac{1}{32a^{28}}$

84. $(2b^{-1})^{-2}(-3b^{-2})^{-3} = (2)^{-2}b^2(-3)^{-3}b^6 = \dfrac{b^8}{4(-27)} = -\dfrac{b^8}{108}$

85. $\left(\dfrac{x^{-2}}{y^3}\right)^{-3} = \dfrac{x^6}{y^{-9}} = x^6y^9$

86. $\left(\dfrac{a^3}{b^{-5}}\right)^{-4} = \dfrac{a^{-12}}{b^{20}} = \dfrac{1}{a^{12}b^{20}}$

87. $\left(\frac{2x^{-2}\cdot 3x^5}{15x^{-8}}\right)^{-2} = \left(\frac{6x^3}{15x^{-8}}\right)^{-2}$
$= \left(\frac{2x^{11}}{5}\right)^{-2} = \frac{2^{-2}x^{-22}}{5^{-2}} = \frac{25}{4x^{22}}$

88. $\left(\frac{2y^{-1}\cdot 5y^7}{20y^2}\right)^{-2} = \left(\frac{10y^6}{20y^2}\right)^{-2}$
$= \left(\frac{y^4}{2}\right)^{-2} = \frac{y^{-8}}{2^{-2}} = \frac{4}{y^8}$

89. $5{,}000 \cdot (20{,}000)^4 = 5 \times 10^3(2 \times 10^4)^4$
$= 80 \times 10^{19} = 8 \times 10^1 \times 10^{19} = 8 \times 10^{20}$

90. $30{,}000 \cdot (20{,}000)^5 = 3 \times 10^4(2 \times 10^4)^5$
$= 96 \times 10^{24} = 9.6 \times 10^1 \times 10^{24}$
$= 9.6 \times 10^{25}$

91.
$(0.00005)^2(2000)^3 = (5 \times 10^{-5})^2(2 \times 10^3)^3$
$= 25 \times 10^{-10} \cdot 8 \times 10^9$
$= 200 \times 10^{-1}$
$= 2 \times 10^2 \times 10^{-1}$
$= 2 \times 10^1$

92. $(0.0006)^2(1000)^6$
$= (6 \times 10^{-4})^2(1 \times 10^3)^6$
$= 36 \times 10^{-8} \cdot 1 \times 10^{18}$
$= 3.6 \times 10^1 \times 10^{-8} \cdot 1 \times 10^{18}$
$= 3.6 \times 10^{11}$

93. $\frac{(10{,}000)^2}{0.000002} = \frac{(1 \times 10^4)^2}{2 \times 10^{-6}} = \frac{1 \times 10^8}{2 \times 10^{-6}}$

$= 0.5 \times 10^{14}$
$= 5 \times 10^{-1} \times 10^{14}$
$= 5 \times 10^{13}$

94. $\frac{(8000)^2}{(0.00001)^3} = \frac{(8 \times 10^3)^2}{(1 \times 10^{-5})^3}$
$= \frac{64 \times 10^6}{1 \times 10^{-15}} = 64 \times 10^{21}$
$= 6.4 \times 10^1 \times 10^{21}$
$= 6.4 \times 10^{22}$

95. $\frac{(0.002)^3(40{,}000{,}000)^3}{(10{,}000)^5}$
$= \frac{(2 \times 10^{-3})^3(4 \times 10^7)^3}{(1 \times 10^4)^5}$
$= \frac{8 \times 10^{-9} \cdot 64 \times 10^{21}}{1 \times 10^{20}}$
$= 512 \times 10^{-8}$
$= 5.12 \times 10^2 \times 10^{-8}$
$= 5.12 \times 10^{-6}$

96. $\frac{(0.0005)^2(10{,}000)}{(500)^4}$
$= \frac{(5 \times 10^{-4})^2(1 \times 10^4)}{(5 \times 10^2)^4}$
$= \frac{25 \times 10^{-8} \cdot 1 \times 10^4}{625 \times 10^8}$
$= 0.04 \times 10^{-12}$
$= 4 \times 10^{-2} \times 10^{-12}$
$= 4 \times 10^{-14}$

R.5 EXERCISES

1. $12x + 8 = 4 \cdot 3x + 4 \cdot 2 = 4(3x + 2)$
2. $18a + 30 = 6 \cdot 3a + 6 \cdot 5 = 6(3a + 5)$
3. $15y^3 - 6y^2 = 3y^2 \cdot 5y - 3y^2 \cdot 2$
$= 3y^2(5y - 2)$
4. $48z^4 - 32z^3 = 16z^3 \cdot 3z - 16z^3 \cdot 2$
$= 16z^3(3z - 2)$
5. $8a^3b^2 + 20a^4b = 4a^3b \cdot 2b + 4a^3b \cdot 5a$
$= 4a^3b(2b + 5a)$
6. $24y^4z^3 + 36y^3z^4$
$= 12y^3z^3 \cdot 2y + 12y^3z^3 \cdot 3z$
$= 12y^3z^3(2y + 3z)$
7. $12x^4 - 20x^3 - 24x^2$
$= 4x^2 \cdot 3x^2 - 4x^2 \cdot 5x - 4x^2 \cdot 6$
$= 4x^2(3x^2 - 5x - 6)$
8. $14y^3 - 21y^2 - 28y$
$= 7y \cdot 2y^2 - 7y \cdot 3y - 7y \cdot 4$
$= 7y(2y^2 - 3y - 4)$
9. $2a^3b - 6a^2b + 6ab$
$= 2ab \cdot a^2 - 2ab \cdot 3a + 2ab \cdot 3$
$= 2ab(a^2 - 3a + 3)$
10. $3w^3z - 12w^2z - 9wz$
$= 3wz \cdot w^2 - 3wz \cdot 4w - 3wz \cdot 3$
$= 3wz(w^2 - 4w - 3)$
11. $4x^3 - 6x^2 = 2x \cdot 2x^2 - 2x \cdot 3x$
$= (2x)(2x^2 - 3x)$
12. $5y^4 - 10y^2 = 5y^2y^2 - 5y^2 2$
$= (5y^2)(y^2 - 2)$
13. $-2x^2 - 6x = (-2x)x + (-2x)3$
$= (-2x)(x + 3)$
14. $-3y^3 - 9y = (-3y)y^2 + (-3y)3$
$= (-3y)(y^2 + 3)$
15. $-5a^5 + 10a^2 = (-5a^2)a^3 - (-5a^2)2$
$= (-5a^2)(a^3 - 2)$

16. $-4b^4 - 12b^2 = (-4b^2)b^2 + (-4b^2)3$
$= (-4b^2)(b^2 + 3)$

17. $-w^3x - w^2x = (-w^2x)w + (-w^2x)1$
$= (-w^2x)(w + 1)$

18. $-zy^3 + zy^2 = (-zy^2)y - (-zy^2)1$
$= (-zy^2)(y - 1)$

19. $x^2 + 8x + 16 = x^2 + 2 \cdot 4x + 4^2$
$= (x + 4)^2$

20. $x^2 + 4x + 4 = x^2 + 2 \cdot 2x + 2^2$
$= (x + 2)^2$

21. $a^2 - 2a + 1 = a^2 - 2 \cdot a \cdot 1 + 1^2$
$= (a - 1)^2$

22. $b^2 - 10b + 25 = b^2 - 2 \cdot 5b + 5^2$
$= (b - 5)^2$

23. $y^2 - 9 = y^2 - 3^2 = (y + 3)(y - 3)$

24. $n^2 - 4 = n^2 - 2^2 = (n + 2)(n - 2)$

25. $9x^2 + 6x + 1 = (3x)^2 + 2(3x) + 1^2$
$= (3x + 1)^2$

26. $25y^2 + 20y + 4 = (5y)^2 + 2 \cdot 5y + 2^2$
$= (5y + 2)^2$

27. $16m^2 - 40mt + 25t^2$
$= (4m)^2 - 2(4m \cdot 5t) + (5t)^2$
$= (4m - 5t)^2$

28. $9s^2 - 24st + 16t^2$
$= (3s)^2 - 2(3s)(4t) + (4t)^2$
$= (3s - 4t)^2$

29. $9x^2 - 16 = (3x)^2 - 4^2$
$= (3x + 4)(3x - 4)$

30. $81a^2 - 25 = (9a)^2 - 5^2$
$= (9a + 5)(9a - 5)$

31. $64n^2 + 48n + 9$
$= (8n)^2 + 2(8n)(3) + 3^2$
$= (8n + 3)^2$

32. $81s^2 - 18s + 1$
$= (9s)^2 - 2(9s)(1) + 1^2$
$= (9s - 1)^2$

33. $25x^2 - 49y^2 = (5x)^2 - (7y)^2$
$= (5x + 7y)(5x - 7y)$

34. $a^2b^2 - y^2 = (ab)^2 - y^2$
$= (ab + y)(ab - y)$

35. $a^2 + 6a + ab + 6b$
$= a(a + 6) + b(a + 6)$
$= (a + b)(a + 6)$

36. $w^2 - 3w + wx - 3x$
$= w(w - 3) + x(w - 3)$
$= (w + x)(w - 3)$

37. $6x^2 - 10x + 3ax - 5a$
$= 2x(3x - 5) + a(3x - 5)$
$= (2x + a)(3x - 5)$

38. $10ax + 5a + 2x + 1$
$= 5a(2x + 1) + 1(2x + 1)$
$= (5a + 1)(2x + 1)$

39. $3y^3 - 4y^2 + 3y - 4$
$= y^2(3y - 4) + 1(3y - 4)$
$= (y^2 + 1)(3y - 4)$

40. $6x^3 - 3x^2 + 10x - 5$
$= 3x^2(2x - 1) + 5(2x - 1)$
$= (3x^2 + 5)(2x - 1)$

41. $8a^3 - 4a^2 + 14a - 7$
$= 4a^2(2a - 1) + 7(2a - 1)$
$= (4a^2 + 7)(2a - 1)$

42. $5t^3 - 10t^2 + 6t - 12$
$= 5t^2(t - 2) + 6(t - 2)$
$= (5t^2 + 6)(t - 2)$

43. $ab - 2b - 3a + 6$
$= b(a - 2) - 3(a - 2)$
$= (b - 3)(a - 2)$

44. $x^2 - xy - 7x + 7y$
$= x(x - y) - 7(x - y)$
$= (x - 7)(x - y)$

45. $x^3 - x^2 + 3 - 3x$
$= x^2(x - 1) - 3(x - 1)$
$= (x^2 - 3)(x - 1)$

46. $ax^2 - 4x^2 + 20 - 5a$
$= x^2(a - 4) - 5(a - 4)$
$= (x^2 - 5)(a - 4)$

47. $x^2 + 5x + 6$
$= x^2 + 2x + 3x + 6$
$= x(x + 2) + 3(x + 2)$
$= (x + 2)(x + 3)$

48. $x^2 + 11x + 30$
$= x^2 + 5x + 6x + 30$
$= x(x + 5) + 6(x + 5)$
$= (x + 5)(x + 6)$

49. $w^2 + 8w + 15$
$= w^2 + 5w + 3w + 15$
$= w(w + 5) + 3(w + 5)$
$= (w + 3)(w + 5)$

50. $u^2 + 19u + 18$
$= u^2 + u + 18u + 18$
$= u(u + 1) + 18(u + 1)$
$= (u + 18)(u + 1)$

51. $v^2 - 2v - 12$
$= v^2 - 6v + 4v - 12$
$= v(v-6) + 4(v-6)$
$= (v-6)(v+4)$

52. $m^2 - 9m - 22$
$= m^2 - 11m + 2m - 22$
$= m(m-11) + 2(m-11)$
$= (m-11)(m+2)$

53. $t^2 - 12t - 28$
$= t^2 - 14t + 2t - 28$
$= t(t-14) + 2(t-14)$
$= (t-14)(t+2)$

54. $q^2 - 4q - 32$
$= q^2 + 4q - 8q - 32$
$= q(q+4) - 8(q+4)$
$= (q-8)(q+4)$

55. $b^2 - 15a + 26$
$= b^2 - 13b - 2b + 26$
$= b(b-13) - 2(b-13)$
$= (b-13)(b-2)$

56. $p^2 - 26p + 25$
$= p^2 - 25p - p + 25$
$= p(p-25) - 1(p-25)$
$= (p-25)(p-1)$

57. $c^2 - 11c + 24$
$= c^2 - 3c - 8c + 24$
$= c(c-3) - 8(c-3)$
$= (c-8)(c-3)$

58. $n^2 - 10n + 21$
$= n^2 - 3n - 7n + 21$
$= n(n-3) - 7(n-3)$
$= (n-3)(n-7)$

59. $2x^2 + 7x + 6$
$= 2x^2 + 3x + 4x + 6$
$= x(2x+3) + 2(2x+3)$
$= (2x+3)(x+2)$

60. $3w^2 + 16w + 5$
$= 3w^2 + w + 15w + 5$
$= w(3w+1) + 5(3w+1)$
$= (3w+1)(w+5)$

61. $15t^2 + 17t + 4$
$= 15t^2 + 5t + 12t + 4$
$= 5t(3t+1) + 4(3t+1)$
$= (3t+1)(5t+4)$

62. $6m^2 + 29m + 20$
$= 6m^2 + 5m + 24m + 20$
$= m(6m+5) + 4(6m+5)$
$= (6m+5)(m+4)$

63. $3n^2 + 16n - 12$
$= 3n^2 - 2n + 18n - 12$
$= n(3n-2) + 6(3n-2)$
$= (3n-2)(n+6)$

64. $4y^2 + 17y - 15$
$= 4y^2 - 3y + 20y - 15$
$= y(4y-3) + 5(4y-3)$
$= (4y-3)(y+5)$

65. $8m^2 + 6m - 27$
$= 8m^2 - 12m + 18m - 27$
$= 4m(2m-3) + 9(2m-3)$
$= (2m-3)(4m+9)$

66. $18p^2 + 9p - 5$
$= 18p^2 - 6p + 15p - 5$
$= 6p(3p-1) + 5(3p-1)$
$= (3p-1)(6p+5)$

67. $8q^2 - 14q + 3$
$= 8q^2 - 2q - 12q + 3$
$= 2q(4q-1) - 3(4q-1)$
$= (4q-1)(2q-3)$

68. $6t^2 - 11t + 4$
$= 6t^2 - 8t - 3t + 4$
$= 2t(3t-4) - 1(3t-4)$
$= (3t-4)(2t-1)$

69. $15z^2 - 19z + 6$
$= 15z^2 - 9z - 10z + 6$
$= 3z(5z-3) - 2(5z-3)$
$= (5z-3)(3z-2)$

70. $10k^2 - 41k + 4$
$= 10k^2 - 40k - k + 4$
$= 10k(k-4) - 1(k-4)$
$= (k-4)(10k-1)$

71. $x^3 - 1 = x^3 - 1^3$
$= (x-1)(x^2 + x + 1)$

72. $y^3 - 27 = y^3 - 3^3$
$= (y-3)(y^2 + 3y + 9)$

73. $a^3 - 8 = a^3 - 2^3$
$= (a-2)(a^2 + 2a + 4)$

74. $b^3 - 1000 = b^3 - 10^3$
$= (b-10)(b^2 + 10b + 100)$

75. $125x^3 - 1 = (5x)^3 - 1^3$
$= (5x-1)(25x^2 + 5x + 1)$

76. $8a^3 - 125 = (2a)^2 - 5^3$
$= (2a-5)(4a^2 + 10a + 25)$

77. $125q^3 - 27 = (5q)^3 - 3^3$
$= (5q-3)(25q^2 + 15q + 9)$

78. $1000b^3 - 343 = (10b)^3 - 7^3$
$= (10b - 7)(100b^2 + 70b + 49)$

79. $27x^3 + 64y^3 = (3x)^3 + (4y)^3$
$= (3x + 4y)(9x^2 - 12xy + 16y^2)$

80. $8h^3 + 125k^3 = (2h)^3 + (5k)^3$
$= (2h + 5k)(4h^2 - 10hk + 25k^2)$

81. $343m^3 + 8n^3 = (7m)^3 + (2n)^3$
$= (7m + 2n)(49m^2 - 14mn + 4n^2)$

82. $a^3b^3 + x^3y^3 = (ab)^3 + (xy)^3$
$= (a\,b + xy)(a^2b^2 - abxy + x^2y^2)$

83. $2x^2 + 8x + 6$
$= 2(x^2 + 4x + 3)$
$= 2(x + 1)(x + 3)$

84. $3x^2 + 6x - 45$
$= 3(x^2 + 2x - 15)$
$= 3(x + 5)(x - 3)$

85. $-2x^3 - 12x^2 - 18x$
$= -2x(x^2 + 6x + 9)$
$= -2x(x + 3)^2$

86. $-4x^4 + 40x^3 - 100x^2$
$= -4x^2(x^2 - 10x + 25)$
$= -4x^2(x - 5)^2$

87. $3a^4 - 3b^4 = 3(a^4 - b^4)$
$= 3(a^2 - b^2(a^2 + b^2)$
$= 3(a - b)(a + b)(a^2 + b^2)$

88. $w^5 - wq^4 = w(w^4 - q^4)$
$= w(w^2 - q^2)(w^2 + q^2)$
$= w(w - q)(w + q)(w^2 + q^2)$

89. $-a^3b - 8b^4 = -b(a^3 + 8b^3)$
$= -b(a + 2b)(a^2 + 2ab + 4b^2)$

90. $-24x^3 + 81 = -3(8x^3 - 27)$
$= -3(2x - 3)(4x^2 + 6x + 9)$

91. $a^3 + 3a^2 - 4a - 12$
$= a^2(a + 3) - 4(a + 3)$
$= (a^2 - 4)(a + 3)$
$= (a - 2)(a + 2)(a + 3)$

92. $x^3 - 5x^2 - 9x + 45$
$= x^2(x - 5) - 9(x - 5)$
$= (x^2 - 9)(x - 5)$
$= (x - 3)(x + 3)(x - 5)$

93. $x^2 - 2x - 12 = 0$
$(x - 6)(x + 4) = 0$
$x - 6 = 0$ or $x + 4 = 0$
$x = 6$ or $x = -4$
The solution set is $\{-4, 6\}$.

94. $y^2 + y - 20 = 0$
$(y + 5)(y - 4) = 0$
$y + 5 = 0$ or $y - 4 = 0$
$y = -5$ or $y = 4$
The solution set is $\{-5, 4\}$

95. $2t^2 + 5t - 3 = 0$
$(2t - 1)(t + 3) = 0$
$2t - 1 = 0$ or $t + 3 = 0$
$t = \frac{1}{2}$ or $t = -3$
The solution set is $\left\{-3, \frac{1}{2}\right\}$

96. $3p^2 - 14p + 8 = 0$
$(3p - 2)(p - 4) = 0$
$3p - 2 = 0$ or $p - 4 = 0$
$p = \frac{2}{3}$ or $p = 4$
The solution set is $\left\{\frac{2}{3}, 4\right\}$

97. $4m^2 - 12m + 5 = 0$
$(2m - 1)(2m - 5) = 0$
$2m - 1 = 0$ or $2m - 5 = 0$
$m = \frac{1}{2}$ or $m = \frac{5}{2}$
The solution set is $\left\{\frac{1}{2}, \frac{5}{2}\right\}$

98. $15w^2 - 8w + 1 = 0$
$(5w - 1)(3w - 1) = 0$
$5w - 1 = 0$ or $3w - 1 = 0$
$w = \frac{1}{5}$ or $w = \frac{1}{3}$
The solution set is $\left\{\frac{1}{5}, \frac{1}{3}\right\}$

99. $r^3 + 5r^2 + 6r = 0$
$r(r^2 + 5r + 6) = 0$
$r(r + 2)(r + 3) = 0$
$r = 0$ or $r + 2 = 0$ or $r + 3 = 0$
$r = 0$ or $r = -2$ or $r = -3$
The solution set is $\{-3, -2, 0\}$

100. $2c^3 - 2c^2 - 4c = 0$
$2c(c^2 - c - 2) = 0$
$2c(c - 2)(c + 1) = 0$
$2c = 0$ or $c - 2 = 0$ or $c + 1 = 0$
$c = 0$ or $c = 2$ or $c = -1$
The solution set is $\{-1, 0, 2\}$

R.6 EXERCISES

1. $\dfrac{b^2 - 16}{b^2 + 8b + 16} = \dfrac{(b - 4)(b + 4)}{(b + 4)(b + 4)} = \dfrac{b - 4}{b + 4}$

2. $\dfrac{2x^2 - 2y^2}{2x^2 - 4xy + 2y^2} = \dfrac{2(x - y)(x + y)}{2(x - y)^2}$
$= \dfrac{x + y}{x - y}$

3. $\dfrac{4x^2+4x-24}{2x^2-18} = \dfrac{4(x+3)(x-2)}{2(x-3)(x+3)} = \dfrac{2x-4}{x-3}$

4. $\dfrac{2a^3+2a^2-40a}{a^3+4a^2-5a} = \dfrac{2a(a^2+a-20)}{a(a^2+4a-5)}$

$= \dfrac{2a(a+5)(a-4)}{a(a+5)(a-1)} = \dfrac{2a-8}{a-1}$

5. $\dfrac{6x^3y^6}{8x^3y} = \dfrac{6x^{3-3}y^{6-1}}{8} = \dfrac{3y^5}{4}$

6. $\dfrac{10a^3b^2}{15ab^4} = \dfrac{2a^{3-1}b^{2-4}}{3} = \dfrac{2a^2}{3b^2}$

7. $\dfrac{-20wz^9}{25w^3z^2} = \dfrac{-4z^{9-2}}{5w^{3-1}} = -\dfrac{4z^7}{5w^2}$

8. $\dfrac{21r^2t}{-28r^5t^3} = \dfrac{3}{-4r^{5-2}t^{3-1}} = -\dfrac{3}{4r^3t^2}$

9. $\dfrac{-2a-2y}{-4a^2+4y^2} = \dfrac{-2(a+y)}{-4(a-y)(a+y)} = \dfrac{1}{2(a-y)}$

10. $\dfrac{-4a^2-12a+40}{-2a+4} = \dfrac{-4(a^2+3a-10)}{-2(a-2)} = \dfrac{-4(a+5)(a-2)}{-2(a-2)} = 2a+10$

11. $\dfrac{-3x^3+3y^3}{-3x^2+3y^2} = \dfrac{-3(x^3-y^3)}{-3(x^2-y^2)} = \dfrac{(x-y)(x^2+xy+y^2)}{(x-y)(x+y)} = \dfrac{x^2+xy+y^2}{x+y}$

12. $\dfrac{2x^2+10x+12}{2x^3+16} = \dfrac{2(x^2+5x+6)}{2(x^3+8)} = \dfrac{(x+2)(x+3)}{(x+2)(x^2-2x+4)} = \dfrac{x+3}{x^2-2x+4}$

13. $\dfrac{4b^2}{21a} \cdot \dfrac{35a^2}{8b^4} = \dfrac{4b^2}{3 \cdot 7a} \cdot \dfrac{5 \cdot 7a^2}{4 \cdot 2b^4} = \dfrac{5a}{6b^2}$

14. $\dfrac{9w^3}{5t^2} \cdot \dfrac{10t^5}{27w^8} = \dfrac{3 \cdot 3w^3}{5t^2} \cdot \dfrac{2 \cdot 5t^5}{3^3w^8} = \dfrac{2t^3}{3w^5}$

15. $\dfrac{6ab^3}{40} \cdot \dfrac{25}{18a^7b} = \dfrac{2 \cdot 3ab^3}{2^3 \cdot 5} \cdot \dfrac{5^2}{2 \cdot 3^2a^7b} = \dfrac{5b^2}{24a^6}$

16. $\dfrac{3xy^3}{15xy} \cdot \dfrac{45xy^2}{18xy^9} = \dfrac{3xy^3}{3 \cdot 5xy} \cdot \dfrac{3^2 \cdot 5xy^2}{2 \cdot 3^2xy^9} = \dfrac{1}{2y^5}$

17. $\dfrac{15x^3}{x^2-2xy+y^2} \cdot \dfrac{x^2-y^2}{5x^7}$

$= \dfrac{3 \cdot 5x^3}{(x-y)^2} \cdot \dfrac{(x-y)(x+y)}{5x^7}$

$= \dfrac{3x+3y}{x^4(x-y)}$

18. $\dfrac{20a^6}{9a^2+12ab+4b^2} \cdot \dfrac{9a^2-4b^2}{4a^3} = \dfrac{20a^6}{(3a+2b)^2} \cdot \dfrac{(3a-2b)(3a+2b)}{4a^3} = \dfrac{5a^3(3a-2b)}{3a+2b} = \dfrac{15a^4-10a^3b}{3a+2b}$

19. $\dfrac{5x+10}{x^2+5x+6} \cdot \dfrac{x^2+6x+9}{10x+30} = \dfrac{5(x+2)}{(x+2)(x+3)} \cdot \dfrac{(x+3)^2}{10(x+3)} = \dfrac{1}{2}$

20. $\dfrac{x^2-x-12}{x^2+x-12} \cdot \dfrac{x^2+4x}{x^2-4x} = \dfrac{(x-4)(x+3)}{(x+4)(x-3)} \cdot \dfrac{x(x+4)}{x(x-4)} = \dfrac{x+3}{x-3}$

21. $\dfrac{4a^5b^4}{a^2-ab} \div \dfrac{24a^8b}{a^2-b^2} = \dfrac{4a^5b^4}{a^2-ab} \cdot \dfrac{(a^2-b^2)}{24a^8b} = \dfrac{4a^5b^4}{a(a-b)} \cdot \dfrac{(a-b)(a+b)}{24a^8b} = \dfrac{ab^3+b^4}{6a^4}$

22. $\dfrac{17x^5y^6}{x^2-y^2} \div \dfrac{51x^5y}{x^2+2xy+y^2} = \dfrac{17x^5y^6}{x^2-y^2} \cdot \dfrac{x^2+2xy+y^2}{51x^5y}$

$$= \frac{17x^5y^6}{(x-y)(x+y)} \cdot \frac{(x+y)^2}{51x^5y}$$
$$= \frac{xy^5+y^6}{3(x-y)}$$

23. $\dfrac{a^2-a-2}{a^2+a} \div \dfrac{a^2-2a}{a^3+3a^2}$
$$= \frac{(a-2)(a+1)}{a(a+1)} \cdot \frac{a^2(a+3)}{a(a-2)}$$
$$= a+3$$

24. $\dfrac{3w^2-3w-18}{6w^2-18w} \div \dfrac{w+2}{2w^2+2w}$
$$= \frac{3(w-3)(w+2)}{6w(w-3)} \cdot \frac{2w(w+1)}{w+2}$$
$$= w+1$$

25. $\frac{8}{3x} + \frac{4}{3x} = \frac{12}{3x} = \frac{4}{x}$

26. $\dfrac{3}{5x^2y} + \dfrac{2}{5x^2y} = \dfrac{5}{5x^2y} = \dfrac{1}{x^2y}$

27. $\dfrac{14b}{7b+1} + \dfrac{2}{7b+1} = \dfrac{14b+2}{7b+1}$
$$= \frac{2(7b+1)}{7b+1} = 2$$

28. $\dfrac{2w^2+1}{w^2+4} + \dfrac{w^2+11}{w^2+4} = \dfrac{3w^2+12}{w^2+4}$
$$= \frac{3(w^2+4)}{w^2+4} = 3$$

29. $\dfrac{1}{x-y} - \dfrac{2x}{x^2-y^2}$
$$= \frac{1(x+y)}{(x-y)(x+y)} - \frac{2x}{x^2-y^2}$$
$$= \frac{x+y-2x}{(x-y)(x+y)}$$
$$= \frac{-1(x-y)}{(x-y)(x+y)}$$
$$= \frac{-1}{x+y}$$

30. $\dfrac{1}{x^2-x-2} - \dfrac{1}{x-2}$
$$= \frac{1}{(x-2)(x+1)} - \frac{1(x+1)}{(x-2)(x+1)}$$
$$= \frac{1-x-1}{(x-2)(x+1)}$$
$$= \frac{-x}{(x-2)(x+1)}$$

31. $\dfrac{4-3w}{2w^2-5w-3} - \dfrac{w}{2w+1}$
$$= \frac{4-3w}{(2w+1)(w-3)} - \frac{w(w-3)}{(2w+1)(w-3)}$$
$$= \frac{4-3w-w^2+3w}{(2w+1)(w-3)}$$
$$= \frac{4-w^2}{(2w+1)(w-3)}$$

32. $\dfrac{t}{3t^2-t-2} - \dfrac{t}{3t+2}$
$$= \frac{t}{(3t+2)(t-1)} - \frac{t(t-1)}{(3t+2)(t-1)}$$
$$= \frac{t-t^2+t}{(3t+2)(t-1)}$$
$$= \frac{2t-t^2}{(3t+2)(t-1)}$$

33. $\dfrac{m}{m^2+m} + \dfrac{5}{m^2+3m}$
$$= \frac{m}{m(m+1)} + \frac{5}{m(m+3)}$$
$$= \frac{m(m+3)}{m(m+1)(m+3)} + \frac{5(m+1)}{m(m+3)(m+1)}$$
$$= \frac{m^2+3m+5m+5}{m(m+1)(m+3)}$$
$$= \frac{m^2+8m+5}{m(m+1)(m+3)}$$

34. $\dfrac{n}{n^2-9} + \dfrac{2}{n^2+3n}$
$$= \frac{n}{(n-3)(n+3)} + \frac{2}{n(n+3)}$$
$$= \frac{n \cdot n}{(n-3)(n+3)n} + \frac{2(n-3)}{n(n+3)(n-3)}$$
$$= \frac{n^2+2n-6}{n(n-3)(n+3)}$$

35. $\dfrac{\frac{1}{2}-\frac{1}{3}}{\frac{5}{4}-\frac{1}{6}} = \dfrac{12\left(\frac{1}{2}-\frac{1}{3}\right)}{12\left(\frac{5}{4}-\frac{1}{6}\right)}$
$$= \frac{6-4}{15-2} = \frac{2}{13}$$

36. $\dfrac{\frac{3}{8}+\frac{2}{3}}{\frac{1}{2}-\frac{1}{4}} = \dfrac{24\left(\frac{3}{8}+\frac{2}{3}\right)}{24\left(\frac{1}{2}-\frac{1}{4}\right)}$
$$= \frac{9+16}{12-6} = \frac{25}{6}$$

37. $\dfrac{\frac{1}{a}+\frac{2}{b}}{\frac{3}{ab}-\frac{1}{ab}} = \dfrac{ab\left(\frac{1}{a}+\frac{2}{b}\right)}{ab\left(\frac{3}{ab}-\frac{1}{ab}\right)}$
$$= \frac{b+2a}{3-1} = \frac{b+2a}{2}$$

38. $\dfrac{\frac{4}{xy}-\frac{3}{xy}}{\frac{2}{x}-\frac{5}{y}}=\dfrac{xy\left(\frac{4}{xy}-\frac{3}{xy}\right)}{xy\left(\frac{2}{x}-\frac{5}{y}\right)}$

$=\dfrac{4-3}{2y-5x}=\dfrac{1}{2y-5x}$

39. $\dfrac{\frac{1}{3t^3}-\frac{5}{6t}}{\frac{4}{9t}-\frac{5}{2t^2}}=\dfrac{18t^3\left(\frac{1}{3t^3}-\frac{5}{6t}\right)}{18t^3\left(\frac{4}{9t}-\frac{5}{2t^2}\right)}$

$=\dfrac{6-15t^2}{8t^2-45t}$

40. $\dfrac{\frac{2}{5m^2}+3}{\frac{1}{10m}-2}=\dfrac{10m^2\left(\frac{2}{5m^2}+3\right)}{10m^2\left(\frac{1}{10m}-2\right)}$

$=\dfrac{4+30m^2}{m-20m^2}$

41. $\frac{3}{x}+\frac{1}{2x}=\frac{1}{6x}+\frac{10}{3}$

$$6x\left(\frac{3}{x}+\frac{1}{2x}\right)=6x\left(\frac{1}{6x}+\frac{10}{3}\right)$$
$$18+3=1+20x$$
$$20=20x$$
$$1=x$$

The solution set is $\{1\}$.

42. $\frac{1}{t}+\frac{2}{3t}=\frac{3}{4t}+\frac{1}{6}$

$$24t\left(\frac{1}{t}+\frac{2}{3t}\right)=24t\left(\frac{3}{4t}+\frac{1}{6}\right)$$
$$24+16=18+4t$$
$$22=4t$$
$$11/2=t$$

The solution set is $\left\{\frac{11}{2}\right\}$.

43. $\dfrac{3}{x-2}-\dfrac{2}{x+2}=\dfrac{1}{x^2-4}$

$$(x^2-4)\left(\frac{3}{x-2}-\frac{2}{x+2}\right)=(x^2-4)\frac{1}{x^2-4}$$

$$3(x+2)-2(x-2)=1$$
$$3x+6-2x+4=1$$
$$x=-9$$

The solution set is $\{-9\}$.

44. $\dfrac{4}{y-3}+\dfrac{6}{y+1}=\dfrac{3y}{y^2-2y-3}$

$$(y-3)(y+1)\left(\frac{4}{y-3}+\frac{6}{y+1}\right)$$
$$=(y-3)(y+1)\frac{3y}{y^2-2y-3}$$

$$4(y+1)+6(y-3)=3y$$
$$10y-14=3y$$
$$-14=-7y$$
$$2=y$$

The solution set is $\{2\}$.

45. $\dfrac{5}{a+5}+\dfrac{7}{2a-3}=\dfrac{4a}{2a^2+7a-15}$

$$(a+5)(2a-3)\left(\frac{5}{a+5}+\frac{7}{2a-3}\right)$$

$$=(a+5)(2a-3)\frac{4a}{2a^2+7a-15}$$

$$5(2a-3)+7(a+5)=4a$$
$$17a+20=4a$$
$$13a=-20$$
$$a=-20/13$$

The solution set is $\left\{-\frac{20}{13}\right\}$.

46. $\dfrac{3}{3m+4}+\dfrac{2}{2m-1}=\dfrac{m}{6m^2+5m-4}$

$$(3m+4)(2m-1)\left(\frac{3}{3m+4}+\frac{2}{2m-1}\right)$$
$$=(3m+4)(2m-1)\frac{m}{6m^2+5m-4}$$

$$3(2m-1)+2(3m+4)=m$$
$$12m+5=m$$
$$11m=-5$$
$$m=-5/11$$

The solution set is $\left\{-\frac{5}{11}\right\}$.

47. Let x represent the number of teachers.

$$\frac{22.4}{1}=\frac{1904}{x}$$
$$22.4x=1904$$
$$x=85$$

There are 85 teachers.

48. Let x represent the number of cups of cereal.

$$\frac{2}{1}=\frac{12}{x}$$
$$2x=12$$
$$x=6$$

Use 6 cups of cereal.

49. Let x represent the number of dogs and $x - 12$ represent the number of cats.

$$\frac{4}{3} = \frac{x}{x-12}$$
$$4x - 48 = 3x$$
$$x = 48$$
$$x - 12 = 36$$

There are 48 dogs and 36 cats.

50. Let x represent the number of cars and $x - 12$ represent the number of trucks.

$$\frac{3}{7} = \frac{x-12}{x}$$
$$3x = 7x - 84$$
$$-4x = -84$$
$$x = 21$$
$$x - 12 = 9$$

There were 21 cars and 9 trucks.

51.

	D	R	T
To	1400	x	$\frac{1400}{x}$
Back	1400	$x+6$	$\frac{1400}{x+6}$

$$\frac{1400}{x+6} = \frac{1400}{x} - 3$$

$$x(x+6)\frac{1400}{x+6} = x(x+6)\left(\frac{1400}{x} - 3\right)$$
$$1400x = 1400(x+6) - 3x(x+6)$$
$$1400x = 1400x + 8400 - 3x^2 - 18x$$
$$3x^2 + 18x - 8400 = 0$$
$$x^2 + 6x - 2800 = 0$$
$$(x-50)(x+56) = 0$$
$$x = 50 \text{ or } x = -56$$

The average speed on the way to Dallas was 50 mph.

52.

	D	R	T
First	600	x	$\frac{600}{x}$
Second	400	$x-10$	$\frac{400}{x-10}$

$$\frac{600}{x} = \frac{400}{x-10} + 2$$

$$x(x-10)\frac{600}{x} = x(x-10)\left(\frac{400}{x-10} + 2\right)$$
$$600x - 6000 = 400x + 2x(x-10)$$
$$600x - 6000 = 400x + 2x^2 - 20x$$
$$-2x^2 + 220x - 6000 = 0$$
$$x^2 - 110x + 3000 = 0$$
$$(x-60)(x-50) = 0$$
$$x = 60 \quad \text{or} \quad x = 50$$
$$x - 10 = 50 \qquad x - 10 = 40$$

First day 60 mph and second day 50 mph or first day 50 and second day 40.

53. Let x be the number in the original group and $x + 4$ be the number in the larger group.

$$\frac{2100}{x} = \frac{2100}{x+4} + 400$$

$$x(x+4)\frac{2100}{x} = x(x+4)\left(\frac{2100}{x+4} + 400\right)$$
$$2100x + 8400 = 2100x + 400x^2 + 1600x$$
$$-400x^2 - 1600x + 8400 = 0$$
$$x^2 + 4x - 21 = 0$$
$$(x+7)(x-3) = 0$$
$$x = -7 \text{ or } x = 3$$

There are 3 students in the original group.

54. Let x be the number in the original group and $x + 4$ be the number in the larger group.

$$\frac{250}{x} = \frac{340}{x+4} + 20$$

$$x(x+4)\frac{250}{x} = x(x+4)\left(\frac{340}{x+4} + 20\right)$$
$$250x + 1000 = 340x + 20x^2 + 80x$$
$$-20x^2 - 170x + 1000 = 0$$
$$2x^2 + 17x - 100 = 0$$
$$(x-4)(2x+25) = 0$$
$$x = 4 \quad \text{or} \quad x = -25/2$$

There are 4 students in the original group.

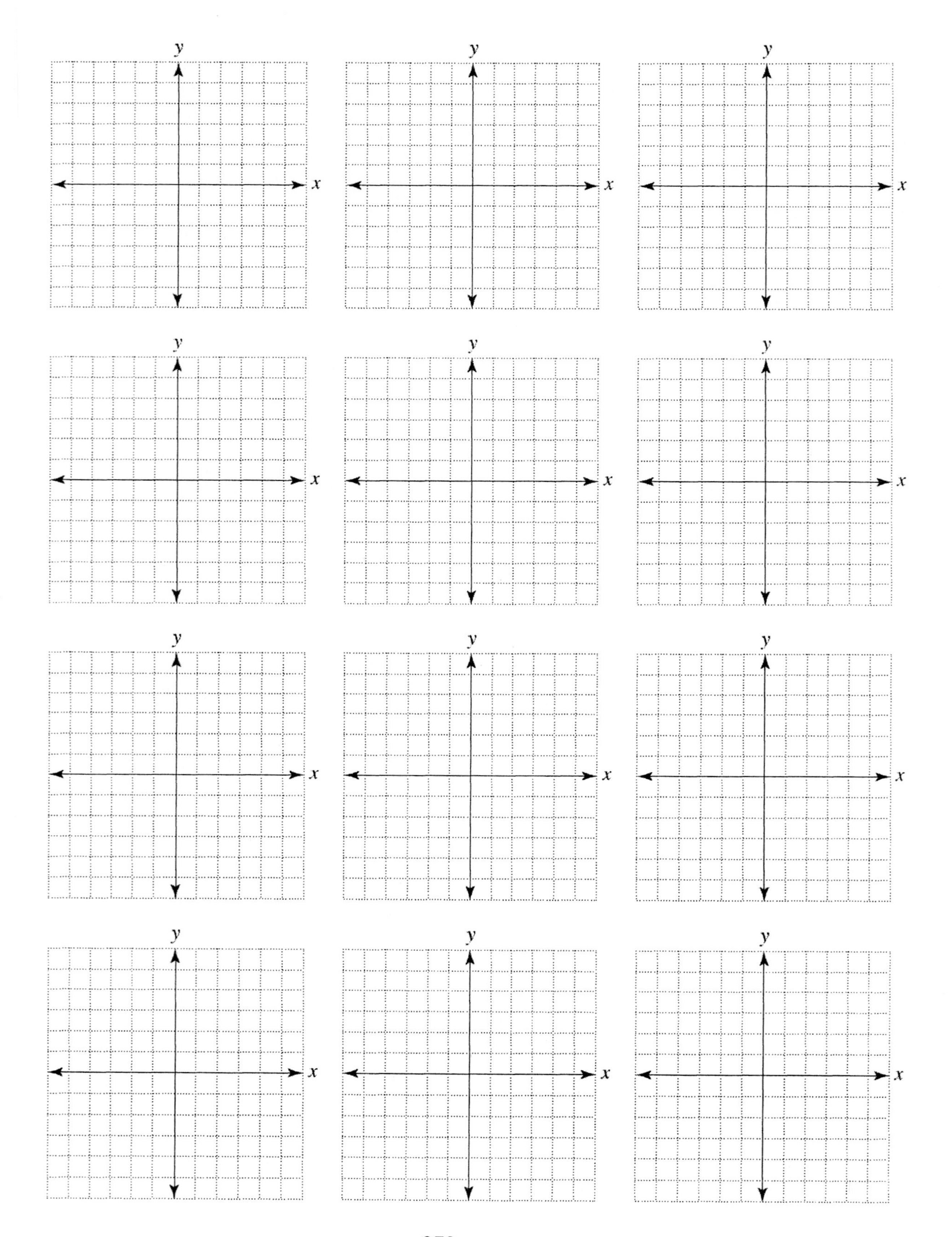
y
x

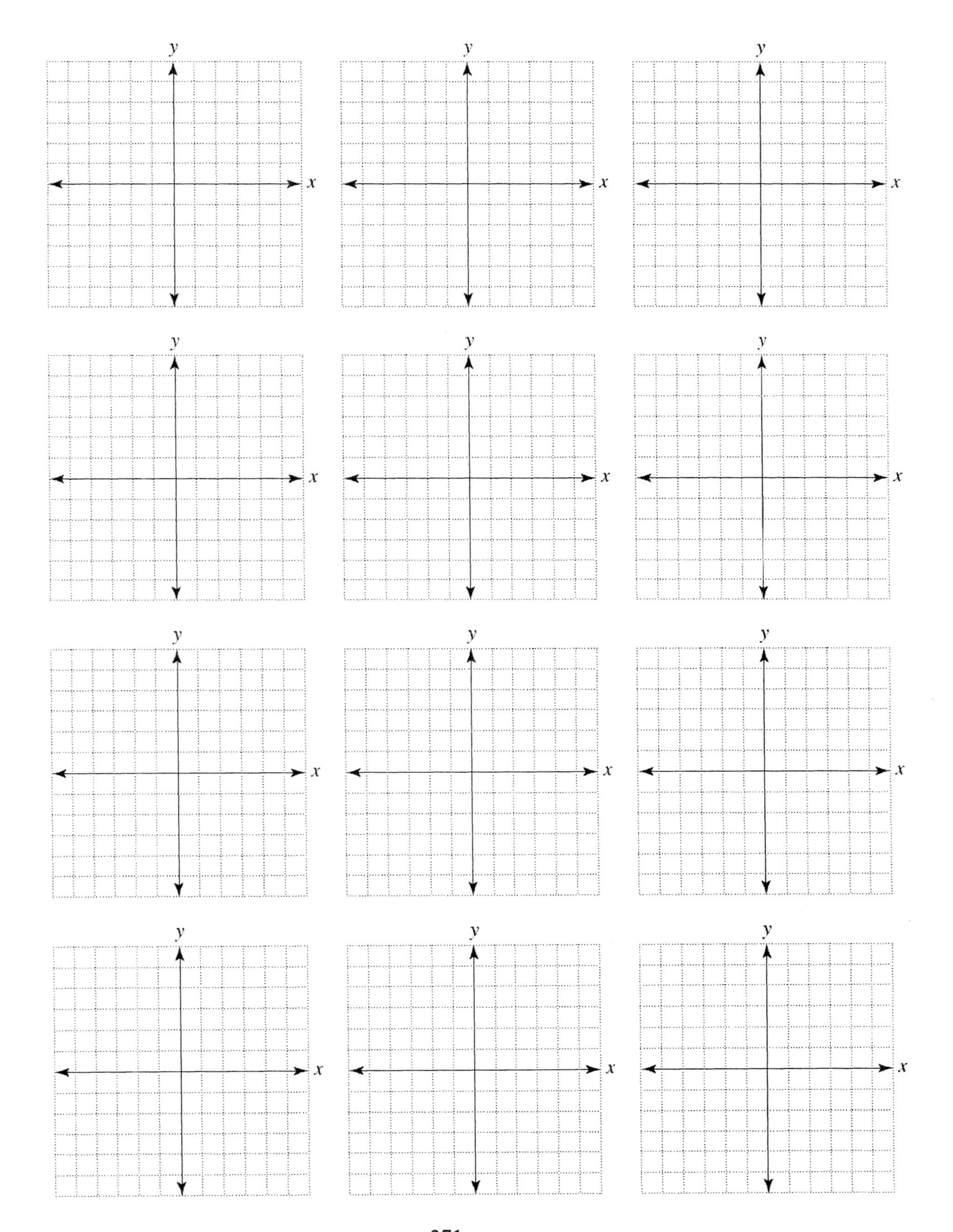

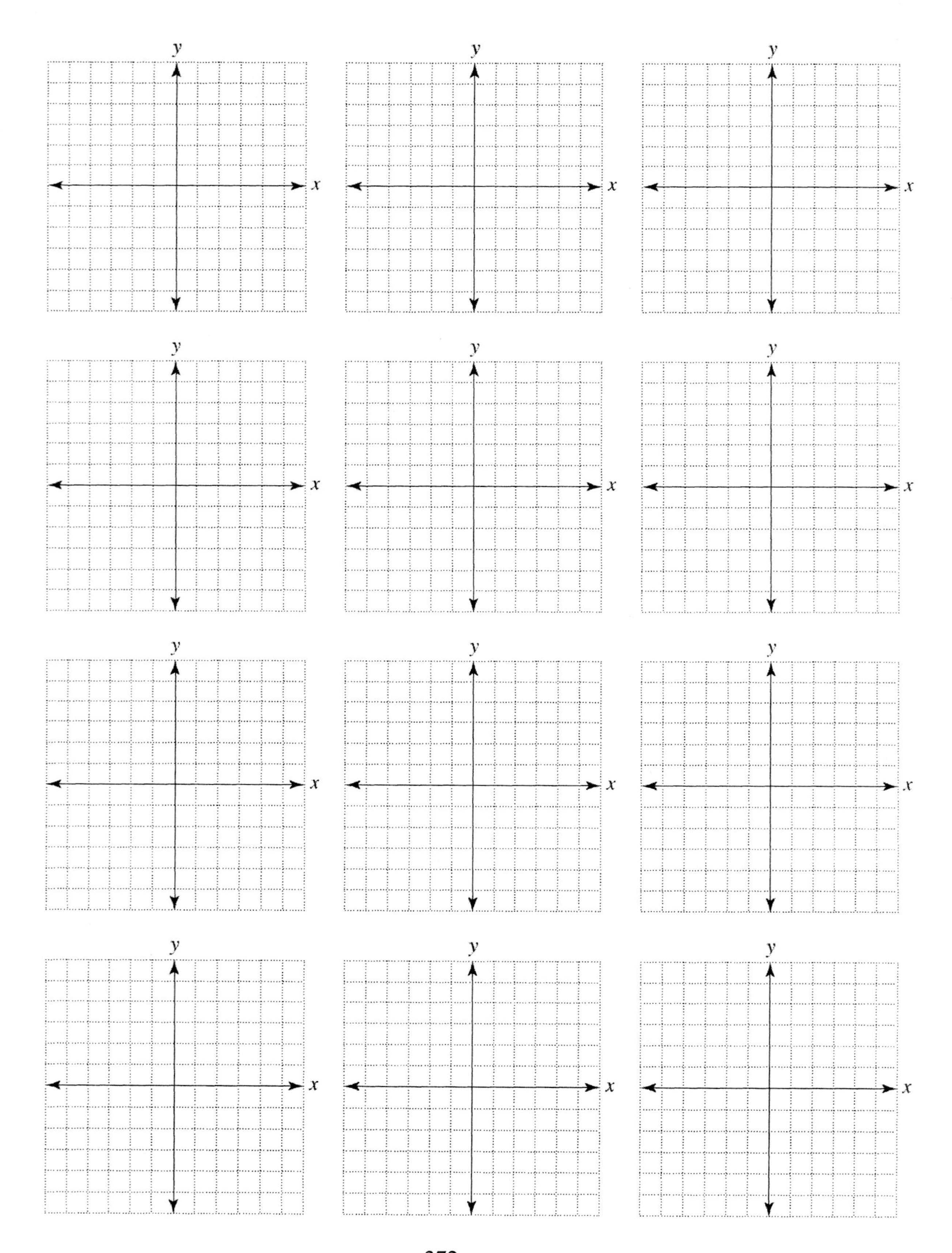

y
x

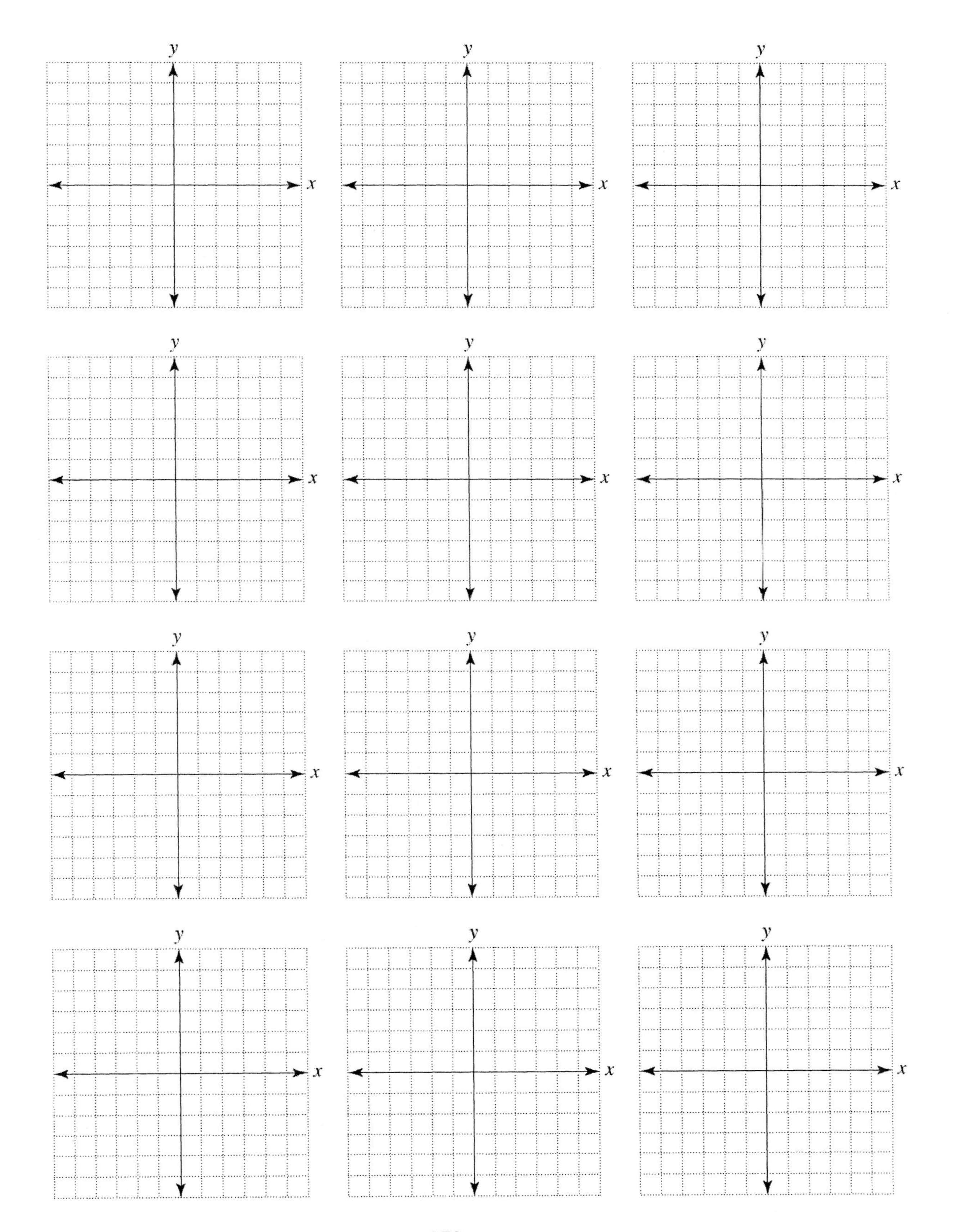

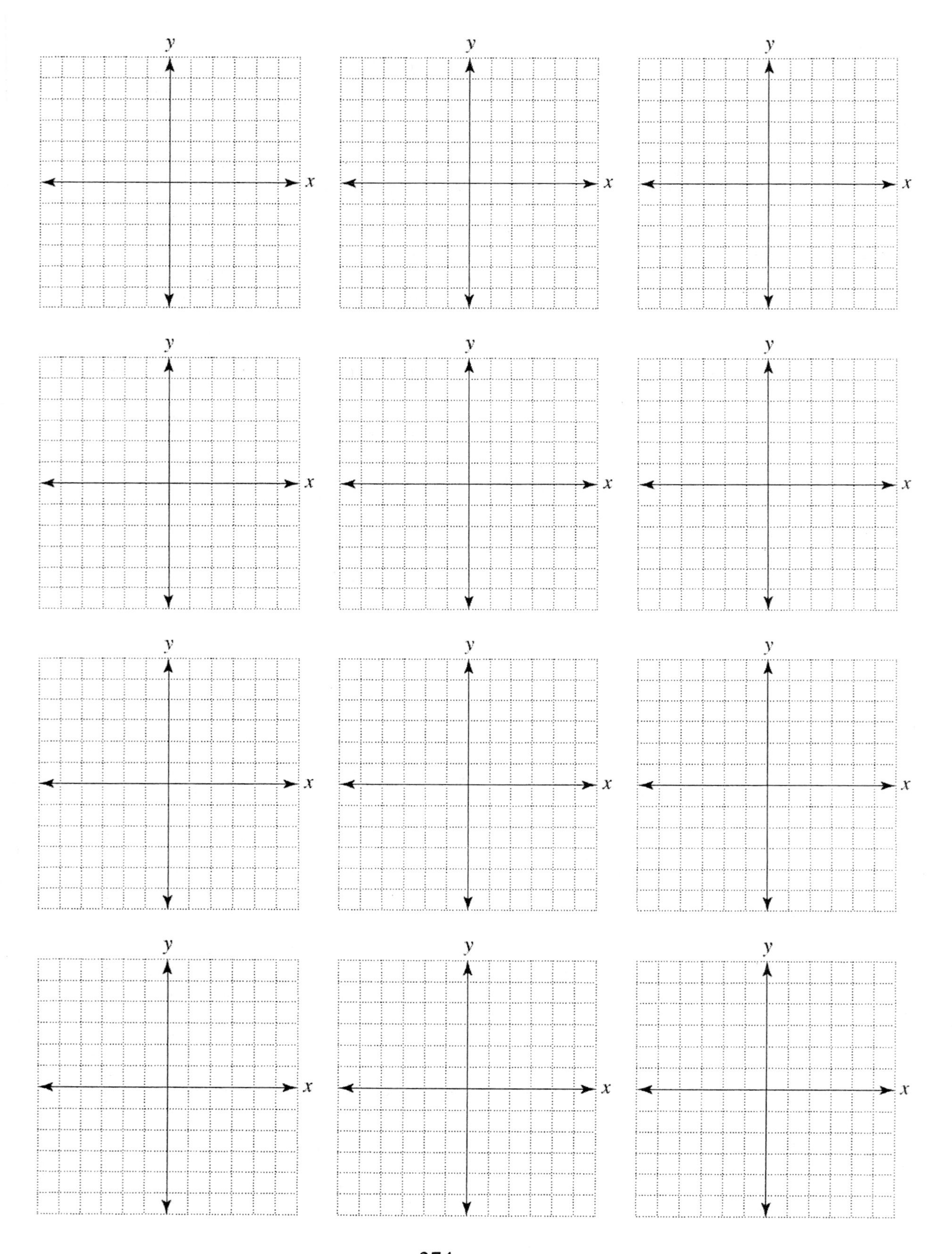
y
x
y
x
y
x
y
x
y
x
y
x
y
x
y
x
y
x
y
x
y
x
y
x

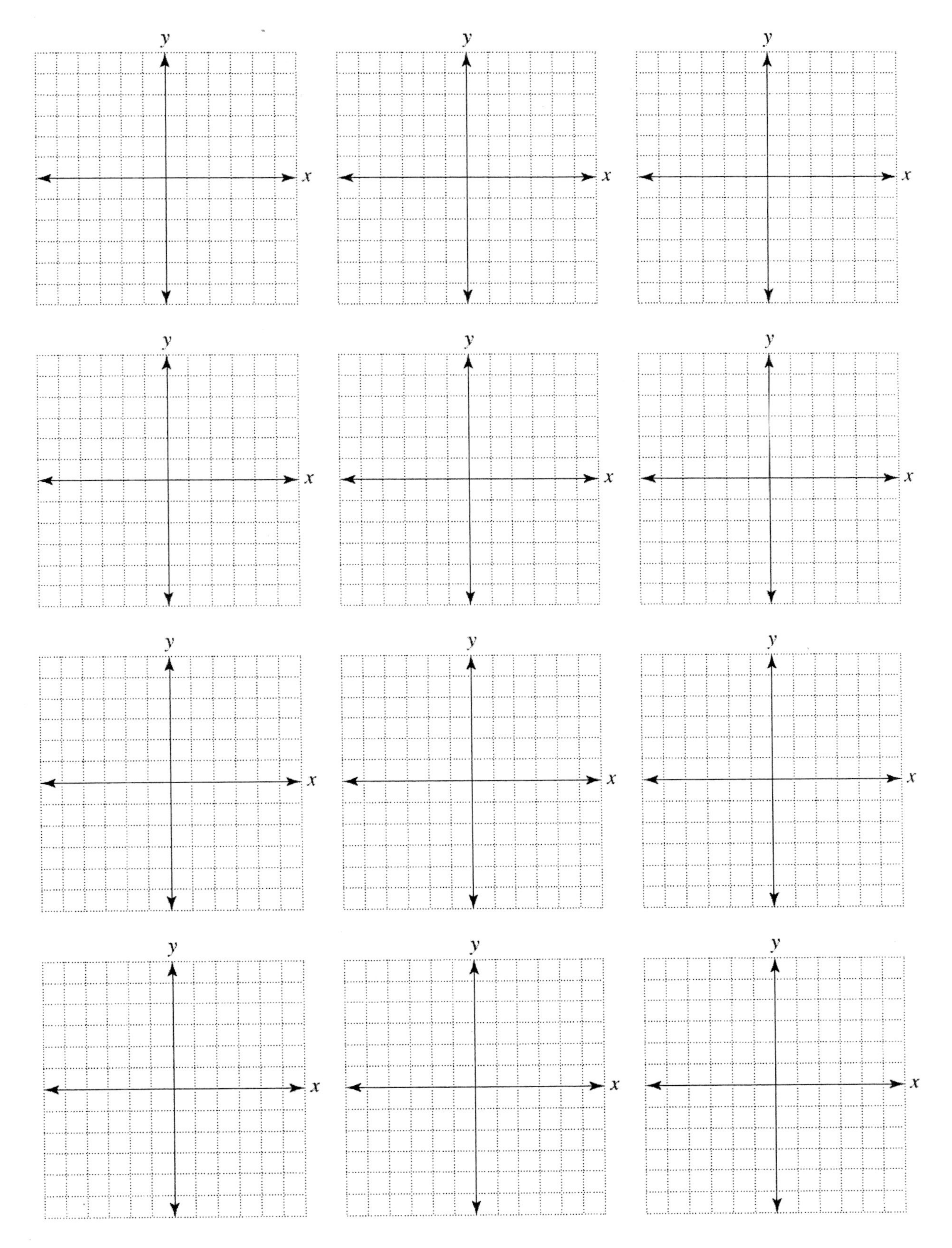
y
x
y
x
y
x
y
x
y
x
y
x
y
x
y
x
y
x
y
x
y
x
y
x

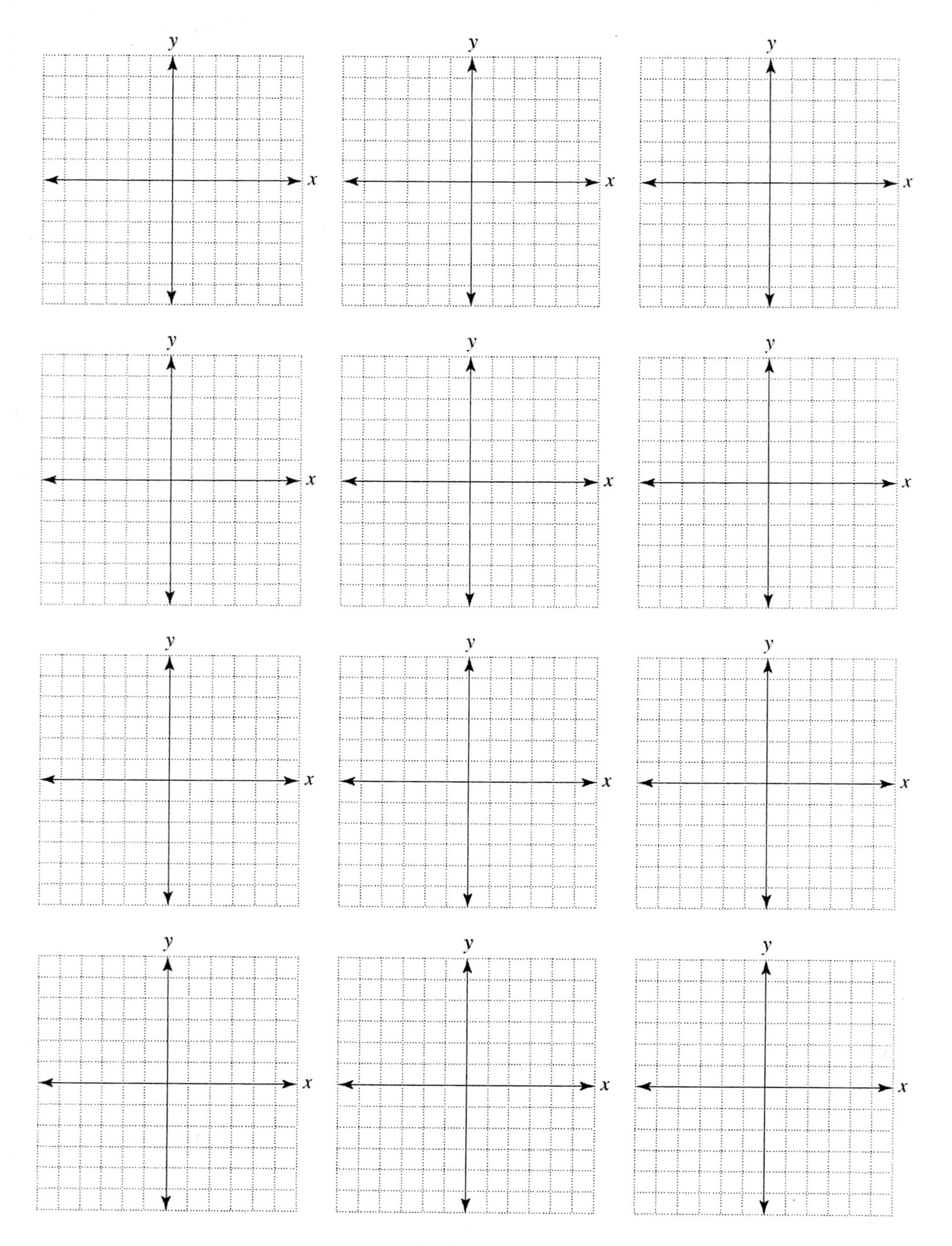

y
x

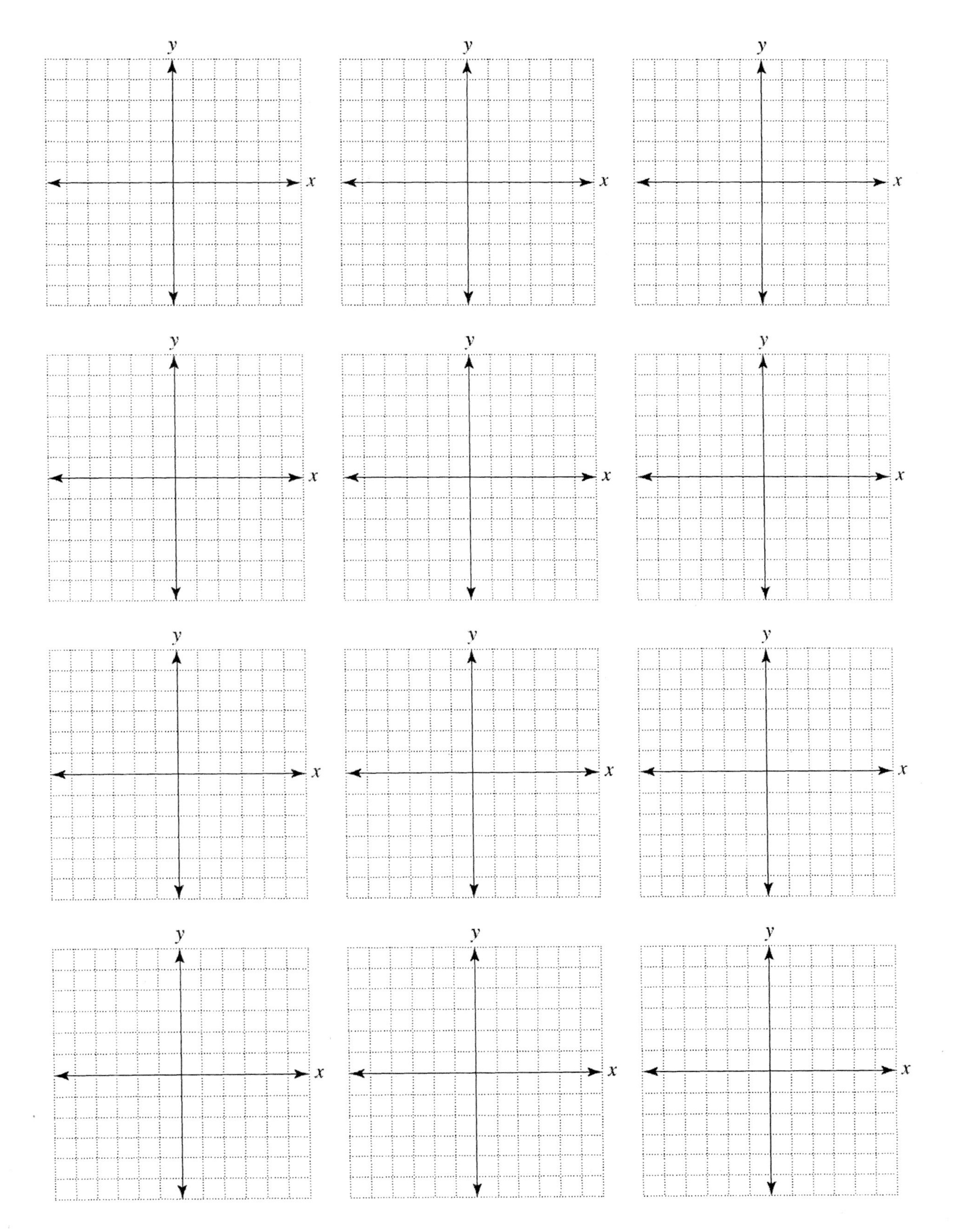
y
x
y
x
y
x
y
x
y
x
y
x
y
x
y
x
y
x
y
x
y
x
y
x

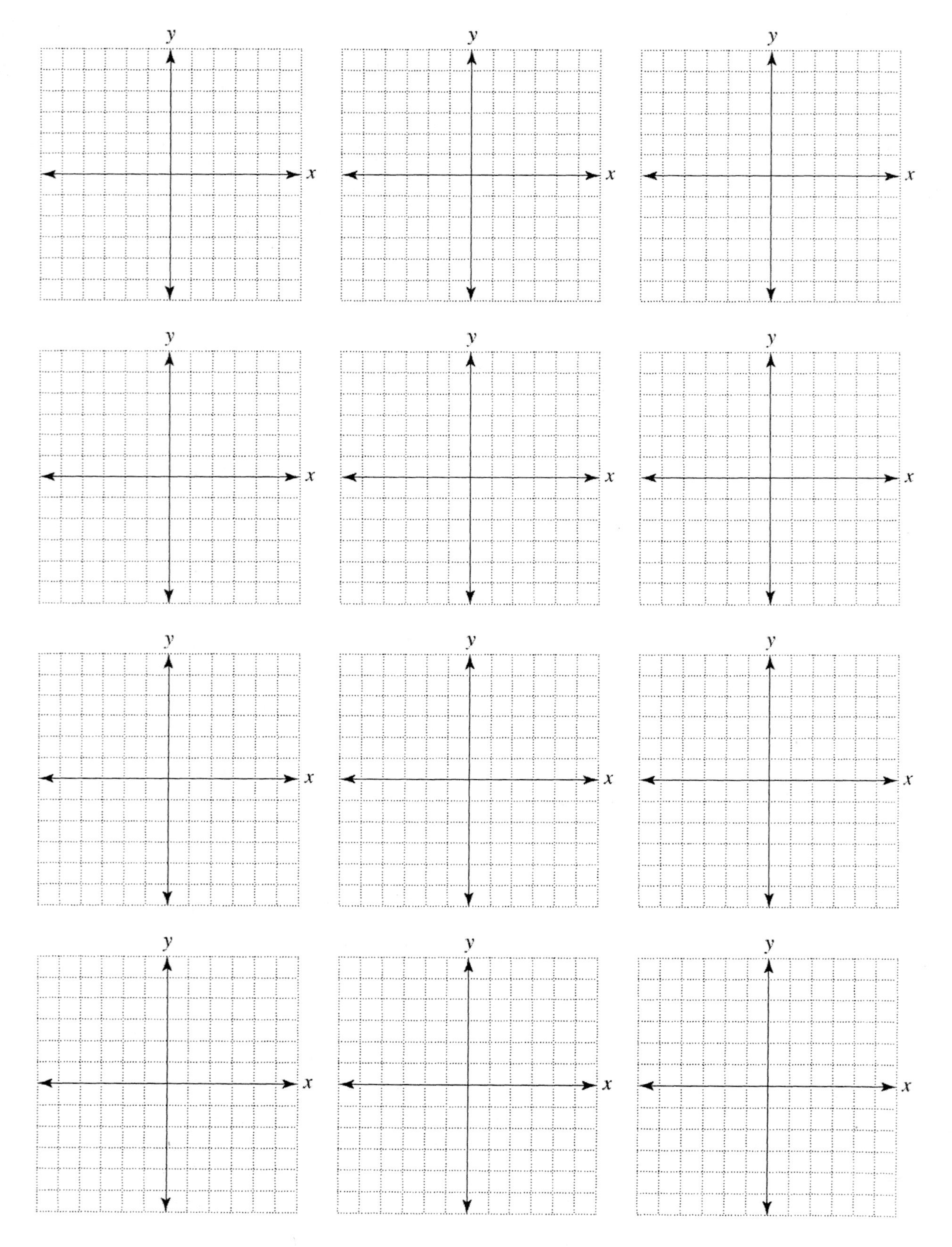
y
x
y
x
y
x
y
x
y
x
y
x
y
x
y
x
y
x
y
x
y
x
y
x

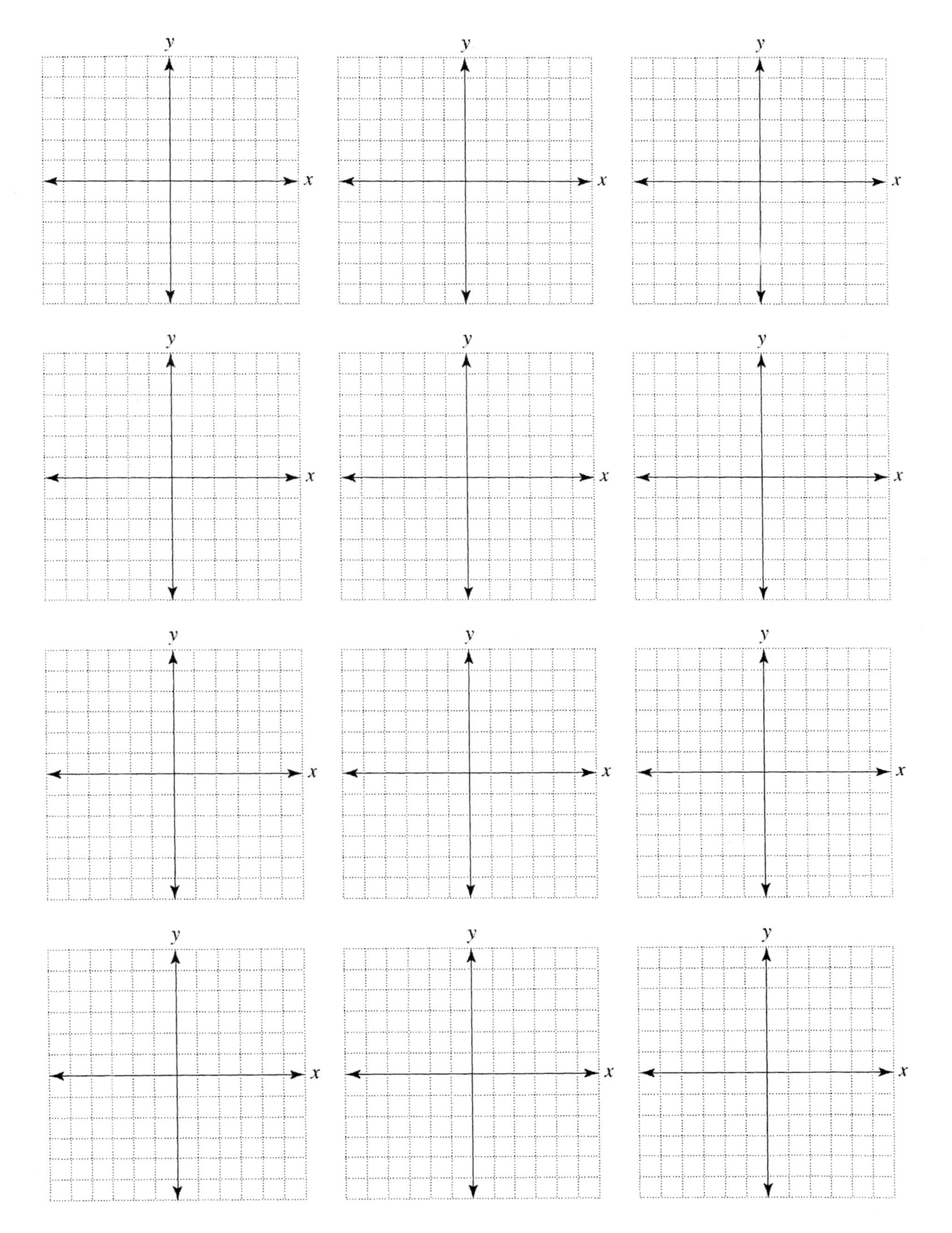
y
x
y
x
y
x
y
x
y
x
y
x
y
x
y
x
y
x
y
x
y
x
y
x